PLANT BIOTECHNOLOGY AND IN VITRO BIOLOGY
IN HET 21ST CENTURY

Current Plant Science and Biotechnology in Agriculture

VOLUME 36

Aims and Scope
The book series is intended for readers ranging from advanced students to senior research scientists and corporate directors interested in acquiring in-depth, state-of-the-art knowledge about research findings and techniques related to all aspects of agricultural biotechnology. Although the previous volumes in the series dealt with plant science and biotechnology, the aim is now to also include volumes dealing with animals science, food science and microbiology. While the subject matter will relate more particularly to agricultural applications, timely topics in basic science and biotechnology will also be explored. Some volumes will report progress in rapidly advancing disciplines through proceedings of symposia and workshops while others will detail fundamental information of an enduring nature that will be referenced repeatedly.

The titles published in this series are listed at the end of this volume.

Plant Biotechnology and In Vitro Biology in the 21st Century

Proceedings of the IXth International Congress of the International Association of Plant Tissue Culture and Biotechnology Jerusalem, Israel, 14–19 June 1998

Edited by

ARIE ALTMAN

Faculty of Agriculture, Rehovot,
The Hebrew University of Jerusalem, Israel

MEIRA ZIV

Faculty of Agriculture, Rehovot,
The Hebrew University of Jerusalem, Israel

and

SHAMAY IZHAR

Agricultural Research Organization, Bet Dagan,
The Ministry of Agriculture, Israel

The Congress was organized in cooperation with:

UNESCO, Paris
The International Atomic Energy Agency, Vienna

SPRINGER-SCIENCE+BUSINESS MEDIA, B.V.

Library of Congress Cataloging-in-Publication Data

DOI 10.1007/978-94-011-4661-6

Printed on acid-free paper

TABLE OF CONTENTS

Section B. Genetic manipulations: Transformation and Gene Expression, Hybridization, Haploidization and Mutagenesis

Section C. Genetic Stability and Instability, Selection and Variability

Section D. Regulation of Primary and Secondary Metabolism

Section E. Model Systems: Cell Cycle, Transport and Signal Transduction

Section F. Biotechnology for Plant Protection: Abiotic and Biotic Stress

Section G. Biotechnology for Crop Improvement: Yield, Quality and Production of Valuable Substances

Section H. Novel Micropropagation Methods

Section I. New Markets and Commercial Applications

Section J. Intellectual Property Rights

EDITORIAL PREFACE

The IXth international congress of the International Association of Plant Tissue Culture (IAPTC), entitled "Plant Biotechnology and In Vitro Biology in the 21st Century", convened in Jerusalem, Israel, June 14-19, 1998. This congress was devoted to all aspects of plant biotechnology, presenting a variety of scientific achievements and techniques: from tissue culture and in vitro methods, to cellular mechanisms, molecular studies and generation of transgenic plants, to commercial applications and intellectual property rights. At the Jerusalem congress, the IAPTC was renamed the International Association of Plant Tissue Culture and Biotechnology (IAPTC&B), to reflect the changes and scope of activities of the association.

Traditional agricultural techniques contributed to remarkable achievements since the dawn of plant and animal domestication, but presently we face several serious limitations. The world's envisaged tripled demand for food, agricultural commodities and natural products cannot be met without environmentally-friendly novel plant biotechnology. The emerging plant biotechnologies, that search into the regulation of genes which are responsible for specific qualitative or quantitative traits, should be able to overcome and complement the constraints of conventional breeding procedures. Plant biotechnology and molecular breeding contribute successfully to production of novel agricultural products and to shortening the breeding period by genome mapping. Most plant biotechnologies are now available for the widespread use in all countries around the globe. They have to be adapted to the specific needs of the local agriculture, the environmental conditions, and the economical constraints. Like any other new revolutionary technique, agricultural biotechnology should be continuously monitored and regulated as to its effects on human welfare and the environment.

The congress brought to Jerusalem over 1, 100 participants from over 60 countries around the globe. The congress scientific program included 6 plenary lectures, 56 concurrent scientific symposia with a total of 240 oral presentations, 6 workshops and over 500 posters. Most of the oral presentations are assembled in this volume. The presentations are arranged in 11 sections, representing the major areas of plant biotechnology.

While we have made every effort to reach uniformity in terminology and style, the published manuscripts remain the sole responsibility of the presenting authors. We wish to express our sincere gratitude to each one of the congress participants and contributors, and all members of the International Advisory Board, the National Scientific Program Committee, and the Organizing Committee. The financial support of national, international and private organizations, and especially UNESCO, Paris and FAO/IAEA Joint Division, Vienna, is gratefully acknowledged. Finally, we would like to acknowledge the very able assistance of Ms. Basia Vinocur in the preparation of this volume, and the helpful cooperation of Kluwer Academic Publishers.

Arie Altman
Shamay Izhar
Meira Ziv

THE INTERNATIONAL ASSOCIATION FOR PLANT TISSUE CULTURE AND BIOTECHNOLOGY (IAPTC&B)

A. Altman
Chairman

A.A. Watad
Secretary-Treasurer

M. Ziv
Journal Editor

IAPTC EXECUTIVE COMMITTEE

A. Altman (Israel)
Chairman

D. Dudits (Hungary)
I. Navarro (Spain)
I.K. Vasil (USA)
A.A. Watad (Israel)
M. Ziv (ex officio, Israel)

ORGANIZING COMMITTEE

A. Altman
Chairman

Y. Eshdat
Secretary

D. Aviv
E. Birnbaum
Y. Gafni
S. Gepstein
S. Izhar
E. Khayat
S. Lavee
H. Lilien-Kipnis
A. Perl
M. Waldman
A.A. Watad
A. Zelcer

NATIONAL SCIENTIFIC PROGRAM COMMITTEE

S. Izhar
Chairman

A. Altman
G. Ben-Hayyim
A. Breiman
R. Fluhr
G. Galili
E. Galun
J. Gressel
J. Hirschberg
R. Levin
O. Reuveni (deceased)
A. Vainstein
M. Ziv

INTERNATIONAL ADVISORY BOARD

M. Abo El-Nil (Kuwait)
J. Aitken-Christie (New Zealand)
A. Atanassov (Bulgaria)
M. Caboche (France)
Z.L. Chen (China)
P. Debergh (Belgium)
M. Elliot (UK)
O.L. Gamborg (USA)
Y.Y. Gleba (Ukraine)
R.B. Goldberg (USA)
M.A. Madkour (Egypt)
W.J. Peacock (Australia)
D. Pental (India)
A. Sasson (France)
J. Schell (Germany)
W.-Y. Soh (South Korea)
M. Van Montagu (Belgium)
L. Willmitzer (Germany)
Z. Xu (China)
Y. Yamada (Japan)
A.M. Zakri (Malaysia)

ACKNOWLEDGEMENTS

The Organizing Committee express their appreciation
to the following institutions and companies for their
support of the Congress.

The International Atomic Energy Agency, (IAEA), Vienna
UNESCO, Paris
Agricultural Research Organization, The Volcani
Center, Israel
Biological Industries Plant Propagation Ltd., Israel
Chief Scientist, Ministry of Agriculture and Rural
Development, Israel
Danziger, "Dan" Flower Farm, Israel
Dekalb Genetics Corporation, USA
DowElanco, USA
DuPont Agricultural Products, USA
German-Israel Fund for Research and International
Development (GIFRID)
Ginosar Tissue Culture Nurseries Ltd., Israel
Hazera Quality Seeds, Israel
Israel Export Institute
Keren Kayemeth, Israel National Fund
Kluwer Academic Publishers, The Netherlands
Makhteshim-Agan Headquarters, Israel
Ministry of Science, Israel
Monsanto Company, USA
Pioneer Hi-Bred International Inc., USA
Rahan Meristem Plant Propagation and Biotechnology
Technion, Israel Institute of Technology
The Hebrew University of Jerusalem, Israel
The Israel Cotton Production & Marketing Board Ltd.
The Peres Institute for Peace
Rhone-Poulenc, France
United States Tobacco Company
Weizmann Institute of Science, Israel

The following Institutions in Israel have also supported the Congress:
El Al Israel Airlines
The Ministry of Tourism
The Jerusalem Municipality

Introduction

THE PLANT AND AGRICULTURAL BIOTECHNOLOGY REVOLUTION: WHERE DO WE GO FROM HERE?

A. Altman
The Otto Warburg Center for Agricultural Biotechnology, Faculty of Agriculture, The Hebrew University of Jerusalem, Rehovot, Israel.
e-mail: altman@agri.huji.ac.il

1. The agricultural scene: Food security and the changing environment

In a world where population growth is outstripping food supply agricultural- and especially plant-biotechnology, needs to be swiftly implemented in all walks of life. The world population is expected to reach 7 billion within 25 years, over 10 billion in the year 2050, while agricultural production is growing at the slower rate of about 1.8 % annually. All human beings depend on agriculture that produce food of the appropriate quality at the required quantities. Since this can no longer be achieved by traditional methods alone, *breeding through plant biotechnology is a necessity*. In the long run the massive and immediate implementation of plant and agricultural biotechnology is more highly crucial than that of medical biotechnology, since more people worldwide die from famine and diseases related to malnutrition than from "modern", western diseases. With "Disaster Management" becoming a central issue in modern sociology research and curricula, *food and food product shortages are the ultimate disaster*. However, unlike most natural disasters, this is one that we can prepare for and even prevent.

Domestication of plants and animals found in the wild, combined with gradual long-term changes in their qualitative and quantitative traits, were the first attributes of agriculture. Domestication, followed by food storage, coincided with the growth of microorganisms. Thus was born classical food fermentation, the earliest known application of biotechnology for the generation of food products. This traditional agriculture now faces several serious limitations:

1. Market limitations: The world is becoming a global village whose free-market rules negate the effectiveness of local pricing policies, and where a dictate of international trade and policies exists. This has greatly affected future developments in agriculture, still the world's largest business.

2. Limitations of natural resources: Global climatic changes (resulting mainly in desertification and salinization), industrialization and urbanization, have reduced land and water availability and caused alarming deterioration of soil, water and air quality.

A. Altman et al. (eds.), Plant Biotechnology and In Vitro Biology in the 21st Century, 1–7.

3. Inherent biological genetic limitations: Although previously highly efficient, the release of new improved genotypes by classical breeding is now too slow to cope with the demands, and is considerably limited by the lack of appropriate "natural" genes that can be introgressed by traditional genetic crosses.

Only two major potential solutions seems to exist for increasing food supply and agricultural commodities, in addition to continuously improving agricultural practices, despite the aforementioned limitations: (1) a search for alternative food sources (e.g., marine or extraterrestrial products), (2) enhanced efficient plant breeding.

2. The need for integration: Combining biotechnology with classical physiology and breeding

Food production, for both quantity and quality, as well as for new plant commodities and products, in developed and developing countries around the globe, cannot rely solely on classical agriculture. Human survival, vis-à-vis a continuous increase in agricultural productivity, depends on the effective *merging of classical breeding with modern plant biotechnology* and the novel tools it provides. The "green revolution", for example, increased wheat production 10-fold in India and several other countries in South east Asia, thereby feeding three times as many people. However, this revolution has already been exploited to its limits, and alternative solutions are required to breed improved crops. Now biotechnology, integrated with classical breeding, is on the verge of creating the "evergreen revolution".

The potential to improve plant and animal productivity and their proper use in agriculture relies largely on newly developed *DNA biotechnology and molecular markers*. These techniques enable the selection of successful genotypes, better isolation and cloning of favorable traits, and the creation of transgenic organisms of importance to agriculture. Together, these generic techniques are both an extension and an integral part of classical breeding, contributing successfully to shortening breeding and selection cycles.

The new plant biotechnology implies the use of recombinant DNA techniques and in vitro cell biology in three major areas:

1. As an aid to classical breeding: This includes the ongoing genome mapping projects, e.g., in Arabidopsis, rice, maize and tomato, combined with the recent activities in functional genomics, proteomics and bioinformatics, and DNA marker-assisted selection. The combined use of these techniques will soon shorten the time required for "classical" breeding and selection cycles.

2. Generation of engineered (transgenic) organisms: In view of the inherent limitations of introgressing new genes by traditional genetic crosses (i.e., lack of appropriate desired genes and crossing barriers), the efficient engineering of plants has already resulted in improved field-grown transgenic plants in several important crops. The result of this impressive development, which began only 18 years ago, have made possible the direct insertion and integration of genes isolated from several organisms, and the creation of novel, and otherwise impossible genetic recombinations.
3. Integration of microorganisms into plant production systems: The biotechnological development of new symbiotic, antibiotic, and antagonistic relationships between plants and microorganisms (fungi, bacteria and insects) using, among other techniques, engineered plants and microorganisms, creates new possibilities. Some of these include biological control of pests, biofertilization and plant growth stimulation, and bio- and phytoremediation.

During the last two decades, these new biotechnologies have been adapted to agricultural practices and have opened *new vistas for plant utilization*. This will continue and intensify in the next decade. Plant biotechnology -- especially in vitro regeneration and cell biology, DNA manipulation and genetic modification of biochemical pathways -- is changing the plant scene in three major areas: (1) growth and development control (vegetative, generative and propagation), (2) protecting plants against the ever-increasing threats of abiotic and biotic stress, (3) expanding the horizons by producing specialty foods, biochemicals and pharmaceuticals. These areas were extensively discussed at the 9th international congress of the IAPTC&B "Plant Biotechnology and In Vitro Biology in the 21st Century", held in Jerusalem in June 1998.

3. Growth control: Vegetative, generative and propagation

The better insight into the control of plant regeneration, morphogenesis and patterns of cell division achieved during the last two decade, is due to three major discoveries: (1) the *totipotency and regeneration ability of plant cells and tissues*, as revealed by cell culture and micropropagation, (2) the elucidation of *genes responsible for hormone production and activation* in plants, (3) active research into the mechanisms and molecular control of the *cell cycle and signal transduction pathways*, in part adopted from previous studies with animal cells, in part unique for plants. These have enabled both the *control and biotechnological manipulation* of vegetative growth, generative patterns (e.g., of flowers and seeds) and of micropropagation.

Vegetative growth: Morphogenetic control mechanisms are still extremely obscure, but the advent of molecular hormone and cell-cycle research is sure to lead to a better understanding of vegetative growth patterns. Thus, the possibility of biotechnologically manipulating *plant growth rate and architecture* can become a reality. For example, potential consequences of controlled auxin overproduction/availability include: adventitious root formation of importance to propagation, cell and organ elongation for biomass production, increased apical dominance of importance to timber production, etc. Controlled cytokinin overproduction/availability can result, among other things, in enhanced bud break -- which is of great importance to plant architecture, branching and compactness -- a desired characteristic for some ornamentals, and delayed leaf and plant senescence. No less important, in this respect, is the potential - as yet not practical - of affecting the orientation and rate of cell division, cell elongation and tissue longevity, by interfering with the cytoskeleton and cell cycle, the synthesis of cellulose and other cell components, and programmed cell death, respectively. A few of these possibilities have already been realized.

Generative development: Flowers, fruits and seeds are extremely important for agriculture. Hence, biotechnological research and development aims to interfere with and control their development and characteristics, and some of the many related studies have already produced practical applications. The major targets in *flower development* are color, scent and senescence. Strategies for the molecular breeding of flower color and scent include over- and underexpression of color (anthocyanins and carotenoids) and scent (volatiles) compounds, with respect to their biosynthesis, cellular transport and targeting. Important targets for controlling *fruit development* include growth, ripening and senescence (as for vegetative growth), color and scent (similar to flowers) and, in addition, flavor -- particularly metabolic control of sugar, acid and other flavor components. Of great importance to fruits are biotechnological strategies for the production of seedless fruits via parthenocarpy (overproduction of auxin), pollen destruction (no fertilization), or arrest of embryo development. The manipulation of *seed development* using biotechnlogical strategies is especially critical, since the seed industry (together with vegetative propagation material) constitutes the germplasm of the future for any type of plant production system. Seeds and vegetative propagules are, practically speaking, packages of genes that form the basis of all advanced and economically viable agricultural industries, both national and private. Biotechniques and molecular strategies are now available for the major seed-based operations: hybrid seed production, generation of artificial seeds (coated somatic embryos), and for the establishment of germplasm banks that may solve some of the biodiversity issues.

Micropropagation: Micropropagation is used routinely to generate a large number of high-quality clonal agricultural plants, including ornamental and vegetable species, and in some cases also plantation crops, fruits and vegetable species. Micropropagation has significant advantages over traditional clonal propagation techniques. These include the potential of combining rapid large-scale propagation of new genotypes, the use of small amounts of original germplasm (particularly at the early breeding and/or transformation stage, when only a few plants are available), and the generation of pathogen-free propagules. This impressive application of the *principles of plant cell division and regeneration to practical plant propagation* is the result of continuous tedious studies in hundreds of laboratories worldwide, many of them in developing countries, on the standardization of explant sources, media composition and physical state, environmental conditions and acclimatization of in vitro plants. Particularly noteworthy are the many recent studies on the molecular of organogenesis and somatic embryogenesis. However, further practical applications of micropropagation, which is also commercially viable, depends on *reducing the production costs* such that it can compete with seed production or traditional vegetative propagation methods (e.g., cuttings, tubers and bulbs, grafting). Techniques that have the potential to further increase the efficiency of micropropagation, but still await further improvements, include: simplified large-scale bioreactors, cheaper automatization facilities, efficient somatic embryogenesis and synthetic seed production, greater utilization of the autotrophic growth potential of cultures, and good repeatability and quality assurance of the micropropagated plants.

4. The need to protect: Abiotic and biotic stress tolerance

The application of molecular genetics and plant transformation to the *diagnosis and control of plant pests* has become one of the major practical success stories of plant biotechnology in the past decade. The availability of dozens of transgenic crop plants which are resistant to a range of insects, viruses and herbicides, as well as to several phytopathogenic fungi and nematodes has been validated under both field and laboratory conditions, and is of great economic importance. Moreover, applying the principles of engineering plants for resistance to these pests to other plants of agricultural importance is now considered routine, although in practice still laborious, especially for new genotypes. Apart from a wider application to additional plants, the real *challenges lying ahead* include: (1) improved expression of the target genes in the plants, especially their spatial and temporal control, (2) the use of wide-spectrum and alternative target genes to circumvent the problem of pest resistance, (3) intensified integration of biological control via the use of selected and engineered microorganisms with a biocontrol potential.

While plant biotechnology has been applied successfully to fighting a large number of pests, this is not yet the case for abiotic stress conditions such as drought, salinity, extreme temperatures, chemical toxicity and oxidative stress. *Drought and salinization are the most common natural causes of lack of food and famine* in arid and semiarid regions, and the most serious environmental threats to agriculture in many parts of the world. Desertification, resulting from overexploitation by the local inhabitants, is often aggravated by regional climatic changes, and results in increased soil erosion and a decrease in land and agricultural productivity. It is estimated that increased salinization of arable land will have devastating global effects, resulting in 30 % land loss within the next 25 years, and up to 50 % in the year 2050. Although more difficult to control and engineer than the usually monogenic traits of resistance to biotic pests and herbicides, the genetically complex response to abiotic stress is globally and regionally far more important. Therefore, *breeding for plant tolerance to drought and salinity stress* should be given a high research priority in all future agbiotech programs. Strategies for the manipulation of *osmotic stress tolerance* in plants might include: expression of osmoprotectants and compatible solutes, ion and water transport and channels, expression of water-binding and membrane-associated dehydrins and other proteins, transcription factors and DNA-binding proteins, etc. Also of specific interest are the intervening stages of stress perception, signal transduction (ABA and others), and protein modification. The discovery of new stress-related genes and the design of stress-specific promoters are equally important.

5. Expanding the horizons: Food, biochemicals and pharmaceuticals

Traditionally, agriculture was targeted to improving the production of plant-derived food, in terms of both quantity and quality. This was also the initial primary target of plant biotechnology. The second phase of plant biotechnology is now gradually being implemented: a shift from the production of low-priced food and bulk commodities to high-priced, specialized plant-derived products. This includes two major categories of biomaterials: (1) direct improvement and modification of specialized constituents of plant origin, and (2) the manufacture in plants of non-plant compounds. Biotechniques, mostly based on the *engineering of metabolic pathways*, are now available for modifying many plant constituents that are *used in the food, chemical and energy industries*. This includes many "primary" metabolites: carbohydrates (starch synthesis, yield and allocation, production of high-amylose or high-amylopectin starch, increased sucrose synthesis for the sugar industry, fructan production, etc.), proteins (improvement of amino acid composition and protein content), oils and fats (ratio of saturated to nonsaturated fatty acids, increased content of specific valuable fatty acids like erucic acid, ricinoleic acid and others). Many other plant constituents are either minor or non-food components, but have specific high- value applications, such as specific fatty acids as an alternative energy source, polysaccharides with heat hysteresis properties and for bioaffinity purification, temperature and salt-resistant enzymes for the food industry, etc. Moreover, the use of plants as *"bioreactors" for the production of "foreign", non-plant compounds* is gaining momentum and may eventually lead to alternative types of agriculture. This includes, for example, production of bioactive peptides, vaccines, antibodies and a range of enzymes -- mostly for the pharmaceutical industry. For the chemical industry, plants can be used to produce, e.g., polyhydroxybutyrate for the production of biodegradable thermoplastics, and cyclodextrins which form inclusion complexes with hydrophobic substances.

6. Supply and demand: Where do we go from here ?

Achievements today in plant biotechnology have already surpassed all previous expectations, and the future is even more promising. The full realization and impact of the aforementioned developments depend, however, not only on continued successful and innovative research and development activities, but also on a favorable regulatory climate and public acceptance. About 12 % of the world's land surface is used to grow crops, and the agricultural area required to support food production -- 0.44 ha / capita in 1961 -- will probably have been reduced to 0.15 ha / capita in 2050. The intensification of agriculture with its aforementioned limitations thus requires enhanced and more efficient plant breeding and the release of economical, high-return and patentable plant-derived products. This cannot be achieved without supporting advanced research and development in biochemistry, physiology, genomics and biotechnology of agricultural plants. Plant scientists now have a central role in society, not unlike their place 300 years ago when Jonathan Swift (1667-1745) stated: "Whoever could make two ears of corn or two blades of grass to grow upon a spot of ground where only one grew before, would deserve better of mankind, and do more essential service to his country, than the whole race of politicians put together."

Selected References

Altman, A., ed. (1998) Agricultural Biotechnology. Marcel Dekker, Inc., New York, ISBN: 0-8247-9439-7. 770 pp.

Altman, A. and M. Ziv (1997) Horticulture biotechnology: a historical perspective and future prospects. In A. Altman and M. Ziv, eds., Horticultural Biotechnology: In Vitro Culture and Breeding, Acta Horticulturae 447: 31-35.

Borlaug, N.E. (1997) Feeding a world of 10 billion people: the miracle ahead. Plant Tissue Culture and Biotech. 3: 119- 127.

Information Systems for Biotechnology (ISB) News Reports (nbiap@vt.edu).

The Bowditch Group Electronic AgBiotech Newsletter (biotechnews@world.std.com).

Plant Biotechnology: Achievements and Opportunities at the Threshold of the 21st Century

Indra K. Vasil
Laboratory of Plant Cell and Molecular Biology, 1143 Fifield Hall, University of Florida, Box 110690, Gainesville, FL 32611-0690, USA

The IXth IAPTC Congress has great symbolism in that it is the last one to be held in the 20th century and the 2nd millenium. As such, it provides a valuable opportunity to pause and to assess the contributions of plant biotechnology to the advancement of science and human welfare, and to gain a perspective for the future. It is a particularly significant time for such an assessment because the first products of plant biotechnology for human use have now been introduced into the marketplace, and there is much debate and concern about the uses and impact of this technology on our lives and the environment.

Plant biotechnology has two important and interdependent components, those of tissue culture and molecular biology. While the beginnings of tissue culture can be traced to the early decades of the 20th century, serious molecular studies of plants did not begin until the early 1970s. The scientific basis for the development of cell and tissue culture systems is derived from the Cell Theory of Schleiden (1838) and Schwann (1839), which states that individual cells within an organism have the "capacity for independent life", and the Darwinian concept of hormonal regulation of plant growth (Darwin and Darwin 1880). During the latter half of the 19th century, these theories were supported by sporadic descriptions of in situ callus and shoot bud formation in wounded tissues and cuttings (see White 1963, Gautheret 1985). Although attempts to culture isolated plant cells and tissues in nutrient solutions were made as early as 1902 (Haberlandt 1902, for translation see Krikorian and Berquam 1969), formal, organized and detailed studies on plant tissue culture did not begin until the early 1930s. These were greatly influenced by the discovery in 1934/1935 of the first naturally occurring plant growth substance, the auxin indole-3-acetic acid, and its effects on plant growth (see Thimann 1977, Gautheret 1985). At the same time Philip White in the United States, and Roger Gautheret and Pierre Nobécourt in France, began their fateful and now famous experiments leading to the unlimited growth of plant roots (White 1934a) and cells in culture (Gautheret 1939, Nobécourt 1939, White 1939), and in vitro organogenesis (the formation of roots in carrot callus by Nobécourt 1939, and shoots in tobacco callus by White 1939). These seminal results,

A. Altman et al. (eds.), Plant Biotechnology and In Vitro Biology in the 21st Century, 9–16.

reported within six weeks of each other in early 1939, were the real beginnings of modern plant tissue culture. During the course of his work on the culture of excised roots from virus infected tomato plants, White (1934b) also made the important observation that the subcultured roots were often virus-free. This was attributed to the absence of the virus in cells of root meristems, a fact which later led Georges Morel (Morel and Martin 1952) to use shoot meristem cultures for the elimination of viruses and micropropagation (Morel 1960), and formed the basis of the present day world-wide micropropagation industry.

The discovery of cytokinins, and the finding that in combination with auxins they modulate shoot morphogenesis (Skoog and Miller 1957), was an important milestone in the development of methods for the regeneration of plants from cultured cells. At about the same time, Jakob Reinert (Reinert 1958) in Germany, and Frederick Steward (Steward et al. 1958) in the United States, described the formation of somatic embryos in callus and cell suspension cultures of carrot, respectively. Although plants could now be regenerated from cultured tissues as well as cell suspension cultures by organogenesis or somatic embryogenesis, unequivocal evidence of the totipotency of plant cells was not provided until 1965, when whole tobacco plants were regenerated from single cells cultured in complete isolation (Vasil and Hildebrandt 1965). Until about 1980, plant regeneration in vitro was limited to model dicotyledonous species, as most legumes, monocotyledons and woody tree species continued to be recalcitrant to sustained growth and regeneration in culture. These problems were eventually overcome by the judicious use of plant growth substances and the culture of immature organ explants for the induction of embryogenic cultures in a wide variety of species, including major crops such as cereals, legumes, cassava, and many woody tree species (see Vasil 1986, Vasil and Thorpe 1994, Thorpe 1995).

The isolation (Cocking 1960) and fusion (Power et al. 1970) of plant protoplasts, and regeneration of plants from them (Takebe et al. 1971), generated much optimism for crop improvement by the production of somatic hybrids. Inspite of much effort, however, no commercially useful novel hybrids of any major crop species have been obtained by protoplast fusion. Nevertheless, protoplasts have proven to be very useful for direct DNA delivery leading to the generation of transgenic plants, and in fundamental studies of promoter function and gene regulation (Vasil et al. 1989, McCarty et al. 1992, Vasil and Vasil 1992, Fehér and Dudits 1994, Hoecker et al. 1995).

The production of haploid plants from cultured anthers (Guha and Maheshwari 1964), and later from microspores, was hailed as a major development toward the rapid production of novel homozygous breeding lines. This procedure, like protoplast fusion, has also not lived upto its oft-stated potential, and has at best produced a few useful varieties in rice and some other crops. Similarly, the contention that in vitro generated variability was novel and useful, and could be exploited to broaden the genetic base of crop species (Larkin and Scowcroft 1981), has been largely discredited and abandoned after long and intensive efforts.

Concurrent with the development of efficient methods for the regeneration of plants from cultured cells, significant advances have been made in the transfer of defined genes into plant cells and the production of transgenic plants. The beginnings of these achievements can be traced to the discovery of the three-dimensional architecture of DNA by Watson and Crick (1953), followed 20 years later by the isolation of restriction enzymes and the production of recombinant DNA (Chung and Cohen 1974). The ability to create recombinant DNA molecules and identify and clone genes, combined with Armin Braun's (1941) pioneering studies on the nature of the crown-gall disease caused by *Agrobacterium tumefaciens*, eventually led to the use of this soil-borne natural vector for the transformation of plants (DeBlock et al. 1984, Horsch et al. 1984, 1985). More recently, the biolistics procedure, developed by John Sanford (Sanford et al. 1987) has proven to be a valuable tool for plant transformation, especially for those species that are not yet amenable to *Agrobacterium*-mediated transformation. Together, these two methods account for a majority of the transgenic plants produced, including many important crop species in which agronomically useful genes have been stably integrated.

These remarkable achievements in plant biotechnology come at an opportune time at the threshold of the 21st century, when humankind is facing catastrophic problems of food security and environmental degradation. The following discussion provides an overview of the enormity of the problems that we face, and how plant biotechnology can play an important role in their resolution (Vasil 1998).

The advent of agriculture nearly 10,000 years ago helped to sustain a steady increase in human population by ensuring a continuing and reliable supply of food. Despite age-old human concerns about balancing population and food supplies, and many population control measures, world population continues to grow at about 1.5% a year, and is projected to reach 8 billion by 2020, and 11 billion by 2050. Almost all of this growth will occur in the already overpopulated, underdeveloped and poorer regions of Africa, Asia and Latin America, which will be home to nearly 90% of the human population.

Food shortages were common during the early and middle years of the 20th century, but the introduction of Green Revolution varieties of wheat and rice, during the mid-1960's in Asia and Latin America, reversed this trend by keeping increases in food productivity slightly ahead of population growth. Remarkably, both China and India, the two most populous nations with chronic food shortages, became net food exporters during this period. Such increases in crop productivity can not be sustained indefinitely, and indeed have begun to decline. The average increase in grain yield of 2.1% a year maintained until 1990, has since declined to only 0.5% a year. At the same time, improved economic conditions in China and India have created a greater demand for better and more varied food products, particularly poultry and meat, which require greater supplies of feed grains (already nearly 35% of world grain production is used for livestock production). Thus it is not surprising that both China and India have become net importers of food in the last few years, and are together projected to import as much as 260 million tons of food grain annually by 2030. As a consequence of the increasing demand and declining productivity, world food reserves currently stand at 48 days - their lowest level since the beginning of the Green Revolution - from a previous high of 77 days. These are well below the minimum required for ensuring food security.

The international agricultural system is operating under considerable stress, and it can be expected to only worsen with food needs of the increasing population. With the current trends of population growth and agricultural production, the demand for food in the most populous parts of the World will double by the year 2025, and nearly triple by 2050. Increases in food productivity of this magnitude can not be brought about in such a short period of time by conventional breeding, or by increasing the amount of arable land, which accounts for 97% of all food production in the world. Indeed, arable land, which is finite and comprises only about 3% of the earth's surface, is rapidly deteriorating and decreasing as a result of water and wind-induced soil erosion, salinization, acidification, over cultivation, compaction, water logging, over grazing, etc. (a net loss of 1.4% of the total top soil each year). These factors, combined with expected increases in population, will actually decrease the global per capita arable land from the current 0.26 hectare to 0.15 hectare by the year 2050 (compared to 0.44 hectares in 1961). In addition, fresh water supplies, essential for modern high-input irrigation agriculture, are becoming severely limited by increased human and agricultural use, and polluted by agricultural run-off and widespread use of agrochemicals.

It is feared that the expected food shortages in the overpopulated parts of the world during the 21st century may lead to widespread social, economic and political unrest, making food security the single-most serious threat to international peace and security. Countries with efficient agricultural systems generally have high living standards, strong economies, lower rates of population growth, and democratic forms of government. Increasing food productivity in a sustainable manner will, therefore, not only provide adequate nutrition to the expanding humanity, but also will help to reduce population growth, protect the environment, promote economic development, and ensure social, political and economic stability.

The only long-term solution to the problem of food security would be to reduce population growth to zero. Admittedly, it is a complex problem involving political, social and economic factors and policies that is outside the remit of this article. In the hope that world population can be stabilized at 11 billion by about 2050, the challenge for the agricultural sector is to double food production by 2025, and triple it by 2050, on less per capita land, with less water, and under increasingly challenging environmental conditions. In reality, food demand may be even larger than the expected increase in population owing to changing dietary habits, particularly in China and India. Therefore, in regions like Africa - which did not benefit from the Green Revolution because of their primitive and inefficient agricultural systems – modern agricultural practices should be introduced to increase food productivity. The feasibility of this has already been demonstrated by the 200-400% increases in crop productivity brought about under the Sasakawa Global 2000 program (Borlaug 1997). In other parts of the world where modern high-input agriculture is already practiced widely, increases in crop productivity like those attained during the Green Revolution may not be feasible in the future, as many of the modern crop varieties are nearing their biological and physical limits of productivity with harvest indices approaching 0.5. Therefore, environmentally benign biotechnological methods, including molecular breeding based on DNA markers, should be used to complement and supplement traditional plant improvement programs to increase crop yields in regions where in spite of advanced agricultural technology and the heavy use of agrochemicals, 42% of crop productivity is lost to competition with weeds, and to pests and pathogens (Oerke et al. 1994). Additional 10-30% post-harvest losses occur due to a variety of factors, especially in the developing countries where storage conditions are poor.

During the 20th century traditional plant breeding has brought about enormous increases in crop productivity, thereby saving millions of hectares of forest and grasslands and wilderness areas, which support biodiversity and vital ecosystems, from being converted into farmland. However, plant improvement by hybridization is slow, and is restricted to a very small gene pool owing to natural barriers to crossability. Advances in plant biotechnology have helped to overcome these barriers by making it possible to transfer defined genes into all major food crops, including cereals, legumes, cassava, potato, and many vegetables and fruits (the market value of transgenic crops is expected to increase from 450 million dollars in 1995 to over seven billion dollars by 2005). The entire global gene pool - whether it be plant, animal, bacterial or viral - has become accessible for plant improvement. The first genes that have been integrated into crop species provide resistance to non-selective and environment friendly herbicides, and many pests and pathogens. Increasingly large acreages of twelve transgenic crops (cotton, canola, maize, soybean, potato, etc.) are being commercially grown for human use and consumption. The global acreage devoted to transgenic crops increased from 7 million acres in 1996 to over 30 million acres in 1997 (James 1997). In 1997, transgenic cotton, soybean, and corn accounted for 18%, 13% and 9%, respectively, of the national acreage in the United States, while 25% of the canola grown in Canada was transgenic.

Farmer acceptance of such crops is high, because of substantial savings in labor and costs related to marked reductions in the use of agrochemicals. Carefully planned introduction and monitoring of such crops, after due consideration and evaluation of possible biological and environmental risks, would greatly help in reducing the enormous crop losses attributed to weeds, pests and pathogens. The use of transgenic crops may indeed have a beneficial effect on the environment by significantly reducing the use of agrochemicals. Other genes for improving crop productivity, and manipulating starch/protein/oil quality and quantity, resistance to stresses such as temperature, drought, and salinity/metal toxicity, are also being isolated and studied. In the foreseeable future, these will be used to produce the second generation of transgenic crops.

Many other beneficial applications of biotechnology are under active development. Production of biodegradable plastics in transgenic plants may lead to substantial reduction in the use of petroleum-based plastics. Good progress is being made in the use of transgenic plants to produce therapeutic proteins and pharmaceuticals, and even edible vaccines. Engineered plants have been shown to be useful in phytoremediation to decontaminate soils containing heavy metals and other toxic substances. Manipulation of photosynthetic efficiency and flowering time, and source/sink relationships, could be used in the future to increase crop yields. The achievement of these objectives will require a much better understanding of the molecular and genetic basis of plant developmental processes, and the discovery of agronomically important genes and the elucidation of their regulatory mechanisms. This task is expected to be greatly facilitated by the international initiatives already underway for the mapping of plant genomes.

In several countries, particularly in Europe, there is considerable opposition to the cultivation and use of transgenic crops as human food. There are serious concerns about the unintended transfer of transgenes to wild and weedy relatives, and possible adverse biological and environmental effects of transgenic crops. These concerns should be carefully considered and addressed in an open, scientific and responsible manner, and strict guidelines should be developed for the field release, monitoring and human use of transgenic plants/products. It is in the interest of the biotechnology community to take an active role in informing the public about the safety and advantages of transgenic crops in order to encourage their acceptance and use, and to reverse the often negative perception of biotechnology. Gene transfer through biotechnology may indeed turn out to be safer and more precise than traditional plant breeding, as it involves the transfer of a very limited number of defined genes with known gene products, rather than the random transfer of thousands of genes with little or no information about their protein products.

Over the next decade a wide variety of transgenic crops will become an important and routine component of agricultural systems in the industrialized countries, for whom they are a luxury rather than a necessity. Transgenic crops will also be used for many industrial and pharmaceutical purposes. Introduction of such crops into the developing countries, where they are needed and will have the most impact, will be slow and largely through the efforts of multinational biotechnology companies, which are establishing biotechnology R & D centers in Brazil, China, India, etc. This is because most developing countries currently lack the scientific and industrial infrastructure to develop and exploit the full extent of these technologies independently. Access of the developing countries to important genes, plasmids/vectors, selectable markers, transformation technologies, etc., is severely limited, because they are expensive and are held by public or private institutions in the industrialized countries. In the longer term, nevertheless, much of the increase in food production to meet the dual challenge of population growth and food demand must of necessity occur in the developing countries on land already under cultivation. If successful, this will impact greatly on their overall economic development, which will help control the relentless increases in population.

International efforts to transfer biotechnology to developing countries have either failed or been largely ineffective. The best long-term solution to this problem would be to improve scientific and technical manpower and infrastructure in the developing countries so that they can themselves make full use of biotechnology. Teaching and training in plant molecular biology and genetics should be made an integral part of high school, college and university curricula. This will initially require collaborations with western academic institutions and international financial assistance. In the past, some of the International Agricultural Research Centers (IARC) have played a critical role in the development of high yielding and superior crop varieties. They should now be encouraged to complement and supplement their breeding programs by adopting methods of molecular breeding, based on DNA-based markers, for more efficient and rapid development of new varieties. At a time when the IARCs are facing substantial budget cuts, it would be inadvisable and impractical to impose any other major biotechnology or molecular biology programs on them. Rather, collaborations in selected priority areas should be encouraged between the IARCs and public and private research groups.

International agriculture in the latter half of the 20th century was dominated by the Green Revolution. Considering the power and many advantages of plant biotechnology, Gene Revolution may well dominate the agriculture of the 21st century. It will be important to remember, however, that plant biotechnology will not be the magic bullet that will solve the problem of food security forever. Rather, if developed and used wisely, it will at best provide a grace period of about 50 years in which to control population growth and finally discover the long-sought road to sustainable development.

References

Borlaug NE (1997) Plant Tissue Cult Biotech 3: 119-127

Chung ACY and Cohen SC (1974) Proc Nat Acad Sci USA 71: 1030-1038

Cocking EC (1960) Nature 187: 927-929

Braun AC (1941) Phytopathology 31: 135-149

DeBlock M, Herrera-Estrella L, Van Montagu M, Schell J and Zambryski P (1984) EMBO J 3: 1681-1689

Darwin C and Darwin F (1880) The Power of Movement in Plants. London

Fehér A and Dudits D (1994) In: Vasil IK and Thorpe TA (eds), Plant Cell and Tissue Culture, pp 71-118. Kluwer, Dordrecht

Gautheret RJ (1939) CR Hebd Seances Acad Sci 208: 118-120

Gautheret RJ (1985) In: Cell Culture and Somatic Cell Genetics of Plants, Vol 2, Cell Growth, Nutrition, Cytodifferentiation and Cryopreservation (Vasil IK, ed), pp 1-59. Academic Press, Orlando

Guha S and Maheshwari SC (1964) Nature 204: 497-498
Haberlandt G (1902) Akad Wiss Wien, Math-Naturwiss Kl, Abt 1 111: 69-92
Hoecker U, Vasil IK and McCarty DR (1995) Genes & Dev 9: 2459-2469
Horsch RB, Fraley RT, Rogers SG, Sanders PR, Lloyd A and Hoffmann N (1984) Science 223:496-498
Horsch RB, Fry J, Hoffmann N, Wallroth M, Eichholtz D, Rogers S and Fraley R (1985) Science 227:1229-1231
James C (1997) ISAAA Briefs No 5. ISAAA, Ithaca, NY
Krikorian AD and Berquam DL (1969) Bot Rev 35: 59-88
Larkin PJ and Scowcroft WR (1981) Theor Appl Genet 67: 443-455
McCarty DR, Hattori T, Carson CB, Vasil V, Lazar M and Vasil IK (1992) Cell 66: 895-905
Morel G (1960) Amer Orchid Soc Bull 29: 495-497
Morel G and Martin C (1952) CR Hebd Seances Acad Sci 235: 1324-1325
Nobécourt P (1939) CR Seances Soc Biol Ses Fil 130: 1270-1271
Oerke E-C, Dehne H-W, Schönbeck and Weber A (1994) Crop Production and Crop Protection. Elsevier, Amsterdam
Power JB, Cummins SE and Cocking EC (1970) Nature 225: 1016-1018
Reinert J (1958) Naturwiss 45: 344-345
Sanford JC, Klein TM, Wolf ED and Allen N (1987) Part Sci Tech 5: 27-37
Schleiden MJ (1838) Müllers Arch Anat Physiol, pp 137-176
Schwann T (1839) Mikroscopische Untersuchungen über die Übereinstimmung in der Struktur und dem Wachstum der Thiere und Pflanzen. No. 176. Oswalds, Berlin
Skoog F and Miller CO (1957) Symp Soc Exp Biol 11: 118-131
Steward FC, Mapes MO and Mears K (1958) Amer J Bot 45: 705-708
Takebe I, Labib C and Melchers G (1971) Naturwiss 58: 318-320
Thimann KV (1977) Hormone Action in the Whole Life of Plants. Univ of Massachusetts Press, Amherst
Thorpe TA, ed (1995) In Vitro Embryogenesis in Plants. Kluwer, Dordrecht
Vasil IK, ed (1986) Cell Culture and Somatic Cell Genetics of Plants, Vol 3, Plant Regeneration and Genetic Variability. Academic Press, Orlando
Vasil IK (1998) Nature Biotechnology 16:399-400
Vasil IK and Thorpe TA, eds (1994) Plant Cell and Tissue Culture. Kluwer, Dordrecht
Vasil IK and Vasil V (1992) Physiol Plant 85: 279-283
Vasil V, Clancy M, Ferl RJ, Vasil IK and Hannah LC (1989) Plant Physiol 91: 1575-1579
Vasil V and Hildebrandt AC (1965) Science 150: 889-890
Watson JD and Crick FHC (1953) Nature 171: 737-738
White PR (1934a) Plant Physiol 9: 585-600
White PR (1934b) Phytopath 24: 1003-1011
White PR (1939) Amer J Bot 26: 59-64
White PR (1963) The Cultivation of Animal and Plant Cells. Ronald Press, New York

PHARMACEUTICAL FOODSTUFFS: ORAL IMMUNIZATION WITH TRANSGENIC PLANTS

Tsafrir S. Mor and Charles J. Arntzen
Boyce Thompson Institute for Plant Research,
Tower Rd. Ithaca, NY 14853, USA

1. Past: The concept of edible vaccines

Vaccination is one of the most cost-effective medical interventions available. The public health triumphs of this century, such as the eradication of smallpox and the near eradication of polio, can be attributed in part to the success of the traditional or "Jennerian" vaccine technology. These effective vaccines consist of either killed or, more often, attenuated strains of the pathogen which are delivered by injection (the Sabine oral polio vaccine being an exception). Unfortunately, however, this approach cannot be extended to many pathogens which are expensive or impossible to culture.

The new generation of biotechnology-derived subunit vaccines holds the greatest promise for global effective reduction of infectious diseases in the next century. Creation of these vaccines involves transferring a gene from human infectious agents, such as viruses or bacteria, into a transgenic host. The transgenic organism then produces a "subunit" of the infectious agent -- a protein which cannot cause disease but which does contain the antigenic fingerprint of the authentic disease-causing agent. In most studies over the last 15 years these subunit vaccines have been purified from the transgenic "production hosts" (such as cultured yeast cells) and injected into vaccinees to induce immunity against a specific disease.

Effective control and possible eradication of epidemic diseases depend on global vaccination plans involving both developed and developing nations. However, not only the manufacturing costs of many vaccines make them prohibitively expensive for the poorer countries of the world, but also the distribution and administration of these vaccines nearly always depend on expensive infrastructure and equipment that limit their use in the third world. A new breakthrough is needed to introduce changes in the way subunit vaccines are produced, distributed and administered to make them available on a global scale.

With this goal in mind, the idea to produce subunit vaccines in edible tissues of transgenic crop plants first emerged (reviewed by Mason, Arntzen 1995; Mor et al. 1998). The attractive feature of transgenic plants is that they combine a cost-effective production system with a safe and efficacious delivery system. Plants are the cheapest source of proteins and potentially also the cheapest source of transgenic proteins. Furthermore transgenic plant technology is fast spreading to developing countries where effective but inexpensive vaccines are most needed. Moreover, being "edible", the vaccines do not need to be purified from the plant tissues thus reducing production costs even further.

Edible plant vaccines are aimed at sensitizing the gut associated lymphoid tissue, which is part of the common mucosal immune system (Mestecky et al. 1997). Responses include the appearance of specific antibodies in blood serum and of specific secretory antibodies in the mucosal secretions of the intestines (and also in saliva, respiratory and reproductive

A. Altman et al. (eds.), Plant Biotechnology and In Vitro Biology in the 21st Century, 17–20.

tract secretions). Secretory antibodies are very important for the prevention of infection by the myriad of microbes that attack at mucosal surfaces (respiratory, enteric, or sexually transmitted diseases). While secretory antibodies can be induced by oral (and nasal) immunization, vaccination by injection rarely induces them. Therefore edible vaccines against mucosal infections are predicted to be efficacious. In addition, becuase they circumvent the need for sterile syringes (scarce in developing countries) they would also be safer than injected vaccines.

Over the last seven years since our group introduced the subject, we, as well as others, have been evaluating the use of transgenic plants as vaccine vectors . Studies using prototype antigens of viral and bacterial origins have verified that immunogenic proteins can be produced in plants and that they can trigger immune responses if the plant tissues which express them are consumed as food. Here will present a summary of research progress concerning the expression of the heat-labile toxin B subunit of *Escherichia coli* (Haq et al. 1995; Mason et al. 1998; Tacket et al. 1998). This research culminated recently with the successful completion of human clinical trials (Tacket et al. 1998), the first government approved use of a pharmaceutical derived from transgenic plants.

2. Present: The proof of concept

Enterotoxigenic *E. coli* (ETEC) and *Vibrio cholerae* are the primary causes of bacterial gastroenteritis. Both bacteria initiate disease by colonizing the intestinal epithelia and both produce enterotoxins, responsible for the diarrheal symptoms. The heat-labile enterotoxin (LT) of ETEC is closely related to the cholera toxin (CT). These toxins consist of an A subunit (LT-A or CT-A), which contains the toxic ADP-ribosylation activity, and five B subunits (LT-B or CT-B), which self-assemble into a doughnut-like pentameric ring structure that has no toxic effect by itself (Sixma et al. 1991). The B subunit pentamer, which has a high affinity for GM_1 gangliosides present on the surfaces of gut epithelial cells, binds the A subunit and facilitates its entry into these cells. There is currently no reliable and effective vaccine against either ETEC or cholera. There is, however, evidence suggesting that oral vaccination can induce the production of mucosal antibodies against LT-B or CT-B, which can neutralize the toxicity of the respective holotoxins by preventing its binding to gut cells (Arakawa et al. 1997; Clemens et al. 1990; Mason et al. 1998).

Several years ago, our group has proved that LT-B could be expressed in edible tissues of transgenic plants and could assemble into pentameric ring (Haq et al. 1995). Most importantly, LT-B expressed in plants could induce the production of both serum and secretory antibodies in mice fed with transgenic potato tubers. The response was specific, as mice fed with control tubers did not exhibit any LT-B-specific antibody response (Haq et al. 1995).

In a subsequent publication from our laboratory (Mason et al. 1998), the degree of protective immunity afforded by oral immunization with plant-derived LT-B was investigated in a surrogate challenge (Richardson et al. 1984). Mice that were fed with potato tubers expressing LT-B and then challenged by gavage of CT accumulated less fluid in their small intestines as compared to the control.

In complementary experiments, Arakawa et al. reported the expression and assembly of CT-B in transgenic potato tubers (Arakawa et al. 1997; Arakawa et al. 1998). As reported in the LT-B experiments (Haq et al. 1995; Mason et al. 1998), the plant-derived CT-B induced significant levels of neutralizing anti-CT-B immunoglobulins in orally immunized mice (Arakawa et al. 1998). In addition, the immune response against CT-B could be boosted by further oral doses (Arakawa et al. 1998). These results suggest that, in addition to being useful for primary vaccinations, edible vaccines may be an efficient means of boosting immune responses after oral or even parenteral vaccination.

These encouraging results provided the motivation to request approval of the U.S. Food and Drug Administration for human clinical trials with transgenic potato tubers expressing LT-B. Approval was granted in 1997

The crucial research adavance which made possible the human clinical trials was the production of adequate levles of expression of LT-B in potato tubers. Our earlier studies had shown only levels of expression of the antigen in the tissue (Haq et al. 1995), which implied that people would have to eat an unreasonably large amount of tuber to receive the desired dose. Two strategies have been used to enhance the expression and accumulation of LT-B. First, it was found that a plant cell localization signal could be fused to the antigenic protein to cause increased cellular accumulation (Arakawa et al. 1998; Haq et al. 1995). More recently, a completely synthetic gene for LT-B was created to give a "plant-optimized" gene. The synthetic gene encodes a protein with the native amino acid sequence, but utilizes codons preferred by plant cells, and lacks problematic sequences such as cryptic plant-specific polyadenylation signals in the bacterial gene (Mason et al. 1998). The protein product of the synthetic gene accumulated to ~0.15% of soluble protein, which was more than 10 fold higher than the best levels obtained previously (Haq et al. 1995). These levels of expression allowed this candidate edible vaccine to be tested in humans, the first human clinical trial of a plant-derived vaccine (Tacket et al. 1998).

In contrast to individuals that consumed untransformed potato tubers, volunteers that consumed raw potato tubers expressing LT-B (0.5-1.0 mg per dose) developed specific anti-LT-B mucosal and systemic immune responses. These responses are comparable to those observed when humans are challenged with 10^9 ETEC bacteria (Tacket et al. 1998). The human clinical trials demonstrate that edible plant vaccines are immunogenic in humans as was previously shown in mice; proving that they can protect humans against a challenge is the next logical step.

3. Future: Practical applications and more challenges

We and others have demonstrated that plants can produce antigenic proteins of various viral and bacterial pathogens and that these proteins can elicit immune responses and even protection by oral and/or parenteral administration. Among those antigens already expressed in plants one finds the hepatitis B surface antigen (Mason et al. 1992; Thanavala et al. 1995), the glycoprotein of rabies virus (McGarvey et al. 1995), the VP1 subunit of foot-and-mouth disease virus (Carrillo et al. 1998), the coat protein of Norwalk virus (Mason et al. 1996), and the LT-B and CT-B subunits of ETEC and *V. cholerae*. (Arakawa et al. 1997; Arakawa et al. 1998; Haq et al. 1995; Mason et al. 1998; Tacket et al. 1998). Progress is being made toward expression of such complex entities as virus like particles of rotavirus (Mor et al., this volume) and the LT holotoxin (Mason et al., unpublished). Furthermore, other mucosal immunogens are currently being evaluated for production in plants, including antigens of pathogens of the respiratory and reproductive tracts.

With the success of the potato LT-B vaccine in the clinical trials, and with the prospective clinical trial of a plant derived recombinant Norwalk virus vaccine, it seems as if the idea of edible plant vaccines has come of age. Although protection assays have only been performed on experimental animals so far, we anticipate that these conclusions will be substantiated by human clinical trials. The usefulness of the concept from a biotechnological point of view is evident by the active interest of various government agencies, not-for-profit organizations and biotechnology companies to pursue commercialization of plant vaccines.

However, there remain some very challenging basic questions to be answered. Will "non-traditional" oral subunit vaccines function at effector sites other than the gut? Is it possible that oral tolerance could develop with food-borne antigens (Garside, Mowat 1997)? Lastly, what is the most appropriate plant tissue to deliver subunit vaccines? Bananas, which are grown in many developing countries seems and the fruit of which is eaten uncooked even by infants, seem to be an especially attractive system (Mor et al. 1998). While many unanswered questions can be posed, we feel that they are implementation issues -- not roadblocks. Vaccines produced in plants offer a new strategy for safe and cost-effective global immunization against infectious diseases.

4. Acknowledgments

We would like to thank our many colleagues whose work we summarized here, including Drs. Hugh Mason, Liz Richter, Gregory May, Kenneth Palmer, Yasmin Thanavalla, John Clements, Mary Estes, Carol Tacket and Myron Levine. Our research is supported by the Thrasher Research Fund, NIH, WHO and NSF. T.S. Mor is supported by a Postdoctoral Award (No. FI-251-97) from The United States – Israel Binational Agricultural Research and Development Fund (BARD).

5. References

Arakawa T et al (1997) Transgenic Res. 6, 403-413.
Arakawa T et al (1998) Nat. Biotechnol. 16, 292-297.
Carrillo C et al (1998) J. Virol. 72, 1688-1690.
Clemens JD et al (1990) Lancet 335, 270-273.
Garside P, Mowat AM (1997) Crit. Rev. Immunol. 17, 119-137.
Haq TA et al (1995) Science 268, 714-716.
Mason HS, Arntzen CJ (1995) Trends Biotechnol. 13, 388-392.
Mason HS et al (1996) Proc. Natl. Acad. Sci. USA 93, 5335-5340.
Mason HS et al (1998) Vaccine 16, 1336-1343.
Mason HS et al (1992) Proc. Natl. Acad. Sci. U. S. A. 89, 11745-11749.
McGarvey PB et al (1995) Bio/Technology 13, 1484-1487.
Mestecky J et al (1997) Behring. Inst. Mitt. 98, 33-43.
Mor TS et al (1998) Trends Microbiol. in press.
Richardson SH et al (1984) Infect. Immun. 43, 482-486.
Sixma TK et al (1991) Nature 351, 371-377.
Tacket CO et al (1998) Nat. Med. 4, 607-609.
Thanavala Y et al (1995) Proc. Natl. Acad. Sci. U. S. A. 92, 3358-3361.

GENETIC CONTROL OF REGENERATION WAS ALTERED DURING ONE-WEEK RIPENING OF IMMATURE MELON COTYLEDONS ON LIQUID/MEMBRANE SYSTEM

Jeffrey Adelberg and Jin Feng Chen
Department of Horticulture
Clemson University, Clemson SC 29634 USA

1. Introduction

Genotypic variation is a large factor affecting the success of regeneration systems in many species, including melon (*Cucumis melo* L.). Genotypic variation obviously has a genetic component, but regeneration response is a phenotype with strong interactive environmental factors. Recently we observed heterosis and cytoplasmic factors combine with Mendelian genetics to create the interactive environmental component for genotypic variation in regeneration of melon (Adelberg et al. 1998). We have followed that work with observations of regeneration following a one-week in vitro "ripening" treatment of immature cotyledons. Regeneration was quantified to describe differences in genotypic expression of regeneration under the influence of explant conditioning from the same three inbreds and cross combinations.

2. Materials and Methods

Self-pollination and reciprocal crosses of greenhouse-grown inbred lines of 'Ogen' ('OG'), 'Rocky Ford Green flesh' ('RFG'), and 'Green Ice' ('GI') were made by hand-pollination. Fruits were harvested between three to four weeks after pollination, when cotyledons filled the seed cavity, and seed coats were still pliable for dissection. Immature cotyledons were aseptically dissected from the seed by cutting off one third of seed at the pointed end eliminating the embryonic axis, while pushing from the blunt end of the seed with forceps forcing the cotyledons to slide out of the seed coat. The two cotyledons were then separated and placed ad axially, six cotyledons per Magenta GA7 vessel (Magenta Corp., Chicago IL). Immature cotyledons were placed on a membrane raft (Osmotek Ltd., Rehovot Israel) on liquid medium (Murashige and Skoog, 1962) containing 3.0% sucrose, 10 μM BA, organic acids and vitamins at pH 5.6 (Adelberg et al. 1994) or 6% sucrose with 10μM BA. This one week of culture prior to placement on agar regeneration media is referred to as one-week ripening. Cotyledons were moved to semi-solid agar regeneration medium (as described above with 0.7% Sigma TC agar).

A. Altman et al. (eds.), Plant Biotechnology and In Vitro Biology in the 21st Century, 21–24.

After five weeks, cotyledons were dissected under aseptic conditions, removing all visible regenerant buds. Pubescent, spherical tissues of light green color formed by periclinal growth on the epidermal surface were distinguished as buds. This experiment was a completely randomized design with six vessel replications (six cotyledon each) for each inbred and reciprocal cross-combination, for the three inbred cross families. The entire factorial was analyzed by ANOVA for both response variables. Interaction of genotype ripening solutions was analyzed by grouping genotypes into three families, of two inbreds and their hybrids. Tukey's Honestly Significant Difference distinguished responses within a family for a particular microenvironment. All data analysis was done in JMP (SAS Institute, Cary NC).

3. Results

Percentage regeneration was not affected by genotype or the ripening treatment used (Table 1). Comparing these data with previous work on conventional agar media, the one-week ripening of tissue on liquid/membrane system did not effect regeneration rate. Approximately 71% of immature cotyledons regenerated on agar media (Adelberg et al. 1998), compared to 78% regenerated following one week transfer from MS with 10 μM BA and 3% sucrose, and 64% of cotyledons regenerated following one week transfer from 6% sucrose solution with 10 μM BA. These differences were not significant and we concluded that this one-week transfer technique results in explants of adequate quality to undergo regeneration.

'Ogen' has been previously described as a recalcitrant genotype (Orts et. al 1987). Following the one-week ripening, 'Ogen' is as equally likely to regenerate as the other two inbreds, 'GI' or 'RFG'. The one-week ripening treatment, therefore can facilitate regeneration of recalcitrant genotypes. When counting numbers of buds per cotyledon, there was a large genotypic difference in regeneration rate (Table 1). The choice of ripening solution did not affect the number of buds regenerated, but the strong interactive effect shows that different genotypes responded to these solutions quite differently.

The inbred that produced the least buds following one-week ripening was 'RFG' (Table 2). Hybrids of 'Ogen' and 'RFG' performed better than either parent, but the heterotic response was significant only following the 6% sucrose ripening treatment. (The reciprocal hybrid was not successfully made that season). When 'Ogen' was crossed with 'GI' , more buds per cotyledon were recovered when 'Ogen' was the maternal parent. Heterosis was significant, but dependent on cytoplasmic factors from 'Ogen'. 'GI', as a parent, also improved the regenerative response of hybrids with 'RFG'. When cotyledons were ripened on media, hybrids from both reciprocal crosses produced more buds per cotyledon then either parent. When ripened on 6% sucrose solution, hybrid cotyledons from both reciprocal crosses produced intermediate numbers of buds to the parents. In the cross of 'GI' and 'RFG', heterosis was only expressed after ripening on media and cytoplasmic factors were not involved.

Table 1. Sum of squares from analysis of variance for the effects of genotype and ripening solutions on the regeneration and shoot growth of melon.

Source of variation	d.f.	Percentage Regeneration	Buds per cotyledon
Genotype (G)	7	2.09ns	21309.16***
Ripening solution (R)	1	0.53ns	86.48ns
G x R	7	5.25ns	4569.60***
Error	316	250.73	41042.17

***=significant at the 0.001 probability level; ns=not significant

4. Discussion

The interaction between genotype and environment played a significant role in determining regenerative ability of explants. In our prior work with the regeneration of melon from the same families of cross-combinations (Adelberg et al. 1998), microenvironment, (e.g. liquid media versus agar) altered the expression of genotypic differences and genetic control. In the present work, the physical microenvironment was similar for both ripening treatments, whereas 6% sucrose has an osmotic pressure of approximately -470 kPA and MS media with 3% sucrose has an osmotic pressure of approximately -450 kPA. The liquid membrane/system does not have a matrix component and mass flow of water across membranes would be nearly equal in the two ripening treatments.

Cotyledons are a commonly used explant source. Immature cotyledons of melon have high regenerative capacity once the cotyledons have enlarged to fill the seed cavity (Adelberg et al 1994). In a ripening fruit at this stage of maturity, embryo fresh weight starts to decrease as dry weight increases (Welbaum and Bradford 1988). Cotyledon tissue at this stage of development is nearly isotonic with the ripening solutions we used. During embryo maturation *in vivo*, sucrose, and to lesser extent minerals, flows from the plant into the embryo. Since embryos were selected from the same fruit and placed in near-isotonic microenvironments, differences in genetic control of regeneration altered during this one-week ripening process was more likely influenced by sucrose concentration or media salts than mass flow of water.

In application, a short physiological conditioning treatment of an explant will alter the way genotypic variation is expressed. Previously recalcitrant genotypes, like OG, may become more regenerative. A genetic component of genotypic variation is heritable, and hybrids of 'OG' regenerate well in this system. Non-uniform regeneration of reciprocal crosses indicate cytoplasmic factors affect heritability of regeneration for many crops, including melon (Adelberg et al 1998; Oridate et al 1992; Molena and Nuez 1995). Physiological conditions during embryo development has a large interactive effect on the expression of cytoplasmic factors affecting regeneration of wheat (Lazar et al 1984; Mathias et al 1986). Heterosis and cytoplasmic factors often modify genetic responses by interactions with the environment. Our research continues to question which environmental factors modify the genetic control of regeneration for melon.

Table 2. Mean performance for regeneration Ogen (OG), Green Ice (GI), Rocky Ford Green, and their reciprocal hybrids.

Genotype	Ripening Solution	Percentage Regeneration	Buds per cotyledon
GI	Media	75a[z]	14.50a
GI x OG		83a	8.06c
OG x GI		100a	29.76b
OG		83a	14.92a
GI	Sugar solution	92a	24.04a
GI x OG		94a	13.89b
OG x GI		100a	24.90a
OG		63a	9.50b
GI	Media	75a	14.50a
GI x RFG		100a	30.67b
RFG x GI		92a	29.64b
RFG		100a	2.56c
GI	Sugar solution	92a	24.04a
GI x RFG		78ab	18.89a
RFG x GI		92a	19.58a
RFG		53b	6.44b
OG	Media	83a	14.92a
OG x RFG		--	--
RFG x OG		100b	19.42a
RFG		100b	2.56b
OG	Sugar solution	63a	9.50a
OG x RFG		--	--
RFG x OG		100b	23.00b
RFG		53a	6.44a

[z]Means separation within families for a particular ripening solution by Tukey's Honestly Significant Difference.

5. References

Adelberg JW et al (1998) In Vitro Cell Dev. Biol. (abstract) 34:45-A.

Adelberg JW et al (1994) HortSci 29:689-691.

Lazar MD et al (1984) Theor. Appl. Genet. 68:131-134.

Mathias et al (1986) Theor. Appl. Genet. 72:70-75.

Molina RV and F Nuez (1995) Plant Cell Tiss. Org. Cult. 43:249-257.

Oridate T et al (1992) Plant Cell Tiss. Org. Cult. 29:27-30.

Orts et al (1987) HortSci. 22:666.

Welbaum G and K Bradford (1988) J. Exp. Bot. 40:1355-1362.

DIFFERENTIAL HORMONAL REQUIREMENTS FOR CLONAL PROPAGATION OF MALE AND FEMALE JOJOBA PLANTS

Veena Agrawal, Surya Prakash and Shrish C. Gupta, Department of Botany, University of Delhi, Delhi-110007, India

1. Introduction

Simmondsia chinensis (Jojoba), a dioecious shrub of arid zone is emerging as an agronomically valuable crop. Its seeds store liquid wax (40-60% by dry weight), similar in properties to sperm whale oil and is thus extensively employed as a high pressure lubricant in heavy machineries, in pharmaceuticals as well as in cosmetic industries all over the globe. Till date, only one paper has appeared on clonal propagation of jojoba by Chaturvedi & Sharma (1989). They have mentioned that male and female plants respond alike in culture. In view of its great economic importance, it is imperative to develop protocols for *in vitro* clonal propagation of elite male and female plants of this taxon. Present paper reports the differential hormonal requirements for *in vitro* clonal propagation of physiologically mature male and female plants.

2. Materials and Methods

2.1 ***Research material :*** The twigs, 8-12 inches long, from 18 to 20-year-old elite male and female plants of jojoba growing at the National Bureau of Plant Genetic Resources (NBPGR) and Central Arid Zone Research Institute (CAZRI), Jodhpur, India, were collected, surface sterilized and cut into nodal segments for inoculation.

2.2 ***Culture medium and culture conditions :*** Various basal media like Gamborg et al's (B_5, 1968), Gresshoff & Doy's (GD, 1973), Murashige & Skoog's (MS, 1962), Knop's (Knop, 1865), Schenk & Hildebrandt's (SH, 1972), and Lloyd & McCown's (WPM, 1981) were tried. The media were also individually supplemented with three different cytokinins such as N^6-benzyladenine (BA); kinetin (Kn) and N^6(2-isopentenyl)adenine (2iP) and auxins like indole-3-butyric acid (IBA), indole-3-acetic acid (IAA), α-naphthalene acetic acid (NAA) and 2,4-dichlorophenoxy acetic acid (2,4-D). The media contained 3% sucrose as carbon source and was gelled with 0.8% agar for shoot proliferation and 0.5% for root induction. The pH of the medium was adjusted to 5.8 prior to autoclaving at 121°C and 15 lb psi pressure for 15 min. The cultures were incubated at 25$\pm$2°C and continuous light of 450-460 $\mu W/cm^2$ intensity emitted by cool and white fluorescent lights (40W Philips).

2.3 ***Analysis of data :*** Results have been expressed as mean $\pm$ S.D. The mean values were calculated on the basis of 24 replicates in each experiment, repeated once. Data were analyzed using analysis of variance (Anova). In each column the means followed by the same letters are not significantly different as calculated by Statistical Analysis System (SAS) at $P < 0.05$.

3. Results

3.1 ***Establishment of shoot proliferating cultures :*** The nodal segments of field grown male and female plants (Fig. 1a) did not elicit any response on all the six basal media tried. Incorporation of BA in MS or B_5 medium induced multiple shoots in very low percentage of explants. Best response of male

A. Altman et al. (eds.), Plant Biotechnology and In Vitro Biology in the 21st Century, 25–28.

explants was recorded at 10 μM BA (Fig. 1b) while for female, 20 μM was optimum (Fig. 1c). Such nodal explants became brown, therefore, further experiments were carried out on microshoots developed on original explants. Nodal segments of the microshoots differentiated multiple shoots in their axils in both male and female plants, but the percentage of responding cultures as well as number of shoots per explant varied significantly at any level of individual cytokinins. Multiple shoots in 80% of male explants developed with an average of 3.5 ± 1.8 shoots per explant on 10 μM BA (Table 1; Fig. 1d) while 100% cultures of the female responded on 20 μM BA, with an average of 4.7 ± 2 shoots per explant (Table 1; Fig. 1g). The average number of shoots at 10 μM BA was comparatively less in female (Fig. 1e). The quantity of callus formed at the basal cut ends was more in male explants (Fig. 1f) in comparison to those in female (Fig. 1g). Both the percentage of explants developing multiple shoots and average number of shoots per explant decreased significantly with increase in the levels of BA in both the sexes, however, their responses differed remarkably (Table 1).

The other cytokinins like Kn and 2iP did not improve the morphogenic response over that of BA neither in male nor in female explants. A maximum of 31% explants induced multiple shoots at 20 μM Kn in female cultures whereas only 6% responded at the same level in male (Table 1). All its other concentrations (1, 5, 10, 30, 40 & 50 μM) failed to induce multiple shoots in male while in female the percentage of explants developing multiple shoots and average shoot number were comparatively much higher (Table 1). Akin to Kn, 2iP too, did not prove effective for inducing multiple shoots especially in male explants. On 50 μM 2iP multiple shoots were induced in 19% cultures with an average of 1.7 ± 0.9 shoots per female explant but at its lower concentrations, the results were insignificant. Incidentally, at the same level (50 μM) the male explants did not develop multiple shoots even in a single culture. Among all the cytokinins tried, 2iP was most effective for better elongation of shoots (Table 1).

3.2 *Rhizogenesis* : Nearly 80% *in vitro* raised shoots induced roots if a pulse treatment of 50 μM IBA for 20 minutes was given to female and 40 minutes to male plant regenerants, prior to their transfer to semi-solid MS medium supplemented with 10 μM IBA + 1 μM BA + 0.5% activated charcoal (Fig. 1h).

3.3 *In vitro hardening and soil transfer* : The plantlets were removed from the medium, washed thoroughly under running tap water and dipped in 0.2% bavistin (fungicide) solution for 15 min. These plantlets were transferred to soil and sprayed periodically with water, covered with polybags to ensure high humidity. The plantlets transferred to soilrite are thriving well for the last five months (Fig. 1i).

4. Discussion

Present investigations clearly reveal that the nodal explants of male and female jojoba plants behave differentially under the same level of cytokinins, when used alone or in combination with auxins in culture media (Prakash et al, unpublished). This is in contrast to the earlier observation (Chaturvedi, Sharma 1989), wherein they have reported that male and female plants of *S. chinensis* responded alike in culture. Requirement of high levels of BA by female explants over the male indicates that the endogenous level of this cytokinin may be low in the former. In males, the optimum response was at 10 μM BA where 80% explants developed multiple shoots, whereas, 20 μM was best for female nodes for differentiating multiple shoots in 100% cultures. Interestingly, at lower levels of BA none of the explants developed multiple shoots in male while in female the response was better (38%). This further shows the differential morphogenic responses of male and female explants at the same level of BA. While comparing the efficacy of three cytokinins (BA, Kn, 2iP), the kinetin and 2iP elicited very poor responses in both male and female explants. However, Kn and 2iP elicited better response in female explants over male, in terms of shoot production and average number of shoots per explant. Our observations of differential hormonal requirements by male and female explants find support from Mehra & Cheema (1985) for a dioecious tree taxon *Populus*

Table 1: Differential morphogenic responses of male and female nodal explants of *S. chinensis* plants on MS medium supplemented with different cytokinins, after two months of inoculation

Cyto-kinins (µM)	Explants developing shoots (%) Female S*	Female M**	Male S*	Male M**	Average number*** of shoots per explant Female	Male	Average length*** of shoots (in cm) Female	Male
BA								
0	36^{c}	0^{e}	0^{d}	0^{e}	0.41 ± 0.6^{g}	0.0 ± 0.0^{e}	1.08 ± 0.4^{e}	0.0 ± 0.0^{e}
1	61^{a}	38^{c}	0^{d}	0^{e}	2.47 ± 0.6^{d}	0.0 ± 0.0^{e}	3.21 ± 1.5^{a}	0.0 ± 0.0^{e}
5	38^{bc}	57^{b}	41^{ab}	33^{b}	2.70 ± 0.7^{c}	2.0 ± 2.0^{b}	2.26 ± 1.0^{b}	1.1 ± 0.6^{b}
10	11^{d}	88^{a}	18^{c}	80^{a}	3.03 ± 0.7^{b}	3.5 ± 1.8^{a}	2.12 ± 1.3^{bc}	1.3 ± 1.4^{a}
20	0.0^{d}	100^{a}	52^{a}	16^{cd}	4.74 ± 2.0^{a}	1.3 ± 0.6^{c}	1.90 ± 1.1^{bcd}	1.2 ± 0.9^{ab}
30	56^{ab}	43^{bc}	52^{a}	19^{c}	2.47 ± 0.7^{d}	1.3 ± 0.9^{c}	1.73 ± 1.2^{cd}	1.1 ± 0.7^{ab}
40	69^{a}	29^{cd}	47^{ab}	6^{ad}	2.20 ± 0.6^{e}	0.9 ± 0.9^{d}	1.77 ± 0.9^{cd}	0.8 ± 0.7^{c}
50	72^{a}	19^{d}	33^{bc}	4^{e}	1.76 ± 0.9^{f}	0.6 ± 0.7^{d}	1.48 ± 0.9^{de}	0.6 ± 0.3^{d}
Kn								
1	52^{bc}	4^{cd}	13^{b}	0^{b}	0.81 ± 0.8^{bc}	0.12 ± 0.0^{c}	1.44 ± 0.9^{b}	1.73 ± 0.8^{a}
5	86^{a}	6^{bcd}	10^{b}	0^{b}	1.40 ± 0.5^{b}	0.14 ± 0.2^{bc}	2.82 ± 1.3^{a}	1.02 ± 0.4^{ab}
10	52^{bc}	14^{bc}	43^{a}	2^{b}	1.60 ± 0.9^{ab}	0.64 ± 0.4^{a}	2.06 ± 0.8^{ab}	1.05 ± 1.0^{ab}
20	59^{bc}	31^{a}	27^{ab}	6^{a}	2.35 ± 1.0^{a}	0.41 ± 0.2^{ab}	1.64 ± 0.8^{b}	0.96 ± 1.2^{ab}
30	66^{ab}	10^{bcd}	24^{ab}	0^{b}	1.24 ± 0.7^{bc}	0.24 ± 0.2^{bc}	2.01 ± 0.8^{ab}	0.85 ± 0.4^{ab}
40	51^{ab}	17^{b}	23^{ab}	0^{b}	0.97 ± 1.0^{bc}	0.24 ± 0.2^{bc}	2.00 ± 1.0^{ab}	1.03 ± 0.5^{ab}
50	66^{ab}	12^{bc}	14^{b}	0^{b}	1.29 ± 0.8^{b}	0.16 ± 0.1^{bc}	1.60 ± 1.2^{b}	0.64 ± 0.5^{ab}
2iP								
1	79^{cd}	0^{d}	87^{ab}	0^{b}	0.83 ± 0.4^{bc}	0.97 ± 0.6^{c}	2.55 ± 1.2^{d}	3.52 ± 3^{e}
5	90^{ab}	2^{cd}	85^{b}	0^{b}	1.02 ± 0.4^{b}	1.12 ± 0.5^{c}	3.77 ± 1.5^{c}	4.23 ± 3^{bc}
10	98^{a}	0^{d}	75^{c}	6^{a}	1.50 ± 0.7^{a}	1.41 ± 0.6^{b}	4.73 ± 2.8^{ab}	4.14 ± 2^{bc}
20	83^{bcd}	8^{b}	75^{c}	0^{b}	1.79 ± 0.9^{a}	0.95 ± 0.5^{c}	5.38 ± 4.4^{a}	4.56 ± 3^{ab}
30	94^{a}	4^{c}	94^{c}	2^{ab}	1.72 ± 0.9^{a}	1.91 ± 0.9^{a}	4.62 ± 3.2^{b}	4.11 ± 4^{bc}
40	85^{bc}	8^{b}	91^{ab}	2^{ab}	1.54 ± 0.7^{a}	1.55 ± 0.7^{b}	4.65 ± 3.2^{ab}	5.24 ± 3^{a}
50	75^{d}	19^{a}	75^{c}	0^{b}	1.70 ± 0.9^{a}	1.41 ± 0.6^{b}	4.65 ± 3.3^{ab}	4.85 ± 4^{ab}

A minimum of 24 cultures were recorded for each treatment and experiment has been repeated once. Mean within the column followed by the same superscript are not significantly different as determined by SAS at $P < 0.05$. S*=single; M**=multiple; ***=Mean±S.D.

ciliata. However, they have found that the female explants responded better at lower levels of cytokinins wherein male did not show any morphogenesis. Dauphin-Guerin et al (1980) while working on another dioecious species, *Mercurialis annua* L. found that the endogenous levels of cytokinins are different in male and female plants. During their investigations they detected free base zeatin only in female while its nucleotides were found in males. In another study, Syzov & Syocev (1970) have recorded high levels of chlorophyll and carotenoid pigments in female especially during flowering season. Further, biochemical work on the quantitative and qualitative nature of the proteins in male and female plants is in progress.

5. Acknowledgements

Financial assistance of the University Grants Commission is gratefully acknowledged. Thanks are also due to Dr. Ashok Mittal for statistical analysis, as well as of Mr. K.J. Rout and Miss Shahnaj Shubhan for help.

6. References

Chaturvedi HC, Sharma M (1989) Pl. Sci. 63, 199-207.
Dauphin-Guerin B et al (1980) Planta 148, 124-129.
Gamborg OL et al (1968) Exp. Cell Res. 50, 151-158.
Gresshoff PM, Doy CH (1974) Z. Pflanzenphysiol. 73, 132-141.
Knop W (1865) Landw. Vers. Stn. 7, 93-107.
Lloyd G, McCown B (1980) Int. Pl. Prop. Soc. Proc. 30, 421-427.
Mehra PN, Cheema GS (1985) Phytomorphology 35, 151-154.

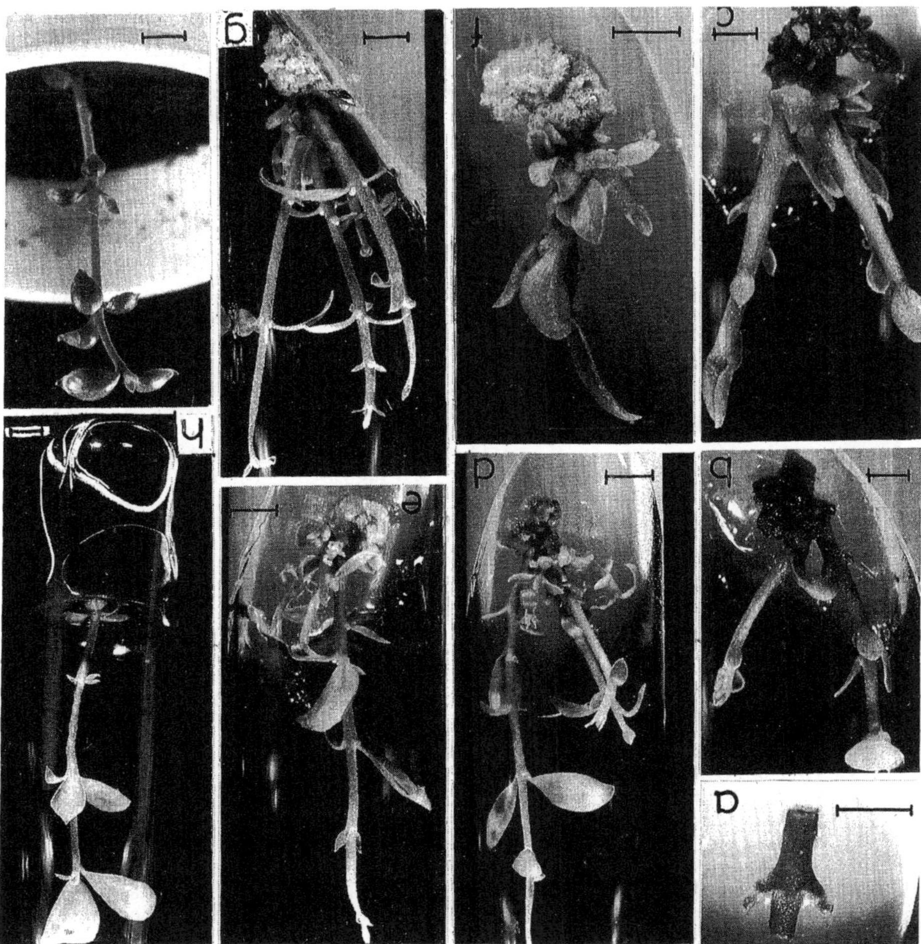

Fig. 1 (a-i) : Differential morphogenic responses in male and female plants of *S. chinensis*. (a), A nodal explant in MS basal medium (control) after 30 days of culture. (b-c), Development of multiple shoots in nodal explant of field - grown plants on MS + 10 μM BA in male (b) and on MS + 20 μM BA in female (c), after 90 days of culture (d-g), Nodal segments of microshoots, at 10 μM BA differentiating multiple shoots in male (d) and very few shoots in female (e); at 20 μM BA showing very low and stunted shoots in male (f) and maximum multiple shoots in female (g), after 60 days of culture. (h), Shoot inducing roots on MS + 10 μM IBA + 1 μM BA + 0.5% AC after pulse treatment of 50 μM IBA for 20 min in female and 40 min in male. (i), one month old tissue - culture raised plantlet growing in soilrite. Bars represent 5 mm.

Murashige T, Skoog F (1962) Physiol. Plant. 15, 473-497.
Schenk RU, Hildebrandt AC (1972) Can. J. Bot. 50, 199-204.
Syzov SS, Syocev MV (1970) Ukr. Bot. Zh. 27, 379-382.

SOMATIC EMBRYOGENESIS IN KOREAN GINSENG (*PANAX GINSENG* C.A. MEYER)

Kwang-Tae Choi, Yong-Eui Choi, Deok-Chun Yang and Ji-Chang Park
Korea Ginseng and Tobacco Research Institute, Taejon, Korea

1. Introduction

Korea ginseng *(Panax ginseng* C.A. Meyer) is perennial medicinal plant. Ginseng plants are troublesome for cultivation and require a period of more than 3 years to produce seeds. At time of seed harvest, zygotic embryos of Korean ginseng are still in an immature globular stage, thus the seeds require stratification and cold treatment for several months. Therefore, tissue culture procedures could contribute to clonal propagation and breeding in ginseng. In many reports of Korean ginseng tissue cultures, the regenerated plantlets from somatic embryos developed into multiple shoots without or inadequate roots, indicating low plant conversion (Butenko *et al.* 1968, Chang and Hsing 1980, Shoyama *et al.* 1988). However, little is known what is the main problem of ginseng plants not forming well-developed roots. The present paper deals with the regenerative ability of somatic multiple and single embryos arising directly from ginseng cotyledons on medium without any growth regulators.

2. Materials and Method

2.1. Induction of somatic embryos

The seeds were stratified in humidified sand to mature for several months at a 10℃ temperature since the zygotic embryos just after harvest were in an immature globular stage (at about 200 μm in length). Cotyledon explants from immature (1 mm in length), midmature (3 mm in length), fully mature zygotic embryo (6 mm in length), and seedling (60 mm in length) of *Panax ginseng* C.A. Meyer were cultured on Murashige and Skoog basal medium (MS, 1962) with 5% sucrose and 0.7% agar for induction of somatic embryos. The seeds were immersed in 70% ethanol for 1 min, then in 1% sodium hypochlorite solution for 1 h, and washed three times with sterile distilled water. After carefully dissecting the zygotic embryos from the seeds, the abaxial sides of the excised

A. Altman et al. (eds.), Plant Biotechnology and In Vitro Biology in the 21st Century, 29–32.

cotyledons were placed on the surface of MS basal medium (Murashige and Skoog 1962) with 5% sucrose, 0.7% agar. The medium was adjusted to pH 5.8 and then was autoclaved at 120℃ for 15 min. Cotyledon explants were cultured in 10×1 cm plastic petri dishes containing 30 ml of medium. The culture room was maintained at 24± 2℃ under 16:8 h photoperiods of 24 μmol $m^{-2}s^{-1}$ intensity under white fluorescent tubes. The frequency of somatic embryo production was determined after 2 months of culture by counting cultured cotyledon explants forming somatic embryos.

2.2. Germination and Plant regeneration

Cotyledonary somatic embryos were transferred to MS medium containing 3×10^{-5} M GA_3 to induce germination of somatic embryos. To investigate the effect of nitrogen compounds on the root and shoot growth of plantlets, the germinating embryos were transferred to modified MS liquid medium with 2% sucrose, with various concentrations and combinations of NH_4NO_3 and KNO_3 in 50 ml liquid medium with filter bridges in 100 ml glass tube. The plantlets were transferred to square plastic pots containing the soil, sand and peat (4:4:3 v/v) in a greenhouse.

3. Results and Discussion

3.1. Development of multiple and single somatic embryos

When the cotyledon explants were cultured on MS agar medium with 5% sucrose, somatic embryos were developed directly near the basal excised region of cotyledon explants (Fig. 1A-B). Somatic embryos were developed into multiple or single state. The frequency of somatic embryo formation was 93% in midmature cotyledons and 66% in mature cotyledons. On the other hand, immature and seedling explants did not produce somatic embryos. On cotyledons from midmature zygotic embryos, most of somatic embryos (96% of total embryos) developed as multiple state, while on cotyledon explants from mature one, 13% of somatic embryos developed into single state. These single embryos that was closely similar to zygotic embryos, developed to normal plantlets.

3.2. Germination and plant regeneration

Seventy three percent of single embryos were developed into normal seedlings with shoot and roots similar to zygotic embryos (Fig. 1E). In contrast, none of multiple embryos produced roots even in some multiple embryos having radicle-like structures. On full-strength MS medium (10.3 mM NH_4^+ + 19.7 mM NO_3^-), root growth of plants was weak, while shoot growth was vigorous. To observe the nitrogen requirement for root growth of plants, the seedlings derived from single embryos were transferred to the liquid media with a filter paper bridge in a glass tube. Nitrogen levels (NH_4NO_3 and KNO_3) of MS media were modified to various concentrations and combinations. On full-strength MS basal medium (10.3 mM NH_4^+ + 19.7 mM NO_3^-), root growth was weak, but shoot growth was

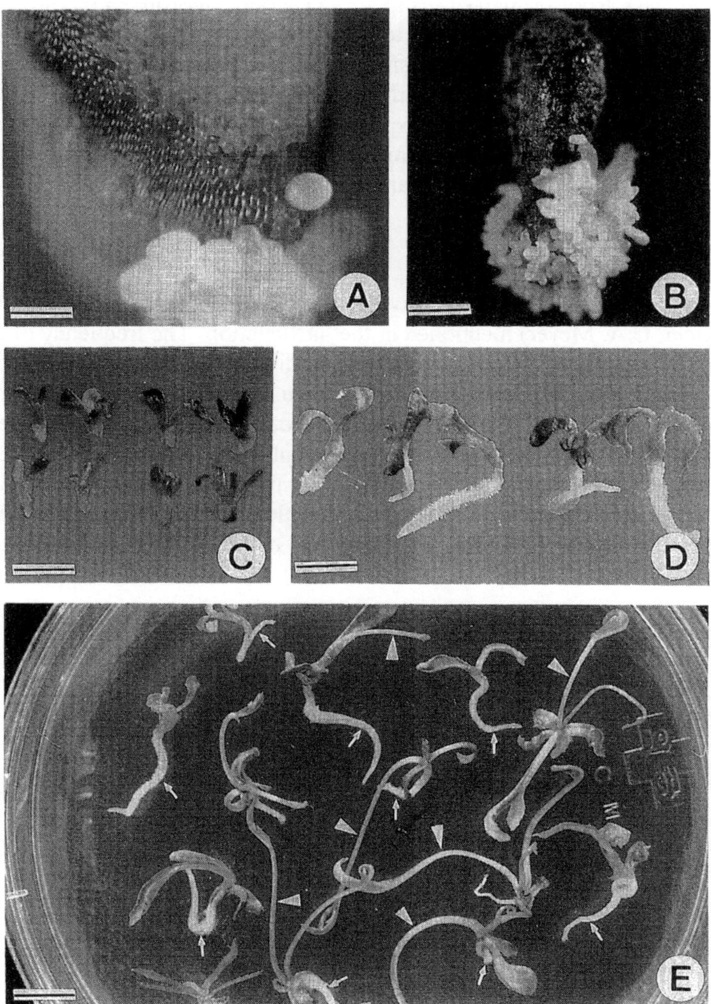

Fig. 1A-E. Plant regeneration from somatic single embryos of ginseng. A: Somatic embryos developed directly from a cotyledon of mature zygotic embryo on MS medium after 1 month. B: Cotyledonary embryos after 2 months. C: Somatic single embryos after separation from the cotyledon explants. D: Germinating somatic single embryos on MS medium with 3×10^{-5} M GA_3. E: Plantlets regenerated from single embryos on 1/2 MS medium (arrows: roots, arrowheads: shoots).

vigorous (Fig. 1E). Root growth of plants was generally vigorous in modified MS media omitting ammonium nitrate. Therefore, to promote the balanced root and shoot growth, all the germinated seedlings were transferred to the half-strength MS medium lacking ammonium nitrate. The ginseng plantlets derived from single embryos have been acclimatized in a greenhouse. In this experiment, we suggest that direct somatic single embryogenesis on hormone-free medium can be applied to an advanced technique for successful plant regeneration from Korean ginseng cotyledon cultures.

4. Conclusion

The present study reveals the regenerative ability of somatic embryos in Korean ginseng (*Panax ginseng* C.A. Meyer) mentioned in the introduction. The frequency of multiple or single embryo formation differed markedly according to the degree of maturity of the cotyledons. Cotyledon explants from midmature zygotic embryos formed multiple embryos, while cotyledons from fully mature zygotic embryos formed single embryos. Somatic single embryos regenerated into normal plantlets with both roots and shoots, while multiple embryos regenerated into only multiple shoots without roots. In the full-strength of MS basal medium, root growth of plantlets derived from single embryos was weak compared to shoots. Deletion of ammonium nitrate of MS medium promoted the root growth of plantlets.

5. References

Butenko RG et al. (1968) Bot. Zh. 7: 906-913.
Chang WC, Hsing YI (1980) Theor. Appl. Genet. 57: 133-135.
Murashige T, Skoog F (1962) Physiol. Plant. 15: 473-497.
Shoyama Y et al. (1988) Planta Med. 54: 155-156.

Levels and *in situ* localization of endogenous cytokinins as chief factors controlling bud regeneration

D. Chriqui[1], A. Azmi[1], A. Guivarc'h[1], A. Chiappeta[3], W. Dewitte[2], J. P. Reynoird[1], E. Boucheron[1] and H. Van Onckelen[2]

[1]University P. M. Curie, Lab. CEMV, 4, place Jussieu, case 150, 4, place Jussieu, F-75252 Paris Cedex 05, France, [2]University of Antwerp, U.I.A., Dept. of Biology, B-2610 Wilrijk, Belgium, [3]Università della Calabria, I-87030 Arcavata di Rende, Italy

Introduction - Fundamental to expected responses when excised plant tissues are placed *in vitro* is the question to which extent explant factors, especially the endogenous hormonal situation, are influencing the receptivity to external stimuli (Cassells et al., 1982). In addition, the main consideration of the recent approaches dealing with the regulation of gene expression by hormones has been the effect of exogenous compounds providing no information on the patterns of distribution of endogenous hormones. This is particularly true for cytokinins which are known to trigger a variety of developmental events both in whole plants and in *in vitro* cultures (Kaminek, 1992; Brzobohaty et al., 1994). Analytical methods to quantify cytokinins have been strongly improved during the recent past years, however, the various approaches to localize cytokinins *in situ* have long been questionable due to their diffusible property. Pioneer works in the area of cytokinin immunocytochemistry (Zavala, Brandon, 1983; Eberle et al., 1987; Sossountsov et al., 1988) stressed the role of aldehyde fixation in the linkage between the cytokinin bases isopentenyladenin (iP), zeatin (Z) and dihydrozeatin (DHZ) and BSA and also emphasized the use of antibodies obtained by immunization with cytokinin ribosides coupled to BSA according to the periodate oxidation method (Erlanger, Beiser, 1964). Following these works, we have developed a new immunocytochemical method focusing on the localization of cytokinin bases which were evidenced to be the active forms (Laloue, Pethe, 1982; Vesely et al., 1994). This method was applied to a range of herbaceous and woody models. The results were checked carefully by several controls. In parallel, the levels of endogenous cytokinins were determined using capillary liquid chromatography-tandem mass spectrometry. A good correlation was obtained between immunolocalization and quantification, allowing conclusions about the patterns of distribution of cytokinins during development and tissue differentiation and during the process of *in vitro* dedifferentiation.

Material and methods - *Nicotiana tabacum* L. cv. SR1 and W38 (wild or harbouring the *bik ipt*, the *cdc2a-gus* or the *cycl-gus* constructs) were grown from seeds in greenhouse (25 °C/ 20 °C, day/night temperature; 90 $\mu Em^{-2}s^{-1}$ light energy). Tobacco pith explants were excised from the middle part of the stem of 3-4 month old plants and placed on Murashige and Skoog's medium (MS) supplemented with 1 $mg.l^{-1}$ NAA and 0.2$mg.l^{-1}$ BA. *Eucalyptus globulus* seedlings and clones of micropropagated

A. Altman et al. (eds.), Plant Biotechnology and In Vitro Biology in the 21st Century, 33–36.

plantlets were grown aseptically on MS alone or supplemented with 0.01 $mg.l^{-1}$ NAA and 0.1 $mg.l^{-1}$ BA prior to hypocotyl excision and transfer *in vitro* (Azmi et al., 1997a) or inoculations with *Agrobacterium tumefaciens*-derived strain (C58 pM 90 / T139 GUS-INT (Azmi et al., 1997b). *Olea europea* juvenile and mature shoot apices were collected on trees grown under natural conditions in Calabria during growth period and winter dormancy.

Immunolocalization of Z, DHZ and iP was performed using affinity purified rabbit antibodies on material fixed for 3 h at 4 °C in a 0.5 % glutaraldehyde-3% paraformaldehyde-PBS mixture, then sectioned (30 - 50 µm) with a vibratome. The floating sections were transferred successively into blocking buffer (0.1 % fish gelatine, 0.5 % BSA, 1 % normal sheep serum, 20 mM glycine, PBS, 3 x 10 mn) then into PBS prior to incubation with primary antibodies in a dilution of 1/100 in blocking buffer supplemented with 0.025 % Tween 20 at 4 °C overnight. After several washes in PBS, they were incubated in the secondary antibody (sheep anti-rabbit IgG) conjugated with alcaline phosphatase for one hour at room temperature, washed with blocking buffer then PBS and rinsed with Tris-HCl buffer. They finally reacted in the presence of nitro-blue tetrazolium and 5-bromo-4-chloro-3-indolyl phosphate for three minutes. Sections were mounted in PBS containing 2 mM EDTA and glycerin and observed immediately. A complete set of controls including incubations without primary antibodies, saturation of the primary antibodies with their antigens and methanolic extraction of the cytokinins prior to incubations was developed. Cytokinin quantification was performed by LC-MS (Witters et al., in press) following purification by a combination of solid phase and immunoaffinity (Redig et al., 1996). The histochemical GUS assay was performed according to Jefferson (1987) modified by the addition of 2 mM K^+ ferricyanide in the incubation medium (Guivarc'h et al., 1996).

Results and Discussion - *Dynamics of cytokinins in tobacco plants during development* - The distribution of cytokinins in cv.SR1 and W38 shoot apices and differentiated organs was analysed by both immunocytochemistry and quantitative tandem mass spectrometry. Immunolabelling was observed for Z, DHZ and iP in vegetative apices and flower buds developing seeds while no signal was detected in prefloral transition apices. This was confirmed quantitatively, with a threefold decrease content of free bases and ribosides, leading to the conclusion that leaf and floral organogenesis were characterized by high cytokinin contents compared to the prefloral phase during which cell division occurred with no organogenesis (Dewitte et al., submitted). Strong signal was also observed in developing leaves and floral organs, in male and female sporogenous tissues. Among the ovular cells, the archesporial cells and later the embryo sac were significantly labelled. At the end of stamen differentiation no labelling was found in mature pollen grains, whereas the vascular bundle of the connective and the cells of the longitudinal slits provided with secondary thickenings were highly labelled. The placenta vascular strands were also heavily stained. After fertilization strong cytoplasmic labelling was observed in the endosperm and in the embryo. In the differentiated parts of stems and leaves, cytokinins were found mainly associated with leaf differentiation and greening, in the internal and external phloem strands, with a heavy labelling in the companion cells, in the vascular cambium, in xylem rays and, in a patchy manner, in pith and cortical parenchyma.

Localization of endogenous cytokinins during pith culture - At the time of excision, pith tissue from transgenic tobacco lines expressing the *cdc2-gus* or the *cyc 1 At-gus* constructs displayed a GUS expression localized sporadically in some cells. Excision and passage *in vitro* led to an overall increase in the expression of GUS activity during the first two days then the expression of both genes became limited to the peripheral cells from which callus formation originated. Dividing cells exhibited a strong GUS positive reaction. A similar pattern of *in situ* localization of endogenous cytokinins was found. While only some pith cells at T_0 were provided with detectable cytokinins, the *in vitro* culture induced an increase in cytokinins in the 3-5 cell layers located near the wound then immunolabelling was found in the peripheral and cambial-like dividing cells. The apoplastic compartment of areas undergoing a process of cell separation and the callus cells redifferentiated into phloem elements were also stained. These results indicated that presence of free cytokinins is associated with many of the processes that occur *in vitro* including cell reactivation and redifferentiation but also that excised tissues are producing *de novo* endogenous cytokinins in relation with wound healing, cell division and cell separation. The presence of cytokinins in cells expressing the *cdc2-gus* construct could be related to the regulation of this gene by cytokinins (Hemerly et al., 1993).

Seasonal variation in cytokinin distribution in olive tree shoot apices - Free cytokinins in this woody model were found to vary considerably according to the season. During winter dormancy, no signal was observed. During spring reactivation, signal became detectable first in the shell-zone of axillary buds, then in the axillary buds themselves and later in the shoot apical meristem and leaf primordia. Interestingly, the labelling was restricted to the meristematic cells and the differentiated parts of the stem were totally devoid of signal. This could be related to the recalcitrancy of differentiated olive tree explants to dedifferentiate.

Cytokinin localization during differentiation and bud regeneration in E. globulus - Eucalypts are highly recalcitrant models to bud regeneration. For *E. globulus*, it was found that only hypocotyls excised from 10-15 days old seedlings were competent for caulogenesis after transfer on MS medium containing 0.01 $mg.l^{-1}$ NAA, 0.2 $mg.l^{-1}$ BAP and 0.2 $mg.l^{-1}$ thidiazuron, the narrow period of competency being closely related to the particular stages of oil gland differentiation (Azmi et al., 1997a). While 2-10 days old hypocotyls were provided with high endogenous cytokinins levels localised in all tissues, the 10-15 days old ones displayed cytokinins restricted to the epidermal and sub-epidermal cells involved in the schizogenic oil gland differentiation, thereafter these cells became fully differentiated and lost both their cytokinin content and their regeneration ability. In this model, presence of endogenous cytokinins in the competent cells thus appeared to be a limiting factor to regeneration. An efficient alternative method using *Agrobacterium tumefaciens* strain 82.139 allowed to regenerate buds on tumors induced on clonal plantlets. This was explained by the high endogenous zeatin and zeatin riboside contents in the shooty tumors compared to the non-shooty tumors induced by the nopaline strain C58 (Azmi et al., 1997b). We found, using a C58 pM90/T139 GUS-INT strain harbouring the wild T-DNA of the 82.139 strain and the 35S-GUS-INT construct, that the shooty tumors were provided by only a few number of transformed cells. Colocalization of cytokinins and transgenic cells in these tumors was also achieved, showing a good correlation between the GUS positive areas and high immunolabelling. As in olive tree, it was also found that cytokinins in mature eucalypts were restricted to the shoot apices and to the vascular strands. This suggested that cell differentiation in these woody models corresponded to a rapid drop in cytokinin bases while herbaceous models retained cytokinins in the differentiated parenchyma and a better capacity for cell reactivation.

Sink effect of endogenous cytokinins - Transgenic tobacco plants expressing the *ipt* gene from *A. tumefaciens* in the axillary buds (Hewelt et al., 1997; Guivarc'h et al., in preparation) were analysed for the *in situ* localization of cytokinins and for the cellular consequences of such a localised expression. Strong immunolabelling was found in the apical part of growing axillary buds of mature plants entering the flowering process. This was associated with very high contents of both phenolics and starch and illustrated the expected involvement of cytokinins both in the regulation of genes of the phenylpropanoid pathway (Deikmann, Hammer, 1995) and in sink formation for carbohydrates (Ehne, Roitsch, 1997). Such an effect of cytokinins on starch accumulation was previously reported as an early consequence of *in vitro* culture on medium containing both sucrose and kinetin and was correlated with transient membrane modifications and sucrose uptake (Brossard-Chriqui and Iskander, 1982). Sucrose unloading and starch accumulation were found to be critical processes in the induction of adventitious root (Brossard, 1977) and shoot meristems (Brossard, 1970). What one may ask therefore is whether the expression of some of the genes considered as cytokinin-dependent are regulated directly by cytokinins or indirectly by sugars accumulated as a result of sink formation by cytokinins. Of particular interest considering all our results together is also to find species- and tissue-specificity of cytokinin distribution. Such patterns as well as the changes in distribution occurring during development and differentiation could explain the variety of responses to exogenous cytokinins that are usually observed among species and explants when placed *in vitro*. We are expecting that the method for *in situ* localization that we developed in the present study will be helpful for new insights in the puzzling problem of cytokinin action alone or combined with auxin, both at the whole plant level and for the control of *in vitro* behaviour of plant tissues.

Acknowledgements - D. Inzé (University of Gent, Belgium) and T. Schmülling (University of Tübingen, Germany) are gratefully acknowledged for the supply of the *cdc2a-gus* and *cycl 1 At-gus* constructs and for the *bik-ipt* tobacco plants, respectively.

References

Azmi A et al. (1997a) Plant Science, 127, 81-90.
Azmi A et al. (1997b) Plant Cell, Tissue Org. Cult., 51, 9-16.
Brossard D (1970) C. R. Acad. Sc., 271, 56-59.
Brossard D (1977) New Phytol., 79, 423-429.
Brossard-Chriqui D, Iskander S (1982) Protoplasma, 112, 217-225.
Brzobohaty B et al. (1994) Plant Mol. Biol., 26, 1483-1497.
Cassells AC et al. (1982) Physiol. Plant., 56, 507-512.
Ehne BR, Roitsch T (1997) Plant J., 11, 539-548.
Deikmann J, Hammer PE (1995) Plant Physiol., 108, 47-57.
Eberle J et al. (1987) Planta, 172, 289-297.
Erlanger, Beiser (1964) Biochem., 52, 68-74.
Guivarc'h et al. (1996) Transgenic Res., 5, 281-288.
Hemerly AS et al. (1993) Plant Cell, 5, 1711-1723.
Hewelt A et al. (1994) Plant J, 6, 879-891.
Kaminek M (1992) Tibtech, 10, 159-164.
Jefferson RA et al. (1997) EMBO J., 6, 3901-3907.
Laloue M, Pethe C (1982) Plant Growth Substances, pp 185-195, Academic Press, NY
Sossountzov L et al. (1988) Planta, 175, 291-304.
Zavala ME, Brandon DL (1983) J. Cell Biol., 97, 1235-1239.
Vesely et al. (1994) Eur. J. Biochem., 224, 771-786.

SOMATIC EMBRYO POLARITY IN *DACTYLIS GLOMERATA* AS DETERMINED BY ELECTRON MICROSCOPY

B.V. Conger and A. Vasilenko
Department of Plant and Soil Science
University of Tennessee
Knoxville, TN 37901-1071 USA

1. Introduction

Somatic embryos initiate and develop from single mesophyll cells in cultured leaf segments of orchardgrass (*Dactylis glomerata* L.) genotype "Embryogen-P" (Conger et al. 1983; Conger, Hanning 1991; Trigiano et al. 1989). Experiments can be designed in which segments from one half leaf can be used for a specified treatment and the corresponding "sister" or "mirror" segments from the other half of the same leaf can serve as a control. This paired half-leaf system was in a space shuttle experiment (STS-64) that flew from 9-20 September 1994. No significant differences in somatic embryogenesis were observed in leaf segments plated 21 d and 14 d prior to launch whereas, a 70% reduction was observed for segments plated 21 h before liftoff (Conger et al. 1995).

Previous results with light microscopy indicated that the first cell divisions take place about 4 d after plating and that approximately 90% of these divisions are periclinal (Trigiano et al. 1989). In the space shuttle experiment, a higher ratio of anticlinal:periclinal first divisions were observed in the 21 h flight treatment compared to the control. Since the first cell divisions would be taking place during the flight, we hypothesize that microgravity affects axis determination and embryo polarity at a very early stage, perhaps the first cell division. The objective of the present investigation was to supplement and corroborate our previous studies and results obtained with light microscopy with transmission electron microscopic (TEM) analyses of early mesophyll cell divisions in cultured orchardgrass leaf segments.

2. Materials and Methods

Tillers were collected from greenhouse grown plants. The outer leaves were removed, and the basal 3 cm portions of the innermost two were excised, split along the midvein, and surface sterilized. These leaf pieces were then cut transversely into six segments approximately 3 mm x 3 mm and plated serially from the basal potion upward onto a

A. Altman et al. (eds.), Plant Biotechnology and In Vitro Biology in the 21st Century, 37–40.

solid Schenk and Hilderbrandt (1972) medium containing 30μM dicamba (3,6 dichloro-2-methoxybenzoic acid), 30 g/L sucrose and 0.8% agar. Incubation was in the dark at 22°C.

Leaf segments were collected at various periods ranging from 5 d to 12 d after plating and fixed with 2.5% glutaraldehyde in 0.1 M phosphate buffer (PB), pH=7.4. They were post fixed in 1% OsO_4 in the same buffer solution. Embedding was in Epon-Araldite resin in flat molds with proper orientation to obtain cross sections of leaf segments and longitudinal axes of the developing somatic embryos. Resin blocks were cut on an LKB ultramicrotome with a diamond knife and 500Å sections were stained with uranyl acetate followed by lead citrate. Stained sections were examined with a Hitachi H-600 transmission electron microscope. The microscope was operated with an electron beam at 75 kV, and the magnification ranged from 2,500x to 15,000x. Serial sections were analyzed for early cell divisions leading to embryo initiation.

3. Results and Discussion

The number and percentage of periclinal to anticlinal first divisions in two-cell stage proembryos as determined by both TEM and light microscopy (LM) are shown in Table 1. Data for LM was derived from various experiments conducted over the past three years. Observations from both types of microscope analyses indicate that approximately 70% of the first cell divisions are periclinal.

Table 1. Number and percentage of periclinal to anticlinal first cell divisions in two-cell stage proembryos as determined by TEM and LM.

Microscopy	No. periclinal	No. anticlinal	Percentage periclinal:anticlinal
TEM	93	41	69:31
LM	840	311	73:27

This is a lower percentage than reported earlier (Trigiano et al. 1989) but higher than that observed in leaf segments plated 21 h prior to launch on an 11 d space shuttle flight. In that case, only about 40% of the first cell divisions were periclinal (unpublished data). Also, a much higher frequency of two-cell stage proembryos were present in the flight compared to the ground controls. This observation suggests that the spaceflight conditions (presumably microgravity) had an inhibitory effect on early cell divisions. Either cells failed to initiate division or discontinued after the first division, especially if it was not periclinal.

Electron micrographs of early cell division are shown in Fig.1. Within 96 h cells become densely cytoplasmic and the first cell divisions are initiated. Chromosomes are aligned at the metaphase plate in the cell in Fig. 1a in preparation for a periclinal division. The three-cell stage embryo in Fig. 1b was formed by two periclinal divisions and polarity is clearly established. Note also, that the embryo is developing independently of the surrounding tissue. A four-cell stage structure formed by an initial anticlinal division followed by two periclinal divisions is shown in Fig. 1c. In this case, polarity has not been established and an embryo may or may not develop directly from subsequent divisions. This structure may give rise to an embryogenic cell mass, termed a proembryonal cell complex by Haccius (1978), in which one or more embryos may initiate and develop. We have observed and reported on this phenomenon previously (Trigiano et al. 1989).

An eight-cell embryo is shown in Fig. 1d. This embryo was formed by two periclinal divisions, followed by two anticlinal divisions and then by two divisions in an oblique plane. Polarity has been established in this embryo; however, the mature structure will probably possess a multiseriate rather than a uniseriate suspensor. Ontogenetic development of orchardgrass somatic embryos, including variation in suspensors, has been published (Gray, Conger 1985).

In conclusion, approximately 70% of the first cell divisions leading to somatic embryo formation from orchardgrass mesophyll cells are periclinal. The plane and direction of this first cell division is probably important in establishing embryo polarity. Im microgravity, divisions may not continue to produce either an embryo or a proembryonal cell complex, whereas, one or the other probably will occur under normal 1-*g* conditions.

4. Acknowledgment

The research was supported in part by NASA grants NAG10-0138, NAGW 3141, NAG10-0221.

5. References

Conger BV, Hanning GE (1991) Crop Sci. 31, 855.
Conger BV et al (1983) Science 221, 850-851.
Conger BV et al (1995) Am. Soc. Gravit. Space Biol. Bull. 9(1), 93.
Gray DJ, Conger BV (1985) Tissue Culture in Forestry and Agriculture, pp 49-57, Plenum Press, New York
Haccius B (1978) Phytomorphology 28, 74-81..
Trigiano RN et al (1989) Bot. Gaz. 150, 72-77.

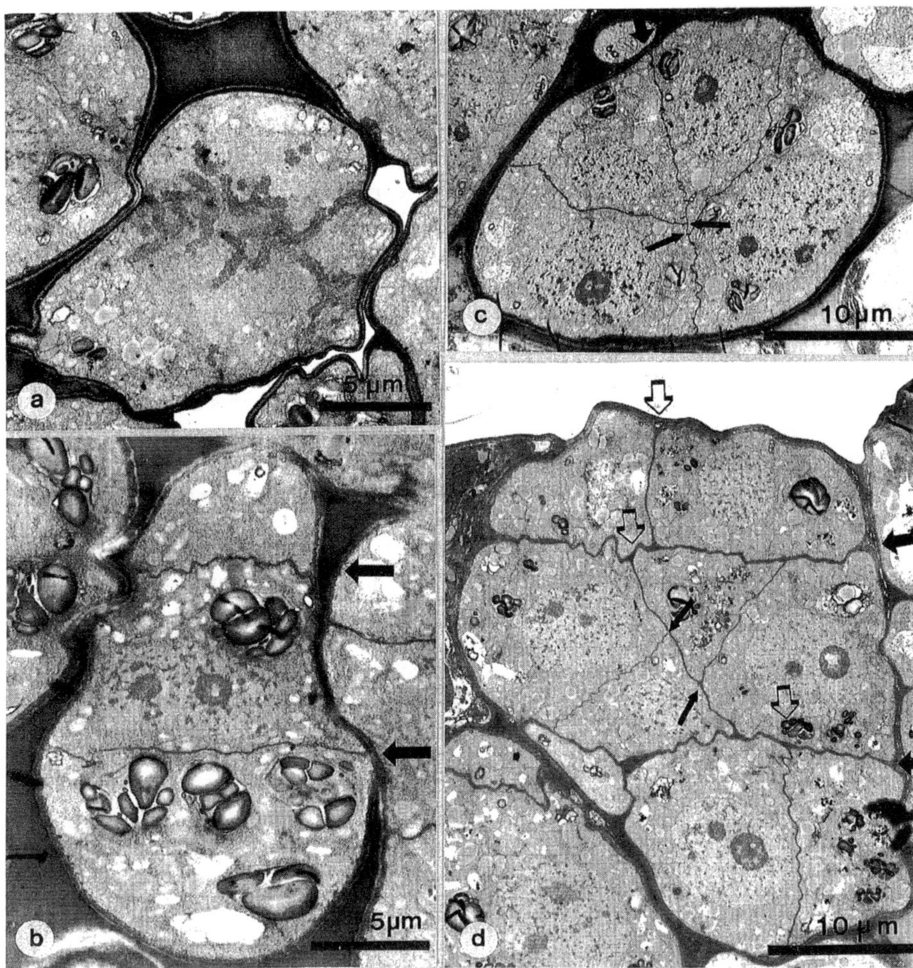

Fig. 1. Electron microscopy of early cell divisions leading to somatic embryo formation from mesophyll cells. (a) Single cell beginning division, note chromosomes on metaphase plate. (b) Three-cell stage formed by two periclinal divisions (arrows). (c) Four-cell stage formed by an anticlinal division (large arrow) followed by two periclinal divisions (small arrows). (d) Eight-cell stage proembryo formed by two periclinal divisions (large black arrows) followed by two anticlinal divisions (open arrows) and then by two oblique divisions (small black arrows).

DUAL EFFECT OF ETHYLENE DURING ROOTING OF APPLE MICROCUTTINGS

Geert-Jan De Klerk, Annie Paffen, Jan Jasik and Vally Haralampieva
Centre for Plant Tissue Culture Research,
PO Box 85, 2160 AB Lisse, The Netherlands

Abstract
During rooting of apple microcuttings, we applied 24h pulses with ACC (1-aminocyclopropane-1-carboxylic acid, an ethylene precursor) or STS (silverthiosulfate, an inhibitor of ethylene action). The timing of the pulse and the auxin concentration in the medium determined the effect (promotive, inhibitory, or no effect). Another major factor was submergence of the basal 4 mm of the stem in the medium: the results indicated that this led to entrapment and accumulation of ethylene in the basal part of the stem (the portion that produces the roots) at high auxin concentration. In stem-segments or slices that are both cultured on top of the medium, ethylene accumulation did not occur. The detrimental effect of auxin on shoot and roots was -at least partly- due to auxin-induced ethylene synthesis.

1. Introduction

Previously we have shown that root formation of apple microcuttings can be dissected into three successive phases (De Klerk et al., 1995; De Klerk, 1995). During the first day after taking the microcutting, certain cells in the stem develop competence to respond to the rhizogenic signal. This period is designated as the dedifferentiation phase. During the following two or three days, these cells become determined to form roots by the rhizogenic action of auxin. During this induction phase, meristemoids of ca. 30 cells are formed. After this, morphological differentiation occurs during which the roots develop. The distinction between different phases each with its own specific requirements, enables to study the role of plant growth regulators in rooting in detail (De

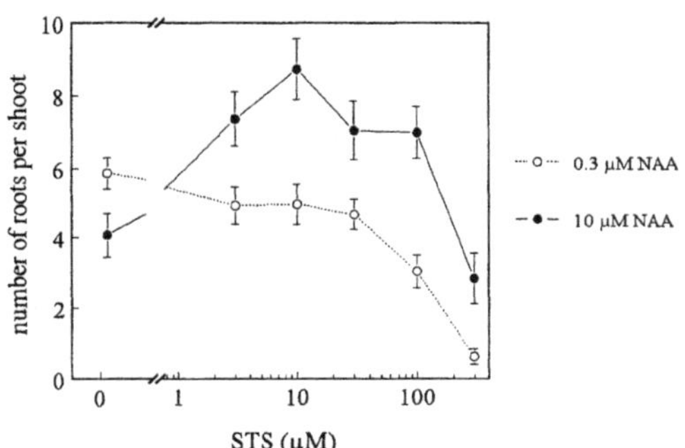

Fig. 1. The effect of STS on rooting of apple shoots cultured for 5 days on medium with 0.3 or 10 µM NAA and increasing concentrations of STS. After 5 days, the shoots were transferred to medium without growth regulators.

A. Altman et al. (eds.), Plant Biotechnology and In Vitro Biology in the 21st Century, 41–44.

Klerk, 1995).

We report here that the contradictory results in literature about the role of ethylene (Mudge, 1988) may be partly explained by differences in ethylene sensitivity among the three phases. We also examined the role of auxin concentration and culture conditions (in particular whether the rooting zone of the stem is emerged or submerged in the medium, cf. flooded plants; Jackson, 1985).

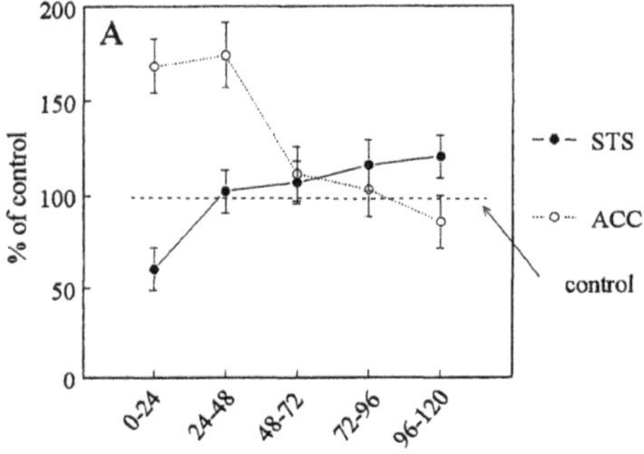

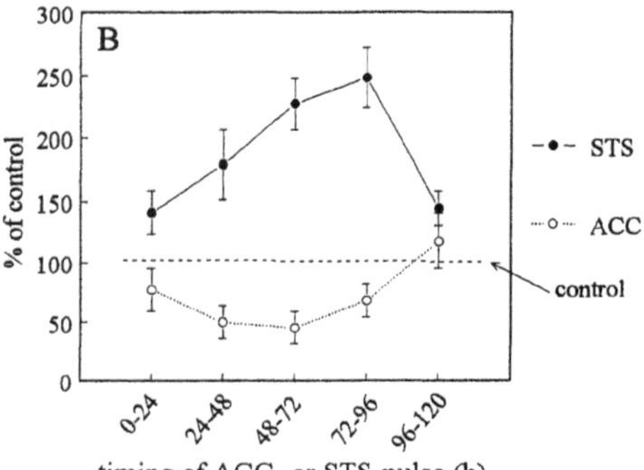

Fig. 2. Effect of 24h pulses with 100 μM STS or 30 μM ACC on rooting of apple shoots. The shoots were cultured on medium with 0.3 μM NAA (A) or 10 μM NAA (B) and the pulses were given at the indicated times. The number of roots is expressed as a percent of the rooting of shoots that had not received a pulse with ACC or STS.

2. Materials and Methods

Shoot production of *Malus* 'Jork 9' was maintained as described previously (De Klerk et al., 1995). Rooting was examined in shoots, 1-cm defoliated stem-segments and 1-mm stem-slices. The rooting procedure has been described previously in detail (De Klerk, Caillat, 1994; De Klerk et al., 1995). Shoots and segments were rooted on 40 ml medium in 2.5 cm high 9-cm Petri dishes (10 shoots or segments per dish). The shoots were put for ca. 4 mm in the medium. The segments were cultured horizontally on top of the medium. The slices were cultured in 1.5-cm high 9-cm Petri dishes on top of 20 ml rooting medium (30 slices per dish). They were with the apical side down on the medium and the dish was incubated upside down. After 5 days culture at 25°C in the dark, shoots, segments and slices were transferred to dishes with hormone-free medium and to the light. Roots were counted 3 weeks after the start of the experiment. In one experiment, from each rooted shoot the length of the longest root was determined. For each determination, 30 shoots, 30 segments or 90 slices were used. In the figures, the means ±SE are shown.

3. Results

For rooting of apple shoots, 10 μM NAA (naphthaleneacetic acid) is supraoptimal and 0.3 μM NAA suboptimal (De Klerk et al., 1997). We first examined the effect of STS (silverthiosulfate, an inhibitor of ethylene action) applied during the full rooting treatment of 5 days. Figure 1 shows that STS stimulated rooting of shoots at 10 μM NAA but had no effect at 0.3 μM. The optimal concentration of STS was 10 μM. At 100 or 300 μM, STS was inhibitory. STS acted by inhibiting ethylene action since an inhibitor of ethylene synthesis, AVG (aminoethoxy-vinylglycine), gave similar results (data not shown). We conclude from this experiment that at 10 μM NAA, but not at 0.3 μM NAA, auxin-induced ethylene synthesis blocked rooting.

We examined when STS acted by giving 24h pulses with STS. At 10 μM NAA, these pulses had little or no effect during the first day, but during the induction period (24-96h) they strongly promoted rooting (Fig. 2B). 24h-Pulses with ACC (1-aminocyclopropane-1-carboxylic acid, a precursor of ethylene) had the opposite effect and strongly inhibited rooting during the induction phase. After 96h, both ACC and STS had little or no effect. At 0.3 μM NAA, 24h pulses with STS or ACC acted only just after the start of the rooting treatment: STS inhibited and ACC promoted rooting (Fig. 2A). After that, both compounds did not have a significant effect.

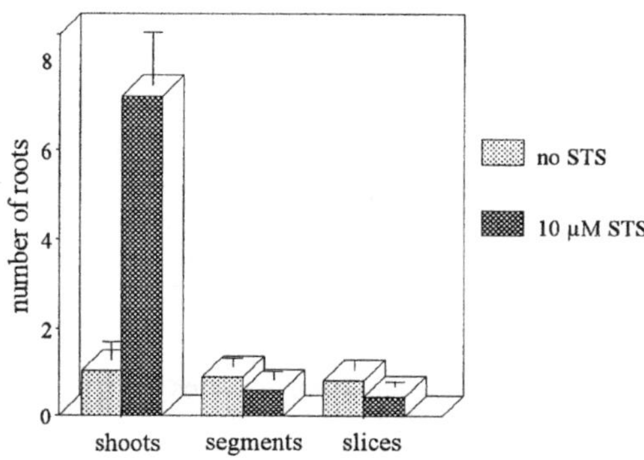

Fig. 3. The effect of STS on rooting of shoots, 1-cm segments and 1-mm slices of apple. The shoots were put for 4 mm into the medium. The segments and slices were on top of the medium. After 5d in the dark with 10 μM NAA with or without 10 μM STS, the explants were transferred to medium without growth regulators.

Stem slices are a simplified system to study rooting (De Klerk, Caillat, 1994). We examined the effect of STS in slices and also in 1-cm stem segments. Both were cultured on top of the medium. Under these culture conditions, STS did not promote rooting (Fig. 3), also not when applied at a range of concentrations.

At 10 μM NAA, the shoots did senesce. This was mediated by ethylene since senescence did not occur when STS had been applied to the medium (not shown). STS not only improved the quality of the shoot but also of the root system as shown by increased root length (Fig. 4).

4. Discussion

Zimmerman et al. noted in 1933 that ethylene increased root formation. After that, ethylene has been reported to inhibit or stimulate rooting, or to have no effect (Mudge, 1988). Here we report two explanations for the variation of effects reported in literature.

First, we found that the effect of ethylene depends on the time of application and the auxin concentration in the medium. Ethylene was promotive during the initial days of the rooting treatment at a low auxin concentration (Fig. 2A). At high auxin concentration this effect was not observed (Fig. 2B). Possibly, ethylene enhanced the sensitivity to auxin as has been reported for flooded plants (Visser et al., 1996) and for derooted sunflower seedlings (Liu, Reid, 1992). During the induction period (24-96h), we observed a strong effect at 10 μM NAA: ACC-pulses were inhibitory and STS-pulses promotive (Fig. 2B). At 10 μM NAA, the tissue probably produced high levels of ethylene that were inhibitory during the induction phase. ACC resulted in more inhibition, whereas the inhibition by ethylene was reduced by the ethylene inhibitor STS.

Second, we have found that the effect of STS depends on culture conditions, viz. submergence vs. emergence of the rooting zone. It is well known from research on flooded plants that physical entrapment of ethylene in submerged plant tissue leads to a high endogenous ethylene level (e.g., Voesenek et al., 1993). This explains why STS promoted rooting in shoots (from which the rooting zone of the stem is submerged in the medium) and not in slices (that are cultured on top of the medium). Thus, the variation of literature-results on ethylene may also be explained by different culture conditions resulting in different levels of trapped ethylene. It should be noted that in studies on tissue culture, accumulation of ethylene in submerged tissues is usually disregarded.

Since ethylene easily escapes from slices, it might be expected that at 10 μM NAA slices show better rooting performance than shoots. This was not found (Fig. 3) because the optimal auxin concentration for slices is lower than that for shoots (De Klerk et al., 1997). The high sensitivity of slices to auxin may be due to a severe wounding response.

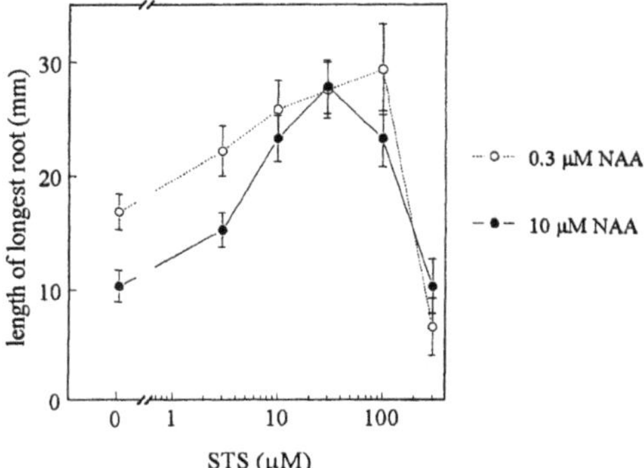

Fig. 4. The after-effect of STS on root growth of apple shoots cultured for 5 days on medium with 0.3 or 10 μM NAA and increasing concentrations of STS. After that, the shoots were transferred to medium without growth regulators.

In the present paper, we report another interesting finding about the effect of auxin during rooting. Our results indicate that at high concentration of auxin, microcuttings produce high levels of ethylene which results in leaf scenescence and poor growth of shoots and roots. When STS is given together with auxin, the microcuttings have an excellent appearance: many long roots are formed, the leaves do not show signs of senescence and the shoots grow well during the rooting treatment. Care should be taken in the use of STS. It has been reported that various morphological characteristics of roots are under the control of ethylene, viz., the formation of aerenchyma (He et al., 1992), the response to mechanical impedance (Zacarias, Reid 1992) and the formation of root hairs (Tanimoto et al., 1995).

References

De Klerk GJ (1995) Current Issues in Plant Cellular and Molecular Biology, pp 111-116, Kluwer Academic Publishers, Dordrecht, The Netherlands.
De Klerk GJ, Caillat E (1994) Adv. Hort. Sci. 8, 15-18.
De Klerk GJ et al. (1995) J. Exp. Bot. 46, 965-972.
De Klerk GJ et al. (1997) Plant Cell Tiss. Or g. Cult. 49, 39-44.
He C-J et al (1992) Plant Physiol. 98, 137-142.
Jackson MB (1985) Ann. Rev. Plant Physiol. 36, 146-174.
Liu JH, Reid DM (1992) J. Exp. Bot. 43: 1191-1198.
Mudge KW (1988) Adventitious Root Formation in Cuttings, pp 150-161, Dioscorides Press, Portland, USA.
Tanimoto M et al. (1995) Plant J. 8, 943-948.
Visser EJW et al. (1996) Plant Physiol. 112, 1687-1692.
Voesenek LACJ et al. (1993) Plant Physiol. 103, 783-791.
Zimmerman W et al. (1933) Contrib. Boyce Thompson Inst. 5, 1-17.
Zacarias L, Reid MS (1992) Physiol. Plant. 86, 301-307.

TIME-LAPSE TRACKING OF ORIGIN AND DEVELOPMENT OF SOMATIC EMBRYOS IN A GYMNOSPERM, NORWAY SPRUCE

L.H. Filonova*, P.V. Bozhkov and S. von Arnold

*Department of Forest Genetics, Uppsala Genetic Center, Swedish University of Agricultural Sciences, Box 7027, S-75007 Uppsala, Sweden; *Permanent address: Institute of Cell Biology and Genetic Engineering, National Academy of Sciences of Ukraine, Zabolotnogo str. 148, 252650 Kiev, Ukraine*

1. Introduction

Understanding the consequence of characteristic morphological stages during embryogenesis is required for any project aimed at isolation of genes which control embryonic pattern formation and post-embryonic development. Besides, selection and classification of embryo mutants, as another prerequisite for the most molecular studies, is impossible without the reference fate map showing correct progression of embryogenesis. Construction of such map in somatic embryogenesis can be achieved by two alternative approaches: (1) using synchronous cell-division systems (e.g., Fukuda et al. 1994), or (2) by time-lapse tracking of development of individual embryogenic protoplasts, cells or multicellular structures (Toonen et al. 1994).

In somatic embryogenesis of gymnosperms, very little is known so far about origin of somatic embryos and their early development. There is much speculation that conifer somatic embryo formation starts directly from either elongated vacuolated or small isodiametric cell with dense cytoplasm; and furthermore embryogenic cell cultures are a complex of early somatic embryos continuously multiplying by cleavage or budding in the presence of auxin and cytokinin (e.g., Attree et al. 1987; Klimaszewska 1989; Tautorus et al. 1990; Durzan et al. 1994; Jasik et al. 1995). This version, however, contradicts with the concept established for angiosperms, which claims that exogenous auxin inhibits somatic embryo formation (Halperin 1966). Other authors hold to the different idea, that somatic embryos arise indirectly, from cell aggregates with low level of differentiation (von Aderkas 1992; Korlach, Zoglauer 1995). There is no question, however, that further individual development of somatic embryos till maturity is promoted by abscisic acid (ABA; Dong et al. 1997).

By the present study we have shown that somatic embryos of Norway spruce are derived from proembryogenic masses (PEMs), which are distinguished from somatic embryos by morphology and pattern of immunocytochemical localization of specific epitope of arabonogalactan proteins (AGP).

A. Altman et al. (eds.), Plant Biotechnology and In Vitro Biology in the 21st Century, 45–48.

2. Materials and methods

Development of more than two thousand single cells and few-celled structures obtained from six embryogenic cell lines of Norway spruce (*Picea abies* L. Karst.) was monitored by time-lapse tracking technique (Golds et al. 1992; Toonen et al. 1994). Initially agarose films contained auxin and cytokinin, followed by their substitution for ABA.

Agarose-embedded samples were taken for light and immunofluorescence microscopy at regular intervals (Baskin et al. 1992; Pennell et al. 1992; Evans et al. 1997).

3. Results

Irrespective of cell line, formation of somatic embryos in agarose films under gradually decreasing levels of auxin and cytokinin was always preceded by growth of less differentiated structure, PEM. PEM structures may be classified under three types according to number of elongated and small cells and degree of polarity. PEM I is a polar structure with a few small cells and only one elongated cell. PEM I grows into PEM II structure by adding new elongated and small cells without disturbing polarity. The latter gradually disappears upon transition from PEM II to PEM III. Increased proliferative activity of PEM III sometime makes this structure more resembling enlarged non-differentiated cell clump rather than polar structure, which can be confused with somatic embryo.

Upon depletion of auxin or after addition of ABA, PEM III transdifferentiates to somatic embryos (for terminology, see Eguchi, Kodama 1993). Constant supply of ABA is required for further development of somatic embryos, which may be either normal (in A-type cell lines) or abnormal (in B-type cell lines; Egertsdotter, von Arnold 1998). ABA-driven development includes stages of early embryogenesis, late embryogenesis and maturation. The duration and progression of histogenesis at each stage is determined by genotype.

The main histological differences between a PEM and somatic embryo are lack of protoderm-like layer and embryonal tube layer in PEMs compared to somatic embryos.

Previously, monoclonal antibody JIM 13-reactive epitope of AGP was detected by immunoblotting of samples of concentrated extracellular proteins from embryogenic cell suspensions of Norway spruce (Egertsdotter, von Arnold 1995). By using immunomicroscopy in the present study, we have found that the patterns of recognition of AGP epitope by JIM 13 are different between cells of PEMs and somatic embryos. Somatic embryos exhibited just very weak localization of antibody-reactive epitope in few cells. On the contrary, PEMs demonstrated abundant localization of the epitope throughout the whole region of small cells. Moreover, the most intensive labelling was found in some cells of PEM III transdifferentiated to somatic embryo.

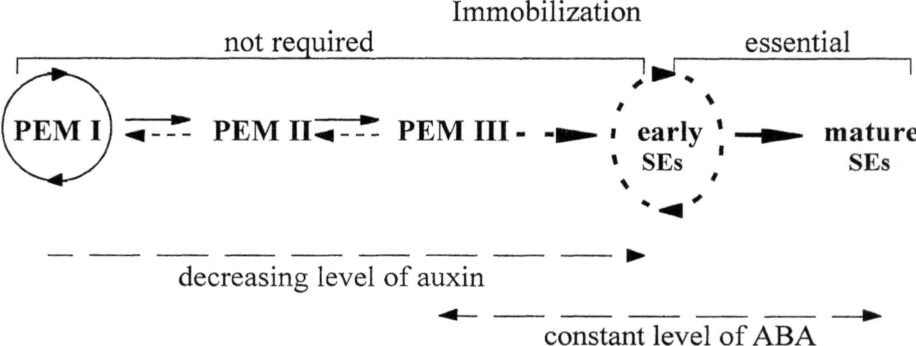

Figure 1. The model of somatic embryogenesis in Norway spruce

4. Conclusions

The pathways of somatic embryogenesis in Norway spruce are summarized in the model shown in Fig. 1. Closed line represents the cyclic process of multiplication at PEM I level by asymmetric division of elongated cells (ADEC). This process predominates under constant supply of auxin and cytokinin in liquid medium. When immobilization is applied or auxin is being gradually depleted, the average level of the whole system can be forwarded biased, to PEM II or PEM III levels (fine lines in Fig. 1). However PEM I structures are always formed from more advanced PEMs by ADEC (fine dotted lines), if only auxin and cytokinin are still present in culture medium (Fig. 1).

The bold dotted line shows trasnsdifferentiation of PEM II to somatic embryos. Those conditions which are inhibitory for PEM multiplication are, contrary, stimulatory for the transdifferentiation (e.g., depletion of auxin, immobilization, increased pH, addition of ABA, etc.). Sometime the process can be reversed back, from somatic embryos to PEM I (not shown). This occurs when the balance of factors which stimulate transdifferentiation is disturbed.

It is tempting to speculate, that PEM dies upon transdifferentiation. This idea is supported by the fact that the presence of AGP epitope recognized by JIM 13 in maize coleoptiles during xylem differentiation has been related to programmed cell death (Schindler et al. 1995).

Somatic embryo level is maintained either by cleavage of early embryos (closed bold dotted line) or through their development to the mature forms (bold line; Fig. 1). ABA is especially required for the latter process, although the balance between two processes is genotype-dependent.

5. Acknowledgements

We are grateful to Dr. Keith Roberts (Department of Cell Biology, John Innes Institute, Norwich, UK) for a kind gift of anti-AGP antibodies. The research was supported by The European Comission FAIR project CT95-0873 and The Swedish International Development Cooperation Agency.

6. References

Attree SM et al. (1987) Plant Cell Repts 6, 480-483.
Baskin TI et al. (1992) Planta 187, 405-413.
Dong J-Z et al. (1997) J. Exp. Bot. 48, 277-287.
Durzan DJ et al. (1994) Int. J. Plant Sci. 155, 677-688.
Egertsdotter U, von Arnold S (1995) Physiol. Plant. 93, 334-345.
Egertsdotter U, von Arnold S (1998) J. Exp. Bot. 49, 155-162.
Eguchi G, Kodama R (1993) Curr. Opin. Cell Biol. 5, 1023-1028.
Evans DE et al. (1997) Cell Biol. Int. 21, 295-302.
Fukuda H et al. (1994) Int. J. Dev. Biol. 38, 287-299.
Golds TJ et al. (1992) J. Plant Physiol. 140, 582-587.
Halperin W (1966) Am. J. Bot. 53, 443-453.
Jasik J et al. (1995) Protoplasma 185, 205-211.
Klimaszewska K (1989) Plant Cell Repts 8, 440-444.
Korlach J, Zoglauer K (1995) Plant Cell Repts 15, 242-247.
Pennell RI et al. (1992) J. Cell Biol. 119, 1371-1380.
Schindler T et al. (1995) Plant J. 7, 25-36.
Tautorus TE et al. (1990) Plant Sci. 67, 115-124.
Toonen MAJ et al. (1994) Planta 194, 565-572.
von Aderkas P (1992) Can. J. For. Res. 22, 397-402.

CONIFER SOMATIC EMBRYO PRODUCTION FROM LIQUID CULTURE

Pramod K. Gupta and Roger Timmis
Weyerhaeuser Technology Center
WTC – G30
Tacoma, WA 98477, USA

1. Introduction

Somatic embryogenesis has been recognized as a technique for large-scale clonal propagation of conifer species. The main advantage of somatic embryogenesis over organogenesis is the possibility of a potentially unlimited production of clones with functional root and shoot poles, and the development of manufactured seeds (Redenbaugh et al., 1987). Further, somatic embryogenesis in a liquid system offers the opportunity for process automation resulting in reduction of manual labor costs. Although numerous conifer species are able to produce somatic embryos on semi-solid medium (Gupta PK, Grob JA, 1995), there are very few reports on somatic embryo development and maturation from liquid cultures (Attree et al., 1995). This paper describes somatic embryo development and maturation of Douglas-fir (*Pseudotsuga menziesii*) from liquid medium.

2. Culture initiation

Embryonal suspensor mass (ESM) is initiated from immature embryos at predome and dome stages (before development of cotyledons) of Douglas-fir. ESM cultures are initiated onto solid medium with auxin and cytokinin. Details of the method and media composition for ESM induction have been published earlier (Gupta et. al.1996). Induction of ESM has not been achieved in liquid medium. Semi-solid medium is necessary for ESM initiation, and lower gelrite concentration (0.1–0.2 %) has been found best for the induction process (Becwar 1996).

2.1 Culture maintenance

ESM cultures are maintained in liquid medium. Maintenance medium contains lower levels of hormones (auxin and cytokinin) than required for initiation because it does not contain activated charcoal (Gupta et al.1994). ESM cultures in this medium multiply by true-to-conifer-type cleavage polyembryony (Gupta PK, Grob JA, 1995). For initiation of liquid culture (suspension culture), 2-3 g ESM are transferred to 250 ml Erlenmeyer flasks with 20-25 ml of liquid medium. Flasks are placed on a rotary shaker (90-110 rpm) in darkness at 23°C. After 7-8 days, old medium is replaced with fresh medium.

A. Altman et al. (eds.), Plant Biotechnology and In Vitro Biology in the 21st Century, 49–52.

Twenty to twenty-five days after establishment of suspension culture, ESM liquid cultures are poured into sterile 100 ml measuring cylinders and allowed to settle for 30 min. The supernatant is discarded, and settled ESM cultures measured for volume. Settled ESM is subcultured in fresh medium at a density of 1:9 (v/v) by transferring 5 ml settled ESM to a 250 ml Erlenmeyer flask containing 45 ml of fresh liquid medium. ESM suspension cultures are maintained by regular weekly subculture.

2.2 Embryo development and maturation

Embryo development and maturation too have been achieved using liquid medium, with the liquid being contained in a supporting matrix such as cheese cloth (Boulay et al. 1988), or polyester pad (Krogstrop 1988, Gupta and Pullman 1991). We use a combination of polyethylene glycol MW 4,000 – 8,000 (10 – 20%), activated charcoal (0.05–1.0%), ABA (2.5–100 mg/L) and GA (1–20 mg/L) in a medium for embryo development and maturation of Douglas-fir (Gupta et al.1994). This liquid development medium is soaked in pads, on the surface of which is placed 1.0 ml settled ESM suspension. High osmolality of the medium and ABA concentration in the pads are maintained by adding a second pad (also soaked with high osmolality medium and ABA) beneath the first (Gupta et al.1996). Fifty to one hundred cotyledonary embryos have been produced per ml of settled ESM suspension of Douglas-fir by this procedure (Gupta et al.1994).

2.3 Germination and establishment in the field

Germination of conifer somatic embryos has not been achieved in liquid medium, but to date has required semi-solid medium. Good quality (zygotic-like) embryos are selected from plates and transferred onto semi-solid medium for germination (Gupta et al.1994). Germination is achieved with 7-10 days incubation in the dark followed by transfer to light. After 6–10 weeks germinants are selected and transferred to soil. Recently, we have produced over 30,000 somatic seedlings of Douglas-fir from a large number of genotypes for clonal field tests. ESM cultures of over 700 genotypes from genetically superior full sib families have been cryostored in liquid nitrogen for the selection of best clones after field testing.

2.4 Bioreactor

Bioreactors offer various advantages over shake flasks due to the possibilities for automation, continuous monitoring and control of growth conditions (agitation, pH, oxygen, and carbon dioxide), larger volume, and maintenance of homogeneous culture (Ammirato and Styer, 1985). ESM cultures of conifers have been grown in bioreactors (Tautorus et al., 1994). For different conifer species, different types of bioreactors have been used (Table 1). At Weyerhaeuser, several ESM lines of Douglas-fir were grown in a 1.5 liter stirred tank bioreactor equipped with low speed, winged, magnetically driven stirrers without cell damage (Timmis et al. 1998). After plating onto the pads imbibed with liquid development medium as described for the shake flask cultures, the embryo yields showed no significant difference from ESM grown in shaker flasks. However, embryos from ESM grown in bioreactors tended to be of larger size and germinated better, although the *smaller* embryos accounted for the improved germination (Timmis et al., 1998).

Table 1: Early stage somatic embryo growth in a bioreactor

1. Airlift-bioreactor	Black spruce *(P. mariana)* Interior spruce (*P. glauca engelmannii*)	Tautorus et al. (1994)
2. Stirred-bioreactor	Black spruce (*Picea mariana*) Interior spruce (*P. glauca-engelmannii*)	
3. Stirred -bioreactor	Monterey pine *(Pinus radiata)*	Smith et al (1994)
4. Stirred-bioreactor (Biostat BF2)	Sitka spruce (*Picea sitchensis*)	Moorhouse et al (1996)
5. Stirred-bioreactor	Douglas-fir (*Pseudotsuga menziesii*)	Timmis et al (1998)

Moorhouse et al. (1996) used a Biostat BF2 (magnetic stirrer base) bioreactor for development of Sitka spruce (Picea sitchensis) somatic embryos. The maturation medium was perfused in the bioreactor to replace the initial proliferation medium in order to induce the development of somatic embryos in a submerged cell culture. However, embryos failed to develop beyond the globular stage. Cotyledonary embryo development has not been achieved in a submerged cell culture bioreactor. A similar perfusion system but using individual plates, was developed by Zhang (1997) using Weyerhaeuser cultures. This was shown to prevent nutrient depletion and improve embryo development.

Cotyledonary embryo development and maturation of White spruce (*Picea glauca*) has been achieved in a bioreactor on flat absorbent pads above the surface of a liquid medium. Medium was continuously supplied to the one end of the pad, while the spent medium exited by gravity from the opposite end (Attree et al. 1994). Development medium was pumped from the reservoir into the bioreactor using a peristaltic pump at the rate of 60 ml per day.

Paques et al. (1992) produced cotyledonary embryo development of Norway spruce (*Picea abies*) in a bioreactor using polyurethane layers in liquid medium. ESM was immobilized in polyurethane layers and placed vertically in liquid maturation medium. The liquid medium was retained in polyurethane layers by capillary action, and replaced frequently with fresh medium. The early-stage embryos in direct contact with liquid medium (submerged) turned brown while those that were not in contact with medium (above the surface of liquid medium) produced cotyledonary embryos.

At Weyerhaeuser, we have used a bioreactor for cotyledonary embryo development of Douglas-fir. In the bioreactor, the medium was supplied semi–continuously from the lower surface of the pads to the developing embryos on the top. Development medium was pumped from the reservoir into the bioreactor until it made contact with the lower surface of the pads. Medium was absorbed in the pads by capillary action, and after few hours, medium was pumped out to the reservoir. This was repeated at regular intervals until mature cotyledonary embryos developed. Higher yields of good quality embryos were produced in this bioreactor.

2.5 Conclusion

This paper shows that it is possible to produce somatic embryos of conifer from liquid medium in a bioreactor. However, different types of bioreactors are currently required for early-stage embryo growth and cotyledonary embryo development and maturation.

3.0 References:

Ammirato PV, Styer DJ (1985) Biotechnology in Plant Sciences – Relevance to Agriculture in the Eighties, pp 161-178, Academic Press, Orlando, USA.
Attree SM et al. (1994) Plant Cell Rep.13, 601- 606
Boulay MH et al. (1988) Plant Cell Rep. 7, 134 –137
Gupta PK et al. (1994) TAPPI Proceedings Biological Science Symposium, pp 35-39, TAPPI Press, USA
Gupta PK et al. (1996) Somatic Cell Genetics and Molecular Genetics of Trees, pp 3-10, Kluwer Academic Publishers, Dordrecht, The Netherlands.
Gupta PK, Grob JA (1995) Somatic Embryogenesis in Woody Plant Vol. 1, pp 81-98, Kluwer Academic Publishers, Dordrecht, The Netherlands.
Gupta PK, Pullman GS (1991) U.S. Patent No. 5,034,326
Krogstrup P (1990) Plant Science 72, 115-123
Moorhouse SM (1996) Plant Growth Regulation 20, 53-56
Paques M et al. (1995) Somatic Embryogenesis in Woody Plant Vol. 1, pp 399 –414, Kluwer Academic Publishers, Dordrecht, The Netherlands
Redenbaugh K et al. (1987) Plant Tissue and Cell Culture, pp 473-494, Alan R. Liss, Inc., New York
Smith DR (1994) TAPPI Proceedings Biological Science Symposium, pp 19-29, TAPPI Press, USA
Tautorus TE et al. (1994) In Vitro Cell. Dev. Bio. Plant 308, 58-63
Timmis R et al. (1998) Abstract In Vitro Cell. Dev. Bio. Plant 34, 21 A
Zhang C (1997) Ph.D. Thesis, University of Minnesota, Minnesota, USA

BASAL NUTRIENT AND HORMONAL REQUIREMENTS FOR DIRECT ORGANOGENESIS OF CEDRUS LIBANI A. RICH

N. Hosseyni, A. Ozan, Z. Kaya
Middle East Technical University, Department of Biological Sciences, Ankara 06531 Turkey

1. Introduction

Toros cedar (Cedrus libani A. Rich) is a native, commercially important tree species in Turkey. Main distribution of Toros cedar occurs in Southern Anatolia, Turkey. Selective cutting of the best trees, lack of management, forest fires have led to the genetic impoverishment of this species in the region.

FAO Panel of Experts listed Cedrus libani as one of the species to be taken under gene conservation program and listed in the first priority species for in situ conservation (FAO, 1977). Since 1961, a considerable effort has been made for intensive management of Toros cedar forests in the Toros Mountains, southern Turkey (Yahyaoglu et al. 1990; Isik, 1990; Lopez et al, 1990; Özdönmez 1990). Recently, the Turkish Forest service has established new plantations, set up seed collection areas, and established clonal seed orchards by grafting to expand the Toros Cedar forests. However, the progress is very slow since good seed years in Toros cedar are rare and periodic (every 3 or 5 years).

The aim of this study was to determine the optimum conditions, the effects of different types and concentrations of plant hormones for in vitro direct organogenesis of Toros Cedar. Here, we report the optimum media composition for development of organogenesis using juvenile and mature explants.

2. Materials and Methods

The seeds were surface sterilized and sown in perlite. Followed by 11 to 15 days of growth excised cotyledons surface sterilized and then placed in media to be tested. For the induction of adventitious bud, shoot and root Murashige & Skoog (MS), Gresshof & Doy (GD) an Mc Cown (MC) salt mixtures were tested. All the salt media used in this study were supplemented with 30 g/l sucrose, 100 mg/l myo-inositol and asparagine, 5 mg/l nicotinic acid and thiamine-HCl and 0.5 mg/l pyrodoxin-HCl, 5 mg/l agar-gel was added to the media. The pH was adjusted to 5.60 or 5.75 before sterilisation. When needed 6-benzylaminopurine (BAP), Naphthalene acetic acid (NAA) and activated charcoal (0.5 g/l) were added prior to sterilization. Activated charcoal was only added to hormone free (HF) media. All the components were purchased from Sigma Chemical Co., St. Louis, MO, USA.

The salt media and hormone combinations for determination of the best concentrations for bud and shoot formation were as following: Medium I contained MS(1/2), BAP (0,2,5,10,20,30 mg/ml), and NAA (0.1 mg/ml); Medium II contained MS(1/2) and BAP (10, 20 mg/ml); Medium III contained MC and BAP (2, 5, 10, 20, 30 mg/l); Medium IV contained MC, BAP (10, 20 mg/ml) and NAA (0.1 mg/ml). Explants were incubated at 25°C under 754 W with 16 hours light photoperiod in media I, II, III

A. Altman et al. (eds.), Plant Biotechnology and In Vitro Biology in the 21st Century, 53–56.

and IV. After 19 days of hormone induction all the explants were transferred into HF media (1/2 strength MC) kept at the same temperature and photoperiod regimes stated above.

For rooting, three months old adventitious shoots obtained from the media II and III were used. The adventitious shoots were incubated in media containing low concentrations of indole-3-acetic acid (IAA) and indole-3-butyric acid (IBA). The rooting media tested were as following. Medium A: MC(1/2), $1.4.10^{-6}$ M IAA or $5.7.10^{-5}$ M IAA; Medium B: MS (1/4), $1.4.10^{-6}$ M IAA or $5.7.10^{-5}$ M IAA; Medium C (1/2), $1.4.10^{-6}$ M IBA or $5.7.10^{-5}$ IBA. The three rooting media contained Boron at a concentration of $3.2x10^{-5}$ M for four weeks then transferred to HF media (Mohammed & Vidaver 1988).

In the case of direct organogenesis with mature explants, plant material was obtained from winter buds of 10 or 15 years old Toros cedar trees according to Lin et al. (1991). For the induction of adventitious bud and shoots the media compositions tested were as following: Medium D: GD (1/2), 1.10^{-6} M Kinetin (KN) and 1.10^{-4} M IAA; Medium E: MS (1/4), 1.10^{-6} M KN and 1.10^{-4} M IAA; Medium F: MC, 1.10^{-6} M KN and 1.10^{-4} M IAA; Medium G: MC and 1.10^{-5} M KN; Medium H: MC and 3.10^{-5} M KN.

After explants were grown for 16 days in induction medium, they were transferred to HF medium containing MC (1/2) and activated charcoal, other media components were the same. The influence of BAP (2 and 5 mg/l) and KN (2 and 5 mg/l) was tested in the medium containing MC basal salt mixture (1/2 strength).

For adventitious root induction only MC salt mixtures with varying amounts of BAP and KN, NAA, 2-4 D (2,4-Dichlorophenoxyacetic acid), IAA, IBA were tested. The rooting of winter buds was carried out as described above for juvenile cotyledonary tissue.

3. Results and Discussion

3.1 *Juvenile Explants.* In both MS and MC media bud induction occurred in the presence of BAP alone. Addition of NAA reduced the bud induction and shoot multiplication on both media. (Table 1). Bud formation was initiated at lower BAP concentrations in MC media. In media containing MC salt mixture and two hormones BAP and NAA together bud induction from the cotyledon was initiated at higher BAP concentrations (Table 2). However, when applied alone at high concentrations BAP exerted inhibitions on adventitious bud formation. The optimum level of BAP for adventitious bud induction and shoot elongation was 2-5 mg/l. The optimum period of culture in the presence of BAP was 19 days.

Shoot multiplication was rapid in media containing 5 mg/l BAP. However, in media supplemented with 2 mg/l of BAP the shoots appeared more healthy. Therefore, 2 mg/l BAP was chosen as the cytokinin concentration for shoot elongation and further cutting and transfer to rooting medium. The HF media for both bud and shoot MC in 1/2 strength and 0.05% activated charcoal was determined as optimum and used for standard shoot elongation medium.

Two months old adventitious shoots were used for rooting experiments. The best results were obtained for rooting in media containing 1/2 strength MC and 1/4strength MS both supplemented with IAA ($1.4x10^{-6}$ M). Percentage of explants with adventitious roots was 28.6 in MC (1/2) and 33.3 in MS (1/4). IBA did not produce rooting at the concentrated tested.

Table1. Effect of various BAP concentrations on the formation of adventitious bud and shoots from juvenile cotyledonary explants growing on media II and III

Media	Basal salt mixture	BAP mg/l	number of explants	% of explants with buds	Standard Error	average number of shoots
III	MC	0	10	0	0	10.0
		2	10	90.0	6.01	4.6
		5	14	92.7	23.21	14.5
		10	15	86.7	12.15	5.7
		20	10	80.0	8.53	3.4
		30	12	58.3	15.75	5.4
II	MS	10	6	83.3	9.19	2.8
		20	6	83.3	7.07	0.7

Table 2. Effect of NAA and various BAP concentrations on the formation of adventitious bud and shoots from juvenile cotyledonary explants in medium I and IV

Media	Basal salt mixture	BAP mg/l	NAA mg/l	Number of explants	% of explants producing advent. buds	Standard Error	Average shoot number
I	MS	0	0	12	0	0	12.0
		0	0.1	13	0	0	12.0
		2	0.1	10	10.0	1.40	2.3
		5	0.1	12	16.6	2.50	6.4
		10	0.1	11	54.4	5.00	5.2
		20	0.1	12	58.3	3.15	8.4
		30	0.1	11	36.4	1.40	2.3
IV	MC	10	0.1	6	33.3	1.41	4.8
		20	0.1	6	50.0	0.70	5.3

3.2 *Mature Explants.* For adventitious bud induction from the mature explants, of the three salt mixtures tested, MC gave the best induction. In the absence hormones growth and elongation of the buds did not occur at all. The optimum time period of culture in the presence of growth hormones was determined to be 16 days.

Effect of hormones KN and IAA on adventitious bud-shoot formation were tested in media containing MC salt mixture. The optimum concentration for KN was 10^{-6} M where bud induction and multiplication from mature explants of Toros cedar occurred. In addition to KN and IAA hormones, BAP and KN individually or in combinations were tested for their bud induction capabilities from winter buds of Toros cedar. BAP and KN when individually supplied to the media stimulated some adventitious needles from callus. However, when they were used in combination no adventitious needle formation occurred (Table3).

Table3. Effect of cytokinin treatments (BAP and KN) on Toros cedar bud multiplication after two months of culturing in MC medium with KN and IAA

Hormone concentration	Number of explants	% of buds surviving and forming shoots	% of explants producing needles
5 mg/l BAP	33	93.9	21.2
2 mg/l BAP	15	93.3	26.6
5 mg/l KN	11	63.6	18.2
2 mg/l KN	10	80.0	20.0
2 mg/l BAP + 2 mg/l KN	15	40.0	0
5 mg/l BAP + 0.5 mg/l KN	8	62.5	0

In this study bud induction was achieved by using cytokinin in the media containing MC basal salt mixture rather than MS salt mixture. The results agree with reports of Villalobos et al. (1985) and Ellis et al. (1991).

In direct organogenesis studies using mature explants MS and MC salt mixtures gave limited adventitious bud induction. Axillary bud induction by tissues from mature conifers has been reported for only a few species (Lin et al.1991; Corchete et al. 1993; Sankara & Venkateswara 1985). The present study demonstrated that although the frequency of regeneration in vitro generally declines with age, the nodal explants from juvenile as well as adult trees responded alike to shoot induction, and there being no significant difference in regeneration frequency. As a result MC formula with vitamin supplements was determined as an optimal basal media for bud induction and development from mature explants. Although limited rooting was obtained by using juvenile tissues as explant, no rooting occurred with mature explants. Shoots induced from juvenile material were more amenable to rooting than those from mature explants.

The results of this study suggested that micropropagation of Toros cedar by utilizing tissue culture technology may be recognized as potentially important tool in reforestation. Progress concerning C. libani research should be further expanded for root induction and acclimatization.

4. References

Corchete M P et al (1993) Plant Cell Reports 12, 534-536.
Ellis D D (1991) Plant Cell Tiss. Org. Cult. 27, 281-287.
FAO (1977) Report of the forth session of the FAO panel of Experts on Forest Gene resources, pp 75, FAO Publications, Canberra, Australia.
Lin Y et al (1991) Plant Cell Tiss. Org. Cult. 26, 161-166.
Mohammed G H, Vidaver W.E 1988. Plant Cell Tiss. Org. Cult. 14, 137-160.
Sankara R, Venkateswara R (1985) Plant Science 40, 51-55.
Villalobos V M et al (1985) Canadian Journal of Botany 63, 2172-2176.
Yahyaoglu Z et al (1990) Cedar Symposium Proceedings, Forest Research Institute, No:59 pp. 325-331, Ankara, Turkey.

AN OVERVIEW OF PROGRESS ON SOMATIC EMBRYOGENESIS IN FOREST TREES

S. Mohan Jain

Department of Plant Production, Crop Husbandry Section, Latokartanonkaari 5, FIN-00014, Box 27, University of Helsinki, Helsinki, Finland.

1. Introduction

An efficient plant regeneration system by tissue culture is a prerequiste for the successful use of biotechnology in forest improvement programmes. Plant tissue culture has a great potential for rapid multiplication of elite tree genotypes for the large-scale plant production (Jain, 1997; Ahuja, 1992). This technology is highly labour intensive and cost effective, and is being widely used in the developing countries (Table 1). Since forest trees are long lived and highly heterozygous, it is important that regenerated plants are genetically uniform of the selected genotypes (Klopfenstein et al, 1997). Commonly, all cell and tissue culture techniques require the growth of tissue and organ culture segments on a suitable nutrient medium that stimulate the development of either callus or direct shoot formation or organogenesis. The regenerated shoots are rooted either *in vitro* or *in vivo*. This approach of plant regneration is termed as micropropagation, which is commercially used for large-scale plant multiplication (Jain, 1997).

Somatic embryogenesis is another highly effective tool for clonal plant multiplication (Jain et al, 1995; Timmis, 1998), and it was for the first time reported in *Daucus carota* (Steward et al, 1958), *Santalum album* (Rao, 1965), *Ilex aquifolium* (Sussex, 1972), and *Picea abies* (Hakman & Von Arnold, 1985; Hakman et al, 1985; Chalupa, 1985). There has been a tremendous progress on somatic embryogenesis and production of somatic seedlings in forestry of almost every commercially important tree belonging to both angiosperms and gymnosperms (Tables 2 & 3; Jain et al, 1995).

2. Advantages and disadvantages of somatic embryogenesis

Somatic embryogenesis offers several advantages in improving forest trees, and they are: a) cost effective large-scale clonal plant production and capable of producing umlimited

A. Altman et al. (eds.), Plant Biotechnology and In Vitro Biology in the 21st Century, 57–63.

Table 1. Tissue culture technology in important woody and fruit trees in some developing countries. TC: tissue culture; M- Micropropagation; SE- somatic embryogenesis

Countries	Tree species	Technology in progress
Bangladesh	Jute, bamboo, coconut, papaya, neem, teak Mahogany	TC:M
China	Banana, Citrus, papaya,, apple, kiwi, Poplar	TC: M,SE
Fiji	Banana, vanilla, taro	TC
India	Banana, apple, cherry, grapes, papaya, citrus, tea, oilpalm, bamboo, poplar, neem, mango, coconut, *Eucalyptus*, coffee, sandalwood	TC: M,SE
Indonesia	Citrus, banana, papaya, bamboo, coconut, oilpalm, cacao	TC: M, SE
Malaysia	Oilpalm, rubber, bamboo, cacao, papaya, banana	TC: M
Nepal	Banana, Citrus, *Eucalyptus, Ficus, Dalbergia, Artocarpus*	TC: M
Pakistan	Datepalm, Citrus, banana	TC: M
SriLanka	Tea, rubber, coconut, banana, pineapple	TC: M
Thailand	Rubber, oilpalm, coffee, mango,lychee	TC: M

number of plants; b) Both root and shoot meristem development occurs in the same step of the process; c) quick and easy scale-up can be achieved via liquid culture e.g. growing of somatic embryos in bioreators and ultimately automation of somatic seed production; d) long-term storage of somatic embryos via cryopreservation and establishment of germplasm banks; e) Development of artificial seeds or somatic seeds; f) encapsulation dehydration - a combination of artificial seed and cryopreservation technologies (Blakesley et al, 1996), and g) Genetic transformation. The advantages of clonal propagation are: 1) commercial exploitation of unique combinations of invaluable genetic traits that can not be reproduced through conventional breeding, and 2) harvesting of uniform material (Timmis, 1998). Merkle (1995) mentioned two major limitations to forest tree somatic embryogenesis: a) low numbers of field-plantable clonal plantlets produced per embryogenic culture, and b) inability to initiate embryogenic culture from mature trees of harwood types. So far, mostly seeds or seedlings of hardwood forest trees have been used as an explant for the induction of somatic embryogenic cultures, and thus propagated material is of unproved genetic value (Merkle, 1995). Both genotypic and somaclonal variation are also limitation factors for clonal propagation in forest trees (Jain et al, 1998).

3. Induction of somatic embryogenesis

Female gametophyte explants have successfully been used to induce somatic embryogenesis in conifers (Jain et al, 1995). Durzan & Gupta (1987) reported that immature embryos of *Pseudotsuga menziesii* excised with female gametophyte still attached via the suspensor system produced higher percentages of embryogenic cultures than those lacking the female gametophyte. Jain et al (1988) obtained three types of calli by culturing mature zygotic embryos of *Picea abies*: a) globular, b) light green and compact, and c) white mucilage. Type a and b calli were non-embryogenic and did not turn into embryogenic cultures (white mucilage) even after several subcultures on the fresh medium. These results indicated the presence of different cell population which differentially responded to the similar growth conditions. In Poplar, somatic embryogenesis is induced from leaf explants either taken from greenhouse- or field-grown plants (Michler & Bauer, 1991). The most common growth regulators such as 2,4-dichlorophenoxyacetic acid (2,4-D) and napthalene acetic acid (NAA) have been used regularly for the initiation of embryogenic cultures, however, in some cases, e.g. in Abies, auxins have been inhibitory and needed a cytokinin, kinetin or benzylaminopurine (BA) (see Jain et al, 1995). Both cytokinin and auxin are needed for the initiation of embryogenic cultures in Picea sps (Jain & Ishii, 1998).

***Table 2.* Somatic embryogenesis in angiosperm woody plants.**

Important selected woody species	
Aesculus hippocastanum	*Betula album, B. pendula*
Camellia japonica, C. sinensis	*Eucalyptus citridora, E. grandi, E. dunni*
Quercus serrata, Q. bicolor, Q. ilex, Q. rubra, Q. suber	*Mangifera indica*
Prunnus sps.	*Citrus* sps.
Salix sps	Oil palm, Date palm
Olive	Kiwi
Avacado	*Santalum album*
Coffea arabica	Walnut
Liriodendron tulipifera	*Castanea dentata*
Magnolia macrophylla	*Robinia pseudoacacia*
Carya illinoensis	*Liquidambar styraciflua*
Ocotea catharinensis	*Malus pumila*
Populus sps.	*Hoheria angustifolia*

The maintenance of somatic embryogenic cultures is usually done on the medium similar to the intiation medium on semi-solid or liquid medium. Gupta & Grob (1995) reported that *Pinus taeda* embryogenic cultures can't be maintained for a long time on the semi solid medium because cultures undergo aging process and often show a decline in embryogenic potential. Chelik and Klimaszewska (1991) separated embryogenic outgrowth as soon as the tissue reaches sufficient mass, otherwise the embryo can be overgrown by callus and lost. Therefore, it is essential to subculture on the fresh medium at 2-3 week interval. Muralidharan & Mascarenhas (1995) maintained spontaneous repetitive somatic embryogenesis in *Eucalyptus citridora* by regular subculturing at 4 week interval for nine years without losing embryogenic potential. The species, genotype and the type of explant are important for the induction of repetitive embryogenesis and the maintenance of embryogenic potential in forest trees (Jain & Ishii, 1998).

Table 3. Somatic embryogenesis in gymnosperm forest trees

Important selected species
Abies alba, *A. balsamea, A. fraseri, A. nordmanniana*
Larix decidua*, L. leptolepis, L. occidentali*
Picea abies*, P. glauca, P. glauca-engelmannii complex, P. glehnii, P. jezoensis, P. mariana, P. omorika, P. pungens, P. ruben, P. sitchensis, P. wilsonii*
Pinus caribaea*, P. elliottii, P. lambertiana, P. nigra, P. patula, P. bungena, P. radiata, P. sylvestris, P. banksiana, P.* taeda, *P. strobus, P. pinaster*
Pseudotsuga menziesii
Sequoia sempervirens
Taxus brevifolia, *T. baccata, T. cupsidata, T. x media*
Cycadales, *Encephalatos sps., Ceratozamia mexicana, C. hildae*

The osmotic potential of the culture medium, and abscisic acid (ABA) are essential for the development and maturation of somatic embryos in conifer trees (Jain et al, 1995). The role of ABA is not useful in maturation of somatic embryos of *Eucalyptus grandis* and *Populus* (Jain & Ishii, 1998). The germination rate of somatic embryos is very much dependent on tree species and genotype. Synchronized germination of matured conifer somatic embryos was accomplished by reducing sugar concentration and addition of activated charcoal in the medium without any growth regulators. The accumulation of lipid reserves and subsequently their utilization is essential during germination. Somatic sedlings of limited number of forest trees have undergone field trials e.g. *Pseudotsuga menziesii* , *Picea abies*, and *Pinus radiata* because of low germination rate of somatic embryos.

4. Commercialization of somatic embryogenesis

Sutton & Polonenko (1998) suggested three major considerations that must be considered in successfully developing a commercial venture for a plant somatic embryogenesis system: a) development of a reliable, high output and cost efficient somatic embryo manufacturing process; b) development of a user-friendly product formulation and biologically stable deivery system; and c) development, maintaenance and service of a market place demand for the somatic embryogenesis products. Cervelli & Senaratna (1995) outlined research and development of plant somatic embryogenesis system that be commonly used for other plants: a) induction of embryogenic cell cultures from zygotic parent material, and preservation and storage of embryogenic material for a long time, b) germination of large number of proembryos through "bulking up" of maintenance somatic embryogenic cultures, c) conversion of proembryos into mature embryos, d) harvesting, e) processing (some degree of drying) and storage, f) germination *in vitro*, g) transplantation into containerized nursery growing mixes, and h) "out-planting" into natural soil systems. Timmis (1998) emphasized reducing the labour requirements for converting somatic embryos into plants.

5. Biochemical and molecular biology of somatic embryogenesis

Seed proteins from a number of conifers have shown that crytalloids are the major storage proteins, which belong to legumin class of storage proteins (Chatthai & Misra, 1998). A comparative study of somatic and zygotic embryos of *Picea abies* showed a similar of storage protein accmulation at the later stage of embryogenesis, however, crystalloid protein began to accumulate during early stage of embryogenesis. Von Arnold et al (1995) demonstrated differences in extracellular proteins between embryogenic callus (type A) capable of producing somatic embryos which can be matured with ABA treatment and type B callus uanable to do so. There were three proteins exclusively found in the medium of growing type A callus and were absent in type B callus medium. One of them with molecular weight 28kda showed around 55% N-terminal similarity to a newly discovered class of antifungal protiens . Also, these two calli differed in peroxidases and chitinases. Minocha et al (1993) showed that embryogenic culures had a higher concentration of free putrecine than spermidine in *Picea abies*. The role of polyamine metabolism seems to be important in the process of somatic embryogenesis (Minocha et al, 1995). Chatthai et al (1997) isolated a novel metallothionein-related cDNA clone PM2.1 and that expressed in somatic and zygotic embryos, haploid tissue, as well as in hormone-and metal-treated seeds and seedlings of dougflas fir. However, the expression of PM 2.1 gene in somatic embryos was dependent upon ABA and osmoticum, and ABA and metal ions in seeds and seedlings.

Molecular markers such as restriction fragment length polymorphism (RFLP), random amplified polymorphic DNA (RAPD), amplified fragment length polymorphisn (AFLP) are needed to identify genetic variants at the early stage of plant development. This is due to long life cycle of forest trees. Molecular marker (RAPDs) analysis indicated identical clonal somatic seedlings of *Picea abies*, *Populus deltoides*, and *Quercus serrata* (Jain & Ishii, 1998). Hashmi et al (1997) showed variation among plants regenerated from embryo callus of peach. Fourre et al (1997) suggested that cytological analysis of somatic seedlings would be useful because they showed cytological variation in somatic seedlings of *Picea abies* and failed to see any variation by RAPD markers. Molecular markers will also be useful in gene mapping, identification of homozygosity, characterization of inbred lines, quantitative trait loci (QTLs), and marker assisted breeding (Schuch, 1991).

6. Genetic transformation of embryogenic cultures

Usually, woody plants are recalcitrant for genetic transformation due to difficulties in plant regeneration. Recently Jain et al (1995) edited a book whic includes reviews on genetic transformation of angiosperms and angiosperm woody plants. Somatic embryogenic cultures have successfully been used in genetic transformation of *Liriodendron tulipifera, Larix* sps, *Picea* sps, *Pinus* sps (Jain & Ishii, 1998) by using either biolistic or *Agrobacteruium*-mediated transformation method. Embryogenic cultures have an advantage of obtaining direct transgenic somatic embryos which can be germinated into transgenic somatic seedlings, avoiding or minimizing somaclonal variation. So far, a few important genes including lignification, *rol*C, resistance against insect and pests, reproductive sterility, drought and herbicide resistant have been introduced in forest trees. Long life cycle and extended vegetative phase of forest trees may hamper the monitoring of transgene expression. Transgene, e.g. may cause abnormality immediately or remain dormant for a long time or even might be lost during the long vegetative phase of the tree species, and therefore, it is rather difficult to predict the behaviour of transgene in the future.

7. References

Ahuja MR (1992) Micropropagation of woody plants, Kluwer Academic Publishers, Dordrecht, The Netherlands.

Blakesley D et al. (1996) Plant Growth Regl. 20, 11-16.

Cervelli R & Seneratna T (1995) Automation and environmental control in plant tissue culture, pp 29-64, Kluwer Academic Publishers, Dordrecht, The Netherlands

Chalupa V (1985) Commun. Inst. For. Cech. 14, 57-63

Chatthai WM & Misra S (1998) Molecular biology of woody plants, Kluwer Academic Publishers, Dordrecht, The Netherlands (in press).

Chatthai WM et al. (1997) Plant Mol Biol, 34: 243-254.

Cheliak WM & Klimaszewska K (1991) Theor. Appl. Genet. 82,185.190.

Durzan DJ & Gupta PK (1987) Plant Sci. 52, 229-235.

Gupta PK & Grob JA (1995) Somatic embryogenesis in woody plants, Vol 1, pp 81-98, Kluwer Academic Publishers, Dordrecht, The Netherlands

Fourre JL et al. (1997) Theor Appl Genet 94, 159-169.

Hakman I & Von Arnold S (1985) J Plant Physiol 121, 149-158.

Hakman I et al (1985) Plant Sci 38, 53-59.

Hashmi G et al. (1997) Plant Cell Rept 16, 624-627.
Jain SM (1997) Plant biotechnology and plant genetic resources for sustainability and productivity, pp 227-233, Academic Press, Austin, USA,
Jain SM & Ishii K (1998) Recent advances in biotechnology for tree conservation and management, International Foundation of Science, Stockholm, Sweden (in press).
Jain SM et al. (1988) *Theor. Appl Genet 76, 501-506*
Jain SM et al. (1995) Somatic embryogenesis of woody plants, Vol 1-3, Kluwer Academic Publshers, Dordrecht, The Netherlands.
Jain SM et al. (1998) Somaclonal variation and induced mutations in crop improvement, Kluwer Academic Publishers, Dordrecht, The Netherlands.
Klopfenstein NB et al. (1997) Micropropagation, genetic engineering, and molecular biology of *Populus.* Gen Tech Rep RM-GTR-297, Fort Collins, Co; USDA, Forest Service, Rockey Mountain and range Expt. Stn. 326 p
Merkle SA (1995) Plant Tiss. Cult. & Biotech. 1, 112-121.
Michler CH & Bauer EO (1991) Plant Sci. 77, 111-118.
Minocha R et al. (1993) Tree Physiol 13, 365-377.
Minocha R et al. (1995) Somatic embryogenesis in woody plants, Vol 1, pp 337-360, Kluwer Academic Publishers, Dordrecht, The Netherlands.
Muralidharan EM & Mascarenhas AF (1995) Somatic embryogenesis in woody plants, Vol 2, pp 23-40, Kluwer Academic Publishers, Dordrecht, The Netherlands
Rao PS (1965) Phytomorph 17, 175-179.
Schuch W (1991) In Vitro Cell Dev Biol 27P, 99-103
Steward FC et al. (1958) Am J Bot, 45, 7095-7098.
Sutton BCS & Polonenko DR (1998) Somatic embryogenesis in woody plants, Vol 4, Kluwer Academic Publishers, Dordrecht, The Netherlands (in press).
Sussex I (1972) Phytomorph 22, 50-59.
Timmis R (1998) Biotechnol Prog 14, 156-166.
Von Arnold S et al. (1995) Somatic embryogenesis in woody plants, Vol 3, pp 17-36, Kluwer Academic Publishers, Dordrecht, The Netherlands.

GENE EXPRESSION DURING SOMATIC EMBRYOGENESIS IN CARROT SUSPENSION CULTURES

A. Komamine [1,2], J. Asami[1], Y. Endo[1], Y. Yamanaka [1], S.Sato [3], M. Smith[4], S. Wada[1], Y. Yanagawa[5], H.Nakagawa[5] and R. Kawahara[4],
[1]Japan Women's University, Mejiro, Tokyo, [2] Research Institute of Evolutionary Biology, Kamiyoga, Setagaya, Tokyo, [3] Mitsui Plant Biotechnology Institute, Tsukuba, [4] RIKEN,Wako, Saitama and Tsukuba, [5] Chiba University, Matsudo, Japan

Somatic embryogenesis in carrot suspension cultures is an ideal system for investigations of the entire process of development as well as a model system for investigations of totipotency expression in plants.
Developmental program of somatic embryogenesis should be regulated by gene expression. In order to investigate gene expression during somatic embryogenesis, a high frequency and synchronous embryogenesis system is required.
We established a high frequency and synchronous embryogenesis from single cells (Fujimura, Komamine 1979; Nomura, Komamine 1985) for investigations of embryogenesis at molecular level in carrot. Using this system, we isolated several genes which were expressed during embryogenesis or specifically at some stages of embryogenesis. In this paper, we report the characterization and expression of these genes during embryogenesis and discuss functions of these genes in embryogenesis.

1. 21D7 gene
21D7 was first identified as a 45-kD carrot antigen. 21D7 protein was detected in embryogenic carrot cultures and meristematic tissues of various species of plants (Smith et al. 1988). 21D7 protein could be detected in competent single cells (State 0), which can divide to form embryogenic cell clusters (State 1) in the presence of auxin, and in State 1 embryogenic cell clusters formed, globular, heart, torpedo shaped embryos. However, when competent single cells were cultured in the absence of auxin, cells became "elongated cells" which could not divide nor differentiate even if they were cultured in the presence of auxin. 21D7 protein could not be detected in competent single cells after transfer to auxin free medium.
21D7 cDNA was cloned by immunoscreening. Using this clone, we confirmed expression of 21D7 gene in growth stages of batch cultures and synchronous cell division culture of *Catharanthus roseus*. The results revealed that 21D7 expression is associated with cell division (Smith et al. 1993). A homologus gene to 21D7 gene, Sun2, was reported in *Saccharomyces cervisiae* (Kawamura et al. 1996). Sun 2 gene

A. Altman et al. (eds.), Plant Biotechnology and In Vitro Biology in the 21st Century, 65–68.

encodes a regulatory subunit of 26S proteasome. Fractionation of carrot and spinach crude extracts showed that 21D7 protein sedimented together with the active 26S proteasomes. The expression of plant 21D7 rescued the yeast *sun* 2 mutant (Smith et al. 1997). It was concluded that 21D7 protein is a subunit of the plant 26S proteasome and plays the important role in cell division which is essential to occur in the early phase of somatic embryogenesis.

2. CEM1

In order to isolate embryo specific genes, we performed differential screening between embryogenic cell clusters cultured in the 2,4-D free medium and those in the 2, 4-D containing medium and 5 independent clones were obtained. One of them, designated CEM1, was revealed to express strongly at proglobular - globular stage in which active cell division occurs, but its expression declined thereafter. Nucleotide sequence and homology search revealed that CEM1 has a high homology with Elongation Factor1 α (70~90% at amino acid level) of eukaryotic species, indicating that CEM1 encodes Elongation Factor 1 α which is an essential protein for elongation of polypeptide in protein synthesis. *In situ* hybridization indicated that CEM1 transcripts were located in actively dividing areas in embryos. It is reasonable that CEM1 is expressed preferentially in actively dividing areas in which active protein synthesis should occur, requiring synthesis of Elongation Factor 1 α (Kawahara et al. 1992).

3. Organ specific genes

Hypocotyl and root specific genes were isolated by differential screening between hypocotyls and roots of carrot seedlings. Four clones were isolated; they were designated CAR3 and 4 (hypocoty specific), and CAR5 and 6 (root specific) and expressed after globular embryos or heart-shaped embryos. Expression of these cDNAs was at very low level in cells which were cultured in the medium containing 2, 4-D, and was strongly suppressed when 2,4-D was added to heart shaped embryos. *In situ* hybridization analysis revealed that CAR4 was expressed in epidermis and in the regions surrounding tracheary elements in torpedo shaped embryos.

4. Homeobox genes

Homeodomains (HDs) are DNA-binding domains that have been well characterized in animals, and HD proteins are thought to be regulators of transcription. To investigate the regulation of gene expression during somatic embryogenesis in carrot an attempt was made to isolate cDNA clones that encode HD proteins, and independent clones (CHB1 through CHB6) were isolated. Transcripts corresponding to CHB1 through CHB6 were expressed at different times during somatic embryogenesis. In particular, transcripts corresponding to CHB2 were expressed in close association with the early development of embryos, indicating that CHB2 may be embryo specific gene. CHB4 and 5 were expressed after torpedo shaped embryos and CHB4 in hypocotyls and CHB5 in hypocotyls and roots, indicating CHB4 and 5 are involved in differentiation of hypocotyls and / or roots during somatic embryogenesis (Kawahara et al. 1995).

5. CEM6

In order to isolate a really embryo specific gene, we performed subtractive differential screening. The cDNA library was constructed from proglobulars. For use as probe in screening, the same cDNA used for library construction was enriched for specific sequences using subtractive hybridization. The cDNA used for subtraction was prepared from suspension cultures 5 days after subculturing in auxin-containing medium. Nine independent differentially expressed cDNA clones were obtained from a screen of 150000 recombinant phages. Northern analysis indicated one of these, CEM6, to be expressed specifically during somatic embryogenesis, i.e. only from globular to torpedo shaped embryos. The amino acid sequence deduced from the nucleotide sequence of the CEM6 cDNA indicates that it encodes a glycine-rich protein containing a hydrophobic signal-sequence like domain. Its early embryo specific expression and sequence characteristics suggest an important role as a protein surrounding cell wall in embryogenesis (Sato et al. 1995). *In situ* hybridization revealed that transcripts of CEM6 were localized in peripheral area of cells in globular to torpedo shaped embryos, but signals of CEM6 began to disappear from root parts of late stage torpedo embryos and no signal could be observed in early seedlings. Antisense CEM6 was introduced to carrot cells in suspension cultures. Transformant cells were selected by kanamycin to obtain stable transformant cell population which was checked by GUS assay, PCR and RT-PCR. From 6 transformant lines obtained, embryogenesis was induced according to Fujimura and Komamine's method (1979). Frequency of embryogenesis was much lower (10～40%) than in the control culture (<90%). Even if embryogenesis was induced, the first appearance of globular stage embryos was much delayed and most formed embryos were polyembryos. Since antisense cDNA cannot suppress completely expression, these findings indicate that antisense CEM6 suppressed embryogenesis, suggesting that CEM6 plays an essential (at least an important) role in somatic embryogenesis.

6. Conclusion

We isolated several genes which were expressed during somatic embryogenesis. They are classified into three categories:

a) Genes involved in cell division

21D7 and CEM6 belong to this category. In somatic embryogenesis, initiation and maintaining of cell division are essential events. 21D7 was found to encode a regulatory subunit of 26S proteasome, functioning to regulate the progress of the cell cycle. Thus expression of 21D7 gene is essential for initiation of cell division and expression of totipotency. CEM1 encodes Elongation Factor1 α which is an essential protein for protein synthesis required to active cell division which occurs in prior to formation of globular embryos.

b) Genes involved in organ formation

In the late stages of embryogenesis, that is, torpedo stage in particular, formation of organs, hypocotyl and root, is initiated. Thus, CAR3, 4 and CHB4 involved in hypocotyl formation, CAR5, 6 and CHB5 involved in hypocotyl and / or root formation were expressed from globular and torpedo stage of embryos. It is interesting that expression of all these genes was suppressed by exogenously supplied auxin which also inhibits embryogenesis.

c) Embryo specific genes

Two genes CHB2 and CEM6, which were expressed specifically only in globular-torpedo stage of embryogenesis, were isolated. CHB2 is one of homeobox genes. CEM6 encodes a glycine-rich protein containing a hydrophobic signal-sequence, may be a glycoprotein. CEM6 transcripts were localized in peripheral cells of torpedo shaped embryos in particular. Introduction of antisense CEM6 suppressed somatic embryogenesis, suggesting CEM6 plays an important role in somatic embryogenesis. Functions of these genes are under investigation.

References

Fujimura T, Komamine A (1979) Plant Physiol. 64, 13-19
Kawahara R et al. (1992) Eur. J. Biochem. 209, 157-162
Kawahara R et al. (1995) Plant Mol, Biol. 27, 155-164
Kawamura M et al. (1996) Mol. Gen. Genet. 251, 146-152
Nomura K, Komamine A (1985) Plant Physiol. 79, 988-991
Sato S et al. (1995) Plant Mol. Biol. 28, 39-46
Smith JA et al. (1988) Planta 174, 462-472
Smith MW et al. (1993) Plant Physiol. 101, 809-817
Smith MW et al. (1997) Plant Physiol. 113, 281-291

PHOTOCONTROL OF SHOOT REGENERATION FROM HYPOCOTYLS OF TOMATO

B. Lercari, S. Moscatelli, E. Ghirardi, R. Niceforo and L. Bertram
Dipartimento di Biologia delle Piante Agrarie, Università di Pisa, Viale delle Piagge 23, I 56124 Pisa, Italy

1. Introduction

Thanks to the manipulation of hormone composition in the medium it has been possible to attribute morphogenic function to specific hormones and to separate spatially and in time the different hormone-dependent steps which are present in the morphogenic process (Christianson and Warnick 1988; Fambrini et al.1996). By contrast, the role of the photomorphogenic pigments (if any) in the discrete steps characterizing the process of morphogenesis *in vitro* is still an open question despite a number of studies showing that light quantity, quality and duration influence both shoot and root formation (Hughes, 1981; Lercari et al. 1986; Economou and Read, 1987; Marcenaro et al, 1994).
In the present paper we investigated the photocontrol of shoot regeneration from hypocotyl explants of an *au* mutant of tomato and its isogenic wild type. The *au* mutant used in this study was isolated from tissue cultures as a result of somaclonal variation (Lipucci di Paola et al., 1988). The etiolated seedling of *au* mutant lacks spectrophotometrically detectable phytochrome, as a consequence of a excision in chromophore synthesis (Terry and Kendrick, 1996). In tomato plants at least five phytochrome genes are expressed (H*au*ser et al., 1997).

2. Materials and Methods

Seeds of the *aurea* mutant of tomato, *Lycopersicon Esculentum* Mill., and of its isogenic wild type cv UC 105, were germinated after disinfection, aseptically on Murashige and Skoog, 1962 (MS) hormone-free medium, to which 30 g/l of sucrose and 7 g/l of Difco Bacto-agar were added. The cotyledons and the root were excised after ten days of culture at 25°C in darkness or under a 16 h photoperiod of white fluorescent light at 50 $\mu mol\ m^{-2}\ s^{-1}$(Philips TL 40 W/ 33 RS fluorescent lamps), unless otherwise specified. At the moment of the excision, the hypocotyl length of the two genotypes was similar bec*au*se the germination of the *au* seeds was slower than that of the wild type seeds, while the subsequent elongation rate of *au* hypocotyls was greater. The excised hypocotyls were

A. Altman et al. (eds.), Plant Biotechnology and In Vitro Biology in the 21st Century, 69–72.

divided into three segments 2.5-3 cm long and transferred to 5 cm diameter Petri dishes containing MS medium without hormones or with indoleacetic acid (IAA), napphtalenacetic acid (NAA), zeatin riboside (Zr), isopentenyladenine(2iP) and benzilaminopurine (BAP), each at three concentrations: 10^{-7}, 10^{-5} and 10^{-4} M. The segments of hypocotyl obtained from apical, middle and basal position on the seedlings were cultured in distinct Petri dishes. Four segments with the same position of origin were cultured on each Petri dish, all segments were placed horizontally on the Petri dish with the same direction of polarity. Hypocotyl explants were grown at 25°C under different light conditions; every week callus, root and shoots formation were recorded.
The red light (0.5, 2.5 and 5.0 µmol $m^{-2}s^{-1}$) was obtained Philips TL 20 W /15 fluorescent tubes filtered with Roscolux red filter N° 27 , far red light (0.5, 2.5 and 5.0 µmol $m^{-2}s^{-1}$) was obtained from Silvania VHO F40T12/232 fluorescent tubes filtered with Roscolux blue filter N° 27, and blue light (0.5, 2.5 and 5.0 µmol $m^{-2}s^{-1}$) was obtained from Philips TL 20W/18 fluorescent tubes filtered with Roscolux blue filter n° 83 (Fig. 1). The values presented are averages ± SD of 8 -12 Petri dishes, i.e. 32 - 48 explants, obtained in two experiments.

3. Results

De-etiolated explants were cultured in presence of white light for different periods of time, then transferred to the dark. The percentage of explants that regenerated shoots was determined after 35 days of culture (end point determination). The experimental results show that the light-dependent morphogenetic ability of de-etiolated wild type hypocotyl is not influenced by their original position on the seedling (Tab. 1). By contrast, there is a significant difference between the morphogenetic potential of *au* explants obtained from different positions on the seedling. Almost half of the apical segments of *au* hypocotyls regenerated shoots, while the regeneration from middle and basal segments was negligible.

Table 1. The percentage of de-etiolated hypocotyls explants that regenerated shoots measured after 35 days of in vitro culture. Mean values±SD are shown.

Segments position	Wild type	Aurea
Apical	64 ±6	52.5 ±6
Middle	50 ±5	0
Basal	55 ±5	0

In the case of etiolated explants from wild type tomato the percentage of middle and basal explants from etiolated seedlings that regenerated shoots was reduced in comparison to that found in de-etiolated explants (compare Tabs 1 and 2). No shoot regeneration was displayed by etiolated *aurea* hypocotyls.

Table 2. The percentage of etiolated hypocotyls explants that regenerated shoots measured after 35 days of in vitro culture. Means values ± SD are shown.

Segments position	Wild type	Aurea
Apical	50 ±6	0
Middle	12 ±3	0
Basal	8 ±2	0

No one of the five hormones added to MS medium was able to eliminate the differences in regeneration capability between the different positions of hypocotyl segments of the aurea mutant. Thus irrespective of the presence of the hormones the regeneration response was null or negligible in the middle and basal segment of the *aurea* mutant (Table 3). No one shoot was regenerated from explants of both the genotypes when cultured in darkness.

Table 3. Effect of the presence of auxins and cytokinins in the culture medium. The percentage of de-etiolated hypocotyls explants that rergenerated shoots was measured after 35 days of culture under a 16 h photoperiod. IAA= indoleacetic acid; NAA= naphtalenacetic acid;BAP= benzylaminopurine;ZR= zeatin riboside; 2iP= isopenteniladenine.

	Wild type			Aurea		
	Apical	Middle	Basal	Apical	Middle	Basal
IAA 0.1μM	39	67	72	63	0	8
IAA 1μM	55	77	87	59	25	0
IAA 10 μM	48	75	72	25	12	12
NAA 0.1μM	90	66	30	50	27	0
NAA 1μM	61	45	30	28	0	0
NAA 10μM	0	0	0	0	0	0
BAP 0.1μM	40	33	40	34	0	0
BAP 1μM	32	21	29	41	0	0
BAP 10μM	0	16	16	0	0	0
ZR 0.1μM	36	30	46	35	16	10
ZR 1μM	55	75	75	25	0	15
ZR 10μM	55	36	50	16	0	0
2iP 0.1μM	50	15	42	37	5	5
2iP 1μM	35	12	22	50	25	O
2iP 10μM	22	29	46	15	0	O

4. Discussion

It appears that in apical segments of wild type tomato the acquisition of the competence and determination for shoots regeneration does not require the presence of light during the seedling growth, whereas the subsequent realization of shoot regeneration requires the presence of light. By contrast in all positions of the au mutant and in middle and basal segments of tomato hypocotyls the presence of light for at least a part of the day is required for all the three steps characterizing the process of shoot regeneration, that is acquisition of competence, determination and finally the realization of shoot regeneration. The fact that the au mutant has a reduced level of phytochrome, strongly support the conclusion that the level of phytochrome plays a crucial role in the determination (position-dependent or not) of the capability of hypocotyl explants of tomato to regenerate shoots. The pivotal role of phytochrome is further supported by: 1) the significative regenerative response induced by daily red light pulses only in apical segments (data not shown); 2) the absence of shoot regeneration when the explants were grown in vitro in darkness even in presence of auxins and cytokinins and 2) histological studies showing that shoot primordia originate from vascular parenchima cells of the hypocotyls explants without an intermediate callus phase (data not shown). Indeed Adam et al, 1994, found that PHY-A genes were maximally expressed in the hook region and in the vascular tissues of the hypocotyl of tobacco.

5. References

Adam E et al. (1994) Plant J. 6, 283-293.
Christianson ML and Warnick DA (1988) Hort Sci. 23, 515-519.
Economou AS and Read PE (1987) Hort Sci. 22, 751-754.
Fambrini M et al. (1996) Plant Sci. 114, 205-214.
H*au*ser B et al. (1997) Planta 201, 379–387.
Hughes KW (1981) Environ Exp Bot 21, 281-288.
Lercari B et al.(1986) Physiol Plant 67, 340-344.
Lipucci Di Paola M et al.(1988) Adv. Hort. Sci. 2, 30-32.
Marcenaro S et al.(1994) Physiol. Plant. 91, 329-333.
Murashige T and Skoog F (1962) Physiol. Plant. 15, 473-497.
Terry MJ, Kendrick RE (1996) Biol. Chem. 271, 21681-21686.

6. Acknowledgements this research was supported by a grant from M.i.P.A., Piano Nazionale Biotecnologie.

FACILITATED INITIATION OF SOMATIC EMBRYOGENESIS IN *ARABIDOPSIS THALIANA* BY MUTATIONS IN GENES REPRESSING MERISTEMATIC CELL DIVISIONS

Andreas P. Mordhorst[1,2], Keete J. Voerman[1], Marijke V. Hartog[1], Ellen A. Meijer[1], Jaques van Went[2], Maarten Koornneef[3], and Sacco C. de Vries[1], [1]Laboratory of Molecular Biology, [2]Laboratory of Plant Cytology and Morphology, [3]Laboratory of Genetics; Department of Biomolecular Sciences, Agricultural University Wageningen, Dreijenlaan 3, 6703 HA Wageningen The Netherlands

Abstract

A very efficient and reproducible system for somatic embryogenesis in Arabidopsis was established by using intact seedlings of the *primordia timing* mutant (*pt*). Embryogenic clusters originated from the enlarged shoot apical meristem (SAM) of the mutant seedlings when germinated in 2,4-D containing liquid media. *pt* somatic embryos had all characteristic embryo pattern elements, but with higher and more variable numbers of cell layers and cells per cell layer. This finding shows that pattern formation can be completed in somatic embryos without the regular cell division pattern seen in zygotic embryos. Embryogenic cell lines were also established from seedlings of other mutants with enlarged SAMs, such as *clavata1* and *clavata3* (*clv*). *pt clv1-4* and *pt clv3-2* double mutants showed additive effects on SAM size and an even higher frequency of seedlings producing embryogenic cell lines. This data suggest that an increased population of noncommitted SAM cells may be responsible for facilitated establishment of somatic embryogenesis in *Arabidopsis*.

Introduction

Plant embryogenesis spans the developmental period that ranges from the zygote to the formation of the mature embryos. In plants, the entire process of embryogenesis can be recapitulated from other cells than the zygote, giving rise to apomictic, androgenetic and somatic embryos (reviewed by Mordhorst *et al.*, 1997). Somatic embryogenesis in the model plant Arabidopsis is an important tool for the analysis of embryogenesis especially because of the availability of many zygotic embryo mutants in this species. Somatic embryos or "embryo-like structures" have been described from a high variety of explant materials (Huang, Yeoman, 1984; Marton, Browse, 1991; Wu *et al.*, 1992; Mathur *et al.*, 1995; Pillon *et al.*, 1996; Luo, Koop, 1997; Mordhorst *et al.*, 1998). Here, we demonstrate the establishment of highly embryogenic cell lines by using the *primordia timing* mutant (*pt*, member of the *hpt/cop2/amp1/pt* complementation group) by germinating seeds in 2,4-D containing medium. Mutant *pt* embryos are characterised by an enlarged shoot apical meristems (SAM) (Mordhorst *et al.*, 1998) from which the initial embryogenic clusters all derived from. Embryogenic cell cultures were also established from seedlings of other mutants with enlarged SAMs, such as *clavata* (*clv*) and *pt clv* double mutants.

Material and Methods

Seeds of the *primordia timing* (*pt*) and the *clavata1* (*clv1-4*) and *clavata3* (*clv3-2*) mutants were kindly provided by I. Vizir (University of Nottingham, United Kingdom) (Vizir *et al.*, 1995) and S. CLARK (University of Michigan, Ann Arbor, Michigan, USA) (Clark *et al.*, 1993; 1995), respectively. For the initiation of cell lines, approximately 30 sterile seeds (Mordhorst *et al.*, 1998) were incubated in 20 ml liquid MS-medium (Murashige, Skoog, 1962) (Duchefa, Haarlem, The Netherlands) containing 2 % w/v

A. Altman et al. (eds.), Plant Biotechnology and In Vitro Biology in the 21st Century, 73–76.

sucrose, 4,5 µM 2,4-D and 10 mM MES at pH 5,8 (MS-4). After a cold pre-treatment of 2 days at 6-8 °C cultures were kept on a rotary shaker (100 rpm) at 25 °C in the light (approx. 3000 lux for 16 h, 8 h dark). Seeds germinated and each seedling developed a callus aggregate. After two weeks the medium was replaced by 20 ml fresh MS-4 medium. After three weeks the percentage of seedlings was determined that developed embryogenic green clusters with a smooth surface in addition to yellowish non-embryogenic callus. Embryogenic cell lines were established by subculturing only the embryogenic clusters from 1 to 3 seedlings in 20 ml B5-4 medium (Gamborg *et al.*, 1968) containing 2 % w/v sucrose, 4,5 µM 2,4-D at pH 5,8. Non-embryogenic lines were obtained by subculturing the yellowish callus in a similar way. Cell lines were maintained by a weekly subculture of approximately 30 - 50 mg embryogenic green clusters in 20 ml B5-4 medium. For the development of somatic embryos parts of aggregates from liquid cultures (approximately 1.5 x 1.5 mm in size) were transferred to agar (0.8 % w/v) - solidified growth regulator free 1/2 MS medium containing 1% w/v sucrose. After 7-10 days single embryos were dissected and transferred to new, vertically oriented 1/2 MS plates. The preparation of semi-thin sections and the determination of the number of cells in the L1 layer of the SAM of mature embryos was described elsewhere (Mordhorst *et al.*, 1998).

Results

Seedlings germinated in liquid medium in the presence of 2,4-D formed yellowish, rough and unorganised callus aggregates from the leaf veins of the entire cotyledons of wild type (L*er*) and *pt* seedlings (Fig. 1 a, b). In addition to this type of callus, another type of SAM-derived callus, characterised by a smooth surface and a bright green colour, developed from *pt* seedlings (Fig. 1b). Subculture of only the embryogenic clusters gave immediately rise to stable (for over 2 years) embryogenic cell lines. New embryogenic clusters were continuously formed on the surface of existing embryogenic clusters leading to large aggregates up to 1.5 cm in diameter that occasionally break apart. After transfer to agar-solidified media lacking growth regulators, mature somatic embryos developed. The morphology of somatic embryos was highly variable. Somatic embryos had fused or separated cotyledon(s) or were even fused to each other (Fig. 1 c,e). In general, embryos were oriented with the basal (root) pole attached to the callus tissue (Fig. 1c) but also the inverse orientation was found (Fig. 1d). To promote germination somatic embryos were separated from each other (fig. 1f). Somatic embryo-derived seedlings were transferred into the greenhouse and developed into fertile plants (Fig. 1g, left plant) even shorter and more bushy than *pt* seed grown plants (Fig. 1g, right plant). The *pt* R_2 generation was indistinguishable from the *pt* plants only propagated through seeds.

Since the embryogenic clusters derived from the enlarged *pt* SAMs, we analysed whether the *clv1* and *clv3* mutants (from which also enlarges SAM had been described by Clark *et al.*, 1995; Running *et al.*, 1995) as well as *pt clv* double mutants were also able to develop embryogenic cell lines. The percentage of seedlings producing embryogenic clusters was correlated with the increased size of the SAM of the original seedling (Table 1). Interestingly in both double mutant additive effects of both single mutants were seen concerning the size of the SAM and in the 'enhanced somatic embryogenesis' phenotype as well. In the case of *clv1-4* and *clv3-2* the amount of embryogenic clusters produced is rather low and cultures contained besides the embryogenic also non-embryogenic aggregates (data not shown).

Because the somatic embryos display differences in morphology to zygotic embryos, it was essential to confirm their internal organisation in order to qualify them as a true somatic embryo. Longitudinal sections through *pt* somatic embryos showed that all pattern elements as seen in the zygotic embryo were present (Fig. 1 h). Hypocotyls of

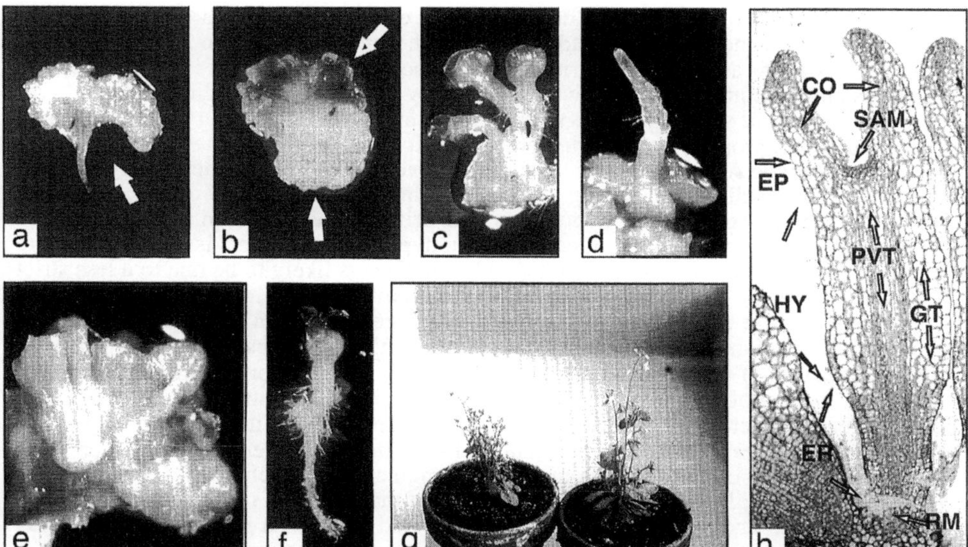

Fig. 1: (a) wild type (L*er*) and (b) *pt* seedlings three weeks germinated in the presence of 2,4-D (the white arrows point to non-embryogenic callus, the outlined arrow to an embryogenic cluster); (c) - (g) development of *pt* somatic embryos, (c) somatic embryos with fused cotyledon(s) and (d) inverted orientation, (e) fused somatic embryos, (f) separated single somatic embryo, (g) somatic embryo-derived (left) and seed-derived (right) *pt* plants in the greenhouse; (h) longitudinal section through somatic embryo; CO, cotyledons, SAM shoot apical meristem, EP, epidermis, PVT provascular strands, HY, hypocotyl, GT, ground tissue, ER, embryonic root, RM root meristem.

Table 1: Number of cells in the L1 layer of SAMs of mature embryos and percentage of seedlings developing embryogenic clusters (% EC) out of the SAM of wild type, *pt, clv1-4, clv3-2* and the respective double mutants .

Genotype	number of cells in L1	n	% EC*
WT (L*er*)	8.2 ± 0.7	6	0
pt	17.8 ± 2.3	8	30
clv1-4	11.9 ± 1.8	7	16
clv3-2	10.3 ± 1.2	6	4
pt clv1-4	21.0 ± 3.2	8	41
pt clv3-2	25.5 ± 4.0	11	42

Values are ±SE.

* Mean of two independent experiments each with two replicates and approx. 30 seedlings per replicate

both wild type (L*er*) and *pt* zygotic embryos are characterised by an invariant number of 3 continuous cell layers of the ground tissue (2 cortex and one endodermis layer) and 8 cell files in the pro-vascular tissue (n = 10). In somatic *pt* embryos, 4.4 ± 1.0 layers of ground tissue and 12.0 ± 5.6 (n = 13) cell files in the pro-vascular tissue were found. In zygotic *pt* embryos developing on plants derived from somatic embryos the numbers of cortex/endodermis cell layers and vascular cell files were identical to seed derived plants again (n = 16). A similar increase cell layers and cells per cell layer was determined in the primary root of *pt* somatic embryos compared to WT (L*er*) and *pt* seed-grown seedlings (data not shown).

Discussion

Two systems for continuous somatic embryogenesis in liquid medium of Arabidopsis are now available. One system employs dissected immature zygotic embryos as starting material (Wu *et al.*, 1992; Pillon *et al.*, 1996; Luo, Koop, 1997; Mordhorst *et al.*, 1998).

The second more efficient system, described here, employs *pt* seeds directly germinated in liquid media. One of the main advantages of the *pt* culture system is the fact that it employs dry seeds as starting material, making laborious dissection of immature zygotic embryos and continuous plant growth unnecessary. Based on the origin of the embryogenic clusters in *pt* seedlings, the enlarged SAM, embryogenic capacity was also found in *clv* mutants, a property not reported previously for these mutants.

Early wild type Arabidopsis zygotic embryo development is characterised by a highly regular pattern of divisions leading to an almost invariant number of cortex cell layers and pro-vascular cell files in hypocotyls and primary roots (Dolan *et al.*, 1993). In somatic *pt* embryos the number of cell layers in hypocotyls and roots is higher and more variable than in wild type and pt zygotic embryos. This is likely to be due to a less strict division pattern in the earlier stages of somatic embryos resulting from the *in vitro* conditions because zygotic embryos from plants regenerated via somatic embryos had the original invariant number of cell layers and cell files. This observation suggests that a regular pattern of embryogenic cell divisions is not required for patterning in somatic embryos. *pt* zygotic embryos are characterised by an enlarged SAM. Because embryogenic clusters developed from the SAM, we propose that it is this property of the *pt* mutant that is responsible for the 'enhanced somatic embryogenesis' phenotype. This is supported by a similar response in the other mutants used such as *clv1-4* and *clv3-2* (Clark *et al.*, 1995; Running *et al.*, 1995). The fact that both *clv* single mutants had less extra SAM cells and both *pt clv* double mutants had more than *pt* single mutants was reflected in their respective embryogenic cluster formation, suggesting even a quantitative relation between both traits. Our interpretation is that with an increased SAM size more undifferentiated or uncommitted cells which are present in immature zygotic embryos and persist in a high number in *pt* and *clv* mutants are able to respond to 2,4-D by initiating a new embryonic program.

Acknowledgements

The authors want to thank I. Vizir and S. Clark for the kind gift of seeds. A. P. M. was supported by grants from the Research School of Experimental Plant Sciences, the Agricultural University Wageningen, the EC programs Human Capital and Mobility, PTP-Biotech and the EC, Life Sciences and Technologies, Biotechnology Program (1994-1998) funding the European Plant Embryogenesis Network. E. A. M. was supported by PTP-Biotech.

References

Clark SE *et al.* (1993) Development 119, 397-418.
Clark SE *et al.* (1995) Development 121, 2057-2067.
Clark SE *et al.* (1997) Cell 89, 575-585.
Dolan L *et al.* (1993) Development 119, 71-84.
Gamborg OL *et al.* (1968) Exp. Cell. Res. 50, 151-158.
Huang BC, Yeoman MM (1984) Plant Sci. Lett. 33, 353-363.
Luo Y, Koop H-U (1997) Planta 202, 387-396.
Marton L, Browse J (1991) Plant Cell Rep. 10, 235-239.
Mathur J *et al.* (1995) Plant Cell Rep. 14, 221-226.
Mordhorst AP *et al.* (1997) Crit. Rev. Pl. Sci. 16, 535-576.
Mordhorst AP *et al.* (1998) Genetics in press.
Murashige T, Skoog F (1962) Physiol. Plant. 15, 473-497.
Pillon E *et al.* (1996) Plant Journal 9, 573-577.
Running MP *et al.* (1995) Methods in cell biology: Plant cell biology, pp. 215-227, California Academic Press, San Diego.
Vizir I *et al.* (1995) Abstract Book 6th Intern. Conference on Arabidopsis Res. pp. 128.
Wu Y *et al.* (1992) Protoplasma 169, 89-96.

Agarose-induced embryoid formation in sunflower protoplasts is triggered by RGD-mediated membrane-matrix adhesion.

Michel Petitprez, Henri Barthou, Christian Brière, and Gilbert Alibert
Laboratoire de Biotechnologie et Amélioration des Plantes, Unité associée INRA, Institut National Polytechnique-Ecole Nationale Supérieure Agronomique, 145 avenue de Muret, F31076 Toulouse cedex, France

1. Introduction

Embedding of plant cells or protoplasts in gelling medium is widely used to improve cell viability and plating efficiency (Smidsrod and Skjak-Braek, 1990). In *Arabidopsis thaliana* (O'Neill and Mathias, 1993) and *Helianthus annuus* (Chanabé et al. 1991), protoplast inclusion in a solid matrix induces changes in division pattern and development. In sunflower, protoplasts cultured in liquid medium divide symmetrically and form loose microcolonies. In contrast, when they are embedded in agarose, most of them divide asymmetrically and display a polar organization at the onset of their development giving rise to compact embryo-like structures (Petitprez et al., 1995). These embryoids progress to the heart-shape stage but do not develop further. Various hypotheses have been proposed to explain the effect of embedding on the fate of protoplasts: (i) limiting oxygen diffusion (Barbotin et al., 1993), (ii) simulating a cell wall, (iii) generating physical constraints that affect the stability of the cells (Asano et al., 1994). We previously showed that an increase of physical constraints such as hydrostatic pressure 0.4-1 MPa had an inhibitory effect on microtubule organization and cell wall synthesis and consequently on cell division (Barthou et al., 1997) but had no effect on division asymetry. Thus, physical constraints cannot solely explain the protoplast division pattern observed in agarose cultures.

Recently, in a study on the cytoskeleton of agarose-embedded protoplasts, Caumont et al. (1997) reported that the agarose matrix interacts with cortical microtubules and stabilizes the cytoskeleton. In animal cells, it is well known that several plasma membrane proteins are involved in linking the cytoskeleton to the extra-cellular matrix, the most prominent and best studied matrix receptors belonging to the integrin family (Hynes, 1992). In plant cells there is an increasing interest in such transmembrane proteins because they could play structural and signaling roles in controlling cell polarity and morphogenesis (Wyatt and Carpita 1993), although clear evidence is still lacking concerning the existence of such components (Reuzeau and Pont-Lezica, 1995). Nevertheless, indirect evidence of the role of integrin-like proteins has been provided in gravisensing, cell growth and polarity. The synthetic tripeptide Arg-Gly-Asp (RGD), which inhibits the binding of integrin to vitronectin or fibronectin inhibits plant cell growth (Schindler et al., 1989), membrane wall adhesions in *Pelvetia* (Henry et al., 1996) and polar axis fixation in *Fucus* (Shaw and Quatrano, 1996). Taken together, these results suggest that adhesion sites could forward extra-cellular signals into the cell via the cytoskeleton. In our system, the agarose matrix could play the role of an artificial wall able to bind transmembrane proteins. In order to test if such connections exist and if they can lead to unequal cell division and subsequent embryoid formation, we tried to prevent protoplast adhesion to the agarose matrix in mechanical (osmotic pressure) and chemical (ProNectin F and RGD peptides) ways.

2. Material and methods

Protoplast isolation and culture. Seeds of *Helianthus annuus* L. var. Emil (Pioneer Hibred International) were cultured as described by Chanabé et al. (1989). Hypocotyls of seven-day-old seedlings were used for isolation of protoplasts which were cultured at a final density of $5x10^4$

A. Altman et al. (eds.), Plant Biotechnology and In Vitro Biology in the 21st Century, 77–80.

mL^{-1} in L4 medium either in liquid medium or embedded in 0.5% low-melting-point agarose (Chanabé et al., 1991). Cultures were maintained 10 days at 25°C in the dark. After this period, they were transferred to the light (25µE.m^{-2}.s^{-1}, 16h illumination). Plating efficiency was assessed at 7 days by counting the division rate and at 14 days by counting microcolonies and embryoids.

Variation of osmotic pressure. Agarose droplets of embedded protoplasts were prepared as described above and allowed to solidify for 1h. They were then submerged with a liquid TLD medium containing 7%, 8% (standard condition) or 10% (w/v) mannitol.

Effect of ProNectin F and RGD peptides. Stock solution of ProNectin F (33µM in water) was diluted to the appropriate concentrations in TLD medium. Heptapeptides (stock solution 5mM) were kindly provided by R.F. Pont-Lezica (UPS-CNRS, Toulouse, Fr). In addition to the YGRGDSP sequence peptide, YGDGRSP was used as control in the same conditions.

Immunocytolocalisation. Protoplasts were fixed for 1h in 2% paraformaldehyde containing 0.03% glutaraldehyde in microtubule stabilizing buffer, treated according to Caumont et al. (1997) and microtubules revealed by a mouse anti β-tubulin.

3. Results

Effect of Osmotic pressure

In order to study the effect of plasma membrane agarose contact, protoplasts were cultured in agarose droplets submerged with a hypoosmotic medium (7% mannitol), a standard medium (8% mannitol) or a hyperosmotic medium (10% mannitol). Swelling of protoplasts following addition of hypoosmotic medium firmly presses the membrane onto the matrix whereas collapsing of the protoplast by addition of hyperosmotic medium reduces protoplast membrane agarose adhesion After 7 days of culture there was no significant difference between the division rates whatever the culture conditions. After 14 days of culture in standard medium (TLD 8% mannitol) the embedded protoplasts that developed gave rise to microcolonies or embryoids with relative proportions of 27% and 73% respectively. When the mannitol concentration of the culture medium was decreased to 7%, the proportion of embryoids increased to 89%. Conversely an increase of the osmotic pressure of the medium to 10% led to a lower proportion of embryoids (69%). When the mannitol concentration was increased up to 12% the formation of embryoids was almost completely suppressed (2,6%). Thus, membrane-agarose contact appears to be necessary for embryoid differentiation.

Effect of ProNectin

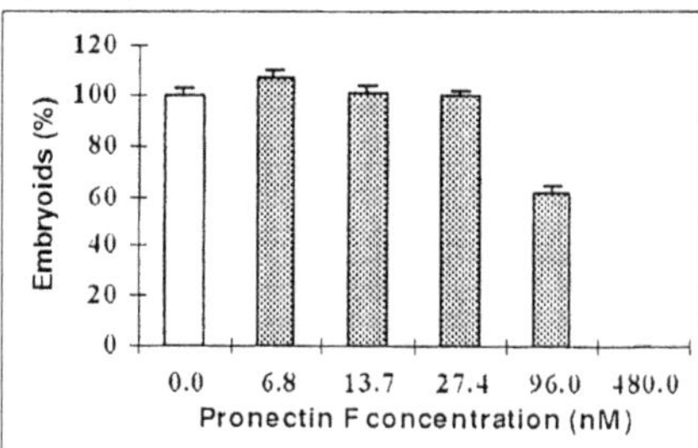

Fig 1. Effect of ProNectin F on the embryoid formation of agarose-embedded protoplasts counted after 14 days

ProNectin F (PNF) is a protein polymer, bearing 13 repeated RGD motifs and is currently used to promote cell attachment on rigid substrates. When agarose-embedded protoplasts are cultured in PNF containing medium the proportion of embryoids was significantly lowered in a concentration-dependent way (Fig 1): low concentration (<27 nM) did not affect embryoid number as compared to the control, whereas inhibition of embryoid formation was highly significant (P=0.01) for 100 nM and complete for 480 nM.

Effect of peptides

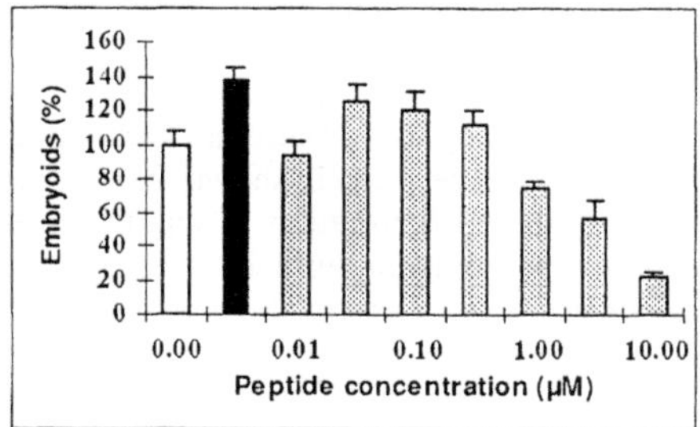

Figure 2. Effect of RGD heptapeptides on the embryoid formation in agarose embedded protoplasts; base 100 refers to the standard conditions. Protoplasts were cultured in the presence of different concentrations of RGD peptides (grey); in the presence of 1 µM DGR peptide (black), or without any added peptide (white). The difference between 1 µM RGD and 1 µM DGR was significant at P=0.01.

The involvement of specific amino acid motifs in protoplast-agarose adhesion was investigated by incubating protoplasts with synthetic heptapeptides containing RGD or DGR motifs. The division rate of protoplasts cultured, either in liquid or after embedding in agarose was not modified by the addition of these synthetic peptides whatever the applied concentration. When agarose embedded protoplasts were treated with 0.05 µM to 0.25 µM RGD peptide the number of embryoids slightly increased (+20%) compared to the untreated control. On the contrary, RGD concentrations over 0.25 µM, inhibited the production of embryoid versus microcolonies (Fig 2). Dose-response analysis indicates that the concentration of peptide required for 50% inhibition (EC_{50}) is around 2 µM. In the case of embedded protoplasts treated with 1 µM RGD peptide, the number of embryoids was significantly reduced (75%) compared to the GDR-treated control (140%) (P=0.01) showing the specific effect of the RGD motif.

Microtubular cytoskeleton modifications

Agarose-embedded protoplasts are characterized by (i) a randomly arranged cortical network of microtubules which gradually organize in long parallel bundles, and (ii) perinuclear microtubules forming a defined basket around the nucleus. When treated with RGD peptide, cortical microtubules appeared as short, sometimes punctuated, microtubules (Fig 3A) in up to half of the protoplasts. In the same way RGD reduces the setting up of the perinuclear microtubules basket (Fig 3B) specially after 48h.

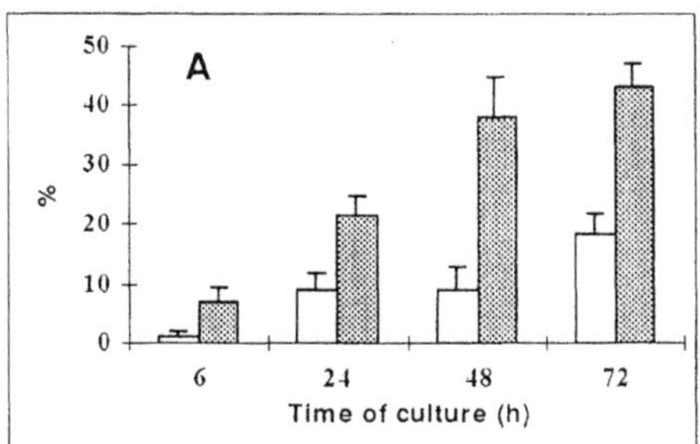

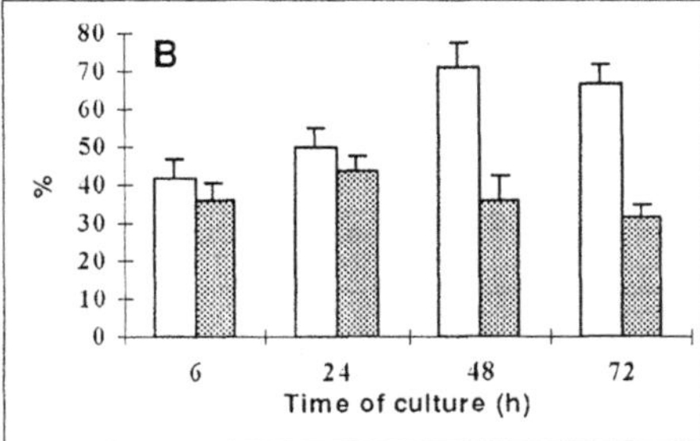

Fig 3. Effect of RGD heptapeptide on the cytoskeleton. Proportion of protoplasts having short cortical microtubules (A) and proportion of protoplasts having perinuclear microtubules (B) in the presence 10µM RGD peptide (grey) or without any added peptide (white).

4. Conclusion

These results suggest that membrane-matrix contacts could be mediated by RGD binding sites of the plasma membrane and transduced by microtubules. In a recent work Maniotis et al. (1997) showed that in endothelial cells mechanical constraints sensed by integrins are forwarded by the cytoskeleton up to the nuclear membrane and the nucleoplasm. These authors suggest that anisotropic constraints could modify the division axis. We have recently shown that agarose embedding of protoplasts changes the perinuclear microtubule organization (Caumont et al., 1997). Such a reorganization could represent a step in the cell signaling process in response to the anchorage of the protoplast membrane to the agarose matrix. Determining the nature and localization of the agarose-membrane anchorage sites, which are responsible for the asymmetric division of embedded protoplasts, provide an exciting goal for the study of the role of the extra-cellular matrix in regulating differentiation and development in plants.

Acknowlegments

The authors would like to express their aknowledgement to Pr. R.F. Pont-Lezica and Dr. H. Canut for their helpful comments . We also thank P. Anson for his technical assistance.

References

Asano Y, Ito Y, Sugiura K, Fujiie A (1994) Improved protoplast culture of Bentgrass (*Agrostis* L.) using a medium with increased agarose concentration. J Plant Physiol **143**: 122-124

Barbotin JN, Nava Saucedo JE, Bazinet C, Kersulec A, Thomasset B, Thomas D (1993) Immobilization of whole cells and somatic embryos: coating process and cell-matrix interactions. *In* K Redenbaugh, ed, Synseeds: application of synthetic seeds to crop improvement, CRC Press, Boca Raton, FL, pp 65-103

Barthou H, Brière C, Caumont C, Petitprez M, Kallerhoff J, Borin C, Souvré A, Alibert G (1997) Effect of atmospheric pressure on sunflower (*Helianthus annuus* L.) protoplast division. Plant Cell Rep **16**: 310-314

Caumont C, Petitprez M, Woynaroski S, Barthou H, Brière C, Kallerhoff J, Borin C, Souvré A, Alibert G (1997) Agarose embedding affects cell wall regeneration and microtubule organization in sunflower hypocotyl protoplasts. Physiol Plant **99**: 129-134

Chanabé C, Burrus M, Alibert G (1989) Factors affecting the improvement of colony formation from sunflower protoplasts. Plant Sci **64**: 125-132

Chanabé C, Burrus M, Bidney D, Alibert G (1991) Studies on plant regeneration from protoplasts in the genus *Helianthus*. Plant Cell Rep **9**: 635-638

Henry CA, Jordan JR, Kropf DL (1996) Localized membrane-wall adhesions in *Pelvetia* zygotes. Protoplasma **190**: 39-52

Hynes RO (1992) Integrins: versatility, modulation, and signaling in cell adhesion. Cell **69**: 11-25

Maniotis AJ, Chen CS, Ingber DE (1997) Demonstration of mechanical connections between integrins, cytoskeletal filaments, and nucleoplasm that stabilize nuclear structure. Proc Natl Acad Sci USA **94**: 849-854

O'Neill CM, Mathias RJ (1993) Regeneration of plants from protoplasts of *Arabidopsis thaliana* L. cv. Columbia (C24), via direct embryogenesis. J Exp Bot **44**: 1579-1585

Petitprez M, Brière C, Borin C, Kallerhoff J, Souvré A, Alibert G (1995) Characterisation of protoplasts from hypocotyls of *Helianthus annuus* in relation to their tissular origin. Plant Cell Tiss Org Cult **41**: 33-40

Reuzeau C, Pont-Lezica RF (1995) Comparing plant and animal extracellular matrix-cytoskeleton connections - are they alike? Protoplasma **186**: 113-121

Schindler M, Meiners S, Cheresh DA (1989) RGD-dependent linkage between plant cell wall and plasma membrane: consequences for growth. J Cell Biol **108**: 1955-1965

Shaw SL, Quatrano RS (1996) The role of targeted secretion in the establishment of cell polarity and the orientation of the division plane in *Fucus* zygotes. Development **122**: 2623-2630

Smidsrod O, Skjak-Braek G (1990) Alginate as immobilization matrix for cells. Trends Biotechnol **8**: 71-78

Wyatt SE, Carpita NC (1993) The plant cytoskeleton-cell-wall continuum. Trends Cell Biol **3**: 413-417

GENE EXPRESSION DIFFERENCES BETWEEN ZYGOTIC AND SOMATIC EMBRYOS MONITORED BY DIFFERENTIAL DISPLAY AND cDNA ARRAY: A POTENTIAL TOOL TO IMPROVE LOBLOLLY PINE SOMATIC EMBRYO QUALITY

Gerald S. Pullman , John Cairney, Nanfei Xu, Xiaorong Feng
Institute of Paper Science and Technology
Forest Biology Group
Atlanta GA 30318

1. Introduction

A continued supply of low cost, high quality fiber is essential for the future success of the U.S. forest products industry. With the worldwide demand for paper expected to increase nearly 50% by the year 2010, efforts are increasing to boost forest productivity by propagating superior trees. Clonal propagation of high value forest trees through somatic embryogenesis (SE) has the potential to rapidly capture the benefits of breeding or genetic engineering programs and to improve raw material uniformity and quality. Clonal propagation research at the Institute of Paper Science and Technology (IPST) currently centers on SE of loblolly pine (*Pinus taeda*). Loblolly pine is dominant on 11.7 million ha. and comprises over half of the standing pine volume in southern U.S. forests due to its fast growth, broad natural range, response to cultural practices, resistance to disease and ice damage, and genetic variability for breeding (Westvaco 1998).

Large scale production of somatic embryos for operational use is occurring for some spruce and firs (Sutton et al. 1997); however, the process depends on species and genotype (Atree et al. 1993, Handley et al. 1994). A few laboratories have reported regeneration of loblolly pine plantlets through SE but the process is inefficient and embryo quality is unsuitable for operational use (Pullman et al. 1994, Becwar et al. 1995). Nevertheless, many forest products companies in the U.S. and the rest of the world recognize that SE will be a key feature of future tree improvement programs and therefore have begun intensive internal SE research.

Developing and improving tissue culture protocols is a lengthy and costly process. The traditional literature-based trial-and-error approach is most often used successfully; however at some point embryo yield and quality improvements become increasingly difficult to obtain. Additional approaches such as studying natural embryo development to mimic the hormonal (Kapik 1994), nutritional, and physical (Pullman 1997) conditions found in vivo or understanding how medium changes over time, such as activated carbon adsorption and pH effects (Van Winkle et al. 1997), can further protocol development. Yet, new ways to improve protocols are still needed.

Changes during embryo development are subtle and difficult to observe visually, but

A. Altman et al. (eds.), Plant Biotechnology and In Vitro Biology in the 21st Century, 81–84.

molecular events are dramatic and conspicuous. A fundamental understanding of gene expression patterns during embryo development would allow more detailed and informative comparison and monitoring of somatic and zygotic embryos. We believe that identifying normal and abnormal gene activity will provide stage-specific markers and hypotheses to improve somatic embryos. In this paper we discuss the use of Differential Display (DD) and cDNA arrays to monitor gene expression over the sequence of zygotic and somatic embryo development. This process provides a new approach which can be used to assay and improve embryo quality.

2. Materials and Methods

2.1 Zygotic and somatic loblolly pine embryos. Loblolly pine cones were collected weekly from a breeding orchard near Lake Charles, Louisiana, and shipped on ice to IPST. Embryos were excised and evaluated for developmental stage (Pullman et al. 1994). Stage 9 embryos were separated by the week they were collected; 9.1 (week 1) 9.2 (week 2) etc. Staged zygotic embryos were sorted into vials partially immersed in liquid nitrogen and stored at -70 °C. Somatic embryos for loblolly pine were initiated as described by Becwar et al. 1995 or with minor modifications. Somatic embryos were grown, selected, and staged as described by Pullman et al. 1994 and stored at -70 °C.
2.2 Molecular Techniques. Differential Display was carried out essentially according to Liang and Pardee (1992). To allow the extraction of mRNA from minute amounts of embryo tissue we developed a magnetic bead extraction method (Xu et al, 1997). cDNA bands from differentially expressed genes were excised from gels, cloned, and their differential expression confirmed using methods modified in our laboratory (Xu et al. 1997, Cairney et al. 1997). The southern blotting procedure described by Cairney et al. (1998) was used to prepare membranes containing quadrate blots for several hundred differentially expressed cDNAs. Resulting autoradiographs were scanned and digitized.

3. Results and Discussion

3.1 Mass isolation of genes differentially expressed in loblolly pine zygotic embryos. We have developed a RNA DD method that is sensitive enough to produce banding patterns from one mid-to late-stage embryo or 10-20 early stage embryos. This technique, which extracts mRNA directly from tissue using oligo(dT) beads, avoids losses inherent in conventional RNA extraction methods, is fast, reliable, and low-cost. Differences in gene expression during development, as well as between somatic and zygotic embryos, can be easily detected. Powered with this new technique, we have used more than 20 sets of primer pairs to perform DD for finely staged zygotic embryos of loblolly pine. We have produced more than 600 DD patterns totaling more than 60,000 bands. Over 600 bands were identified which increased or decreased during embryo development. From these, we have isolated and sequenced 417 cDNA clones. This is among the largest collections of cDNAs related to plant embryogenesis.
3.2 Detection of gene expression by micro-array assay. The expression of our cloned DNA in loblolly pine embryos needed to be confirmed. Northern blotting is usually used to estimate the level of transcript, but requires a few μg of RNA for each hybridization .

Since we have several hundred probes, such an approach would be impractical given the amount of tissue available. We have taken a reverse approach, spotting cloned cDNAs on a piece of membrane as individual dots and using mRNA from the embryo as probe to hybridize to the cDNAs on the membrane. As little as 10 ng of mRNA is needed and can be obtained from one to several embryos depending on stage. In our protocol, the mRNA is converted to cDNA and the cDNA is labelled and amplified using PCR. The expression of all the cDNAs on the membrane is determined by one hybridization. The similarities as well as differences in the expression of those genes are clearly visualized when different embryo materials are examined.

Currently, we have performed high density array Southerns for both somatic and zygotic embryos at all the developmental stages. Our dot array Southern data indicates that gene expression of our late stage somatic embryos resembles middle stage zygotic embryos; many transcripts present during late zygotic embryogenesis are absent in somatic embryos and late stage somatic embryo gene expression patterns resemble the patterns of middle stage zygotic embryos.

Cairney et al. 1998 have discussed how this gene expression information may be used to improve the process of somatic embryogenesis. As shown in Figure 1, the high density array Southerns allows rapid evaluation of embryos subjected to protocol changes. Following the expression of a known gene permits inferences about metabolism and is very valuable in developing media-improvement hypotheses. Further, detailed gene expression studies may help by providing an understanding of the timing and location of gene expression (e.g. in situ hybridization). The isolation of key genes also gives us the ability to monitor the expression of these genes as stage specific markers and allows protocol variations to be quickly evaluated.

4. References

Attree SM, Fowke LC (1993) Plant Cell Tiss. Org. Cult 35, 1-35.

Becwar, MR, Pullman GS (1995) Forestry Sci. **44**, 287-301.

Cairney J et al. (1997) TAPPI Biol. Sci. Sym. 85-91, ISBN 0-89852-695-7.

Cairney J et al. (1998) Applied Biochemistry and Biotechnology (in press).

Handley et al. (1994) TAPPI Biol. Sci. Sym. 47-55, ISBN 0-89852-930-1.

Kapik R (1995) Tree Physiol. 15, 485-490.

Liang P, Pardee AB (1992) Science **257**, 967-971.

Pullman GS, Webb DT (1994) TAPPI Biol. Sci. Sym. 31-34, ISBN 0-89852-930-1.

Pullman GS (1997) TAPPI Biol. Sci. Sym. 41-48, ISBN 0-89852-695-7.

Sutton BD et al. (1997) TAPPI Biol. Sci. Sym. 191-194, ISBN 0-89852-695-7.

Xu N et al. (1997) Plant Mol. Biol. Reporter, 15, 377-391.

Van Winkle SC, Pullman GS (1997) TAPPI Biol. Sci. Sym. 49-56, ISBN 0-89852-695-7.

Westvaco Company (1998) Forest Focus 22, 10-13.

5. Acknowledgments. The authors thank the Member Companies of IPST and the Georgia Consortium for Pulp and Paper Competitiveness for support.

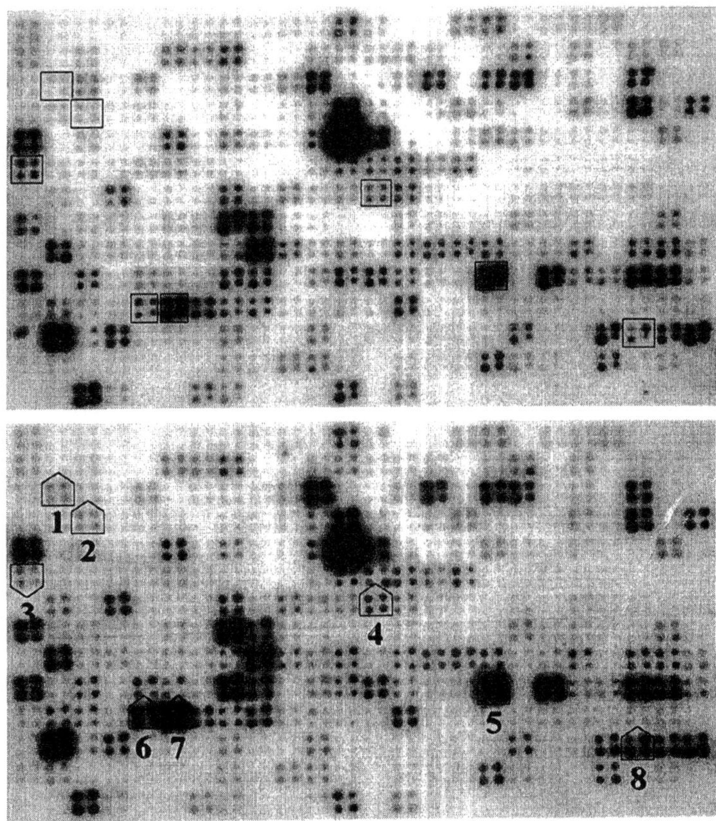

Figure 1. Detection of gene expression by high density array Southern hybridization for loblolly pine genotype 333 after 12 weeks on two maturation media. Top, 5.2 mg/L ABA; bottom 10 mg/L ABA. Arrows up indicate increased gene expression the 10 mg ABA treatment; arrow down, expression lower in 10 mg ABA treatment. Squares in top panel mark the corresponding spots marked in the bottom panel. Gene 1 (LPS064), expression is usually higher in ZE than in SE; 2 (LPS092) expressed in late ZE; 3 (LPZ049) is starch synthase, higher level in ZE; 4 (LPZ091) LMW heat shock protein, found in late stage ZE; 5 (LPZ202) lea gene (late embryo abundant); 6 (LPZ215) higher level in late ZE; 7 (LPZ216) lea gene; 8 (LPZ270) 70S heat shock protein, found in late ZE. A lower level of #3 means a decreased synthesis of starch in 10 mg ABA treatment. All the others bring the expression closer to ZE.

A TOBACCO KNOTTED-1 RELATED HOMEOBOX GENE FOR INVESTIGATING *IN VITRO* SHOOT FORMATION

C.M. Ramage and R.R. Williams
School of Land and Food, The University of Queensland, Brisbane, Queensland 4072, Australia

1. Introduction

Determining the mechanisms by which plants initiate and regulate meristem development has major implications for a wide variety of recalcitrant species. The variation of individual components of the plant tissue culture environment, particularly plant growth regulators, has marked effects on *in vitro* developmental pathways. However, little is known of the links between these components and molecular changes in meristem initials. Molecular markers would contribute to a better understanding of these developmental pathways.

Recently, insights into the formation of the shoot meristem have come from studies of the "Knotted" class of homeobox genes, containing an evolutionarily conserved 180bp DNA motif (McGinnis et al., 1984). Genes containing this motif have been shown to control a variety of developmental programs, acting as biological switches to control the fate of cells during morphogenesis (Hayashi and Scott, 1990; Jackson et al., 1994).

Knotted homeobox genes have been identified in a number of monocot and dicot species and the similarity in their amino acid sequences suggests a fundamental role in the process of shoot meristem formation. Knotted type homeodomains can be divided into two classes. Class 1 genes (eg. Kn1 and STM) share high sequence homology within the homeodomain and are expressed primarily in meristem enriched regions such as the shoot and floral primordia and also in the unexpanded stem but not in leaves or roots. (Kerstetter et al., 1994). Class 2 genes on the other hand are expressed in varying levels in all tissues.

The discovery of these genes have led us to ask the question "Could a knotted gene be used as a functional early detection marker for *in vitro* studies of shoot formation?" To address this we have used a PCR based cloning approach to isolate a 389bp fragment of a knotted like gene from *Nicotiana tabacum* var. zz100 (TobH1). We are currently examining expression of this gene during tobacco leaf disc organogenesis.

A. Altman et al. (eds.), Plant Biotechnology and In Vitro Biology in the 21st Century, 85–88.

2. Procedure

Two degenerate oligonucleotide primers were designed by reverse translation of highly conserved regions of the published *Arabidopsis* shootmeristemless (STM) protein (Long et al., 1996). These oligonucleotides were used to PCR amplify a 389bp fragment from 1st strand cDNA derived from *in vitro* grown *N. tabacum* vegetative shoot meristem RNA (Figure 1.). The fragment was cloned into a pGEM® T-easy vector and sequenced on both strands by the dideoxynucleotide chain-termination method using an automated sequencing system (Applied Biosystems model 373A ver. 1.2.1). Sequence analysis of the cDNA and prediction of the amino acid sequence was carried out using ClustalW (1.4) and PHYLIP 3.5c (provided by J.Felsenstein, Department of Genetics, University of Washington, Seattle, WA).

RNA gel blot analysis was performed using total RNA isolated from various tissue types using Trizol™(Life Technologies) after the method of Chomczynshi and Sacchi (1987). Ten micrograms of total RNA were electrophoresed on 1.4% (w/v) native agarose/TBE gels using the procedure of Liu and Chou (1990). Following electrophoresis, gels were soaked in 20x SSC and RNA transferred to an N+ Hybond nylon membrane (Amersham) in 20xSSC overnight. After transfer, RNA was immobilised onto the membrane by UV irradiation (Stratagene UV stratalinker 2400) and the membrane hybridised using the procedures of Brown and Mackey (1997). The 389bp fragment was ^{32}P labelled by random primed extension (Amersham labelling kit) and exposed overnight on a Phosphor imaging plate (Molecular Dynamics).

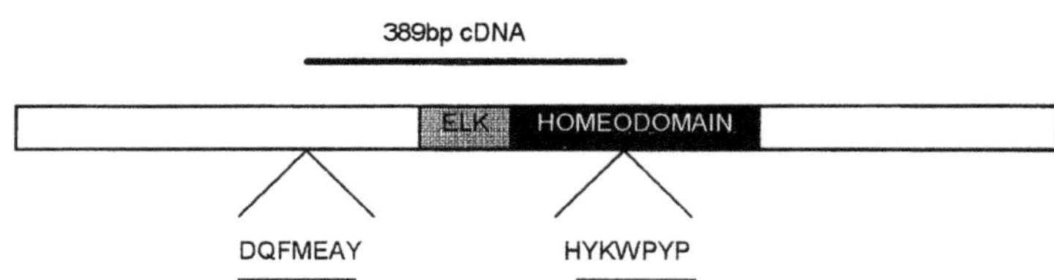

Figure 1. Degenerate oligonucleotides were designed based on two conserved amino acid blocks from the published shootmeristemless protein sequence (Long et al., 1996). Arrows indicate the direction in which the primers were designed.

3. Results & Discussion

The TobH1 fragment isolated encodes a predicted protein of 132 amino acids long and contains a region of the consensus homeodomain motif. Immediately adjacent to, and upstream of the homeodomain, is the conserved ELK domain (Vollbrecht et al., 1991). Based on the classification of Kerstetter et al., (1994) and the sequence similarity with other plant homeobox genes (Figure 2), it is reasonable to suppose that TobH1 is a class 1 Knotted gene.

The cDNA sequence of TobH1 has 98% similarity and 100% amino acid identity to NTH15, a recently isolated Knotted like gene from *N. tabacum* cv. Samsun NN (Tamaoki et al., 1997). Discrepancy between these sequences may reflect variation between the Samsun and zz100 varieties and/or errors during reverse transcription. Alternatively, they may represent distinct genes from a small gene family. Knotted-like genes from several plant species are members of small gene families, with variable 3' and 5' regions and conserved homology within the homeobox and ELK domains. Tobacco is an alleotetraploid, therefore we would expect a number of copies in this genome. We are currently doing 5' and 3' RACE to obtain a full length sequence of TobH1 and to further characterise this fragment.

```
TOBH1 DQFMEAYCEMLTKYEQELSKPFKEAMVFLSRIECQFKALTLTSSSESVAALGEAIDRN...GSSEEEVDV 70
NTH15 ----------------------------------------------------------...---------
LET6  -----------------------------------------APN-SHES-----M---...---D-----
STM   -----------V-----------------Q-V-----S-S-S-P-SFSGYGET-----N.N--------M
SBH1  ---------------------L----L--Q-------N--IS--D...F-SN-GG---...-----D--L
KN1   -------H---V-FRE--TR-LQ---E-MR-V-S-LNS-SISGR-.....-RNILS....S-----DQEG
OSH1  -------H---V--RE--TR-LQ---E--R-V-T-LNT-SISGR-.....-RNILS...-------DQEG

                       ---------------------
TOBH1 NNG.....FIDPQAEDQELKGQLLRKYSGYLGSLKQEFMKKRKKGKLPKEARQQLLDWWTRHYKWPYP 138
NTH15 ---.....------------------------------------------------------------
LET6  ---.....--------R--------------------------------------V---L--I-----
STM   ---.....-V------R------------------------------------------S--------
SBH1  H-......M-------RD--------------------------------------E--N--------
KN1   SG-ETELPEV-AHGV-----HH--K------S-----LS--K--------------S--DQ-------
OSH1  SG-ETELPE-DAHGV-----HH--K------S-----LS--K-------D------N--EL-------
```

Figure 2. Deduced coding sequence of the tobacco TobH1 fragment aligned with the coding sequences of other Knotted class 1 plant homeobox proteins *N. tabacum*, NTH15 (Tamaoki et al., 1997); *Lycopersicon esculentum,* LeT6 (Janssen et al., 1998); *Arabidopsis thalania*, STM (Long et al., 1996); *Glycine max*, SBH1 (Ma et al., 1994); *Zea mays*, Kn1 (Vollbrecht et al., 1991); *Oryza sativa,* OSH1 (Matsuoka et al., 1993). Identical residues are indicated by dashes. Amino acids in bold indicate degenerate priming sites. The extent of the homeodomain is marked with a box and the ELK domain by a dashed line.

An RNA gel blot was used to determine the size of the full length TobH1 transcript and expression in various tissues. A single band, 1.6Kb, is detected in RNA from developing stems (Figure 3.). A weaker band of the same size, from vegetative shoot apices, is also evident but no signal is produced from roots or leaves. This expression pattern is the same as for NTH15, strongly indicating that TobH1 and NTH15 are the same gene.

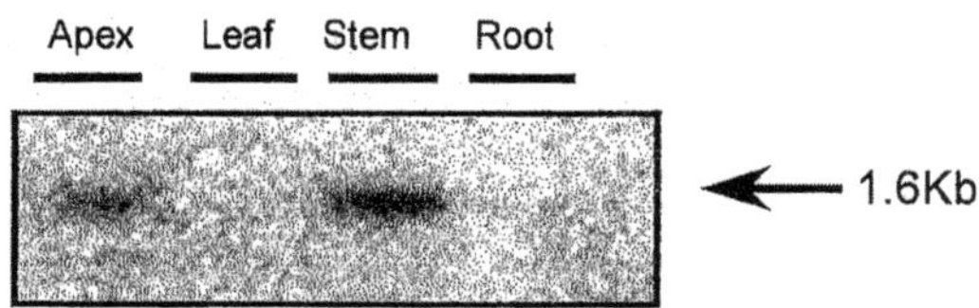

Figure 3. RNA gel blot analysis for TobH1. Total RNA from different tissue types. Spectrophotometry and a representative RNA gel stained with ethidium bromide was used to determine equal amounts of RNA prior to blotting and hybridisation. Each lane consists of 10μg total RNA. Blots were probed with ^{32}P labelled TobH1 cDNA.

4. Conclusion

We have used a PCR based cloning approach to isolate a 389bp fragment of a knotted like gene from *in vitro* grown *N. tabacum* var. zz100. Based on the conservation of sequence identity and expression pattern, TobH1 is most likely part of the NTH15 knotted gene. This fragment is being used to address the question "Could a knotted gene be used as a functional early detection marker for *in vitro* studies of shoot formation?"

5. Acknowledgments

We would like to thank Dr. B. Carroll and his research group for the use of their facilities and taking time to explain the complexities of molecular biology. We would especially like to thank Mr. D. Harrison for his help with phosphor imaging. This research was supported by University of Queensland collaborative research project number 97/QUALR3018G.

6. References

Brown T and Mackey K (1997) In, Current Protocols in Molecular Biology suppl. 37 4.91-4.916.
Chomczynshi P and Sacchi N (1987) Analytical Biochemistry 162, 156-159.
McGinnis W et al., (1984) Nature 308, 428-433.
Hayashi S and Scott MP (1990) Cell 63, 883-894.
Jackson D et al., (1994) Development 120, 405-413.
Janssen B et al., (1998) Plant Molecular Biology 36, 417-425.
Kerstetter R et al., (1994) Plant Cell 6, 1877-1887.
Liu YC and Chou YC (1990) Biotechniques 9(5), 558.
Long JA et al., (1996) Nature 379, 66-69.
Matsuoka M et al., (1993) Plant cell 5, 1039-1048.
Tamaoki M et al., (1997) Plant Cell Physiology 38(8), 917-927.
Vollbrecht E et al., (1991) Nature 350, 241-243.

IN VITRO CLONAL PROPAGATION OF ALEPPO PINE (*PINUS HALEPENSIS* MILL.)

T. Tzfira, A. Ya'ari, O. Yarnitzky, J. Riov and A. Altman
The Kennedy-Leigh Centre for Horticultural Research and The Otto Warburg Center for Biotechnology in Agriculture, The Hebrew University of Jerusalem, Rehovot 76-100, Israel.

1. Introduction

For more than fifty years *Pinus halepensis* has been the major planted forest tree throughout Israel, and it occupies an important place, together with *P. brutia*, in other Mediterranean regions. Its high tolerance to dry climate and its ability to grow on poor marginal soils (Weinstein 1989) has made it a preferable species for reforestation in Israel. In recent years, however, its importance has been diminished because of its susceptibility to *Matsucoccus josephi*, which causes death of both young seedlings and very old trees (Wilcocx 1979; Liphschitz, Mendel 1989). Recently it was reported that some trees, possibly hybrids of *P. halepensis* and *P. brutia*, exhibit an extensive growth and high resistance to the pest (Korol et al 1995). Such resistant trees can be a good source for aforestration if they can be vegatetivly propagated. We assessed the regeneration potential of both mature embryos and adventitious fascicular shoots, for the clonal propagation of *P. halepensis*.

2. Materials and Methods

Adventitious shoot induction on mature embryos. Free-pollinated seeds of *P. halepensis* were imbibed for 48 h in running tap water, and then surface-sterilized as previously described (Tzfira et al 1996). The apical part of the seed was cut with a no. 11 scalpel blade, and the seed coats and endosperm were removed with forceps, revealing the mature embryo. The embryos were placed either in normal (cotyledons up) or inverted orientation (cotyledons down) on a Z-36 regeneration medium (Ya'ari 1998) in petri dishes. Three weeks after they had been placed in the inverted orientation, the embryos were replanted in the normal orientation on the same medium. The regeneration medium was supplemented with 0.3 mg/l Conjugate-G (conjugate of "phenoxy-auxin" with glycine, kindly provided by Prof. J. Riov, The Faculty of Agriculture, Rehovot, Israel) and 2.25 mg/l BA. Regenerated shoot clusters were separated from the embryo after six weeks, and their basal end was dipped for 6 h in 1 μM IBA. The shoot clusters were then cultured in a 1:1 vermiculite:peat mixture for rooting and hardening.
Fascicular shoot induction on seedlings. Three-year-old greenhouse-grown plants of *P. halepensis* were pruned and sprayed four times, at weekly intervals, with 200 mg/l BA or

A. Altman et al. (eds.), Plant Biotechnology and In Vitro Biology in the 21st Century, 89–92.

TDZ dissolved in 0.1% (w/v) BB5™ surfactant. The resulting BA-induced fascicular shoots (about 5-7 cm in length) were excised, surface-sterilized in 2% $NOCl_2$ for 15 min, followed by a 2-min immersion in 50% EtOH and 0.1% BB5™ surfactant, and then washed twice in sterile water. For rooting, the basal cut surface of the shoots was dipped in 1μM BSA-IBA (conjugate of BSA with IBA, provided by Prof. J. Riov) for seven days, or explants were directly injected with wild-type *Agrobacterium rhizogenes* (strain LBA9365) bacterial suspension followed by transfer to vermiculite:calcar mixture.

3. Results and Discussion

Mature embryo cultures. A week after placing the embryos in the normal orientation in the medium, the cotyledons opened, turned green and became thicker. Soon after, several knobs were visible on the cotyledons (Fig 1A), marking the beginning of primordial bud organization. Four weeks after planting, 1-mm long needles regenerated, marking the development of adventitious buds. The average number of adventitious buds was six per embryo, and was accompanied by callus production, mainly from the root apex, which was imbedded in the media. Microscopic observations reveled that adventitious buds regenerated mainly on cotyledons which were in direct contact with the media. It was previously reported that regeneration in other *Pinus* spp. (e.g. *P. canariensis*, Martinez Pulido et al 1990 and *P. radiata*, Aitken et al 1981) occurred exclusively on cotyledons which were in direct contact with the media. Although regeneration in *P. halepensis* occurred also from other cotyledons, it was suggested that better contact between the medium and the cotyledons might enhance bud regeneration. Thus, a comparison was made between mature embryos placed either in inverted or normal orientation in the medium. In both cases, the cotyledons opened, turned green, became thicker and produced knobs on them. However, although the number of knobs in the inverted embryos was much higher than in the normally oriented embryos (about several dozens), they failed to further develop into buds if left inverted in the medium. This may suggest that the direct contact between the medium and the cotyledons, which was found to highly enhance knob production, might inhibit their future development into buds. Re-planting in normal orientation, after three weeks in the inverted position, allowed the development of buds (Fig 1B), with an average of 40 adventitious buds per embryo.

Culture of inverted mature embryos to enhance adventitious bud formation in Himalayan pines was also reported (Agrawal et al 1991), and we further studied whether the direct contact between the cotyledons and the medium or the embryos' polarity was the main factor for improved regeneration in *P. halepensis*. Mature embryos were planted in either the normal or inverted orientation in petri dishes, which were also placed in the normal or inverted orientation. Regeneration of a large number of adventitious buds was observed only in embryos planted in the inverted orientation followed by replanting in the normal orientation, regardless of the orientation of the petri dish. Thus, it can be concluded that direct contact between the medium and the cotyledons, rather than polarity, is the crucial factor for bud regeneration in *P. halepensis*.

Individual regenerated shoots were separated from the cultured embryo after six weeks, but did not survive on various media and growth-regulator combinations. However, when two to five bud clusters were cultured in the dark on basic GD (Gresshoff, Doy 1974) medium the survival rate reached 95%, yet with no root regeneration or shoot elongation. The importance of shoot elongation for rooting of *Pinus* adventitious shoots was emphasized in several works (Amerson et al 1988; David 1982). However, the use of different media with various growth regulators did not enhance shoot elongation (except for needles). However, rooting of these adventitious shoot clusters, following treatment with 1 μM IBA for six hours and planting in 0.8% agar in petri dishes or in 1:1 moistured vermiculite:peat mixture in 0.5 L plastic boxes was beneficial. The average rooting percentages was 16% and 85% in agar or vermiculite:peat mixture, respectively. Rooted shoot clusters (Fig 1D) was then transferred to a fog chamber for hardening and further growth and development.

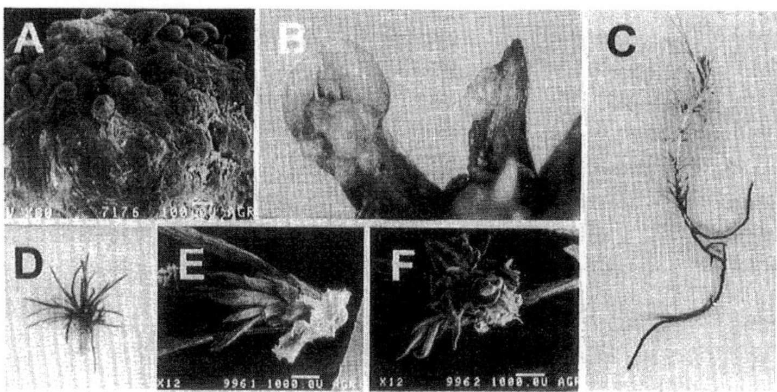

Fig 1. Clonal propagation of *Pinus halepensis* from mature embryos and fascicular shoots. (A) Early primordial organization on the cotyledon of a mature embryo. (B) Adventitious bud regeneration from cotyledons. (C) Adventitious root regeneration at the basal end of an adventitious shoot. (D) Rooted adventitious shoot cluster. (E) Well-developed fascicular shoot regenerated following BA-treated plants/treatment (F) Stunted fascicular shoot regenerated from TDZ-treated seedlings.

Fascicular shoot induction and rooting. Cytokinin treatment of 3-year-old plants increased the number of fascicular buds induced almost fourfold, as compared with control seedlings (which were only pruned). Most of the regenerated buds developed from bracyblasts and were of two types: (1) well-developed, long, single shoots carrying several whorls of long juvenile needles (Fig 1E) in BA-treated and control plants, and (2) stunted shoots carrying dense and short juvenile needles (Fig 1F) in TDZ-treated plants. Although in the latter case a large number of shoot primordia were also observed at the bracyblast basal region, most of these primordia failed to develop and resulted in an average of only two adventitious shoots per bracyblast. The potential of TDZ treatment for inducing premordia formation is impressive, if accompanied by primordia development. However, due to the lack of further growth of the dense structure, all attempts at efficient surface-sterilization and further *in vitro* manipulation of these shoots failed.

BA-originated fascicular shoots were treated in several ways with free auxins, resulting in less than 10% rooting. Dipping the shoots for a period of one week in a 1μM BSA-IBA solution resulted in 57% rooting after seven weeks of culture in vermiculite:calcar mixture. Fascicular shoots cultured on agar-based medium or vermiculite alone failed to produce roots even after a long BAS-IBA treatment. The advantage of auxin conjugates over free auxins was demonstrated in several other cases (Riov 1993; Felker, Clark 1981) and was attributed to the slow release of active auxins. Furthermore, the inability of treated shoots to root on agar or vermiculite media emphasized the importance of media aeration for the rooting of *P. halepensis*. Improved rooting, due to the use of aerated and porous media, was also demonstrated in other *Pinus* species, e.g. *P. brutia* x *P. halepensis* (Scaltsoyiannes et al 1994) and in *P. radiata* (Horgan, Holland 1989).

The use of *A. rhizogenes* to improve rooting of cuttings was demonstrated in several works (e.g. Strobel 1989; Patena et al 1988), including *Pinus* (McAffe et al 1993). Several strains and different inoculation techniques were assessed for rooting of BA-originated fascicular shoots, and direct injection of LBA9365 bacterial suspension to the shoot base was found most effective (e.g. Fig 1C). The rooting percentage of infected shoots was similar to that of shoots treated by IBA-BSA. The resulting two-year-old chimeric plants, which originated from fascicular shoots and represent a single *P. halepensis* clone, are now at their third year in the greenhouse.

4. References

Agrawal et al (1991) Biotech. Lett. 1, 14-19.
Aitken J et al (1981) Can. J. For. Res. 11, 112-117.
Amerson et al (1988) In: Genetic Manipulation of Woody Plants. pp. 117-137. (Hanover JW and Keathley DE, eds) Plenum Publishing Corporation, New York.
David A (1982) In: Tissue Culture in Forestry. pp. 72-108. (Binga JM and Durzan DJ, eds) Martinus Nijhoff, Dr. W. Junk Publishers, The Hague.
Felker P and Clark PR (1981) J. Range Manage. 34, 466-468.
Gresshoff PM and Doy C (1972) Planta 107, 161-170.
Horgan K and Holland L (1989) Can. J. For. Res. 19, 1309-1315.
Korol L et al (1995) Silvae Genet. 44, 186-190.
Liphschitz N and Mendel Z (1989) Can. J. Bot. 67, 2692-2703.
Martinez Pulido C et al (1990) Can. J. For. Res. 20, 1200-1211.
McAffe BJ et al (1993) Plant Cell, Tissue and Organ Culture 34, 53-62.
Patena L et al (1988) Acta Hort. 227, 324-329.
Riov J (1993) Acta Hort. 329, 284-288.
Scaltsoyiannes A et al (1994) Ann. Sci. For. 51, 171-182.
Strobel GA (1989) In: Adventitious Root Formation in Cuttings. pp. 10-115. (Davis TD et al, eds) Dioscorides Press, Portland, OR.
Tzfira T et al (1996) Plant Cell Rep. 16, 26-31.
Weinstein A (1989) For. Ecol. Manag. 26, 215-225.
Wilcox H (1975) La'Yaran 25, 49-51 (in Hebrew).
Ya'ari A (1998) M.Sc. Thesis, The Hebrew University of Jerusalem, Israel.

NORWAY SPRUCE SOMATIC EMBRYOGENESIS: ENDOGENOUS LEVELS OF PHYTOHORMONES DURING SOMATIC EMBRYO DEVELOPMENT

M. Vágner[1,4], Z. Vondráková[1], J. Špačková[1], M. Cvikrová[1], J. Eder[1], H. Lipavská[2], J. Albrechtová[2], H. Svobodová[2] and I. Macháčková[3]

[1]Institute of Experimental Botany, Prague, Czech Republic; [2]Charles University, Fac. Sci., Prague, CR; [3]De Montfort University, Norman Borlaug Inst. Plant Sci., Inst. Exp. Bot., Prague, CR, [4]E-mail: vagner@ueb.cas.cz

1. Introduction

Somatic embryogenesis offers a tremendous potential for the micropropagation of conifers with a relatively long reproduction cycle and low fructification frequency, and those in which traditional micropropagation by axillary budding is ineffective. The Norway spruce represents an excellent example of this. Moreover, the Norway spruce is one of the most abundant conifers in Northern and Central Europe whose population is endangered by pollution and insects in some areas. The multiplication of particular genotypes which have a higher degree of resistance to such unfavourable environments is of great current concern. The potential advantages of production of plants through somatic embryogenesis over cutting of plants derived from organogenesis in *P. abies* are an almost unlimited number of plants obtained from one donor plant within a short period of time, the possibility of cryopreservation of embryogenic cell lines, the future prospect of reducing labour and embling costs resulting from the fact that the technique is amenable to automation, and finally, the possibility of rejuvenating mature trees identified as elite genotypes by this method.

Since 1985, when somatic embryogenesis of *P. abies* was reported for the first time (Chalupa 1985, Hakman et al. 1985), extensive progress has been made. Growing empirical experience enabled the production of mature somatic embryos, the regeneration of plants, and the successful transfer of emblings *ex vitro*. Relatively abundant data are available concerning the use of exogenous plant growth regulators in the manipulation of demanded developmental steps of somatic embryo initiation, growth and maturation. In contrast, both in angiosperms and even more in gymnosperms, there is still very limited information on the endogenous levels of plant hormones during both the zygotic and somatic embryo development, and in consequence, on the definition of their function in somatic embryogenesis.

The aim of the work was to describe the changes in endogenous phytohormones during Norway spruce somatic embryo development and maturation, and to study the possible effect of polyethyleneglycol on these patterns. Moreover, detailed anatomical analysis of developing embryos was made to describe ontogenetic events coinciding with the changes of particular phytohormone levels.

A. Altman et al. (eds.), Plant Biotechnology and In Vitro Biology in the 21st Century, 93–96.

2. Material and Methods

2.1. In vitro cultures: Embryogenic culture of *Picea abies* L. (Karst.), genotype AFO 541 was a generous gift from dr. Bercetche (AFOCEL, France). The cultures were grown on media according to Gupta and Durzan (1986), solidified with 0.75 % (w/v) agar (Sigma), and the pH was adjusted to 5.8 ± 0.05 prior to autoclaving. The maintenance medium was supplemented with 5 μM 2,4-D, 2 μM kinetin, 2 μM BAP (all Sigma), and 30 $g.l^{-1}$ sucrose (Lachema, C.R.). The medium for development and maturation of somatic embryos lacked auxin and cytokinins. 20 μM ABA (Sigma) was used to trigger development of the embryos. ABA was used in combination with none, 3.75 or 7.5 % (w/v) polyethyleneglycol 4000 (PEG) (Fluka). The PEG solution was autoclaved separately and mixed with the remaining part of the medium after autoclaving. All organic substances except sucrose were also diluted separately, the pH was set to 5.8 ± 0.05, and then the solution was filter-sterilised and added to the cooled autoclaved media. The embryogenic cultures were maintained by weekly subculturing to the fresh media in Magenta vessels (Sigma) or Drigalski plates (diameter 100 mm and height 35 mm), containing 40 ml of medium. Cultures were kept in darkness at 24 ± 1°C.

2.2. Phytohormonal analyses: The embryogenic cultures were sampled according to the time elapsed after the transfer to the maturation medium. In order to eliminate the effect of fluctuations in endogenous phytohormone levels during subcultivation, the cultures were sampled between the 5th and the 7th day of the subcultivation period. Embryogenic suspensor mass (ESM) collected from at least 3 vessels was gently wiped out with filter paper, weighed and immediately frozen in liquid nitrogen. In later stages of maturation, the developing somatic embryos were manually separated from the remaining ESM, and both parts were weighed and analysed separately. Individual phytohormones were extracted and analysed by the methods listed below:

IAA - HPLC with fluorimetric detection (Eder et al., 1988),
ABA - GC-ECD (Vágner et al. 1998),
cytokinins – HPLC and ELISA (Macháčková et al. 1993),
polyamines – HPLC (Slocum et al. 1989),
ethylene – GC-FID (Vágner et al. 1998).

3. Results and Discussion

ABA in the internal medium was shown to be necessary for proembryo development. Neither cessation of plant growth regulators, nor the inclusion of PEG evoked complete embryo development alone or in combination. The ABA level in the ESM grown on proliferation medium was very low, close to the detection limit. After inclusion of 20 μM ABA to the external medium, endogenous level of ABA rose up markedly for two weeks (Fig. 1), and then continuously declined even several weeks before the omission of ABA from the medium. ABA levels in somatic embryos were lower compared to remaining ESM. Interruption of the supply of ABA after 52 d resulted in more accelerated decrease of endogenous ABA in mature somatic embryos, however, a considerable level of ABA in somatic embryos was still present after 12 d of cultivation on ABA-free medium. An addition of PEG increased endogenous ABA accumulation only if expressed on FW basis (results not shown), however, the differences in ABA accumulation were not significant on DW basis.

The ESM grown on proliferation medium contained relatively low endogenous IAA concentration (Fig. 2). The levels of IAA remained stable for two weeks after the passage to the maturation medium. A dramatic increase of IAA level was recorded in developing somatic embryos. The endogenous level of IAA decreased in the later stages of embryo maturation. The presence of PEG in the medium resulted in the decrease of the peak of IAA and its shift to the left on the time axis, and an accelerated decrease of IAA in of mature somatic embryos. Total cytokinins (sum of Z, ZR, iP and iPA) decreased within one week after its passage to the maturation medium and then remained relatively steady. The effect of PEG on cytokinin levels was not significant. iP represented the most abundant cytokinin during ESM growth, whereas iPA became prevalent in later stages of somatic embryo development (results not shown). The endogenous content of free polyamines reached its maximum on the 3rd and 28th day after the passage to the maturation medium (Fig. 3). In both cases, significantly higher content of free polyamines was recorded in the PEG-supplemented medium. In the late maturation stage, free polyamines were replaced by their soluble conjugates (results not shown). Ethylene emanation decreased after transfer of ESM to the maturation medium and negatively correlated with the concentration of PEG in the medium (Fig. 4). Minimal production of ethylene was recorded after 4 weeks. The subsequent continuous rise of ethylene production should rather be due to the ESM than mature embryos (in this case the culture was not divided into the developing embryos and remaining ESM as was done in the phytohormonal analyses above).

Presented description of endogenous phytohormonal levels during somatic embryo development and maturation is considered as the first step of this research. Further research will focuse on the modulation of these levels in close relation to the desired physiological and developmental events during somatic embryo development and maturation.

4. Acknowledgement

This work was supported by grant No. 522/96/K186 (GACR) and by COST 822 Project.

5. References

Chalupa V (1985) Commun. Inst. For. Cech. 14: 57-63.

Eder J et al. (1988) Physiology and Biochemistry of Auxins in Plants, pp. 389-390, Academia Praha and SPB Acad. Hague.

Gupta et al. (1986) In Vitro Cell Dev. Biol. 22: 685-688.

Hakman I et al (1985) J. Plant. Physiol. 121: 149-158.

Macháčková I et al. (1993) Physiol. Plant. 87: 160-166.

Slocum RD et al (1989) Plant Physiol. 89: 512-517.

Vágner M et al (1998) Adv. Hort. Sci. 12: 11-18.

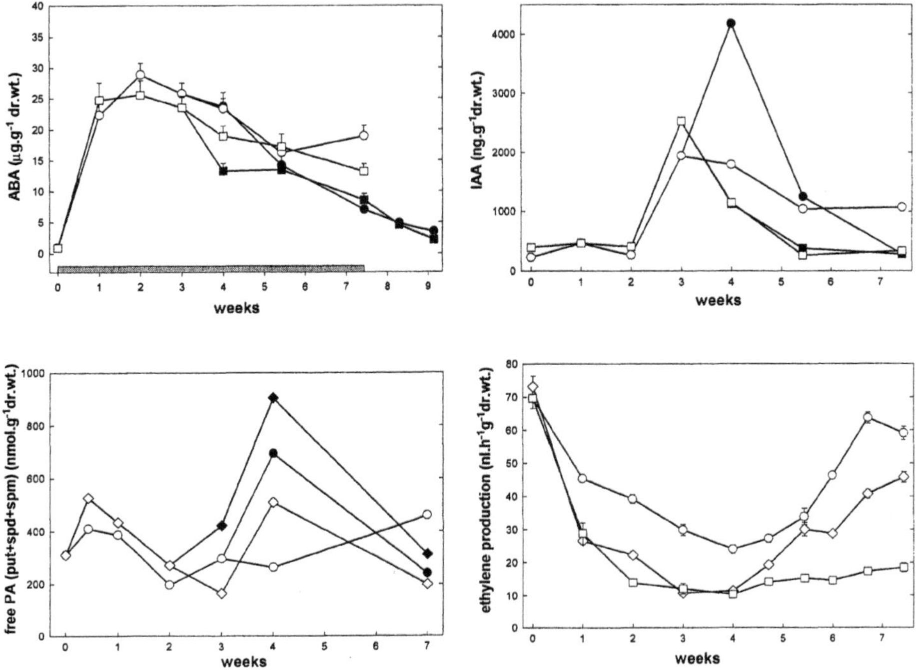

Fig.1-4: Endogenous content of ABA (Fig.1), IAA (Fig.2), free polyamines (Fig. 3), and production of ethylene (Fig.4) during Norway spruce somatic embryo development and maturation. Horizontal bar in Fig.1 indicates duration of ABA treatment. X-axes: weeks on maturation medium. Circles – medium supplemented with 20µM ABA, diamonds - 20µM ABA + 3.75 % PEG, squares - 20µM ABA + 7.5% PEG, open symbols – ESM, closed symbols – developing somatic embryos.

MODIFICATIONS OF LEAF MORPHOGENESIS INDUCED BY INHIBITION OF AUXIN POLAR TRANSPORT

Zhi-hong Xu* and Di-an Ni
National Laboratory of Plant Molecular Genetics, Shanghai Institute of Plant Physiology, Chinese Academy of Sciences, Shanghai 200032, China
(*Present mail address: Chinese Academy of Sciences, 52 Sanlihe Road, Beijing 100864, China)

1. Introduction

Auxin is an unique hormone with polar transport among those well-known plant hormones. It has been shown that auxin polar transport plays an important role in a number of physiological process (such as gravitropism, phototropism, early development of floral bud formation, etc.) (Goldsmith,1977; Okada et al,1991). Our previous work(Liu et al,1993) already demonstrated that auxin polar transport is essential for establishment of bilateral symmetry in early development of zygotic embryos. In *Brassica juncea*, auxin polar transport inhibitors induced the formation of a collar cotyledon, instead of two opposite cotyledons on the globular embryos cultured. The fused cotyledon also was observed in *Arabidopsis* mutant *pin-1* with impaired auxin polar transport(Liu et al, 1993). The present study aims at revealing whether it also affects pattern formation of leaf morphogenesis in shoot formation from leaf explants cultured and sterile seedlings.

2. Materials ans Methods

Auxin polar transport inhibitors used in this study are 9-hydroxy-fluorene-9-carboxylic acid (HFCA), 2,3,5-triiodobenzoic acid (TIBA) and trans-cinnamic acid (CA).

The leaf explants of tobacco (*Nicotiana tobacco* L. cv. Ge Xin No1) were cultured on MS medium with 2 mg/l BA, supplemented with different concentrations of auxin polar transport inhibitors. Effects on shoot formation and leaf morphogenesis were observed during the culture. Other two species, *Crassula arboresiens* and *Orychophragmus violaceus*, also were studied with the same method.

In order to observe the effect of auxin polar transport on leaf morphogenesis of seedlings, the sterile seedlings of tobacco and *Brassica chinensis* L.(cv. Ji-mao-cai) were used as materials, that were obtained from the seeds germinated on MS supplemented with auxin polar transport inhibitors.

3. Results

3.1. Effects of auxin polar transport inhibitors on leaf morphogenesis in shoot formation from the cultured leaf explants

A. Altman et al. (eds.), Plant Biotechnology and In Vitro Biology in the 21st Century, 97–99.

The tobacco leaf explants in culture were readily induced to form shoots on MS medium with cytokinin. On auxin transport inhibitor (HFCA, TIBA or CA) contained medium, all three inhibitors tested induced to form the adventitious shoots with trumpet-shaped leaves and/or fused leaves, though the most effective concentration varies with different inhibitors used (Ni et al, 1996). The highest frequency of trumpet-shaped leaf formation could reach up to 82% (based on number of the explants with trumpet-shaped leaf in total number of explants cultured) and over 15% (based on number of adventitious shoots with trumpet-shaped leaf in total number of the shoots examined), when they were cultured on the medium with 7.5 mg/l TIBA (see Table 1). The similar effect was observed in the cultures of the leaf explants of *Crassula arboresiens* and *Orychophragmus violaceus*.

Table 1. Effects of auxin polar transport inhibitors at different concentrations on frequencies of the formation of adventitious shoot with trumpet-shaped leaf in the culture of tobacco leaf explants

Inhibitor treatment	Altered/Total*	Percentage (%)
control	0/1273	0
1 mg/L TIBA	13/425	3.06
4 mg/L TIBA	16/261	6.13
7.5 mg/L TIBA	51/337	15.13
2 mg/L HFCA	35/738	4.74
7.5 mg/L HFCA	46/584	7.88
20 mg/L HFCA	56/769	7.28
10 mg/L CA	11/612	1.80
20 mg/L CA	13/533	2.35
40 mg/L CA	17/824	2.06
60 mg/L CA	16/552	2.90
80 mg/L CA	43/986	4.36
100 mg/L CA	53/633	8.37

*Number of adventitious shoots with trumpet-shaped leaves/total number of adventitious shoots examined.

3. 2. Effects of auxin polar transport inhibitors on leaf pattern formation of seedlings

Two auxin transport inhibitors, TIBA and HFCA, were tested in the germination medium. The seedlings of tobacco and *Brassica chinensis* obtained on the media with 15 or 30 mg/l TIBA or HFCA also formed a trumpet-shaped or fused leaf at the 5th or 6th leaf position. The frequencies of the seedlings with such abnormal leaf could reach to 13.5% and 25.5% respectively in tobacco and *Brassica*. After the seedlings with trumpet-shaped leaf were transferred on 1/2 MS medium, normal shoots could be developed from inside the center of the trumpet-shaped leaves in some seedlings, and the others swelled at the base and could not grow normally.

Histological observations of tobacco seedlings showed that while there are only one main vascular bundle and several minor ones in normal leaves, two main vascular bundles exist in the fused laeves, and several vascular bundles of more or less the same size in the trumpet-shaped leaves.

In conclusion, the results described above demonstrated that auxin polar transport is essential for the bilateral symmetry pattern formation, not only for cotyledons during early embryogenesis, but also for leaves in shoot formation.

4. Acknowledgments

The research project was supported by China National Natural Science Foundation (No 39670078) and National Laboratory of Plant Molecular Genetics, Shanghai Institute of Plant Physiology. The authors wish to thank Prof. Zhen-au Xia, Dr. Yong-ning Chen, and Mr. Ling-jian Wang for their taking part in the project.

5. References

Goldsmith MHM(1977) Ann. Rev. Plant Physiol. 28,439-478.
Liu CM et al(1993) Plant Cell 5,621-630.
Ni DA et al(1996) Acta Bot. Sinica 38,867-869.
Okada K et al(1991) Plant Cell 3,677-684.

Quantitative Analysis of Douglas-fir Somatic Embryogenesis

Chun Zhang, Chetan Gadgil and Wei-Shou Hu
Department of Chemical Engineering and Materials Science,
University of Minnesota,
Minneapolis, MN 55455, U.S.A.

Introduction

Douglas-fir (*Pseudotsuga menziesii*) is one of the world's most commercially important timber trees. Due to the increasing demands for forest products and the decreasing woodland area, there is a need for increasing wood production on existing lands. The use of elite tree clones is an effective way of increasing productivity. Somatic embryogenesis is considered to be an economically viable method of mass propagation of such elite clones (Aitken-Christie *et al.*, 1995).

Somatic embryogenesis in Douglas-fir involves 3 stages: a maintenance stage, a singulation stage and a maturation stage. The number of cells with embryogenic potential is increased in the maintenance stage; the transfer into ABA singulation stage initiates embryo development, while cotyledonary embryos are formed in the maturation stage. The final stage is traditionally carried out in Petri dishes (Gupta *et al.*, 1995).

Previous efforts in this area have focused on an understanding of the embryogenesis process, both *in vivo* and *in vitro*. Embryo classification is carried out in an arbitrary manner and depends on the human operator carrying out the classification process.

We have developed a perfusion reactor for embryo maturation. In this paper, we present a quantitative study of the kinetics of somatic embryo development in the perfusion reactor. The embryos are classified into their developmental classes using an automated classifier, which eliminates errors due to operator biases. We have also formulated a defined medium to replace the casamino acids that were originally used. The results of embryo development in such a medium are also presented.

Materials and Methods

Cell line and suspension culture

Suspension cultures of Douglas-fir *(Pseudotsuga menziesii)* were kindly provided by Weyerhaeuser Corporation. The composition of the medium used in the developmental stages and details of the procedures used have been described elsewhere (Taber *et al.*, 1998).

A. Altman et al. (eds.), Plant Biotechnology and In Vitro Biology in the 21st Century, 101–104.

Perfusion reactor

A membrane perfusion reactor was developed for Douglas fir maturation cultures. The reactor consisted of two glass chambers that were separated by a cellulose membrane with 0.45 μm pore size (Millipore, Bedford, MA). The lower chamber was filled with the maturation medium so that the membrane was fully in contact with the liquid medium. The working volume of the reactor was approximately 33 ml. 2 mL of cell suspension from week 3 ABA singulation culture was poured on the membrane evenly. The two glass chambers of the reactor were then sealed with a clamp and kept in the dark at 24±1 °C. 6 ml fresh medium was perfused into the reactor every day. An equal amount was withdrawn from the reactor to maintain a constant working volume. The medium used was the normal maturation medium except no activated charcoal was included.

Neural-Network Classifier

The images of embryos were obtained and processed as previously described (Chi *et al.*, 1994). A commercial neural network software (Ward System Group, Frederick, MD) was used in constructing all neural network classifiers. A hierarchical decision tree was used to classify the embryos into three normal classes (S1, S2 and S3 embryos) and one abnormal class. Node #1 (the first branching point) of the decision tree separated the embryo population into normal and abnormal embryos. The normal embryos were then further separated into three individual classes at node #2.

In this pattern recognition system, two neural network classifiers were used (node #1 and #2). Both of these two neural classifiers used a three-layer feedforward fully connected structure. The input layer consisted of 19 neurons corresponding to 10 size-related and 9 Fourier features obtained from a Fast Fourier Transfer of embryo contours. The output layer contained 2 and 3 neurons for the classifiers at node #1 and #2, respectively. 30 and 25 neurons were used in the hidden layer of the two classifiers. In this study, a total of 660 pre-classified embryos were used as examples to train the neural network classifiers. Backpropagation learning algorithm was used to adjust weighing factors of neural network classifiers.

Fisher discriminant analysis

Fisher discriminant analysis uses a linear combination of a set of variables of observations that maximize the distance of mean values between groups while minimize the within-groups variations for classification. It can be proved that the linear combination that maximizes the distance between the means of the various groups is the eigenvector β_m corresponding to the largest eigenvalue of the matrix $\mathbf{W}^{-1}\mathbf{B}$, where **B** is the between-groups sum of squares matrix and **W** is the matrix of the sum of the within-groups sums of squares matrices.

A commercial statistical software (IMSL, Inc., Houston, Texas) was used to calculate β_m.

Results and Discussion

Neural Network Classifier

The results of the classification of the embryos as normal and abnormal embryos are shown in Table 1. Of the 271 normal embryos used for testing, the classifier was able to correctly classify 267. However, the accuracy of classification of abnormal embryos was only 69%.

The classifier for the second node (classification of the normal embryos as S1, S2 and S3 embryos) was trained with 580 pre-classified normal embryos. 271 normal embryos that had not been used in the classification were used to test the accuracy of the classifier. Table 2 shows the results of the classification. The elements along the diagonal indicate the correctly classified embryos. The classifier was able to classify the embryos with an accuracy of more than 80% in all cases.

Table 1. Node 1 Results

Human Operator		*Neural net Classifier*	
	Total	*Normal*	*Abnormal*
Normal	271	267	4
Abnormal	32	10	22

Table 2. Node 2 Results

Operator		*Neural Net Classifier*		
	Total	*S1*	*S2*	*S3*
S1	66	54	11	1
S2	136	6	118	12
S3	65	0	10	55

Perfusion Reactor Development

The maturation medium for batch culture contains activated charcoal. The use of activated charcoal is not desirable as its quality varies from batch to batch and hence introduces an uncontrollable variable in the medium composition. Charcoal is thought to be beneficial because it absorbs harmful metabolites or inhibitors that are secreted by the cells. This effect can be achieved by replacing the medium periodically. A perfusion reactor was constructed to enable such a medium replacement.

The yield of mature S2 and S3 embryos from the perfusion reactor and control Petri-dish cultures is shown in Fig 1. Use of the perfusion reactor increased yields of mature S3 embryos by 125% compared to control cultures. Thus the perfusion reactor achieved greatly superior yields than the control cultures.

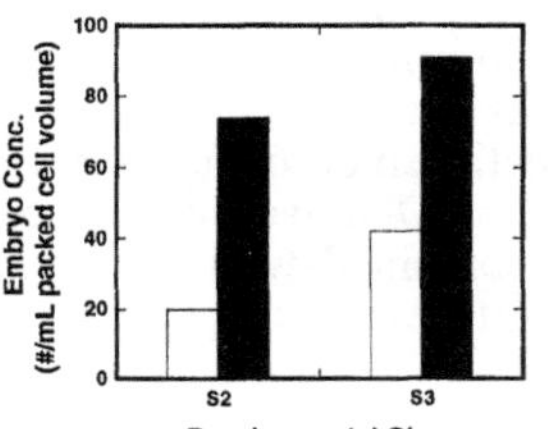

Fig.1 Perfusion Reactor Yield

Besides charcoal, the maturation medium also contains casamino acids, whose composition varies depending on the batch that is used. In order to remove this drawback, we replaced the casamino acids with a mixture of amino acids. It was observed that the defined medium could successfully replace casamino acids without any significant decrease in the yields of mature embryos.

The embryos obtained from perfusion culture were not only more numerous, but also visibly more mature. However, it is desirable to obtain a quantitative description of the degree of maturity. To this end, a Fisher

Discriminant analysis was carried out and the Fisher discriminant corresponding to each embryo was computed. The distribution of these values is shown in Fig 2. It is seen that the index (referred to as Maturation Index) can provide a quantification of the developmental state, with increasing values corresponding to a more mature stage. To quantify the effect of the perfusion reactor on embryo development, the maturation indices for the embryos obtained from the perfusion culture and control cultures were determined (Fig. 3). It is seen that the maturation index of perfusion culture embryos, shows a narrower distribution, implying a more uniform quality of embryos.

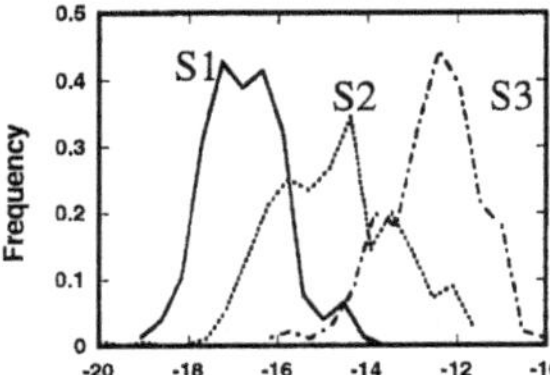

Fig. 2. Distribution of Maturation index (M.I.)

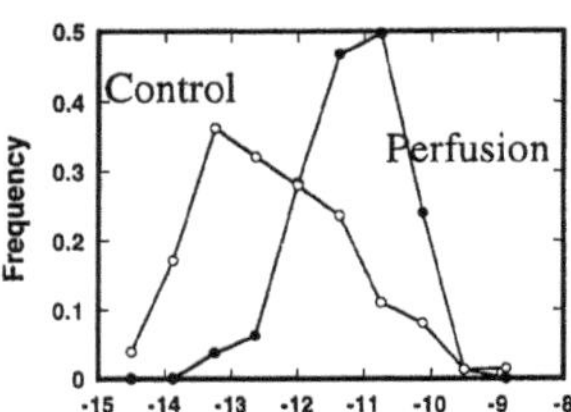

Fig 3. Distribution of M.I. for batch and perfusion embryos

Conclusion

We have used of an automated neural network-based classification system for classifying Douglas-fir somatic embryos into their developmental classes. We have also demonstrated the accuracy of such classifiers.

In order for an accurate study of the embryogenesis process, a defined medium is necessary. We have developed a defined medium to replace the casamino acids used in the maturation medium. The use of a perfusion reactor eliminates the need for charcoal in the developmental medium. The perfusion reactor also results in higher yields of mature embryos.

We have developed a statistical classifier based on Fisher Discriminants to quantify the developmental state of Douglas-fir somatic embryos. Using this, we have demonstrated that the perfusion reactor results in a more uniform and more mature embryo population compared to control batch cultures.

Somatic embryogenesis is potentially an effective means of mass-propagation of Douglas-fir. The pattern recognition system that we have developed, combined with the defined medium and perfusion reactor, will help in quantitative studies of somatic embryo development.

References

Aitken-Christie, J., Kozai, T., and Smith, M.A.L. (1995). "Automation and Environmental Control in Tissue Culture," Kluwer Academic Press, Dordrecht.

Chi, C-M., Vits, H., Staba, E.J., Cooke, T.J. and Hu, W-S. (1994) Morphological kinetics and distribution in somatic embryo cultures. Biotechnol. Bioeng., 44, 368-378.

Gupta, P.K., Timmis, R.R., Timmis, K.A., Carlson, W.C., and Welty, E.D.E. (1995). Somatic embryogenesis in Douglas-fir (Pseudotsuga menziesii). In "Somatic embryogenesis in woody plants." (S. Jain, P. Gupta and R. Newton, eds.), Vol. 3, pp. 303-313. Kluwer Academic Publishers, Dorfrecht

Taber, R.P., Zhang, C., and Hu, W.-S. 1998. Kinetics of Douglas-fir *(Pseudotsuga menziesii)* somatic embryo development. Can. J. Botany (in press).

GENETIC AND MOLECULAR ANALYSIS OF MALE FERTILITY AND CYTOPLASMIC DNA VARIATION IN INTERSPECIFIC *SOLANUM* SPP. SOMATIC HYBRIDS

T. Bastia[1], N. Scotti[1], L. Monti[1], E.D. Earle[2] and T. Cardi[1]
[1] CNR-IMOF, Res. Inst. for Vegetable and Ornamental Plant Breeding, via Universita' 133, 80055 Portici, Italy, and [2] Dept. of Plant Breeding, Cornell Univ., Ithaca, NY 14853-1902, USA

1. Introduction

Somatic hybridization is a useful technique for gene transfer between *Solanum* species. By contrast with other methodologies, it is also a powerful method to create new nuclear-cytoplasmic organelle assortments and induce variability in organellar DNA (Earle 1995). In potato, male sterility and several important traits are controlled by cytoplasmic genes or their interaction with nuclear ones. However, the genetic control of these traits is not well known. In some interspecific sexual hybrids of *Solanum* spp., nuclear factors from the wild species interact with cytoplasmic factors of the cultivated one to cause male sterility (CMS) (Kaul 1988). Genetic analysis has revealed that one or few nuclear genes are involved in this interaction; in contrast the cytoplasmic factors involved have not yet been characterized. As a matter of fact, while the chloroplast genome of some *Solanum* species has been extensively studied, only recently has the organization of the mitochondrial genome of *Solanum tuberosum* been examined, and very little is known about other *Solanum* species.

Somatic hybrids regenerated in two protoplast fusion experiments between *S. tuberosum* (*tbr*) and the sexually incompatible wild species *S. commersonii* (*cmm*) were all male sterile except for one (Cardi et al. 1993a; Carotenuto, Bastia 1995). In this paper we report some results relative to the characterization of the male sterility observed in the somatic hybrids, and to the molecular analysis of the variability generated in the organellar genomes as a consequence of protoplast fusion and regeneration.

2. Materials and methods

The two population of somatic hybrids (SH and FPN4) between *cmm* (accessions Cmm1 and Cmm3o, respectively) and *tbr* (dihaploid clones Svp11 and Atl9, respectively) were produced as previously described (Cardi et al. 1993b; Carotenuto, Bastia 1995).

A. Altman et al. (eds.), Plant Biotechnology and In Vitro Biology in the 21st Century, 105–108.

Pollen production was analyzed by shaking the anthers with a battery-powdered vibrator and pollen viability was determined by staining with 1% acetocarmine. Southern blots were prepared with total DNA and analyzed with chloroplast and mitochondrial probes.

3. Results and discussion

No segregation of the the fertility phenotype was observed in plants derived from selfing the male fertile hybrid SH9A or crossing the male sterile hybrid SH9B (regenerated from the same callus) with SH9A (Table 1). Moreover, 345 plants derived from crossing 15 other sterile somatic hybrids with SH9A were all male sterile (data not shown). On the other hand, male fertility was restored in 5 out of 14 plants obtained by crossing SH9B with Raritan, a *S. tuberosum* variety known to possess nuclear genes responsible for the restoration of male fertility in male sterile interspecific *Solanum* spp. sexual hybrids.

Table 1. Pollen production in plants derived from selfing the male fertile somatic hybrid SH9A and from crossing the male sterile somatic hybrid SH9B with SH9A and with the variety Raritan

Cross	No. of plants analyzed	Pollen production	
		none/poor	medium/abundant
SH9A x SH9A	472	0	472
SH9B x SH9A	184	184	0
SH9B x Raritan	14	9	5

These results suggest that the male sterility of *cmm* (+) *tbr* somatic hybrids is due to interactions between nuclear genes from the wild species with cytoplasmic (mitochondrial?) genes from the cultivated one, as hypothesized in other interspecific sexual hybrids in *Solanum* spp. In fact, triploid hybrids obtained from sexual crosses between a tetraploid *cmm* clone (as male parent) with a dihaploid *tbr* clone were all male sterile (Novy, Hanneman 1991); in contrast triploid hybrids deriving from reciprocal crosses of the same parents showed good pollen production and stainability (Carputo et al. 1995). Further, it can be hypothesized that the single fertile somatic hybrid either had the intact mitochondrial genome of the wild species or had lost the *tbr* mitochondrial regions involved in nuclear-cytoplasmic interaction through mitochondrial DNA rearrangements *in vitro*.

When the two fusion parents (Cmm1 and Svp11) were tested for RFLP polymorphism in cp or mt DNA, we found a relatively low cpDNA polymorphism: of the four cpDNA probes tested in combination with 8-9 restriction enzymes, only the pStB153 probe in combination with *Bam*HI showed polymorphism. A higher variability was found in the mitochondrial genomes of the two species. However, some mtDNA regions, such as *cox3*, *rrn26*, and *rps3-rpl16*, were highly conserved, since no polymorphism was found even with 8-10 restriction enzymes (Table 2).

Table 2. Number of enzyme x probe combinations which revealed polymorphism between Cmm1 and Svp11

	No. of Probes	No. of Enzyme/ Probe	Total no. of E x P combinations	
			analyzed	polymorphic
mtDNA	14	4-10	81	22
cpDNA	4	8-9	33	1

On the basis of these results, 1 cpDNA and 9 mtDNA probes were tested in Southern analysis using a sample of 31 Cmm1 (+) Svp11 male sterile somatic hybrids and the male fertile clone SH9A.

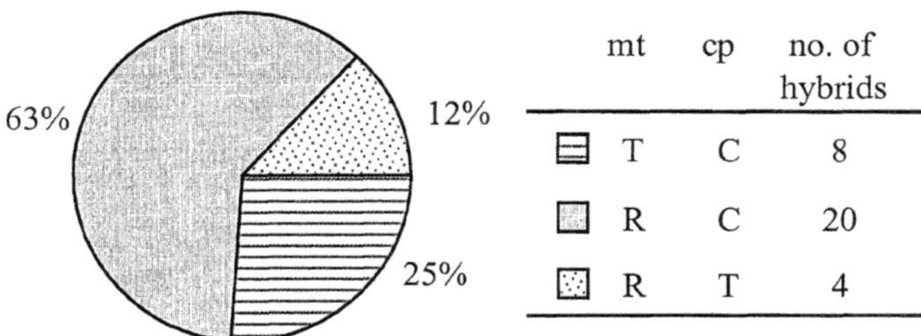

Figure 1. Percentage of hybrids showing different combinations of fusion partner-specific mitochondrial (mt) and chloroplast (cp) loci. C = *cmm* pattern; T = *tbr* pattern; R = rearranged pattern

As reported in Fig. 1, the hybrids showed a large variability at the cytoplasmic level. Most of them (75%) had the mitochondrial genome rearranged (some regions from Cmm1 and others from Svp11) and the chloroplasts from either Cmm1 (63%) or Svp11 (12%). The remaining 25% had inherited the mitochondria from *tbr* and the chloroplast from *cmm*. Other combinations of organelle genomes were not obtained. These results confirm that somatic fusion is an efficient method to generate genetic variability at the cytoplasmic level.

When the *tbr* fusion partner was pre-treated with a herbicide to bleach the in vitro plantlets (SH population), 88% of the hybrids showed the pattern of the wild species. In contrast, chloroplasts of the two species were transmitted randomly when neither fusion parent was pre-treated (FPN4 population). This behaviour may be due to a preferential sorting-out of plastids derived from the bleached fusion partner.

The frequency distribution of male sterile somatic hybrids with 0-9 mtDNA *cmm*-specific loci (Fig. 2) indicates that most of them have a prevalence of *tbr* mtDNA fragments, while the fertile one (SH9A) has a prevalence of *cmm* mtDNA regions (7/9). Although no strict correlation between any mitochondrial genome region and male sterility has been found so far, our results are in agreement with the hypothesis that a rearrangement in the mitochondrial genome of the male fertile somatic hybrid has determined the loss of the *tbr* mitochondrial region involved in the interaction with nuclear genes of the wild species and then in male sterility.

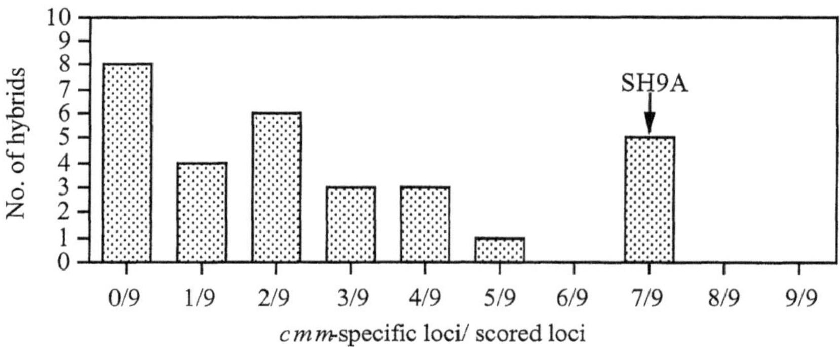

Figure 2. Number of male sterile somatic hybrids with 0-9 mtDNA *cmm*-specific loci. The mtDNA composition of the male fertile hybrid (SH9A) is also indicated

4. Acknowledgements

Contribution no. 156 from CNR-IMOF. Research partially supported by the Italian Ministry of Agriculture, Project "Biotecnologie Vegetali", and by NATO, CRG 960133

5. References

Cardi T et al. (1993a) Theor. Appl. Genet. 87, 193-200.

Cardi T et al. (1993b) Am. Potato J. 70, 753-764.

Carotenuto N, Bastia T (1995) pp 145, XXXIX Annual Meeting of the Italian Society of Agricultural Genetics (SIGA), Vasto Marina, Italy, September 27-30.

Carputo D et al. (1995) Euphytica 80, 123-129.

Earle ED (1995) The Molecular Biology of Plant Mitochondria, pp 557-584, Kluwer Academic Publishers, Dordrecht.

Kaul MLH (1988) Male sterility in higher plants, Springer-Verlag, Berlin.

Novy RG, Hanneman RE, Jr. (1991) Am. Potato J. 68, 151-169.

TRANSGENIC TROPICAL MAIZE WITH *cryIAb* AND *cryIAc* GENES FROM MICROPROJECTILE BOMBARDMENT OF IMMATURE EMBRYOS

N. Bohorova, W. Zhang, P. Julstrum, S. McLean, B. Luna, R.M. Brito, L. Diaz, M.E. Ramos, R. Diaz, P. Estanol, R.V. Ordonez, M. Pacheco, C.R. Castillo, M. Salgado, D. Hoisington

CIMMYT, Lisboa 27, Ap.Post. 6-641, 06600 Mexico D.F., Mexico.

INTRODUCTION

Maize has been one of the prime targets for genetic manipulation in monocotyledonous grains. The first demonstrations of progress were the successful production of transgenic plants by microprojectile bombardment (Klein et al. 1989; Fromm et al. 1990; Gordon-Kamm et al. 1990; Genovesi et al. 1992; Walter et al. 1992; Frame et al. 1994; Wan et al. 1995, Brettschneider et al. 1997); then successful hygromycin (Walters et al. 1992), bialaphos (Spencer et al. 1990), and glyphosate (Howe et al. 1992) selection of stable transformants, and recent *Agrtobacterium* - mediated gene insertion in maize plants (Ishida et al. 1996). Most studies on maize transformation have utilized genotypes adapted to temperate zones (Fromm et al. 1990; Gordon-Kamm et al. 1990; Walter et al. 1992; Armstrong et al. 1995) and plants regenerated from these lines were shown to transmit the recombinant DNA to their progeny. Little or no attention, however, has been focused on the transformation potential of maize germplasm and inbred lines adapted to tropical and subtropical regions. Production of genetically transformed plants depends both on the ability to integrate foreign genes into target cells and the efficiency with which plants are regenerated from genetically transformed cells. Embryogenic calli and plant regeneration were obtained from 50% of tropical and subtropical lines, 87% of midaltitude lines, and 75% of highland lines tested (Bohorova et al. 1995) and type II callus with high potential for plant regeneration from tropical maize was produced (Prioli and Silva 1989; Carvalho et al. 1997). These studies serve as the basis for developing transgenic technology for maize inbreds adapted to tropical conditions. As part of a UNDP-sponsored project to develop insect resistant tropical maize germplasm using genetic engineering, we developed the techniques to transform elite CIMMYT tropical, subtropical, and midaltitute inbreds and to stably integrate and express insecticidal proteins in tropical and subtropical maize germplasm.

MATERIALS AND METHODS

Preparation and culture of immature embryos

Tropical maize inbred lines CML67, CML72, CML216, CML323, and the hybrids from these lines were used in the experiments. Immature embryos (1.0-1.5 mm) were aseptically removed from the kernels and placed scutellum up on the initiation medium. N6C1 medium was used for embryogenic callus initiation and maintenance (Bohorova et al. 1995). For callus initiation and maintenance, the cultures were incubated in darkness at 28°C and embryogenic tissue was subcultured every 21 days. Plants were regenerated from embryogenic calli by transferring tissue to Magenta containing MSR medium and then transferred to MSE medium for root formation (Bohorova et al. 1995). The cultures were incubated in a growth culture room at 28°C with a light/dark photoperiod of 16:8 hours .

Plant expression vectors

In order to study which genes are most appropriate for the generation of transgenic plants, we screened the toxic activity of native isolates of *Bacillus thuringiensis* (*Bt*) and specific

A. Altman et al. (eds.), Plant Biotechnology and In Vitro Biology in the 21st Century, 109–113.

cryI proteins against four major tropical maize pests: corn earworm (CEW), fall armyworm (FEW), southwestern corn borer (SWCB), and sugarcane borer (SCB) (Bohorova et al. 1996; Bohorova et al. 1997). Based on these results, we included plasmids containing synthetic *cryIAb* and *CryIAc* genes (kindly supplied by Dr. I. Altosaar, Canada) and *bar/gus* (kindly supplied by Dr. S. Maddock; Pioneer Hi-Bred, U.S.A.) carrying the ß-glucuronidase (*gus*) and the selectable *bar* genes under the control of the cauliflower mosaic virus (CaMV) 35S promoter.

DNA coating and bombardment
For microprojectile bombardment, prewashed 50 µl aliquots of gold particles (40 mg gold particles suspended in 1 ml distilled water using a procedure from Bio-Red Instruction Manual) were coated with 5 µl plasmid DNA (1 µg/1µl) on ice. Fifty µg sterile aliquots of 2.5M$CaCl_2$ were stored at 4°C and mixed with 20 µl of 0.1M spermidine in a microfuge (1.5 ml) and added to the solution of particles with DNA. The mixture was vortexed for 3 minutes at room temperature and centrifuged in a microfuge for 1 min. The supernatant was removed and discarded. The DNA particles were washed with 240 µl ethanol (75%), resuspended in 240 µl absolute ethanol, and 3-5 µl of the suspension was spread onto the center of each macrocarrier, air-dried, and bombarded. The DNA for bombardment was adjusted to a concentration of 1 µg/µl in TE buffer (1mM Tris, pH 7.8; 0.1 mM Na_2EDTA and is stored at -20°C). Fifty fresh embryos (aseptically removed from maize caryopses or precultured on callus initiation medium for 4 days) were placed on disposable petri plates (100 x 15 mm) with 10 ml N6C1 medium and arranged in a circle about 2 cm in diameter in the center of the plates. For osmotic pre-bombardment treatments, the material was kept on N6C1 medium with 12% maltose for 4 hours. Bombardments were performed using the Bio-Rad PDS-1000 helium-driven biolistics particle delivery system. Each plate of tissue was bombarded twice at a rupture pressure of 900, 1100, or 1350 to 1500 psi. Different bombardment parameters were evaluated: particles with different diameters: 0.4-1.0 µm (gold powder, spherical, Aldrich Chem. Co.) and 1.0 µm, 1.5 µm, 1.9 µm (Bio-Red gold powder); particles per bombardment 14-900 µg/shot; particle suspension volume per bombardment (3-5 µl); DNA per bombardment (5-20 µl); and number of shots per target (1-2 shots). GUS activity was detected histochemically as described by McCabe (1988).

Selection of transformed embryos/calli and recovery of transgenic plants
Selection of transformed cells was achieved using either phosphinothricine (PPT) (Sigma) or bialaphos (B) (Meiji Seika Kaisha, Ltd., Yokohama, Japan). Concentration ranges from 1 to 10 mg/l of PPT or bialaphos were tested with non-transformed calli to derive a more efficient selection system. Two different approaches were followed for the selection of transformants. In the first approach, the embryos or calli were transferred to the N6C1 medium with all of the medium components as described by Bohorova et al. (1995), or N6C1 medium containing 1 mg/l PPT/bialaphos (N6C1B1), and cultured in darkness for 7 days. In the second selection experiment, we cultured the bombarded materials on the N6C1 medium, but excluded both L-proline and casein hydrolysate (N6C2), or added to N6C2 medium 1 mg/l PPT/bialaphos, and then kept the materials for 7-14 days in darkness. The duration of the selection process was 50-75 days on the media with 1, 3, 5, and 10 mg/l PPT/bialaphos, depending of the genotype used, refreshing the selection medium every 2 weeks. All cultures were kept at 28°C under dark conditions in the growth culture chamber.

Plant regeneration, Basta™ testing and leaf bioassay
Plantlets developed on regeration medium were transferred to soil in environmentally controlled growth chambers and greenhouse conditions, and grown for further analyses. Plants were further selected by painting the fifth or sixth leaf near the tip of the youngest fully extended leaf with 2% Basta™ solution containing 0.1% Tween 20. All Southern positive plants for *cry* genes were crossed with the respective CML lines to produce T1 seeds. From each event we tested at least 20 plants in T1 progeny. Three infestations of 15

neonate SWCB larvae (per infestation) were applied during the test period. Plants were scored for feeding damage 2 weeks after the final infestation using 1-9 scale (1 best; 9 worst), and were divided into two classes: resistant and susceptible.

Southern hybridization
For molecular confirmation, genomic DNA was isolated from leaf samples of each putative transformant as well as the untransformed tropical lines. Plant genomic DNA was extracted from freeze-dried young leaf tissue (0.5 g) of primary transformants and progeny plants according the procedure described by Hoisington et al. (1994).

RESULTS

Optimization of DNA delivery parameters
Highly regenerable embryos were bombarded with gold particles coated with plasmid DNA (*gus/bar*) to optimize delivery of DNA-coated microprojectiles and co-trasformed with *cry*IAb and *cry*IAc gene constructs. Different microprojectile bombardment parameters were evaluated using transient ß-glucoronidase (*gus*) assays. In our case, the highest number of transient *gus* expression events were observed with embryogenic immature embryos. For bombardment of 20 petri plates, each with rings consisting of about 50 embryos, the following procedures were used: 1) fresh isolated immature embryos (1-1.5 mm) or 4-day precultured immature embryos were obtained; 2) 4 hour pre-bombardment osmotic treatment of the embryos was performed on the N6C1 medium with 12% maltose; 3) 50 µg gold particles (1.0 µm in diameter) coated with 5 µl plasmid DNA (2.5 µl DNA from two different plasmids) were put on ice and add a mixture of 20 µl spermidine and 50 µl $CaCl_2$; 4) vortex the mixture using a vortex shaker with a multiple platform head for 3 min at room temperature; 5) the mixture was centrifuged in a microfuge for 1 min at 13,000 x g and the supernatant was removed and discarded; 6) the solution was resuspended in 240 µl absolute ethanol twice; 7) five µl of suspension were pipetted onto the center of a macroccarrier previously positioned in a macrocarrier holder and air-dried; 8) two bombardments were applied with 1,350 psi rupture discs, using particle densities of 30 µg/shot; 9) target materials were positioned approximately 8 cm below the microprojectile stopping plate (3rd shelf from bottom).

Major differences in transient gene expression were not observed between the particle diameters and densities. An average number of 60 to 100 transient *gus*-signals per embryo was recorded from maize embryos using particle densities of 14, 30, and 75 µg/shot. Each blue spot of *gus*-expressing cells had a densely stained blue central core, with a less densely stained surrounding region. With particle densities of 150, 300, 600, and 900 µg/shot fewer blue spots appeared, but more diffused blue color was detected on the surface of the embryos or callus.

Acceleration pressure has been optimized and for compact calli established from CML72 and CML216, 1350 psi is used, and for friable tissue, such as that originating from CML67 embryos, 1,100 psi is used.

Selection and regeneration of transgenic plants
Two protocols were used for selection with bialaphos. The first experiment included the bombardment of immature embryos or 4 day-old precultured from immature embryos. In this experiment, cells were grown on callus initiation medium (N6C1) for 7 days post-bombardment or on N6C1 with 1 mg/l bialaphos for 7-10 days. The proliferating explants were subsequently transferred to N6C1 medium with 5 or 10 mg/l bialaphos, depending on the genotypes used. Four weeks later, the difference between the resistant explants and the control explants became more pronounced. Resistant colonies appeared at 6-8 weeks. All calli of non-transformed tissue of CML67 died within a few days on selection medium with 5 mg/l bialophos, but transformed embryogenic calli and somatic embryos of CML216 developed on the selection medium with 10 mg/l bialaphos.

To have an effective phosphinothricin selection it is important to omit not only glutamine from the selective medium, but also several other amino acids. For this reason, in the second experiment we excluded both L-proline and casein hydrolysate from the selective medium (N6C2) and increased the concentration of bialophos to 10 mg/l. In this experiment, after bombardment embryos/calli were plated immediately onto medium with the selective agent bialaphos using a concentration of 1 mg/l. They were kept on this medium for 1 week, then transferred to a selection medium with 5 or 10 mg/l bialaphos for 4 more weeks. To regenerate the selected embryos/calli to plantlets, a low concentration of bialophos or PPT was used to obtain development of transgenic fertile plants. The concentration of 5 mg/l bialaphos in the regeneration medium and 1-3 mg/l in rooting medium resulted in recovery of transformed plantlets. Within 4-5 months putative transformed maize plants were transferred to soil and grown under greenhouse conditions. All the plantlets were healthy and successfully grew into fertile mature plants.

Analysis of T0, T1, and T2 plants

As reported previously, we chose the constructs in order to have effective herbicide selectable markers containing *bar* and *gus* genes to co-transform with *cryIAb* and *cryIAc* genes. This allowed us to screen putative transgenic plants by Basta™ bioassay. Most of T0 plants were morphologically normal pollinated in two direction crosses. Plants showing moderate to high resistance to Basta™ herbicide were further checked using Southern blot analyses. The T0 plants regenerated from nontransformed calli were susceptible to the herbicide treatment, as expected.

All T0 plants were analyzed for the presence/ absence of introduced genes by DNA hybridization. Three transgenes *bar*, *gus*, and *cry* were used as probes to hybridize with genomic DNA. All 49 transgenic plants had different independent transformed events, which were blotted and probed with *cryIA* fragments. Southern hybridization analysis confirmed that these 49 plants were derived from 18 independent transformation events because they all showed different hybridization patterns. For these independently produced transgenic plants the estimated copy number of the intact fragments varied from approximately 1 to 10.

For phenotypic confirmation, a simple insect bioassay was performed with leaf samples from each transgenic plant carrying the *cryIAb* gene, already confirmed by Southern blot. For example, from 367 bombarded immature embryos in experiment #77 (lab list), 50 embryos were recovered after selection, producing 119 plants that were transferred to the greenhouse. Seventeen plants were classified as fully resistant to Basta™, none were resistant to SWCB in T0 progeny, and none displayed resistance in T1 progeny. The 17 plants originated from 3 different embryos belonging to 3 different events that carried *bar*, *gus*, and *cryIAb* genes. Analysis of the other event (#4) indicated that only 13 plants out of 211 were hybridized with *bar*, *gus*, and *cry* genes and presented resistance to herbicide in T0 progeny; none of these plants expressed resistance to the Southwestern corn borer in T0 and T1 progeny.

Resistance to larvae of Southwestern corn borer presented only in plants from event #16 (lab. list), which carried *bar*, *gus*, and *cryIAc* genes. In T0 progeny, transgenic plant showed resistance to SWCB larvae and the plants in T1 progeny segregated 19:17 (resistant to susceptible), close to the expected Mendelian ratio. All Southern positive plants for *cryIAc* gene were crossed with the respective inbred lines. Twenty seedlings were obtained and analyzed using insect feeding assays with leaf samples. Segregation data of T2 progeny indicated that none of the plants presented resistance to Basta™, and the segregation ratio for resistance to the SWCB was 6:14 (resistant to susceptible).

Discussion

In our experiments on developing transgenic technology for tropical and subtropical maize, we focused on optimizing biolistics-mediated gene delivery into cell cultures; applying enhanced gene expression vectors for efficient selection of transformed tissue; and using various parameters and selection regimes for recovery of transgenic maize. Our goals were to establish a replicable plant genetic transformation system for tropical maize, to introduce gene(s) into intact cells, and to have them adequately express in maize plants. Expression of *Bt* genes in maize, resulting in excellent protection against European corn borer in the field, was advanced when Koziel et al. (1993) transformed elite cultivars of maize with truncated *cryIAb* gene and with the expression of an insecticidal protein in the progeny of 173 transgenic corn events screened by Armstrong et al. (1995). With the use of transformation techniques to obtain new cereal crop products, we expect genetically improved tropical maize containing *Bt* genes to play a vital role in integrated pest management in tropical areas. The new products should provide breeders with another resource for enhancing the genetic composition of their germplasm. However, the actualization of genetically modified plants' potential, particularly tropical maize upon which so many of the world's poor rely, depends on a broad array of non-technical issues, especially, intellectual property rights and related regulatory developments (patent structures and guides for deployment of genetically modified organisms).

Acknowledgments

The authors would like to thank David Poland and the group from CIMMYT publications for preparing this manuscript. Thanks to Dr. I. Altosaar, University of Ottawa, Canada, and Dr. Sh. Maddock, Pioneer Hi-Bred Int., USA for kindly providing the constructs for maize transformation, and Dr. Sarah Fennell now working on genetic analysis of transgenic maize. The work to develop transgenic maize is supported by UNDP.

References:

Armstrong CL et al (1995) Crop Sci. 35:550-557.
Bohorova NB et al (1995) Maydica, 40/3 p. 275-281.
Bohorova NB et al (1996) Entomophaga, v.41, 2, 153-165.
Bohorova NB et al (1997) Journ of Econ Entom Micr. Control.90 (2): 412-415.
Brettschneider R et al (1997) Theor Appl Genet 94:737-748.
Carvalho CHS et al (1997) Plant Cell Reports 17:73-76.
Frame BR et al (1994) The Plant Journal 6:941-948.
Fromm ME et al (1990) Bio/Technology 8:833-839.
Genovesi D et al (1992) In vitro Cell Dev Biol 28, 124 A
Gordon-Kamm WJ et al (1990). The Plant Cell 2:603-618.
Hoisington D et al (1994). Lab.Prot.:AMGL. Mexico, D.F. CIMMYT.
Howe AR et al (1992) In vitro Cell. Dev. Biol. 28(3):124 A.
Ishida Y et al (1996) Nuture Biotechnology v.14, June, 745-750.
Klein TM et al (1989) Plant Physiol. 91, 440-444.
Koziel M et al (1993) Bio/technology 11, 194-200.
McCabe E et al (1988) Bio/technology, vol. 6,923-926.
Prioli LM,Silva WJ (1989) Rev Bras Genet 12:553-566.
Spencer TM et al (1992) Plant Mol Biol 18: 201-210
Walter DA et al (1992) Plant Mol Biol 18:189-200.
Wan Y et al (1995) Planta 196: 7-14.

BYPASSING JUVENILE STAGE IN GENETIC TRANSFORMATION OF CITRUS PLANTS

M. Cervera, J. Juárez, A. Navarro, J.A. Pina, N. Duran-Vila, L. Navarro, L. Peña.
Dept. Plant Protection and Biotechnology, Instituto Valenciano de Investigaciones Agrarias (IVIA), Apartado Oficial 46113, Moncada, Valencia, Spain

1. Introduction

Improvement of woody fruit species through genetic engineering will have limited applications unless tissue from mature plants of elite varieties can be readily transformed. Transformation and regeneration of these plants are usually restricted to juvenile tissues (Cervera et al. 1998a), which exhibit greater regenerative potential (Durzan 1990) and greater susceptibility to transformation than mature ones (Smith 1920). Regenerated plants from juvenile tissues will have juvenile characters and several years will be needed before horticultural and commercial traits of the transgenic plants could be evaluated. Development of transformation procedures that could bypass the juvenile stage would greatly reduce the time involved in improving woody fruit trees by genetic engineering.
Sweet orange (*Citrus sinensis* L. Osb. cv. Pineapple) has been chosen as a model woody fruit tree plant to develop a transformation procedure for mature tissue of a selected cultivar. Citrus is the most extensively grown fruit crop worldwide and sweet orange accounts for approximately 70% of the citrus total production. Up to 20 years may be needed for sweet oranges to lose juvenile characters and bring into production. We describe here a method that allows genetic transformation and regeneration of mature transgenic sweet orange plants.

2. Material and Methods

2.1. Source and preparation of plant material

Buds collected from *C. sinensis* L. Osbeck cv. Pineapple maintained in a screenhouse of the virus-free Germplasm Bank Collection of the IVIA were grafted on seedlings of *C. volkameriana* Ten & Pasq. at the greenhouse (18-27°C). Newly elongated mature shoots were allowed to develop one, two or three flushes, that were used separately as the source of mature tissues. Stem pieces (20 cm long) were stripped of their leaves and thorns and disinfected. One-cm-long internodal stem segments in a semi-hardened stage were cut transversely from the stem pieces.

2.2. Regeneration of mature material

The regenerative potential of the three first flushes from the newly elongated mature plants as well as material from juvenile plants was compared. Internodal stem segments were cultured horizontally in MSB3 medium (MS salts, White vitamins, 3 mgl^{-1} BAP, 3%

A. Altman et al. (eds.), Plant Biotechnology and In Vitro Biology in the 21st Century, 115–118.

sucrose and 1% agar, pH 5.7) (Peña et al. 1995a), maintained in the dark for 8 weeks at 26°C and 60% RH and then exposed to a 16 h photoperiod under an illumination of 45 $\mu Em^{-2}s^{-1}$ with no changes in temperature and humidity conditions. Regeneration frequency was evaluated three weeks later as the number of regenerating explants per total number of cultured explants.

2.3. Bacterial strain and vector

A. tumefaciens EHA105 containing the binary plasmid p35SGUSINT (Vancanneyt et al. 1990) was used as vector system for transformation. *A. tumefaciens* EHA105 is a disarmed derivative of *A. tumefaciens* A281 (Hood et al. 1993), which had been previously established as supervirulent in this species (Cervera et al. 1998b). Two gene cassettes in the T-DNA, 35S-*uidA*(GUSINT)-35S and NOS-*nptII*-NOS, served respectively as reporter and selectable marker genes. Bacterial suspensions were prepared as described in Peña et al. (1995a,b).

2.4. Transformation of mature material

Only the first flush from mature plants was used in two independent transformation assays. Internodal 1-cm-long stem segments from these shoots were inoculated and cocultivated on feederplates for 3 days. Feederplates were prepared by pipetting 2 ml of 6 to 7-day-old tomato cell suspensions on the surface of 25 ml of TCS solid medium (MS salts plus tomato vitamins, 3% sucrose, 2 mgl^{-1} IAA, 1 mgl^{-1} 2-ip, 2 mgl^{-1} 2,4-D, pH 5.7 (Cervera et al. 1998a) with a sterile Whatman 5 paper on the top in 10 x 1.5 cm (diameter x height) plates. Following cocultivation, the explants were blotted dry and transferred to MSB3 medium containing 100 mgl^{-1} kanamycin, 250 mgl^{-1} cefotaxime and 250 mgl^{-1} vancomycin. The plates were maintained in the dark for 15 days at 26°C and then transferred to a 16 h photoperiod in the conditions above mentioned. Explants were subcultured to fresh medium every 4 weeks. Regenerated shoots were harvested from the stem segments and excised in two pieces. The shoot basal ends were assayed for GUS activity (Jefferson 1987) and the remaining portions were shoot-tip grafted *in vitro* onto Troyer citrange (*C. sinensis* L. Osbeck x *P. trifoliata* L. Raf.) seedlings, as previously described (Peña et al. 1995a,b). After three weeks, the *in vitro* grafted plantlets were again grafted in the greenhouse onto vigorous 5-month-old seedlings of Rough lemon (*C. jambhiri* Lush).

2.5. DNA analysis

DNA was isolated from leaves according to Dellaporta et al. (1983). For Southern analysis, 15 μg of *Eco*RI, *Hin*dIII or *Pst*I-digested DNA samples were separated on 0.8% agarose gels, blotted to nylon membranes (Hybond-N+, Amersham) and probed with DIG (Boehringer Mannheim) labeled coding regions of *uidA* or *nptII* genes following the supplier instructions.

3. Results and Discussion

3.1. Invigoration of mature tissue through grafting on juvenile rootstocks

Previous experiments in our laboratory (unpublished results) had confirmed the limited regenerative potential of explants from aged mature citrus plants. Thus, to increase the regeneration ability of mature explants, buds from adult trees were grafted on vigorous seedlings. Regeneration from stem segments from the first, second and third flushes of newly-grafted invigorated mature sweet orange plants was evaluated in comparison to regeneration from stem segments from juvenile plants. Results indicated that explants from the first and second flushes produced similar regeneration frequencies, significantly higher than that of the explants from the third flush (Table 1). The first flush of the adult plants was selected as the source of tissue for genetic transformation experiments.

3.2. Application of transformation process on tissue from invigorated sweet orange plants
The transformation procedure which had been optimized for regeneration of transgenic juvenile sweet orange plants (Cervera et al. 1998a) was applied to explants from the first flush of invigorated mature plants. This process included the inoculation of explants with the supervirulent *A. tumefaciens* strain EHA105, a 3-day cocultivation period on tomato feederplates and selection on 100 mgl^{-1} kanamycin, exposing the explants to darkness during the initial 15 days of selection. After 2-5 months on MSB3 selective medium, regenerated shoots 0.2-0.5 cm high (Fig. 1A) were excised from the explants, assayed for GUS activity and *in vitro* grafted on Troyer citrange seedlings (Fig. 1B). In two different experiments, 211 shoots were recovered from 294 inoculated mature explants. Thirteen shoots (6.1%) were GUS-positive. Transformation frequencies were comparable in both experiments, approximately three fold lower than that obtained for juvenile material (results not shown). It is probable that aging decreased the susceptibility of plant cells to *Agrobacterium* infection. All of the putatively transgenic and control plantlets were again grafted at the greenhouse on juvenile Rough lemon seedlings.
High frequency regeneration of escapes was obtained. Whereas GUS+ shoots regenerated in all cases from callus from the cambium, escapes regenerated directly from the explants with almost no callus formation. Observation of serial thin sections under the microscope (results not shown) suggested that competent cells for transformation were cambial dedifferentiated (actively dividing) cells.

3.3. Confirmation of transgenic nature and mature ontogenic stage of sweet orange plants
Southern blot analysis confirmed the presence of stable integrated *uidA* and *nptII* gene cassettes into the plants' genome in a variable number of copies (Fig. 2).
The putative mature transgenic sweet orange plants showed morphology and growth habits of an adult plant, as compared to control mature plants. In fact, whereas juvenile plants showed a pronounced thorniness, transgenic mature plants were almost thornless, similar to the mature plants from which the explants were taken for transformation. After fourteen months in the greenhouse, the transgenic and control plants started to flower, confirming their mature nature. Both flowers and immature fruits from the transgenic plants showed a dark blue color after overnight incubation in X-Gluc (Fig. 3A, 3B).
These results confirmed the maintenance of the ontogenic mature stage of the invigorated mature plants as well as the transgenic plants. Interestingly, transgenic events kept their epigenetic mature state even after a process of dedifferentiation, induction and redifferentiation, necessary to shift the cells to a competent stage for transformation. Therefore, we have directly transformed and regenerated mature tissues of citrus, bypassing the juvenile stage. This process greatly shortens the period of time until flowering and bearing fruits and decreases the time to achieve horticulturally acceptable characteristics by years. It could be very valuable to apply this transformation and regeneration system to transgenic plant production of other woody fruit species.

4. References

Cervera M et al (1998a) Transgenic Res. 7, 51-59.
Cervera M et al (1998b) Physiol. Mol. Plant Pathol. 52, in press.
Durzan D (1990) Plant Aging. Basic and Applied Approaches, vol 186, pp 19-25, NATO ASI Series.
Dellaporta SL et al (1983) Plant Mol. Biol. Rep. 4, 19-21.
Hood EE et al (1993) Transgenic Res. 2, 208-218.

Jefferson RA (1987) Plant Mol. Biol. Rep. 5, 387-405.
Peña L et al (1995a) Plant Cell Rep. 14, 616-619.
Peña L et al (1995b) Plant Sci. 104, 183-191.
Smith EF (1920) An introduction to bacterial diseases of plants, pp 413-472, WB Saunders Company, London.
Vancanneyt G et al (1990) Mol. Gen. Genet. 220, 245-250.

TABLE 1. Effect of the flush on the regeneration frequency from *Citrus sinensis* mature internodal explants and comparison to juvenile explants.

Material		Regeneration frequency[1] (%)[2]	Mean of shoots per explant ± SE[3]
Juvenile		53 (39)	4.3 ± 0.4 a
Mature	1st flush	23 (38)	2.1 ± 0.4 b
	2nd "	18 (38)	1.4 ± 0.6 b
	3rd "	5 (39)	1.7 ± 1.3 ab

[1] Regeneration frequency (%) defined as the number of regenerating explants per the total number of cultured explants.
[2] Figures in brackets represent the total number of evaluated explants after discarding contaminated cultures.
[3] Different letters indicate significant differences at 0.5% level using Student's t.

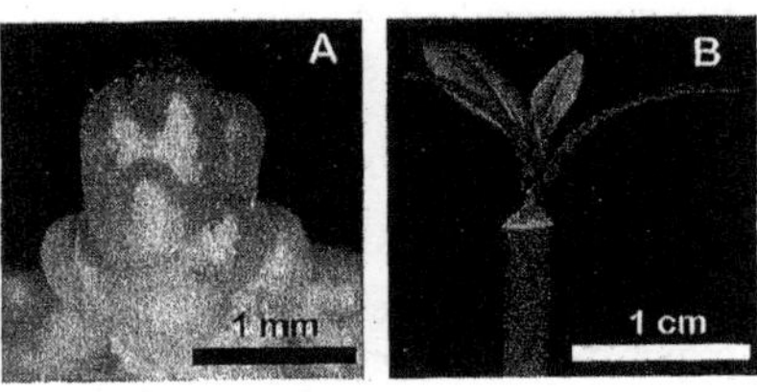

FIGURE 1

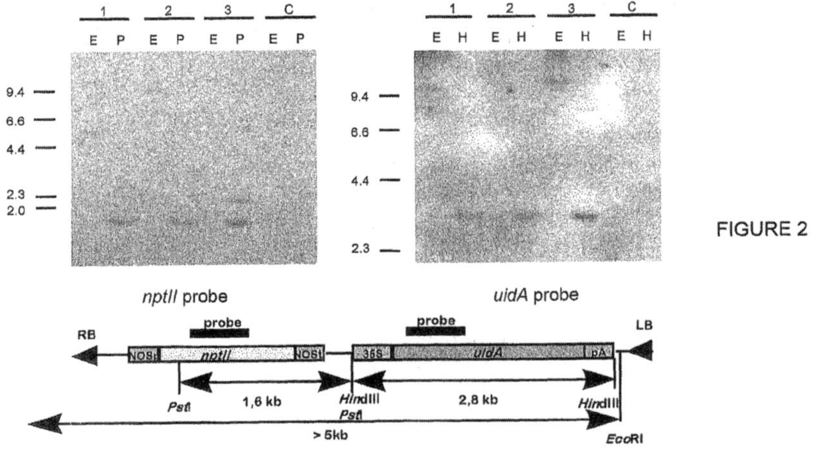

FIGURE 2

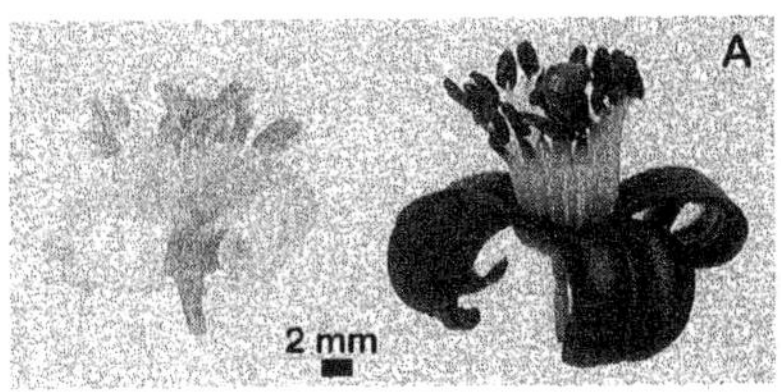

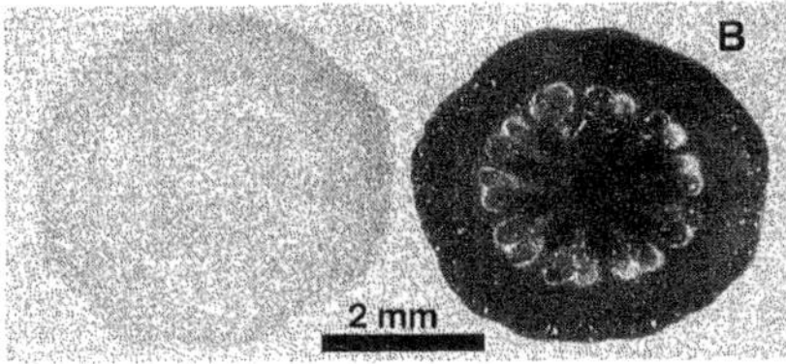

FIGURE 3

GREEN HAPLOID PLANT PRODUCTION IN DURUM WHEAT BY VARIOUS HAPLO-METHODS

H. Chlyah, S. Cherkaoui, N. Saidi, O. Lamsaouri, M. Mdarhri-Alaoui, H. Benkirane, O. Chlyah, O. Amail

University Mohammed V, Science Faculty, Biology Department, B.P. 1014, Rabat, Morocco

1. Introduction

Haploid plant production in durum wheat (*Triticum turgidum* ssp. *durum* = *T. durum*), and subsequent chromosome doubling to constitute homozygous diploid plants, can be very useful in breeding programmes for improvement of this species, in particular for salt and drought tolerance.

Anther culture (androgenesis), although it has led to some green haploid plant formation in bread wheat,gave almost total albino plant regeneration in durum wheat (Chlyah, Saidi 1991 ; Saidi et al. 1997), except for a few particular genotypes (Cattaneo et al. 1991, Ghaemi et al. 1995), and therfore cannot yet be used.

To attempt to overcome this handicap of albino plant regeneration observed in anther culture, various physiological, chemical and environmental factors have been tested ; anther culture of plants produced by backcrossing a *T.aestivum-T. durum* hybrid to *T. durum* has also been carried out.

Other haplo-methods have also been used: unfertilized ovary culture (gynogenesis) and wide crosses with maize or *Hordeum bulbosum*.

2. Haplo-methods

2.1. Androgenesis (anther culture)

Anther cultures (microspores in the uninucleate stage) of sixteen cultivars of durum wheat and one F1 hybrid were compared. A pretreatment of wheat plants with a chemical hybridizing agent (CHA), which prevents pollen differentiation, was applied in one experiment. A cold pretreatment (3°C) of excised wheat spikes was maintained for 0, 2, 4-6, 8 and 15 days and anthers then cultured on four induction media : C17, BPTG, P2 and N6 (see Saidi et al 1997). All contained sucrose (9%), agar (8 or 10gl-1) and 2,4-dichlorophenoxyaceticacid (2,4-D) at 2mg.l-1 and all but C17 also contained kinetin (0.5mg.l-1). Embryos produced were transferred to one of three regeneration media : R9, C17 or N6 which contained various regulatory substances and amino compounds, sucrose (2 or 3%) and agar. Anther culture of plants produced through backcrosses after an initial *T. aestivum-T. durum* cross were done in optimal conditions

A. Altman et al. (eds.), Plant Biotechnology and In Vitro Biology in the 21st Century, 119–122.

as defined in the preceding experiments .

2.2. Unfertilized ovary culture (gynogenesis)

Eight durum wheat genotypes were tested. As in anther culture, spikes were cold pretreated for 0, 5,10,15 or 20 days. Ovaries (excised when microspores in anthers were in uni- or binucleate stages) were cultured on MS medium containing various auxins and cytokinins, sucrose (90g l-1) and agar (8g l-1) (Mdarhri-Alaoui et al 1998). Three light-dark conditions were tested : 1- continuous darkness, 2- 2-5 weeks darkness, then 16h photoperiod ; 3- continuous light (16h photoperiod). Calli formed were transferred to a MS medium without regulatory substances, then to a R9 medium containing 1mg l-1 of both indolyl acetic acid (1AA) and kinetin.

2.3. Wide-crossing

Nine genotypes of durum wheat were crossed with 8 varieties of maize or with *H. bulbosum*. Emasculation of durum wheat spikes, pollination, 2,4-D treatments of florets and detached tillers, and culture procedures for rescued haploid embryos have been described elsewhere (Saidi et al. 1998).

Chromosome counts in regenerated plants were done on squash preparations of root tip cells after staining with the Schiff reagent. Chromosome doubling of haploid plants was done by a colchicine (0.25%) treatment for 4-5h at room temperature.

3. Results and discussion

3.1. Androgenesis in durum wheat cultivars : problem of albinism

In order to optimize androgenic embryo formation and regeneration in durum wheat, various factors were studied (see Saidi et al. 1997 for complete data). All 17 genotypes studied were capable of embryo formation in the range of 3.2-25.4 embryos per 100 anthers, but after shoot formation on a regeneration medium (16 combinations of growth and trophic substances) and over 60000 anthers cultured, only 2 green haploid plants were obtained, all others were albino.

Optimization of other factors (use of a CHA treatment, 8-day cold pretreatment of spikes, C17 induction medium, transfer of 1mm long 21 day old embryos) resulted only in improved rates of embryo formation and/or albino regeneration, not regeneration of green plants.

3.2. Androgenesis in plants from 3 back-cross generations after a *T. aestivum* x *T. durum* cross

In an attempt to transfer the capacity for green androgenic plant formation *of T. aestivum* to *T. durum*, we used the following protocol of crosses :

T. aestivum x T. durum	→	F1 (2n=35)
F1 x *T. durum*	→	BC1 (2n=28-35)
BC1 x *T. durum*	→	BC2 (2n=±28)

BC2 x *T. durum* → BC3 (2n=28).

Two cultivars of aestivum wheat : 'Nesma' (a normal genotype) and 'Verry's'='Tilila' (genotype containing the 1BL/1RS translocation from rye) were each crossed with 4 genotypes of durum wheat. Anthers from BC1, BC2 and BC3 plants from crosses with 'Verry's' produced, in general, a percentage of green haploid plants twice as big as that obtained from the cross with 'Nesma'. In both cases, the number of green plants decreased from BC1 to BC3, while the number of albino plants increased. Doubled haploid green plants produced through this method from the BC3 generation had 28 chromosomes like durum wheat, while retaining some capacity for green plant formation through androgenesis This method involved a lengthy crossing procedure and seemed mainly successful in the presence of the 1BL/1RS translocation

3.3. Unfertilized ovary culture : callus and green plant regeneration

Using this method, effects of genotype and two environmental factors were studied. The genotype effect of durum wheat was again important : only 5 of the 8 cultivars studied produced calli and 4 of these also produced green plants (Mdarhri-Alaoui et al 1998). A cold pretreatment (5 – 10 days) of spikes was necessary for organogenesis. The light regime was also important : a period of darkness (2-5 weeks) followed by light (16h photoperiod) was necessary for green plant formation ; ovaries cultured in continuous darkness or continuous light (16h photoperiod) gave no results. Compared with androgenesis, this technique had the advantage of producing 100% green haploid plants. However, the presence of callus, even reduced, could be a source of genetic variability.

3.4. Wide crossing method (chromosome elimination) : haploid embryo and plant formation

This method of haploid production with maize or *H. bulbosum* as pollinators, has been successful in aestivum wheat for many years and wide crossing with maize has recently been applied to durum wheat (O'Donoughue, Bennett 1994a, Savaskan et al 1997, Saidi et al 1998). So far in the literature, no green haploid plants have been reported after *T. durum-H. bulbosum* crosses although fertilization was shown to occur (O'Donoughue, Bennett 1994b).

3.4.1. Durum wheat x maize

- Wheat genotypic effect. All nine genotypes of durum wheat showed ovary enlargement in approximately 75% of florets and all produced some embryos : for about 100 pollinated florets (all maize genotypes grouped), numbers of embryos varied significantly from 1.9 to 8.1 depending on the cultivar. After embryo rescue, subsequent germination to form at least one green haploid plant was observed for each wheat genotype. However, rates of plant formation varied widely (3- 33.5 plants per 100 embryos). These differences seemed related to the quality of the embryo at the time of its rescue.

- Maize genotypic effect. A comparison of mean green plant formation per100 embryos for 8 cultivars of maize (all wheat genotypes grouped), showed values from 16.4% to 3%. Although most crosses produced embryos, some were incapable of germination to form green plants. The best pollinator was able to hybridize with 7 of the 9 wheat genotypes, the worst with only 2 genotypes.

3.4.2. Durum wheat x *H. bulbosum*
Although all 9 wheat genotypes (same genotypes as in preceding crosses with maize) tested showed at least 50% ovary development, only 5 produced embryos at rates from 2.3% to 6.7% and only embryos of 4 genotypes were able to germinate and produce one or more green haploid plants. Therefore, although it has been demonstrated that this cross could result in haploid green plants, incompatibilities with several wheat genotypes has reduced the value of *H. bulbosum* as pollinator.

4. Conclusion

Whereas androgenesis through anther culture remained useless for our durum wheat cultivars because of widespread albinism, it was possible from backcross generations after aestivum-durum crosses, particularly if the aestivum parent had the 1BL/1RS translocation. Unfertilized ovary culture was relatively simple and produced all green haploid plants but a callus phase before organogenesis has so far been unavoidable. Wide crossing with *H. bulbosum* and maize produced green haploid plants directly from embryos, and the cross with maize was the most effective for a wide range of durum genotypes.

5. Acknowledgements

We thank the Francophone Agency for Higher Education and Research (AUPELF-UREF) for financing this research.

6. References

Cattaneo M et al (1991) J. Genet. Breed. 45, 369-372.
Chlyah H, Saidi N (1991) Amélioration des Plantes pour l'Adaptation aux Milieux Arides, pp 138-148, John Libbey Eurotext, Paris.
Ghaemi M et al (1995) Cereal Res. Comm. 23, 215-222.
Mdarhri-Alaoui M et al (1998) C. R. Acad. Sci. Paris, Life Sciences 321, 25-30.
O'Donoughue L, Bennett MD (1994a) Theor. Appl. Genet. 89, 559-566.
O'Donoughue L, Bennett MD (1994b) Theor. Appl. Genet. 87, 673-680.
Saidi N et al (1997) Plant Cell Tiss. Org. Cult. 51, 27-33.
Saidi N et al. (1998) Can. J. Bot. in press.
Savaskan C et al. (1997) Plant Breed. 116, 299-301.

EVALUATION OF GIBBERELLIN 20-OXIDASE AND *ROLC* GENES FOR DWARFING ORNAMENTAL PLANTS

I.S. Curtis[1], M.R. Davey[1], P. Hedden[2], A.L. Phillips[2], D.A. Ward[2], S.G. Thomas[2], K.C. Lowe[1] and J.B. Power[1]

[1]Plant Science Division, School of Biological Sciences, University of Nottingham, University Park, Nottingham NG7 2RD, UK, and [2]IACR-Long Ashton, Department of Agricultural Sciences, University of Bristol, Long Ashton BS41 9AF, UK

Abstract

A gibberellin 20-oxidase gene from pumpkin and the *rolC* gene from *Agrobacterium rhizogenes* were introduced into *Solanum dulcamara* using *A. tumefaciens*-mediated delivery. Southern hybridisation confirmed the presence of transgenes in kanamycin-resistant regenerants. The phenotypes of transgenic plants indicated significant alterations in development, especially in relation to stature. Dwarfism in transgenic *S. dulcamara* is discussed in relation to GA metabolism.

1. Introduction

Chemical growth retardants, such as daminozide, are commonly used in commercial practice for modifying the morphology of ornamental species. Many of these compounds act by inhibiting the biosynthesis of plant hormones, particularly the gibberellins (GAs). An alternative strategy, involving genetic manipulation of hormone biosynthesis, would be less harmful to both producers and the environment. An *Agrobacterium*-mediated transformation system has been developed (Curtis *et al.*, 1998) to transfer genes with dwarf-inducing potential, namely, a gibberellin 20-oxidase (GA 20-ox) gene from seed of pumpkin (*Cucurbita maxima*) and the *rolC* gene from *Agrobacterium rhizogenes*, into *Solanum dulcamara* (Bittersweet, Woody Nightshade), which is grown as an ornamental in its variegated form. GA 20-ox catalyses several steps in GA biosynthesis producing, predominantly, biologically inactive products (Lange *et al.*, 1994). The over-expression of GA 20-ox may thus reduce the content of active GAs. Expression of the *rolC* gene induces dwarfism by an uncertain mechanism, although it has been suggested that it affects the metabolism of cytokinins (Nilsson *et al.*, 1993; Fais *et al.*, 1996) and/or GAs (Nilsson *et al.*, 1993). Studies on the introduction of genes affecting dwarfism into *S. dulcamara*, may provide a model for altering morphology of other plants, particularly ornamentals, and could lead to reductions in the use of chemical growth retardants.

2. Materials and Methods

2.1 Bacterial strains and plasmids

The binary vector pLARS121, carrying CaMV 35S.GA 20-ox and *nos.npt*II.Tg7 genes between right and left T-DNA borders, and pCV002 with the CaMV 35S.*rolC* and *nos.npt*II.*nos* genes (Schmülling *et al.*, 1988), were introduced into *A. tumefaciens* strain GV3101. *Agrobacterium tumefaciens* strain LBA4404 carrying ORF12 (the *rolC* gene with its own promoter; rolC.*rolC*) and the *nos.npt*II.*nos* gene in pMOG23 (Sijmons *et al.*, 1990), together with the supervirulent pTOK47 (Jin *et al.*, 1987), was also used.

2.2 Production of transgenic plants

Transgenic plants of *S. dulcamara* were produced using the leaf disc transformation system described by Curtis *et al.* (1998).

A. Altman et al. (eds.), Plant Biotechnology and In Vitro Biology in the 21st Century, 123–126.

2.3 NPTII ELISA

An NPTII ELISA Kit (5 Prime → 3 Prime, Inc., Boulder, USA) was used to detect and to quantify NPTII protein in crude plant extracts as described by Curtis *et al.* (1995).

2.4 Southern hybridisation

Genomic DNA was extracted from young fully expanded leaves of control and transformed plants (Dellaporta *et al.*, 1983). DNA (10 μg) from GA 20-ox transformed plants was digested with *Hin*dIII; DNA from *rolC* transformants was restricted with *Eco*RI. Digested DNA was electrophoresed, blotted and membranes probed with digoxigenin (DIG)-labelled GA 20-ox or *rolC* probes (McCabe *et al.*, 1997).

2.5 Quantification of GA

Material from shoot apices, stems and leaves (5-6 plants/construct) was harvested from transgenic and non-transformed plants and freeze-dried. GA concentration was determined by GC-MS using [17-$^{2}H_2$]GAs as internal standards (Croker *et al.*, 1990).

2.6 Phenotypes of regenerated plants

Measurements (in cm) were made, 50 d after transfer of plants to the glasshouse, for height, internodal length, maximum lamina length and width, petiole length and the number of stems from the basal region. The concentrations of chlorophylls *a* and *b* and total carotenoids (all μg ml^{-1} mg^{-1} leaf) were determined spectrophotometrically (Lichtenthaler, 1987). The extent of leaf wrinkling was assessed, based on arbitrary units from low to severe. The time (d) taken to anthesis was recorded following transfer of plants to the glasshouse. Pollen viability (%) was determined by fluorescein diacetate (FDA) staining (Curtis *et al.*, 1996), for grains from freshly dehisced anthers (4 flowers/plant). The length (cm) of petals (5 flowers/plant) was measured from fully opened flowers.

2.7 Statistical analyses

Data is presented as mean ± standard error of the mean. Statistical significance between mean values was assessed by Student's *t*-test (Bishop, 1971).

3. Results

3.1 Transformation efficiency

Shoots were regenerated from *Agrobacterium*-inoculated leaf disks; uninoculated explants formed shoots only on medium lacking kanamycin sulphate. There was a significant increase in the mean number of shoots from explants inoculated with *Agrobacterium* carrying the GA 20-ox ($P<0.01$) or rolC.*rolC* ($P<0.001$) genes, compared to explants inoculated with *Agrobacterium* carrying the 35S.*rolC* gene (48 and 18 plants/30 explants, respectively). No significant difference was found between the number of shoots from explants inoculated with the GA 20-ox or rolC.*rolC* genes.

3.2 ELISA detection of NPTII expression and molecular characterisation of regenerated plants

Non-transformed plants had undetectable levels of NPTII protein. Shoots, which rooted on 50 mg l^{-1} kanamycin sulphate, were positive for NPTII protein; for GA 20-ox plants (n = 10), rolC.*rolC* plants (n = 17) and 35S.*rolC* plants (n = 6), the respective values were 2.9 ± 0.5, 2.3 ± 0.3 and 3.0 ± 0.8 ng NPTII protein mg^{-1} total protein. Southern analysis of DNA extracts from GA 20-ox, rolC.*rolC* and 35S.*rolC* transformed plants (n

= 4, 5 and 5, respectively) revealed 1-2 copies of the transgenes per genome. DIG-labelled probes failed to hybridise to genomic DNA from non-transformed plants.

3.3 GA, pigments and phenotypic analyses of transformed plants

The concentration of the active gibberellin (GA_1) in plant tissues showed marked differences between transgenic and non-transformed plants. The concentrations of GA_1, in stems of GA 20-ox, rolC.*rolC* and 35S.*rolC* transformed plants were 15%, 41% and 24%, respectively, of the concentration in non-transformed plants. A reduction in the mean concentration of GA_1 was observed in apices (73% of control value) and leaves (35% of control value) for GA 20-ox plants. In stems of the GA 20-ox transformed plants, there were also reductions in the concentrations of all members of the 13-hydroxy GA pathway, except for the inactive tricarboxylic acid, GA_{17}, which accumulated in the transformants. Pigment concentrations of leaves and data relating to phenotypic differences between transgenic and non-transformed plants is summarised in Table 1.

Table 1. Analysis of transgenic and non-tranformed (control) plants of *Solanum dulcamara*

Character	Control (n = 5)	GA 20-ox (n = 10)	rolC.*rolC* (n = 17)	35S.*rolC* (n = 6)
Chlorophylls *a, b*	0.73 ± 0.11	1.45 ± 0.09***	1.16 ± 0.07***	0.41 ± 0.05**
Carotenoids	0.15 ± 0.02	0.27 ± 0.01***	0.21 ± 0.01***	0.07 ± 0.01***
Plant height	51.80 ± 3.60	21.70 ± 2.17***	32.41 ± 1.70**	14.50 ± 0.67***
Internode length	2.91 ± 0.26	1.55 ± 0.18**	2.20 ± 0.16	0.27 ± 0.02***
Lamina length	7.13 ± 0.48	6.00 ± 0.51	4.57 ± 0.11***	2.61 ± 0.11***
Leaf width	5.77 ± 0.39	4.26 ± 0.27*	3.63 ± 0.11***	1.61 ± 0.08***
Leaf wrinkling	none	none	severe	severe
Petiole length	3.18 ± 0.20	2.49 ± 0.09**	2.70 ± 0.07*	0.49 ± 0.03***
No. of stems	1.60 ± 0.40	1.10 ± 0.10*	6.18 ± 0.74***	7.83 ± 0.70***
Days to anthesis	89.40 ± 1.08	87.90 ± 1.22	102.00 ± 3.92***	130.00 ± 2.73***
Petal length	0.88 ± 0.02	0.80 ± 0.01**	0.65 ± 0.01***	0.30 ± 0.00***
Pollen viability	95.00 ± 1.67	98.30 ± 0.21*	60.18 ± 0.42***	62.00 ± 1.18***

Values are mean ± s. e. m., n = number of plants analysed, *P<0.05, **P<0.01, ***P<0.001 compared with corresponding mean control values.

The GA 20-ox and rolC.*rolC* transformed plants contained significantly (P<0.001) more pigments compared to controls, whilst the 35S.*rolC* plants had significantly (P<0.01) lower levels of chlorophyll *a*, chlorophyll *b* and carotenoids compared to GA 20-ox and rolC.*rolC* transgenic and non-transformed counterparts. All transgenic plants exhibited a significantly (P<0.01) reduced height compared to controls, as reflected by a reduced internodal length. Both rolC.*rolC* and 35S.*rolC* transformants exhibited reduced leaf area with severe leaf wrinkling compared to non-transformed plants. In terms of growth habit, all rolC.*rolC* and 35S.*rolC* transformants developed significantly (P<0.001) more stems than control plants; GA 20-ox plants had single stems similar to controls. Plants transformed by both *rolC* constructs reached anthesis significantly (P<0.001) later compared to GA 20-ox and non-transformed plants. All transgenic plants exhibited significantly (P<0.01) smaller petal lengths compared to controls, with pollen viability in rolC.*rolC* and 35S.*rolC* plants being significantly (P<0.001) lower than in controls and the GA 20-ox transformed plants.

4. Discussion

Dwarfism in *S. dulcamara* has been achieved by the integration and expression of GA 20-ox and *rolC* genes. Expression of the pumpkin GA 20-ox gene reduced the concentration of the major active GA, GA_1, in apices, stem and leaf tissues, probably accounting for reduced plant height and internodal lengths. GA 20-ox gene expression also increased pigmentation in leaf and stems, with an increase in chlorophyll *a*, chlorophyll *b* and carotenoids. The principal consequences of expression of the *rolC* gene, under its own or a 35S promoter, were dwarfism, leaf wrinkling, a bushy habit and a reduction in leaf area, flower size and fertility. Similar changes have been observed in *Nicotiana tabacum* transformed with the *rolC* gene (Schmülling *et al.*, 1988). In addition, over-expression of the *rolC* gene with the CaMV 35S promoter enhanced dwarfism in *Solanum*. Although the function of the *rolC* gene is unclear (Fais *et al.*, 1996), the present results from the GC-MS analyses revealed alterations in GA concentrations. However, it is unclear whether such changes contributed to the dwarf plant phenotype observed, or whether dwarfism was a consequence of other biochemical changes. Overall, this study provides a baseline for future extension of this technology to commercially-important ornamental plants, particularly Chrysanthemums.

5. Acknowledgements

This work was supported by MAFF under OC 9407. IACR-Long Ashton receives grant-aided support from BBSRC. The authors thank S.J. Croker, R. Lepe and S. Magnus for GC-MS analyses.

6. References

Bishop ON (1971) *Statistics for Biology*, 2nd Edn., Longman Group Ltd., London, UK.

Croker SJ, Hedden P, Lenton JR and Stoddart JL (1990) Plant Physiol. 94, 194-200.

Curtis IS, He C, Power JB, Mariotti D, de Laat AMM and Davey MR (1996) Plant Sci. 115, 123-135.

Curtis IS, Power JB and Davey MR (1995) *Plant Gene Transfer and Expression Protocols, Methods in Molecular Biology*, Vol. 49, (Jones H. ed.), pp 149-159, Humana Press, Totowa, NJ, USA.

Curtis IS, Power JB, Hedden P, Phillips A, Ward DA, Lowe KC and Davey MR (1998) *Tree Biotechnology: Towards the Millennium*, (Davey MR, Alderson PG, Lowe KC and Power JB, eds.), Nottingham University Press, Nottingham, UK (in press).

Dellaporta SL, Wood J and Hicks IB (1983) Plant Mol. Biol. Rep. 1, 19-21.

Fais M, Strnad M, Redig P, Dolezal K, Hanus J, Van Onckelen H and Schmülling T (1996) The Plant J. 10, 33-46.

Jin S, Komari T, Gordon MP and Nester E (1987) J. Bacteriol. 169, 4417-4425.

Lange T, Hedden P and Graebe JE (1994) Proc. Natl. Acad. Sci. USA 91, 8552-8556.

Lichtenthaler HK (1987) Methods in Enzymol. 148, 350-382.

McCabe MM, Power JB, de Laat AM and Davey MR (1997) Mol. Biotechnol. 7, 79-84.

Nilsson O, Moritz T, Imbault N, Sandberg G and Olsson O (1993) Plant Physiol. 102, 363-371.

Schmülling T, Schell J and Spena A (1988) The EMBO J. 7, 2621-2629.

Sijmons PC, Dekker BMM, Schrammeijer B, Verwoerd TC, Van den Elzen PJM and Hoekema A (1990) Bio/Technol. 8, 217-221.

AGROBACTERIUM-MEDIATED TRANSFORMATION OF *PHASEOLUS* BEANS

W. Dillen, M. Zambre, J. De Clercq, A. Goossens, J. Kapila, E. Vranová, M. Van Montagu and G. Angenon
Laboratorium voor Genetica, Departement Genetica, Vlaams Interuniversitair Instituut voor Biotechnologie (VIB), Universiteit Gent, K.L. Ledeganckstraat 35, B-9000 Gent, Belgium

1. Introduction

The grain legumes are an important group of crop species, belonging to the family of the *Fabaceae*. They are grown for human and animal nutrition all over the world. Examples include soybean (*Glycine max*), peanut (*Arachis hypogaea*), pea (*Pisum sativum*), faba bean (*Vicia faba*), lentil (*Lens culinaris*), cowpea (*Vigna unguiculata*), mung bean (*Vigna radiata*), chickpea (*Cicer arietinum*), and common bean (*Phaseolus vulgaris*).

Phaseolus beans are the major source of protein for millions of people, mostly with low incomes, in Central and South America and in Africa. The genus comprises approximately 200 species of which five are cultivated (*P. vulgaris*, *P. acutifolius*, *P. lunatus*, *P. coccineus* and *P. polyanthus*). Most of the dry bean production is confined to *P. vulgaris*. This species is cultured in over 70 countries in temperate, sub-tropical and tropical climates. It is therefore of capital importance that the crop be improved from the nutritional and agronomic point of view. Increasing the sulfur amino acid content and enhancing the tolerance for pests and diseases are the main challenges. In this respect, transgenic plants would be of great help to breeders.

A few steps are essential in the production of transgenic plants. First, foreign DNA has to be delivered to cells. This step poses no problems for *Phaseolus* because, in addition to direct gene delivery, *Agrobacterium* infection can readily be achieved (e.g. McClean et al. 1991). Second, cells in which the foreign genetic material has been stably integrated have to be selected, and plants have to be regenerated such cells. This step is problematic in grain legumes in general and in *Phaseolus* in particular.

The lack of *in vitro* regeneration capacity has prompted investigators to devise 'regeneration-independent' approaches. Such an approach has met with success (Russell et al. 1993). Seedling apical meristems were submitted to particle bombardment and after stimulation of axillary shoot proliferation, chimeric plants were obtained. Putative germline transformants were identified by histological ß-glucuronidase assays. Plants that passed the transgenes on to the progeny were recovered at a frequency of 0.03% of the harvested shoots. The transgenic plants contained multiple and sometimes

A. Altman et al. (eds.), Plant Biotechnology and In Vitro Biology in the 21st Century, 127–130.

incomplete copies inserted at one locus. The transgenes were stably inherited through many generations. The method was successful for one cultivar (Seafarer) but not for three other cultivars (Bountiful, Carioca, and A295). Using the same strategy, Aragão et al. (1996) obtained germline transformants of cultivar Olathe. It was hypothesized (Aragão, Rech 1997) that at least two features render a genotype suitable for transformation through meristem bombardment, namely (1) a good ability for axillary bud formation and (2) a proper apex morphology, i.e. a meristem that is not covered by leaf primordia. The observation that genotype-dependent factors such as apex morphology and shoot proliferation competence are important, temper the expectation that meristem bombardment will provide a genotype-independent approach towards transformation. In addition, the unfavorable integration patterns associated with biolistic gene delivery necessitate the production of a large number of transgenics which, in view of the low efficiencies, is hard to achieve. Nevertheless, the method clearly allows the production of transgenic plants in cases where regeneration pathways more complex than axillary shoot proliferation (e.g. somatic embryogenesis, organogenesis from callus) are not available.

2. Results

Ideally, transgenic plants would be produced by means of *Agrobacterium*-mediated gene transfer, followed by selection of and regeneration from transformed tissues. We anticipated that a procedure involving shoot formation from callus would suit this purpose. With the exception of one study (Mohamed et al. 1993), regeneration from callus in *P. vulgaris* has however not been reported. Strikingly, the two genotypes for which this type of regeneration was achieved had been constructed from interspecific crosses of *P. vulgaris* with *P. acutifolius* (the tepary bean), suggesting that their competence for regeneration derives from the latter species. In a screening of a range of *P. acutifolius* genotypes, regeneration capacity was indeed encountered (Dillen et al. 1996). Green nodular callus with a competence to regenerate was established from floral or vegetative buds or from pedicels of *P. acutifolius* using a combination of TDZ and IAA. Of twelve genotypes tested, seven were responsive in that they produced morphologically normal shoots. Of six genotypes, fertile regenerants were established in the greenhouse. The regeneration capacity seemed to be restricted to the subspecific group *P. acutifolius* var. *acutifolius* rather than *P. acutifolius* var. *tenuifolius* and occurred in wild as well as in cultivated genotypes. We have found that several *P. acutifolius* x *P. vulgaris*-derived common bean cultivars also harbor a capacity for regeneration, be it to a lesser extent than *P. acutifolius*. On the other hand, screening of a wide range of wild *P. vulgaris* accessions revealed little regeneration potential (Zambre et al. 1998a). We therefore decided to exploit the regeneration competence of *P. acutifolius* to obtain transgenic plants.

Callus of one (wild) *P. acutifolius* genotype (NI576) was co-cultivated with *Agrobacterium tumefaciens* strain C58C1Rif^R containing the helper plasmid pMP90 (Koncz, Schell 1986) and the binary plasmid patarc3-B1b. This plasmid contains between the T-borders the neomycin phosphotransferase II (*nptII*) gene under control of the nopaline synthase (*nos*) promoter and the octopine synthase 3' termination and polyadenylation signals, the *Escherichia coli* ß-glucuronidase (*uidA*) gene (Jefferson 1987) with the potato *st-ls1* intron (Vancanneyt et al. 1990) under the control of the CaMV 35S promoter and the *nos* 3' processing and polyadenylation signals, and, in addition, a genomic fragment presumably coding for the *P. vulgaris* arcelin-5a seed storage protein (Goossens et al. 1995). This protein is thought to be involved in the resistance of a wild *P. vulgaris* accession against the Mexican bean weevil (*Zabrotes subfasciatus*), a major post-harvest pest of dry common beans. After selection on the antibiotic geneticin and regeneration, a transformant was produced. Its transformed status was confirmed by molecular and genetic analysis of the progeny (Dillen et al. 1997a). The transgenes segregated in a three to one fashion, indicating integration of the T-DNA at a single locus. Seeds were harvested and tested for the presence of arcelin. The relative concentration of arcelin-5a was in the order of 20% of total seed protein. To our knowledge, this is the first documented demonstration of *Agrobacterium*-mediated transformation of a member of the genus *Phaseolus*. The procedure has been repeated several times by now and many more transformants have been generated.

The method is currently being optimized for several parameters. It was found that relatively low temperatures during co-cultivation (22°C) greatly improved T-DNA transfer (Dillen et al. 1997b). This observation may be of relevance for plant transformation in general and for the transformation of tropical species in particular. For these species co-cultivation is often performed at the (higher) temperatures that maximally support *in vitro* regeneration.

Transgenic *P. acutifolius* NI576 plants may be used to introgress transgenes into *P. vulgaris* by interspecific hybridization. Noteworthy in this respect is the reported high cross-hybridizing ability of *P. acutifolius* N576) with *P. vulgaris* (Haghighi et al. 1984; Haghighi, Ascher 1988).

We are currently pursuing two goals. The first is to obtain transgenic plants of a domesticated rather than a wild *P. acutifolius* genotype. This is of practical value, e.g. for the production of engineered seed proteins. Transgenic callus lines of two domesticated genotypes have been established. Shoot production from these lines has been initiated and the first plantlets for progeny analysis are expected to become available soon. The second objective is to achieve direct transformation of *P. vulgaris* with *Agrobacterium*. To this end, hybrid-derived *P. vulgaris* genotypes are employed. Rather than relying on pedicel explants from greenhouse-grown plants, as reported before (Mohamed et al. 1993), we succeeded in producing regeneration-competent callus from *in vitro*-germinated seeds (Zambre et al. 1998b). With this method, shoot production no longer depends on environmental conditions experienced by explant donor plants, the most probable cause for our poor success in reproducing the results of Mohamed et al. (1993). We found that regenerable callus of *P. vulgaris* can be transformed with *Agrobacterium* as efficiently as *P. acutifolius* callus. The selection for transformed callus lines and subsequent regeneration of plants is under investigation.

3. Conclusions

Transgenic plant production in *P. vulgaris* remains problematic. So far, two alternatives are available. One approach, namely particle bombardment of meristems, has been applied repeatedly with success (Russell et al. 1993; Aragão et al. 1996; Aragão, Rech 1997) and allows the direct introduction of genes into (some) *P. vulgaris* cultivars. It is however very laborious and inherently contains the drawbacks of the biolistic approach, i.e. complex and unpredictable plasmid integration patterns. The other approach (Dillen et al. 1997a) involves plant regeneration from callus and uses *Agrobacterium* for gene delivery. Transgenic plants can be produced more efficiently and, for practical purposes, less plants need to be generated as integration patterns obtained with *Agrobacterium* are generally more favorable. However, so far, the approach is only successful for *P. acutifolius*. As such, the introduction of transgenes into *P. vulgaris* would require a lengthy backcross program, including interspecific hybridization.

In the near future, further progress with both approaches is expected to result in more feasible systems for transformation of *P. vulgaris*. With respect to the meristem transformation approach, the use of *Agrobacterium* in combination with bombardment (Brasileiro et al. 1996) is interesting. As for the regeneration-based approach, the use of regeneration-competent (hybrid-derived) *P. vulgaris* genotypes may allow direct transformation of the crop with *Agrobacterium*.

4. References

Aragão FJL et al. (1996) Theor Appl Genet 93, 142-150.
Aragão FJL, Rech EL (1997) Int J Plant Sci 158, 157-163.
Brasileiro ACM et al. (1996) J. Amer. Soc. Hort. Sci. 121, 810-815.
Dillen W et al. (1996) Plant Sci. 118, 81-88
Dillen W et al. (1997a) Theor. Appl. Genet. 94, 151-158.
Dillen W et al. (1997b) Plant J. 12, 1459-1463.
Goossens A et al. (1995) Plant Physiol 109, 722.
Haghighi K et al. (1984) Ann Rep Bean Improv Coop 27, 42-43.
Haghighi KR, Ascher PD (1988) Sex Plant Reprod. 1, 51-58.
Jefferson RA (1987) Plant Mol. Biol. Rep. 5, 387-405.
Koncz C, Schell J (1986) Mol. Gen. Genet. 204, 383-396.
McClean P et al. (1991) Plant Cell Tissue Org. Cult. 24, 131-138.
Mohamed MF et al. (1993) J. Am. Soc. Hortic. Sci. 118, 158-162.
Russell DR et al. (1993) Plant Cell Rep. 12, 165-169.
Vancanneyt G et al. (1990) Mol. Gen. Genet. 220, 245-250.
Zambre M et al. (1998a) Ann. Rep. Bean Improv. Coop. 41, 119-120.
Zambre M et al. (1998b) Plant Cell Rep. 17, 626-630.

CONTRIBUTIONS OF PROTOPLAST FUSION TO IMPROVEMENT OF BRASSICA CROPS

E. D. Earle[1], T. Cardi[1], M.H. Dickson[2], L. N. Hansen[1], D. W. Heath[1], J.-P. Ren[2] and M. Sigareva[1]
[1]Dept. of Plant Breeding, Cornell University, Ithaca, NY 14853 USA
[2]Dept. of Hort. Sciences, NY Ag. Exp. Station, Geneva, NY 14456 USA

1. Introduction

In earlier days of plant tissue culture, protoplast fusion provided the only method of introducing or combining genes somewhat beyond the limits of sexual reproduction. Protoplast fusion is now rather overshadowed by plant transformation, a very powerful method that permits transfer into plants of genes from any taxonomic group. In view of the success of gene transfer, is protoplast fusion still of interest and value?

Protoplast fusion retains some advantages over gene transfer, including the ability to alter cytoplasms for practical use or study and the ability to transfer single or multiple genes not yet isolated as DNA sequences. Moreover, regulatory, legal, and social constraints on fusion-derived plants are generally less than for transgenic plants. Protoplast fusion can claim some success in broadening the genetic base of crop plants, increasing access to genes from sexually incompatible plant species, and facilitating improvement and transfer of cytoplasms. These concepts are illustrated by several examples from work on Brassica vegetable and oilseed crops in our lab.

2. Materials and Methods

Protoplasts were isolated from leaves of *in vitro* grown plantlets, fused with polyethylene glycol, and cultured in a feeder layer system. Pretreatments with iodacetate and, in some cases, gamma-irradiation were used to prevent division of unfused protoplasts or eliminate the nuclear DNA of one fusion partner. Procedures for cultures and assays are described in detail in the references cited.

3. Results and Discussion

3.1. Resynthesis of amphidiploid *B. napus* from selected diploid progenitors

B. napus is an amphidiploid that contains the nuclear genomes of two diploid species, *B. rapa* and *B. oleracea*. Fusion of protoplasts from *B. rapa* and *B. oleracea* lines selected for their nuclear-encoded traits or organellar genomes (or both) can create new

A. Altman et al. (eds.), Plant Biotechnology and In Vitro Biology in the 21st Century, 131–134.

and possibly improved types of *B. napus*. These can be used either to broaden the genetic basis of *B. napus* itself or as starting material for backcrosses with *B. rapa*.

Manipulation of fatty acid levels in oilseeds

Heath and Earle (1995) fused protoplasts from *B. rapa* and *B. oleracea* accessions reported as high in erucic acid (22:1; useful for industrial purposes) after extensive screening of germplasm collections. Fifty somatic hybrids were recovered from 167 calli in a fusion experiment. Seeds from these plants were assayed for fatty acid composition after greenhouse and field trials. Most had erucic acid levels similar to the mean of the fusion partners; however, several had stably higher levels than either partner grown under similar conditions. Analogous results were obtained from fusions in which the partners were selected for low levels of linolenic acid (18:3; preferred for cooking oil) (Heath, Earle, 1997). Thus it is possible to enhance genetic variability through fusion and to fix desirable characters from the selected fusion partners.

Targeted alterations in fatty acid composition have also been achieved through gene transfer methods; however, fusion-derived *B. napus* showed additional beneficial characters. Some improvements came from one fusion partner (large seed size from *B. rapa*), while others were not apparent in either partner (better resistance to powdery mildew; less lodging and pod shattering). Thus multiple agronomic improvements may be achieved from a single fusion combination.

Resistance to softrot disease

B. napus was also resynthesized by fusion of diploid progenitors selected for tolerance to softrot, a bacterial disease caused by *Erwinia carotovora* ssp. *carotovora*.(Ren 1998). About 750 accessions and breeding lines of *B. rapa* and 30 of *B. oleracea* were screened for resistance. Protoplasts from 3 selected *B. rapa* lines were fused with "Shogun" broccoli, and 12 somatic hybrids were recovered. Most of them were more resistant to *E. carotovora* than the fusion partners (especially *B. rapa*) and also than commercial *B. napus* (Westar) and *B. napus* resynthesized from unselected partners. Resistance appears to be controlled by several additive genes that can be combined. Progeny lines obtained by self pollination or backcrosses of the somatic hybrids with the *B. rapa* partners also showed enhanced resistance. Some of the backcross plants were much more resistant than the *B. rapa* partner. These results suggest that resynthesis of *B. napus* followed by backcrossing can allow transfer of genes for disease resistance (or other traits) from *B. oleracea* to *B. rapa* crops.

3.2. Development of *B. oleracea* vegetables with enhanced disease resistance

Crucifers that are difficult to cross sexually with *B. oleracea* may be resistant to diseases that damage *B. oleracea* vegetables. Somatic hybridization of the resistant line with susceptible *B. oleracea* can be followed by backcrosses with a vegetable for possible recovery of resistant crops. Rapid-cycling *B. oleracea* (Crucifer Genetics Cooperative #3-1) was selected as the fusion partner for such experiments because of its excellent regeneration from protoplasts (Hansen, Earle 1994).

Resistance to *Xanthomonas campestris* pv. *campestris* (causal agent of blackrot)

A *B. carinata* line (previously described as *B. napus*) with virtual immunity to *X. campestris* was the resistance donor in fusions that produce somatic hybrids resistant to blackrot (Hansen, Earle, 1995). Pollination with broccoli and embryo rescue produced one backcross progeny, also resistant. Through additional backcrosses and selections, resistant plants with the morphology and DNA content of broccoli were recovered (Earle et al. unpublished). Resistant plants often produced fewer resistant progeny than expected; however, a 1997 field trial in Cortland, NY gave encouraging results. Two self-pollinated lines had almost 100% plants rated 1 on a 1-5 scale, while the two broccoli backcross lines scored 2.8 or 4.5. The best materials are being distributed to seed companies. A more extensive field trial, including cabbage lines, is in progress.

Resistance to *Alternaria* disease

Sinapis alba was used as a source of resistance to *Alternaria brassicicola* and *A. brassicae*. Initial fusions yielded resistant somatic hybrids, but no progeny were obtained (Hansen, Earle 1997). Further fusions produced 17 resistant somatic hybrids, from which five resistant broccoli backcross progeny were obtained by embryo rescue (Sigareva, unpublished). Pollination of one of them with broccoli gave >60 progeny. Some were more resistant than broccoli (though less than the previous generations) and were also male and female fertile. These plants are being advanced in hopes of producing broccoli with stable enhanced tolerance to *Alternaria*. Fusion-derived lines are also being compared to transgenic broccoli expressing a *Trichoderma harzianum* endochitinase gene, some of which show tolerance to *Alternaria* (Mora unpublished).

3.3. Intraspecific and interspecific transfer of male sterile cytoplasms in *B. oleracea*

Direct transfer of cold-tolerant Ogura CMS from broccoli to cabbage

Organelle assortment after fusion has eliminated chloroplast-related cold sensitivity from Ogura CMS Brassica lines (e.g., Walters et al. 1992). After cold-tolerant Ogura CMS became available, a further question was whether protoplast fusion enabled faster conversion of elite fertile lines to CMS than sexual backcrosses.

To address this question, fusion-mediated transfer of the improved Ogura CMS from broccoli to fertile cabbage was attempted. Studies of regeneration of 40 cabbage lines from leaf protoplasts showed high genotype specificity: 12 failed to regenerate; 8 had 10% or less shoot formation from colonies; 4 had >40% response (Sigareva, Earle 1997). Cybridization experiments in which the broccoli protoplasts were gamma-irradiated (usually 30 krad) produced 280 diploid CMS cabbage plants from three lines (Sigareva, Earle 1997). CMS plants were identified by a PCR assay using primers for the Ogura CMS-related mitochondrial DNA region. Presence of the expected 500 bp PCR band accurately predicted male sterility of flowering plants in the field. CMS cabbage plants were ready for transfer into soil within 8 months after the fusion, much faster than by sexual backcrosses of biennial cabbage. However, this benefit is limited to cabbage lines with at least some degree of regenerability.

Transfer of Anand CMS to *B. oleracea*

Access to the cold-tolerant Ogura CMS is limited by patent control, so an alternate CMS for hybrid seed production in *B. oleracea* vegetables would be useful. The "Anand" CMS available in *B. rapa* (Crucifer Genetics Cooperative #1-31; apparently derived from *B. tournefortii* via *B. juncea*) was transferred to rapid cycling *B. oleracea* by cybridization experiments (Cardi, Earle 1997). In PCR assays using primers based on the radish *atp6* gene, CMS cybrids have a strong 570 bp band absent in fertile plants (Cardi, Earle 1998). The 22 diploid CMS cybrids recovered showed mitochondrial DNA rearrangements and varied substantially in their flower morphology and female fertility. Backcrosses of the best materials have produced CMS broccoli, cauliflower, and cabbage lines being tested by seed companies for possible commercial use.

4. Conclusions

These studies show that horticultural improvements in Brassica crops can be obtained by a variety of protoplast fusion strategies. Efficient regeneration systems are essential since large numbers of fusion-derived plants may be required in order to obtain some that have the desired features. Cybridization is an effective method for direct transfer of desirable organelle-encoded phenotypes into regenerable lines. Backcrosses of somatic hybrids often prove to be harder than obtaining the hybrids. Reliable assays are important to select appropriate fusion partners, identify fusion products, and monitor specific traits of interest. Cooperation with breeders and seed companies helps ensure movement of laboratory-derived materials toward agricultural use.

5. References

Cardi T, Earle ED (1997) Theor. Appl. Genet. 94, 204-212.
Cardi T, Earle ED (1998) Eucarpia Cruciferae Newsl. 20, 53-54.
Hansen LN, Earle ED (1994) Plant Cell Reports 13, 335-339.
Hansen LN, Earle ED (1995) Theor. Appl. Genet. 91, 1293-1300.
Hansen LN, Earle ED (1997) Theor. Appl. Genet. 94, 1078-1085.
Heath DW, Earle ED (1995) Theor. App. Genet. 91, 1129-1136.
Heath DW, Earle ED (1996) Plant Cell Reports 15, 939-944.
Heath DW, Earle ED (1997) Euphytica 93, 339-343.
Ren JP (1998) Ph.D. Thesis, Cornell University, Ithaca, NY.
Sigareva M, Earle ED (1997) Theor. Appl. Genet. 94, 213-220.
Walters TW et al. (1992) Plant Cell Reports 10, 624-628.

6. Acknowledgements

This work was supported in part by a grant from the Vegetable and Flower Seed Permanent Research Fund and by the Cornell Center of Advanced Technology in Biotechnology. We thank Reeds Seeds for conducting the blackrot field trials.

EXPRESSION OF A CHIMERIC GRAPEVINE STILBENE SYNTHASE GENE IN STABLE WHEAT TRANSFORMANTS

Sebastian Fettig[1] and Dieter Hess[1]
[1]Universität Hohenheim, Institut für Pflanzenphysiologie -260-, Emil-Wolff-Str. 25, D-70593 Stuttgart, Germany

1. Introduction

We have transferred a chimeric stilbene synthase (*sts*)-gene into wheat, the most important human food crop. Stilbene synthases are producing stilbene-type phytoalexins like resveratrol, a substance with fungicidal potential (Langcake, Pryce, 1976). *Sts*-genes from groundnut (Hain *et al.*, 1990) and grapevine (Hain *et al.*, 1993) have already been transferred to other plants. Tolerance against fungal infections was improved in tobacco (Hain *et al.*, 1993) and possibly in rice (Stark-Lorenzen *et al.*, 1997). In tobacco and petunia, overexpression of *sts*-genes led to a substrate competition between stilbene and chalkon synthases causing male sterility (Fischer *et al.*, 1997).

In addition, resveratrol seems to be responsible for some beneficial effects of red wine on human health, e.g. inhibition of platelet aggregation (Bertelli *et al.*, 1995), vasorelaxing activity (Chen, Pace-Asciak, 1996), or cancer chemopreventive activity (Jang *et al.*, 1997).

In the experiments to be described here we used embryogenic scutellar callus as a target for particle bombardments. Eighteen German spring wheat cultivars had been investigated. 'Combi' and 'Hanno' were found to be most suitable for callus induction, regeneration (Viertel *et al.*, 1998) and biolistic gene transfers (Iser *et al.*, in preparation).

The first step in producing wheat with improved fungal resistance, male sterility, and/or positive effects on human health has been done.

2. Material and Methods

2.1. Callus culture and particle bombardment

Induction and culture of embryogenic scutellar callus derived from immature embryos were performed as described (Viertel *et al.*, 1998). After a culture of 3 weeks the embryogenic callus was subcultured for bombardments. The plasmid pStil 2 (see results) was cotransferred with the selectable marker plasmids pAHC 25 (Christensen, Quail, 1996) or pDM

A. Altman et al. (eds.), Plant Biotechnology and In Vitro Biology in the 21st Century, 135–138.

302 (Cao *et al.*, 1992), respectively. The mixture was precipitated on gold particles (1 μm diameter, BioRad, final amount 250 μg per shot) using the standard procedure provided by BioRad. Particle delivery was performed using the device PDS 1000/He (BioRad). After 2 days, the calli were transferred to fresh medium for one week. For selection, phosphinothricin was added in the second week at a concentration of 1 mg l^{-1}, then raised weekly to 2.5 and finally 5 mg l^{-1}. After 4 weeks, the embryogenic calli were subcultured on regeneration medium R1 (Ahuja *et al.*, 1982) containing 5 mg l^{-1} phosphinothricin. Plantlets converted from embryoids were treated as described (Viertel *et al.*, 1998)

2.2. *Southern* blot-analysis

For *Southern* blot analyses, extraction of genomic DNA, digestion (for enzymes see Fig. 1), separation, transfer to nylon membranes, hybridisation and exposition to film material was carried out as described (Hess *et al.*, 1990). The *sts*-probe consisted of the *sts*-coding region isolated as a 1.9-kb *Hin*DIII/*Bam*HI fragment from pStil 2.

2.3. Reverse transcriptase (RT)-PCR

Total RNA was isolated from leaves of transgenic and control plants. Reverse transcription and cDNA amplification was carried out using the Boehringer Titan-One-Tube RT-PCR-kit in a Pharmacia LKB Gene ATAQ Controller. Primers were chosen to be complementary 165 bp upstream and 332 bp downstream the 356 bp intron of the *sts*-gene. The reaction consisted in a reverse transcription step at 50°C for 30 min followed by a denaturation step at 94°C for 2 min and 40 cycles each of 90 s at 94°C, 10 s at 57°C and 30 s at 68°C.

2.4. Reversed-phase (RP) HPLC-analysis

Leaves of transgenic and control plants were ground in liquid nitrogen and treated with 1 N HCl for one hour on a boiling waterbath. Subsequently, the samples were extracted with diethyl-ether. The ether-phase was removed, dried and the residue redissolved in methanol. All these steps were carried out under light protection. RP-HPLC was performed on a Beckman 344 M high performance liquid chromatograph using a Nucleosil C18 column, H_2O-acetonitrile as eluent (dotted line). For detection of resveratrol, a Shimadzu RF-530 fluorescence monitor was used (excitation λ: 330 nm, emission λ: 374 nm).

3. Results and discussion

The coding region of a *sts*-gene from grapevine (Hain *et al.*, 1993) was linked to the ubiquitin-promoter from maize. This promoter is constitutively expressed at high levels in monocotyledonous plants (Christensen and Quail, 1996). As strong inducible promoters for wheat are not available at present a constitutively expressed *sts*-gene was used. Furthermore, high *sts*-activities and an additional male sterility in transgenic tobacco using the constitutive *sts*-strategy were described (Fischer *et al.*,1997).

The resulting plasmid pStil 2 was used for co-transfers with the selectable marker plasmids pAHC 25 (containing the *uidA*- and the *bar*-gene), and pDM 302 (containing the *bar*-gene), respectively.

The highest transformation efficiency (0.29%) was found in one experiment with 'Combi'. Taking together all bombardment experiments, 8 transformed 'Combi' T_0-plants and one 'Hanno' T_0-plant regenerated from 7794 'Combi' and 4244 'Hanno' bombarded scutellar calli. From control calli exposed to selective pressure no regenerant plants were obtained. All transferred genes were showing co-segregation in the T_1. In 3 progenies the segregation was in a mendelian manner.

Integration of the genes transferred was analysed by *Southern* blotting. All 8 'Combi' plants and the 'Hanno' plant were found to be transgenic for the *bar*-gene and, if used, for the *uidA*-gene. Seven of the 8 'Combi' T_0-plants and the 'Hanno' T_0-plant were found to be co-transformants due to the additional integration of the *sts*-gene (Fig. 1). These 8 plants will be followed further

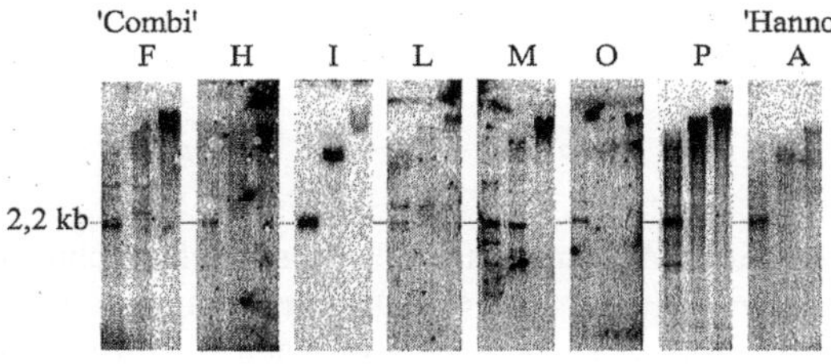

Figure 1. *Southern* blot analysis of seven 'Combi' T_0-plants and one 'Hanno' T_0-plant. Left lanes: genomic DNA digested with *Pst*I/*Bam*HI for the *sts*-coding region (2.2 kb). Middle lanes: genomic DNA digested with *Sac*I which possesses a singular recognition site at the end of the *sts*-coding region. Therefore, for each integrated copy one hybridisation signal is to be expected if no rearranged copies were present. Right lanes: Undigested genomic DNA. Hybridisation signals in the high molecular DNA indicate absence of free plasmid DNA.

High activity of the selectable marker/reporter genes was found only in T_0-plants containing the selectable marker/reporter gene governed by the ubiqutin promoter, with the exception of 'Combi' T_0-plant H. In all plants containing the selectable marker gene governed by the actin promoter, activity of the selectable marker/reporter genes was repressed just as in previous wheat transformation experiments (Hess *et al.*, 1990, Hess, 1996, Srivastava *et al.*, 1996, Viertel *et al.*, 1997). In 3 of the transgenics a methylation of the *sts*-gene was found.

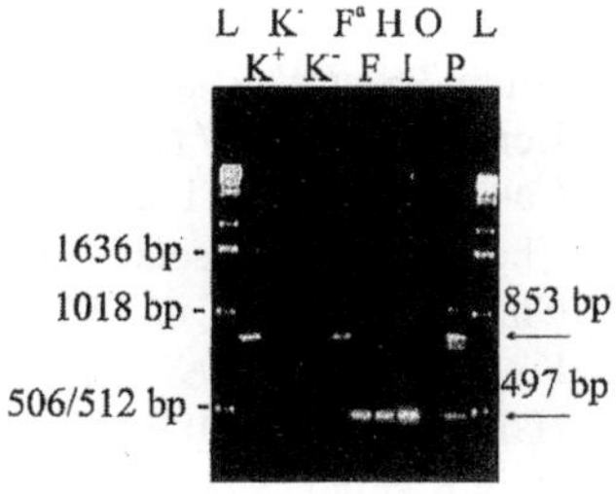

Figure 2. RT-PCR analysis of T_0-plants. The 853 bp amplification product of the *sts*-DNA was found in RT-PCR assays of plasmid-DNA (lane K^+), of genomic DNA of T_0-plant F (lane F^d) and of, as impurity, RNA of T_0-plant P. The 497 bp amplification product of reverse transcribed *sts*-mRNA was found RT-PCR assays of T_0-plants F, H, I and P. Lanes K^-: negative controls.

Expression of the *sts*-gene was shown on mRNA-level by RT-PCR (Fig. 2). S*ts*-mRNA was found in 4 of the seven *sts*-transgenic 'Combi' T_0-plants. From the other 3 'Combi' T_0-plants, and from the 'Hanno' T_0-plant no *sts*-mRNA could be detected.

Activity of the *sts*-gene was confirmed in T_1-plants by determination of resveratrol, which had been released from its glycosides, using HPLC and mass spectrometry. From all control plants a small peak with the same retention time as the synthetic resveratrol was detected. Also the mass of resveratrol (228.2 g mol^{-1}) was found in controls. The occurrence of resveratrol in some *Poaceae* has already been described (Powell *et al.*, 1994). Therefore, it can not be excluded that wheat contains endogenous resveratrol as well. Its concentration was low and could be clearly distinguished from that of transgenics. Since control plants were never showing signals neither in *Southern* blot analyses nor in RT-PCR experiments, the homology of the supposed wheat enzyme to the grapevine enzyme has to be low.

Coming back to the qualities to be expected from *sts*-containing plants, in our experiments there was no evidence for **male sterility** due to an overexpression of the *sts*-gene. Since resveratrol was accumulated in a glycosidic form, a possible **fungal resistance** still has to be confirmed. Concerning possible **effects on human health**, *sts*-gene transfers using endosperm specific promoters are initiated. The first step in achieving the last mentioned qualities has been done. At present, it is not yet possible to consume beneficial resveratrol without increasing the alcohol level - just by enjoying red wine together with resveratrol-white bread.

4. Acknowledgements

We thank F. Scheyhing for construction of pStil2, A. Harmsen and B. Rösingh for technical assistance, I. Klaiber, Institut für organische Chemie, Universität Hohenheim, for mass spectrometry analyses and I. Bertling, Dept. Hort. Sci., University of Natal, South Africa, for language improvements.

5. References

Ahuja PS *et al* (1982) Z Pflanzenzüchtg 89, 139-144
Bertelli AAE *et al* (1995) Intl J Tiss React 17, 1-3
Cao J *et al* (1992) Plant Cell Rep 11, 586-591
Chen CK, Pace-Asciak CR (1996) Gen Pharmacol 27, 363-366
Christensen AH, Quail PH (1996) Transgen Res 5, 213-218
Fischer R *et al* (1997) Plant J 11, 489-498
Hain R *et al* (1990) Plant Mol Biol 15, 325-335
Hain R *et al* (1993) Nature 361, 153-156
Hess D *et al* (1990) Plant Sci 72, 233-244
Hess, D (1996) *In vitro* haploid production in higher plants 2, pp 375-392, Kluwer Academic Publishers, Dordrecht, The Netherlands
Jang MS *et al* (1997) Science 275, 218-220
Langcake P, Pryce RJ (1976) Physiol Plant Pathol 9, 77-86
Powell RG *et al* (1994) Phytochem 35, 335-338
Srivastava, V *et al* (1996) Theoret Appl Genet 92, 1031-1037
Stark-Lorenzen P *et al* (1997) Plant Cell Rep 16, 668-673
Viertel K *et al* (1997) Vortr Pflanzenzüchtg 38, 17-39
Viertel K *et al* (1998) J Plant Physiol 152, 167-172

PROSPECTS FOR THE ISOLATION OF GENES CONTROLLING TREE-SPECIFIC TRAITS BY USING A TRANSPOSON TAGGING APPROACH

M. Fladung

Federal Research Centre for Forestry and Forest Products
Institute for Forest Genetics and Forest Tree Breeding
Sieker Landstr. 2, 22927 Grosshansdorf, Germany

1 Introduction

The application of modern molecular technologies previously developed on crop plants to forest tree species offers new opportunities for studies on gene regulation and gene expression. In trees traits are of interest which are associated with tree-specific characteristics, such as fruit quality and flower sterility in fruit trees, and wood properties in forest trees. So far, very little is known about genes controlling these traits, and only few genes have been cloned. By means of molecular tools different RAPD-, RFLP- or AFLP-markers can be identified corresponding to specific wood properties. Another strategy is the transposon tagging approach using endogenous transposable elements (Walbot 1992) which have been detected in a number of annual plant species (Döring and Starlinger 1986, Gierl and Saedler 1992), but not in trees so far.

However, the application of transposable elements (TE's) is not only restricted to those species that contain well characterised transposons, but has also been extended to species in which none or weakly characterised endogenous transposons are known. Genetic transfer of isolated and well-characterised TE's in association with reporter genes indicating transposon excision has been extended to establish a gene tagging system also in these plants (Baker et al. 1986, Chuck et al. 1993).

The advantage of using phenotypic or morphological marker genes as reporter genes is that their expression can be easily observed at all times during the life cycle of the transgenic plant. The expression of the marker gene is expunged when the TE is inserted between the promoter and the coding sequence of the marker gene. Transgenic plants carrying such constructs are phenotypically similar to untransformed controls when the TE is not excised. After excision of the TE, the plant phenotype is altered based on the action of the morphological marker gene in the transgenic plant. Several bacterial genes well known to influence plant phenotype in transgenic plants were successfully used as reporter genes for transposon excision, e.g. *ipt* (Estruch et al. 1991), *rolC* (Spena et al. 1989, Jones et al. 1992) and *iaaL* (Spena, unpublished).

A. Altman et al. (eds.), Plant Biotechnology and In Vitro Biology in the 21st Century, 139–142.

The present work describes the transfer of constructs carrying the maize transposable element *Ac* into the genome of aspen and hybrid aspen. Two reporter gene systems were employed for transformation. In one system the *rolC* gene was under control of two different promoters, while in the second the *iaaL* gene expression was controlled by one promoter.

2. Material and Methods

Transformation of aspen, bacterial strains and plasmids

Two different aspen clones (*Populus tremula* L.; W52, Brauna 11) and one hybrid clone (*P. tremula x tremuloides* Michx.; Esch5) were used for transformation experiments (Fladung et al. 1997). The *Agrobacterium tumefaciens* strains GV3101 (nopaline type; intermediate virulence) and LBA4404 (octopine type; low virulence were employed for transformation experiments (Fladung and Ahuja 1996). The plasmid pPCV002-*Ac*-CaMVC, for example, carries the *rolC* gene under control of the 35S-promoter of the cauliflower mosaic virus, with insertion of a transposable element *Ac* from *Zea mays* between the 35S promoter and the *rolC* coding region. As a plant selection marker the *npt-II* gene leading to kanamycin resistance was used in each case.

DNA and RNA extraction, PCR, Southern, northern and western analysis

Extraction of DNA, PCR, Southern and northern analysis were performed as described by Fladung and Ahuja (1995) and Fladung et al. (1997). The PCR reaction was performed with *rolC*, *Ac*, *iaaL* and *npt-II* specific 20- to 24-mer primer pairs (Fladung et al. 1997). Non radioactive labelling of the *rolC*, *Ac*, *iaaL* and *npt-II* specific probes using DIG-dUTP (Boehringer Mannheim) was done in a PCR reaction (Fladung and Ahuja 1995). Isolation of total protein for western analysis, SDS-Polyacrylamide gel electrophoresis, western blotting and non-radioactive detection was performed as described by Kettig (unpublished) and the manual of the 'Western-Light Plus™ Chemiluminescent Detection System' (Boehringer Ingelheim Bioproducts). Rabbit antiserum containing polyclonal antibodies against the *rolC* protein was kindly provided by F. Salamini, Max-Planck Institute for Plant Breeding, Cologne, Germany.

3. Results and Discussion

Insertion of the transposon, *Ac,* between the 35S- or the rbcS-promoter, and the coding region of one of the reporter genes *iaaL* or *rolC* causes inactivation of this gene. Therefore, transgenic aspens carrying one of the 35S-*Ac-iaaL*, 35S-*Ac-rolC* or rbcS-*Ac-rolC* constructs grew normally and were phenotypically not distinguishable from untransformed controls. The presence of the *Ac*-transposon in the genome of these transgenic aspen was confirmed by PCR using highly specific primer pairs and Southern analysis (Fladung and Ahuja 1997, Fladung et al. 1997).

Excision of *Ac* during leaf histogenesis restored *iaaL* or *rolC* activity. However, this event was only visible in the *Ac-rolC* carrying transgenics as light-green spots on darker-green leaf background but not in the *Ac-iaaL* carrying ones. As revealed by PCR analysis, out of 62 independent 35S-*Ac-rolC* and rbcS-*Ac-rolC* transgenic aspen transformants obtained 59 carried both *Ac* and *rolC* gene sequences. The remaining three clones revealed a *rolC*-specific but no *Ac*-specific band (without *rolC* phenotype). Southern blot analysis confirmed the presence of *rolC* and *Ac* specific fragments in the remaining 59 independent clones (Fladung et al. 1997).

Molecular evidence for *Ac* excision was provided by PCR analysis as well as in Southern, northern (Fladung and Ahuja 1997) and western experiments. In PCR analysis, bands of about 1.5 kb were obtained (length of the 35S-*rolC* gene construct), when a specific primer pair was added to the PCR reaction (Fladung and Ahuja 1997). Such amplification products of 1.5 kb in size were obtained from the light-green sectors only (phenotypic indicating *rolC* function), but not in the dark-green leaf parts (length of 35S-*Ac-rolC* approximately 6 kb). In northern experiments, a specific *rolC* transcript was found in the light-green sectors of the leaves, but no signal could be detected in the dark-green ones. Thus, using leaf color as a phenotypic marker the activity of the transposon *Ac* can be followed in rbcS-*Ac-rolC* and 35S-*Ac-rolC*-transgenic aspen as shown for 35S-*Ac-rolC* transgenic tobacco (Spena et al. 1989) or rbcS-*Ac-rolC* transgenic tomato plants (Jones et al. 1992).

Sequencing the transition promoter to reporter gene of PCR-fragments which were obtained from light-green sectors only revealed precise excisions of the TE with some exceptions (Figure 1). So far, out of 19 analyses made three cases were found of which two show modifications in the nucleotide sequence of the residual flanking sequences.

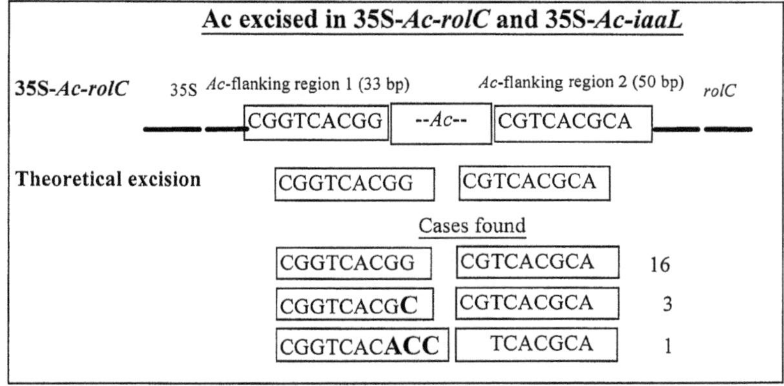

Figure 1: The transposon *Ac* is integrated between the 35S-promoter and the *rolC* or *iaaL* gene flanked by cloning sequences. Following excision three different cases were found of remaining sequences.

Evidence for re-integration of the *Ac* element into the genome of the aspen plants has been provided by Southern-analysis. Employing different restriction enzymes fragments of known sizes are expected when *Ac* is excised or not excised when using labeled *rolC*- or *iaal*-specific probes (Table 1). Rehybridizing the blots which reveal *Ac* excision with *Ac*-labeled probe demonstrates the re-integration of the element into a new position in the genome of the transgenic aspen. The application of the Inverse-PCR technique allows sequencing of genomic regions flanking the transposon in the new genomic context. In a single plant it is possible to isolate many of such re-integration events. Therefore, sequence informations of flanking regions of many individual re-integration events provide the feasibility to characterize larger parts of the genome. The sequence can be used to screen gene banks for homology to known genes or to look for typical features specific for plant genes. Alternatively, cloned plant fragments can be employed as molecular markers to identify their position in gene maps and their linkage to already mapped genes.

Restriction	Restriction site				Fragment size (bp)			
enzyme	35S	*Ac*	*rolC/iaaL*	*npt*-II	*Ac* not excised		*Ac* excised	
					rolC	*iaaL*	*rolC*	*iaaL*
EcoRV	yes	no	yes		5800	6200	1300	1700
NcoI	yes	no	no	yes	7200	10000	2700	4500

Table 1: Expected sizes of *rolC* or *iaal* labeled fragments cases when *Ac* is excised or not excised in Southern analysis using two different restriction enzymes

4. Acknowledgements

This study was supported by the Deutsche Forschungsgemeinschaft, Bonn. The technical assistance of Olaf Novitzki and Gabi Bentien, and greenhouse work of Hans Seuthe and Willi Holzwart are greatefully acknowledged. I thank Dr. T. Schmülling (University Tübingen, Germany), Dr. A. Spena (Max-Planck-Institute for Plant Breeding, Cologne, Germany) and Dr. J.D.G. Jones (John Innes Centre, Norwich, UK) for supplying the *Agrobacterium* strains.

5. References

Baker B et al (1986) Proc. Natl. Acad. Sci. USA 83, 4844-4848.

Chuck G et al (1993) Plant Cell 5, 371-378.

Döring HP, Starlinger P (1986) Ann. Rev. Genet. 20, 175-200.

Estruch, JJ et al (1991) Science 254, 1364-1367.

Fladung M, Ahuja MR (1995) Biotechniques 18: 3-5.

Fladung, M, Ahuja, MR (1996) Transgenic Organisms and Biosafety, Horizontal Gene Transfer, Stability of DNA and Expression of Transgenes, pp 275-281, Springer Verlag Berlin, Heidelberg.

Fladung, M, Ahuja MR (1997) Plant Mol. Biol. 33, 1097-1103.

Fladung M, Kumar S, Ahuja MR (1997) Trans. Res. 6, 111-121.

Fladung M, Muhs, H.J., Ahuja MR (1996) Silv. Genetica 45, 349-354.

Gierl A, Saedler H (1992) Plant Mol Biol 19: 39-49.

Jones JDG et al (1992) Proc Royal Soc Edinburgh 99B, 107-119.

Spena A et al (1989) The Plant Cell 1, 1157-1164.

Walbot V (1992) Ann Rev Plant Physiol 43: 49-82.

Sweet Potato *(Ipomoea batatas* L.*)* Biotechnology: Progress and Perspectives.

R. Garcia, R. Moran, D. Somontes, J. Mena, E. Pimentel, Z. Zaldua, A. Lopez, M. Garcia.
Centro de Ingenier□a Genetica y Biotecnologia. Camagüey. P.O.Box 387, C.P. 70100. Camagüey, Cuba. Fax: (53) 322- 61587. Tel: (53) 322- 61295.

INTRODUCTION

Although sweet potato is one of the most important plants around the world (FAO, 1989), biotechnological work on it has lagged behind. In the last years few groups have reported their experiences in sweet potato molecular biology manipulation to increase its nutritional quality (Lopez et al. 1996) and to give it pest resistance (Newel et al. 1995). Others have been working on protoplast isolation and regeneration (Sihachakr and Ducreux, 1987), and regeneration and transformation from roots, petioles, stems and leaves (Gosukonda et al. 1995; Carswell and Locy, 1984). The most important factors affecting the evolution of the " in vitro" culture, regeneration and transformation responses of sweet potato is the genotype worked on (Gosukonda et al. 1995).

The sweet potato weevil *(Cylas spp.)* is the major biological antagonist of sweet potato worldwide. The improved cultivars delivered do not show a stable performance for different ecological conditions; most of them depend on soil structure or the physiological and botanical features of the cultivar rather than on the genetic production of chemical defenses against the pest (Sutherland, 1986). The lack of any strong conventional genetic improvement programs to obtain pest resistance make sweet potato an important target to be modified by biotechnological tools. Here, we show our experiences on sweet potato biotechnology and our strategies to improve this important crop.

MATERIALS AND METHODS

Regeneration and transformation conditions

Leaves and stems were taken from young plants, from Jewel cultivar, grown in glass culture tubes containing 7 ml of MS medium. The "in vitro" plants are kept during six weeks at 25°C in a 12/12 photoperiod, 66% relative humidity. We have studied more than 151 combinations of growth regulators including: Indol-3-acetic acid (0, 0.25, 0.5, 0.75, 1.0, 1.25, 2.0); Naphtalen-acetic acid (NAA) at 0.1 ; 0.5; 1.0; 2.0; Zeatin riboside (0.11, 0.22 , 0.5, 0.75, 1.0); Kinetin (0.1, 0.5, 1.0, 2.0); 6-benzilaminopurine (BAP) at 0.25, 0.5, 1.0, 2.0, and 3.0; Paclobutrazol (PPP) at 0.1; 0.5; 1.0; 2.0 and PPP at 0.0; 1.0; 2.0 ; 3.0, (all in $mg.l^{-1}$).

The direct shoots regeneration efficiency calculated as follows: RE= (RS/ TNE) x 100. Were RE, represents the regeneration efficiency of direct shoots in percent (%); BR, total number of regenerated shoots; and TNE total number of tested explants.

The term regeneration frequency was introduced to determine the functional shoot or root emission per explant. It was calculated as follows: RF= NRP/ TNE. Were RF, is the regeneration frequency per explant; 5321., is the number of rooted functional shoots; and TNE, represent the total number of assayed explants.

Agrobacterium mediated transformation procedure

Agrobacterium tumefaciens strain C58C1-pGV2260 (Str^r; Spc^r; Rif^r; Amp^r) was used for plant transformation. Binary plasmid pDEBTT carrying the *Bacillus thuringiensis* var. *tenebrionis* delta-endotoxin and the *npt* II genes.

Several factors affecting the transformation efficiency has been evaluated in our studies such as: *Co-cultivation medium; Co-culture time; Use of vir gene stimulators; Co-culture conditions; Regeneration medium.*

A. Altman et al. (eds.), Plant Biotechnology and In Vitro Biology in the 21st Century, 143–146.

All the regeneration media tested were supplemented with kanamycin at 45 mg l^{-1} as selectable. The frequency of transformation (TF), also termed % of kanamycin resistant plants, was determined as the number of shoots that rooted on the propagation medium containing Kanamycin (KRS) per the total number of explants in an experiment (TNE). It is expressed as follows: TF= (KRS/ TNE).

Biological tests

Some of the kanamycin resistant clones and one untransformed were selected to perform the first biological test for sweet potato weevil resistance. Under controlled environmental conditions, 12 plants per clone were planted in an earth chamber. Established parameters for a good tuber yield were provided. Between each plant, highly infected tubers were planted, in order to guarantee pest population as inoculum.

Damages were evaluated 120 days after planting, considering the parameters: infestation percent (I), and pondered degree of infestation (PDI). Both of them, following a traditional methodology established in Cuban agriculture by using the procedure of Thousend and Heuberge, 1948, they were calculated as follows: $I=\Sigma(a.b).100/D\ (n)$; $PDI=\Sigma(a/n)\ (b)$;

where: a= Values of scale of damages. (From 1 to 4); b= Number of tubers per value of the scale, N= Total number of tubers, D= 4 (Maximum value of the scale).

In the experiment under controlled conditions, results of parameters I and PDI, were statistically compared using Duncan test for variance analysis both between treatments (clones and control plants) and replicates.

RESULTS AND DISCUSSION

The stem explants were able to form non-embryogenic calli on the cut side of the explants but early organogenic shoots appear first than these calli and they were always regenerated from the meristematic area placed in the nodal section. The best shoot emission for the **Jewel** cultivar were obtained on several media. Remarkably, auxin influence to induce direct organogenesis was observed to be greater than that of cytokinins. It was usual to find more stimulation for non embryogenic calli formation when the explants were treated with cytokinins (even at low concentrations) than with auxins. In some cases shoots were achieved by giving moderated auxin treatments (IAA 0.5 mgl^{-1}; NAA 0.1 mgl^{-1}). There is not any previous reference of using auxin based stimulation alone to induce plant regeneration on sweet potato.

Table 1. Regeneration efficiency on different regeneration media for Jewel cultivar. **BN-3 medium**, MS + Sucrose 3% + BAP (0.1) + NAA (1.0); **PN-2 medium**, MS + Sucrose 3% + PPP (1.0 mgl^{-1}) + NAA (1.0 mgl^{-1}); **MPM medium**, MS + Sucrose 3% + IAA (0.5) in $mg.l^{-1}$.

MEDIUM	Direct shoots (%)	Indirect shoots	Regeneration frequency
BN-3	68.9	100	0.7
PN-2	85.00	100	0.85
MPM	200.00	100	3

A high regeneration frequency was obtained on several combination of NAA and Paclobutrazol. However, there was a inhibitory effect on the normal shoot development when Paclobutrazol concentration was over 1.5 mgl^{-1}.

Taken into account that leaves behaved better than stem in all our regeneration experiments we chosen this explant source as target tissue for the next steps of the project.

Long term cultures in liquid medium very much affected the physiological and morphological stage of the explant during the transformation step. Although, it was proved the liquid medium gives more possibilities for the full interaction between the target explant and the bacterium it was better to co-cultivate the explants not for long periods in liquid medium. When the explants were co-culture in liquid medium for more than 36 hours 100 percent of them died after planting on the regeneration medium supplemented with kanamycin at 45 mg l^{-1}. The optimum time for co-cultivation in liquid was established for 24 hours; for this period of co-cultivation the 100 percent of the explants kept green and formed kanamycin resistant roots.

Table 2. Transformation frequency on three different media.

MEDIUM	Km^r shoots (%)	Km^r roots (%)	TF
BN-3	30.2	76.8	93.25
MPM	55.3	100	100
PN-2	45.3	82.3	100

Long term culture of the explants with the bacterium were possible on solid medium. However, some few transgenic roots and shoots were obtained (1% and 3% for the best treatment).
No important influence of acetosyringone on the transformation efficiency was found. The same number of transgenic shoots were obtained, both when the acetosyringone was present in the medium and when it was not.

Biological test results
According to parameters I and PDI, two transgenic clones, C27 and C1, have shown the least weevil infection. In this experiment, the damages in tubers of C27 and C1 were at least 5 times and almost twice lower than untransformed plants, respectively.

Table 3. Evaluation of transgenics sweet potato clones for resistance to sweet potato weevil (*Cylas formicarius sp. elegantulus*) under controlled conditions, 120 days after planted. **C:** Untransformed control plants belonging to **Jewel** cultivar.

Clone	I	PDI.
C-27	2.59	0.104
C-1	8.01	0.319
C	14.09	0.591

By Dellaporta´s protocol for plant DNA extraction (Dellaporta et al. 1983), a high quality material was obtained in all cases. In some kanamycin resistant clones, bands at the same size of the one from the positive control, were amplified by P.C.R. The same results were obtained after hybridization of P.C.R products with the selected probe. This fact confirmed the successful transformation with the B.t.t. toxin gene. Fig. 1.I shows the results of the Southern hybridization of the P.C.R. products from 2 of the clones selected among several that were positive to this test. Neither by P.C.R., or by P.C.R. product hybridization, the bands were detected in untransformed control plants.
Western blot, performed with proteins from the clones C27 and C1, showed that CryIIIA toxin expression was detected on the expected molecular weight in both, and there was not any band in the lane of untransformed control plants (data not reported). Despite these results, the toxin expression was visually estimated no higher than 0.005% of the total protein content, even for the clone C27 that exhibited higher levels.
There is coincidence between the expression levels of the CryIIIA toxin and the mRNA detection. Regarding the low toxin expression detected, some papers have described the weak expression of the B.t. endotoxins in transformed plants carrying the native bacterial genes. Numerous factors could affect the expression of heterologous genes in transgenic plants, such as the codon usage, the A+T

content, the molecular context at the translational start, mRNA sequence and structure and some others (Ely, 1993).
Hybridization of total DNA digestion with KpnI for clone C27 showed a single band migrating at a size between 4.3 kb and 6.5 kb bands of the molecular weight marker. In the case of the digestion with EcoRV of the same clone's DNA, two bands were observed. One of them migrated at a size of around 4 kb and the other one, of around 6 kb. No bands were observed in lanes corresponding to digestion with the same enzymes of untransformed control plants' DNA (data not reported) These results confirmed the integration of the *cry*IIIA gene into the genome of this clone, thus its transgenic character. This pattern allowed us to suggest that the gene was inserted in a single copy.

CONCLUSIONS

Our progress and results on sweet potato biotechnology and molecular biology to confer pests resistance have demonstrated that biotechnological manipulation of this tropical crop is a suitable tool to improved its agronomic performance. Nowadays, we are strongly working in building a synthetic deltaendotoxin gene to increase the expression levels in the plant tissues and to grow up the resistance against weevils. Transgenic plants with the novel synthetic genes are almost to be generated and will be tested under field condition on the second half of 1999.

REFERENCES

Barton KA et al (1987) Plant Physiol. 85, 1103-1109.
Blakesley D et al (1996) Plant Cell Reports 15, 873- 876.
Carswell GK, Locy RD (1984) Plant Cell Tiss. Org. Cult. 3, 229- 236.
Dellaporta SL et al (1983) Plant Molecular Biology Reporter 1, 19-21.
Ely S (1993) *in: Bacillus thuringiensis*, An Environmental Biopesticide: Theory and Practice., 1993, pp.115., John Wiley & Sons, Ltd., **(Ed.)**
FAO Production Year Book. 1989. Food and Agriculture Organization, Rome.
Gosukonda RM et al (1995) HortScience 30, Suppl 5, 1074- 1077.
L⌝pez A et al (1996) Biotecnolog□a Aplicada 13, 265- 270.
Murashige T et al (1962) Physiol. Plant 15, 473-497.
Newel CA et al (1995) Plant Science 107: 215-227.
Sihachakr D et al (1987) Plant Cell Rep 6, 326- 328
Sutherland JA (1986) Trop. Pest Manage. 32, 304- 315.
Thousend CR et al. (1948) Plant Disease Report, 340-343.

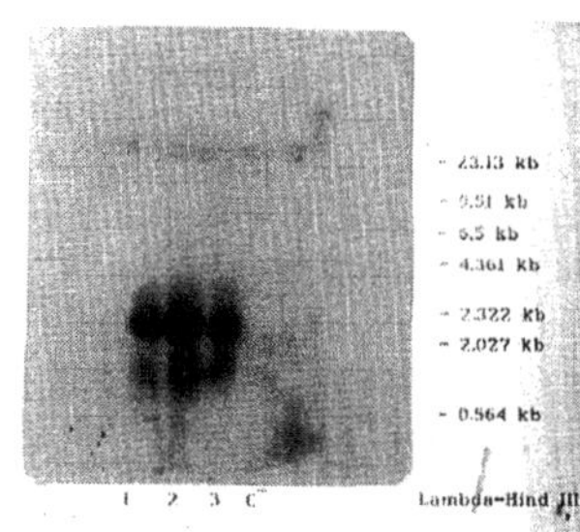

Figure 1- Molecular analysis of transgenic plants.
(I) ***Southern blot of the PCR products.*** **Lane 1-** DNA from Btt., **Lane 2-** DNA from clone C1, **Lane 3-** DNA from clone C27, **Lane 4-** DNA from untransformed control plants.

Cytoplasmic diversity caused by mitochondrial (mt) DNA dynamics and mt gene expression in Petunia

H. Hauschner, V. Yesodi, S. Izhar, Y. Tabib, N. Firon

Introduction

Mitochondria (mt), which are specialized organelles whose primary function is the conservation of oxidatively derived energy and its utilization for ATP synthesis, produce a large proportion of the energy that cells require and contain their own DNA. Plant mt genomes are large in size and complex in sequence organization. The rapid structural mt DNA changes observed in plants (Gray 1992) may cause duplication of genes and generation of pseudogenes and novel open reading frames (*orfs*) as well as re-shuffling of genes and their regulatory sequences. These phenomena cause diversity that may have evolutionary advantages during stress conditions. The study of plant mt genome evolution and genome expression may be facilitated by the analysis of mt mutants. One of the few mt mutations characterized in plants so far is the cytoplasmic male sterility (CMS) mutation (Hanson, Conde 1985), that is usually manifested by the abortion of normal pollen development and is commercially used for the efficient production of hybrid seed.

Tissue culture conditions were found to affect mtDNA reorganization and were thus used in order to try and generate novel mtDNA mutations in *Petunia*. Prolonged maintanance in tissue culture of cells derived from CMS *Petunia* produced indeed a novel mt mutation (Yesodi et al 1997a). We have used the mt mutant mentioned above, as well as a CMS mutant, in order to study the molecular mechanisms that are responsible for the generation of mt mutations (Soferman-Avshalom *et al* 1993, Yesodi *et al* 1995, Yesodi *et al* 1997a). Our results indicate that these two mutants, which were found to contain novel chimeric *orfs*, were generated by the mechanism of mtDNA homologous recombination. As a result of such mtDNA reorganization, diversity in the 5' region of subunit 3 of the NADH dehydrogenase (*nad3*) enzymatic complex has been created, affecting the transcription of this gene. Diversity in the 5' and 3' non-coding sequences of the α subunit copies of the mt ATPase enzymatic complex (*atpA*), in both male fertile and CMS *Petunia,* has also been detected (Yesodi et al 1997b). These changes were indeed found to affect *atpA* transcription in the two cytoplasms.

A. Altman et al. (eds.), Plant Biotechnology and In Vitro Biology in the 21st Century, 147–150.

Methods

Plant material: The following *Petunia* lines were used in this study: *P. parodii* L. S. M. line 3699, *P. hybrida* Hook Vilm lines 3704 and 11734 (cv. Rosy Morn), all with normal (F) cytoplasms, as well as CMS *P. parodii* line 3688 with an (S) cytoplasm (isonuclear with line 3699) and the CMS *P. hybrida* lines 11795 and 11736 with (S) cytoplasms (isonuclear with lines 3704 and 11734, respectively).

Isolation of total and mt DNA, DNA transfer and hybridization, mt DNA library construction and DNA sequence analysis: Were performed as previously described by Yesodi *et al.* (1997a, 1997b).

Total and mt RNA preparation, RNA transfer and hybridization: Total RNA was isolated from young leaves using the Tri-Reagent method (Molecular Research Center, Ohio). Mt RNA was isolated from a mt-enriched fraction of cell suspension cultures, using the Tri-Reagent method. 50-60 μg RNA were loaded on a agarose-formaldehyde gel and blotted to a nylon membrane (Qiagen). *AtpA* and *nad3* probes were prepared as described in Yesodi *et al.* (1997a, 1997b).

Results and Discussion

The homologous recombination mechanism plays a major role in creating cytoplasmic diversity in Petunia

The maternally inherited CMS phenotype in *Petunia*, has long been attributed to the presence of the mt chimeric *S-Pcf* sequence (Young & Hanson 1987). The *S-Pcf* locus consists of three *orf*s that are co-transcribed (Hanson *et al.* 1989). The first *orf*, *Pcf* detected only in CMS *Petunia*, contains parts of the *atp9* and *cox2* genes and an unidentified reading frame, *urf-s* (Fig. 1). The second and third *orf*s contain the single-copy *nad3* and ribosomal protein S12 (*rps12*) sequences, respectively. More recently it has been suggested that the sequence of events leading to the generation of *S-Pcf* might have involved introduction of an *urf-s* low abundance mt DNA, via homologous recombination, into the main mt genome 5' to *nad3*, at a region where a low-homology *urf-s* related sequence was detected (Yesodi *et al.* 1995). Recently, the only mutation detected so far in *S-Pcf* has been characterized (Yesodi *et al.* 1997a). In cell line R13-138, which was generated from a CMS somatic hybrid that was kept in tissue culture for seven years, a change in the first *orf* of *S-Pcf* has been characterized: the *atp9* sequence has been lost, while exon 1 of the normal copy of the *cox2* gene (including the original ATG sequence) and its adjacent 5' sequence, have been introduced. The data suggest that this reorganization was the consequence of a homologous recombination event involving part of the *cox2* coding region. In the two mt mutants of *Petunia* described above, the sequence 5' to the *nad3* and *rps12* genes has been altered, suggesting the possibility that different promoters are regulating the expression of these genes in the mt mutants compared to wild type *Petunia* (Fig. 1). Sequencing of the 5' region of the *nad3* gene in male fertile *Petunia*, resulted in the detection of a truncated *atpA* copy. Homologous recombination through a progenitor *atpA* sequence is assumed to have caused the rearrangement in the 3' portion of the *atpA orf* and the generation of the truncated *atpA* gene (Yesodi *et al* 1997b). The differences in the 5' and 3' non-coding sequences of the *atpA* copies in male fertile and CMS *Petunia* suggest the possibility of differences in the regulation of the expression of the *atpA* gene in these two cytoplasms.

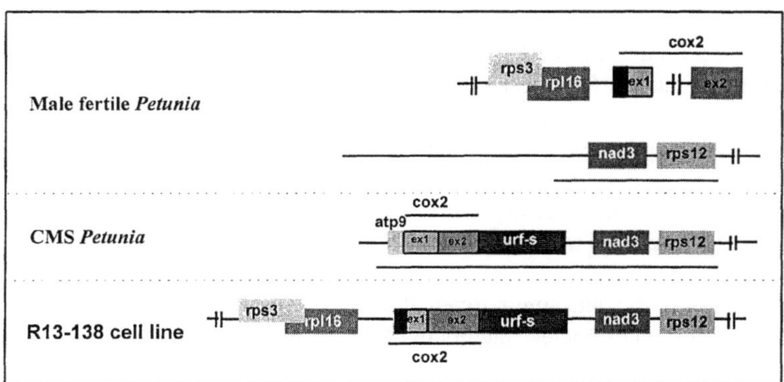

Figure 1: Alteration in the 5' region of the *nad3* and *rps12* genes in two mt mutants of *Petunia*

Diversity in transcription of the mt *atpA* and *nad3* genes

Northern blot analysis of mt RNA derived from cells, kept in suspension culture and hybridized with an *atpA* probe, indicated the presence of 1.0-kb, 2.0-kb, 2.3-kb, 2.6-kb, 4.0-kb, 4.6-kb transcripts in male fertile *P. parodii* and 2.0-kb, 2.6-kb, 4.4-kb, 6.2-kb transcripts in CMS *P. parodii*. A 1.8-kb *atpA* transcript was detected in male fertile *P. hybrida* while in the isonuclear CMS *P. hybrida* line two transcripts, 1.8-kb and 4.6-kb were detected. The 1.8-kb and 4.6-kb transcripts were also detected in both male fertile and CMS *P. hybrida* RM. In mt RNA of cell line R13-138 (derived from fusion of CMS *P. parodii* and male fertile *P. hybrida* cells) the 1.8-kb *atpA* transcript was detected. These different transcripts, which may be the result of either different transcription initiation sites or processing of mt RNA, indicate the existence of diversity in *atpA* transcription. Such diversity in mt gene expression may be advantageous for the plant under stress conditions. The results indicate that there are differences in *atpA* transcription between isonuclear male fertile and CMS *Petunia*. The results also indicate differences in *atpA* transcription between male fertile *P. parodii* and male fertile *P. hybrida* as well as between CMS *P. parodii* and CMS *P. hybrida* lines, suggesting that the nuclear genome affects the transcription of the mt *atpA* gene. It should be mentioned that in the case of (F) *P.parodii* and (F) *P.hybrida* the differences in *atpA* transcription may be due to differences between the two (F) cytoplasms rather than due to the effect of the nuclear genome (although at the DNA level very few differences between the sequence of the *atpA* copies in the two (F) cytoplasms were detected, Yesodi *et al.* 1997b). In order to resolve this question, we are currently doing the following crosses: *P.parodii* X *P.hybrida* and *P.hybrida* X *P.parodii.*

Northern blot analysis of mt RNA, hybridized with an *nad3* probe, gave a "smeary" pfofile, suggesting that the *nad3* transcripts may be degrading rapidly. Similar results were obtained using six different RNA extracts. The same blots were then used for hybridization with *atpA.* In mt RNA derived from suspension culture cells of CMS *P. hybrida* RM a major 2.5-kb and two additional 1.4-kb and 2.0-kb transcripts were apparent, while in the isonuclear (F) line 1.3-kb and 1.6-kb transcrips were detected, indicating differences in *nad3* transcription between CMS and (F) cytoplasms.

There were differences in the intensity and the sharpness of the *nad3* hybridizing transcripts between the three *Petunia* nuclear backgrounds, suggesting that the nuclear background may affect either the synthesis or the stability of the *nad3* transcript. The highest intensity hybridizing *nad3* transcripts were obtained using the *P. hybrida* RM nuclear background.

We are thus planning to use the data summarized above as a starting point for the detection and characterization of nuclear factors that control plant mt gene expression in both wild type and mutant *Petunia* lines.

References

Gray M (1992) Int. Rev. Cytol. 141, 233-357.
Hanson, M. R. *et al* (1985) Int. Rev. Cytol. 94, 213-266.
Hanson, M. R. *et al* (1989) Oxford Survey Plant Mol. Cell Biol. 6, 61--85.
Soferman-Avshalom, O. *et al* (1993) Theor. Appl. Genet. 86, 308-311.
Yesodi, V. et al (1995) Mol. Gen. Genet. 248, 540-546
Yesodi, V. et al (1997a) Mol. Gen. Genet. 255, 106-114.
Yesodi, V. et al (1997b) Curr. Genet. 32, 348-357.
Young, E. G., Hanson, M. R. (1987) Cell 50,41-49.

SOMATIC HYBRIDS OF SOLANUM TUBEROSUM AND RELATED SPECIES

John P. Helgeson and Geraldine T. Haberlach
Agricultural Research Service, United States Department of Agriculture, Department of Plant Pathology, University of Wisconsin, Madison, WI 53706

Introduction. Wild species related to potato have many interesting disease resistances and other characters that could be useful in potato breeding. Often, however, these species are difficult or impossible to cross with potato. For this reason the asexual combining of genomes by somatic hybridization appears to many workers to be one means by which the sexual incompatibilities might be bypassed. Somatic hybrids of *Solanum tuberosum* and related species have been sought for over 20 years with various degrees of success. Hybrids that have been produced between *S. tuberosum* and other *Solanum* spp.include :*S. tuberosum + S. acauli* (38) , *S. tuberosum+ S. berthaulti (31), S. tuberosum+S. brevidens* (1,6 ,*29), S. tuberosum+S. bulbocastanum (5,13,15,35), S. tuberosum +S. chacoense (21), S. tuberosum +S. circaeifollium* (26), *S. tuberosum +S. commersonii(8,20), S. tuberosum + S. etuberosum (27), S. tuberosum + S. nigrum* (7), *S. tuberosum+S. papita (19), S. tuberosum + S. phureja (30), S. tuberosum +S. pinnatisectum(*33), *S. tuberosum +S. polyadenium (*Hunt and Helgeson, unpublished), *S. tuberosum+ S. torvum (18).*

Development of somatic hybridization techniques. Many different techniques need to be successful if one is to obtain somatic hybrids. First, it is necessary to have a method for producing viable protoplasts. With potato, these procedures began to be successful in the mid 1970's. Shepard and Totten, for example, based some of their procedures on the results obtained with tobacco to develop isolation and regeneration schemes for potato (32) .We analyzed a large number of wild species for protoplast regeneration before settling on a procedure (11). Subsequently, fusion procedures with polyethylene glycol (1) or electrofusion (27,34) have proven to be generally useful for obtaining hybrids. Selection of hybrids from amongst unfused cells can present some difficulties, however the problem is really not too difficult. Often, the somatic hybrids exhibit "Hybrid vigor" and can be selected on that basis. Thus either the shoots will grow more rapidly or be earlier to differentiate from the calli. Alternatively, if one knows the differentiation requirements for the two species, the media can be manipulated in sequence so that only hybrid plants are obtained (3). It is also possible to use various color stain differences, including various fluorescent and vital stains or cells containing and lacking chloroplasts (17) can be used to select hybrids. Finally, it is possible to transform the parent to give differing antibiotic resistances to such compounds as kanamycin or hygromycin and then select cells on media containing both of the antibiotics (25) .

A. Altman et al. (eds.), Plant Biotechnology and In Vitro Biology in the 21st Century, 151–154.

The techniques developed by many workers have resulted in the substantial array of available hybrids listed above Only hybrids between two *Solanum* species are listed there. However, hybrids between various dihaploid *S. tuberosum* lines (2) as well as wider hybrids, e.g. between potato and tomato (Lycopersicon esculentum) have also been produced. Also, a number of workers have sought to selectively modify either the cytoplasmic of genetic compounds of the cells prior to fusion. If, for example, one inactivates most of one nucleus before fusion the polyploid problem of symmetrical fusions, which result in the summation of the two genomes into a single entity, may be eliminated. However, in this paper, we cover only symmtrical fusions.

The rationale for carrying out somatic hybridizations. The natural curiosity that one may have for studying the events that accompany the interaction of the different genomes of two sexually incompatible species is sufficient justification for carrying out the experiments. However, there are also some very practical results that can come from somatic hybrids of two species for which sexual hybridization has proven impossible. In particular, enhanced disease or stress resistances of the fusion products can result. Traits from non tuber bearing species may be transferred in to tuber-bearing lines (*e.g.* PLRV resistance from *S. brevidens* (12) or PVY resistant from S. *etuberosum* (28). The pioneering efforts of Binding and his collaborators (7) to transfer herbicide resistance from *S. nigrum* into potato is another example of a possible practical result. Also, the synthesis of a "good" tetraploid somatic hybrids from two selected diploids is a long-cherished goal of Wenzel and his co-workers (*e.g.* 9).

Studies at Madison to obtain new disease resistances. The work in the author's laboratory has focused on fusions of potato with diploid, 1EBN species including *S. brevidens, S. bulbocastanum. S. commersonii. S. etuberosum and S. polyadenium.* These are generally sexually incompatible even with diploids extracted from *S. tuberosum.* However, they often are reported to have disease resistances that might be highly useful if incorporated in potato varieties. The general scheme used in our research is to obtain seeds of a reportedly disease resistant accession of the wild diploid species, surface sterilize the seeds, grow plantlets up *in vitro* and then to test seedlings for resistance to the particularly disease. At this point we develop the procedures for obtaining protoplasts from the selected seedling. This saves a considerable amount of trouble as even different seeds from a given accession can have very different regeneration rates from protoplasts (11). Seedlings selected for fusion work are then maintained *in vitro* so that all key parent lines can be retained for future comparisons with hybrids.

Both PEG fusions (1) and electrofusions (27) have been used successfully in our laboratory, although the latter procedure has been used by preference if the numbers of protoplasts obtained from one or the other of the parents is relatively low. Much of our earlier work was on somatic hybrids between *S. brevidens,* a wild South American diploid species that forms no tubers and either dihaploid or tetraploid potato lines. In our hands, the hexaploid somatic hybrids were easier to cross than the tetraploid hybrids (10) and hybrids with a highly fertile tetraploid such as a selected clone of PI 203900 were much more fertile than hybrids with less fertile varieties such as Russet Burbank.

Somatic hybrids between *S. brevidens,* or its related non-tuber - bearing species, *S. etuberosum,* and potato proved to have a number of very useful disease resistances which, generally, can be passed on to further generations by sexual crosses. The resistances include those to viruses PLRV and PVY (12,28). Some of the lines also appeared to be very resistant to *Verticillium dahlii* and *Alternaria solani* (Helgeson, personal observation). Somatic hybrids between *S. brevidens* and *S. tuberosum* PI203900 were found to be highly resistant of tuber soft rot caused by *Erwinia* spp. (4) and to ring rot (Cornebacterium michigensis, German and Helgeson, unpublished data). Recently, we have obtained somatic hybrids between *S. commersonii* and potato that are highly resistant to bacterial wilt caused by *Pseudomonas* (now *Ralstonia) solanacearum* (20,22).

Randomly amplified polymorphic DNA primers (RAPDs) and restrictionfragment length polymorphism(RFLP) probes were found to be very useful for proving that the plants under study were actually somatic hybrids, for determining that whole sets of wild-species chromosomes were actually present in the hybrids and for examining the extent of interaction between the two different genomes (23,24,27,36, 37).

The discovery of resistance to *Phytophthora infestans* (late blight) in progeny of somatic hybrids between *S. bulbocastanum* and potato has resulted in a change of direction in our latest efforts. While testing for early blight resistance in a field that received no fungicide treatments, the experiment was overtaken by an epidemic of late blight caused by the US 8 (A2 mating type) of *P. infestans.* Although all of the standard U. S. cultivars in the experiment were killed, several somatic hybrids and their sexual progeny show little or no infection (15) . These results were subsequently confirmed in experiments at 8 different locations around North America (16). Efforts to map the resistance gene(s) that come from *S. bulbocastanum* are underway in the laboratory.

Conclusions. Somatic hybridization between various *Solanum* ssp. has been attempted in many different laboratories in many different countries. It is clear that fertile somatic hybrids can be obtained and that characters "trapped" by this procedure can be passed on to sexual progeny. Thus it is now possible to tap a whole group of genetically distinct 1 EBN plants that previously had presented considerable difficulty to breeders. There is a cost in doing so, however. The rearrangement of the beneficial traits that accompanies the crosses can often present real problems in terms of poor tuber traits amd other plant characteristics. Thus, the somatic hybridization procedures - that now have been around for more than 20 years - have not been as useful as might have been originally envisioned. However, since some extremely valuable resistances to diseases such as ring rot , bacterial wilt and late blight can be obtained by these means one can expect that in the future there will be continual progress in obtaining disease resistant plants with the aid of this technique. The improved plants , which no longer require extensive pesticide treatments for survival, can be expected to be very valuable given the world-wide wish for decreasing pesticide use.

References

1. Austin S et al (1985) Plant Sci. 39, 75-82.
2. Austin S et al (1985) Theor. Appl. Genet 71, 172-175.
3. Austin S and JP Helgeson (1987) Plant Molecular Biology, Plenum pp 209-222.
4. Austin S et al (1988) Phytopathology 78, 1216-1220.
5. Austin S. et al (1993) American Potato Journal 70, 485-495.
6. Barsby T et al 1984 Plant Cell Reports 3, 165-167.
7. Binding H et al (1982) Theor. Appl. Genet 63, 273- 277.
8. Cardi T et al (1993) American Potato Journal 70, 753-764.
9. Deimling S et al (1988) Plant Breeding 101, 181-189.
10. Ehlenfeldt M and JP Helgeson (1987) Theor . Appl. Genet . 73, 395-402 .
11. Haberlach G et al (1985) Plant Science 39, 67-74.
12. Helgeson JP et al (1986) Plant Cell Reports 3, 212-214.
13. Helgeson JP et al (1988) Plant Cell Tiss. Org. Cult. 12, 185-187.
14. Helgeson JP et al (1993) American Potato Journal 70, 437-452.
15. Helgeson JP et al (1995) American Potato Journal 72, 629.
16. Helgeson JP et al (1998) Theor. Appl. Genet (in press).
17. Hunt G and JP Helgeson (1989) Plant Science 60, 252-257.
18. Jadari R et al (1992) Euphytica 64, 39-47.
19. Kaendler C et al (1996) Theor. Appl. Genet 92, 455-462.
20. Kim H et al (1993) Korean Society of Plant Tissue Culture. 20, 337-344.
21. Komarnitskii IK et al (1980) Dok.Akad. Nauk Ukr. SSR 4, 75-77.
22. Laferriere, L et al (1997) In Bacterial Wilt Disease, Springer, PP 316-320.
23 .McGrath JM et al (1994) Theor. Appl. Genet 88, 917-924.
24. McGrath JM et al (1996) Genetics 142, 1335-1348.
25 Masson J et al (1989) Theor. Appl. Genet78, 153-159 1989.
26. Matteij WM et al. (1992) Theor. Appl. Genet 83, 459-466.
27. Novy R and JP Helgeson (1994) Theor . Appl. Genet 89, 775-782.
28. Novy R and JP Helgeson (1994) Theor . Appl. Genet 89,783-786.
29. Pehu E et al (1989) Theor . Appl. Genet 78, 696-704.
30. Puite K et al(1986) Plant Cell Reports 5, 262-265.
31. Serraf I et al (1991) Plant Science 76, 115-126.
32. Shepard J and R Totten (1977) Plant Physiology 60,313-316.
33. Sidorov VA et al (1987) Theor . Appl. Genet 74, 364-368.
34. Templaar M et al (1987) Plant Science 48, 99-106.
35. Thieme R et al (1997) Euphytica 97, 189-200.
36. Williams CE et a .(1990) Theor. Appl. Genet 80, 545-551.
37. Williams CE et al (1993) Genetics 135,1167-1173.
38.Yamada T et al (1997) Breeding Science 47, 229-236.

A NEW AND VERSATILE AGROBACTERIUM-BASED PLANT TRANSFORMATION VECTOR

R. Hellens, N. Joyce and P. Mullineaux
John Innes Centre, Norwich Research Park, Colney, Norwich, NR4 7UH, United Kingdom

1. Introduction

The transformation of plant species for both experimental and biotechnological applications is most often achieved using *Agrobacterium tumefaciens*-mediated procedures (Ellis 1994). Binary Ti plasmids were developed primarily in the 1980s and have proved very useful in routine plant genetic manipulation work. However, the full potential for the versatility of binary Ti vectors has not been fully realised. The current designs available in the public domain are becoming increasingly obsolete as the demands and range of plant molecular biology increases. Current vectors tend to be in excess of 9kb in size, of low copy number in *E.coli*, and lack versatility in both cloning site and marker gene selection. In this proceedings paper we report the construction of a new vector Ti system, pGREEN, with allied selectable marker and reporter genes which addresses all of the above problems.

2. The pGREEN binary Ti plasmid system

pGREEN is a binary Ti vector which has been designed to be small, versatile, high copy number in *E.coli* and low copy number in *Agrobacterium*. The basic vector, pGREEN0000, is 3.23kb, has a colE1 replication origin (ori) for *E.coli* and use the *nptI* gene for selection of kanamycin resistance in both bacterial species (Fig.1). The *lacZ'* gene provides α-lac complementation (blue/white colony selection) when DNA fragments are inserted into the restriction enzyme sites derived from the pBluescript polylinker. The left and right borders (LB and RB) are flanked internally by blunt cutting *HpaI* and *StuI* sites respectively. The basic pGREEN vector contains no selection for plant transformation, but a number of selection and marker cassettes have been designed such that they can be easily cloned into the LB or RB providing the researcher with unprecedented amount of flexibility in the configuration of the T-DNA. Figure 2 shows the 7 CaMV 35S promoter-based selectable marker and reporter gene cassettes. In addition, there are 7 equivalent nos-based cassettes available. All restriction sites duplicated in the pGREEN T-DNA have been removed from the cassettes by site-directed mutagenesis.

A. Altman et al. (eds.), Plant Biotechnology and In Vitro Biology in the 21st Century, 155–158.

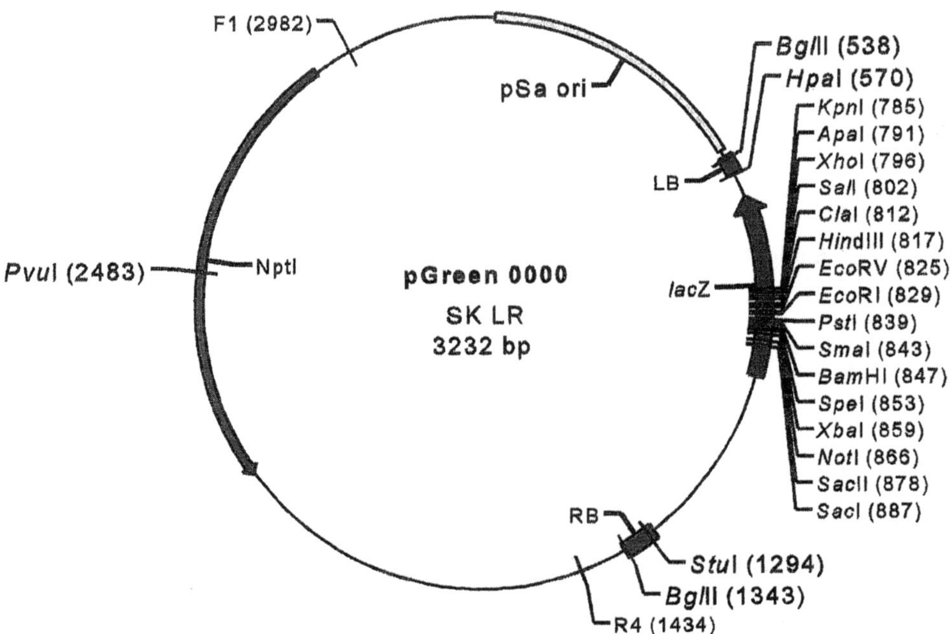

For replication in *Agrobacterium*, the pSa replication locus has been used. In order to reduce the size of the pGREEN plasmid, the pSa-RepA gene, which acts *in trans* on the pSa ori, has been put onto a second plasmid which is compatible with pGREEN in *Agrobacterium*. This second plasmid confers tetracycline resistance and provides a multiple cloning site for further developments currently being contemplated (see section 5). Because the pGREEN plasmid requires the repA function, selection for the pGREEN plasmid (kanamycin resistance) automatically selects for the repA plasmid (pJICSa_Rep; Fig. 3) so tetracycline does not need to be included in the culture media.A range of *Agrobacterium* strains have been transformed with pJICSa_Rep to simplify *Agrobacterium* transformation with pGREEN, these include LBA4404, GV2260, AGL1, EHA105 and LBA9402 (*A.rhizogenes* strain).

3. Plant transformation with pGREEN

The following plant species have been transformed with versions of pGREEN containing selectable marker and reporter genes; peas, tobacco, *Arabidopsis*, cabbage, oil-seed rape and potato. These experiments have also included transformation experiments with various genes inserted in the polylinker. In all cases, the pGREEN vectors produced transgenic plants with

Figure 2

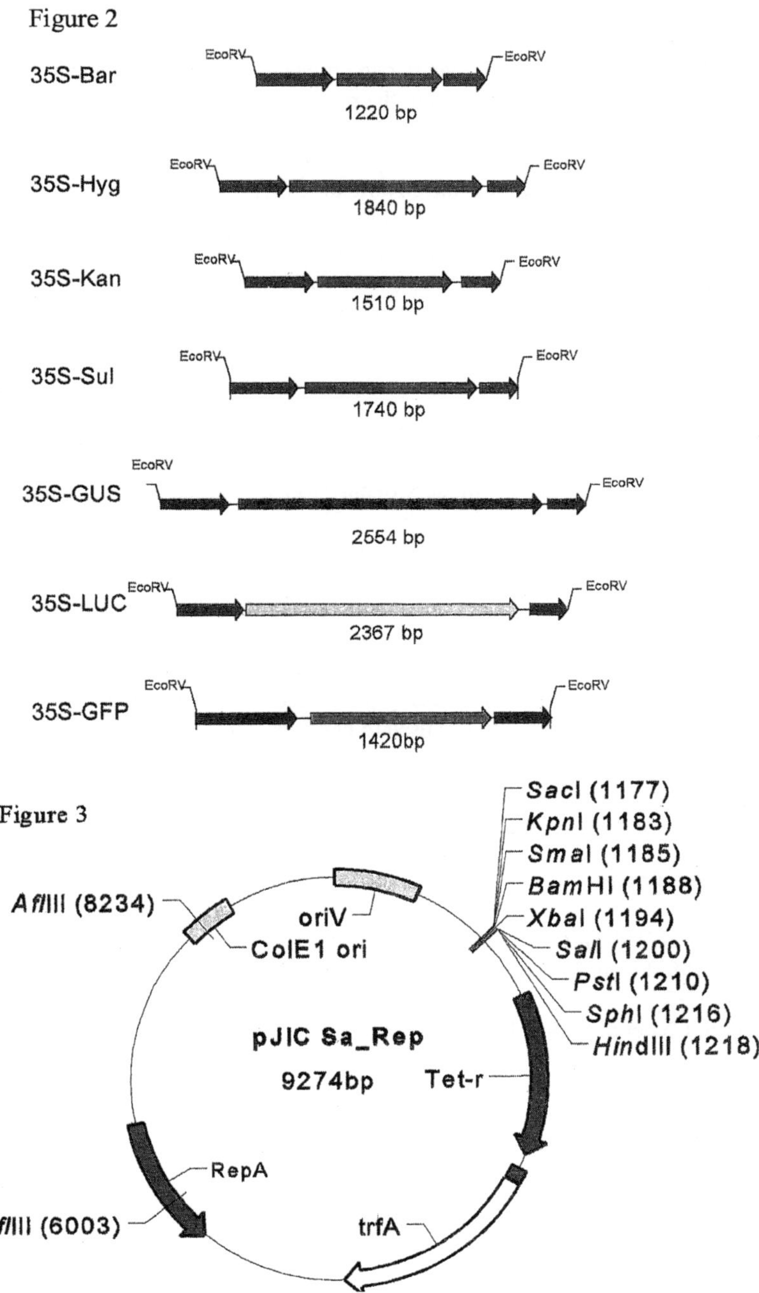

Figure 3

at least the same efficiency as the Ti vector normally used to transform the species in question. For example, the performance of pGREEN0049 (LB nos-*aph3'II*-nos[+], RB 35S-*luc*-CaMV polyA[-]) in transforming potato compared with the equivalent pBIN19 construct showed that there was no difference in the performance of the two binary Ti plasmids (E.A. Edwards, personal communication). Equally, the use of pGREEN0229 (LB nos-*bar*-nos[+]) harbouring a 35S-GUS gene containing an artificial intron (R. Hellens; unpublished data) gave 7 stable transformed pea lines from 200 explants (a frequency of ca. 0.4%; S. Bean, personal communication) which is typical for this transformation procedure using a *bar* gene-containing binary Ti plasmid (Bean et al 1997).

While analysis of integrated DNAs is not complete, so far we have uncovered nothing unusual compared with plants transformed with earlier binary Ti vectors.

4. pGREEN Website

A website has been prepared which provides all the sequence information, plasmid/marker gene permutations, techniques and order forms. The address is http://www.uea.ac.uk/nrp/jic/infoserv/depart/appgen/pgreen/a_hom_fr.htm.

5. Further developments

pJICSa-Rep can be used as a platform for further improvements in transformation technology. For example, a version of pJICSa_Rep has been produced which possesses its own T-DNA. This will permit co-transformation experiments to be conducted in conjunction with pGREEN, both plasmids being co-resident in the same *Agrobacterium* strain. Alternatively, pJICSa_Rep could be used to harbour extra copies of virulence (*vir*) genes for development of "super-hypervirulent" strains of *Agrobacterium* and/or as the basis for developing binaryTiBAC (Bacterial Artificial Chromosome) vectors (Hamilton et al 1996). Such vectors would be much more convenient to handle than those currently available.

6. References

Bean, S et al (1997) Plant Cell Rep.16, 513-519.

Ellis JR (1993) Plant Molecular Biology LABFAX , pp253-279, BIOS Scientific Publishers, Oxford, UK.

Hamilton CM et al (1996) Proc. Natl. Acad. Sci. USA 93, 9975-9979.

Inheritance of Transgenes in Cereal Plants

Antje Kluth, Dirk Becker, Horst Lörz
Institute of General Botany, AMP II, University of Hamburg,
Ohnhorststraße 18, D-22609 Hamburg, Germany

Abstract

Based on efficient *in vitro* culture and regeneration systems, we have developed reliable protocols for direct DNA uptake into barley, triticale, oat, maize and wheat. Our present experiments are aiming at the transfer of genes of agronomic interest. However, transgenic crop plants will only be of value if their engineered phenotype is faithfully transmitted through subsequent generations in a predictable manner. Therefore, we investigated the inheritance and expression of transgenes at the molecular level in transgenic wheat plants. In most of our independent lines, we observed a mendelian inheritance pattern. Gene inactivation was observed in a few lines containing multicopy-integrations of the transgene. We investigated non-mendelian inheritance in a single-copy transgenic wheat plant line. Transmission rate amounted 1:1 instead of the expected 3:1-ratio. Furthermore, non of the analyzed plants was homozygous for the transgene. Several sources for these phenonemons were investigated in detail.

Introduction

In the last few years the microprojectile mediated gene transfer (Sanford et al. 1987) has become the most often used transformation method for a wide range of cereal crops. Existing problems like somaclonal variation of regenerants from long term *in vitro* cultures could be overcome by directly targeting primary tissues. Up to now, scutellar tissue of rice (Christou et al. 1991), maize (Koziel et al. 1993; Brettschneider et al. 1997), hexaploid triticale (Zimny et al. 1995) and wheat (Weeks et al. 1993; Vasil et al. 1993; Becker et al. 1994), immature inflorescences of tritordeum (Barcelo et al. 1994), barley microspores (Jähne et al. 1994) and leaf base segments of oat (Gless et al. 1998) have been used successfully to obtain fertile transgenic plants via particle bombardment.
For optimization of the bombardment process and for selection of stably transformed plants we most often used pDB1, a plasmid containing the visualizable marker gene *uidA* under the control of the rice actin1 promoter and the selectable marker gene *pat* under the control of the CaMV 35S promoter. Bombarded tissues were selected on phosphinotricin containing medium. Putative transgenic plants were analyzed for the expression of the

A. Altman et al. (eds.), Plant Biotechnology and In Vitro Biology in the 21st Century, 159–164.

marker gene by BASTA spraying. The presence and inheritance of the transgene was analyzed by BASTA treatment of progeny plants, Northern- and Southern hybridizations over successive generations.

Several independ lines were analyzed. We observed a mendelian pattern of inheritance in most of the cases. Genetic analysis showed that in most of the lines the transgene inserted at one, in a few cases also at more loci in the genome. The integration pattern was in most cases very complex. Nearly 55% of the transgenic lines contained between 2 and 5 copies of the transgene. Only in a few cases transgenic plants contain a single copy or more than 5 integrated genes.

All lines which showed inactivation of the transgene contained multicopy-integrations. Loss of BASTA resistance was associated with a hypermethylation of the *Msp*I/*Hpa*II and *Sal*I site in the coding region of the *pat* gene. *Pat* specific mRNA was not detectable, indicating a transcriptional regulation of expression.

The subject of our investigations reported here was a single-copy transgenic wheat plant line which showed a non-mendelian segregation pattern and lacks homozygous plants. Detailed analyses were done with respect to several sources of segregation distortion.

Prerequisites for commercial use of transgenic crop plants are the stable expression of the transgene and its faithful inheritance. Both could be observed in many genetically engineered plants (Christou et al. 1989, Potrykus et al. 1987). Though there are several reports about instable expression of transgenes and their non-mendelian inheritance. From classical breeding and genetics several factors are known which are influencing expression and segregation. Additional factors in the production of transgenic plants are the integration locus and the number of integrated copies which both are accidential and might have some influence on expression and /or inheritance of the transgene (for example: Mittelsten-Scheid et al. 1991; Meyer et al. 1992; Matzke et al. 1993, 1994).

In most of the reported cases loss of transgene activity was caused by transgene inactivation on different expression levels. Transcription is inhibited by *cis-* or *trans-*inactivation, hypermethylation and/or alteration of chromatine structure. Cosuppression causes a posttranscriptional inactivation (reviews: Finnegan, McElroy 1994; Flavell 1994; Meyer 1995; Kumpatla et al. 1998).

Beside these phenonemons there are mechanisms which lead to a true non-mendelian inheritance of the transgene as there are gene conversion, influences of gametophytic genes or self-incompatibility genes, integration into an essential gene, maternal or paternal inheritance or an aberrant karyotype.

Results and discussion

Segregation of the *pat* gene in T2

Segregation of the *pat* gene was analyzed on the phenotypic and on molecular level in progeny of 11 self-pollinated, single-copy *pat*-positive T1-plants. 1400 plants were treated with BASTA by spraying or by covering single leaves respectively, if the plants were subjected to further analysis. Transmission rate of BASTA resistance was on average 1:1, devation amounted 0.52. None of the analyzed T1-plants was homozygous for BASTA resistance.

Gene silencing

Transgene silencing caused for example by cosuppression, *cis*- or *trans*-inactivation is one of the most often reported reasons for loss of transgene expression while the transgene remains in the genome. Both mechanisms of silencing are reversible in the following generations (Meyer, 1995; Finnegan, McElroy, 1994; Kumpatla et al., 1998). To examine the wheat plant line with regard to these phenomenons 50 BASTA resistant and BASTA sensitive T2-plants were analyzed for the presence of the *pat* gene by Southern blot analysis. In all BASTA resistant plants the *pat* gene could be detected, the BASTA sensitive plants showed no hybridization signals. As loss of transgene expression was due to loss of the transgene itself gene-inactivation could be excluded as a reason for instability of transgene expression in the examined wheat plant line.

Analysis of the karyotype

Techniques for producing transgenic plants can easily result in transgenics that possess aberrant karyotypes (Karp 1993). Integration of the transgene into a chromosome that was not in a diploid but in a triploid or tetraploid state would cause a deviation from normal segregation pattern like Matzke and coworkers (1994) observed in an aneuploid transgenic tobacco line. For detection of possible chromosomal aberrations each three metaphases of 10 T2-plants and a non-transgenic control were subjected to a karyotype analysis. Chromosome number was 42 in all cases according to the normal chromosome set in hexaploid wheat. Therefore aneuploidy was not the source of non-mendelian inheritance in this plant line.

Analysis of nuclear and ptDNA

Integration of the transgene into the plastom or the mitochondriom might result in a non-mendelian fashion of inheritance. Usually inheritance of plastidal and mitochondrial DNA in wheat is maternal (Miyamura et al 1987; Corriveau, Coleman 1988). Two methods were used to analyze the plant line with regard to these phenonemon: reciprocal backcrosses and Southern blot analysis of nuclear and ptDNA for presence of the *pat* gene.

Reciprocal backcrosses of transgenic plants with wildtype plants should give first evidence for a possible preferential maternal or paternal inheritance of the transgene. Each 25 wildtype x transgenic- and transgenic x wildtype progeny plants were analyzed by Southern blot analysis. Segregation rate of the pat gene amounted 0.6:1 (*pat*-positive:*pat*-negative) for each direction of the backcrosses indicating that transmission mode was not influenced maternally or paternally.

To examine this possibility on the molecular level ptDNA and ncDNA of transgenic plants were analyzed by Southern blot analysis. In ptDNA the *pat* gene could not be detected. NcDNA showed strong signals after hybridization with a *pat* probe indicating transgene integration into nuclear DNA.
According to these results integration of the transgene into the organelles was not the reason for the observed non-mendelian inheritance.

Anther culture

Until now no plant that was homozygous for the transgene could be observed in the examined line. This could be connected with the integration of the transgene into an essential gene which has to be available in at least one intact allele. The expected segregation ratio would be 2:1 in this case (*pat*-positive:*pat*-negative) like Peng and coworkers (1995) assumed for a transgenic rice plant line which showed a 2:1-segregation rate and was missing homozygosity.
To investigate the plant line with regard to lethal homozygosity anther culture was used as a technique to produce homozygous plants in an artificial way. Several *pat*-positive regenerants were obtained derived from different T1-plants indicating that integration of the transgene caused no lethal homozygosity. The plants were phenotypically normal. Flow cytometry revealed that some of the regenerants doubled their chromosome number spontaneously. Progeny of these fertile plants was analyzed for BASTA resistance and presence of the *pat* gene. They were uniformly *pat*-positive and BASTA resistant according to the homozygous state of the parental plants.

Flanking sequences

After exclusion of transgene inactivation, alteration of the karyotype, lethal homozygosis and transgene integration in the plastom or the mitochondriom as causes for non-mendelian inheritance and missing of homozygosity a further point to investigate were positioning effects. Transmission of the transgene would be disturbed not only if the transgene integrated into an essential gene but also if it was linked to a self-incompatibility locus or a gene playing a role in gamete contest or if it was integrated into a region that was influenced by a recombinational hot spot. Especially repetitive DNA and rDNA seem to have the tendency to keep their original state (Peterhans et al. 1990; Dvorak et al 1987; Lassner et al. 1986). Sequencing the flanking DNA should give first hints for the characteristics of the transgene´s surroundings.
Using LM-PCR a 800bp fragment of genomic DNA flanking the CaMV 35S promoter was amplified. Cloning and sequencing of this fragment are in progress.

Materials and methods

All employed materials and methods are described in detail in the diploma thesis of Kluth (Hamburg 1997).

References

Barcelo P, Hagel C, Becker D, Martin A, Lörz H (1994) Transgenic cereal (Tritordeum) plants obtained at high efficiency by microprojectile bombardment of inflorescence tissue. The Plant Journal 5(4): 583-592

Becker D, Brettschneider R, Lörz H (1994) Fertile transgenic wheat from microprojectile bombardment of scutellar tissue. The Plant Journal 5(2): 299-307

Brettschneider R, Becker D, Lörz H (1997) Efficient transformation of scutellar tissue of immature maize embryos. Theor Appl Genet 94: 737-748

Christou P, Ford TL, Kofron M (1991) Production of transgenic rice (*Oryza sativa* L.) plants from agronomically important Indica and Japonica varieties via electric discharge particle acceleration of exogenous DNA into immature zygotic embryos. Bio/Technology 9: 957-962

Christou P, Swain WF, Yang NS, McCabe DE (1989) Inheritance and expression of foreign genes in transgenic soybean plants. Proc Natl Acad Sci USA 86: 7500-7504

Corriveau JL, Coleman AW (1988) Rapid screening method to detect potential biparental inheritance of plastid DNA and results for over 200 angiosperm species. Amer J Bot 75: 1443-1458

Dvorak J, Jue D, Lassner M (1987) Homogenization of tandemly repeated nucleotide sequences by distance-dependent nucleotide sequence conversion. Genetics 116: 487-498

Finnegan J, McElroy D (1994) Transgene inactivation: Plants fight back! Bio/Technology 12: 883-888

Flavell RB (1994) Inactivation of gene expression in plants as a consequence of specific sequence duplication. Proc Natl Acad Sci 91, 3490-3496

Gless C, Lörz H, Jähne-Gärtner A (1998) Transgenic oat plants obtained at high efficiency by microprojectile bombardment of leaf base segments. J Plant Physiol 152: 151-157

Jähne A, Becker D, Brettschneider R, Lörz H (1994) Regeneration of transgenic microspore-derived, fertile barley. Theor Appl Genet 89: 525-533

Karp A (1993) Are your plants normal? - Genetic instability in regenerated and transgenic plants. Agro-Food-Industry Hi-Tech May/June: 7-12

Kluth A (1997) Analyse der nicht-mendelnden Vererbung eines Transgens in Weizen (*Triticum aestivum* L.). Diploma thesis, Universität Hamburg, Germany

Koziel GM, Beland GL, Bowman C, Carozzi NB, Crenshaw R, Crossland L, Dawson J, Desai N, Hill M, Kadwell S, Launis K, Lewis K, Maddox D, McPherson K, Meghji MR, Merlin E, Rhodes R, Warren GW, Wright M, Evola SV (1993) Field performance of elite transgenic maize plants expressing an insecticidal protein derived from *Bacillus thuringiensis*. Bio/Technol 11: 194-200

Kumpatla SP, Chandrasekharan MB, Iyer LM, Li G, Hall TC (1998) Genome intruder scanning and modulation systems and transgene silencing. Trends in plant science 3: 97-104

Lassner M, Dvorak J (1986) Preferential homogenization between adjacent and alternate subrepeats in wheat rDNA. Nuc Ac Res 14: 5499-5512

Matzke MA, Neuhuber F, Matzke AJM (1993) A variety of epistatic interactions can occur between partially homologous transgene loci brought together by sexual crossing. Mol Gen Genet 236, 379-386

Matzke AJM, Neuhuber F, Park Y-D, Ambros PF, Matzke MA (1994) Homology-dependent gene silencing in transgenic plants: epistatic silencing loci contain multiple copies of methylated transgenes. Mol Gen Genet 244, 219-229

Meyer P, Linn F, Heidmann I, Meyer H, Niedenhof I, Seadler H (1992) Endogenous and environmental factors influence 35S promoter methylation of a maize A1 gene construct in transgenic petunia and its colour phenotype. Mol Gen Genet 231, 345-352

Meyer P (1995) Variation of transgene expression in plants. Euphytica 85: 359-366

Mittelsten-Scheid O, Paszkowski J, Potrykus I (1991) Reversible inactivation of a transgene in Arabidopsis thaliana. Mol Gen Genet 228, 104-112

Miyamura S, Kuroiwa T, Nagata T (1987) Disappearence of plastid and mitochondrial Nucleoids during the formation of generative cells of higher plants revealed by fluorescence microscopy. Protoplasma 141: 149-159

Peng J, Wen F, Lister R, Hodges TK (1995) Inheritance of *gusA* and *neo* genes in transgenic rice. Plant Mol Biol 27: 91-104

Peterhans A, Schlüpmann H, Basse C, Paszkowski J (1990) Intrachromosomal recombination in plants. EMBO J 9: 3437-3445

Potrykus I, Paszawski J, Saul MW, Negrutiu I, Shillito RD (1987) Direct gene transfer to plants: facts and future. In: Green CE, Somers DA, Hackett WP, Biesboer DD (eds) Plant tissue and cell culture: 289-302. New York: Alan R Liss Inc

Sanford JC, Klein TM, Wolf ED, Allen N (1987) Delivery of substances into cells and tissues using a particle bombardment process. J Part Sci Technol 5: 27-37

Vasil V, Srivastava V, Castillo AM, Fromm ME, Vasil IK (1993) Rapid production of transgenic wheat plants by direct bombardment of cultured immature embryos. Bio/Technol 11: 1553-1558

Weeks JT, Anderson OD, Blechl AE (1993) Rapid production of multiple independent lines of fertile transgenic wheat (*Triticum aestivum*). Plant Physiol 102: 1077-1084

Zimny J, Becker D, Brettschneider R, Lörz H. (1995) Fertile, transgenic *Triticale* (x *Triticosecale* Wittmack). Molecular Breeding 1: 155-164

STUDIES ON WHEAT PROLYL ISOMERASE IN TRANSGENIC PLANTS

I. Kurek[1], R. Dulberger[1], P. Christou[2], A. Breiman[1]
[1] Department of Plant Science, The George S. Wise Faculty of Life Sciences, Tel Aviv University, Israel; [2] The John Innes Centre, Norwich, UK.

INTRODUCTION

Peptidyl prolyl isomerases catalyse the interconversion between cis and trans forms of the peptide bond preceding proline residues in proteins. Three distinct families of prolyl isomerases have been identified: the cyclophilins, the FKBPs and the parvulins (Schmid 1997; Dolinski, Heitman 1997). Cyclophilins and FKBPs have attracted attention as the cellular receptors for the immunosuppresants cyclosporin A, FK506 and rapamycin, but this role in drug action is distinct from the enzymatic activity. Members of the three prolyl isomerase families are highly conserved, abundant and expressed in multiple cellular compartments, suggesting that these enzymes play a critical role in cell physiology. Because prolyl isomerases can accelerate slow protein refolding steps *in vitro*, it has been speculated that this is what they do *in vivo*. A number of proteins physically associated with different prolyl isomerases such as kinases and transcription factors have been identified but the physiological significance of these associations remains to be established. The prolyl isomerases have been implicated in various cellular functions (independent of their prolyl isomerase activity) such as DNA degradation activity and RNA maturation.

Recently, plant prolyl isomerases have been shown to be involved in basic cellular processes: for example the *Arabidopsis* mutant *PAS1* mutated in the FKBP70 was characterised by its aberrant embryo and root morphology (Vittorioso et al. 1998) and the Cyp40 was shown to interact with a thylakoid protein phosphatase (Fulgosi et al. 1998).

We have isolated and characterised two wheat prolyl isomerases belonging to the FKBP family (Blecher et al. 1996; Kurek et al. 1998). The wheat FKBP73 is expressed in young and dividing tissues and not affected by stresses, whereas the wheat FKBP77 is induced by heat stress. The two proteins exhibit 85% sequence identity, possess a similar structure with three PPIase FKBP12- like domains, a calmodulin domain and a tetratricopeptide (TPR) domain .

In order to further study the function of these proteins, the wheat FKBP73 and FKBP77 were expressed in *E.coli* and the enzymatic activity was analysed. It was found that only the first PPIase domain exhibits the enzymatic activity, and that the TPR domain is involved in interaction with the heat shock protein HSP90 (Reddy et al. 1998).

To obtain further insight into the biological role of the FKBP domains we produced transgenic rice plants which express the individual domains of the FKBP73.

MATERIALS and METHODS

Ten to fifteen-day-old immature seeds from *Oryza sativa* cultivar *Indica* (M12) were bombarded by the direct DNA transfer system based on electric discharged particle acceleration (ACCEL^{TM}) technology as previously described (Christou et al, 1991). The plasmid pWRG4515 (pUC19 backbone, CaMV35S pro :: *hmr* :: NpA + rice Gt3 pro :: *gus* :: NpA); (Christou, Ford 1995) was cotransformed with the pAL76

A. Altman et al. (eds.), Plant Biotechnology and In Vitro Biology in the 21st Century, 165–168.

vector (a modified pFC7: *ubiquitin* pro + intron :: polylinker :: NpA); (Lonsdale et al. 1995) containing: (a) The full wFKBP73; (b) wFKBP77; (c) wFKBP73 PPIase 1,2,3 domains; (d) the wFKBP73 engineered with the FLAG epitope at the N-terminus and (e) wFKBP73 in which the first domain was exchanged with the 2nd domain. The DNA preparation, the isolation of immature embryos, particle bombardment and plant regeneration were performed as previously reported (Christou et al. 1991). Transformants were selected on hygromycin at 50 mg/l. The first leaf of each transgenic plant was analysed by western blot decorated with the polyclonal antibodies raised against the recombinant wFKBP73 (anti FKBP73) (Blecher et al. 1996), or with the polyclonal antibodies, raised against the synthetic peptide from the C- terminus of the wFKBP77 (anti HS24C) (Kurek et al. 1998).

The plants were grown in controlled greenhouse conditions with 65% humidity at 28°C day and 25°C night with 12 h light regime. Six plants bearing panicles were transferred to a growth chamber with the same conditions except for 25°C day and night. To examine flowers, panicles sampled from 12 plants (six from the greenhouse and six from the growth chamber) were brought into the laboratory within jars with water. To excise the flower, the tips of the lemma and palea were cut off and the two bracts were then forcibly separated from each other. For examination with the scanning electron microscope (SEM), floral parts were fixed in FAA (formalin-acetic acid-alcohol) or in 3% glutaraldehyde in 0.3 M phosphate buffer and rinsed in buffer. The material was then dehydrated through an ethanol series. This material as well as fresh pollen were coated with gold and viewed in a JEOL 35 SEM.

Viability of pollen of nondehiscent anthers was examined with 1% aniline blue in lactophenol or using the Alexander method (1969). Viability of released pollen grains was estimated with 1% aniline blue or with the tetrazolium test (Hauser, Morrison 1964; Stanley, Linskens 1974). Stigmas were examined for presence of pollen grains by staining with a mixture of light green and acid fuchsin.

RESULTS and DISCUSSION

Rice plants expressing the wFKBP73, the three PPIase (1,2,3) domains without the TPR and calmodulin motifs, the wFKBP73 in which the 1st PPIase domain was exchanged with the second (2,2,3) and plants expressing the wFKBP77 were obtained (Table1). Whereas no sterile plants were detected among the plants expressing the FKBP73, 15% of the plants expressing the domains 2,2,3 were sterile and all plants expressing the PPIase domains 1,2,3 without the TPR and calmodulin domains were sterile (Table1). In order to determine the anatomy of sterility, flowers were examined with the light microscope and with the SEM.

The flowers were found to be tightly enclosed by the coriaceous paleae. Extrusion of anthers did not occur but in a few spikelets (see below). Excised mature anthers were shriveled (Fig 1), variously asymmetrical, and non dehiscent. In the pollen samples viewed in the SEM, only aborted grains were present (Fig 2). The percentage of non-stainable pollen ranged from 76% to 100% with a mean of 93%. An ovary and two feathery stigmas were present in all flowers. However, no pollen grains were found on stigmas of mature flowers. Of the 12 plants observed, only one plant displayed two or three anthers extruded from each of three spikelets. In these anthers the percentage of non-stainable pollen ranged from 25% to 52% with a mean of 41.5%. Germinated as well as ungerminated pollen grains were present on two of the seven stigmas examined. For comparison, normal anthers and pollen from wild rice plants were examined (Figs. 3, 4). In these plants 96% of the pollen grains were stainable.

In the transgenic rice plants expressing FKBP73 PPIase 1,2,3 domains, male sterility is expressed at various sites, stages of development and functions of the male reproductive system: 1. arrest of normal differentiation and functioning of anthers as obvious from their abnormal appearence; 2. inhibition of normal development of pollen grains, as evident from the high percentage of aborted (non-stainable) pollen in the vast majority of flowers; 3. abscence of a functional dehiscence mechanism (prevents deposition of pollen on the stigmas) resulting in (explains) the absence of pollen grains on the stigmas, and 4. defective filament elongation preventing extrusion of anthers.

The results presented in this paper indicate that overexpression of the truncated FKBP73 missing the TPR and calmodulin motifs may be associated with anther malformation. It is possible to conceive an imbalance in the interaction of FKBP with other proteins such as HSP90 (since the TPR responsible for this interaction is missing) which may be the trigger for basic cellular defects. Further studies may reveal the molecular link between the overexpression of the truncated FKBP expressing the three FKBP12-like domains and the male sterility.

Table 1. OVER EXPRESSION OF WHEAT FKBPs IN T_0 TRANSGENIC RICE

Gene	Total	Fertile[a]	Sterile[b]	% Sterility
FKBP73	4	4	0	0
FKBP73 FLAG	6	5	1	16
FKBP73 PPIase 1,2,3 dom	**20**	**0**	**20**	**100**
FKBP73 2,2,3	14	11	3	21
FKBP77	7	6	1	15
Total	**51**	**26**	**25**	

a. The number of seeds collected from the fertile plants ranged from 120 to 200 seeds per plant with a mean of 156 seeds per plant

b. The number of seeds collected from the sterile plants ranged from 6 to 0 seeds with a mean of 0.66 seeds per plant.

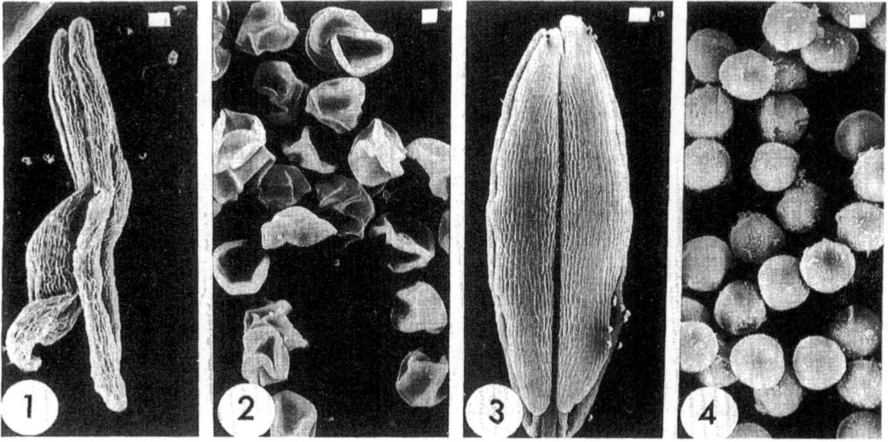

Figures 1-4
Scanning electron micrographs of anthers and pollen grains of rice plants. **1,2** Transgenic rice. **3,4** Wild rice. **1.** Abnormal anther excised from closed mature spikelet of a transgenic rice plant. **2.** Aborted pollen grains. **3.** Normal undehisced anther of wild rice excised from a spikelet prior to anthesis. **4.** Pollen grains from undehisced anther. Figs. **1, 3** Bars = 100 μm. Figs. **2, 4** Bars = 10 μm.

REFERENCES

Alexander MP (1969) Stain Technol. 44, 117-122

Blecher O et al (1996) Plant Mol. Biol. 32, 493-504

Christou P et al (1991) Bio/Tech. 9, 957-962

Christou P, Ford TL (1995) Transgenic Research 4, 44-51

Dolinski K, Heitman E (1997) Guidebook to Molecular Chaperones and Protein Folding Catalysis, pp 359-369, Oxford: Oxford University Press

Fulgosi H et al (1998) EMBO J. 17, 1577-1587

Hauser EJP, Morrison JH (1964) Am. J. Bot. 51, 748-753

Kurek, I et al (1998) Plant Physiol. (in press)

Lonsdale DM et al (1995) Plant Mol. Biol. Rep. 13, 343-345

Rassow J et al (1995) Trends in Cell Biol. 5, 207-212

Reddy RE et al (1998) Plant Physiol. 118, 1395-1401

Schmid FX (1997) Molecular Chaperones in the Life Cycle of Proteins, pp 361-389 New York, Basel, Hong Kong: Marcel Dekker Inc.;.

Stanly RG, Linskens HF (1974) Pollen: biology, biochemistry and management. Springer, Berlin, Heidelberg, New York.

Vittorioso P et al (1998) Mol. Cell Biol. 18, 3034-3043

GA-induced gene expression in petunia flowers

Y. Leitner-Dagan, A. Izhaki, G. Ben-Nissan, A. Borochov and D. Weiss.
The Kennedy Leigh Centre for Horticulture, The Hebrew University of Jerusalem, Rehovot 76100, Israel.

Introduction

The plant hormone gibberellin (GA) plays an essential role in a wide range of developmental processes including petal growth and pigmentation (Huttly, Phillips 1995). We have previously shown that GA is essential for *Petunia hybrida* flower development (Weiss, Halevy 1989). When detached petunia corollas were grown *in vitro* in a sucrose medium, they elongated and became pigmented only in the presence of GA_3 (Weiss et al. 1992). Analysis of endogenous GAs indicated the presence of biologically active GAs, including GA_1 and GA_4, in the anthers and corollas at the stages of rapid elongation and anthocyanin accumulation (Weiss et al. 1995). The effect of the hormone on corolla pigmentation was due to the activation of the anthocyanin biosynthetic genes such as chalcone synthase (*chs*) (Weiss et al. 1992). However, the effect of the hormone is not exclusive to these genes; it also promotes the expression of genes from general metabolic pathways, such as triosephosphate isomerase (*tpi*) (Ben-Nissan, Weiss 1995). Recently, we have cloned (Ben-Nissan, Weiss 1996) the petunia homologue (*gip*) of the tomato *gast1* (Shi et al. 1992). *Gip* is expressed in corollas during cell elongation and is induced specifically by GA_3. Expression analysis of the gene provided evidence supporting its possible role in GA-induced corolla elongation (Ben-Nissan, Weiss 1996).

Although many GA-induced genes have been cloned and characterized, very little is known about the early steps in GA-signal transduction. Genetic analysis of GA signal transduction revealed several mutants, including those exhibiting decreased and increased signals (Swain, Olszewski 1996). Mutation in *Arabidopsis* at the *spindly* (*spy)* locus creates a phenotype resembling the wild type treated with exogenous GA and therefore suggests that the *spy* product acts as a repressor of GA signal transduction in *Arabidopsis*. The *spy* gene was cloned and sequence analysis suggested that the SPY protein is an *O*-GlcNAc transferase (Jacobsen et al. 1996).

In the present work we used pharmacological and molecular approaches to study the transduction of the GA signal in petunia corollas. We analyzed the effect of GA_3 and inhibitors of signal transduction on the expression of three different genes: *gip* and *chs* that are induced by GA_3, and *tpi* that is slightly promoted by the hormone. Furtheremore, we

A. Altman et al. (eds.), Plant Biotechnology and In Vitro Biology in the 21st Century, 169–172.

cloned the petunia homologue of the *Arabidopsis spy* gene and began to analyze its expression in the flower.

Results and Discussion

Ca^{2+}, calmodulin and protein dephosphorylation are required for GA-induced gene expression

To study the possible role of calcium in GA signal transduction, we used several Ca^{2+} antagonists, including the calcium chelator BAPTA and the calcium-channel blocker LaCl. Neither inhibitor affected the induction of the gene by GA_3. On the other hand, the intracellular calcium-channel blocker, ruthenium red (RR), completely blocked the effect of the hormone (Fig.1). RR inhibited the slight promotion of *tpi* expression induced by GA_3, but not its basal expression. These results suggest that the effect of RR is specific to the GA signal, and that Ca^{2+} from intracellular sources such as endoplasmatic reticulum, but not from the apoplast, is involved in the transduction pathway of the signal. The calcium ionophor A23187 is widely used to increase Ca^{2+} level in the cytoplasm (Christie, Jenkins 1996). Although we found that blocking intracellular calcium channels inhibited GA_3-induced gene expression, we did not succeed in mimicking the effect of the hormone by treatment with the calcium ionophor. Similar results were obtained by Gilroy (1996) in barley aleurone cells, where inhibition of cytoplasmatic Ca^{2+} level prevented GA stimulation of α-amylase secretion, but mimicking GA-induced changes of calcium concentrations did not mimic GA action. Since Ca^{2+} is involved in the transduction of many different signals, it is reasonable to assume that it acts in conjunction with some other signaling components to induce the specific response (Bowler et al. 1994). It is also possible that GA enhances Ca^{2+} in a particular subcellular microdomain (Knight et al. 1996) and the calcium ionophor cannot initiate this effect.

Ca^{2+} regulates the activities of many proteins via the activation of the calcium-binding protein, calmodulin. In barley aleurone protoplasts, calmodulin is involved in GA-induced α-amylase protein secretion, but not in GA-induced α-amylase gene transcription (Gilroy 1996). We used the well-characterized calmodulin antagonist, W-7, and showed that in petunia corollas, it inhibits GA_3-induced gene expression (Fig.1). We also tested the effect of W-5, a close structural analog of W-7 that is less effective as a calmodulin inhibitor. W-5 had a smaller effect than W-7, and higher concentrations of W-5 were required to inhibit gene expression.

Protein phosphorylation and/or dephosphorylation play an important role in stimuli-response coupling. Kuo et al. (1996) found that okadaic acid (OA), a potent inhibitor of protein phosphatase 1 and 2A, blocks GA-induced α-amylase gene expression in wheat aleurone cells. Our study showed similar results: OA completely blocked the induction of *gip* and *chs* genes by GA_3 (Fig.1). Since inhibition of protein phosphatases inhibited GA-signal transduction, it is reasonable to expect that inhibition of protein kinases would replace GA and induce, on its own, the expression of the genes by itself. We used two protein-kinase inhibitors, staurosporine and H-7, which affect a broad range of serine/threonine and tyrosine kinases (Tamaoki 1991), but neither induced *gip* and *chs* expression in the absence of the hormone. Our results indicate a possible involvement of protein kinases in GA signal transduction in petunia corollas. Although staurosporin did not inhibit the induction of *chs* and *gip* by GA_3, H-7 partially inhibited the effect of the hormone. Thus, it is possible that both protein phosphorylation and dephosphorylation are involved in GA-signal transduction in petunia corollas.

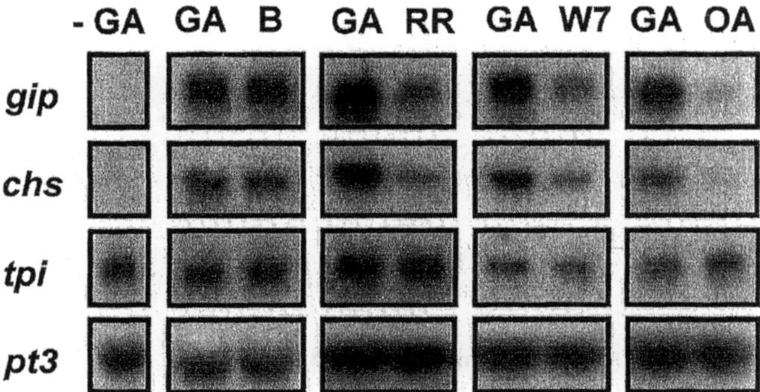

Figure 1. Detached corollas were grown in sucrose solution for 24 h and then transferred to fresh sucrose solution (-GA) or to sucrose solution with 10 μM GA_3 (GA) with or without 2.5 mM BAPTA (B), 5 μM RR, 50 μM W7 or 3 μM okadaic acid (OA). After 6 h RNA was extracted and analyzed for *gip*, *chs* and *tpi* expression. All the blots were re-hybridize with the ribosomal *pt3* probe.

The petunia homologue of the Arabidopsis spy gene: isolation and characterization

Recently a GA-signal transduction repressor (*spy*) has been cloned from *Arabidopsis*. (Jacobsen et al. 1996). To clone the petunia homologue of the *spy* gene, we screened an anther-specific cDNA library with the *Arabidopsis spy* probe. A screening of 50,000 plaques resulted in one positive clone. After several rounds of purification, a single plaque was selected and the insert, later called *spyp*, was sequenced. The *spyp* cDNA is 3513 bp long, bearing an open reading frame coding for 932 amino acids. The deduced protein sequence exhibited 83% identity with the *Arabidopsis* SPY (Fig. 2).

Spyp expression was analyzed in the corolla at different developmental stages. Ploy(A)+ RNA analysis revealed the accumulation of a 3.6-kb transcript in the corolla. Relatively high levels of expression were found at the early stages of corolla development (stages 1 and 3), before rapid elongation and the accumulation of anthocyanin took place. At a later stage (stage 5), during corolla expansion, *spyp* expression declined. *Gip* mRNA, on the other hand, began to accumulate at stage 3, reaching a maximum level at stage 5. Since *gip* expression is a good indication of the presence of GA in the tissue (Ben-Nissan, Weiss 1996), these results suggest that *spyp* expression decreases when GA level increases. However, application of GA_3 to detached corollas had no effect on *spyp* transcript level.

```
A.SPY   MAWTERDVENGKESDSLGNNGFLKGVQSSSDSKGSPVRISPVKKSFEGKDAIT 53
        |  .: : :.::|...:.:||| .| .|||.| |    :|| :|  :|.|.:.
SPYP    M..VGLEDDTERERSPVVENGFSNGSRSSSSSAGV...LSPSRKVTQGNDTLS 48

A.SPY   YANILRSRNKFVDALAIYESVLQKDSGSIESLIGKGICLQMQNMGRLAIESFA 103
        ||||||.||||.||||:||.:|:||| .:|. |||||||| ||.|.||:::|.
SPYP    YANILRARNKFADALALYEAMLEKDSKNVEAHIGKGICLQTQNKGNLAFDCFS 98

A.SPY   EAIKLDPQNACALTHCGILYKDQGRLVEAAESYQKHL 143
        |||:|||:|||||||||||.|::|||||||||||| |
SPYP    EAIRLDPHNACALTHCGILHKEEGRLVEAAESYQKAL 138
```

Figure 2. Alignment of the petunia SPY (SPYP) N-terminal amino-acid sequence with this of SPY from *Arabidopsis* (A. SPY).

References

Ben-Nissan G, Weiss D (1995) J. Plant Physiol. 147, 58-62.

Ben-Nissan G, Weiss D (1996) Plant Mol. Biol. 32, 1067-1074.

Bowler C et al. (1994) Cell 77, 73-81.

Christie JM, Jenkins GI (1996) Plant Cell 8, 1555-1567.

Gilroy S (1996) Plant Cell 8, 2193-2209.

Huttly AK, Phillips AL (1995) Physiol. Plant. 95, 310-317.

Jacobsen SE et al. (1996) Proc. Natl. Acad. Sci. USA 93, 9292-9296.

Knight H et al. (1996) Plant Cell 8, 489-503.

Kuo A et al. (1996) Plant Cell 8, 259-269.

Shi L et al. (1992) Plant J. 2, 153-159.

Swain SM, Olszewski NE (1996) Plant Physiol. 112, 11-17.

Tamaoki T (1991) Methods in Enzymology, pp. 340-347, Academic Press, San Diego, CA.

Weiss D, Halevy AH (1989) Planta 179, 89-96.

Weiss D et al. (1992) Plant Physiol. 98, 191-197.

Weiss D et al. (1995) Plant Physiol. 107, 695-702

ENGINEERING THE PLASTID GENOME: PROBLEMS AND POTENTIAL

Pal Maliga
Waksman Institute, Rutgers, The State University of New Jersey, 190 Frelinghuysen Road, Piscataway, NJ 08854-8020, USA

1. Introduction

The plastid genome of higher plants is a 120-kb to 160-kb double-stranded DNA present in 1,900 to 50,000 copies per leaf cell. To obtain genetically stable transplastomic lines every one of the plastid genome copies (ptDNA) should be uniformly altered in a plant. Transformation is accomplished through the following steps: (i) introduction of the transforming DNA, encoding antibiotic resistance, by the biolistic process or PEG treatment; (ii) integration of the transforming DNA by two homologous recombination events and (iii) elimination of wild-type genome copies during repeated cell divisions on a selective medium. As integration of foreign DNA always occurs by homologous recombination, plastid transformation vectors contain segments of the plastid genome to target insertions to specific locations. The plastid vectors also contain a marker for selection. Useful, non-selectable genes are cloned next to selectable marker genes, with which they are introduced into the plastid genome. Homology-directed genome manipulations have included introduction of point-mutations and deletion of targeted genes.

Plastid transformation in a higher plant was first reported in tobacco (*Nicotiana tabacum*) by Svab et al. in 1990. Although the technology is eight-years old, as of the meeting date (June, 1998), it has been extended only to the model species *Arabidopsis thaliana* (Sikdar et al. 1998). Progress to date, and the reasons for the delay in applying this technology is the topic of this review.

2. The bottlenecks of the technology

DNA Delivery. Efficient DNA delivery to plastids is a rate-limiting step of plastid transformation. Since the particles are not aimed and they penetrate only a few cell layers, most cells in a bombarded sample are never hit by a DNA-carrying particle. Bombardment of tobacco leaves with the helium-driven Biorad Biolistic Gun (and selection for *aadA*) on average reliably yields one transplastomic line per sample (Svab and Maliga 1993). However, in exceptional cases more than ten transplastomic lines may be obtained (Zora Svab, unpublished) indicating, that through a better control of biomaterials and/or DNA delivery this efficiency could be readily increased by a factor of

A. Altman et al. (eds.), Plant Biotechnology and In Vitro Biology in the 21st Century, 173–176.

ten. PEG-mediated DNA uptake is potentially more efficient than the biolistic protocol since during PEG treatment every protoplast is exposed to DNA. PEG treatment is reported to yield up to 2-4 transplastomic lines per 10^5 treated protoplasts (Golds et al. 1993; O'Neil et al. 1993; Koop et al. 1996). Higher efficiency obtained using PEG treatment may justify the significant labor input required to handle protoplasts.

Marker genes and tissue culture systems. Plastid transformation is routinely and efficiently practiced only in tobacco (*N. tabacum*). Even in tobacco, our protocol (Svab and Maliga 1993) may be practiced only with c.v. Petit Havana, or other cultivars with a similar, prolific shoot regeneration response to cytokinin treatment. Furthermore, the transformation efficiency depends on the selective drug, and the particular gene conferring resistance. Spectinomycin is a better choice for selection than kanamycin, as selection for spectinomycin (*aadA*) resistance yields on average one transplastomic line per bombarded leaf (Svab and Maliga, 1993), whereas usually only a single transplastomic kanamycin resistance line (*neo* or *kan* gene) is recovered from 25 bombarded leaf samples (Carrer et al. 1993; Carrer and Maliga, 1995). However, selection for spectinomycin resistance by mutant 16S rRNA genes is a lot less (50 to 100 x) efficient than by *aadA* (Svab and Maliga, 1993). Efficient selection demands selective propagation of the rare transformed plastid without killing the cell. The efficiency of spectinomycin selection could be due to protection of the entire cell by a few *aadA*-expressing transformed genome copies after DNA delivery. The marker genes which have this property should be either very highly expressed or, alternatively, a small amount of the gene's product should meet the demand. Since drug sensitivity is specific to taxonomic groups, e.g. cereals are naturally resistant to spectinomcyin, testing of additional marker systems is essential for extension of the technology to other crop plants.

Gene expression technology. Tissue- and development-specific expression of nuclear transgenes can be obtained with the appropriate promoters. For the transplastomic technology to be competitive, it should also offer the benefit of tissue-specific transgene expression. The natural source of expression signals are those of native plastid genes. Much more is known about gene expression in chloroplasts, the plastid type differentiated for photosynthesis, than about gene expression in non-green plastids.

Plastid gene expression is regulated at both transcriptional and post-transcriptional (mRNA stability, translation) levels. Transcription of plastid genes is by two distinct RNA polymerases: the eubacterial-type, multi-subunit, plastid-encoded plastid RNA polymerase (PEP) and the phage-type, nucleus-encoded plastid RNA polymerase (NEP) (Maliga 1998). The σ^{70}-type PEP promoters, such as those driving the expression of the photosynthetic *psbA* and *rbcL* genes are suitable for the expression of transgenes in leaves (Staub and Maliga 1993; 1994; Shiina et al. 1998). In contrast, the *rrn* operon PEP promoter is suitable for transgene expression in all tissue types, including embryogenic tissue culture cells (Silhavy et al. 1998b). Transcription by the PEP in the dark is reduced four- to ten-fold. High levels of mRNA can be maintained in the dark through stabilization of the mRNA by incorporating the *rbcL* 5' untranslated region in the chimeric constructs (Shiina et al. 1998).

Promoters for the NEP fall in two classes: Type I NEP promoters are contained in ~15 bp upstream (-14 to +1) of the transcription initiation site (+1), whereas Type II NEP promoters are located downstream (-5 to +25), with no sequence conservation between the two (Sriraman et al. 1998). Most NEP promoters are inactive in chloroplasts, and could be identified only in non-photosynthetic plastids (Hajdukiewicz et al. 1997; Kapoor et al. 1997; Hübschmann and Börner 1998; Silhavy and Maliga 1998a). This is good news: these NEP promoters should then function in some non-photosynthetic plastid type. Indeed, one of these, the Type I *rpoB* promoter, is highly active early on during chloroplast development (Silhavy and Maliga 1998a). However, the Type I *clpP* promoter (PclpP-111) is constitutive in rice and maize, actively transcribing the *clpP* protease subunit gene in leaves and in embryogenic cells (Silhavy and Maliga 1998a,b). The tobacco Type II *clpP* promoter (PclpP-53) is also constitutive (Sriraman et al. 1998). It appears therefore, that NEP promoters are highly regulated, and may become a valuable source of assorted tissue-specific plastid promoters.

An alternative to native plastid promoters are artificial expression systems. An artificial system based on the expression of plastid transgenes from a T7 promoter by a nucleus-encoded, plastid-targeted T7 RNA polymerase has been reported (McBride et al. 1991). It is not clear yet, whether or not T7 RNA polymerase expression interferes with plastid function.

3. Advantages of plastid transgenes

Extending plastid transformation to other crop plants is not trivial, as discussed above. However, the significant advantages associated with incorporating transgenes in the plastid genome warrant further efforts.

High level protein expression. Rubisco, or ribulose-1,5-bisphosphate carboxylase, the most abundant protein on earth, makes up to 65% of the total soluble protein in leaf extracts (Ellis 1979). The large subunit of Rubisco is the product of the plastid *rbcL* gene, a testament of the plastid's capacity for protein synthesis, and probably an indicator of the highest protein levels which may be obtained from plastid transgenes. Initial success in this area is expression in tobacco chloroplasts of neomycin phosphotransferase (1%; Carrer et al. 1993), β-glucuronidase (2.5 %; Staub and Maliga 1993) and the *Bacillus thuringiensis* insecticidal protoxin (3-5%; McBride et al. 1995).

Expression of pathways from polycistronic mRNAs. The plastid translation machinery is prokaryotic in nature, and will efficiently translate polycistronic mRNAs (Staub and Maliga, 1995). Expression of multiple genes from a polycistronic message would simplify engineering of pathways, since several genes functioning in different steps of a given pathway may be expressed as one transcription unit. This would be more practical than individually fine-tuning the expression of multiple nuclear genes, and combining them by crossing. Given the natural variability in its size we expect, that the plastid genome may be readily enlarged by up to an additional 50 kb of DNA, possibly encoding 25 to 35 new genes. The largest increase obtained thus far was about 10 kb (2 x 5 kb) in the repeated region (McBride et al., 1995).

Containment through the lack of pollen transmission. Containment is secured by the lack of pollen transmission of plastids in most crop plants, including maize, rice, wheat, cotton and oilseed rape. If integrated into the plastid genome, herbicide resistance genes will not be transmitted by pollen to related weeds outside the field. Transplastomic plants and their seed progeny, however, will be uniformly resistant to the herbicide due to uniform maternal inheritance of plastids. An example is expression of a glyphosate resistance gene in tobacco plastids (Daniell et al. 1998)

Lack of transgene position effect and gene silencing are additional advantages of incorporating transgenes in plastids. These will be discussed elsewhere.

4. Acknowledgements. Original research cited here was supported by grants from the National Science Foundation, The Rockefeller Foundation and Monsanto Co. to PM.

5. References

Carrer H et al (1993) Mol Gen Genet 241, 49-56.
Carrer H, Maliga P (1995) Biotechnology 13, 791-794.
Daniell H et al. (1998) Nature Biotechnology 16, 345-348.
Ellis RJ (1979) TIBS 4, 241-244.
Golds et al (1993) Biotechnology 11, 95-97.
Hajdukiewicz PTJ et al (1997) EMBO J 16, 4041-4048.
Hübschmann T, Börner T (1998) Plant Mol Biol 36, 493-496.
Kapoor S et al (1997) Plant J 11, 327-337.
Koop et al (1996) Planta 199, 193-201.
Maliga P (1998) Trends Plant Sci 3, 4-6.
McBride K et al (1991) Proc Natl Acad Sci USA 91, 7301-7305.
McBride K et al (1995) Biotechnology 13, 362-365
O'Neil et al (1993) Plant J 3, 729-738.
Shiina T et al (1998) Plant Cell in press.
Sikdar SR et al (1998) Plant Cell Reports 17, in press.
Silhavy D, Maliga P (1998a) Curr Genet 33, 340-344.
Silhavy D, Maliga P (1998b) Curr Genet 34, 67-70.
Sriraman P et al (1998) Nucleic Acids Res in press.
Staub J, Maliga P (1993) EMBO J 12, 601-606.
Staub J, Maliga P (1994) Plant J 6, 547-553.
Staub J, Maliga P (1995) Plant J 7, 845-848.
Svab Z et al (1990) Proc Natl Acad Sci USA 87, 8526-8530.
Svab Z, Maliga P (1993) Proc Natl Acad Sci USA 90, 913-917.

ELECTROFUSION OF FIELD PEA AND *LATHRYUS* PROTOPLASTS FOR HYBRID DEVELOPMENT

J.S. McCutchan[1,2], P.J. Larkin[2], P.A. Stoutjesdijk[2], E.C.K. Pang[1] and P.W.J. Taylor[1]

[1]Department of Crop Production, Institute for Land and Food Resources, University of Melbourne, Parkville, Victoria 3052, AUSTRALIA and [2]CSIRO Plant Industry, GPO Box 1600, Canberra, ACT 2601, AUSTRALIA

Introduction

Ascochyta blight is currently the most severe fungal disease of field pea (*Pisum sativum* L.) in Australia. *Lathryus sativus* (grasspea) has been found to contain the highly sought-after resistance genes for aschochyta blight, though attempts have failed at introducing useful traits from this species into field pea through conventional crossing (Pang, Brown, 1993). Protoplast isolation from field pea has been described for commercial cultivars (Lehminger-Mertens, Jacobsen, 1989; Puonti-Kaerlas, Eriksson, 1988) but never for a wild genotype. A protocol for obtaining protoplasts from *L. sativus* has not been published previously. This paper describes protocols for establishing a culture system for field pea protoplasts from a wild genotype, the establishment of shoot, callus and suspension cultures of *L. sativus*, and subsequent protoplast isolation. Protocols for electrofusion are also described.

Pea protoplast culture

Seeds of an Ethiopian field pea accession were surface sterilised and the zygotic embryos implanted into 'Sorbarods' (Sigma) with hormone-free liquid MS (pH 5.8) (Murashige, Skoog, 1962) in sterile culture vessels. Seedlings were germinated at 24 °C in 16h light/8h dark conditions ('standard conditions'). Tissue was finely cut, then digested at 25 °C overnight in darkness in enzyme solution (630 mOsm kg^{-1}, pH 5.8) including 1.5% Cellulase Onozuka R-10 and 0.5% Macerozyme R10. The protoplast/enzyme mixture was filtered through a series of stainless steel sieves (328, 110 and 50 µm) and washed three times in a salt/glucose washing osmoticum (660 mOsm kg^{-1}). Protoplasts were centrifuged at 110g for 10 minutes between washes.

Isolated protoplasts were cultured at a density of 10^5 protoplasts ml^{-1} in calcium alginate beads (Larkin et al, 1988) surrounded by liquid KM8p containing 1.0 mg l^{-1} 2,4-D and 0.2 mg l^{-1} zeatin (pH 5.6). Cultures were maintained on an orbital shaker

A. Altman et al. (eds.), Plant Biotechnology and In Vitro Biology in the 21st Century, 177–180.

under standard conditions. Liquid KM8p medium surrounding the beads was replaced with fresh medium at seven days. Over subsequent weeks this was replaced progressively with KM1, KM2 and KM3 (Larkin et al, 1988). After approximately four to five weeks, colonies were released from the calcium-alginate matrix using a 50 mM sodium-citrate/5 mM Tris solution (pH 5.8) (Larkin et al, 1988). In an attempt to produce somatic embryos or shoots, calli were plated onto somatic embryo induction media (Loiseau et al, 1995; Kysely, Jacobsen, 1990; Lehminger-Mertens, Jacobsen, 1989) or shoot induction media (Bohmer et al, 1995).

Protoplast yield was consistently high (2×10^7 protoplasts g^{-1} tissue), protoplasts were uniform in size and showed cytoplasmic streaming (Fig 1a). Division was first observed 6-10 days after isolation (Fig 1c), and reached a maximum frequency of 90% (McCutchan et al, 1997). Somatic embryos developed after nine months on medium containing 1 mg l^{-1} 2.4-D (Lehminger-Mertens, Jacobsen, 1989b) (Fig 2a) and shoot primordia developed after seven months on 10 μM TDZ (Fig 2b). However, these regeneration responses were rare in this genotype.

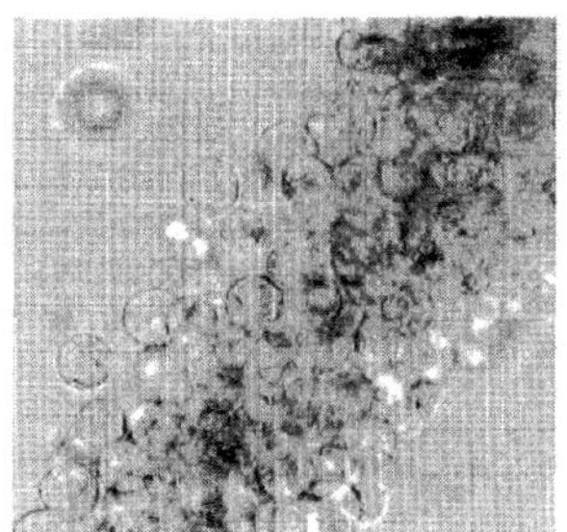

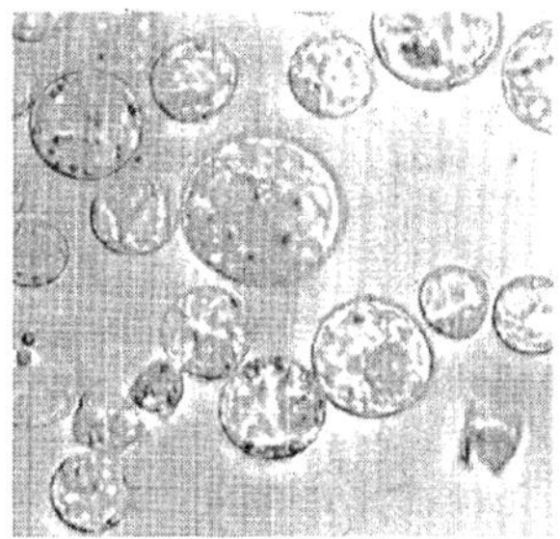

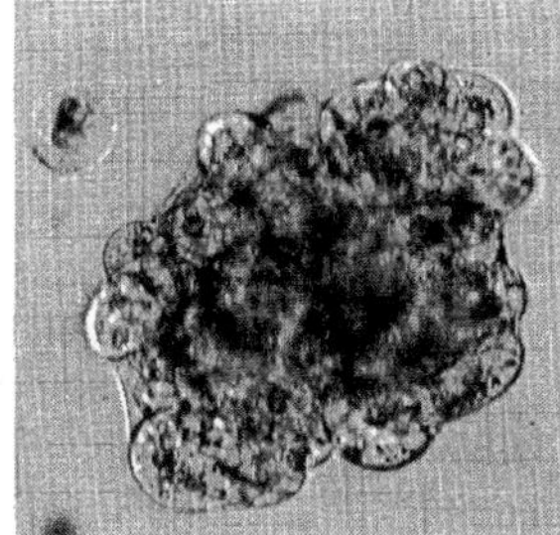

Figure 1 (a) Pea protoplasts released during digestion (b) Purified *Lathyrus sativus* protoplasts (c) Pea cell division at day 20

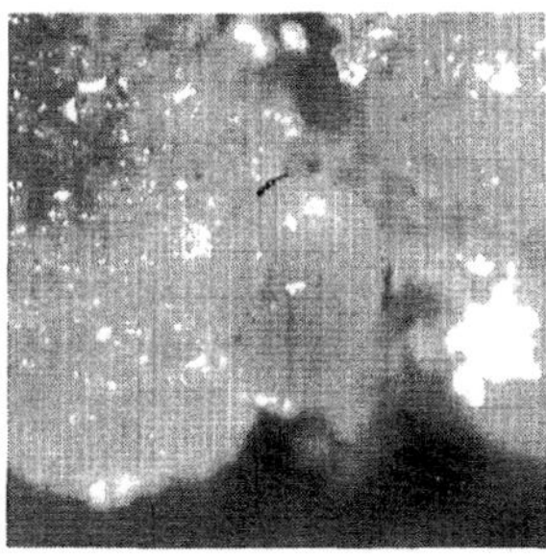

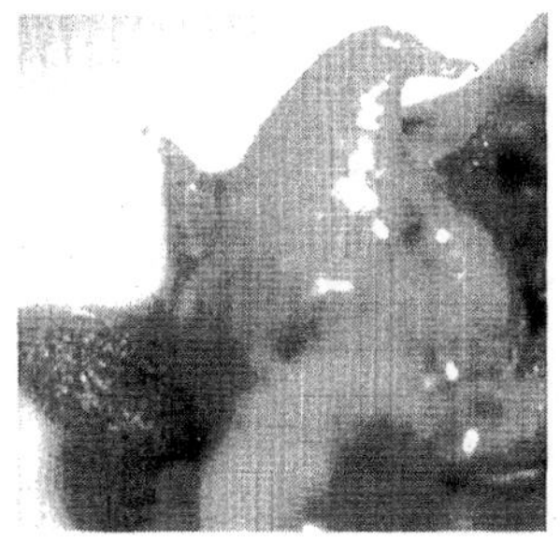

Figure 2 (a) Somatic embryo and (b) shoot primordia of *P.sativum*

Lathyrus protoplasts

Seeds of *Lathryus sativus* were surface sterilised before transfer to solid MS medium containing 20 g l^{-1} sucrose (pH 5.9) for germination under standard conditions. Shoot cultures were initiated by excising apical shoots and transferring to RL medium supplemented with 210 ug l^{-1} IAA (Phillips, Collins, 1984). Sections were excised and transferred to B5 medium supplemented with 200 mg l^{-1} casein hydrolysate and a range of concentrations of 2,4-D and kinetin. Cultures were maintained at 24 °C and 27 °C, and callus was scored qualitatively, and subcultured every three to four weeks.

Friable callus was transferred to liquid B5 medium supplemented with 1 mg l^{-1} 2,4-D and 0.1 mg l^{-1} kinetin to initiate suspension cultures. Cultures were maintained in conical flasks on an orbital shaker at 110 rpm. Shoot tips and young leaves were excised from shoot cultures, and sliced finely in enzyme solution. Alternatively, suspension culture was centrifuged for 10 min at 100 g to give a packed cell volume, which was resuspended in enzyme solution. The enzyme solution contained 1.0% (w/v) cellulysin, 0.5% driselase, 0.5% pectinase and 0.1% pectolyase (600 mOsm kg^{-1}, pH 5.8). Tissue was digested overnight at 25 °C in darkness under stationary conditions (shoot culture explant) or on a 25 rpm orbital shaker (suspension culture explant). The protoplast-enzyme mixture was purified as for pea, with the inclusion of an intermediate wash step in which protoplasts were purified by centrifuging through a Percoll-glucose stepwise density gradient (Li, 1993).

Lathryus sativus shoot cultures grew rapidly, with some degree of root development and green leaves. Callus established from these shoot explants grew rapidly, displaying friable callus that was light in colour. After 16 weeks in suspension, dividing cell suspension cultures were established.

Protoplast pretreatments and electrofusion

Pea protoplasts were suspended in 20 ml wash solution and pre-treated with 3 mM iodoacetamide (IOA) at 4 °C for 30 minutes. Protoplasts were washed three times in electrofusion buffer (EFB) to remove excess ions. EFB consisted of 0.53M glucose, 1 mM $CaCl_2.2H_20$ and 545 μM MES (pH 5.8). *Lathyrus* protoplasts were suspended at 2 x 10^5 ml^{-1} in EFB and pretreated with 30 kRads gamma irradiation from a ^{60}Co source. Field pea and *Lathyrus* protoplasts were mixed in equal quantities at a density of 2 x 10^5 ml^{-1} in EFB, and dispensed in 600 μl aliquots into the wells of a 25-well electrofusion plate. Electrofusion was carried out at a range of values, optimising five parameters; pre-pulse AC alignment voltage (170-230 V cm^{-1}), ramping down time (1-11 s), DC pulse voltage (1200-1400 V), post-pulse AC alignment voltage (30-90 V cm^{-1}) and post-pulse AC alignment time (10-50 s). Pre-pulse AC alignment time was kept constant at 30 s, and DC pulse length was kept constant at 15 μs. Heterokaryons were cultured in calcium alginate beads as described earlier.

Electrofusion is currently underway for optimisation. Heterokaryons have been obtained at approximately 5% frequency, as a proportion of the initial number of protoplasts (Fig 3). It is well documented that cell growth at low densities, a condition which can be expected to be imposed following protoplast fusion experiment, is promoted by other cells in the immediate vicinity (Lowe et al, 1996). A nursing system (pea protoplasts in agarose droplets) was established, which will be utilised for the recovery of heterokaryons from future fusion experiments (Fig 4).

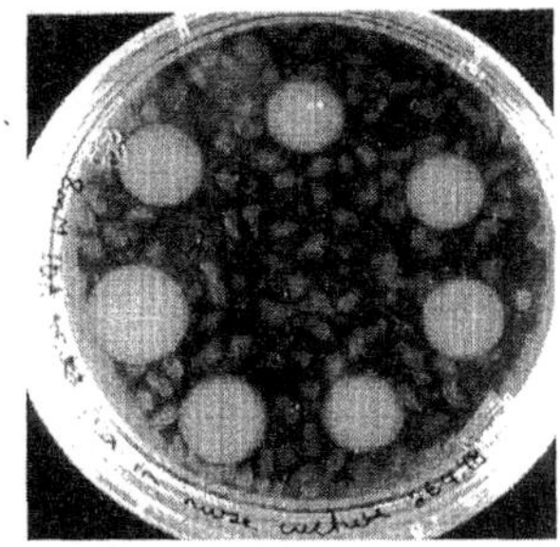

Figure 3. Nurse culture system

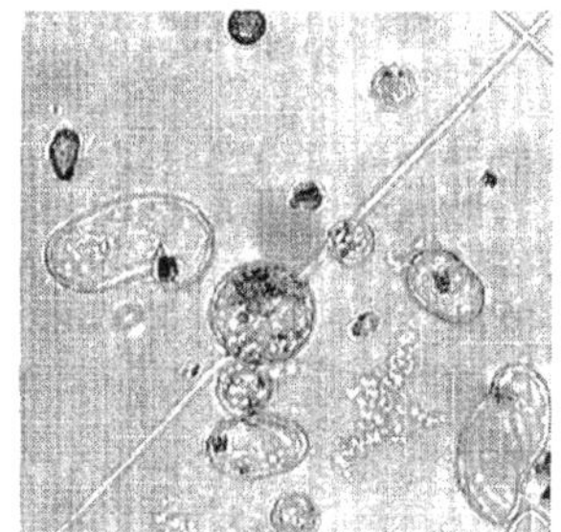

Figure 4. Pea-*Lathryus* heterokaryon

After establishment of the suspension culture, protoplast yield was consistently high. Protoplasts were uniform in size and showed cytoplasmic streaming (Fig 1b).

References

Bohmer P et al (1995) Plant Cell Rep. 15, 26-29.

Glimelius K (1988) Progress in Plant Protoplast Research, pp. 159-168. Kluwer Academic Publishers.

Kysely W, Jacobsen HJ (1990) Plant Cell Tiss. Org. Cult. 20, 7-14.

Larkin PJ et al (1988) Plant Sci. 58, 203-210.

Lehminger-Mertens R, Jacobsen HJ (1989) Plant Cell Reports 8, 379-382.

Li Y (1993) PhD thesis, Australian National University.

Loiseau et al (1995) Plant Cell Tiss. Org. Cult. 41, 267-275.

Lowe et al (1996) Plant Tiss. Cult. Biotech. 2(4), 175-186.

McCutchan JS et al (1997). Proc. Vth meeting IAPTC (Aust. branch), pp. 103-107.

Murashige T, Skoog, F (1962) Physiol Plant 15, 473-97.

Pang E, Brown J (1993) Field Pea Breeding Workshops, Horsham, Australia.

Phillips GC, Collins GB (1984) Handbook of Plant Cell Culture, pp. 169-210. New York: Macmillan.

Puonti-Kaerlas J, Eriksson T (1988) Plant Cell Rep. 7, 242-245.

An efficient protocol for somatic embryogenesis and its use in developing transgenic tea (*Camellia sinensis* (L.) O. Kuntze) for field transfer

Tapan Kumar Mondal, Amita Bhattacharya, Anil Sood and Paramvir Singh Ahuja
Division of Biotechnology, Institute of Himalayan Bioresource Technology,
Palampur 176061, H. P. India

Introduction

With the increasing popularity of tea (*Camellia sinensis* (L.) O. Kuntze) as an important beverage drink throughout the world, the demand for high quality and high yielding varieties is also rising. Therefore, genetic improvement of tea through biotechnological means has become a necessity. Improvement via *Agrobacterium tumefaciens* mediated transformation holds potential. However, there is an urgent requirement for an efficient *in vitro* system of regeneration for its successful implementation.

Inspite of several reports on tea somatic embryogenesis the following problems required attention:

i) Proper stage of explant for high frequency induction of somatic embryogenesis,
ii) asynchrony of secondary embryogenesis,
iii) maturation and precocious germination, and
iv) achieving field transfer.

The present work resolves the above constraints and utilizes the resultant protocol of somatic embryogenesis for developing transgenic tea.

Materials and Methods

Seeds of *Camellia sinensis* were collected from the Institute's Experimental Farm during different seasons and surface sterilized following the method of Sood *et al.*, 1993). De-embryonated cotyledon slices were inoculated on 1/2 MS (Murashige and Skoog, 1962) medium supplemented with 2.5 mg/l NAA and 0.2 mg/l BAP.

While the same medium with 2.0 mg/l BAP and 0.2 mg/l IAA was used for normal embryo development, MS medium modified with reduced nitrates and sulfates was used for secondary somatic embryogenesis.

A. Altman et al. (eds.), Plant Biotechnology and In Vitro Biology in the 21st Century, 181–184.

Embryos were subjected to different parameters like desiccation, chilling at 4°C for 15, 30, 45, 60 days, sucrose (3-9%), alternative sugars and auxin inhibitors, ABA (0-62.5 mg/l) and GA_3 (0.5-5.0 mg/l) (pers. comm., 1998). Plantlets from germinated embryos were transferred to soil in Hikko-trays and were maintained in the poly-house.

This protocol for somatic embryogenesis was then used for *Agrobacterium tumefaciens* mediated transformation using disarmed hyper-virulent strains like EHA105 and LBA4404 with p35SGUSINT and nptII genes.

The effect of different levels of Kanamycin (0-100 µg/ml) was studied on further growth of somatic embryos for the selection of transformants. Also, the effect of different levels of bactericidal antibiotics like carbenicillin, cefatoxime and sporidex on the bacteria as well as on the regeneration potential of the somatic embryos was studied.

Bacteria were cultured overnight at 28°C and 150 rev./min in liquid YEMB (pH 7.0) containing 50µg/ml Kanamycin. Bacterial cells were pelleted at 6000rpm/4°C/20 minutes and disolved in liquid embryo induction medium.

Bacteria at mid-log phase containing 10^9 cells/ml were used to infect globular somatic embryos. Different parameters as given below were studied for the optimization of the transformation protocol.

i) size of embryos (0-1 mm, 2-5 mm, >5 mm)
ii) preculture period (0, 1, 2, 3, 4, 5 days)
iii) method of injury (pricking, glass wool, sterilized sand)
iv) period of co-cultivation (0, 1, 2, 3, 4, 5, 6, 7 days)
v) pH of co-cultivation medium (5.2, 5.3, 5.4, 5.5, 5.6, 5.7, 5.8 and 5.9)
vi) inducing agents (100µM sucrose solution, acetosyringone)

Based on the optimized parameters, 0-1 mm entire embryos were soaked in petri-dish containing sucrose solution and over night grown bacterial solution (1:9 ml dilution) for 20 minutes with gentle shaking. Embryos were blotted on a sterile filter paper and then co-cultivted on secondary embryogenesis medium at 26-28°C. After 5 days, the embryos were first selected on 25µg/ml and then on 50 and 100 µg/ml Kanamycin. While transient expression of GUS was measured following the method of Jefferson, (1987) after 5 days, the Kanamycin resistant transformants were subjected to molecular characterization through PCR using reverse and forward primers of gus and nptII genes.

gus
5'-GGT-GGG-AAA-GCG-CGT-TAC-AAG-3'
5'-TGG-ATC-CCG-GCA-TAG-TTA-AA-3'

nptII
5'CCA-TCG-GCT-GCT-CTG-ATG-CCG-CCG-T-3'
5'AGG-CGA-TAG-AAG-GCG-ATG-CGC-TGC-3'
PCR amplifications were carried out in 50μl containing 1 μl DNA solution, 250 μM dNTPs, 25 ng nptII primers and 25ng gus primers, 2 units of Taq DNA polymerase and 1x buffer (10 mM Tris-Hcl (pH 9), 1.5mM $Mgcl_2$,50mM Kcl and 0.01% gelatin). Reaction was subjected to 30 cycles of 1 min at 92°C, 1 min at 55°C and 2 min at 72°C. Amplified DNA was detected by ultraviolet light after electrophoresis on 1% (w/v) agarose ethidium bromide gels.

Results and Discussion

The cotyledon explants from the seeds of *C. sinensis* were maximally responsive during the months of November and December when 66.7% of the explants from mature seeds only, responded (Mondal *et al.* 1998). Normal development of all stages of embryos i. e. globular, heart and cotyledonary shapes were observed on the embryo development medium.

While 10% germination was observed when embryos were chilled for 45 days, 40% germination was observed in 3% sucrose. Maximum germination was, however, observed at 111 mM Maltose and 20μM trans-cinnamic acid. GA_3, ABA and desiccation was inhibitory to normal germination (pers.comm., 1998). The inhibitory role of ABA and GA_3 in carrot somatic embryogenesis was reported by Vasil, (1996). 70% embryos germinated and 50% of them survived in the field.

While Kanamycin level at 25 μg/ml arrested the growth of somatic embryos, 35 μg/ml killed the nontransformed embryos within 30 days. Of the three bactericidal antibiotics, carbenicillin and sporidex can kill *A. tumefaciens* effectively at the level of 250 mg/l and 400mg/l respectively. At these concentrations both the antibiotics specially sporidex enhanced secondary embryogenesis. Cefatoxime at 250 mg/l was toxic to normal embryo development. That, cefatoxime is toxic to regeneration of explants has also been reported by Tosca *et al.*, 1996 in *Camellia x williamsii*

Intact uninjured embryos (1-2 mm) without wounds were highly susceptible to *Agrobacterium* mediated transformation when they were not pre-cultured as in tamarillo (Atkinson *et al.*, 1993). Wounding probably inhibits regeneration in tea due to oxidation of phenolics at wound sites. Inhibitory effect of wounding was reported by Ducrocq *et al.*, 1994 in *Datura inoxia*. Maximum transient GUS expression of 40-48% was observed when sucrose and not Acetosyringone was used as an inducing agent and when embryos were co-cultivated for 5 days on a medium of 5.8 pH.

PCR analysis from 5 independent transformants out of 18 Kanamycin resistant events show that gus and nptII genes have been stably integrated (Fig. 2).

Reference

Atkinson G R, Gardner C R (1993) Plant Cell Repo. 12 .347-351.

Ducrocq *et al.* (1994) Plant Mole Biol. 25 . 995-1009.

Dodd W A (1994) Int. J. Trop. Agric. Vol XII 3&4. 212-247.

Jefferson *et al.* (1987) EMBO J. 6. 3901-3907.

Murashige T., Skoog F (1962) Physiol. Plant 15. 473-497.

Sood *et al.* (1993) J. Plantation Crops 21 (Supplement). 295-300.

Tosca *et al.* (1996) Plant Cell Repo. 15 541-544.

Vasil I K (1996) Plant Cell Tiss. Org.Cult.

Vieitez A M (1994) Somatic embryogenesis in woody plants. Vol 2. Angiosperm, pp 235-276. Kluwer Academic Publishers. Dordrecht. The Netherlands.

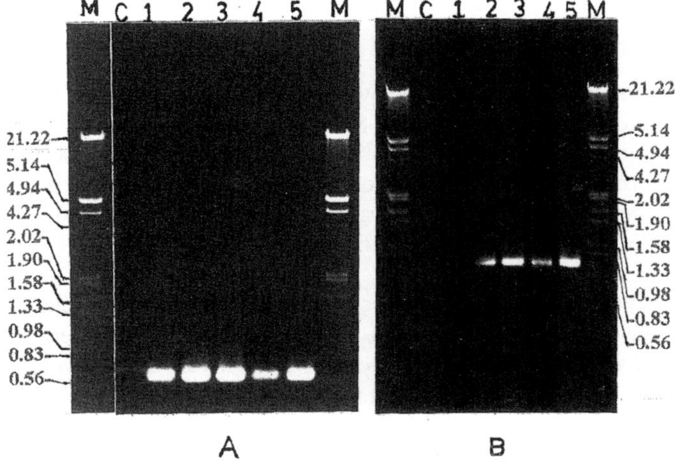

Polymerase chain reaction analysis to detect the presence of gus and nptII gene in transgenic tea

A: PCR amplification of a 693 bp fragment of nptII gene
Lane M: Eco R1/Hind III digest of λ DNA
Lane C: Non transformed tea
Lane 1-5: Independent transformant of tea

B: PCR amplification of a 650 bp fragment of gus gene
Lane M: Eco R1/Hind III digest λ DNA
Lane C: Non-transformed tea
Lane 1-5 : Independent transformant of tea

Regeneration of adventitious shoots in process of genetic transformation

Teresa Orlikowska
Research Institute of Pomology and Floriculture, Skierniewice, Poland

The basis of genetic transformation is the fast and stable introduction of foreign DNA into the genoms of single cells, followed by repeated divisions, and finally regeneration of the whole plants from the progeny of these cells. Regeneration is therefore a very important element of the transformation process and knowledge of regeneration from desirable plant genotypes should precede transformation experiments. Such information can identify the most productive organs and tissues, the proper physiological status of donor plants, and appropriate hormone treatments for efficient regeneration. An effective regeneration system is characterized by an unlimited supply of sterile explants and the ability to maintain long-lasting regeneration from single cells.

Such regeneration systems have been developed for safflower (Orlikowska, Dyer, 1993) and gerbera. Primary explants were induced to regenerate on medium containing 0.5 mg-l thidiazuron (TDZ) and 0.1 mg-l naphthaleneacetic acid (NAA); after 3 days explants were transferred to regeneration medium containing 0.1 mg-l TDZ + 0.1 mg-l NAA (safflower) or 0.5 mg-l benzyadenine + 0.01 mg-l NAA (gerbera). Cotyledons and first leaves of 3 to 11 day-old safflower seedlings were used, although cotyledons gave more stable and predictable results than leaves probably because of better surface contact with the medium and hormone stimuli. After removal of the first leafy buds, organogenic calli were obtained from the remaining tissues. The average yield of regenerated buds obtained within 4-5 months from one cotyledon exceeded 200.

For gerbera, the youngest three or four folded leaves with the petioles (total lenght 2-5 mm) from 3-week-old shoot cultures were used. Primary buds were obserwed to regenerate directly from petiole bases within 2 to 4 weeks. These buds and the lamina were removed and the petioles transferred to fresh medium where organogenic calli with adventitious shoots were produced. This method has been successful for several months with increasing effectiveness (Fig. 1).

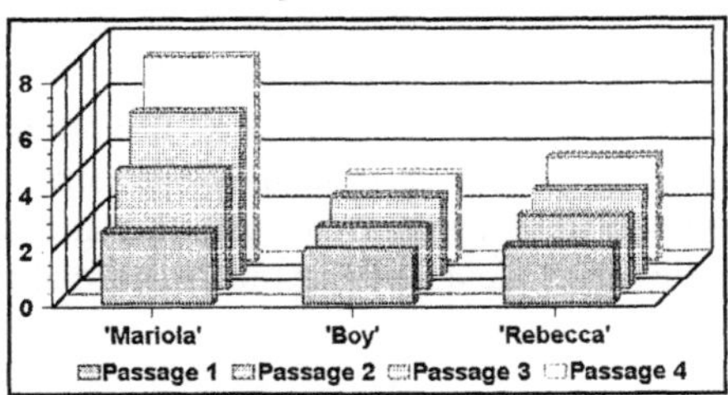

Fig. 1. Number of gerbera shoots per initial explant in four passages. Percentages of regenerated calli in subsequent passages were: 'Mariola' - 91, 99, 99, 100; 'Boy' - 49, 59, 65, 67; 'Rebecca' - 45, 52, 56, 59.

A. Altman et al. (eds.), Plant Biotechnology and In Vitro Biology in the 21st Century, 185–188.

The use of an optimized protocol for regeneration and high transitional expression levels of reporter genes in primary explants do not necessarily guarantee that a sufficient number of independent transgenic plants will be obtained. Some of the reasons for low numbers of expressing transgenic plants and some possible remedies are discussed below.

Low transfer rate of T-DNA. The number of transformed cells in cocultivated or bombarded tissue is much lower than the number of cells competent to regenerate or able to acquire regeneration competence. However, the number of transformed cells can be increased by optimizing T-DNA transfer conditions, such as adjusting inoculum density and duration of cocultivation, using a more agressive bacterial strain, or adopting a different plasmid construct. In the case of biolistic delivery, the type and density of projectiles and addition of DNA stabilizers can be experimentally optimized.

Low number of cells responding to hormonal stimuli. The primary reason for low rates of regeneration is that the explant does not contain enough regeneration-competent cells or cells that can acquire competence during tissue culture. There are several published approaches to increase the numbers of such cells, including:

- use the most productive explants from young seedlings (Zhu et al., 1996), from cultures of a proper age (De Bond et al., 1996) or from organs originated from adventitious shoots (Rugini, Muganu, 1998),
- injuring explants by removal of epidermis (Dong et al., 1993), puncturing (De Jong et al., 1993), scrapping or squashing (Nowak et al., 1997), cutting (Sriskandarajah et al., 1994), bombardment (Knittel et al., 1994), sonication ((Zhang et al., 1994), vortexing with carborundum (Cheng et al., 1996) or shaking with glass beads (Greyburg, Vicks, 1995),
- preculturing explants before transfer of T-DNA on auxin and/or cytokinin medium (Krens et al., 1996).

In our experiments squashing of gerbera shoot tips and scrapping petioles resulted in an increase in the number of transformed shoots per explant. It was especially important for the recalcitrant cultivar 'Boy' (Nowak et al., 1997), in which regeneration was increased to 80% and the mean number of transgenic shoots per initial explant to 0.5, as compared to 30% and 0.04 shoots for the control, respectively.

Transformation competence is not equivalent to regeneration competence. Occasionally, explants may show very high reporter gene (eg. GUS) expression after cocultivation or bombardment but after some time expression levels fall dramatically and the final yield of transgenic shoots is very low or absent. This phenomenon is thought to result from T-DNA being transferred to superficial or wounded tissues only while regeneration is only possible from differrent tissues or cell types (Mukhopadhyay et al., 1992). Therefore, preliminary experiments correlating transient expression levels (preferably from a GUS reporter gene containing an intron and a constitutive promoter) with the most regenerable tissues can often provide good direction towards optimizing the entire regeneration scheme.

Regeneration is suppressed by wounding and/or *Agrobacterium* infection stress. The process of tissue wounding and infection by *Agrobacterium* is often accompanied by plant cell hypersensitive reactions, ethylene biosynthesis, and death of infected and surrounding cells. There are several remedies which may decrease these detrimental effects on transformation and regeneration efficiencies:

- preculturing of explants before cocultivation can separate wound stress from infection stress (De Jong et al, 1993),

- ventilation tissue culture vessels and addition of anti-ethylene agents such as Co^{2+} ions, proline, and AVG can decrease ethylene production and accumulation in the explant environment and adding silver nitrate can reduce susceptibility of tissues to ethylene effects (Biddington, 1992),

- adding antioxydants such as cysteine, citric and ascorbic acids, and phenol traps such as dithiothreitol and polyvinylpolypyrrolidone to the medium can reduce the toxicity of plant phenolic compounds (Perl et al, 1996 and literature cited).

Silver nitrate has been recommended to improve regeneration from rose leaves (Dubois, de Vries 1995). However, in our experiments on rose, silver nitrate added together with cefotaxime after cocultivation, decreased shoot regeneration and instaed promoted growth of non-regenerative callus (Orlikowska et al., 1996). We also found that silver nitrate added alone to medium not containing any organic forms of nitrogen could limit *Agrobacterium* growth after cocultivation for up to 20 days. Therefore, cefotaxime could be omitted from the medium during the critical period required for induction of regeneration. Silver nitrate decreased regeneration of gerbera when continuously present in the medium, but when included in the medium sporadically (one passage during every 3-4 passages), it temporarily restored regeneration in the presence of cefotaxime and kanamycin (T. Orlikowska, unpubl.).

Antibiotics can affect plant regeneration efficiency. The widely used antibiotics carbenicillin and cefotaxime can influence regeneration from plant tissues and their effects vary across plant genotypes (Nauerby et al, 1997). This effect can be explained by the fact that carbenicilline has been shown to be metabolized to auxin-like compounds in plant tissues (Holford, Newbury, 1992). Another aid for rapid elimination of *Agrobacteria* without reducing regeneration efficiency was a short vacuum infiltration of cocultivated tissues with MS medium of pH 3.0 followed by infiltration with antibiotics (Hammerschlag et al., 1997).

In our experiments carbenicillin slightly stimulated regeneration of safflower (Orlikowska et al., 1995) and cefotaxime decreased regeneration and multiplication of gerbera and rose (T. Orlikowska, unpubl.).

Toxic effects of selective agents. Selective agents are lethal not only to non-transformed cells but can also be harmful to transformed cells that do not rapidly and actively detoxify the compound. Therefore, the initiation of selection and concentration of the agent should be chosen carefully. If the regeneration system is not very efficient, a low selection pressure can be applied initially followed by increasing selection pressure when buds are formed. When selection with 5 mg-l kanamycin was applied directly after cocultivation, direct regeneration from gerbera explants was completely lost and transformed calli became visible on explants after one month (T. Orlikowska, unpubl.).

A related phenomenon observed is that, after cocultivation, dying cells insulate transformed cells from hormonal, nutritional and selective stimuli of the medium and secrete toxic metabolites. To avoid such negative effects on cell division, regeneration and selection, healthy tissues should be separated from necrotic areas as soon as practicable.

During selection it can be very helpful if toxic effect of the agent are visible as early as possible. It is not the case of kanamycin, which causes gradual bleaching of veins and lamina. For safflower, geneticine was more effective than kanamycin, because it caused darkening of cut surfaces of non-transformed explants within 1 week (Orlikowska et al., 1995). The herbicide phosphinotricin (PPT) was most effective for gerbera and rose because almost no nontransformed tissue survived selection. Necrotic spots on initial leaves were visible within

1 week and nontransformed shoot tips with young leaves became necrotic within 2 weeks (T. Orlikowska, unpubl.). However, it is important to remember that differrent medium components can increase or decrease susceptibility of cells to PPT (Escandon, Hahne, 1991). In one case PPT was shown to stimulate embryogenesis (Hebert-Soule et al., 1995). PPT can also be used to reselect putative transformed plants in the greenhouse by leaf painting assays. Non-recalcitrant genotypes safe time and money. Plant regenerability is highly dependent on genotype. In species with a short juvenile period, transformation and regeneration are often more straightforward for particular cultivars or plant types. Therefore, it can be more efficient in the long run to transform the regenerable cultivars and then transfer the gene to recalcitrant types through hybridization, rather than attempting to adapt methods for recalcitrant genotypes.

Use a transformation system that does not depend on regeneration *in vitro*. Future advances in transformation technology and techniques will probably permit widespread use of methods which are not based on *in vitro* regeneration. For example, pollination using pollen that was transformed with T-DNA by bombardment (Van der Leede-Plegt et al., 1996), vacuum infiltration of bacteria or plasmids into whole plants (Bechtold et al., 1993), and infiltration of organs such as meristems, embryos or spikes (Sautter et al., 1995) all have potential to avoid regeneration difficulties and such associated problems as somaclonal variation and chimeric transgenic plants.

Acknowledgement

The author is indebted to Dr. William E. Dyer for critical review of the manuscript.

References

Bechtold N et al. (1993) C.R. Acad. Sci. Paris, Sci. de la Vie 316, 1194-1199
Biddington NL (1992) Plant Growth Regulation 11, 173-187
Cheng Y-H et al. (1996) Plant Cell Reports 16, 127-132
De Bondt et al. (1996) Plant Cell Reports 15, 549-554
De Jong et al. (1993) Plant Cell Tissue Organ Culture 32, 263-270
Dong JZ, McHughen A (1993) Plant Science 88, 61-71
Dubois LAM, de Vries DP (1995) Gartenbauwisseschaft 60, 249-253
Escandon AS, Hahne G (1991) Physiol. Plant. 81, 367-376
Greyburn WS, Vick BA (1995) Plant Cell Reports 14, 285-289
Hammerschlag FA et al. (1997) J. Am. Soc. Hort. Sci. 122, 758-763
Hebert-Soule et al. (1995) Plant Cell Reports 14, 380-384
Holford P, Newbury HJ (1992) Plant Cell Reports 11, 93-96
Knittel et al. (1994) Plant Cell Reports 14, 81-86
Krens et al. (1996) Plant Science 116, 97-106
Mughopadhyay A et al. (1992) Plant Cell Reports 11, 506-513
Nauerby et al (1997) Plant Science 123, 169-177
Nowak E et al. (1997) Biotechnologia 4 (39), 27-37 (In Polish with English summary)
Orlikowska TK, Dyer WE (1993) Plant Science 93, 151-157
Orlikowska TK et al. (1995) Plant Cell Tissue Organ Culture 40, 85-91
Orlikowska TK et al. (1996) J. Appl. Genet. 37A, 122-125
Perl et al. (1996) Nature Biotechnology 14, 624-628
Rugini E, Muganu (1998) Plant Cell Reports 17, 581-585
Sautter C et al. (1995) Euphytica 85, 45-51
Sriskandarajah S et al. (1994) Plant Cell Tissue Organ Culture 36, 317-329
Van der Leede-Plegt LM et al. (1996) Acta Hortic. 430, 529-530
Zhu X et al. (1996) Plant Cell Reports 16, 71-75
Zhang LJ et al. (1991) Bio/Technology 9, 996-997

GENERATION OF TRANSGENIC CARNATION PLANTS WITH NOVEL CHARACTERISTICS BY COMBINING MICROPROJECTILE BOMBARDMENT WITH *AGROBACTERIUM TUMEFACIENS* TRANSFORMATION

M. Ovadis, A. Zuker, T. Tzfira, A. Ahroni, E. Shklarman, G. Scovel, H. Itzhaki, H. Ben-Meir and A. Vainstein
The Kennedy-Leigh Centre for Horticultural Research and The Otto Warburg Center for Biotechnology in Agriculture, The Hebrew University of Jerusalem, Rehovot 76-100, Israel.

1. Introduction

As one of the major contributors to the cut-flower market and a commercial leader in terms of number of stems sold worldwide (Jensen, Malter 1995), carnation (*Dianthus caryophyllus* L.) has been an important target for the breeding of new varieties with novel characteristics. Although new carnation varieties are continuously being produced through classical breeding, their high heterozygosity and limited gene pool, and a lack of knowledge regarding their genetic makeup, severely restrict such breeding programs (Woodson 1991). Thus, the possibility of genetically transforming carnation, as well as other cut-flower species, via direct gene transfer is quite attractive (reviewed by Zuker et al. 1998).

Although genetic transformation of carnation cultivars has been reported, and despite the great progress and interest in gene transfer to this crop, the actual transformation procedure is considered routine in only a few laboratories (Zuker et al. 1998). We have developed a unique and efficient transformation protocol which has resulted in the regeneration of non-chimeric transgenic carnation plants of several cultivars, and we have used this procedure to generate carnations with novel characteristics.

2. Materials and Methods

Plant material and tissue culture. Non-rooted cuttings of greenhouse-grown carnation cultivars were surface-sterilized as previously described (Zuker et al. 1995), and their leaves and shoot apices removed. The various *in vitro* media (Table 1) were all based on Murashige and Skoog (1962) basal medium (MS) supplemented with 30 g/l sucrose, adjusted to pH 5.8, and solidified with 8 g/l agar.

Microprojectile bombardment-mediated wounding of plant tissue. Stem explants were wounded in a Biolistics PDS 1000/He system using tungsten particles prepared as described by Zuker et al. (1995). The explants were bombarded twice at 1500 psi with the launching plate placed 9 cm away from the tissue.

Agrobacterium strains. *A. tumefaciens* strains AGLO carrying the binary vectors pCGN7001 or pAM*rolC* and EHA105 carrying pKIWI105 were grown in liquid Luria-Broth (LB) medium supplemented with 0.0192 mg/l acetosyringone and the appropriate antibiotics. PCGN7001 and pKIWI105 carried *uidA* (GUS), pAM*rolC*

A. Altman et al. (eds.), Plant Biotechnology and In Vitro Biology in the 21st Century, 189–192.

appropriate antibiotics. PCGN7001 and pKIWI105 carried *uidA* (GUS), pAM*rolC* carried the *rolC* gene, and all three plasmids carried *nptII* genes. Bacteria were harvested by centrifugation at 12000*g* for 2 min and the bacterial pellet, after resuspension to an OD_{550} of 0.5 in liquid co-cultivation media, was used for the transformation.

Table 1. Media compositions for genetic transformation of carnation.

Medium	Additives (mg/l)
Co-cultivation	5 NAA, 0.0192 As
SI-T1 (regeneration from stem explants)	0.1 NAA, 1 TDZ, 300 Cb, 100 Kn
SI-B1 (regeneration from leaf explants)	0.1 NAA, 1 BAP, 300 Cb, 100 Kn
Elongation and rooting	0.1 NAA, 0.1 GA, 200 Cb, 100 Kn
Propagation	0.1 NAA, 0.1 GA

As, acetosyringone; BAP, 6-benzylaminopurine; Cb, carbenicillin; GA, gibberellic acid; Kn, kanamycin; NAA, α-naphthalene acetic acid; TDZ, 1-phenyl-3(1,2,3-thiadiazol-5-yl)-urea.

Genetic transformation and regeneration of transgenic plants. Wounded explants were inoculated with bacterial suspension for 10 min. Following a co-cultivation period of 3 days in the dark followed by 2 days in the light on co-cultivation media, three primary nodes were sectioned into 2- to 3-mm slices and were subjected to the first regeneration and selection cycle on SI-T1 medium. After 10 days, axillary shoots that had developed from existing buds were excised, and the explants were cross-sectioned and replanted on fresh SI-T1 medium. After 2 weeks, the regenerated adventitious shoot clusters were excised from the stem explants and their leaves were pulled off and cultured on SI-B1. The transgenic nature of shoots regenerated on the SI-B1 second regeneration and selection medium was confirmed through both histochemical (Stomp 1992) and molecular (Zuker et al. 1995) analyses. Transgenic plants were transferred to an elongation and rooting medium, then to the greenhouse.

3. Results and Discussion

Genetic transformation of carnation. Bombardment-mediated wounding followed by inoculation with *Agrobacterium* led to reproducibly high gene-transfer frequencies as evidenced by the transient GUS expression assay in stem explants. Co-cultivation with *Agrobacterium* for 3 days in the dark followed by 2 days in the light was previously reported to increase transformation frequency in cv. White Sim stem explants, relative to other light/dark regimes (Ahroni 1996). These conditions were applied to five different carnation varieties using two different *Agrobacterium* strains and in all cases, 80-90% of the inoculated explants expressed GUS reproducibly.

Histochemical analyses of adventitious shoots regenerated from stem explants in the first selection cycle revealed chimeric GUS expression in most of them. These plantlets failed to survive when further cultured under kanamycin. Thus, a second regeneration and selection step was required to both eliminate chimeric plants and allow the regeneration of stable transgenic shoots.

In the second selection cycle, leaves from the chimeric plantlets generated in the first selection cycle served as the source for adventitious shoot regeneration. Histochemical analysis of randomly chosen plants generated following the second selection cycle, at different developmental stages, revealed a stable, non-chimeric pattern of GUS expression. Further analyses of the transformed plants revealed integration of both *uidA* and *nptII* genes into the plant genome, as well as their inheritance by their progeny.

Using the aforementioned optimal conditions, i.e. wounding of stem explants *via* bombardment, co-cultivation with *Agrobacterium* under the specified dark/light regime, and two cycles consisting of different regeneration/selection methods, an overall stable transformation efficiency of ca. 2 independent transgenic plants per 10 stem explants was achieved. Moreover, almost no escapees were generated and ca. 90% of all the plantlets generated following the second selection cycle expressed marker genes.

Transgenic carnation plants expressing the* rolC *gene. The market demand for cut flowers with novel traits is the main driving force behind the production of new varieties. Hence, genetic engineering approaches are most attractive, as novel traits can be immediately employed (reviewed by Mol et al 1995; Zuker et al 1998). We evaluated the ability of the *rolC* gene from *Agrobacterium rhizogenes* to genetically alter and improve horticultural traits in transgenic carnation plants. Although its precise mode of action in plants is still not clear (Faiss et al. 1996), it has been successfully used to produce transgenic plants with altered morphology (Nilsson, Olsson 1997). The phenotypic implications of the *rolC* gene include reduced apical dominance, male sterility, dwarfness induction, and extensive rooting (Nilsson, Olsson 1997).

Expression of *rolC* under the constitutive 35S CaMV promoter led to several advantageous morphological alterations in transgenic carnation. *In vitro*, *rolC*-transgenic lines of cv. White Sim exhibited increased axillary bud breakage, reduced apical dominance and shortened internodes as compared to control plants. Unique and interesting phenomena were observed in non-detached leaves of *rolC*-transgenic carnation plants. Shoot regeneration occurred, albeit at a low frequency, from the distal area of leaves on plants grown *in vitro*, even on hormone-free media (Fig. 1A). It should be noted that previously, the well-documented regeneration capacity of carnation leaves had only been reported from the leaf base following tissue culture of detached leaves in the presence of hormones (van Altvorst et al. 1995).

rolC-transgenic carnation plants exhibited increased axillary bud breakage and development when grown *ex vitro* under standard commercial greenhouse conditions (Fig. 1B). Transgenic plants also exhibited delayed flowering, as compared with non-transformed plants, with some clones exibiting higher vegetative shoot length. One of the practical implications of increased bud breakage and development is the possibility of obtaining a higher number of stem cuttings from a given mother plant. Indeed, when comparing the number of stem cuttings obtained from *rolC*-transformed and non-transformed plants of the same age, we found the yield to be three times higher for the former. Furthermore, stem cuttings obtained from *rolC*-transgenic plants were better developed, as inferred from their higher dry weight and length.

Stem cuttings from *rol*C-transgenic plants also exhibited better rooting ability as compared with those from their non-transformed counterparts following treatment with commercial rooting powder: *rol*C-transgenic stem cuttings rooted in a shorter time and produced more roots. Furthermore, while the use of commercial rooting powder was almost obligatory for the rooting of non-transformed stem cuttings, *rol*C-transgenic cuttings exhibited good rooting even without it (Fig. 1C). Improved rooting enables reduction in the culturing time of stem cuttings as well as better plant establishment and growth. Other morphological assessments, such as yield of *rol*C-transgenes in the greenhouse, are currently under investigation.

Fig 1. Morphological traits in 35S-*rolC*-transgenic carnation plants. (A) *In planta* spontaneous shoot regeneration from the distal area of non-detached leaves (arrow). The *rolC*-transgenic carnation plantlet was cultured on hormone-free MS media. (B) Increased axillary bud breakage and delayed flowering in 2-month-old *rol*C-transgenic (t) versus control (c) carnation plants. (C) Root production in 2-week-old *rol*C-transgenic (right) and non-transformed (left) stem cuttings grown in hormone-free perlite.

4. Acknowledgments

The authors wish to thank Shemi Ltd. (Shdema, Israel) for providing the plant materials. This research was supported by Grant US-2233-92 from BARD — the United States-Israel Binational Agricultural Research and Development Fund, by the Ministry of Agriculture, by the Ministry of Science and by the Association of Israeli Flower Growers.

5. References

Ahroni A (1996) M.Sc. thesis (The Hebrew University of Jerusalem, Israel).
Faiss M et al (1996) Plant J. 10, 33-46.
Jensen MH and Malter AJ (1995) World Bank Technical Paper 253, 144-146.
Murashige T and Skoog F (1962) Physiol. Plant. 15, 473-497.
Nilsson O and Olsson O (1997) Physiol. Plant 100, 463-473.
Stomp AM (1992) in GUS Protocols, SR Gallagher (ed), Academic Press, San Diego, CA, pp. 103-113.
van Altvorst AC et al (1995) Plant Cell Tiss. Org. Cult. 40, 151-157.
Woodson WR (1991) HortScience 26, 1029-1033.
Zuker A et al (1995) Sci. Hort. 64, 177-185.
Zuker A et al (1998) Biotech. Adv. 16, 33-79.

SEXUAL TRANSFER OF TRANSGENE (*BAR*) INTO NON-TRANSFORMED WHEAT GENOTYPES AND CELL-LEVEL SELECTION OF THE MARKER GENE IN MICROSPORE- AND ANTHER CULTURE

J. Pauk[1], R. Mihály[1,2], R. Hänsch[3], J. Schulze[3] and R.R. Mendel[3]
[1]Cereal Research Non-profit Company, 6701 Szeged, P.O.B. 391, Hungary
[2]University of Agricultural Sciences, 2103 Gödöllô, Páter K. 1., Hungary
[3]Botanical Institute, TU, 38106 Braunschweig, Humboldtstr. 1., Gerrmany

1. Introduction

Fertile transgenic wheat (*Triticum aestivum* L.) plants have been produced during the past few years (Vasil et.al 1993, Weeks et al. 1993, Becker et al., Nehra et. al 1994). At the end of 20th century the fundamental discoveries made in the field of molecular biology and cell and tissue culture have initiated a biological revolution in plant breeding and agricultural production. The first genetically engineered food crops have been introduced into the market in some countries (Nehra et al 1995), but several questions have been under discussion. The fate of transgene is an up-to-date question in plant genetic and breeding research.

Earlier, six independent fertile transgenic lines of spring wheat bearing selectable marker gene *bar* had been obtained. This paper gives a report on the sexual transfer, selection and use of the *bar* gene in non-transformed genotypes up to development of homozygosity in the offspring generations.

2. Material and Methods

2.1. Plant material

Donor plants of CY-45 spring wheat (*Triticum aestivum* L.) line which was previously selected for it's high response to somatic tissue culture was used. The wheat line was grown in fitotron chamber and in greenhouse and immature embryos were excised, cultured, bombarded, bialaphos selected and regenerated as described earlier by Vasil et al. (1993). Crossing (emasculation, pollination) and embryo cultures were made according to protocols in cell and tissue culture handbooks. The microspore and anther culture were made using our previous published methods (Pauk et al. 1992, Poulimatka et al. 1997).

2.2. Transgenic experiments

To optimize the transformation system, transient expression experiments were carried out using pAHC25 plasmid containing *UidA* reporter gene and *bar* marker gene. In the stable transformation experiments, pAHC 20 plasmid containing the selectable *bar* gene for resistance to the herbicide (Finale 14 SL 150g/l ammonium glufosinate made by AgrEvo

A. Altman et al. (eds.), Plant Biotechnology and In Vitro Biology in the 21st Century, 193–196.

Gmbh.) was used. The selectable *bar* gene encodes the enzyme phosphinothricin acetyltransferase (PAT) that inactivates phosphinothricin (PPT), the active ingredient of the herbicide by acetylation. Plasmid DNAs were precipitated, absorbed to gold particles and delivered to target tissue using a DuPont PDS-1000/He device.

3. Results

3.1. Development of homogenous transgenic wheat lines

Six transgenic spring wheat lines (106-3a, 116, 117, 124, 128, 129) were obtained from transformation experiments of bombarded immature embryo-derived calli using pAHC20 plasmid molecule, followed by bialaphos selection. All of the R_0 regenerants were identified as transgenic plants using PAT assay, RNA-RNA hybridazation and herbicide spray.

Five of the R_0 six lines produced complete fertile heads while one line (106-3a) was partially fertile. The offspring of the complete fertile transgenic (T) lines (116, 117, 124, 128, 129) were used in the sexual genetransfer experiments. The T_1 generation of 4 transgenic lines (116, 124, 128, 129) showed a 3:1 segregation in PAT assay. The line 117 showed close to 3:1 segregation ratio. The PAT positive lines were self-pollinated and the subsequent generation was checked by direct herbicide spray. In the T_2 generation, the four complete fertile lines (116, 124, 128, 129) expressed a homogenous herbicide resistance. The line 117 was homogenous. Later the T_2 homogenous herbicide resistant plants were used as male parent in the crossing experiments.

3.2. Sexual transfer of the *bar* transgene

For the breeder, the ability to transfer foreign gene and its Mendelian inheritance is a very important aspect. In crossing experiments the homogenous T_2 transgenic line 124 was crossed with different non-transgenic females. The isolated hybrid embryos from the crosses were tested in medium containing bialaphos (Table 1.). The result of this experiment showed the dominant character of *bar* gene. All the hybrid embryos germinated onto 5mg/l bialaphos medium expressed resistance, while the control variety (Góbé) showed sensitivity to the bialaphos as expected.

In the next experiment different transgenic lines (male) were tested on a wild type (Gar. non tansgenic) variety (Table 2). The sexual transfer of transgene in case of line 116, 124, 128 was good, all the hybrid embryos expressed resisitance on the bialaphos embryo culture. However, two thirds of the hybrid embryos of transgenic line 117 were sensitive to the bialaphos. It indicated that the line 117 was not homogenous.

3.3. Selection of the transgene in anther and microspore culture

To develop the cell-level selection system of bar gene, anther and microspore culture experiments were made using transgenic lines and bialaphos selection. T_2 PAT resistant

Table 1. Sexual transfer of transgene (*bar*) in wheat by non-transgenic x transgenic crossing. T Hybrid embryos were tested in bialaphos included embryo culture.

Genotype or combination	Tested embryos	Germinated embryos on 5 mg/l bialaphos	Fertile plants
Control variety:			
Góbé	12	0	0
Crossing comb.:			
RK-1	9	8	8
RK-2	20	20	20
RK-3		5	5
RK-4	2	2	2
RK-5	12	11	11
Total:	48	46	46

Table 2. Sexual transfer of transgene (*bar*) in wheat by non-transgenic x transgenic crossing. Test of different transgenic lines (male) with Gar. non-transgenic female var.) in embryo culture.

Genotype or combination	Tested embryos	Germinated on 5 mg/l bialaphos	Transplanted in soil	Bialaphos sensitive
Gar. non-trans.	12	0	0	12
Gar. x 116	9	9	9	0
Gar. x 117	6	6	2	4
Gar. x 124	12	12	12	0
Gar. x 128	9	9	9	0
Total:	48	36	32	16

Table 3. Bialaphos selection for *bar* gene from anther culture of different *bar* transformed wheat lines. Number of embryoids on the different bialaphos included medium.

Transgenic wheat lines	Plated anthers onto P-4mf medium	Bialapohs content in medium		
		0 mg/	5 mg/l	after 2-week 5 mg/l
116	3 x 50	175	140	147
117	3 x 150	340	331	392
124	3 x 100	356	157	298
128	3 x 50	138	162	226
129	3 x 50	98	137	112
Total:		1107	927	1175

transgenic lines were used as donors in anther culture (Table 3.). Three treatments were applied, in the first treatment, bialaphos was not used in the anther culture, in the second treatments the medium was supplemented with bialaphos added into liquid mediun at time of isolation while in the third treatment, bialaphos supplement was added 2 weeks after isolation. In all the treatments, there was no significant difference in embryoid production. Similar results were obtained in microspore culture, but the 3 mg/l bialaphos gave the best selection result.

3.4. Production of transgenic haploids and doubled haploids in anther culture

The haploids and spontaneous diploids regenerated from selective anther and microspore culture experiments expressed total herbicide resistance to Finale 14SL. Via chromosome doubling of transgenic haploids, homogenous diploid (DH) wheat lines were obtained. The stability of *bar* foreign gene in doubled haploid wheat genome is under study.

4. Conclusions

The fertile T_1 generation expressed 3:1 segregation ratio for bar^+/bar^- plants. The transformed foreign gene (*bar*) was stable in the wheat genome after two self pollinations and PAT assays. It is therefore, possible to transform transgene successfully via sexual way into non-transgenic genotypes. Cell-level selection of *bar* gene was possible in anther and microspore culture. The regenerated haploids and spontaneous diploids are herbicide resistant. Finally, by using *in vitro* haploid technique (anther and microspore culture), it's possible to produce homogenous *bar* resistant wheat lines by cell-level selection.

5. Acknowledgements

The autors wish to thank Dr. G. Schwarz, Dr. K. Schledzewski, M. Csôsz, K. Talpas, É. Kóta and V.N. Gathaara for their excellent assistance. The funding by BMF/OMFB and OTKA through research grants is highly appreciated.

6. Reference

Becker D et al (1994) Plant Jour 5: 299-307.
Nehra NS et al (1994) Plant Jour 5:285-297.
Nehra NS et al (1995) Plant Breeding Abstracts 65: 803-808.
Pauk J et al (1991) Plant Breeding 107: 18-27.
Puolimatka M (1997) Cer. Res. Comm. 24: 393-400.
Vasil V et al (1993) Bio/technol 11: 1553-1558.
Weeks JT et al (1993) Plant Physiol 102:1077-1084.

PROTEASES FOR THE IMPROVEMENT OF PLANT TRANSFORMATION

Avihai Perl, Pierre Coutos Tevenot[1], Sahar Nachman, Gollop Rachel, Elyasi Rafi and [1]Michel Boulay. The Department of Fruit Tree Breeding and Molecular Genetics, The Volcani Center, P.O.Box 6, Bet-Dagan, 50250, Israel, and [1]Moet&Chandon, Oenologie, 20 Avenue de Champagne, B.P. 140, F-51333, Epernay.

Embryogenic cell lines of *Vitis vinifera* cv. Ugni Blabc (UB), co-cultivated with different strains of *Agrobacterium tumefaciense* undergo necrogenesis leading subsequently to cell death. Antioxidants that were previously utilized in grape transformation (Perl et al. 1996) were found to have only a limited effect blocking this necrogenesis. We assumed that this necrogenesis is due to a hypersensitive like response (HR) of the embryogenic lines to *Agrobacterium tumefaciens.* If a putative proteinaceous elicitor is responsible for inducing the HR-like response in UB embryogenic cell lines, then treatments with certain proteases might reduced this necrogenesis (Wei et al 1992). Ion channels were recently reported to initiate signal transduction in higher plants including during bacterial attack (Ward et al 1995). Channel blockers like lanthanum and k-252a were previously shown to inhibit the HR response of tobacco cells elicited by harpin - a protein excreted by several bacteria during the pathogenesis process (Wei et al 1992; He et al 1993; 1994; Bauer et al 1995). The plant growth regulator, jasmonic acid (JA) has been shown to induce a similar HR-like necrogenesis when applied to the growth of our embryogenic cell lines (Perl et al in preparation). Aspirin or related benzoic compounds were shown to repress gene expression activated by JA (Pena-Cortes et al. 1993; Doherty et al. 1988). In animal systems, almost all aspirin-like drugs inhibit the cylooxygenase activity of the enzyme prostaglandin endoperoxidase synthase (Smith and Marnett). Prostaglandins and the phytohormone JA display some similarity in both chemical structure and biosynthetic pathways (Needleman et al. 1996). Interestingly, inhibitors of lipoxygenase will attenuate mammalian inflammatory responses and some plant responses to wounding or pathogen attack (Preisig and Kuc 1987; Stelzig et al. 1983; Pena-Cortes et al. 1993). Thus, we tried to inhibit necrogenesis and subsequently improve transient GUS expression by interfering in three different stages in which this HR-like response might be modulated. We studied the effects of: 1. Proteases. 2. Ion channel blockers, and 3. Jasmonic acid biosynthesis inhibitors, on reduction of necrogenesis and improvement of transformation efficiencies.

Material and Methods

Transformation procedure Intact explants were co-cultivated with *A. tumefaciens* strains EHA 105 harboring the plasmid PCT160. This plasmid carries the coding sequence of *uidA* reporter gene interrupted by the IV2 intron (Eches et al 1986). 2-4 g of UB embryogenic calli were co-cultivated (10 min. at room temperature) with 10

A. Altman et al. (eds.), Plant Biotechnology and In Vitro Biology in the 21st Century, 197–200.

ml suspension of 5-8x10^6/ml *A. tumefaciens.* The suspension was removed from the dish and the embryos were transferred to a Nitsch and Nitsch (1969) NN Basal Medium with or without a booster cocktail for 48 h at 25°C under darkness. Intact 14-days old tobacco seedling were vacumm infiltrated with *Agrobacterium* for 10 min, followed by co-cultivation of 48 h. on solid NN medium. After washing the seedling with NN medium supplemented with 400 mg/l claforan, the seedling were maintained on wet paper filter for 48 h. GUS staining was performed and blue spots or sectors were counted.

Protease treatment: Embryogenic explants were co-cultivated for 48 h with Agrobacteria in NN medium supplemented with 5-20 mg/l of a filter-sterilized protease type XIV originating from *Streptomyces griseus*. Calli were assayed for positive GUS staining 2 days after co-cultivation and for necrotic embryos 7 days after co-cultivation.

Transformation boosting cocktails: Tobacco seedlings and/or embryogenic cells of grapes were incubated for 1 h before contact with Agrobacterium in NN liquid media supplemented with either: Forskolin (0.025-0.2 mM); α-Aminoadipic acid (0.01-0.2 mM); Lanthanum nitrate (0.002-0.01 mM) and Piroxicam (0.05-0.6 mM).

Results and Discussion

Effect of protease treatment on necrogenesis. When embryogenic cell line of UB were co-cultivated with Agrobacterium in the presence of the protease, we observed a dramatic reduction in calli necrogenesis from 97% (without protease treatment) to only 8% (Data not shown). These results indicates that indeed a proteinaceous fraction is responsible for the elicitation of the necroting response of this grape cultivar to the presence of Agrobacterium.

Effect of compounds activating the signal transduction pathway on GUS expression

We assumed that activating the signal transduction pathway prior to co-cultivation might reduce transformation efficiencies. Tobacco seedlings were pretreated for 1 h with either Forskolin or α-Aminoadipic acid. It was evident that Forskolin pretreatment reduced GUS expression from 86% in nontreated seedlings to 11% in seedlings treated with 0.2 mM Forskolin (Fig 1). α-Aminoadipic acid had even a stronger effect and reduced GUS expression to 2% (Fig. 2). Forskolin is a diterpene isolated from the Indian plant Coleus and is an activator of adenylate cyclase leading to increased in cAMP levels and subsequently in the activity of protein kinase A. L-α-Aminoadipic acid is known to selectively increase intracellular concentrations of free Ca^{++}.

Effect of ion channel blockers on necrogenesis

We assumed that blocking the signal transduction pathway might result with an increase in GUS expression following grape-*Agrobacterium* interaction. Before co-cultivation the embryogenic explants were subcultured for 1 h. in fresh NN medium supplemented with Glyburide. The data shows an reduction of cell necrogenesis in UB embryos from 98% (without pre-treatment) to 29%. Best results were obtained when the UB calli were incubated with 0.05 mM Glyburide (Fig 3). The efficiency of GUS staining was increased from 3% in control calli to 18% in the presence of 0.02 mM Glyburide (Fig 4). Glyburide is a sulfonyurea that selectively blocks

ATP-sensitive K^+ channels. Reduced K^+ conductance causes membrane depolarization and influx of Ca^{2+} through voltage sensitive Ca^{2+} channels.

The effect of Prostaglandin biosynthesis inhibitors on necrogenesis

UB embryogenic calli were pre-treated with Piroxicam for 1 h. before co-cultivation. The data indicate that this pretreatment reduced necrogenesis. The best results were obtained with 0.6 mM Piroxicam that reduced cell necrogenesis from 98% (in controls) to 26% (Fig 5). Piroxicam in mammalians inhibits the cylooxygenase (COX) activity of the enzyme PGH and may act in plants as a blocker of JA biosynthesis by blocking the conversion of 13-HPLA to 12-oxo-PDA (Pena-Cortes et al 1993).

Conclusions

This present study demonstrated methods for increasing the efficiency of transformation of certain cultivars by minimizing the necrotic response of these culture to a putative elicitor originating from Agrobacterium tumefaciens. We utilized pharmaceuticals able to either block ion channels or inhibit JA biosynthesis or by proteases able to degrade protein elicitors. The overall increase in efficiency of GUS expression is not restricted to grapes but is also applicable to plants in general including monocotyledonous species (Perl et al in preparation).

Acknowledgments

This research was supported by NCRD Israel; BMFT Germany; Israeli Ministry of Agriculture; and The United States-Israel Binational Agricultural Research and Development Fund.

References

Bauer et al (1995) Molecular Plant Mic. Interc. 8, 484-491.
Coutos-Thevenot P et al (1992) *Champagne Moet&Chandon patent* No. 92FR-005724 920512
Creelman RA et al (1992) Proc. Natl. Acad. Sci. 89, 4938-4941.
Doherty HM et al (1988) Physiol. Mol. Plant Pathol. 33, 377-384.
Eckes P et al (1986) Mol. Gen. Genet. 199, 216-224.
He SY et al (1993) Cell 73, 1255-1266.
He SY et al (1994) Molecular Plant Micr. Interc. 7, 289-292.
Li X et al (1992) Plant Mol. Biol. 20, 1037-1048.
Needleman P et al (1986) Annu. Rev. Biochem. 55, 69-102.
Nitsch JP (1969) Phytomorphology 19, 389-404.
Pena-Cortes et al (1993) Planta 191, 123-128.
Preisig CL et al (1987) Plant Physiol. 84, 891-894.
Smith WL Marnett LJ (1991) Biochim. Biophys. Acta 1083, 1-17.
Stelzig DA et al (1983) Plant Physiol. 72, 746-749.
Thompson CJ et al (1987) EMBO J. 9, 2519-2523.
Vancanneyt G et al (1990) Mol. Gen. Gent. 220, 245-250.
Ward M et al (1995) The Plant Cell 7, 833-844.
Wei R et al (1992) Science 257, 85-88.

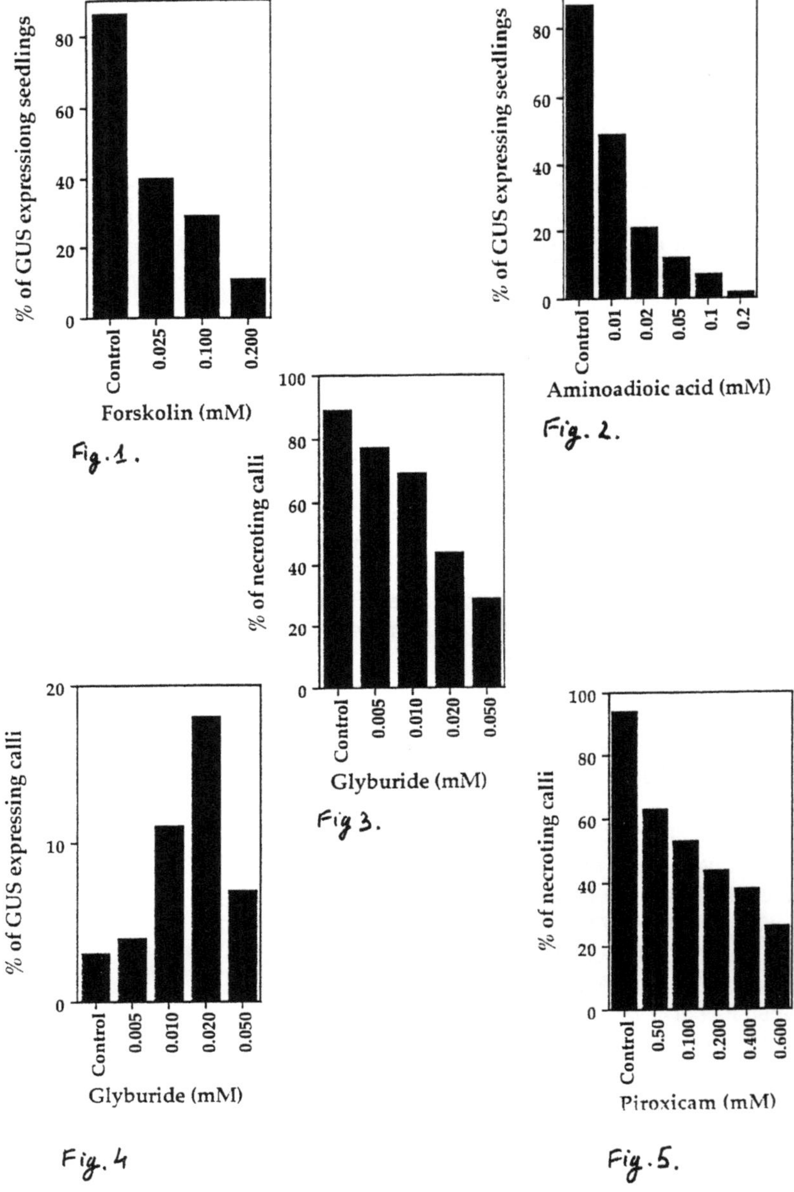

% of GUS expressiong seedlings
80
60
40
20
0
Control
0.025
0.100
0.200
Forskolin (mM)
Fig. 1.
% of GUS expressing seedlings
80
60
40
20
0
Control
0.01
0.02
0.05
0.1
0.2
Aminoadioic acid (mM)
Fig. 2.
% of necroting calli
100
80
60
40
20
0
Control
0.005
0.010
0.020
0.050
Glyburide (mM)
Fig 3.
% of GUS expressing calli
20
10
0
Control
0.005
0.010
0.020
0.050
Glyburide (mM)
Fig. 4
% of necroting calli
100
80
60
40
20
0
Control
0.50
0.100
0.200
0.400
0.600
Piroxicam (mM)
Fig. 5.

PLANT REGENERATION FROM UNINUCLEAR MICROSPORE SUSPENSION CULTURES OF *Aesculus hippocastanum* L.

Ljiljana Radojevic, Nevena Marinkovic, Sladjana Jevremovic and Dusica Calic
Institute for Biological Research "Sinisa Stankovic", 11 060 Belgrade, Yugoslavia

Abstract

Androgenesis induction in suspension culture of uninuclear microspores of *Aesculus hippocastanum* L. was achieved in MS liquid medium supplemented with MS mineral solution (Murashige, Skoog, 1962), 5% sucrose, mixture of vitamins (Radojevic, 1991), l-proline (Pro), casein hydrolysate (CH, 200 mgL^{-1}, each), 2,4 - dichlorphenoxyacetic acid (2,4-D) and 6- furfurylaminopurine (KIN, 4.5 μM, each). Formation of androgenic embryos (AE) in different stages of development ***i.e.*** globular (GE), heart – shaped (HE) and embryo with cotyledons (AECo) were obtained in liquid and / or solid MS medium 4 weeks after suspension plating. The germination (100%) of AECo was achieved in agar medium supplemented with 400 mgL^{-1} l-glutamine (Glu), 0.1% activated charcoal (AC) and 0.5 % polyethylene glycol (PEG). The best plant conversion rate of AECo (11%) was obtained after the successive transfer from agar M_2 (M_0 + AC 0.5 %) medium to solid N medium without hormones.

1. Introduction

The actual genetic improvement of trees requires a different strategy when resorting to *in vitro* methods. The perspective of raising embryos of haploid origin from microspores *in vitro* through anther and / or suspension culture by androgenesis thus offers many foreseeable advantages. The application of these methods for the propagation and improvement of *Aesculus hippocastanum* (Radojevic, 1995) offers the obtainment of many valuable genotypes.
In this report, for the first time, we present our results of recent investigation on induction, formation and plant regeneration from uninuclear microspores of *Ae. hippocastanum* in suspension cultures.

2. Material and methods

2.1. Plant material

Anthers were collected from horse chestnut tree (*Aesculus hippocastanum* L., 2n=40, age 100 years), growing in the Botanical Garden "Jevremovac" in Belgrade. Flower buds (4-5 mm long) were surface sterilized (absolute ethanol, 10 min) and anthers were dissected out.

A. Altman et al. (eds.), Plant Biotechnology and In Vitro Biology in the 21st Century, 201–204.

2.2. Suspension initiation and maintenance

Hundred anthers with uninuclear microspores were cultured in 100 cm^3 liquid MS medium for androgenesis induction (Radojevic, 1991) supplemented with: Pro, CH (200 mgL^{-1}, each), 2,4-D and KIN (4.5 μM, each) at horizontally shaker 4 weeks (wks) in the dark at 25 ± 1 ^{0}C.
The suspension were subcultured every 4 wks by addition of fresh MS medium of the same composition as used for induction of androgenesis.
After 30 days, microspore suspensions were filtrated through sieve (200μ) and plated by Bergmann technique (1960) on the same MS medium with 2.4-D and KIN (0.04 and 4.5 μM, respectively).

2.3. Plant regeneration

For further development of AECo into plantlets, we undertaken the study of the influence of amino acid ***i.e.*** l-Glu and some additives, such as, AC, PEG (Type 8000, product "SIGMA") alone, or in combination with abscisic acid (ABA) on the germination and plant conversion.
Media for germination of AECo (M_0 , M_I– M_{10}): M_0 = MS mineral solution + mixture of vitamins (Radojevic, 1991) + 2% sucrose + 0.7 % agar + Glu 400 mgL^{-1}; M_1 = M_0 + AC 0.1 %; M_2 = M_0 + AC 0.5 %; M_3 = M_0 + AC 1.0 %; M_4 = M_0 + PEG 0.5 %; M_5 = M_0+ PEG 2.5 %; M_6 = M_0 + PEG 5.0 %; M_7 = M_0 + AC 0.1% + PEG 0.5 %; M_8 = M_0 + AC 0.1 % + PEG 5.0 %; M_9 = M_0 + AC 0.1 % + ABA 9.5 μM and M_{10}= M_0 + AC 0.1% + ABA 9.5 μM + PEG 5.0 %.
Medium for plant conversion was MS mineral solution with mixture of vitamins (Radojevic, 1991), 2% sucrose, 0.7% agar, without hormones. Cultures were grown under the fluorescent tubes, "TESLA" of 20 W, 4500 ^{0}K and incandescent light, at 16 h light / 8h dark at 25 ± 1 ^{0}C.

2.4. Scanning electron microscopy

For scanning electron microscopy samples from suspension after 8 wks, were placed on clean stubs (directly from petry dishes) and examined in JOEL JSM Type 35 scanning microscope of different magnification.

2. Results

Uninuclear microspore suspension cultures of *Ae. hippocastanum* were established in MS liquid medium with 2,4-D and KIN (4.5 μ M, each). Seven days after the plating, microspore division occurred and two, three and four-celled embryogenic pollen grains appeared (Figs.1-3.). The lager aggregates ***i.e.*** "micro", embryogenic calli and group of AE at different stages of development (Figs. 4 and 5.) were formed on MS medium with 2,4-D and KIN (0.04; 4.5 μM, respectively).

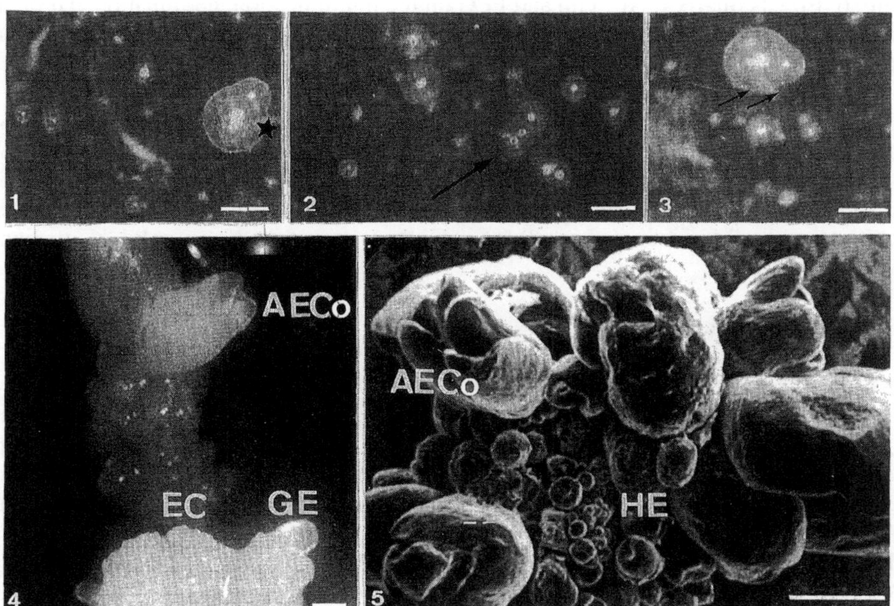

Figs.1-3. Developing two (aster), three (arrow) and four-celled embryogenic pollen grains (2 arrows) after 1 wk in culture. Figs. 4 and 5. Embryogenic complex (EC) after 4 wks (Fig. 4.) and scanning electron microscopy micrograph of EC with GE, HE and AECo after 8 wks in culture (Fig.5.). Figs. 1-5. Bar = 1000 μm.

After the cultivation of AECo on M_0 and M_1-M_{10} media during period of 30-60 days, the germination and, also, the multiplication have occurred (Fig.6.)

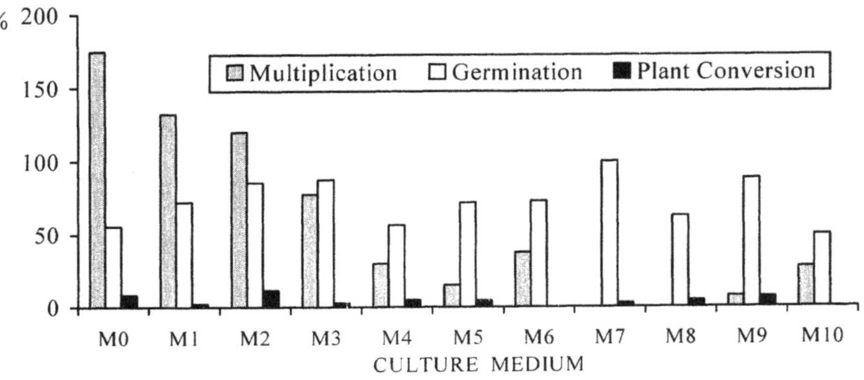

Fig. 6. Effects of ABA, AC, Glu and PEG on the multiplication, germination and plant conversion of AECo *Aesculus hippocastanum* L. obtained in microspore suspension culture.

The germination (100%) of AECo was obtained on M_7 medium containing AC (0.1%) and PEG (0.5%). Increase of AC (from 0.1 to 0.5 %) and PEG (from 0.5 to 5.0 %) stimulate the germination of AECo from 72 to 82 % and from 56 to 73 %, respectively. The best plant conversion rate (11%) was obtained when AECo, after cultivation on M_2 medium with 0.5% AC, were transferred on solid N medium without hormones.

3. Discussion

As it was reported earlier, auxin (2,4-D) and kinetin (4.5 μM, each) were necessary factors for the induction of androgenesis in anther culture of horse chestnut (Radojevic, 1991, 1995). This balance of 2,4-D and KIN had the same effect on androgenesis induction in suspension microspore cultures of same species (Radojevic et al., 1996), while the germination of AECo, required the presence of ABA (9.5 μM), AC (0.5%) and PEG (5.0 %).

The role of ABA, AC and PEG in media for the maturation and germination of somatic embryos of *Ae. hippocastanum* was recently confirmed (Capuana and Debergh, 1997).

Application of this technique represents an original approach for the obtainment of haploids in horse chestnut, as well as in other woody plants.

5. Acknowledgement

This work was supported by The Ministry for Science and Technology of Serbia, grant N^0 03E21.

6. References

Bergmann L. (1960) J. Gen. Physiol. 43: 841-851.

Capuana M. , Debergh P.C. (1997): Plant Cell, Tiss. and Org. Cult 48, 23-29.

Murashige T. , Skoog F. (1962) Physiol. Plant. 15, pp. 473-497.

Radojevic L. (1991) Biotechnology in Agriculture and Forestry Vol. 16, Trees III, pp. 111-141, Springer –Verlag, Berlin, Heidelberg.

Radojevic L. (1995) Somatic Embryogenesis in Woody Plants, Vol. 2, pp. 409-444, Kluwer Academic Publishers, Dordrecht. The Netherlands.

Radojevic L. et al. (1996): 10th FESPP Congress, Septembar 9-13, Florence, Italy, p. 39.

SOMATIC CELL HYBRIDIZATION FOR TRANSFER OF DISEASE RESISTANCE IN *BRASSICA*

U. Ryschka, G. Schumann, E. Klocke, P. Scholze, R. Krämer
Institute for Breeding of Vegetable, Medicinal and Aromatic Plants
Neuer Weg 22/23, D-06484 Quedlinburg, Germany

1. Introduction

Cabbage (*Brassica oleracea* var. *capitata* L.) which is cultivated in many countries of north and south climatic zone on more than 2 million ha, is an important vegetable crop. The yield of cabbage cultivation is limited by diseases and pests. The most important diseases are caused by pathogenic fungi. Clubroot (*Plasmodiophora brassica*) is spread all over the world and causes high losses, particularly on humid acid soils. In many other *Brassicaceae* black leg (*Phoma lingam*) becomes more important to cabbage cultivation, and also *Alternaria* (*A. brassicae*, *A. brassicicola*) is among the most important diseases reducing yields in cabbage. Heavy destruction in cabbage fields is caused by turnip mosaic potyvirus.
This is the reason why introduction of genes for disease resistance is one of the several important goals in cabbage breeding. The genes for resistance are only available in species distantly related to cabbage. Although sexual hybridization between *B. oleracea* var. *capitata* and other wild relatives has been achieved in a few cases, such crosses are extremely difficult to produce, and depend on the genotypes evolved in the cross. Where interesting genes have been identified and isolated, they can be transferred by transformation, but for most traits the genes have not been identified. Somatic hybridization offers an alternate method for the transfer of unidentified genes and polygenic traits. By fusing protoplasts from two species, nuclear as well as organellar genetic information from both parents can be combined in one step. The present investigation reports intergeneric somatic hybridization between *B. oleracea* var. *capitata* as recipient and *Barbarea vulgaris*, *Matthiola incana*, and *Raphanus sativus* as donor, which was primarily attempted to increase the variability and to introgress desirable resistance genes.

2. Material and methods

2.1. Plant materials

The *Brassica oleracea* var. *capitata* used in these experiments was the cultivar 'Korso'. *Barbarea vulgaris*, *Matthiola incana* and *Raphanus sativus* genotypes carried resistance against *Plasmodiophara, Phoma, Alternaria* that had been selected in a lot of accessions.

A. Altman et al. (eds.), Plant Biotechnology and In Vitro Biology in the 21st Century, 205–208.

2.2. Protoplast isolation and irradiation

Resistance donor leaf protoplasts were isolated from *in vitro* grown plants as described (Ryschka et al. 1996). Etiolated hypocotyl protoplasts of *B. oleracea* var. *capitata* were isolated from 5-day-old seedlings grown in dark at 26 °C an MS medium (Murashige and Skoog 1962) without plant growth regulators. The enzyme mixture was the same as used for the mesophyll protoplasts but macerozym R-10 (0.3 %) was used instead of driselase, Mkc-hemicellulase and Pectinase. Prior to the fusion the mesophyll protoplasts of the resistance donors were irradiated with a X-ray dose of 92 Gy.

2.3. Protoplast fusion and culture

The fusion of the protoplasts was conducted according Ryschka et. al. (1996). About $2x10^6$ protoplasts were used for one fusion experiment. The protoplast culture medium consisted of B5 mineral salts and vitamins (Gamborg et al. 1968) and the following components: 45.5 g/l mannitol, 45.5 g/l sorbitol, 2.5 g/l glucose, 125 mg/l ribose, 150 mg/l casein hydrolysate, 875 mg/l $CaCl_2 \cdot 2H_2O$, 0.5 mg/l 2,4-D (2,4-dichlorophenoxyacetic acid), 0.2 mg/l NAA and 0.2 mg/l BAP (6-benzylaminopurine) [modified according to Kartha et al. 1974].
After 3-5 d cultures were transferred in diffuse light (0.65 W/m^2, 16 h photoperiod) and the osmotic pressure of the medium reduced (addition of 500 μl culture medium containing 20 g/l sucrose instead of mannitol and sorbitol, no additional $CaCl_2 \cdot 2H_2O$). A similar dilution of the medium followed 3-4 d later. Within 14 d cell colonies were plated on solid medium (B5 salts and vitamins, 150 mg/l casein hydrolysate, 30 g/l sucrose, 8 g/l Difco agar and plant growth regulators identically to the by protoplast culture medium). 5 weeks after protoplast isolation the callus tissues were transferred to medium in which B5 salts were replace by MS salts (containing 0.02 mg/l NAA and 2 mg/l BAP). 2 weeks later regeneration medium with MS components, 150 mg/l casein hydrolysate, 8 g/l agar, 1 mg/l BAP and 0,04 mg/l GA_3 (gibberellic acid) was used. Shoots bud developed at 25 °C and 6.5 W/m^2 light were transferred to MS medium without hormones and than to MS medium with 0.2 mg/l NAA.

2.4. Molecular plant characterization

Small *in vitro* plants were investigated for their hybrid character with RAPD (random amplified DNA) analyses. For this reason total DNA was isolated and amplified according to Dorokhov and Klocke (1997). For more information about new combinations in the hybrid genomes southern hybridization with various mitochondrial, nuclear and chloroplast specific probes was carried out. the probes were digoxigenin-labeled. The DIG-system for filter hybridization was applied as described by the manufacturer Boehringer Mannheim/Germany.

2.5. Resistance analysis

Resistance to *Alternaria brassicae, A. brassicicola* and *Phoma lingam* were tested according to Ryschka et al. (1996). For tests with clubroot a resting spore suspension (10^6 spores/ml) was used. Inoculum was extracted from diseases roots of Chinese cabbage by electric homogenizer, filtered and centrifuged as described by Williams (1966), and then diluted and quantified with a haemocytometer.

Single plants, planted in 5 cm pots, were inoculated with 2 μl of the suspension. Disease assessment was carried out seven weeks after inoculation.

3. Results and discussion

Two independent fusion experiments have been carried out in which donor protoplasts of *Barbarea vulgaris* and *Matthiola incana* were exposed to X-ray and fused with the recipient protoplasts of *Brassica oleracea* var. *capitata*. The mesophyll protoplasts of *Matthiola incana* were instable and collapsed after fusion. In this cases hypocotyl protoplast were used for the somatic hybridization, but the fusion products could be not distinguished from the parental protoplasts soon after fusion. Three independent fusion experiments have been carried out with the parents *Raphanus sativus* and *Brassica oleracea* var. *capitata*. The fusion frequency was between 10-18 %. After 2-4 months of culture were a total of 44 somatic hybrids by the fusion between *Brassica oleracea* var. *capitata* + *Barbarea vulgaris*, 29 between *B. oleracea* + *Matthiola incana* and 264 between *B. oleracea* + *Raphanus sativus* produced. Hybridity of fusion products was established on the basis of morphology and DNA analyses.

A great deal of variation in the morphology was observed among the somatic hybrids. It was possible to differentiate the hybrid plants according to their morphology to those similar to *Brassica oleracea* and those showing the intermediate morphology between *B. oleracea* and the resistance donor. The hybrids with intermediate morphological exhibit traits like hairs, changed shape of leaves and flowers. Sometimes flower morphology varied within the same plant.

With the RAPD-PCR analysis it was tried to verify the hybrid character in early stages of plant development. In comparison to the DNA-fingerprint from each parent the regenerated plants gave different amplification fragment patterns. The RAPD-pattern of the hybrids *B. oleracea* + *M. incana* had a lot of common fragments with *B. oleracea* but not typical fragments from *M. incana*. However there were differences in this kind that the regenerated plants showed new fragments or some fragments from *B. oleracea* failed. Somatic hybrids of white cabbage with *B. vulgaris* revealed a RAPD-fingerprint similarly to one or another of the parent's, but in each case it distinguished from there with the presence of new fragments or the absence of typical parental fragments. The source of the amplification products by RAPD-analysis is mainly the nuclear DNA, so the RAPD-fingerprint give only a information about these part of total DNA. With digoxigenin-labeled probes it was demonstrated that some of the hybrid plants had new combinations in their mitochondrial DNA.

Somatic hybrids of white cabbage with *Barbarea vulgaris* showed a different pattern of resistance. With the exception of 8 plants from 44 hybrids, resistance to Phoma was found in all regenerants. Resistance to *Alternaria brassicicola* was expressed in 10 plants together with *Phoma* resistance. Five plants showed good resistance to *Alternaria, Phoma* and *Plasmodiophora*. In some tests sensibility of *Alternaria* to environmental and host induced factors resulting in variation of symptom expression could be confirmed. As tests showed, there are good chances to find resistance to *Alternaria* in somatic hybrids from combinations *B. oleracea* var. *capitata* with *Barbarea vulgaris* as well as to clubroot in plants produced by hybridizations between white cabbage and *Matthiola incana*, respectively. Resistance to *Alternaria* was found also in somatic hybrids with *Raphanus*, although *R. sativus* were shown as being susceptible. This indicates that, after protoplast fusion, new resistance principles could be formed of which the relationship is not yet clear.

To transfer single, isolated and cloned genes into plants, methods such as *Agrobacterium* system or direct gene transfer could be used. However, if the goal is to transfer several genes or if the genes of interest are unidentified somatic hybridization may be a more efficient technique.
Asymmetric protoplast fusion provides a method for combining genomes of distantly related species facilitating gene flow from wild species to *Brassica oleracea* var. *capitata*. It also offers a unique opportunity of producing novel combinations of nuclear and cytoplasmic genes that are not possible through conventional methods (Flick et al. 1983; Pelletier et al. 1983, Kirti et al. 1992).

References

Dorokhov, D.B.; Klocke, E. (1997): A rapid and economic technique for RAPD analysis of plant genomes. Russian J of Genetics, 33: 358-265

Flick, C.E.; Bravo, J.E.; Evans, D.A. (1983): Organelle segretation fallowing plant protoplast fusion. Trends Biotechnol. 1: 90-93

Gamborg, O.L.; Miller, R.A. and Ojima, K. (1968): Nutrient requirements of suspension cultures of soybean root cells. Exp. Cell Res. 50: 151-158

Kartha, K.K.; Michayluk, M.R.; Kao, K.N.; Gamborg, O.L. and Constabel, F. (1974): Callus formation and plant regeneration from mesophyll protoplasts of rape plants (*Brassica napus* L. cv. Zephyr). Plant Sci. Lett. 3: 265-271

Kirti, P.B.; Narasimhulu, S.B.; Prakash, S.; Chopra, V.L. (1992): Somatic hybridization between *Brassica juncea* and *Moricandia arvensis*. PlantCell Reports 11: 318-321

Murashige, T. and Skoog, F. (1962): A revised medium for rapid growth and bioassays with tobacco tissue cultures. Physiol. Plant 15: 473-497

Pelletier, G.; Primard, C.; Vedel, F.; Chetrit, P.; Remy, R.; Rousselle, P.; Renard, M. (1983): Intergeneric cytoplasmic hybridization in Cruciferae by protoplast fusion. Mol Gen Genet 191: 244-250

Ryschka, U.; Schumann, G.; Klocke, E.; Scholze, P.; Neumann, M. (1996): Somatic hybridization in Brassicaceae. Acta hort. 407: 201-204

Williams, P.H. (1966): A system for the determination of races of *Plasmodiophora brassicae* that infect cabbage and rutabaga. Phytopatology 56: 624-626

EXPRESSION OF *ARABIDOPSIS THALIANA* ENDO-1,4-ß-GLUCANASE (*cel*1) IN TRANSGENIC POPLAR PLANTS.

Ziv Shani, Mara Dekel, Galit Tsabary, Christian Sig Jensen, Tzvi Tzfira, Raphael Goren, Arie Altman and Oded Shoseyov*. The Kennedy Leigh Centre for Horticultural Research and The Otto Warburg Center for Agricultural Biotechnology. The Faculty of Agriculture, The Hebrew University of Jerusalem, P.O. Box 12, Rehovot 76100, Israel.

* Corresponding author

ABSTRACT

The isolation of an elongation-specific endo-1,4-ß-glucanase-*cel*1 from *Arabidopsis thaliana* was made possible by the fact that considerable homology exists between different endo-1,4-ß-glucanase (EGase) genes from different plants. The *A. thaliana cel*1 cDNA gene was cloned and found to encode a 54-kDa protein. Northern blot analysis of *cel*1 suggested its developmental regulation. RNA transcripts were undetectable in fully expanded leaves as well as at the basal internode of flowering stems. However, a strong transcript signal was detected in the elongating zone of flowering stems. Transgenic poplar plants expressing *cel*1 gene under CaMV 35S promoter, had significantly longer internodes as well as longer fiber cells. These results further substantiate the link between *cel*1 expression and plant cell elongation.

INTRODUCTION

Plant cell elongation is a fundamental process with primary importance in plant-tissue development. Cell elongation requires relaxation of the rigid primary cell wall (Carpita, Gibeaut 1993, Cosgrove 1993, Fry 1988, Roberts 1994). Several mechanisms for this relaxation have been suggested, including the activities of endo-xyloglucan transferase (Nishitani, Tominaga 1992), xyloglucan endotransglycosylase (Fry et al. 1992) and expansins (McQueen-Mason, Cosgrove 1995). EGase has been suggested to play an important role in the elongation process (Shoseyov, Dekel-Reichenbach 1992, Verma et al. 1975) . Most of the EGase genes isolated to date have been studied in relation to fruit ripening and abscission zones. Wu el al. 1996 cloned the EGase gene from pea and showed its expression to be induced by auxin in elongating epicotyls. The *A. thaliana* endo-1,4-ß-glucanase (*cel*1) gene, the mRNA of which is expressed in young, fast-growing tissues, has been recently cloned (Shani et al. 1997).

The progressively increasing consumption of forest products, the steady deterioration of natural forests due to man's activities, and global ecological changes may be leading us to ecological disaster (Dean et al. 1997). In recent years, Populus species have acquired an important place in the basic and applied research of woody plants. The successful transformation of different tree species and the development of molecular breeding methods for woody plants, together with increased demand for wood in the timber, pulpwood and paper industries should lead to extensive, accelerated molecular breeding programs. Structural modification of plant tissues via genetic engineering requires specific promoters that can drive gene expression in the specific tissue of interest in its early developmental stage. Since *Arabidopsis thaliana* has long been regarded as a model system for plants (Goldberg 1988), it is reasonable to explore, whether the data gathered from this tiny annual can be extrapolated to forest trees.

A. Altman et al. (eds.), Plant Biotechnology and In Vitro Biology in the 21st Century, 209–212.

In the present work, we describe the expression pattern of *A. thaliana cel*1 elongation-specific promoter, and the effect of cel1 cDNA expression under CaMV-35S promoter on growth phenotype of transgenic poplar plants

MATERIAL AND METHODS

Plant material and isolation of plant nucleic acids

Arabidopsis thaliana cv. Columbia plants were grown at 24°C under a 16-h photoperiod, using cool-white fluorescent light (50-60 µE m^{-2} S^{-1}). Aspen (*Populus tremula*). Plantlets were maintained by *in vitro* propagation of stem segments in 90 mm vials in the growth room.

Northern blot analysis

A 768-bp DNA probe, base 399-1167 accession # X98544 generated from the *cel*1 cDNA clone was used for northern blot analysis. In each experiment, 40-50 plants were used to extract total RNA from the following tissues: fully expanded leaves, basal internode of the flowering stems and elongating zone of flowering stem.

Construction of transgenic plants expressing cel1 *cDNA under CaMV-35S promoter*

The *cel*1 cDNA was cloned into the binary vector pBI121 (Clontech Laboratories, Inc., Palo Alto, CA). The construct was mobilized into disarmed EHA105 *Agrobacterium tumefaciens* by the freeze-thaw method (Holsters et al.1978). Aspen plant stem segments were transformed as described previously (Tzfira et al. 1997). Transgenic plants were selected on kanamycin. Plants regenerated from independent transformation events, as confirmed by Southern blot analysis were analyzed.

RESULTS AND DISCUSSION

RNA transcript levels of cel1 *in different tissues.*

Northern blot analysis of *cel*1 was carried out using the 768-bp *cel*1 cDNA fragment as a probe (Figure1). RNA transcripts were undetectable in fully expanded leaves, as well as at the basal internode of flowering stems. However a strong transcript signal was detected in the elongating zone of flowering stems of normal plants. These results suggest that Cel1 is involved in *A. thaliana* cell elongation.

Transgenic plants expressing cel1 *cDNA under CaMV-35S promoter*

Transgenic poplar plants expressing cel1 cDNA under CaMV-35S promoter were analyzed. Plants that were positive for Southern and Northern blots were examined under microscope for average length of fiber cells as compared with control wild type plants (Fig 2). Average length of internodes was examined in transgenic plants compared to the control (Fig.3). The results clearly show that expression of *cel*1 cDNA affected growth of transgenic poplar plants.

REFERENCES

Carpita NC, Gibeaut DM (1993) Plant J 3, 1-30.
_Cosgrove DJ (1993) Plant Physiol 102, 1-6.
Dean JFD et al (1997) Engineering Biotechnology, vol 57, pp. 1-44. Springer-Verlag, Berlin, Heidelberg.
Goldberg RB (1988) Science 240, 1460--1467.
Fry SC (1988) Longman Scientific & Technical, New York.
Fry SC et al (1992) Biochem J. 282, 821-828.
Holsters M et al (1978) Mol Gen Genet 163, 181--187.
McQueen-Mason SJ, Cosgrove DJ (1995) Plant Physiol 107, 87-100.
Nishitani K, Tominaga R (1992) J Biol Chem 267, 21058-21064.
Roberts K (1994) Curr Op in Cell Biol 6, 688-694.
Shani, Z et al (1997) J. Plant Mol.Biol. 34, 837-842.
Shoseyov O, Dekel-Reichenbach M (1992) Acta Hort 329, 225-227.
Tzfira T et al (1997) Plant Mol Biol Rep 15, 219--235.
Verma DPS et al (1975) J Biol Chem 250, 1019-1026.
Wu SC et al (1996) Plant Physiol 110, 163-170.

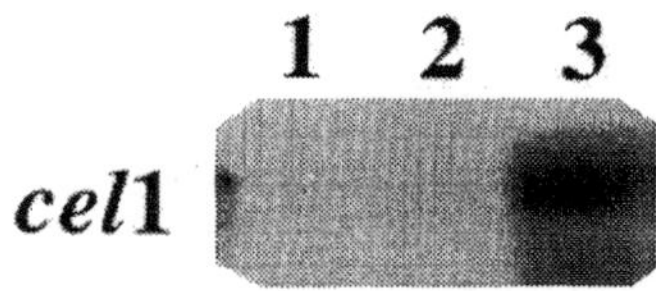

Figure 1. Northern blot analysis of *cel*1. 1. Fully expanded leaf. 2. The basal internode of the flowering stem. 3. Elongating zone of the flowering stem.

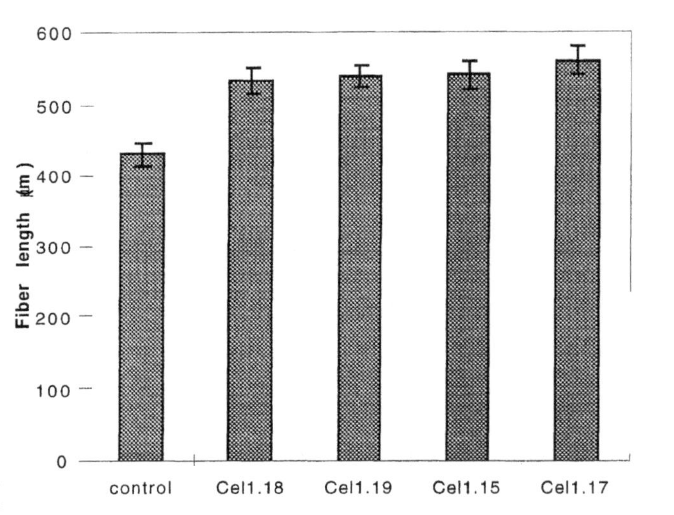

Fig. 2. Average length of fiber cells of several lines of transgenic poplar plants expressing cel1 under CaMV-35S promoter compared with wild type.

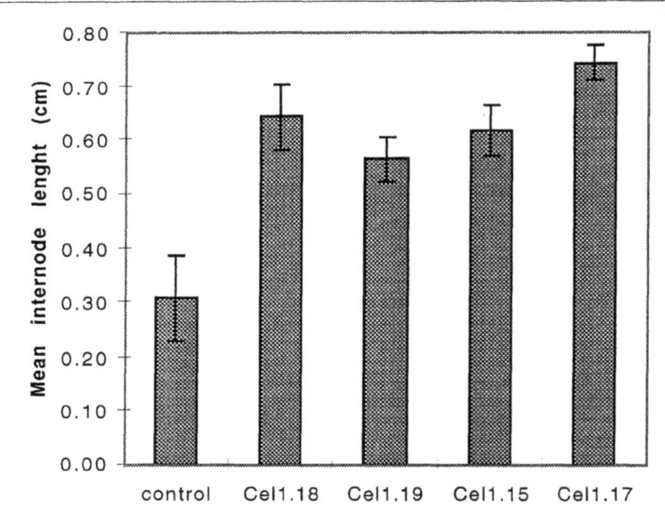

Fig. 3. Average length of internodes of several lines of transgenic poplar plants expressing cel1 under CaMV-35S promoter compared with wild type.

CELLULOSE BINDING DOMAIN INCREASES CELLULOSE SYNTHASE ACTIVITY IN *ACETOBACTER XYLINUM*, AND BIOMASS OF TRANSGENIC PLANTS.

Ziv Shani[1] Etai Shpigel[1,2], Levava Roiz[1] , Raphael Goren[1] Basia Vinocur[1], Tzvi Tzfira[1], and Arie Altman[1], Oded Shoseyov[1]*. [1]The Kennedy Leigh Centre for Horticulture Research and The Otto Warburg Center for Agricultural Biotechnology, The Faculty of Agriculture, The Hebrew University of Jerusalem, P.O. Box 12, Rehovot 76100, Israel.
[2] CBD-Technologies Ltd. Kiryat Weizmann, P.O.Box 156 Nes-Ziona, 74101 Israel.

* Corresponding author.

ABSTRACT

Recombinant cellulose-binding domain (CBD) was found to modulate the elongation of different plant cells *in-vitro*. At low concentrations (0.01-1 μg ml^{-1}), CBD enhanced elongation of *Arabidopsis thaliana* L. roots. At high concentrations (100-500 μg ml^{-1}), CBD dramatically inhibited root elongation in a dose-responsive manner. Maximum effect on root hair elongation was at 100 μg ml^{-1}, whereas root elongation was inhibited at that concentration. Using *Acetobacter xylinum* L. as a model system, CBD was found to increase the rate of cellulose synthase in a dose-responsive manner, up to fivefold as compared with the control. Electron microscopy examination of the cellulose ribbons produced by *A.xylinum*, showed that CBD treatment resulted in a splayed ribbon composed of separate fibrillar subunits as compared with a thin and uniform ribbon in the control. Expression of *cbd* modulated the growth of transgenic plants. Biomass production was significantly higher in selected clones as compared with the control.

INTRODUCTION

Endogenous regulation of cell elongation appears to be dominated by cell-wall mechanics. This process is a result of the interaction between internal turgor pressure and the mechanical strength of the cell wall (Steer, Steer 1989). XG is bound to cellulose microfibrils in the cell walls of all dicotyledons and some monocotyledons (Roberts 1994). The XG bound to the cellulose microfibrils cross-links the cell-wall framework. Plant-cell expansion requires the integration of local wall-loosening and the controlled deposition of new wall materials.

The gram-negative bacterium *Acetobacter xylinum* has long been regarded as a model of cellulose synthesis, mainly because it separates between the cellulose microfibril synthesis and cell wall formation (Ross et al. 1991). Cellulose synthesized by *A. xylinum* is produced as separate ribbons composed of microfibrils, thus potential interactions with other polysaccharides do not exists as in plant cell wall. Since polymerization and crystallization are coupled processes in cellulose synthesis in *A. xylinum,* interference with the crystallization results in the acceleration of polymerization (Benziman et al. 1980). Some cellulose-binding organic substances can also alter cell growth and cellulose-microfibril assembly *in*

A. Altman et al. (eds.), Plant Biotechnology and In Vitro Biology in the 21st Century, 213–218.

vivo. Direct dyes, carboxymethyl cellulose (CMC) and fluorescent brightening agents (FBAs, e.g. calcofluor white ST) prevent microfibril crystallization in *A.xylinum*, thereby enhancing polymerization. These molecules bind to the polysaccharide chains immediately after their extrusion from the cell surface, preventing normal assembly of microfibrils and cell walls (Haigler 1991).
Shoseyov and Doi (1990) isolated a unique cellulose-binding protein (CBPA) from the cellulolytic bacterium *Clostridium cellulovorans* L.. This major subunit of the cellulase complex was found to bind to cellulose, but had no hydrolytic activity and was essential for the degradation of crystalline cellulose. The *cbpA* gene has been cloned and sequenced (Shoseyov et al., 1992). Using PCR primers flanking the cellulose-binding domain gene, the latter was successfully cloned into an overexpression vector that enabled us to overproduce the 17-kDa CBD in *Escherichia coli.* The recombinant CBD exhibits very strong affinity to cellulose (Goldstein et al. 1993).

MATERIALS AND METHODS

Expression and purification of CBD

Overexpression and purification of recombinant CBD was obtained in *E.coli* BL21 (DE3) harboring the pET-CBD plasmid (Goldstein et al., 1993).

Seed germination

Arabidopsis thaliana seeds were washed in 70% ethanol for 1 min, followed by five washes with distilled water. About 100 seeds per treatment were soaked in 1 ml of distilled water containing different concentrations of CBD or BSA in 2 cm diameter, 10 cm long glass culture tubes. The tubes were placed in a growth chamber at 25°C under a 16/8 h of light/dark photoperiod. At different time intervals or after 3 days the lengths of the shoot, root and longest representative root hair were measured in each seedling. The examinations were conducted in three populations of 30 seedlings per treatment.

The effect of CBD on cellulose synthesis in Acetobacter xylinum.

Acetobacter xylinum strain ATCC 23769 was kindly donated by the laboratory of Prof. Moshe Benziman at The Hebrew University of Jerusalem. Cells were grown for 24 h under constant shaking at 30°C in a medium consisting of 0.5% bactopeptone, 0.5% yeast extract, 2% glucose and 0.3% K_2HPO_4 pH 6, containing 1.5 Unit/ml *Trichoderma viride* L. cellulase (Fluka). The cells were harvested by centrifugation and washed twice with pre-cooled phosphate buffer (50 mM NaH_2PO_4 pH 6). The bacterial pellet was resuspended in phosphate buffer to a concentration of 2 mg ml^{-1} dry weight (2.5 $O.D_{600}$=1 mg ml^{-1}). One ml reaction mixtures were placed in 20 ml scintillation vials containing 0.8 mg cells ml^{-1} phosphate buffer. Cellulose synthesis was initiated by the addition of 40 mM glucose (D-[U-^{14}C] glucose, Amersham International Plc, England) at a specific activity of 40,000 cpm μmol^{-1} and was conducted for 1-2 h at 30°C with constant shaking. $^{14}CO_2$ formed was trapped in coverless 1.5 ml tubes containing 0.2 ml 1 M NaOH placed in the reactions vial. The reaction was stopped by the addition of 0.1 ml of HCl 0.5 M to the bacterial suspension and was further incubated for 15 min. One hundred and fifty μl of the NaOH solution containing the trapped $^{14}CO_2$ were transferred to scintillation tubes. The cells and the cellulose were transferred to 1.5 ml tubes, centrifuged and washed three times with water. The cells were lysed by mixing with 0.2 N NaOH 1% SDS; cellulose was recovered on a GF/A filter (Whatman,Maidstone, England) washed with 15 ml of water to remove radioactive background and dried in an oven at 60°C. Filters and NaOH containing trapped $^{14}CO_2$ were counted in a scintillation counter using Opti-fluor (Packard, Meriden, USA) scintillation liquid for glucose incorporation (cellulose synthase activity) and respiration respectively.

Electron microscopy was conducted by placing a copper grid on top of a drop of the appropriate solution at room temperature. The cellulose synthesis reaction contained 0.5 mg ml^{-1} dry weight cells in phosphate buffer, 40 mM glucose with or without CBD at a concentration of 300 μg ml^{-1}. The reaction was incubated for 30 min and then stopped with 2.5% glutardialdehyde for 30 min, washed three times with water and dried. The grids were negatively stained with 1.5% PTA (phospho-tungstic acid) and examined with a Jeol 100 CX electron microscope operating at 80 kV.

RESULTS AND DISCUSSION

The effect of CBD on plant growth in vitro

A. thaliana seedlings were grown in the presence of different concentrations of CBD. Figure 1A shows that low concentrations (0.01-1 μg ml^{-1}) of CBD caused an approximately 30% increase in root elongation relative to the control. Higher CBD concentrations caused a significant inhibition of root elongation in a dose-responsive manner. At lower CBD concentrations, no significant effect on root elongation was observed. However, at 100 μg ml^{-1}, CBD had the opposite effect on root hair elongation, as compared to its effect on root elongation, an almost twofold increase in root hair elongation was observed at this CBD concentration (Fig. 1B). Only at the highest concentration (500 μg ml^{-1}) the effect of CBD on root and root-hair elongation was a similar dramatic inhibition (Fig. 3).

CBD is shown to modulate the elongation of various plant tissues. We propose that the elongation effect of CBD is driven by its ability to bind to cellulose and prevent the normal assembly of microfibrils and consequently cell wall. The inhibitory effect of CBD can be explained by steric hindrance of the cellulose fibrils by excess amounts of CBD, which block the access of enzymes and other proteins that modulate cell elongation via loosening of the rigid cellulose-fibril network. This hypothesis is supported by Nevins, who prevented auxin-induced elongation with anti-β-D-glucan antibodies (Hoson, Nevins 1989) or with antibodies specific to cell-wall glucanases (Inouhe, Nevins 1991).

The effect of CBD on cellulose synthesis in Acetobacter xylinum.

Resting cells of *Acetobacter xylinum* were allowed to synthesize cellulose in phosphate buffer containing radioactive glucose and different concentrations of CBD or calcofluor (as positive control) and BSA (as negative control) for 1 h or for the indicated time. Cellulose synthase activity was determined as glucose incorporation.

Figure 2 shows the effect of CBD at different concentrations (10 - 500 μg ml^{-1} , 0.6 - 30 μM) compared with 1 mM calcofluor and 100 μg ml^{-1} BSA (1.5 μM). CBD increased glucose incorporation in a dose-responsive manner by up to fivefold at 500 μg ml^{-1}. Calcofluor increased the rate by twofold while BSA had no effect. It is evident that polymerization and crystallization are coupled reactions in cellulose synthesis in *Acetobacter xylinum* bacteria (Benziman et al. 1980). CBD enhances incorporation of radioactive glucose in *A. xylinum* by interference with the crystallization process. Our hypothesis is supported by Haigler's (1991) review, in which dyes and fluorescent brightening agents that bind to cellulose alter cellulose microfibril assembly *in vivo*. Modifications in cell shape were observed when red alga (Waaland, Waaland 1975) and root tips (Hughes and McCully, 1975) were grown in the presence of dyes. It is now evident that these molecules can bind to the cellulose chains immediately upon their extrusion from the cell surface of prokaryotes and eukaryotes (Haigler, Brown 1979, Benziman et al. 1980, Haigler et al. 1980, Brown et al. 1982) and prevent crystal-structure formation (Haigler, Chanzy 1988). In addition, the rate of cellulose polymerization has been shown to increase up to fourfold in the presence of dye (Benziman et al. 1980). Crystallization has been proposed to be the bottleneck in this coupled reaction, and its prevention to result in accelerated polymerization.

Electron-microscopy examination of the cellulose ribbons produced by *A. xylinum*, showed that CBD treatment resulted in a splayed ribbon composed of separate fibrillar subunits as compared with a thin and uniform ribbon in the control (not shown). The effect of CBD as observed by electron microscopy is comparable to the effect of CMC (carboxy methyl cellulose) rather than to the effect of calcofluor (Haigler 1991); in both cases the cellulose ribbon only splayed. The effect of CBD on cellulose synthase activity was higher than the effect of CMC, and was comparable and even higher than that of calcofluor. The different effects of CBD, CMC and calcofluor can be attributed to the differences in their molecular weights and their affinities to cellulose. CMC (90 kDa) can only prevent the normal association of larger fibrillar subunits, and therefore hardly alter crystallization, whereas the small molecule calcofluor prevents the glucan chain association immediately after its initiation. CBD is somewhere in-between the two molecules: on the one hand it is not small enough to prevent the association of very small fibrils as done by calcofluor, but on the other hand its high affinity to cellulose makes it an efficient cellulose-intercalating agent, which leads to an up to fivefold increase in cellulose synthesis rate.

Biomass production of transgenic plants expressing CBD

Tobacco f1 plants of p35SC1 (CBD expressed under 35S promoter) and control plants transformed with pBI121 (Clontech Inc) were germinated on plates and grown for 4 weeks. Selected clones expressing CBD were transferred to fresh medium and grown for additional 3-4 weeks in the growth room and then transferred to the green house for 3-4 weeks. Dry weight of the plants was measured. The results indicate that CBD-transgenic plants produce significantly more biomass as compare to the control (Fig.3).

REFERENCES

Benziman M et al., (1980) Proc Natl Acad Sci 77, 6678-6682.
Brown RMJ et al., (1982) 218, 1141-1142.
Cresti, A. Tiezzi, Eds, Sexual Plant Reproduction. Springer-Verlag, Berlin, pp. 89-97.
Goldstein et al., (1993) J Bacteriol 175, 5762-5768.
Haigler CH (1991), *In* CH Haigler, PJ Weimer, eds, Biosynthesis and Biodegradation of Cellulose. Marcel Dekker, Inc, New York, Basel, Hong Kong, pp 99-124.
Haigler C, Brown RM Jr. (1979) J Cell Biol 83, 70.
Haigler CH, Chanzy H (1988) J Ultrastruct Mol Struct Res 98, 299-311.
Haigler C et al., (1980) Science 210: 903-906.
Hoson T, Nevins DJ (1989) Plant Physiol 90, 1353-1358.
Hughes J, McCully ME (1975) 50, 319-329.
Inouhe M, Nevins DJ (1991) Plant Physiol 96, 426-431.
Roberts K (1994) Curr Opin Cell Biol 6, 688-694.
Ross P, Mayer R, Benziman M (1991) Microbiological Reviews 55, 35-58
Shoseyov O, Doi RH (1990) Proc Natl Acad Sci USA 87, 2192-2195.
Shoseyov et al., (1992) Proc Natl Acad Sci USA 89, 3483-3487.
Steer MW, Steer JM (1989) New Phytol 111, 323-358.
Waaland SD, Waaland JR (1975) Planta 126, 127-138.

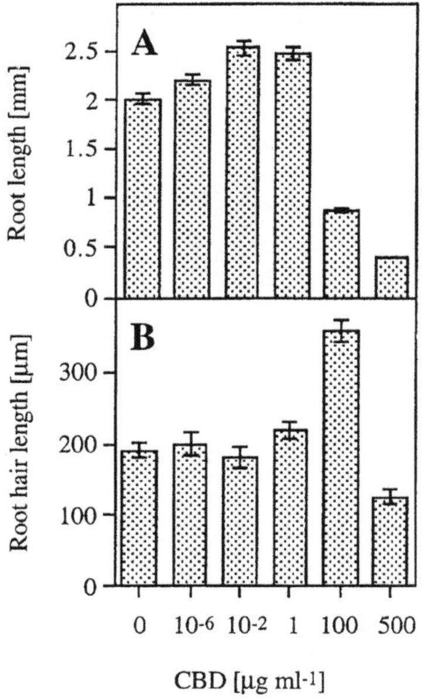

Figure 1. The effect of CBD on root (A) and root hair (B) length. Bars represent standard error.

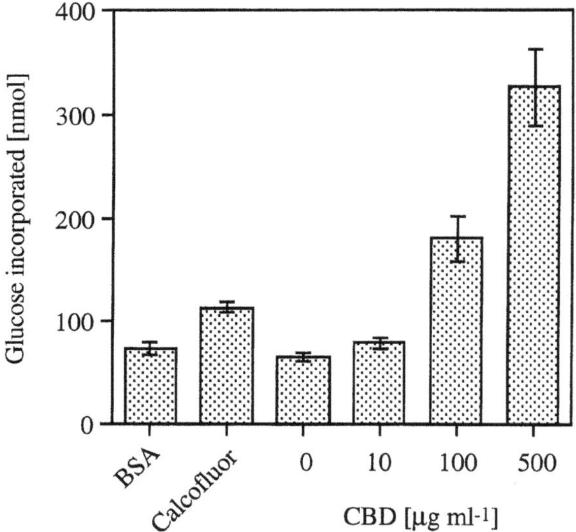

Figure 2. The effect of different concentrations of CBD, 1 mM calcofluor (as a positive control) and 100 μg ml^{-1} BSA (as a negative control) on cellulose synthase activity in *Acetobacter xylinum*.. Bars represent standard errors.

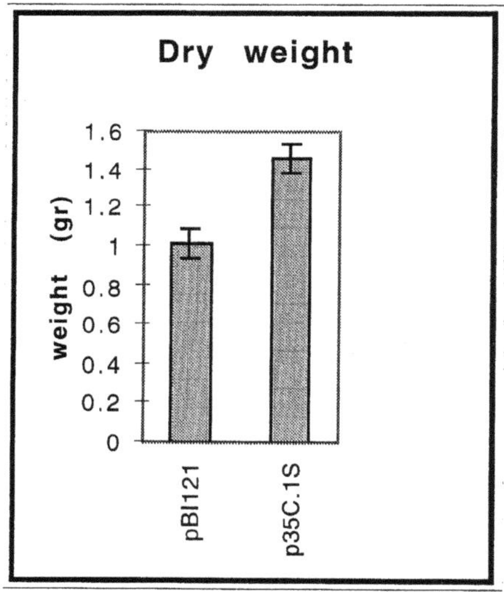

Figure. 3. Biomass production of transgenic tobacco plants expressing CBD (p35SC1)

GENE EXPRESSION IN PLANT-ENDOMYCORRHIZAL FUNGAL COMMUNICATION

O.Shaul, R. David, G. Sinvany, I. Ginzberg, H. Badani, S. Wininger, B. Ben Dor, N. Ovdat, N. Atzmon and Y. Kapulnik.
Agronomy and Natural Resources Dept., Inst. of Field and Garden Crops. ARO, The Volcani Center, Bet Dagan 50250, Israel.

1. Introduction

Vesicular arbuscular mycorrhizal (AM) fungi have been recognized as obligate symbionts of about 80% of the land plants in most ecosystems. The symbiosis is biotrophic and is normally mutualistic; the compatible interactions are based on nutritional transfer between the symbiotic partners. Changes in plant gene expression during the early stages of the mycorrhizal symbiosis have been described (van Rhijn et al., 1997), as well as low-level or transient defense responses (Bonfante-Fasolo, Perotto 1990; David et al., 1998; Gianinazzi-Pearson et al., 1995; Harrison, Dixon 1994; Kapulnik et al., 1996; Lambais, Mehdy, 1993; Volpin et al., 1994; 1995).

The mechanism governing the modulation of plant-defense-related gene expression during AM colonization is not understood. In the present study, we demonstrate the ability of *Glomus intraradices* to down-regulate the expression of chemically induced pathogen-related (PR) genes and the potential involvement of the plant cytokinin in this process.

2. Materials and Methods

Studies were conducted with seeds of *Nicotiana tabacum cv. Xanthi nc* grown under axenic conditions in 0.5-kg pots containing autoclaved sand, and treated as specified by Volpin et al. (1994). Semi-synchronous infection by AM fungi was performed by applying 300-500 spores of *G. intraradices* NPI strain 26 in a layer 4 cm below the soil surface, prior to sowing. Control plants were treated in the same way except for the application of spores; no fungal colonization was detected in the controls during the experiments.

The cytokinins, kinetin and 6-benzylaminopurine (BAP) (Sigma, MO, USA) were applied to leaves by infiltration at a concentration of 10^{-5} M, as described by Klement (1963). In experiments where the synthetic chemical, 2,6-dichloroisonicotinic acid (INA) was used to induce the expression of PR proteins, the treated plants received a single application of a solution containing 50 ppm of the chemical by watering or by foliar spray 4 days prior to sampling (Vernooij et al., 1995).

Protein extraction and Western blot analysis were carried out as previously described (David et al., 1998). RNA extraction and Northern blot analysis were preformed as described previously (Sambrook et al., 1989)

3. Results

Expression of PR proteins in mycorrhizal and non-mycorrhizal tobacco roots

Seventy-two per cent of the inoculated tobacco roots were colonized 5 weeks after sowing. Mycorrhizal and non-mycorrhizal plants showed the same growth rate pattern as reflected by root and shoot biomass (data not presented). Western blot

A. Altman et al. (eds.), Plant Biotechnology and In Vitro Biology in the 21st Century, 219–222.

analysis was used to investigate the expression of the PR proteins. PR-1a was not expressed constitutively in tobacco roots (Fig. 1 - INA), therefore, in order to examine the effect of AM fungus on PR-1a expression, plants were treated with the chemical INA to induce PR-1a. As shown in Fig 1, PR-1a protein accumulated to a high level in INA-treated, non-mycorrhizal control roots, but could not be detected in INA-treated mycorrhizal roots, indicating a suppression of the induction of that gene, by the mycorrihzal fungus. The same blot was tested for the expression of the 32-kDa basic chitinase (PR-3) and the basic glucanase (PR-2).

In contrast to PR-1a, the basic PR-3 is constitutively expressed in tobacco roots (Fig 1, and Neale et al., 1990). In mycorrhizal roots, however, no accumulation of this PR-3 was observed (Fig. 1, -INA). When plants were treated with INA, significant accumulation of the basic chitinase protein was observed in the control plants (Fig. 1,+ INA, -AM), but the level of this protein in INA-treated mycorrhizal plants was significantly lower (Fig. 1, + INA, +AM). The level of PR-2 was not affected by the mycorrhizal fungus colonization (Fig. 1).

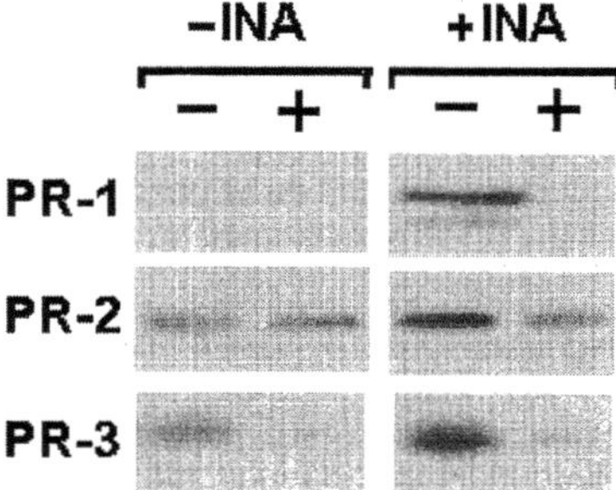

Fig. 1. Effect of INA application on the levels of PR-1a, PR-2 and basic PR-3 in mycorrhizal (+) and non-mycorrhizal (-) tobacco roots. Roots of 35-day-old plants were sampled 96 h after INA application, and their proteins were analyzed by Western blot and probed with the PR- specific antibodies.

Cytokinin affects basic chitinase and PR-1a accumulation

In early attempts to evaluate the cytokinin level in tobacco plant roots, we found that mycorrhizal roots contained twice the quantity of ZR cytokinin-like molecules as the non-mycorrhizal control (data not presented). Western blot analysis was used to investigate the PR-1a, acidic chitinase and basic chitinase levels in tobacco leaves, following cytokinin application. Infiltration of water or a mixture of cytokinins (at 10^{-5} M) into the first expanded tobacco leaf was followed 48 h later by INA application (foliar spray) to induce the PR proteins. Four days after the INA application, significantly lower levels of PR-1a, basic chitinase (32 kDa) and acidic chitinase (28 kDa) accumulation were observed in cytokinin-treated (+) leaves than in the non treated control (-) (Fig. 2). The level of acidic chitinase accumulation was affected to a lesser extent by the cytokinin application than that of basic chitinase (Fig. 2).

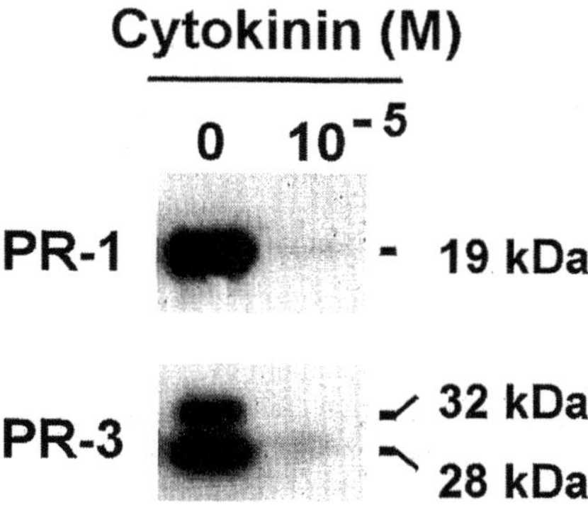

Fig. 2. Effect of cytokinin application on the levels of PR-1a and basic PR-3 in tobacco leaves. Leaves were treated with INA 48 h after the infiltration of the cytokinin mixture and were sampled for protein analysis 96 h later. The Western blot was probed with antibodies against PR-1 and PR-3.

4. Discussion

Mycorrhizal fungal colonization in tobacco roots resulted in significant modifications in gene expression and protein accumulation, compared with non-mycorrhizal roots (presented results and David et al., 1998); we demonstrated a down-regulation of basic chitinase and PR-1a. While coordinated regulation of gene transcription and protein level has also been observed in mycorrhizal bean roots (Lambais, Mehdy, 1993), the present study considers the suppression not only of a constitutively expressed PR protein (basic PR-3), but also of the chemically induced PRs (PR-1a and basic PR-3).

A direct application of cytokinins to tobacco leaves has resulted in reduced accumulation of chitinase and PR-1a proteins (Fig. 2; Shaul et al., personal communication). Similarly, Shinshi et al. (1987) reported a significant decrease of chitinase, and of its mRNA level in cultured tobacco, following elevation of cytokinin in the growth medium. It is tempting to speculate that the higher ZR cytokinin levels found in mycorrhizal plants may be involved in the alteration of the gene expression in mycorrhizae. Cytokinin may serve as a signal molecule in the mycorrhizal system, or act as a growth factor involving in plant development and symbiotic organ development. In fact, a number of cytokinin-like molecules has been observed previously in some monocotyledonous (Allen et al., 1980) and dicotyledonous (Dixon et al., 1988; van Rhijn et al., 1997) plants colonized by AM fungi.

High levels of cytokinin reduce plant resistance to fungal invasion and delay the induction of the hypersensitive reaction (HR) (Beckman, Ingram, 1994; Haberlach et al., 1978). Chitinase and glucanase activities and their transcript levels have been found to be significantly reduced in tobacco tissues, following alterations in the auxin/cytokinin balance (Felix, Meins, 1986). These data reinforce the interpretation that the suppression of the host plant defense response and gene expression observed during the early stages of colonization by *G. intraradices* is an active process, which is of important to the successful establishment of the symbiosis mediated by cytokinin. Changes in hormonal levels of mycorrhizal roots appear to present a new class of secondary plant responses to AM fungus colonization.

Acknowledgment

Part of these investigations were supported by the German-Israel Foundation for Scientific Research & Development, project number:G-0388-199.12/94.

REFERENCES.

Allen M F et al (1980). *Can. J. Bot.* 58: 371-374.
Beckman KB, Ingram DS (1994). Physiol. Molec. Plant Pathol. 44, 33-50.
Bonfante-Fasolo P, Perotto S (1990). In: Electron Microscopy Applied in Plant Pathology, pp 265-275, Springer, Berlin, Heidelberg, New York Tokyo,
David R et al. (1998) MPMI, In Press.
Dixon RK et al. (1988) Tree Physiology 4, 9-18.
Felix G, Meins F Jr. (1986). Planta 167, 206-211.
Gianinazzi-Pearson V et al. (1995). Can. J. Bot. 73, S526-S532.
Haberlach G T et al (1978). *Plant Physiol.* 62: 522-525.
Harrison M J, Dixon RA (1994). Plant J. 6, 9-20.
Kapulnik Y et al. (1996). New Phytol. 133, 59-64.
Klement, Z. (1963) Nature 199, 299-300.
Lambais MR, Mehdy MC (1993) Mol. Plant-Microbe Interact. 6,75-83.
Neale AD et al. (1990) Plant Cell 2, 673-684.
Sambrook J et al. (1989) Molecular Cloning. A Laboratory Manual, 2nd edition. Cold Spring Harbor Laboratory, Cold Spring Harbor, NY.
Shinshi H et al. (1987) Proc. Natl. Acad. Sci. USA 84, 89-93.
van Rhijn P et al (1997) Proc. Natl. Acad. Sci. USA 94, 5467-5472 .
Vernooij B et al. (1995) Mol. Plant-Microb. Interact. 8, 228-234.
Volpin H et al. (1994) Plant Physiol. 104, 683-689.
Volpin H et al. (1995) Plant Physiol. 108, 1449-1454.

DIRECTED FERTILIZATION TO BREED PLANTS FOR TOXIC IONS RESISTANCE DURING POLLINATION

Boris Tokarev and Larisa Lozovsky
Center for Desert Agrobiology
J. Blaustein Inst. for Desert Research
Ben-Gurion University of the Negev, Sede Boqer 84990, Israel

Summary

Pollen tubes growing in the pistil during fertlization are affected by toxic ions, which influence the pollen tube growth rate. Since the tubes of different genotypes have varying susceptibility to the stress ions, the growth of these tubes can be inhibited to different degrees, leading to selective fertilization. It is possible to provide directed fertilization and to achieve some desired changes in the next generation, by influencing the pollen tubes during pollination. In our experiments with nitrate reductase deficient chlorate resistant barley lines, we demonstrate that pollination by F_1 heterozygote pollen under chlorate treatment of the acceptor chlorate-resistant plants significantly changes the segregation in the F_2 generation, with regard to nitrate reductase activity. In the progeny of a control variant of pollination {F_1 (Chlo18 x Viner) x Chlo 18}, the fraction of the plants which are heterozygous for the *nar 1* recessive allele with a normal NRA is close to half of the whole progeny, in conformance with Mendel's law. But for progeny obtained by pollination under chlorate toxic action, the NRA distribution is significantly distorted in favor of the recessive homozygotes which are deficient in NRA. The fraction of the progeny with normal NRA decreases down to 1/7 of the whole F_2 population tested. These results show that there is strong selective pressure which reduces fertilization by pollen tubes carrying a normal *Nar* allele. In other words, they have NR activity and are thus susceptible to toxic chlorate action. We have, therefore, shown that the presence of toxic ions in acceptor plants can act as a fertilization screen, concentrating resistant genes in the progeny. This approach can be used, for example, to select rare resistant mutations.

A. Altman et al. (eds.), Plant Biotechnology and In Vitro Biology in the 21st Century, 223–226.

1. Introduction

When he opened The International Conference on Biology and Ecology of Pollen, D. Mulcashy named a gametophyte phase of plant life "the forgotten generation". But it is in this generation that the most rapid and significant genetic changes occur. (Mulcashy, 1985) For practical application of pollen selection, the following key facts should be stressed.

First, the character of the gene expression in pollen is postmeiotic. Second, there is a selection effect among different pollen genotypes on the next sporophyte generation, demonstrating that gametophytic genes are expressed in sporophyte generation as well as *vice versa*. These genes can be selected both at diploid sporophyte and haploid gametophyte levels. The third and the forth of the main factors are the pollen gaploidy and the huge number of pollen grains, produced by the single sporophyte plant. The above mentioned features permit using the gametophyte systems at the same distinguishing levels as those of bacteria or fungi populations.

It has been shown that up to 80% of the sporophyte genes, controlling different key metabolic processes, are expressed in gametophytes as well, and these values are in very good agreement for different plants. (Tanksley et al., 1984; Sari-Gorla et al., 1994); The susceptibility of the growth rate of the pollen tubes to toxic ions is as variable as that of sporophytes. The presence of different ions in a pistil during pollination can cause drastic changes in the competition between genetically different pollen tubes and can provide for selective fertilization.

In other words, the fraction of ovules fertilized by the pollen tubes resistant to the stress action of the toxic ions will be increased. So the next sporophyte generation will be enriched by plants with improved resistance to the stresses. An analogous approach can be used for selection of the genetic lines with changed patterns of mineral nutrients metabolism. NRA deficient lines of barley are examples of such lines (Tokarev, Shumny, 1977).

In practical applications, the toxic ions in question could be introduced directly into the pistils of the recipient plants by the application of their solutions, or by growing the recipient plants on nutrient solutions containing the corresponding selective agent. It is also possible to select the pollen grains germinating *in vitro* on media with selective agents. Such selected pollen grains can then be transferred to the recipient stigma to obtain their progeny. The validity of both such approaches is shown experimentally (Tokarev et al., 1990). Pollen can be taken from the donor plants treated with mutagens.

This simple idea opens wide new prospects for genetic approaches to the study of the fundamental questions of mineral nutrition and plant resistance to soil stress. The approach may be used as a basis for developing a powerful new experimental tool for the identification and selection for features of mineral nutrition and toxic stress resistance expressed in the huge pollen population.

The numerous data supporting the feasibility for such an approach were collected over the last decade. For example, it has been shown that a gametophyte selection permits salt resistant genotypes in the progeny to be revealed after hybridization of Licopersicon with Solanum (Sacher et al., 1983).

In our model experiments, the possibilities of pollen selection were demonstrated and a feasible experimental technique was developed, using a mutant line of barley deficient in nitrate reductase activity (NRA) Chlo 18. This mutant was selected from the cultivated variety, Viner, on the basis of its chlorate resistance after mutagenic treatment. It has virtually no NR activity as a result of a biochemical point mutation (Tokarev, Shumny, 1977).

Chlorate ClO_3 was widely used in agricultural practice as a system herbicide before introduction of organic synthetic herbicides. It was found that ClO_3 itself is not toxic for plants, but being similar to nitrate NO_3, it can be reduced by nitrate reductase (NR) to ClO_2 which in turn has strong phytotoxic action.

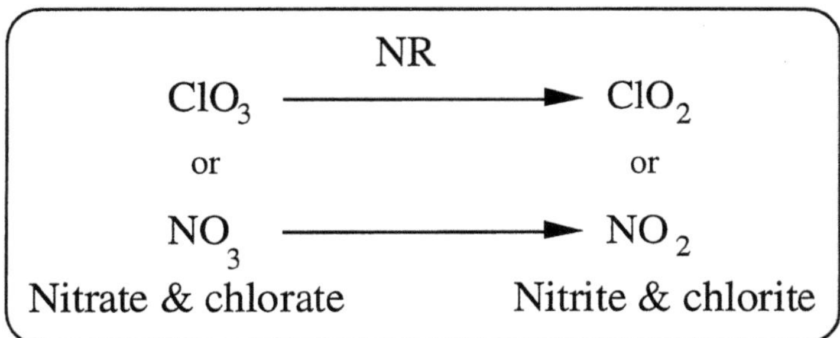

Figure 1. Reduction of chlorate and nitrate by nitrate reductase.

2. Materials and Method

We used as a source of pollen, the F_1 hybrids, (Chlo 18 x Viner), having the normal NRA, and, as the pollen acceptor plants, Chlo 18. a chlorate resistant NRA deficient line. The acceptor plants were emasculated, and a week before pollination, their upper leaves were embedded in the test tubes with 0.1% solution of $KClO_3$.For the control pollination we used emasculated plants without chlorate treatment. Seeds obtained by the artificial pollination were germinated.

Seedlings were grown for 10 days. Their first leaves were used for determination of NRA *in vivo,* Plants with the NRA less than 0.1 of the normal NRA level for a wild type, were considered as NRA deficient homozygotes.

3. Results

The results of the described experiments are presented in Figure 2. We can see that in the progeny of a control variant of pollination a fraction of the progeny heterozygous for a *Nar* 1 recessive allele having normal NRA, is close to a half of the whole progeny, as can be expected from the general theoretical considerations. But for progeny obtained by the pollination under potassium chlorate toxic action, the NRA distribution is significantly distorted in a favor of the recessive homozygotes, which are deficient in their NRA. The fraction of the progeny having normal NRA decreases down to 1/7 of the whole F_2 population tested. These results offer evidence that there was strong selective pressure preventing the fertilization by the pollen tubes carrying a normal *Nar* allele—in other words, having NR activity and thus susceptible to chlorate toxic action.

We can therefore conclude that there was strong selective pressure preventing the fertilization by the pollen tubes carrying a normal *Nar* allele, i.e., having normal NRA and so, susceptible to chlorate toxic action.

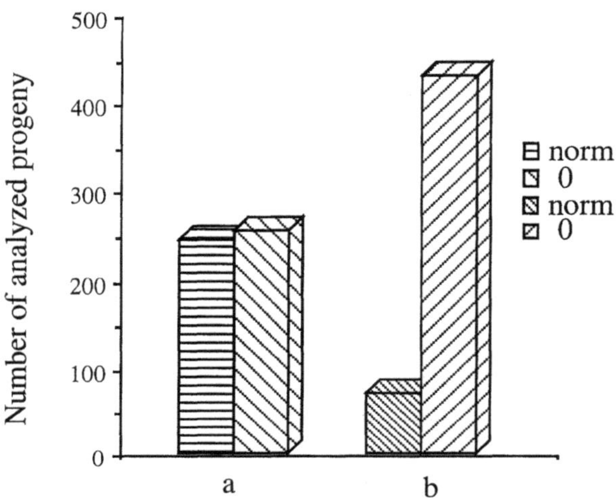

Fig. 2. Effect of chlorate treatment during pollination on the NRA distribution in the Segregations of backross progeny {F_1 (Chlo 18x Viner)} x Chlo 18.} (a): Control pollination. (b): pollination under chlorate treatment.

4. References

Mulcashy D. (1985) The Biotechnology and Ecology of Pollen. In The Proceedings of International Conference University of Mass. Amherst, USA. Preface 7–11.

Sari-Gorla M. et al. (1994) Theoretical and Applied Genetics 88(6-7): pp. 812–817.

Sacher R. F. et al. (1983) Pollen biology and implication for plant breeding: Eds. Mulkashy D.L. and Ottaviano, p. 329.

Tanksley S. et al. (1984) Science, 213, p. 453.

Tokarev B. (1988) The Problems of Application of Microgametophyte Selection in the Plant Breeding, pp. 86–92, Zernograd, Russia.

Tokarev B., Shumny V. (1977) Genetica. (Rus) 13, 2, 2097–2103.

Tokarev B. et al. (1990). Arabidopsis Inf. Service, 1990, 27, 9–12.

IDENTIFICATION OF GENES EXPRESSED IN ROOTS AND NODULES OF *LOTUS JAPONICUS*

K.J. Webb[1], L. Skøt[1], S. Mizen[1], M. Paniske[2], B. Jørgensen[1] and M. Nicholson[1]
[1]Institute of Grassland and Environmental Research, Aberystwyth, SY23 3EB, UK,
[2]The Sainsbury Laboratory, John Innes Centre, Colney, Norwich, NR4 7UH, UK.

1. Introduction

Lotus japonicus is a small, perennial legume that has several important features that make it a suitable subject for a promoter trapping approach to gene isolation. It has a small diploid genome of 12 chromosomes and prolific, autogamous seed production. We have highly inbred lines and can routinely and reliably generate large numbers of genetically transformed plants (Handberg, Stougaard, 1992). This species also offers the important agronomic features of symbiotic interactions with *Rhizobium* species and arbuscular mycorrhiza fungi and secondary products that are lacking in species such as *Arabidopsis* and rice.

In our search for symbiotic genes, we have taken a promoter-trapping approach (Webb et al., 1996) using a version of the β-glucuronidase (GUS) reporter gene lacking a promoter. This gene can only be expressed, and therefore detected, when it inserts into a native plant promoter. In parallel, we established three other populations of *Lotus japonicus* plants: non-transformed regenerants, transformants with selectable markers only and EMS mutants. Primary regenerants, primary transformants and their selfed progeny were also screened for any mutations in their symbiotic interactions with *Rhizobium*..

2. Procedure

Plants were regenerated or genetically transformed from seedlings of a highly inbred line (S_8) of *Lotus japonicus* (Regel) Larsen cv. Gifu B-129-S9 as described by Handberg and Stougaard (1992). Seedlings were transformed using *Agrobacterium tumefaciens* LBA 4404 with either pLX412 (coding for hygromycin resistance) or pΔ*gus*Bin19 (coding for kanamycin resistance with a promoterless GUS, as described by Topping et al., 1991). Primary transformants with putatively tagged genes were grown in hydroponic culture, while seedlings were grown in

A. Altman et al. (eds.), Plant Biotechnology and In Vitro Biology in the 21st Century, 227–229.

vermiculite. For infection with *Rhizobium loti* (strain NZP2238), all plants were grown in nitrogen-free nutrients. For infection with the arbuscular mycorrhiza fungus *Glomus intraradices*, plants were grown as described by Wegel et al., (1998). Tissues with putatively tagged genes were stained with 5-bromo-4-chloro-3-indolyl-β-D-glucuronide (X-gluc) for GUS activity.

3. Results and Discussion

We have identified several plants that are affected in their symbiotic interactions from two out of the four populations of *Lotus japonicus*. No equivalent plants or mutations were seen in primary regenerants and transformants or in their progeny.

In our **EMS mutagenesis programme,** we identified fourteen symbiotic mutants after inoculation with *Rhizobium loti*, ten of which were stable in the M_3 generation. Five mutants were defective in nodulation and five in nitrogen fixation. One nodulation mutant was also defective in its interaction with arbuscular mycorrhiza fungi. Infection attempts of the mutant with both *Glomus intraradices* and *Gigaspora maritima,* were aborted soon after the fungi penetrated the rhizodermis. Host cell death was observed at the infection site, suggesting the induction of a defence response in the mutant roots, possibly triggered by an inappropriate recognition of the symbiotic fungus. The wild-type gene probably functions as an inhibitor of plant defence responses (Parniske et al., 1998).

In our **promoter-trapping programme,** we have successfully tagged genes expressed in developing nodules and in root hairs of *Lotus japonicus* immediately after challenging with *Rhizobium*. In total, we identified seven transformants expressing GUS in nodules.

The functions of the these tagged genes are unknown at present but their expression patterns in each of the seven independently transformed plants suggest that they are distinct from each other and from genes previously cloned from nodules. In some plants, these genes were also expressed in other tissues, such as vascular tissues, connective tissues of anthers and in germinating embryos. Expression of these genes in embryos suggests they have a fundamental role in plant development.

All seven tagged transformants have between one and three copies of T-DNA in their genomes. Using inverse PCR, T-DNA border fragments from two of the transformants have been cloned. One of these transformants (T90), looks particularly interesting since both nodules and root hairs were GUS positive after inoculation with *Rhizobium*. This plant had only one copy of the T-DNA, suggesting that the cloned border fragment represented the gene responsible for GUS activity.

The border fragment from T90 (containing approximately 0.9kb of *Lotus* DNA) was sequenced. It contains TATA box-like elements close to the T-DNA border, which would be consistent with it having promoter activity. The border fragment was used a probe to isolate a 12.5 kb genomic clone from a library of *L. japonicus* cv. Gifu wild-type DNA. Nucleotide sequencing of this clone is currently in progress, and will assist in elucidating the role of this tagged plant gene in symbiosis.

We plan to determine whether these tagged genes are also involved in general plant defence responses by challenging tagged plants with other symbionts and pathogens. Preliminary experiments revealed GUS activity in stem explants of transformant T554 following infection with *Agrobacterium rhizogenes* and in arbuscles in roots of transformant T90 after inoculation with the arbuscule containing plant cells in progeny of transformant T90 after inoculation with *Glomus intraradices.*

Further characterisation of these nodulation mutants and promoter-trapped transformants will provide valuable information for identifying critical processes involved in both nodule formation and arbuscle formation. At the molecular level, we have already cloned one gene – our next step is to confirm its function.

4. References

Handberg K, Stougaard J (1992) *Lotus japonicus*, an autogamous, diploid legume species for classical and molecular genetics. Plant Journal 2, 487-496.

Parniske M, Webb KJ, Martini I, Genre A, Bonfante P (1998) Analysis of *Gigaspora* and *Glomus* infection sites in symbiotic mutants of *Lotus japonicus*. International Congress of Mycorrhiza (ICOM). Upsala, Sweden.

Topping JF, Wei W, Lindsey K (1991) Functional tagging of regulatory elements in the plant genome. *Development,* **112**, 1009-1019.

Webb KJ, Nicholson MM, Jorgensen B, Mizen S, Skøt L (1996) T-DNA mutagenesis in *Lotus japonicus.* Emerging Model Legume Systems: Tools and Recent Advances. Knoxville, Tennessee July 12-14, 1996. Abstr. no. 5.

Wegel E, Schauser L, Sandal N, Stougaard J, Parniske M (1998) Mycorrhiza mutants of *Lotus japonicus* define genetically independent steps during symbiotic infection. MPMI (in press).

5. Acknowledgements

The authors thank David Hughes for invaluable technical support, Dr Jens Stougaard (Denmark) for providing *Lotus japonicus* seed and Dr Barry Scott (New Zealand) for providing *Rhizobium loti* strain NZP2238. IGER is grant aided by BBSRC. BJ and MN were funded by BBSRC PMB phase II.

Agrobacterium-MEDIATED STABLE TRANSFORMATION OF BARLEY (*Hordeum vulgare* L.)

H. Wu, A. C. McCormac, M. C. Elliott and D-F. Chen
The Norman Borlaug Institute for Plant Science Research, De Montfort University, Scraptoft, Leicester LE7 9SU, U. K.

1. Introduction

Advances in tissue culture methodology and development of new, more potent T-DNA vectors have rendered major cereal crops, including rice (Hiei et al. 1994), maize (Ishida et al. 1996), barley (Tingay et al. 1997) and wheat (Cheng et al. 1997) amenable to *Agrobacterium*-mediated DNA delivery. The type and quality of starting materials have proved to be of critical importance. In the case of barley, it was necessary both to pre-wound the immature embryos (IEs) and also to remove the embryo axis, in order to achieve a good level of recovery of stable transformants. The use of IEs iposes a particularly rigorous quality requirement. *Agrobacterium*-mediated transformation of barley has, to date, been reported only for the variety Golden Promise and it is likely that more recalcitrant varieties will demand adaptation of the methodology. In order to address these issues, alternative barley tissue-culture systems have been used to prepare targets for *Agrobacterium*-mediated transformation.

2. Materials and Methods

2.1. Plant materials, bacterial strains and plasmid vectors

Three spring varieties (Golden Promise, Dissa, and Delibes) and four winter varieties (Intro, Puffin, Igri, and Pastoral) were grown in a growth chamber at 12^0C, 50%-70% humidity and 16 h photoperiods of 25W/m^2. For winter varieties, a vernalization treatment was carried out at 6^0C for 8 weeks. IEs 1.0-2.0 mm long were used for callus induction. Calli were transferred to liquid M-1 medium (Wu et al. in press) for the establishment of the IE derived suspension (IED). Anthers containing microspores at the mid/late uninucleate stage were harvested for shed microspore culture (Wu et al. in press). The microspore-derived embryoids/proembryoids (E/P) were transferred to M-1 liquid medium for the establishment of microspore-derived suspension (MSD) or for co-cultivation directly. A set of new binary vectors in a supervirulent *Agrobacterium* strain, EHA101(pBECKS) (McCormac et al. 1997) was used.The T-DNA vectors contained visually-detectable marker genes (*gusA* , *Cl/Lc* or *sgfp-S65T*), and also herbicide resistance gene *(bar)*. The *Agrobacterium* culture was detailed in McCormac et al. (1998).

A. Altman et al. (eds.), Plant Biotechnology and In Vitro Biology in the 21st Century, 231–234.

2.2. Co-cultivation, selection and regeneraton
IEs, suspension cultures (IED and MSD) and E/P were immersed in bacterial suspensions for 15-20 minutes and then transferred to M-1solid medium or MN6 medium (Wu et al. in press) for co-cultivation at 25^{0}C for 4-5 d. They were subsequently transferred to M-1 medium supplemented with 5 mg l^{-1} PPT (for the first round of selection) or 10 mg dm^{-3} (for the second round of selection) and cefotaxime at 250 mg dm^{-3}. Calli were subcultured onto fresh selection medium at intervals of 3 weeks. After at least two rounds of selection, surviving calli were transferred to FHG medium (Hunter 1988) for regeneration.

2.3. GUS histochemical assay and Southern blot analysis
GUS activity was assayed according to the histochemical method of Jefferson (1987), 20% methyl-alcohol was added to the assay solution to eliminate endogenous GUS activity. Southern blot analysis was carried out according to standard protocols.

3. Results and Discussion

3.1. Transient gene expression
After the inoculation of EHA101 (pBECKS.red) to IEs, all three spring varieties and four winter varieties displayed red cells as the result of expressing the foreign *C1/Lc* genes. The frequency of T-DNA delivery (the number of IEs displaying at least one red cell out of the total number of IEs) varied from variety to variety, so did the number of pigmented cells on each IE (see Table 1). The cells displaying the transformed phenotype were typically associated with a limited region of the embryo, being almost exclusively located on the surface of the scutellum structure which was immediately adjacent to the coleoptile/coleorhiza axis. Only in very few (0.01% of the total) red cells was the reporter expressed in cells on the opposite surface of the scutellum. This type of cell/tissue specific targeting was also observed when using pBECKS.GUS/intron and pBECKS.sGFP. The application of acetosyringone to the inoculating bacterial culture did not change the localization of transformation. IEs inoculated on M-1 or MN6 showed comparable patterns of transformation; however, incubation on MN6 medium gave higher and more consistent numbers of red cells. No cell proliferation was observed in the targeted regions.

Table 1. Transient expression of *C1/Lc* genes when IEs of different varieties were targeted by EHA101 (pBECKS.red).

Cultivar	IEs with red spots/total	Max. spots/IE	Average spots/IE
Delibes	40/58 (64.7%)	188	32
Dissa	26/39 (66.7%)	205	22
Golden Promise	58/66 (87.8%)	485	48
Igri	13/38 (34.2%)	17	5
Intro	42/64 (65.6%)	110	19
Pastoral	28/56 (50.6%)	50	16
Puffin	29/61 (47.5%)	97	25

3.2. Selection of Transformants
Calli derived from cells co-cultivated with EHA101 (pBECKS.GUSint/bar) were selected on solid M-1 medium supplemented with 5 mg l^{-1} PPT. When suspension cultures (IED or

MSD) were used, effective elimination of non-transformd cells required completion of the second round of PPT at10 mg l^{-1}. Mosaic staining for GUS was first detected 6 weeks after co-cultivation. After 3 months of selection, histochemical detection revealed uniformly stained callus pieces. The numbers of calli surviving this selection are summarised in Table 2. It was noteworthy that the prolonged tissue culture regimes resulted in a loss of the capacity to regenerate plants from the transformed callus lines.

Table 2. Transformed barley cell lines and plantlets obtained after *Agrobacterium*-mediated transformation of suspension cultures (MSD and IED) and E/P.

Variety	Starting material	Nature of transformed material	No. resistant callus lines/ plantlets showing GUS (%of inoculated cell clusters)
Dissa	IED	Callus lines	13 (1.9%)
Igri	MSD	Callus lines	3 (0.4%)
Dissa	E/P	Plantlets	3 (2.0%)

When E/P were directly co-cultivated with EHA101(pBECKS.GUSint/bar), and a shortened period of *in vitro* culture was employed, regeneration of transformed plantlets was achieved. Southern blot analysis of one such plantlet confirmed the presence of transgenes (Fig.1).

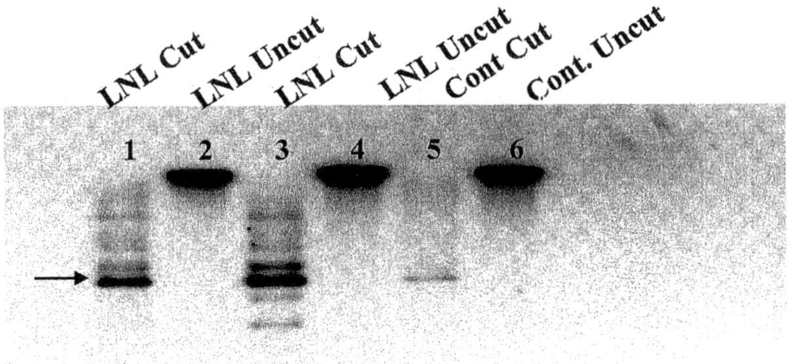

Fig 1. Southern blot analysis of a PPT-resistant callus line (**a,f**) and PPT- resistant plantlet (**d,i**) with probes for the T-DNA-contained *bar* (**a-e**) or *gusA* (**f-j**) genes. Control samples are shown in panels **b,c, g** and **h**. Predicted *Agrobacterium* bands are shown in panels **e** and **j**

3.3. Southern blot hybridization

A number of independently transformed lines of the barley callus as well as one plantlet were analysed by Southern blot hybridization. Fig 1 shows the banding patterns of a transformed callus line (A47-2-6) and a plantlet (A87-1-b) . Bands which hybridized strongly to probes for the *gusA* and *bar* genes were detected. No evidence of major rearrangments of the transgenes was found after use of restriction enzymes which released predictably sized fragments from the T-DNA restriction of the T-DNA at a single site showed that the probed genes had been disassociated from the original binary plasmid and inserted elsewhere. These observations support the conclusion that *Agrobacterium*-mediated transformation had resulted in the insertion of the T-DNA fragment into the barley genome. Different banding patterns, which varied in the size and the number (from 1 to 5) of fragments hybridizing to T-DNA probes, were observed after independent transformation events.

4. Summary

Barley IEs were co-cultivated with a supervirulent *Agrobacterium* strain [EHA101(pBECKS)]. The T-DNA vectors contained visually-detectable reporter genes (*gusA*, *C1/Lc* or *sgfp-S65T*). The IEs of all seven barley varieties displayed transient T-DNA expression. However, all three reporters showed that T-DNA delivery was preferentially targeted to those regions which were characterised by a poor capacity for callus induction. Cells of suspension cultures which had been co-cultivated with EHA101 (pBECKS GUSint/bar) were able to proliferate under selection regimes and a number of PPT resistant callus lines, which expressed the co-transferred *gusA* gene, were isolated. When E/P were used as starting materials, and a shortened period of *in vitro* culture was employed, transformed plantlets were regenerated. Southern analysis of one such plantlet confirmed the presence of transgenes.

5. Acknowlegements

Dr. Wu is funded through the UK MAFF, Crop Molecular Genetics Programme.

6. References

Cheng M et al. (1997) Plant Physiol. 115, 971-980
Hiei Y et al. (1994) Plant Journal 6, 271-282
Hunter CP (1988) Ph. D. Thesis, Wye College, University of London
Ishida Y et al. (1996) Nature Biotechnology 14, 745-750
Jefferson RA (1987) Plant Mol Biol Rep 5, 387-405
McCormac AC et al. (1997) Molecular Biotechnology 8, 199-213
McCormac AC et al. (1998) Euphytica 99, 17-25
Tingay S et al. (1997) Plant Journal 11, 1369-1376
Wu H et al. Plant Cell Tiss. Org. Cult. (in press)

Characterization of a pollen specific gene and its functional analysis in transgenic plants

George J. Wullems, Floyd, R.A. Wittink, Peter de Groot, Jan Derksen and Jan A.M. Schrauwen.
Department of Experimental Botany, Research Group Molecular Plant Physiology.
University of Nijmegen, The Netherlands.

Introduction

Gene regulation during pollen development

Although the morphological and chemical aspects of pollen development have been studied in great detail, knowledge of the underlying molecular processes is relatively limited. The developmental events leading to anther formation and pollen release are precisely timed and regulated (Koltunow *et al.*, 1990; Scott *et al.*, 1991). These processes occur in a strict, chronological order that correlates with floral bud size. This opens an easy way to collect relative homogenous populations of pollen at discrete stages of development.

The high transcriptional and translational activity during pollen development leads to a considerable increase in the total amount of RNA, mRNA and dry weight from the late unicellular microspore stage up to mature pollen. A detailed study of the stage-related expression of mRNAs in the gametophyte during four developmental stages showed that some mRNAs are only present in a single stage. Some other mRNAs are present during more than one stage but their concentrations vary. The latter transcripts either accumulate up to maturity, or reach their highest concentration already before anthesis (Schrauwen *et al.*, 1990; Bedinger and Edgerton, 1990). In addition, there are also RNAs present at a constant level during all stages investigated.

Pollen formation can be divided in two developmental periods, with different sets of genes expressed (Mascarenhas, 1990). Transcripts of the 'early' genes appear soon after meiosis and are reduced or undetectable in mature pollen. Transcripts of 'late' genes are first detected after microspore mitosis and continue to accumulate as pollen matures. Progress in the isolation of pollen expressed genes has been made by differential screening of cDNA libraries from both anthers and pollen. In this way genes have also been characterised with a less distinct temporal expression pattern than the 'early' and 'late' genes (Twell *et al.*, 1993).

The knowledge of genes involved in pollen development has been focused on 'early' sporophytic genes which are almost all expressed in the tapetum, and 'late' genes which are maximally expressed in mature pollen.

A. Altman et al. (eds.), Plant Biotechnology and In Vitro Biology in the 21st Century, 235–239.

Expression of the late pollen specific gene ntp303 during development and pollen tube growth.

The late pollen-specific gene *ntp303* (Weterings *et al.*, 1992) was isolated by differential screening from a cDNA library prepared with RNA from mature pollen. The gene is fully characterised concerning its sequence, its promoter composition and its spatial and temporal expression kinetics (Weterings, 1993). Northern blot analysis confirmed that the *ntp303* RNA is only detectable in bicellular and germinating pollen.

In tobacco pollen, transcription of *ntp303* starts after pollen mitosis I and continues during pollen tube growth with a strong accumulation of the messenger during maturation. *Ntp303* is regulated concerted with a group of other genes in pollen of tobacco. Their transcripts appear after the first haploid mitosis and accumulate during the maturation of the microgametophyte (Schrauwen *et al.,* 1990). Northern blot analysis in our lab have proven that the pattern of *ntp303* gene expression during in vitro ripening of pollen is similar to the expression pattern of *ntp303* during in vivo maturation (van Herpen and de Groot, pers. Comm.). This shows that the expression of the gene is regulated by the haploid gametophyt itself, without any interaction with sporophytic factors. In addition, pulse labelling experiments *in vitro* RNA labelling during germination have shown that the gene is also actively expressed in the growing pollen tube. (Weterings *et al.*, 1992). This active transcription of a pollen specific gene during germination is in contrast to previous reports on an anther specific clone from *Lycopersicon esculentum* (*lat52*), which suggest that no transcription occurs during the first hours of pollen tube growth (Twell *et al*,1989; Ursin *et al.*, 1989).

In situ hybridisation experiments showed that transcription starts at mid-bicellular pollen stage and was only detectable in the vegetative cell. *Ntp303* transcripts are still present near the tip of pollen tubes 72 hours after pollination of the pistil (Reijnen *et al.*, 1991; Weterings *et al.*, 1995). These data indicate a function of the gene during pollen tube growth or during the process of fertilisation.

Dissection of the *ntp303* promoter for the *cis*-regulatory elements was assayed by 5'deletion analysis and the micro-projectile facilitated transient expression technique. These experiments showed that the minimal pollen-specific promoter is located between -103 and -51 nucleotides upstream of the transcriptional start site. In this region two hexameric sequences AAATGA, of which one was inverted, proved to have transcription activating properties. Using gel mobility shift assays, we demonstrated that this minimal functional promoter interacts with a leaf nuclear GT-1 binding activity. These data suggest that the ubiquitous transcription factors are involved in the pollen-specific expression of the gene (Hochstenbach *et al.*, 1996).

The tissue specificity of the NTP303 protein is in agreement with the pollen-specific expression of the *ntp*303 gene as established earlier, but the kinetics of the protein appearance differs from transcription kinetics (Weterings *et al.*, 1992). Whereas the transcripts were detectable from the early bi-cellular stage of pollen development , the protein was detectable not earlier than one hour after germination and accumulated during pollen tube growth. This suggests a repression of translation during maturation.

In conclusion, gene *ntp*303 encodes a 69 kD glycoprotein, whose synthesis is differentially regulated during pollen maturation and pollen tube growth.

Immuno-localization of the NTP303 protein

Electron microscopic immuno-localisation studies showed that the NTP303 protein is located in the vegetative, and not in the generative membranes of maturating pollen grains, in the pollen tubes and in callose plugs. On protein gel blots however, the NTP303 protein remained below the detection level during pollen maturation but increased tremendously above this level during pollen tube growth, *in vivo* as well as *in vitro*. The discrepancy in time of protein appearance as revealed by immuno-localisation and protein gel blot analysis can be explained by the specific accumulation of NTP303 protein in the membranes of the pollen tube. In this respect it is interesting to compare the late pollen gene *ntp*303 with the class of late pollen genes described earlier by Mascarenhas (1993). These genes show an accumulation of transcripts during maturation which are translated early during germination (Mascarenhas, 1975; Mascarenhas and Mermelstein, 1981). The ntp303 gene fits in the class of late pollen specific genes.

The NTP303 protein is present in the microgametophyte from bicellular pollen to germinated pollen tubes, but not in the pollen tube-tip. The protein is located in the vegetative membrane of the vegetative cell including the membrane surrounding the generative cell and sperm cells. Besides the location of NTP303 at the vegetative plasma membranes, the protein was also present in callose plugs. However, because callose and NTP303 do not co-localise argues against an involvement of NTP303 with callose biosynthesis.

The location of NTP303 in all vegetative membranes except the pollen tube tip and its strong accumulation during development and pollen tube growth raised the question of its particular function. As NTP303 is strictly expressed in pollen and homologous transcripts are found in many species, its function must relate to a pollen-specific event.

In conclusion, we have shown that the late pollen gene ntp303 is translated during pollen development, germination and pollen tube growth in a very specific manner. The protein is specifically present in the plasma membrane of the vegetative cell, the vegetative cell that envelopes the gametes and in callose plugs.

Loss of function analysis of the pollen specific gene ntp303

To study the function of the gene *ntp*303, both an anti sense and co-suppression approach were used to examine the effect of abolishing the production of NTP303 on pollen development and pollen tube growth. Therefore, the *ntp303* promoter has been used to express both anti sense and additional sense messengers of the *ntp303* gene in transgenic plants. The *ntp303* promoter has a similar activity as the *lat52* promoter used in comparable studies (Lee *et al.*, 1996; Muschietti *et al.*, 1994; Xu *et al.*, 1995). However, *ntp303* is strictly expressed in pollen whereas *lat52* also exhibit some activity in the endosperm (Twell, 1995). Based on the reported gametophytic expression of *ntp303*, 1:1 transmission of the linked kanamycin-resistance in seeds of self-pollinated progeny of primary transformed plants was used to select transgenic plants. Progeny analysis using *ntp303* transformed plants as male and wild type plants as female revealed that transgene constructs were not transmitted to the progeny, suggesting that NTP303 is involved in effective fertilisation.

To ascertain this relation between cause and effect, single copy plants were chosen for further analysis. Some multicopy plants were also selected to include transgenic plants, which might exhibit a more severe pollen phenotype. However, no relation was found between the number of transgenic inserts and an effect on pollen phenotype or seed set.

Within the selected plants the amount of *ntp303* messengers in pollen grains was reduced to 75 or 50 % of the wild type level. A reduction of 50% in *ntp303* messengers is the expected level for a single copy insertion of a post-meiotic activated gene. So, half of the pollen population should have an abolished NTP303 production. The fact that the plants could be selected on bases of their non-Mendelian segregation pattern indicates that *in vivo* the function of NTP303 is affected. As transformed pollen is capable of *in vitro* germination, *ntp303* is apparently not involved in hydration or not important for the initial germination. So, NTP303 is involved in a pollen event that occurs during travelling the style or is accomplished in the fertilisation itself. To discriminate between a possible retarded pollen tube growth and an affected gamete delivery it is necessary to obtain seeds that result from transformed pollen only. A pollen selection procedure or pollen embryogenesis resulting in homozygous transgenic plants could be helpful to answer this (Touraev *et al.*, 1995).

In conclusion, the results indicate that the transgenic pollen, in which the *ntp303* gene is silenced, do not take part in effective fertilisation since the linked kanamycin-resistance marker is not transmitted to the offspring plants through the male germ line.

References

Bedinger PA, Edgerton MD (1990) Plant Physiol 92:474-479

Hochstenbach R, De Groot P, Jacobs J, Schrauwen J, Wullems G (1996) Sex Plant Reprod 9:197-202

Koltunow AM, Truettner J, Cox KH, Wallroth M, Goldberg RB (1990) Plant Cell 2: 1201-1224

Lee, H.S., Karunanandaa, B., McCubbin, A., Gilroy, S., and Kao, T.H. (1996) Plant J. 9, 613-624.

Mascarenhas,J.P. and Mermelstein, J. (1981) *Acta Soc. Bot. Polon.* 50, 13-20

Mascarenhas JP (1990). Ann Rev Plant Phys Mol Biol 41:317-338

Mascarenhas, J.P. (1993). *Plant Cell* 5, 1303-1314.

Muschietti, J., Dircks, L., Vancanneyt, G. And McCormick, S. (1994). *Plant. J.* 6, 321-338.

Reijnen WH, Vanherpen MMA, De Groot PFM, Olmedilla A, Schrauwen JAM, Weterings KAP, Wullems GJ (1991). Sex Plant Repr 4:254-257

Schrauwen JAM, De Groot PFM, Vanherpen MMA, Vanderlee T, Reynen WH, Weterings KAP, Wullems GJ (1990). Planta 182:298-304

Scott R, Hodge R, Paul W, Draper J (1991). *Plant* Science 80:167-191

Toeraev, A., Fink, C.S., Stöger, E., and Heberle-Bors, E. (1995).. PNAS 92, 12165-12169.
Twell D, Wing R, Yamaguchi J, McCormick S (1989). Mol Gen Genet 217:240-245
Twell D, Patel S, Sorensen A, Roberts M, Scott R, Draper J, Foster G (1993). Sex Plant Reprod 6:217-224
Ursin VM, Yamaguchi J, McCormick S (1989) Plant Cell 1:727-736
Weterings K, Reijnen W, Vanaarssen R, Kortstee A, Spijkers J, Vanherpen M, Schrauwen J, Wullems G (1992). Plant Mol Biol 18:1101-1111
Weterings K, Reijnen W, Wijn G, Vandeheuvel K, Appeldoorn N, Dekort G, Vanherpen M, Schrauwen J, Wullems G (1995) . Sex Plant Repr 8:11-17
Wittink F.R.A. (1998). Thesis University of Nijmegen.
Xu, H., Knox, B., Taylor, P.E., and Singh, M.B. (1995).. PNAS 92, 2106-2110.

DETECTION OF ECONOMICALLY IMPORTANT VARIABILITY IN MICROPROPAGATION

A. C. Cassells[1], S. M. Joyce[1], R. F. Curry[1] and T. F McCarthy[2],
[1]Department of Plant Science and [2]Department of Biochemistry, National University of Ireland, Cork, Ireland

1. Introduction

Variability in micropropagated plants includes random phenotypic variation ('somaclonal variation') and developmental variation. In the former, the progeny may require grading and if the variants are above an arbitrary percentage, the client may reject the batch. In the latter, plant maturation may be altered in the population with consequent economic losses (George, 1993). While the presence, elimination or partial elimination of intracellular biotic contaminants, chimeral breakdown and, or rearrangement and unstable loci may explain some of the variability encountered, these problems should be avoided by testing of the plants prior to clonal propagation (Stage 0 in micropropagation; see Cassells, 1997). A generalization is that the appropriate choice of cloning strategy will also avoid somaclonal variation (George, 1993) However, even using nodal cloning, economically important variability in the form of epigenetic variation can be encountered (Cassells et al. 1997; 1998). Since the whole batch may be uniformly affected, e.g. 'hyperjuvenile' (developmentally retarded), this represents economically important variability for the industry.
Micropropagators frequently lack knowledge of the crop and variety which is based on trials i.e. large scale cultivation and inspection for disease symptoms and genetic aberrations. Often, there are constrains on the use of nodal culture and they are obliged to use axillary bud proliferation or adventitious regeneration. In axillary bud proliferation it is difficulty to be confident with multiple re-cyclying that the cloning has not drifted to a mixed pathway of axillary bud proliferation and adventitious regeneration as has been suggested in strawberry where genetic instability increases with time in culture (Boxus, pers.comm.). In potato, where nodal culture is used commercially, there are restrictions on the number of subcultures/time in culture (Cassells 1997). Literature surveys show that for a given species/genotype, different authors have used a very wide range of media and culture conditions which affect microplant morphology and thus may be influencing crop yield characteristics (Cassells et al. 1998).
Here, the use of flow cytometry, RAPD analysis of DNA and of bisulphite-modifed DNA, and image analysis, for the detection of genetic and epigenetic variability of potato microplants and their progeny is described and compared with results obtained with image analysis (Cassells et al., 1997; 1998).

2. Procedure

2.1 Materials And Methods

A. Altman et al. (eds.), Plant Biotechnology and In Vitro Biology in the 21st Century, 241–244.

2.1.1 Tissue culture

Nodal culture and adventitious regeneration of *Solanum tuberosum* 'Golden Wonder' was as described in Curry and Cassells (1998).To induce morphological variation at the microplant stage, the control nodes were grown in agar-solidified medium and the treatment batches were grown in polyurethane foam impregnated with liquid medium (Cassells and Walsh, 1998). In both cases, the cultures were grown in Magenta vessels with Suncap lids (Sigma Chemical Co., Dublin, Ireland).

2.1.2 Genome analysis

RAPD analysis was as described in Cassells et al. (1997) and Joyce et al. (1998); bisulphite modification of DNA was as described in Joyce et al. (loc. cit.). Flow cytometry was as described in Curry and Cassells (1998).

2.1.3. Morphological assessment..

Image analysis of leaf shape parameters as described in Cassells et al. (1997; 1998). Morphological assessment was based on the UPOV guidelines.

3. Results

Flow cytometry detected the presence of aneuploids and polyploids in a random population of 50 adventitious regenerants from potato internode explants. The euploid value for the DNA content of potato using mung bean as a standard (Curry and Cassells, 1998) was 3.102 +/- 0.097 pg DNA. 4 anueploids with DNA content ranging from 3.95 to 4.19 were detected; as were 2 polyploids with DNA contents of 6.3 and 6.2 pg, respectively. As tissues may contain cells in the G2 phase of cell division, plants were only deemed polyploid when a peak of c. 12.5 pg was detected (Curry and Cassells, loc. cit.).

RAPD analysis was carried out using random primers to distinguish between juvenile and mature potato leaves and between the different phenotypes (vitromorphs) produced in agar and foam culture. No differences were detected using random primers. RAPD analysis of DNA and of bisulphite treated DNA from microplant leaves and mature leaves from the glasshouse and agar- and foam-grown potato microplants was also carried out. RAPD analysis of bisulphite modified DNA, however, detected band polymorphisms between juvenile and mature leaves and less evident differences between the vitromorphs. These differences were more obvious when customised primers with CG and CA tails were used, especially so for the primers with CG-tails at their 3' ends (Fig 1).

4. Discussion

A prerequisite for the selection of the techniques used here was that they be cost effective. RAPD- and image-analysis and flow cytometery, have low consumable costs and do not require high labour skills once set up. DNA based techniques involving DNA restriction (e.g. AFLPs) or the design of specific primers were rejected on cost (Seal, 1997) and degree of difficulty, respectively.

Flow cytometry can be used to monitor genome stability at the callus and microplant population level using polyploidy or aneuploidy as a marker (Curry and Cassells, 1998); RAPDs are capable of detecting somaclonal variation at the microplant population level (e.g. Cassells et al. 1997). RAPDs using random primers may detect random variability in the population but fail to detect individual morphological variants (Cassells et al. 1997) so these techniques cannot be used to discard aberrart phenotypes at 100% efficiency, but may be used to reduce the number of aberrants below the economic threshold.Flow cytometry cannot detect epigenetic chage that does not involve large changes in DNA content. Here it has been shown that RAPD analysis of bisulphite modified DNA has the potential to detect epigenetic differences based on changes in methylation. The principle of the treatment is that bisulphite modification of DNA converts cytosine to uracil but does not modify methylated cytosine (Frommer et al. 1992). Consequently, after modification, strands may be non-complementary if they were unmethylated, and DNA from different developmental stages where the differences are associated with different levels of methylation, will show asymmetric band amplification. Image analysis is capable of detecting genetic variability at the population level based on increase in the standard deviation of polygenic traits e.g. leaf shape but requires character stabilization before measurements can be made i.e. leaf expansion (Cassells et al. 1997). Guard cell measurements can also be used to detect variability in ploidy but also measurements must be carried out on an expanded leaf.

Micropropagators are aware that changes in media and the culture environment affect the morphology of microplants and have the problem that they cannot predict the consequences of optimizing the *in vitro* stages for the quality of the microplants produced. Previously, it has been shown that image analyses can detect epigenetic variability and determine relative maturity of plants from leaf shape parameters in potato and pelagonium (Cassells et al. 1997; 1998). The limitation is that microplants, samples of production, have to be grown to the first expanded leaf stage before this evaluation can be made and a reference population or images derived from same where the microplants are grown in a controlled environment, must be available.

Further work is underway to test the generality of the conclusion from the comparison of the methods described here and to evaluate the potential of modified DNA RAPDs analysis to predict relative maturity of microplant originating from different protocols based on relative numbers of bands present after bisulphite treatment.

5. References

George EF (1993) Plant Propagation by Tisuue Culture, Exegetics, Basingstoke, UK.
Cassells AC (1997) Plant and Microbial Contamination in Micropropagation pp. pp 1-14, Kluwer, Dordrecht, The Netherlands
Cassells AC et al. Angew. Bot. 71, 125-130
Cassells AC et al. (1998) Potato Seed Production by Tissue Culture (Ph. Boxus et al. eds.) CEC, Luxembourg (in press).
Curry RF, Cassells AC (1998) Methods in Molecular Biology: Plant Culture Protocols (RH Hall ed.), Humana Press, New York (in press).
Cassells AC, Walsh C (1998) Proc. COST 822 WG1 Meeting, Berlin. CEC, Luxembourg (in press).
Frommer et al. (1992) proc. Natl. Acad. Sci. USA 89, 1827-1831
Joyce SM et al. (1998) Pottao Seed Production by Tissue Culture (Ph Boxus et al. eds.), CEC, Luxembourg (in press).
Seal S (1997) Plant and Microbial Contamination in Micropropagation (AC Cassells ed.) pp. 123-130, Kluwer, Dordrecht, The Netherlands

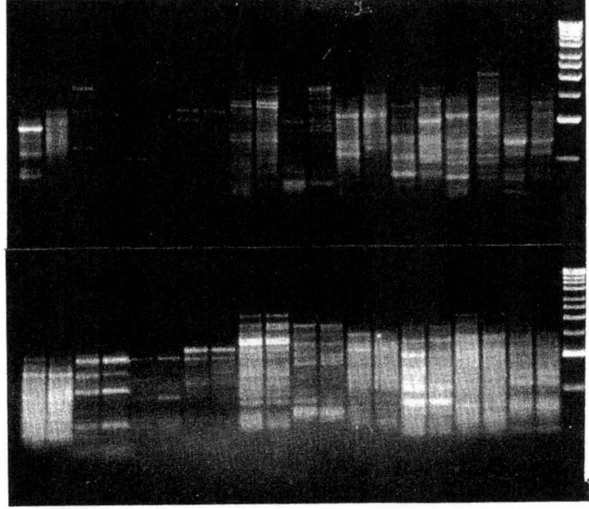

Figure 1. (*top*) Band polymorhisms between bisulphite modified DNA from juvenile and mature potato leaves. The DNA samples are compared in paired lanes using primer with CG-tails at their 3' ends; first 10 lanes; 3'CA primer tails lanes;12-20. Lane 21 is a 1 kb DNA reference ladder. The primers used for lanes 1-10 were 5'AGGCGGCACG3'; 5'CATACGGGCG3'; 5'GACGCTATCG3'; 5'GTTCGCTCCG3' and 5'GAGCTGGCG3'; for lanes 12-20 were 5'TCCACGGGCA3', 5'CCTGGGTCCA3', 5'AGGCCGGTCA3', 5'CAGCGGGTCA3' and 5'TCGTGGCACA3', respectively.
(*bottom*) Band polymorphisms for bisulphite modified DNA from in vitro leaves from agar and foam grown microplants, respectively. Details as above.

ISOLATION OF TISSUE CULTURE-INDUCED POLYMORPHISMS IN BANANAS BY REPRESENTATIONAL DIFFERENCE ANALYSIS.

C.Cullis[1] and K. Kunert

NovoMark Technologies LLC, Cleveland, OH, USA
[1]Department of Biology, Case Western Reserve University, Cleveland, OH 44106 USA

1. Introduction

Tissue culture is an essential component of plant biotechnology and genetic manipulation. The appearance of variation in cultures and regenerated plants (somaclonal variation) is an unavoidable consequence of this process. The appearance of plants that are not true-to-type through the in vitro propagation is a continuing concern. This concern has limited the use of in vitro propagation of elite germplasm in some cases.

A number of molecular tools have been investigated to develop molecular markers which will aid in the detection and minimization of somaclonal variants. Included in these technologies are restriction fragment length polymorphism (RFLP), random amplified polymorpic DNAs (RAPDs) and amplified fragment length polymorphisms (AFLPs) and simple sequence repeat polymorphisms (SSRs). Each of these has had some utility, but the drawback of all of them is that they tend to look at a very small fraction of the genome for any one particular reaction. Therefore a large number of primers or probes have to be tested in order to find useful markers.

The experiments described in this paper are based on the premise that there is a labile fraction of the genome which will be altered whenever a somaclonal variant is observed (Cullis 1977). That is, there are sites within the genome which are especially labile and may be altered, often without any specific phenotypic variation occurring. The isolation of such regions will enable the use of these sites to monitor the progress of the variation occurring, and also to estimate the degree to which phenotypic variation may be found in the regenerants from culture. Therefore two different types of markers may be available, one which is indicative of genomic variation occurring, and the second which is specific to a given phenotypic change. Both of these markers have their utility and both should be detectable by the approach described here.

A. Altman et al. (eds.), Plant Biotechnology and In Vitro Biology in the 21st Century, 245–248.

The test system we have chosen is the banana crop since this crop poses some special problems in attempts to improve the characteristics of popular varieties. The dessert banana is a triploid and is sterile so it cannot be improve by conventional breeding. The plants are propagated by a tissue culture process which generates off-types at a rate which increases with time in culture. The genetic basis for these off-types is unknown and only one potential molecular marker for a single off-type (dwarfism) has been described (Damasco et al 1996). Thus a general method which can be used to generate a series of useful markers would be valuable in bananas and may point the way to obtain similar useful markers in other vegetatively propagated plants.

2. Methods

2.1 Representational Difference Analysis (RDA)

Representational difference analysis belongs to a class of DNA subtractive technologies (Lisitsyn et al 1993). The basic concept is that a comparison between two DNAs can be made by using one of hem in excess in a hybridization reaction to remove all the sequences held in common between the two DNAs. Therefore, what remains are those sequences which differ between the two. However, for complex DNAs, it is necessary to reduce the complexity of the hybridizing mixture for the subtractions to be efficient. The reduction in the complexity can be achieved by any method which reproducibly generates a subpopulation of the genome. Obviously, when a representation is being used, only some part of all the differences existing between the two DNAs will be detected.

RDA depends on the use of the polymerase chain reaction (PCR) to amplify very low concentrations of DNA sequences obtained without the need to start with prohibitively large amounts of material. The methodology is based on altering the ends of the DNA sequences so that only certain combinations can be amplified, so that a steady depletion of common sequences occurs. The depletion is done by hybridizing using increasingly large amounts of the driver (the source of the sequences to be eliminated) with reducing amounts of he tester (the source of the sequences of interest to be isolated). After each round of hybridization and amplification, only those sequences which have formed a double stranded fragment, with both strands arising from the tester, will be amplified. The subtraction product is then subjected to further rounds of amplification until no further change in the subtracted fraction occurs. At this stage the subtraction fragments are inserted into a plasmid vector for further identification and characterization.

The representations have been generated by restriction digests with various enzymes. To generate fragments which have structural polymorphisms, restriction enzymes unaffected by the methylation status of the DNA have been employed, while to isolate possible modification polymorphisms methylation sensitive restriction enzymes have been used.

The subtractions can be performed as a one-on-one (one DNA used as the driver and one used as the tester) or one on many (one DNA used as the tester and many different DNAs mixed as the driver). The power of the many-on-one subtraction is that a polymorphism common to a series of lines, but different from an original line can be isolated in a single step.

2.2 Plant material.

The initial banana subtractions were performed using DNAs isolated from the cultivars Williams and an identified, characterized off type which had arisen through in vitro propagation. The initial off-type was one that showed the masada and chlorotic phenotypes. The subtractions were performed using BamHI and HpaII representations. The subtractions were performed in both directions, that is, using the off-type DNA as the tester and Williams DNA as the driver and the reverse, Williams DNA as the tester and the off-type DNA as the driver. The difference clones are used in hybridizations with DNAs isolated from Williams, a variety of off-types derived from Williams through tissue culture, Grand Nain (GN) and dwarf types derived from GN.

2.3 Results.

Difference clones were obtained from four separate subtractions using the BamHI and HpaII representations. Using the BamHI representations, subtraction products were obtained when either of the DNAs (Williams or the off-type) was used as the tester. The clones obtained from these subtractions are currently awaiting further characterization. The HpaII subtractions have been more extensively characterized. A number of the difference clones have been hybridized with labeled amplicons from the tester and driver DNAs to identify those which are genuine difference clones and those which are effectively escapes from the subtraction. Two of these cloned sequences have been characterized further.

2.3.1 Plasmid cHpa15

When this plasmid was labeled and hybridized to HpaII digested DNAs from Williams and a variety of off types polymorphisms were apparent. Three bands were observed in the Williams digest, and zero (in the Williams off-types) or one (in the GN off-type). This indicates that in the Williams off-types there appears to be a possible deletion of this sequence, and that all the five off types so far tested show a polymorphism for this sequence. This may not be surprising as the tested off-types were all characterized as masadas or chlorotic or both. However, it is promising that the subtraction has yielded a fragment which is altered in all of the tested off-types, as hypothesized earlier.

2.3.2 Plasmid cHpa5

When this plasmid was labeled and was hybridized to HpaII digested DNAs from Williams, GN and a variety of off types, polymorphisms were apparent. There was a very faint signal with Williams DNA and a significantly stronger one with GN. The normal GN gives a series of bands and some residual hybridization at the origin. However, two different dwarf GN lines give a much weaker signal in the lower molecular weight bands and a varying signal at the origin (one being stronger and the other being weaker). The data for the Williams off-types is somewhat different. In general there was a stronger signal at the origin, and three of the off-types had two novel low molecular weight bands.

3. Conclusions

The preliminary evidence reported here indicates that RDA is a useful technique for identifying polymorphisms arising during the process of tissue culture. The sequences identified so far appear to be variable in all the off-types tested and so appear to have a use in identifying the occurrence of change during in vitro propagation.

A method for identifying specific alterations associated with a particular off-type could include a two stage process. Initially a set of variable sequences common to many off-types could be identified by using a pooled driver DNA fraction subtracted from a single tester. This has been shown to be useful in other systems (OH and Cullis unpublished). Then a pool of independently arising similar off-types can be used as the driver, with, added to this pool, the sequences already identified which commonly alter. In this way a small subset of polymorphisms, one, or some, of which should be involved in the generation of the observed phenotype would be isolated.

Polymorphic markers such as those described above have a number of uses. The commonly varying sequences could also be used to determine the fraction of the regenerants which are likely to be variant by checking a representative sample of the regenerants for alterations. They could also be used to monitor the stability of the in vitro cultures and to rapidly determine the effect of media changes on the stability of those cultures. Finally, the characterization of the molecular nature of the polymorphisms should shed some light onto the mechanisms by which somaclonal variation arises.

4. References

Cullis CA (1977) Heredity 38, 129-154.
Damasco O et al (1996) Plant Cell Rep. 16, 118-122.
Lisistyn NA et al (1993) Science 259, 946-951.

GENETIC STABILITY ANALYSIS OF CRYOPRESERVED *PRUNUS* FERLENAIN ROOTSTOCK BY RAPD AND AFLP

B.HELLIOT[1], D.MADUR[2], M.BRISON[1], E.DIRLEWANGER[2], MT. de BOUCAUD[1]

[1] LPCV, Univ. Bordeaux 1, Av. des Facultés, 33405 TALENCE, FRANCE

[2] UREFV, INRA, BP81, 33883 VILLENAVE D'ORNON Cedex, France

Prunus is a genus of the Rosaceae family, including 77 species with a wide range of fruit trees, ornamentals and also rootstocks (e.g. plum, peach, apricot, cherry and almond) which are economically important in the temperate climate regions. Threatened by natural aggressions (diseases and climat) and by human activities (agricultural technologies and deforestation), the biodiversity of the *Prunus* germplasm must be preserved. Cryopreservation is an interesting strategy for the conservation of genetic ressources, which allows to overcome some of the serious limitations encountered by the traditional strategies (field, seed or *in vitro* collections). However, few works have been done to check the genetic stability of the regenerated material after cryopreservation.

For this study, we decided to compare two different cryopreservation procedures : i) **the slow-freezing procedure** which is a rapid method but in which cryoprotection of meristems is realized with a solution containing 12.5% of the mutagenic suspected DMSO, ii) **the encapsulation-dehydration procedure** which is more robust and time consuming. In case of cryopreservation of *Prunus* Ferlenain meristems, both techniques gave 50-60% of regrowth rate. Moreover, in order to avoid any somaclonal variations occurring during *in vitro* culture preceding the freezing, we decided to start a new *in vitro* collection from a mother plant selected in the INRA orchard. From the never micropropagated *in vitro* plantlets, called **Vi**, we excised meristems that we separated in three subsets to obtain : i) *in vitro* shoots, called control-control **(CCi)**, obtained just after the meristem culture, ii) *in vitro* shoots grown from encapsulated-dehydrated meristems either frozen (frozen bead, **Fbi**) or not (control bead, **Cbi**), iii) *in vitro* shoots grown from frozen meristems by the slow-freezing procedure (frozen, **Fi**) or from PVS-2 cryoprotected meristems (control, **Ci**). The index "i" refers to the shoot Vi from which meristems were excised. The genetic stability of the cryopreserved material has been checked with two different molecular biology techniques. We used first the **R**andom **A**mplified **P**olymorphism **DNA** (RAPD), a simple, rapid and common tool used for plant characterization and then the **A**mplified **F**ragment **L**ength **P**olymorphism (AFLP) which is more reproducible, reliable and commonly used for plant fingerprinting. Both strategies allow to cover large regions of the genome.

For the RAPD study, we utilized 42 selected 10-mer oligonucleotide primers on 78 initially tested, to compare 59 vitroplantlets regenerated from cryopreserved meristems either by the slow-freezing procedure or by the encapsulation-dehydration procedure with 103 positive controls (mother plant, Vi, CCi, Ci, Cbi) and one different genotype (Fereley rootstock). For the AFLP study, we used 9 different primer combinations to compare 35 vitroplantlets regenerated after

A. Altman et al. (eds.), Plant Biotechnology and In Vitro Biology in the 21st Century, 249–250.

cryopreservation, either by the slow-freezing procedure or by the encapsulation-dehydration procedure with 17 positive controls (mother plant, Vi, CCi, Ci, Cbi) and with two different genotypes (Fereley and Avifel rootstocks).

First, both molecular biology techniques we used, allowed us to characterize each genotype with many genetic markers among the 85 RAPD fragments and 565 AFLP fragments. Moreover, we didn't observe any difference between the patterns of cryopreserved material, either by the slow-freezing procedure or by the encapsulation-dehydration procedure and between the patterns of the frozen samples by the two procedures and the non frozen samples. We can, first conclude that DMSO seems to have no mutagenic effect on the plant genome. Then, temperature variations occuring during the freezing and thawing steps seems also to have no effect on the genetic stability. Thus, cryopreservation by the slow-freezing procedure, which is rapid technique, can be used in the future.

Cryopreservation seems to have no negative effect on the genetic stability of the frozen meristems and appears for that reason to be a very attractive tool to protect the biodiversity of the *Prunus* species. In fact, we can now consider the creation of a *Prunus* germplasm bank, maybe in Bordeaux, where is already located the European *Prunus* DataBase.

BANANA GERM-PLASM IMPROVEMENT AT RAHAN MERISTEM

Eli Khayat
Rahan Meristem, Kibbutz Rosh Hanikra, 22825, Israel. E-mail: rahan_bio@rahan.com

1. Introduction

The high incidence of "off-types" produced by banana meristem culture is a major concern to commercial growers. At the same time, somaclonal variation is an important tool for the improvement of banana germplasm. Banana plant propagators and researchers have learned, over the years, to reduce the frequency of "off-types" to a manageable level. The majority of the banana "off-types" in the "Cavendish" variety appear as 'dwarf' or 'giant'. It has been demonstrated that both are associated with sensitivity to gibberellic acid (Cote et al., 1990). However, the specific genes responsible for these mutants have not been identified so far. Several gibberellin insensitivity mutants have been isolated and characterized in other plant species, but none of these have been associated with somaclonal variation. Unlike "Cavendish" types, an array of phenotypic variations was reported in plantains. Most variations in plantains were related to the morphology of the inflorescence and fruit (Vuylsteke et al., 1991). Thus, it seems that the highly frequent types of variation are genotype specific.

2. Field Selection

The components of yield in bananas are bunch weight and the number of bunches per hectare. Due to somaclonal variation, individual clones often differ with respect to yield and fruit quality. The deviation within a micropropagated population of clones was used for explant selection.

The selection scheme was initiated with three hundred mats (three plants per mat) of 'Grand Nain' banana plants which were propagated by meristem culture from six mother clones according to Cronauer and Krikorian (1984). After three months of ex-vitro hardening, the plants were transferred to the Western Galilee Banana Experimental Station. Bunch bearing stems (ratoons) were selected according to practices of banana cultivation in the Western Galilee (Israel). During the six years of the experiment, bunches were weighed and data collected (Table 1).

A. Altman et al. (eds.), Plant Biotechnology and In Vitro Biology in the 21st Century, 251–254.

Table 1

		Clone				
Parameter	**5 -1**	**6 - 6**	**37- 5**	**42- 5**	**17- 1**	**Control**
Bunch wt. (kg)	37.82	37.47	35.68	37.44	33.44	30.47
Bunches/ mat/ year	2.83	2.83	3.00	2.83	2.33	1.74
Calculated yield (t/ ha /yr)	**88.19**	**87.38**	**88.10**	**87.31**	**64.22**	**43.59**

Based on the results of the above experiment, the best clones were multiplied by meristem culture as mentioned above. The performance of these clones in total yield, though differing from each other, was consistently above the control population.

3. Involvement of retro-transposon in somaclonal variation

The high frequency of deviation from the original clone occurring during the in vitro culture suggests that the tissue culture process promotes genomic changes. However, the mechanism of somaclonal mutations in bananas is unknown. In recent studies Hirochica et al. (1996), provided clear evidence that extensive duration of rice cell culture activated retro-transposing elements. Furthermore, the tobacco retro element Tto 1 undergoes transcriptional activation by wounding as well as by methyl jasmonate (Takeda et al., 1998). The structural features as well as transcriptional activation of retro-elements resembles retroviruses. Under normal conditions they remain dormant and upon activation they are transcribed, reverse transcribed to cDNA molecules and reintegrated in new loci in the genome. We have examined the activation of retro-elements in the banana genome by extensive in vitro conditions, which induced abnormal phenotypes.

Somaclonal variants with both high and low sensitivity to GA were generated after extensive duration in tissue culture. Both 'off-types' were detectable by a relatively simple bioassay developed in our laboratory (table 2). The results obtained by this assay provide evidence to the hypothesis that 'Dwarf' and 'Giant' phenotypes are related to GA sensitivity. Sandoval et al. (1995) have shown that in addition to the difference in sensitivity, partitioning between the different GA metabolites differed between 'Dwarf', 'Giant' and the normal phenotypes.

Table 2
Influence of GA_3 on "internode" length of in-vitro normal and mutant banana plantlets. Plantlets were grown for 5 weeks on GA-containing or hormone-free media.

Clone / Treatment	**Normal (cv. Grand Nain)**	**Dwarf (G. Nain mutant)**	**Extra dwarf mutant**
Control (hormone free)	14.2 b **)	10.8 c	9.6 d
GA_3 (10 mg/L)	21.7 a	14.9 b	9.6 d

Using degenerate primers taken from published sequence of rice 'Tos 17' (Hirochica et al. 1996), we have isolated a putative 344 bp retro-transposon homologue from the *Musa* genome.

Fig. 1

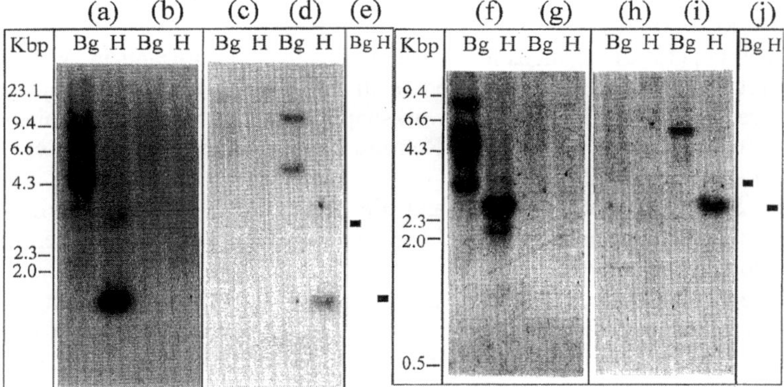

We have used the 344 bp PCR fragment to probe retro-elements on a Southern blot, comparing cut and uncut DNA from a soma-clonal mutant and the original pre-cultured mother meristem. All uncut DNA samples hybridized to the probe. In the lane in which the

DNA of the pre-cultured meristem was loaded the probe hybridized to two fragments (fig. 1), while the LNL mutant hybridized to at least four additional fragments. The addition of bands on the Southern blot indicates a propagation of the retro-element in the LNL mutants. At this point, the number of integrated copies represented per band is unclear. However, in the LNL the hybridized signal was intensified (see arrow in fig. 1). This may indicate multiplication of retro-elements in a close proximity to the original retro sequence. A cascade form of retro-elements has been reported in other species (San Miguel et al., 1996). This phenomenon may explain the high frequency of a single phenotypic variation. If a gene associated with GA sensitivity resides in close proximity to the original retro-sequence and the distribution of insertions is biased to short distances, we expect a high rate of mutants involving GA sensitivity. We are currently investigating this hypothesis.

4. List of references

Cote, F., Sandoval, J., Marie, Ph. And Auboiron E., (1993). Variations in micropropagated bananas and plantains. Literature Survey, Fruits 48:11-20.

Cronauer, SS and Krikorian AD (1984). Multiplication of *Musa* from excised stem apices. Annals of Botany 53:321-328.

Hirochika H., Sugimoto K., Otsuki Y., Tsugawa H. and Kanda M., (1996). Retrotransposons of rice involved in mutations induced by tissue culture. Proc. Natl. Acad. Sci. USA 93, pp.7783-7788.

Reuveni, O., (1990). Methods for detectionof somaclonal variants in 'Williams' bananas. In: Jarret R. (ed), Identification of genetic diversity in the genus *Musa,* pp. 108-113. Proceedings of the international workshop held at Los banios, Philippines INIBAP.

SanMiguel P. et al. (1996). Nested Retrotransposons in the Intergenic Regions of the Maize Genome. Science 274: 765-768

Takeda S., Sugimoto K., Otsuki H. and Hirochika H. (1998). Transcriptional activation of the tobacco retrotransposon *Tto 1* by wounding and methyl jasmonate. Plant Molecular Biology 36: 365-376

Sandoval, J., Kerbelec, F., Cote,F., and Doumas, P., (1995). Distributions of endogenous gibberellins in dwarf and giant off-types banana (*Musa acuminata* AAA cv. 'Grand Nain') plants from in vitro propagation. Plant Growth Regulation 17:219-224.

VuylstekeD., Swennen R., and De Langhe E. (1991). Somaclonal variation in plantains (*Musa* spp., AAB group) derived from shoot-tip culture. Fruits: 46 (4) 429-439.

CYTOCHIMERA DISSOCIATION THROUGH SHOOT-TIP CULTURE OF MIXOPLOID BANANAS

N. S. Roux[1], J. Dolezel[2] and F.J. Zapata-Arias[1]
[1]Plant Breeding Unit, FAO/IAEA Agriculture and Biotechnology Laboratory, Seibersdorf, Austria. [2]Laboratory of Molecular Cytogenetics and Cytometry, Institute of Experimental Botany, Olomouc, Czech Republic.

1. Introduction

Induced mutation techniques are particularly important for *Musa* species where there is limited sexual reproduction that could generate genetic variation, the basis for selection. *In vitro* techniques have created new opportunities to induce mutations in vegetatively propagated crops; shoot-tip culture has been used in combination with mutation induction in banana and plantain (Novak et al., 1987). As in other biotechnology approaches such as genetic transformation, *in vitro* mutation induction techniques applied to *Musa* improvement have a major limitation: the treatment of multicellular meristems results in a high degree of chimerism. The chances of a mutated cell growing out into a sector or layer and manifesting itself depend on its position within the apex as well as its growth rate as compared with the surrounding (non-mutated) cells (van Harten, 1997). To dissociate chimeras, it is often suggested to propagate irradiated propagules into the fourth vegetative cycle, M_1V_4 but basic studies have not been carried out to verify this postulation. It is difficult to monitor gene and chromosome mutations, but genome mutations (changes in ploidy level) can be detected easily using flow cytometry. The aim of this study was to analyse the dynamics of cytochimera (mixoploidy) dissociation during several cycles of shoot-tip culture.

2. Procedure

2.1. *Plant material*

The banana cultivar 'Grande Naine' is a triploid *Musa acuminata* (AAA) clone of the Cavendish subgroup. The clone used was obtained from the Musa Germplasm Transit Center of the International Network for the Improvement of Banana and Plantain (INIBAP), Katholieke Universiteit Leuven (KUL), Belgium. 'Grande Naine' is an important edible cultivar for export. It is also the most widely micropropagated cultivar.

A. Altman et al. (eds.), Plant Biotechnology and In Vitro Biology in the 21st Century, 255–258.

2.2. *Culture conditions and treatment*

Conditions for the micropropagation of bananas were the same as described by Novak et al. (1989). 700 single shoot-tip meristems, 3 to 5 mm long, were isolated from actively growing cultures (14 days) and dissected longitudinally. 0.1 M colchicine solution was filtered-sterilized (0.22 μm) and added to the liquid medium to reach a final concentration of 5.0 mM; the treatment period was 48 hours (van Duren et al., 1996). After treatment, the meristem tips were washed three times with sterile distilled water and transferred to semi-solid propagation medium. In handling the treated shoot-tips of Grande Naine (AAA), a small piece of each shoot was isolated aseptically at each transfer for ploidy screening.

2.3. *Flow cytometric analysis of the ploidy level*

Thirty days after treatment, the ploidy of the surviving shoots was assessed by flow cytometry. Samples for flow cytometry were prepared according to Dolezel et al. (1994 and 1997) using a two-step procedure (Otto, 1990). Diploid *Musa acuminata* was used as internal standard. DAPI-stained nuclei were analyzed with a PA-I flow cytometer (Partec, Germany).

3. Results and discussion

3.1. *Effect of the colchicine treatment*

At the first subculture, thirty days after treatment, only 10% of the explants survived (Fig. 1). Of the surviving explants, 81% were mixoploids, mainly 3n+6n.

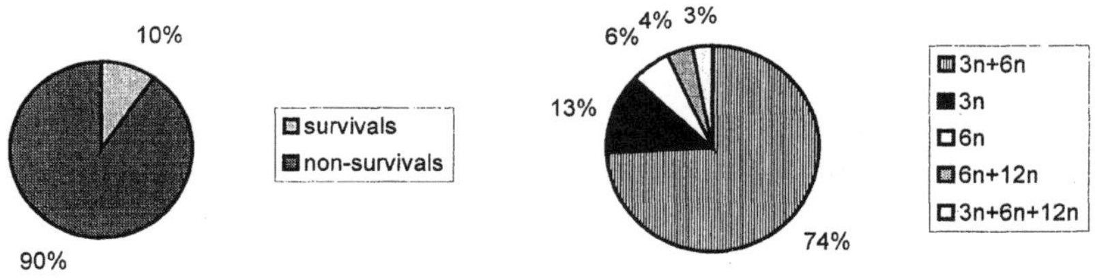

A. **B.**

Fig.1. Effect of colchicine *in-vitro* shoots of 'Grande Naine':
A. Survival of treated explants. **B.** Ploidy of surviving explants

3.2. *Mixoploidy dissociation*

The aim of the study was to induce a high percentage of mixoploidy (3n+6n) and to analyse how many subcultures would be necessary to dissociate the cytochimeras completely into triploids (3n) and hexaploids (6n). From the 51 induced mixoploid shoots obtained in M_1V_0 , 10 were selected and coded (GN-I,GN-II, ...GN-X). These shoots were subcultured every 30-40 days and the ploidy of each single shoot was monitored by flow cytometry before transfer to a fresh semi-solid propagation medium.

The descendent of the ten mixoploid shoots were screened until the fourth generation (M_1V_4). The proportion of mixoploidy decreased from 100% to 67% after one subculture, to 48% after the second, to 39% after the third and to 29% after the fourth (Fig. 2). The percentage of triploids increased slowly during the two first subcultures and then rose rapidly, to reach 42% at the fourth generation probably because of the higher multiplication rate of triploids. The highest percentage of hexaploids (39%) was obtained in M_1V_2.

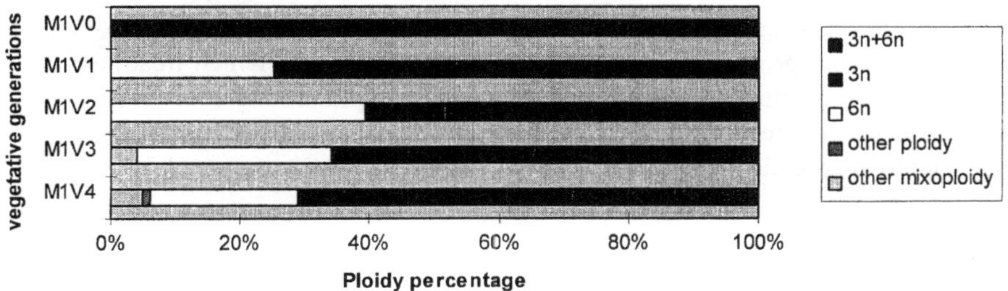

Fig. 2: Chimeral dissociation of mixoploids during 4 successive vegetative generations.

As can be observed from Fig. 3, there was a great variation in ploidy distribution from clone to clone in M_1V_4 generation. All shoots from 2 (GN-II and GN-X) of the 9 treated clones (one clone was contaminated and discarded) remained chimeric even after four subcultures. It is interesting to note that these two clones had also the lowest multiplication rate. Generally the clones could be classified into 3 groups according to their proportion of mixoploid shoots (100%, 30-50% and 10-15%). We selected one clone of each category for further propagation. Preliminary results showed that the proportion of mixoploid shoots did not decrease significantly in M_1V_5 and M_1V_6.

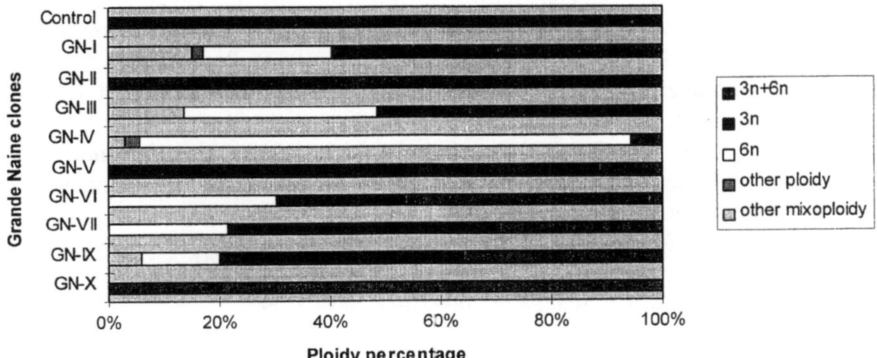

Fig. 3: Frequency of triploids, hexaploids, and mixoploids regenerated from 9 mixoploid banana 'Grande Naine' cultures and propagated until M_1V_4 generation.

4. Conclusions

Our results indicate that through shoot-tip culture, complete chimerism dissociation is not possible even after 6 subcultures. The fact that during the three first vegetative generations the proportion of chimerism was reduced and then tended to stabilize could be caused by the mericlinal chimeras changing into periclinal ones. It is difficult to extrapolate from the specific mode of action of colchicine in doubling the chromosome number, on the effects of other physical or chemical mutagenic agents which are harder to detect. Nevertheless, before concluding that a shoot or a plant is an homohistont, several additional propagation cycles are necessary no matter which treatment is used. One way to dissociate chimerism faster is by using a rapid propagation system. It would be even better to develop a single cell origin system such as an adventitious bud propagation technique or somatic embryogenesis as was mentioned by van Harten (1997).

5. Acknowledgements

The is paper is dedicated to the memory of Dr. O. Reuveni.

This work was funded by a Joint FAO/IAEA/BADC (Belgian Agency for Development and Cooperation) Coordinated Research Project. The authors wish to thank colleagues from the FAO/IAEA Plant Breeding and Genetics sub-Programme for their valuable suggestions and careful reading of the manuscript.

6. References

Dolezel, J., Dolezelova, M. and Novak, F. J. (1994). Nuclear DNA amount in diploid bananas (*Musa acuminata and M. balbisiana*). Biol. Plant. 36: 351-357.

Dolezel, J., Lysak, M.A., Van den Houwe, I., Dolezelova, M. and Roux, N. (1997). Use of flow cytometry for rapid ploidy determination in *Musa* species. Infomusa 6: 6-9.

Novak, F.J., Donini, B., Hermelin, T. and Micke, A. (1987). Potential for banana and plantain improvement through *in vitro* mutation breeding. ACORBAT Memorias VII Reunion. CATIE, Turrialba, 67-70.

Novak, F.J., Afza, R. van Duren, M. and Omar, M.S. (1989). Mutation induction by gamma irradiation of *in vitro* cultured shoot-tips of banana and plantain (*Musa* cvs). Trop. Agric. (Trinidad) Vol. 67 No. 1: 21-28.

Otto, F. (1990). DAPI staining of fixed cells for high-resolution flow cytometry of nuclear DNA. In: Methods in Cell Biology, Vol. 33. Darzynkiewicz, Z., Crissman, H.A. eds., pp. 105-110, Academic Press, Inc., New York.

Van Duren, M., Morpurgo, R., Dolezel, J. and Afza, R. (1996). Induction and verification of autotetraploids in diploid banana *(Musa acuminata)* by *in vitro* techniques. Euphytica 88:25-34.

Van Harten, AM. (1997). Mutation Breeding in Vegetatively Propagated Crops. 15th IAEA/FAO interregional training course on advances in technologies for induced mutations in crops. 20 May-27 June 1997, Seibersdorf, Austria, 92 pp.

MOLECULAR AND FIELD ANALYSIS OF SOMACLONAL VARIATION IN TRANSGENIC PLANTS

F. Sala[1], A. Arencibia[1,2], S. Castiglione[1], P. Christou[3], Y. Zheng[1,4], Y. Han[5]
[1] Depart. of Biology, Univ. of Milano, Via Celoria, 20133, Milano, Italy; [2] C.G.E.B., P.O. Box 6162, Ciudad Habana, Cuba; [3] John Innes Centre, Norwich Res. Park, Norwich NR4 7UH, UK; [4] Depart. of Biology, Northeast Normal Univ. Changchun, P.R. China; [5] Inst. of Forestry, Chinese Academy of Forestry, Beijing 100080 China.

1. Introduction

Transgenic plants are expected to integrate a foreign gene in an otherwise unmodified genome. Indeed, the foreign gene is stably integrated in the genome of selected plants and is transmitted through the sexual progeny (Bellini et al., 1992). On the other hand, there is no reason to exclude that genomic changes may have occurred in transgenic plants and that these determine visible or cryptic phenotypic and metabolic changes. In fact, all transformation protocols foresee cell culture in the dedifferentiated state before plant differentiation and this has been demonstrated to be a powerful mutagenic treatment (Walbot, Cullis, 1983 and 1985). Changes in phenotypic traits, accompanied by decrease in agronomic potential have already been verified in transgenic rice (Shuh et al., 1993) and potato plants (Dale, McPartland, 1992). In general, it is normal practice to disregard transgenic plants with undesirable agronomic traits and to select those that show the highest expression of the transgene and are apparently normal for other traits.

It is expected that a better understanding of the phenomenon may help to reduce the hazards connected with it. Thus, we have asked the following questions: How frequent are genomic changes in transgenic plants? Are they inherited in the sexual and clonal progeny? Is their frequency correlated with the transformation procedure and/or with the integration of the foreign gene? Is there a correlation between genomic changes and changes in agro-industrial traits? Can somaclonal variants be interest within the selection schemes of transgenic populations?

2. Materials and Methods

Tools for genome analysis were RFLP, RAPD, AFLP, RAMP and SSR (Bao et al., 1996; Wang et al., 1996, Arencibia et al., 1998; Karp et al., 1998). Plants were: (a) the Indica-type rice (*Oryza sativa* L.), cv Chinsura Boro II, in the case of protoplast transformation (Bao et al., 1996), the indica-type cv. Thaibonnet and the Japonica-type rice cv. Lido and Carnaroli for particle bombardment and cell electroporation (Arencibia et al., 1998), (b) poplar (*Populus nigra* L.), a cv. from Baku, Azerbaijan, transformed with *Agrobacterium tumefaciens* (Wang et al., 1996), and (c) sugarcane (*Saccharum spp.*), cv. Ja60-5, transformed by cell electroporation (Arencibia et al., 1997).

3. Results and discussion

Transgenic populations. In the last 3 years we produced transgenic populations of rice, poplar and sugarcane by using different transformation protocols (protoplast treatment, *Agrobacterium tumefaciens* infection, particle acceleration and electroporation). In the case of poplar (Wang et al., 1996, Zheng et al., unpublished) and sugarcane (Arencibia et al., in preparation) agronomic traits have also been analysed.

Evaluation of genomic changes in the transgenic plants. For a quantitative comparison of the molecular data gathered with by RFLP, RAPD, AFLP, RAMP and SSR analysis

A. Altman et al. (eds.), Plant Biotechnology and In Vitro Biology in the 21st Century, 259–262.

within the different transgenic populations, the polymorphism index (P.I.) has been calculated; this is defined as the per cent of polymorphic DNA sequences evidenced within each population. Although this value depends on different factors, such as amplitude of the analysed population, genome complexity and type of analysed DNA sequence (*e.g.*: microsatellite vs. conserved sequences), nevertheless it is a useful to compare the extent of genomic changes in different transformation experiments. Table 1 summarised the results produced when analysing the genome of the transgenic populations.

Table 1. *Polymorphism index* (**P.I.**) *in transgenic plant populations.* (**P.I.** *is calculated as the per cent of polymorphic DNA sequences evidenced with each molecular tool.*

Exp. n°	Plant species / Transformation approach	n° of individuals in the population	Tool for DNA analysis	P.I	Reference
1	Rice / Protoplasts treatment	18	RAPD	**19.4**	Bao et al. (1996)
2	Rice / Electric discharge particle bombardment	36	RAPD AFLP RAMP	**0.0** **1.1** **1.6**	Arencibia et al. (1998)
3	Rice / Cell electroporation	15	RAPD AFLP RAMP	**0.0** **0.9** **2.1**	Arencibia et al. (1998)
4	Poplar / *A. tumefaciens*	17	RFLP RAPD	**27.9** **77.3**	Wang et al. (1996)
5	Sugarcane/ Cell electroporation	42	RAPD AFLP RAMP	**0.0** **1.3** **4.1**	Arencibia et al. (1997) and manuscript in preparation

In the case of rice (Table 1, exp. 1, 2 and 3), the P.I. values clear show the large variability in the genome of transgenic plants produced from protoplast, as compared with the relative stability of the plants produced with the other two approaches. Fig. 1A, reporting the UPGMA dendrograms based on the Dice similarity index (Castiglione et al., 1993) produced on the bases of the RAPD analysis shows that: (i) the embryogenic cell cultures cluster in nearby positions with the highest dissimilarity from control plants; (ii) the first generation transgenic plants, as well as the two successive self-pollination progenies, have a modified and dissimilar genomic structure which turn out to be closer to that of non-transformed plants, as compared with cultured cells.

In the case of poplar, in which a *B.t.*-toxin gene conferring insect resistance has been integrated in the genome by *A. tumefaciens* treatment, the best performing clones had been analysed for transgene integration and expression as well as for DNA changes (Wang et al., 1996). The P.I. values of Table 1 (exp. 4) show the occurrence of extensive DNA changes. The UPGMA dendrogram of Fig. 1 shows, surprisingly, that transgenic clones cluster in groups of complete, or almost complete, homology as if genomic changes had not occurred at ramdom, but following specific pattern. This cannot be explained at the present stage of knowledge, since the assumption that clones having the same genomic constitution could have been obtained as multiple regenerants from a unique event of transformation is ruled out by the agronomic data presented below.

As for sugarcane, where a *B.t.*-toxin gene conferring resistance to the borer (*Diatraea saccharalis* Fab.) attack had been integrated into the genome and transgenic clones were selected on the bases of GUS-staining and insect-resistance (Arencibia et al., 1997), the verified P.I. values (Table 1, exp. 5) showed very limited genomic changes.

Evaluation of agronomic and morphological changes in transgenic plants. No large scale field evaluation has yet been performed in the case of the rice transgenic populations. However, it was noticed that, in the case of protoplast transformation, most transgenic plants were apparently normal, while others showed somaclonal variation (albinism, sterility, dwarfism). In the other two cases, regenerated rice plants showed, in small scale field trials, no apparent agronomic changes as compared to controls.

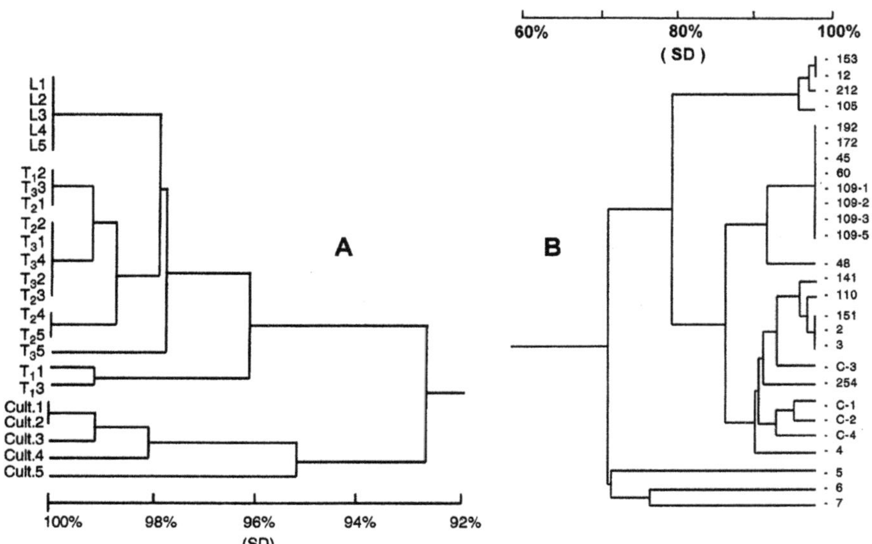

Fig. 1. *UPGMA dendrograms showing relationships among genomes. In* (**A**)*, rice leaves (L), cultured cells (Cult.), transgenic plants (T_1) and their progenies (T_2 and T_3) and, in* (**B**)*, original* P. nigra *clone (2, 3), control plants regenerated without transformation (C). Poplar plants belonging to different species (4,5,6,7) and transgenic poplar clones (all other samples). Data are expressed as per cent of similarity among the tested genomes. Details are in Bao* et al. *(1996) and Wang* et al. *(1996).*

In the case of poplar, transgenic plants have been propagated by cuttings and are being evaluated in field trials in China (at Manas, Xinjiang). Changes were recorded in leaf morphology in some of the transgenic clones (Wang et al., 1996) and are carried on in adult flowering plants. These changes are not random, but have been classified into three distinct classes. Leaves of each class are similar to those of known poplar accessions. Furthermore, clones that map in identical position in the dendrogram of Fig. 1 (for instance, clones 192, 172, 45, 60, 109) are characterised by different leaf shape. The cause of this is not know.

In the case of sugarcane, a population study in field trials under artificial borer inoculation showed large variance for all studied characters in the transgenic population (Table 2). Phenotypic variance was almost double in the transgenic population as compared with that within the C1- and the C2-control populations. This may be related with the differential expression of the *tcryI(A)b* gene and, consequently, with different levels of plant protection..

When considering the 5 elite insect-tolerant sugarcane clones now under field trial in Cuba, it was found that out of the 39 morphologic, agronomic and industrial traits accepted for the definition of a sugarcane variety, 33 traits were identical to those of the original genotype, while 6 morphological traits showed difference. Surprisingly, these variant traits were common to all 5 selected clones. An explanation for the observed convergent somaclonal variation is not available, but origin from a unique transformation event can been ruled out.

Table 2. *Agronomic traits related to borer resistance recorded in a sugarcane field trial after 25 days of insect infection. The analysed populations were: (**A**) transgenic plants, (**B**) plants from in vitro culture (C1-control) and (**C**) plants from the original germoplasm (C2-control).*

Sugarcane Population	Tested plants (number)	Dead stalks ** (%)	Stalk height ** (m)	Stalk diameter** (cm)	Stalks per stool ** (n°)	Affected internodes ** (%)
Transgenic	42	49.1±4.2 (744.96)	1.19±0.1 (0.38)	2.33±0.09 (0.39)	9.93±0.5 (11.04)	23.18±3.2 (450.02)
C1-control	23	56.9±3.5 (355.20)	0.96±0.08 (0.19)	2.04± 0.1 (0.19)	10.04±0.6 (10.09)	24.15±3.7 (393.83)
C2-control	41	53.3 ±2.9 (324.37)	0.94±0.07 (0.17)	2.10±0.1 (0.25)	10.95±0.4 (7.27)	18.98±2.6 (256.17)
Bartlett´s test		7.977(*)	7.157(*)	4.510(NS)	1.693 (NS)	3.013 (NS)

* - Significant for P<0.05 in the Bartlett's variance homogeneity test; (NS) - Not significant in the Bartlett's variance homogeneity test; ** - Population mean and standard error (in brackets: variance)

The in vitro cell culture as the major cause of genomic disturbance. In the case of exp. 2, 3 and 5 of Table 1, control plants were regenerated after the in vitro cell culture steps used for transformation, but in the absence of exogenous DNA and, consequently, without the selective agent. The molecular analysis showed that DNA changes carried by the transgenic plants were quantitatively equivalent to those recorded for these control plants.

In conclusion, we have demonstrated that phenotypic and genomic changes associated with the cell culture may occur in transgenic rice, poplar and sugarcane plants. Plant transformation through particle bombardment and cell electroporation have been shown to produce the lowest levels of genomic changes. This has implications for the management of the transgenic population in field trial. We point that biological activity of the foreign gene, as well as, somaclonal variation must be simultaneously evaluated in populations studies.

4. References

Arencibia et al. (1997) Mol. Breeding 3, 247-255

Arencibia et al. (1998) Mol. Breeding 4, 99-109.

Bao et al. (1996) Transgenic Res. 5, 97-103.

Bellini et al. (1992) Plant Science 82, 193-200.

Dale PJ et al. (1992) Theor. Appl. Genet. 84, 585-591.

Karp et al., eds. (1998) Molecular tools for screening biodiversity: plants and animals. Chapman and Hall, U.K.

Schuh W et al. (1993) Plant Sci. 89, 69-79.

Walbot V, Cullis CA (1983) Plant Mol. Biol. Rep. 1, 3-11.

Walbot V, Cullis CA (1985) Ann. Rev. Plant Physiol. 36, 367-396.

Wang et al. (1996) Transgenic Res. 5, 289-301.

Comparative analysis of genomic DNA methylation status and field performance of plants derived from embryogenic calli and shoot meristematic cultures

Shibo Zhang[1], Shoahua Zhang[1], Myeong-Je Cho[1], Phil Bregitzer[2] and Peggy G. Lemaux[1]

[1] Department of Plant and Microbial Biology, University of California, Berkeley CA 94720

[2] USDA-ARS, Aberdeen ID 83210

1. Introduction

Genomic instability has been widely observed in *in vitro*-derived plants; it is proposed to result from the break-down of normal cellular control (Phillips *et al*, 1994). Following DNA introduction into scutella directly or embryogenic callus derived from the scutella of immature embryos (IEs), numerous transgenic barley lines were produced (Wan and Lemaux, 1994; Lemaux *et al*, 1996). Evaluation of these lines revealed severe reduction in agronomic performance in field tests (Bregitzer *et al.*, 1998) and instability of the transgene and its expression (Bregitzer *et al.*, unpublished data). We propose that poor performance may be in part due to *in vitro*-induced genetic or epigenetic variability. Plants derived from alternative culture methods were analyzed to study the effects of these methods on genomic stability. Plants were regenerated from tissue obtained using methods of standard embryogenic culture (Wan and Lemaux 1994), modified embryogenic callus (Cho *et al.*, 1998) and cultured shoot meristems. Initial results with plants derived from the various methods will be compared relative to methylation polymorphism and field performance.

2. Materials and Methods

2.1 *In vitro* culture and plant regeneration

Three donor plants, each derived from the same single seed-derived line of barley cv. Golden Promise (GP) were grown in a controlled-condition growth chamber (Lemaux *et al.* 1996). After ca. 3 months, IEs were isolated from each donor plant and cultured using three *in vitro* culture methods (Fig. 1). In the standard embryogenic callus induction method (SECI; Wan and Lemaux 1994), a total of 30 IEs were randomly harvested on 3 different days from the 3 donor plants and plated and maintained for 3 months on callus-induction medium (CIM), MS + 2.5 mg/L 2,4-D; callus from each IE was maintained as an individual line. In a modified embryogenic callus-induction method (MECI, Cho *et al.*, 1998), 20-30 IEs were similarly isolated and cultured on DBC1 medium (MS + 2.5 mg/L 2,4-D, 0.01 mg/L BA, 5.0 µM $CuSO_4$) for the first month and transferred to DCB2 (MS + 2.5 mg/L 2,4-D, 0.1 mg/L BA, 5.0µM Cu SO_4) until regeneration. In the shoot meristem method (SMC, Zhang *et al.*, 1997), 20 mature embryos were isolated from the donor plants, germinated on MS medium and shoot apices isolated and cultured on MPM medium (MS +500 mg/L casein hydrolysate + 0.5 mg/L 2,4-D + 2.0 mg/L BA). R_0 plants were regenerated from 1- and 3-month old cultures from SECI and MECI and 3- and 6-month old cultures from SMC by transferring tissue to FHG medium; 1-2 plants from each line were transferred to the greenhouse to produce R_1 seeds. As control material, ten mature seeds from the 3 donor plants were randomly collected, germinated, and grown in the greenhouse.

A. Altman et al. (eds.), Plant Biotechnology and In Vitro Biology in the 21st Century, 263–267.

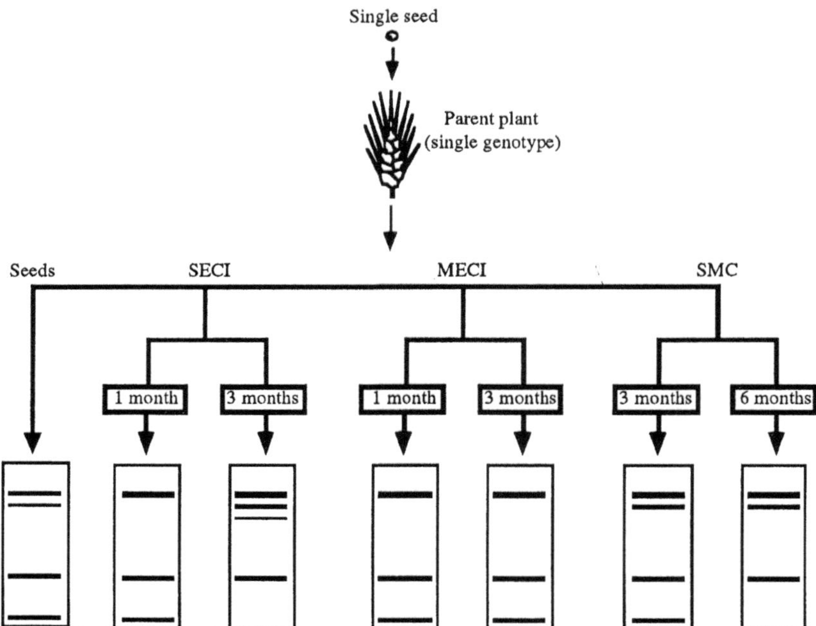

Figure 1. Schematic Representation of Methylation Polymorphism Experiment

2.2 Genomic DNA methylation analysis

DNA from leaf tissue of greenhouse-grown plants was made (Cone 1987), digested with *Hpa*II, *Msp*I, transferred to Zeta-Probe GT membrane (Bio-Rad Laboratories, Hercules, CA), and hybridized with barley probes, ABG 010, ABG 498, ABG602, ABC 155, ABC 256, *hor* 1&2; (Grain Genes: http://http://wheat.pw.usda.gov/probes/) using manufacturer's protocol. After washing, the blot was exposed to Kodak BioMax MS film.

2.3 Field testing

R_1 seed from each individual R_0 plant was used for testing at two sites, Aberdeen ID (irrigated) and Tetonia ID (dryland), 3 replicates at each location. For each tissue culture-derived line, the plot consisted of 8 individual plants planted alternatively with multiple-plant hills of wheat, alternating rows of single barley plants and wheat. Plots were designed with 1 or 2 wheat hills; each barley plant was surrounded by 4 wheat hills, providing relatively equal competition for each barley plant. Heading date, plant height, yield and 100-seed weight were determined on a single plant basis and compared to measurements made on plants of the single plant-derived, uncultured GP control. Means, calculated on a per plot basis, were used for statistical analyses; data was randomized as a randomized complete block design.

3. Results

3.1 *In vitro* culture and plant regeneration

One month after initiation, 26 callus lines, obtained from the 30 IEs of GP using SECI, produced R_0 plants; 19 callus lines produced R_0 plants at 3 months. One or two plants were regenerated from each line. After 1 month, 30 callus lines were obtained from MECI material; plants were regenerated from callus of each line. Thirty callus lines produced R_0 plants after 1 month and at 3 months, 13 lines gave rise to R_0 plants. For SMC, all 20 cultured shoot apices yielded axillary shoot meristems that proliferated (Zhang *et al.*, 1997).

Because of the relatively slow rate of *in vitro* proliferation of the axillary shoot meristematic domes in the first 2 months, compared to the growth of callus, plants were regenerated from SMC at 3 and 6 months.

3.2 Genomic DNA methylation analysis

Genomic DNA methylation patterns were analyzed in DNA from R_0 plants derived from the 3 treatments, using 6 DNA probes (3 from genomic DNA, 3 from cDNA) and the methylation-sensitive enzymes *Hpa*II and *Msp*I. In the 8 control plants analyzed, there were differences in the degree of methylation change among the 6 DNA probes when DNA was digested with *Hpa*II. Probing with Hor 1&2 was variable among the plants; the other 5 probes yielded relatively stable patterns among the 8 control plants. Because of the variability of the hybridization patterns with Hor 1&2, this probe was not used in the analysis of the plants.

The results using the 5 probes on the plants from 1-month-old SECI and MECI were that methylation pattern changes occurred in 32.7 % of SECI plants and 31.1 % of MECI plants. Only 4 % of the plants from the 3-month old SMC had methylation polymorphisms. Sixty-four percent of the plants from 3-month old SECI cultures had polymorphisms; those from MECI remained similar to MECI levels at 1 month, *i.e.* 32.7% (1 month) versus 33.3 % (3 months). In the plants from the 6-month-old SMCs, the percentage of plants with polymorphism increased to 24.0 %. A hybridization pattern using probe ABG 010 on *Hpa*II-digested DNA is shown in Fig. 2 A, B. Both hypo- and hyper-methylation changes are observed, depending on the DNA probe used, as exemplified in Figure 2. When *Msp*I was used, no hybridization pattern changes were observed in any of the plants tested from all three treatments (data not shown).

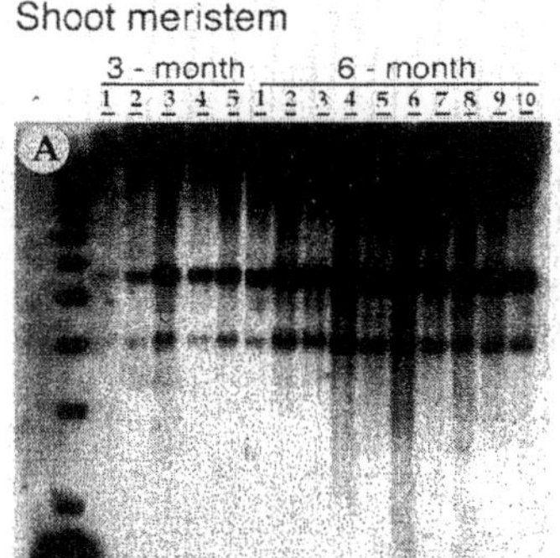

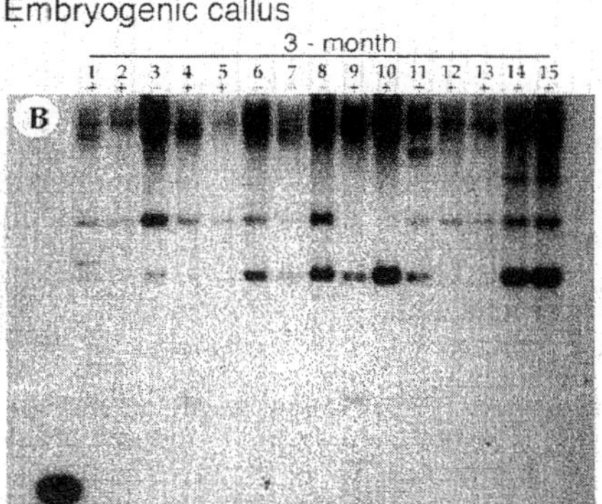

Figure 2. DNA hybridization analysis of methylation polymorphism in regenerated plants.

To determine if methylation pattern changes in R_0 plants were heritable, DNA from R_1 plants from the various methods was restricted with *Hpa*II and analyzed with 2 probes (ABG 498, ABG 010). The DNA hybridization pattern from the progeny of control plants was identical to that of parental plants. In R_1 plants of SECI-3, altered methylation patterns were 100 % heritable in some R1 plants analyzed; in others, new bands were observed

3.3 Field test

Heading Date - The heading date of tissue culture-derived plants was not statistically different than that of the controls. Tissue culture-derived plants were generally shorter and plant-to-plant variability was higher, as compared to control plants.

Height (cm) - Average height of control plants was 62.3 cm for the first and second regeneration dates, respectively. Average plants heights for other materials were: SECI = 60.8*/59.3*; MECI = 62.0/59.5*; SMC = 62.9/60.6* (* = statistically significant difference).

Yield (gm/plant) - Average yield of control plants was 41.7. For the first and second regeneration dates, respectively average plant yields were: SECI =29.6*/ ; MECI = 33.4*/26.8*;SMC = 37.8*/30.8*.

100-seed weight (gm/100 seeds) - Average 100-seed weight of control plants was 3.5. For the first and second regeneration dates respectively, average 100-seed-weights were: SECI = 3.4/3.1*; MECI = 3.4/3.4; SMC = 3.5/3.2*.

4. Discussion

Nearly all successful and reproducible transformation procedures utilize tissues cultured for varying lengths of time *in vitro*. Cultured somatic tissues accumulate genetic and epigenetic changes, termed somaclonal variation (SCV). This leads to the possibility that SCV, an inherent characteristic of *in vitro* cultured cells and transformation processes, might impact the application of genetic engineering to plant improvement. Earlier studies have documented the phenotypic manifestations of SCV in tissue-culture-derived barley lines (Ullrich *et al.* 1991; Bregitzer and Poulson 1995; Kihara *et al.* 1997). Our preliminary results confirm that the accumulation of changes occurs in a time-dependent manner. That is, the number of changes rises with increasing time in culture. In addition, the extent of methylation polymorphism changes depending on the nature of the cultured tissue. The number of changes in plants derived from shoot meristematic cultures was the lowest of the three methods, at both the earlier time point (1 month for SECI and MECI; 3 months for SMC) and the later time point (3 months for SECI and MECI; 6 months for SMC).

The differences in the number of changes seen in plants derived from the three methods might be due to differential effects of the various *in vitro* proliferation processes on genomic stability. The dedifferentiation, which occurs during the formation of embryogenic callus tissue from IEs, might destabilize the genome to a greater extent than the formation of proliferating meristematic domes from axillary meristems. This explanation is supported by the methylation analysis of the 1- and 3-month-old plants from MECI. *In vitro* proliferation of this tissue switched from an embryogenic type of callus tissue at 1 month to a more shoot meristematic growth at 3 months (Cho *et al.* 1998); tissues maintained under the SECI regime remained as embryogenic tissue. The conversion of the MECI tissue to a shoot meristematic mode of growth after 1 month might have stabilized the genome resulting in no further changes in numbers of methylation pattern changes. The conclusion from our genomic methylation analysis is consistent with the that observed in maize plants derived from embryogenic callus also from the immature scutellum (Kaeppler and Phillips, 1993); DNA methylation patterns are changed significantly in plants derived from embryogenic callus.

It has been proposed that methylation changes are a major factor in SCV (Phillips *et al.* 1994). Many examples of heritable changes of methylation patterns have been documented, suggesting that these epigenetic changes will impact the breeding programs used in the development of transgenic germplasm. Tissue culture-derived changes in agronomic performance and malting quality have been shown to be heritable through multiple generations, *e.g.* performance changes noted as early as the R_2 regeneration persist at least until the R_4 generation (Bregitzer and Poulson, 1995; Bregitzer *et al.* 1995; Bregitzer *et al.* 1998). Our preliminary analyses also indicate that in barley the *in vitro* process can result in heritable genomic changes in DNA methylation patterns that are correlated with compromised field performance. In addition our preliminary analyses indicate that the use of shoot meristematic cultures or the conversion of embryogenic cultures to a meristematic growth mode decreases methylation polymorphism and reduces SCV.

Modern cereal cultivars owe their superior performance characteristics to complex allelic interactions which, though poorly understood, can be easily disturbed by changes in the genome. The uncontrolled generation of SCV during the *in vitro* culturing needed to identify transformed tissue is likely to lead to undesirable performance characteristics. The use of tissues maintained in a more meristematic state might lessen this impact. The additional stresses of the transformation process also likely further compromise and impact the plants deriving from transformed tissues (Bregitzer *et al.* 1998). The observed correlation between methylation polymorphism and SCV can be used to identify and ameliorate these effects.

5. References

1. Bregitzer, P., Halbert, S.E., Lemaux, P.G. (1998) Theor. Appl. Genet.96: 421-425.
2. Bregitzer, P., Poulson, M. (1995) Crop Sci. 35: 1144-1148.
3. Bregitzer, P., Poulson, M., Jones, B.L. (1995) Cereal Chem. 72: 433-435.
4. Cho, M.-J., Jiang, W., Lemaux, P.G. (1998) Plant Science, in press
5. Cone, K. (1989) Maize Genetics Cooperation News Letter 63: 68.
6. Kaeppler, S.M., Phillips, R.L. (1993) Proc. Natl. Acad. Sci. USA 90: 8773-8776.
7. Kihara, M., Takahashi, S., Funatsuki, H., Ito, K. 1996. In: Proc. VII Int. Barley Genet. Symp. & V Int. Oat Conf., Saskatoon., 517-19.
8. Lemaux, P.G., Cho, M.-J., Louwerse, J., Williams, R, Wan, Y. (1996) Bio-Rad US/EG Bulletin 2007: 1-6.
9. Phillips, R.L., Kaeppler, S.M., Olhoft, P. (1994) Proc. Natl. Acad. Sci. USA 91: 5222-26.
10. Ullrich, S.E., Edmiston, J.M., Kleinhofs, A., Kudrna, D.A., Maatougui, M.E.H. (1991) Cer. Res. Commun. 19: 245-260.
11. Wan, Y., Lemaux, P.G. (1994) Plant Physiol 104: 37-48.
12. Zhang, S., Williams-Carrier, R., Jackson, D., Lemaux, P.G. (1998) Planta 204: 542-549.

INDUCTION OF POLYPHENOL OXIDASE IN *SEMPERVIVUM* L.

Veronika Abram, Marjan Donko, Andreja Štepec
Department of Food Science and Technology, Biotechnical Faculty,
University of Ljubljana, Jamnikarjeva 101, 1111 Ljubljana, Slovenia

1. INTRODUCTION

Sempervivum L. plants belong to the family of *Crassulaceae* (Swart, 1991), the subfamily *Sempervivoideae* (Stevens et al., 1992). They used to grow mainly in Southern Europe all the way from Spain to Asia Minor, but nowadays they are widespread in other parts of Europe also. *Sempervivum* is an evergreen plant with meaty juicy leaves. Plants can survive prolonged periods of drought and they are known to have CAM metabolism (Winter, Smith, 1996).

Phenolic compounds are secondary metabolites and they give a fruit or vegetable a certain flavour, colour and palatability. They are considered important compounds in the plants defense mechanism against herbivores and micro-organisms, and enzymatic browning reactions (Sapers, 1993).

Polyphenol oxides (PPO) has been found in animals, plants and micro-organisms (Steffens et al., 1998). The enzyme can catalyze two types of reactions: hydroxylation of monophenols and/or oxidation of o-dihydroxyphenols to o-quinones. Polyphenol oxidase from various plants can have either only one or both activities. It is a metalloprotein with two Cu ions in the active site of the enzyme (Martinez, Whitaker, 1995), where two conserved domains have been found. This enzyme is located mainly bound to the membranes of the plastids, although a free soluble form has been found too. Some investigators consider this free form of polyphenol oxidase as a form that appears after activation of the inactive membrane bound polyphenol oxidase (Steffens et al., 1998). A great variation in the molecular weight of polyphenol oxidase has been reported. The molecular weight from some sources can reach from 40-45 up to 59-65 or 70-72 kDa.

The inducibility of polyphenol oxidase activity can be caused by biotic or abiotic wounding. It is known that application of jasmonic acid or its volatile derivative - methyl jasmonate, can increase the level of polyphenol oxidase and protein inhibitors in

A. Altman et al. (eds.), Plant Biotechnology and In Vitro Biology in the 21st Century, 269–272.

some plants (Thaler et al., 1996). Here, we report the increased activity of polyphenol oxidase in *Sempervivum* L. by methyl jasmonate.

2. PROCEDURE

2.1. Materials and Methods

Reagents. All reagents used were of analytical or HPLC grade purity.

Plant material. *Sempervivum tectorum*, L. plants were from the Botanical Garden in Ljubljana, while *Sempervivum* for the induction experiments were collected in the field.

Extraction of polyphenols. Extraction procedures for phenolic compounds were performed following the methods of Abram, Piretti (1995).

Induction of polyphenol oxidase. *Sempervivum* plants collected in the field were immediately transferred to containers by using the soil from the area. In each container 6 plants of approximately the same size were planted and left to grow in the laboratory for 1 day before the experiment with induction was started. The optimum time of induction was determined with the plants grown for 3, 6, 9, 12, 24 and 48 hours in the presence of a 10 % ethanol solution of methyl jasmonate. Control plants were grown in the presence of 95 % ethanol for the same period of time.

Isolation of polyphenol oxidase. Polyphenol oxidase was isolated and partially purified according to the procedure of Sanchez-Ferrer et al. (1989) and Donko, Abram (1997).

Concentration of proteins. Proteins were determined by the method of Bradford.

Polyphenol oxidase activity. Polyphenol oxidase activity was determined using 4-methyl catechol as a substrate (Sanchez-Ferrer et al., 1989).

SDS-PAGE electrophoresis. Electrophoretic separation of proteins in the final supernatant (sup4) were performed on a 3 % of concentration and a 10 % of separation gel following the method of Laemmli.

Cytochemical determination of polyphenol oxidase. Localisation of polyphenol oxidase in fresh, untreated leaves of *Sempervivum* was done by the method of Vaughn (1987).

3. RESULTS AND DISCUSSION

Following the above extraction procedures we isolated 0.07 % of oligomeric and 0.13 % of polymeric polyphenols from fresh leaves of *Sempervivum tectorum,* L. We found that the 3 main oligomeric constituents could be flavone glycosides. After acidic hydrolysis of the oligomeric polyphenols, TLC on Silica gel-60 and HPLC we confirmed that kaempferol was the unique aglycone of the oligomeric polyphenolic fraction in this plant. Purification of the polymeric polyphenols by ion exchange chromatography and Haslam degradation was carried out to determine the constituents of the procyanidins detected after the Bate-Smith reaction. The products of Haslam degradation were analyzed and by HPLC we identified (–)-epigallocatechin and (–)-epigallocatechin-3-gallate as the major components of the procyanidins in the polymeric polyphenol fraction (Abram, Piretti, 1995).

When different stress was used to induce PPO in *Sempervivum* plants, we found that a 10 % (v/v) ethanol solution of methyl jasmonate after 24 hours was the most effective method of inducing PPO activity in the final supernatants. It increased the total specific activity approximately 6 times, while drought for 33 days was not effective. The polyphenol oxidase activity remained almost unchanged. Drought as a stress was not effective in our experiment, but *Sempervivum* is known as a plant which can sustain water deficiency very well. The mechanism of induction by mechanical wounds is considered to be the same as in the case of methyl jasmonate. Wounded plants synthesize a certain amount of jasmonic acid or its derivatives and when our results were compared to the results of Thipyapong et al. (1995) we concluded that both responses were very similar. They obtained a 1.7 times greater polyphenol oxidase activity after mechanical wounds, while in our experiments mechanical wounds increased the total specific activity 1.3 times.

Induction of polyphenol oxidase activity with a 10 % (v/v) ethanol solution of methyl jasmonate after different time intervals is presented in Fig. 1. An increase of total polyphenol oxidase activity was observed after 6 hours of induction and reached its maximum value after 24 hours. When the change of polyphenol oxidase activity per hour was calculated, we obtained the greatest increase after 6 hours.

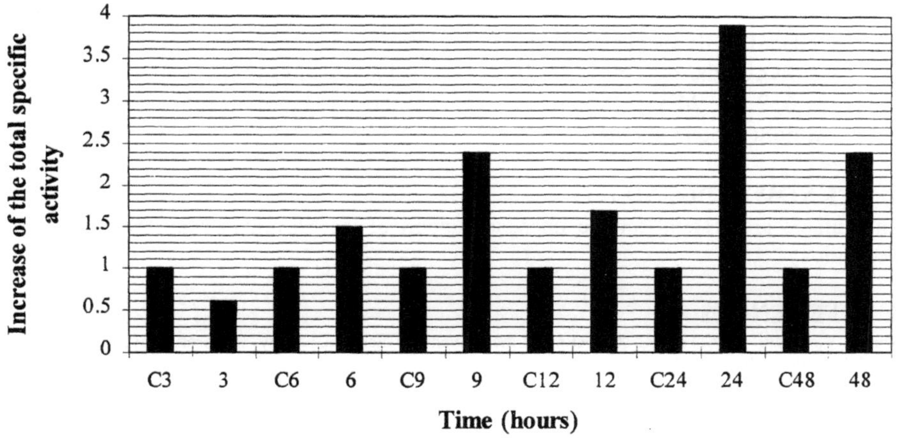

Fig. 1. The increase of the total specific activity in *Sempervivum* after different time of growth in the presence of a 10 % (v/v) ethanol solution of methyl jasmonate. C0, 3, 6, 9, 12, 24, 48 - control plants after different times of induction; 3, 6, 9, 12, 24, 48 - plants grown in the presence of a 10 % (v/v) ethanol solution of methyl jasmonate.

The electrophoretic separation on SDS-PAGE gel of the proteins from the final supernatants (sup4) from plants grown for different times in the presence of a 10 % (v/v) ethanol solution of methyl jasmonate is presented in Fig. 2. From the electropherogram the most intensive line was the one with an M_r of about 60 kDa, which can correspond to the molecular weight of PPO, according to the other published results.

Cytochemical localisation of polyphenol oxidase was performed on the leaves of fresh, uninduced plants in order to find how high the constitutional level of PPO is in *Sempervivum*. From the electron microscopic picture we did not see any polyphenol oxidase, which agrees well with the results of biochemical tests.

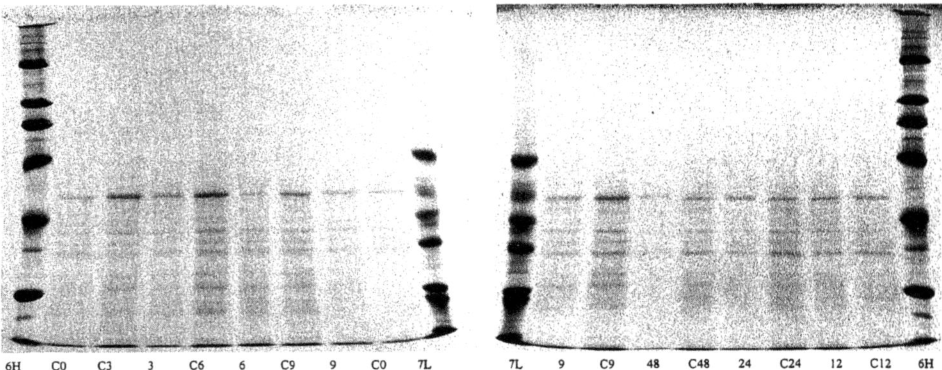

Fig. 2. SDS-PAGE electrophoresis of proteins from the final supernatants of plants grown in the presence of a 10 % (v/v) ethanol solution of methyl jasmonate after 3, 6, 9, 12, 24 and 48 hours. C-control plants, 6H and 7L high and low molecular weight protein standards.

4. ACKNOWLEDGEMENT.

The authors are grateful for the financial support of the Slovenian Ministry of Science and Technology.

5. REFERENCES

Abram, V., Piretti, M.V.(1995) 1 st Meeting of the Slovenian Chemical Society. Portorož, Slovenija, Sept. 28th-Oct, 1st, Abstracts, 86.

Donko, M., Abram, V. (1997) *Stress of Life:* 1-5 July, Budapest, Hungary: Abstracts, 12.

Martinez, M.V., Whitaker (1995) J.R. *Trends Food Sci. & Technol.*, 6, 195-200.

Sanchez-Ferrer, A., Villalba, J., Garcia-Carmona F.(1989) *Phytochemistry*, 28 (5), 1321-1325.

Sapers, G.M. (1993) *Food Technol.*, 47 (10), 75-81.

Steffens, J.C., Harel, E., Hunt, M.D.,Thipyapong, P. (1998) In Polyphenols 96, pp. 223-250, INRA,Paris, France.

Stevens, J.F., Hart, H.T., Malingre, H.H., Malingre, T.M. (1992) *Phytochemistry*, 31(11), 3917-3924.

Swart, D. (1991) Kakteen und andere Sukkulenten, 42 (1), 20-23.

Thaler, J.S., Stout, M.J., Karban, R., Duffey, S.S. (1996) *J. Chem. Ecol.*, 22 (10), 1767-1781.

Thipyapong, P., Hunt, M.D., Steffens, J.C. (1995) *Phytochemistry*, 40 (3), 673-676.

Vaughn, K.C. (1987) In CRC Handbook of Plant Cytochemistry, Vol. 1. pp.159-162. CRC Press, Boca Raton, USA.

Winter, K., Smith, J.A.C. (1996) In: Crassulacean acid metabolism, pp. 1-13, 350-352, Springer-Verlag, Berlin, Germany.

T-DNA and "gain of function" tobacco mutants with altered threonine metabolism

R. Amir[1*], H. Karchi[1], L.Yang[1], A. Perl[2], and Gad Galili[1]

[1] Department of Plant Sciences, The Weizmann Institute of Science, Rehovot 76100, Israel;
[2] Department of Fruit Tree Breeding and Molecular Genetics, Agricultural Research Organization, Bet Dagan, Israel. [*] Present address: Department of Plant Molecular Biology, Migal, Kiryat-Shmona 10200, Israel.

Abstract
A transferred DNA (T-DNA) tagging vector, containing four enhancers of the 35S gene promoter (Walden, 94), was used to obtain "gain of function" tobacco mutants with altered threonine metabolism. From 150 million transferred protoplasts, 17 plants were regenerated whose growth was resistant to a high level of threonine and to its toxic analog hydroxynorvaline. The majority of these plants contained a single T-DNA insert, genetically co-segregating with the threonine resistance. The mutants consisted of two categories: threonine overproducers and threonine non-overproducers. The overproducer mutants were probably connected with regulation of threonine biosynthesis while the non-overproducers mutants may be results of altered threonine sequestration.

Introduction
Threonine is an essential amino acid produced by plants. Due to its major nutritional quality, it is important to understand its regulatory mechanism (Galili, 95).

The biosynthetic pathway of threonine is via the aspartate family. Part of this pathway is common to other amino acids of this family, lysine, methionine and isoleucine (Fig 1.). Threonine biosynthesis is regulated by feedback inhibition of two enzymes in its pathway, aspartate kinase (AK) and homoserine dehydroginase (HSD) (Frankard, 92; Galili, 95). Up to now, work on threonine has concentrated mainly on the biochemical level (biosynthesis and catabolism). Little is known about other factors involved in the metabolism of threonine, such as its transport and its incorporation into proteins.

One of the best approaches to studying the factors involved in its metabolism is through genetics. In this study, we applied the strategy of activation T-DNA tagging, with the aim of not only generating mutants modified in threonine metabolism, but also isolating the genes involved. Activation T-DNA tagging involves the use of a T-DNA, containing the transcriptional enhancer sequence of the 35S RNA promoter of cauliflower mosaic virus, cloned as a tetramer near the right T-DNA border (Hayashi, 92; Walden, 94). Following insertion of such T-DNA into the plant genome, it is to be expected that flanking plant genes will come under the influence of the transcriptional enhancer and be overexpressed, producing a dominant mutation. This allow direct selection for a specific phenotype from amongst a population of T-DNA generated primary transformants. To generate a large number of primary transformants, transformation is carried out by protoplast/*Agrobacterium* co-cultivation.

Plants are sensitive to a high level of threonine because it causes feedback inhibition of AK and HSD, leading to methionine and isoleucin starvation (Bryan, 80) Hence, to find mutations with a gain of function in threonine metabolism, we looked for mutants resistant to a high level of threonine.

A. Altman et al. (eds.), Plant Biotechnology and In Vitro Biology in the 21st Century, 273–276.

Material and Methods

Plant material and bacterial strains

Tobacco plants (*Nicotinia nicoterium*) was used in this study. The T-DNA tagging vector pPCVICEn4HPT, kindly provided by Rick Walden, was used for tagging (Walden, 94). In brief, between the T-DNA border sequences, the vector contains a hygromycine-resistance gene linked to the nos promoter, an *Escherichia coli* plasmid origin of replication and an ampecilin-resistant gene, as well as the enhancer sequence of the 35S RNA promoter of CaMV, cloned as a tandem tetramer at the right border sequence. The *Agrobacterium* host for plant transformation was GV3101. Control experiments involved the culture of the protoplasts together with *Agrobacterium* (harboring pPCVHPT), containing the hygromycin resistant cassette, but lacking the multiple enhancer sequences (Hayashi, 92).

Protoplast isolation and *Agrobacterium* co-cultivation

Mesophyl protoplasts were isolated from 6-8 week-old sterile tobacco plants (Negrutiu, 87). In all the experiments, 5 x10^{7} protoplasts were co-cultivated with A*grobacterium*, containing pPCVICEn4HPT. Tagging by co-cultivated protoplasts was carried out when the protoplasts had divided, and had an average of 4-6 cells each, as described by Fritze (95). Two days after co-cultivation, the protoplasts were embedded in agarose containing hygromicin and threonine.

The calli were transferred to MS medium, containing 2 mM threonine to maintaining the threonine selection plus NAA and BAP to induce shoot formation and hygromycin. Shoots were transferred to Nitch medium, containing hygromycin and 5 mM threonine for rooting. Untransformed plants could not form roots on this level of threonine. For a positive control for these experiments we used transgenic plants, overexpressing the bacterial AK and thus overproducing threonine (Shaul & Galili, 92). These transgenic plants are resistant to a toxic level of threonine and to its toxic analog, DL-β-hyydroxynorvaline (HNV).

Genetic analysis of threonine-resistant plants

Each of threonine resistant plants was subjected to Southern analysis, using the 1.9 Kb *Hind III* fragment of the hygromycin-resistant gene as a probe. The T0 plants were self-crossed and/or crossed with wild-type plants, and at least 100 seeds were assessed for inheritance of the hygromycin marker and phenotypic changes. Hygromycin-resistant plants were self-crossed, and the progeny assessed for homozygous T2 plants. Finally, the protoplasts were isolated from the T2 homozygote and the heterozygote and proved able to form a callus in the presence of a selective level of threonine, thereby confirming the stability of the original phenotypic selection.

Measurements of free amino acid levels

Mutant plants were grown for about two months in the greenhouse. The third to fifth leaves (counting from the upper apex) were removed and ground in liquid nitrogen. Free amino acids were extracted from a sample of the frozen leaves, as described by Shaul & Galili (92).

Results and Discussion

Isolation of threonine resistance mutants

The experiment was initiated with 5×10^7 protoplasts for each level of threonine selection: 1.5, 2.5 or 3.0 mM (Table 1). It was found that 2.5 and 3.0 mM threonine were toxic to untransformed protoplasts, while 1.5 was a moderate level. The transformation frequency with *Agrobacterium*, containing the active tagging vector, was judged to be approximately 15% . Ten weeks after protoplast isolation, calli formed.

Table 1- From protoplasts to plants. The selection program and results of regeneration on different threonine levels.

Stages	mM threonine in media	1.5	2.5	3.0
Protoplasts	1.5; 2.5; 3.0	5×10^7	5×10^7	5×10^7
Protoplast infection	1.5; 2.5; 3.0	8×10^6	8×10^6	8×10^6
Calli formation	2	>150	54	12
Shoot formation	2	>300	126	0
Root formation	5	6	11	0

In order to tag genes that expressed at different developmental stages in the plants, we chose to work also with a moderate level of threonine. Thus, although calli can regenerate to shoots on 1-5 mM threonine, we added only 2 mM threonine to the medium. Over 150 calli were formed from the selection initiated on 1.5 mM threonine, with 54 and 12 calli from 2.5 mM and 3.0 mM, respectively (Table 1). These calli were removed to a solid MS medium, containing standard amounts of auxin and kinetin to stimulate shoot formation. No shoots regenerated from calli initiated from 3.0 mM threonine. In all, 17 mutants succeeded in rooting on Nitch medium containing 5 mM threonine (Table 1). These mutants were transferred to the greenhouse for further analysis.

We also tried to select gain of function mutants on HNV (a toxic analog of threonine). Two different experiments (each of them beginning with 4×10^6 protoplasts) were initiated. Although resistant calli were formed on 0.1-0.3 mM HNV, they failed to regenerate to shoots after removal to MS medium. Untransformed protoplasts failed to survive under such conditions.

Genetic analysis of the mutants

Plants were grown to maturity and segregation analysis was carried out by self-crossing, or crossing with untransformed individuals, followed by testing for segregation of the hygromycine-resistant gene in the progeny. Twelve of the mutant plants segregated in a 3:1 ratio, indicating that the T-DNA was inserted into a single locus.

Southern analysis was carried out, using genomic DNA isolated from leaf tissue of plants representing each of the different threonine resistance lines. Digesting the plant DNA with EcoRI, a restriction site not present in the T-DNA, allowed us to estimate the size of the genomic DNA flanking outside of the T-DNA. The sizes of hybridizing fragments in the different mutant liens led us to suspect that each of the recovered individuals was genetically unique and resulted from an independent transformation event.

Phenotypic characteristics of the mutants.

The hygromycin-resistant lines of T1 were examined for the threonine level in their leaves, and for the ability of the protoplasts to form a callus in the presence of selective levels (4 -7 mM) of threonine in the medium. This was performed on R-plants (resistant to hygromycin) and on S-plants (sensitive to hygromycin), both types the progenies of T0 plants. The R-plants varied in their extent of resistance to threonine, but were all more resistant than S-plants or wild type plants, most of whose protoplasts died under this condition. The difference between the R plants and the others was more clear-cut when HNV was added to the protoplast media. The protoplasts from wild type and/or S plants died within ten days, whereas those of the mutants survived and formed calli. Differences between R and S plants were also observed when the seeds were germinated on medium containing 0.2 mM HNV: wild type and S seedlings suffered and exhibited a yellow color, whereas the R seedling grew well and remained green.

In seven of the 17 mutants, leaves from R-plants possessed 2-5 fold higher free threonine than S plants. This trend was also observed in T2 plants. The mutants can therefore be divided into two categories: threonine overproducers and those that do not overproduce threonine. Overproduction is probably connected with regulation of threonine synthesis. Threonine overproduction can occur, for example, when AK or HSD isozymes - which are not sensitive to a high level of threonine - were enhanced by enhancer activity, resulting (paradoxically) in overproduction of threonine in such mutants. The absence of overproduction may be the result of threonine uptake, transport, or its incorporation into proteins with a high threonine content, or other causes. With the gene tagging approach we employed, there is no way of knowing in advance what precise role each tagged genes may play in threonine metabolism. However, it can be predicted that any gene whose deregulated expression reduces threonine toxicity may be tagged. We intend to identify the genes rescued from our mutants, in order to gain a better understanding of threonine metabolism.

Acknowledgments

We are indebted to Ms. Inbal Neta, Ayelet Nir and Merav Ziv for excellent technical assistance. This research was supported by grants from MINERVA Foundation and the Leo and Julia Forchheimer Center for Molecular Genetics. GG is an incumbent of the Bronfman Chair of Plant Sciences.

References

Bryan JK (1980) The Biochemistry of Plants, pp 403-452, Academic Press, New-York.
Frankard V et al (1992) Plant Physiology 99, 1285-1293.
Fritze K et al (1995)The Plant Journal 7, 261-271.
Galili G (1995) The Plant Cell 7, 899-906.
Hayashi H et al (1992) Science 258, 1350-1352.
Negrutiu I et al (1987) Plant Molecular Biology 8, 363-373.
Shaul O, Galili G (1992) Plant Physiology 100, 1157-1163.
Walden R. et al (1994) The Plant Journal 13, 4729-4736.

ACYLATED ANTHOCYANINS FROM CARROT CELL CULTURES: BIOSYNTHESIS AND SPECIFICITY OF THE ACYLATION REACTIONS

D.C. BAKER[1] J.L. CLARK[1] and D.K. DOUGALL[2].
Departments of Chemistry[1] and Botany[2], University of Tennessee,
Knoxville, TN 37996-1100, USA.

1. INTRODUCTION

Some of the naturally occurring anthocyanins have a greatly increased color retention at near neutral pH's compared to others. This increased color retention is an advantage when these compounds are to be used to color food, neutraceuticals and pharmaceuticals. The anthocyanins with increased color retention typically have several cinnamic acid residues attached and removal of these markedly decreases the color retention of the pigment (Dangles et al. 1993). There is an anthocyanin with one cinnamic acid residue attached (Alatanin C) which also has increased color retention at neutral pH (Yoshida et al. 1991). Alatanin C has the acyl group attached to one end of a ß-1,6-linked disaccharide chain attached to the 3-position of the anthocyanidin ring. Available information suggests that the disaccharide linkage between the chromophore and the cinnamic acid is important in the color retention. A similar arrangement is also found in a number of the anthocyanins produced by carrot cell suspension cultures (Baker et al. 1994; Gläβgen et al. 1992) and these also have increased color retention at near neutral pH relative to their non-acylated counterparts (Dougall et al. 1997).

When carrot cell cultures are treated with the cinnamic acids which are found on their anthocyanins, the proportion of the anthocyanin acylated with that acid increases. When related cinnamic acids are provided to the cultures, new anthocyanins acylated with the provided acid are produced (Baker et al. 1994, Dougall 1989). This observation led to the idea that a series of anthocyanins, differing only in the acyl group, could be produced by providing specific acids to the carrot cell cultures and that this series of anthocyanins could be used to examine the role of the acyl group in the color retention.

2. PROCEDURES

The cultures, feeding experiments, isolation and characterization of anthocyanins are described in Baker et al. (1994) and Dougall et al. (1998). The extracts used to synthesize glucose esters were prepared and assayed as described by Halaweish and Dougall (1990) with the 1 mM acid in the assay mixture. The neutral esters were separated from the starting materials as described by Leznicki and Bandurski (1988) and the products then examined by HPLC on a C18 column with a methanol/water gradient. The extracts used for the acyl transfer reaction were prepared and assayed as described by Gläβgen and Seitz (1992) except where noted. The preparation of glucose esters will be described elsewhere.

A. Altman et al. (eds.), Plant Biotechnology and In Vitro Biology in the 21st Century, 277–280.

3. RESULTS AND DISCUSSION

3.1 Preparation of acylated anthocyanins.

We have fed 15 cinnamic and benzoic acids to anthocyanin producing wild carrot cell cultures, one at a time, and obtained 17 anthocyanins differing only in the acyl group. The extra 2 were anthocyanins acylated with 4-chlorobenzoic acid and 4-trifluoromethylbenzoic acid which were produced metabolically from the corresponding cinnamic acids that had been provided (Dougall et al. 1998). We were unable to detect the presence of anthocyanin acylated with the corresponding benzoic acid in deliberate searches for them, after feeding 3,4,5-trimethoxycinnamic acid and 4-methoxycinnamic acid to the cultures. We did obtain anthocyanins acylated with these benzoic acids when they were provided to the cultures. We concluded that these cultures have only a low specificity for the acid they use to acylate anthocyanins.

3.2 Specificities of Enzyme Catalysed Reactions.

The low specificity for the acid used to acylate anthocyanins led us to examine the ability of crude tissue extracts to activate the acids and transfer them anthocyanins. In carrot, in contrast to the other known cases where acyl groups are transferred from coenzyme A esters to anthocyanins (Fujwara et al. 1997 and references cited therein), the intermediates are ß-D-glucose esters. Halaweish and Dougall (1990) showed that extracts of carrot cells could carry out the glucose ester synthesis with five acids and Gläβgen and Seitz (1992) demonstrated the transfer reaction with glucose esters of the three cinnamic acids found on carrot anthocyanins using crude extracts of carrot cells. In view of the acylation with benzoic acids in addition to cinnamic acids and the wide range of acids that the cells could use, the ability of the extracts to esterify and transfer these acids were examined.

3.2.1. Specificity of glucose ester synthesis.

The specificity of Sephadex treated extracts prepared as described by Halaweish and Dougall (1990) for the acids used to synthesize glucose esters was examined using UDP-^{14}C-glucose. Table 1 shows that the formation of glucose esters was demonstrated with 16 acids.

Table 1. Glucose ester formation from UDP-^{14}C-glucose and acids.

Acid	picomoles of product/2 hrs
3,4,5-trimethoxycinnamic acid	77 ± 14 (3)
3,4,5-trimethoxybenzoic acid	73 ± 7 (3)
4-trifluoromethylcinnamic acid	47 ± 6 (4)
4-trifluoromethylbenzoic acid	30 ± 7 (4)
4-chlorocinnamic acid	37 ± 1 (2)
4-chlorobenzoic acid	13 ± 9 (3)
4-hydroxycinnamic acid	63 ± 3 (2)
4-hydroxybenzoic acid	5 ± 0.4 (3)
4-hydroxy-3,5-dimethoxy-cinnamic acid	71 ± 1 (2)
4-hydroxy-3,5-dimethoxy-benzoic acid	103 ± 3 (2)
cinnamic acid	10 ± 1 (2)
4-nitrocinnamic acid	10 ± 1 (2)
4-hydroxyphenylpropionic acid	16 ± 1 (2)
4-methoxycinnamic acid	37 ± 12 (2)
3-methoxycinnamic acid	33 ± 8 (2)
2-methoxycinnamic acid	23 ± 6 (2)

We could not detect ester formation with 4-dimethylaminocinnamic acid or 4-hydroxyphenylacetic acid and this was not pursued further. In the cases of reactions with 3,4,5-trimethoxycinnamic and 4-chlorocinnamic acids, separate and non radioactive incubation mixtures gave products which cochromatographed on HPLC with the corresponding synthetic 1-ß-D-glucose esters. The data in Table 1 indicate that extracts of carrot cells catalysed the formation of glucose esters with many acids that were used by cell cultures to acylate carrot anthocyanins. The low levels of activity with 4-hydroxybenzoic, and 4-dimethylaminocinnamic acids are difficult to interpret at this time. It is possible that the concentrations of acid in the assay were too low for maximum product formation.

UDP-Glucose was the best nucleotide substrate for the reaction with 3,4,5-trimethoxycinnamic acid while TDP-glucose gave about one quarter of the product obtained with UDP-glucose. No product was formed with ADP-, CDP-, or GDP-glucose.

3.2.2. Specificity of the acyl transfer reaction.

The specificity of the transfer reaction was examined initially using crude extracts prepared as described by Gläßgen and Seitz (1992). The semiquantitative summary of the results is given in Table 2.

Table 2. Concentration of acylated anthocyanin in the transfer reaction mixture after 2 hours incubation with crude enzyme preparations and various 1-D-glucose esters at 1 mg/mL.

	Acyl Group	Concentration µM
1)	3,4,5-trimethoxycinnamoyl-ß	39
2)	3,4,5-trimethoxybenzoyl-ß	0
3)	4-chlorocinnamoyl-ß	53
4)	4-chlorobenzoyl-ß	15
5)	4-chlorobenzoyl-α	0
6)	4-hydroxycinnamoyl-ß	6
7)	4-hydroxybenzoyl-ß	0
8)	cinnamoyl-ß	21
9)	benzoyl-ß	no standard
10)	3-phenylpropionyl-ß	no standard

No Standard = new peak was present in the HPLC trace from this treatment but it could not be quantitated because the authentic compound was not available.

Of the 9 compounds with the ß-configuration which were tested, all the cinnamic acid esters were substrates for the transfer reaction. The rate of transfer of 4-hydroxycinnamoyl radical from the glucose ester to anthocyanin appeared to be very low. When the transfer was measured with several concentrations of the 4-hydroxycinnamoyl ester, the concentration of product was doubled when the ester concentration was 10 mM ie three times that used in Table 2.

The specificity of the extract for glucose esters was examined using 3-*O*-{ß-D-glucopyranosyl-(1→6)-[ß-D-xylopyranosyl-(1→2)-]ß-D-galactopyranosyl}cyanidin (CyaGXG) as an acceptor. When the specificity for the acceptor was examined using the 3,4,5-trimethoxycinnamoyl- or 4-hydroxycinnamoyl- ester, the rate was approximately 50% higher with 3-*O*-{ß-D-glucopyranosyl-(1→6)-ß-D-galactopyranosyl}cyanidin (CyaGG) at 0.44 mM than it was with CyaGXG at 3.4 mM. No new anthocyanin was detected when 3-*O*-[ß-D-xylopyranosyl-(1→2)-ß-D-galactopyranosyl]cyanidin (CyaGX) or 3-*O*-ß-D-galactopyranosylcyanidin (CyaG) were used as potential acceptors.

Two of the four benzoates, namely the 4-chlorobenzoyl- and the benzoyl-ester were also transferred. The α-anomer of 4-chlorobenzoate was not transferred indicating that the transfer reaction is specific for the ß-anomers of glucose esters. 3,4,5-Trimethoxybenzoyl- and 4-hydroxybenzoyl- esters did not transfer. Transfer of these acyl groups was not obtained with extracts made from tissue cultured for 4 to 8 days or when the pH of the assay buffer was varied from pH 6.0 to pH 8.0. 3,4,5-Trimethoxybenzoyl-1-ß-D-glucose inhibited the transfer of 3,4,5-trimethoxycinnamoyl-1-ß-D-glucose by 25% when present at 22 mM but was not inhibitory at 6.7 mM. The 4-hydroxybenzoyl-1-ß-D-glucose inhibited the transfer of 3,4,5-trimethoxycinnamoyl-1-ß-D-glucose by 95% at 8.4 mM and completely at 28 mM. With 0.46 mM 3,4,5-trimethoxycinnamoyl-1-ß-D-glucose in the presence of 10 mM 4-hydroxycinnamoyl-1-ß-D-glucose, the total acylated anthocyanin formed was unchanged but 25% of it was acylated with 3,4,5-trimethoxycinnamic acid. This suggests that both compounds were substrates for the same enzyme.

Our results show that for the majority of acids which are used by carrot cell cultures to acylate anthocyanins, the pathway is via the 1-ß-D-glucose esters. We have not been able to demonstrate the transfer of two benzoic acids. We suggest that in these two cases, the simplest interpretation is that the glucose esters of these compounds are used but that our methods of making cell free extracts have not retained the transfer activity with these two compounds ie there is a second transfer enzyme for which they are substrates. The specificity of the transfer reaction for the anthocyanin component, together with the fact that CyaGG and acylated CyaGG's are at the best only minor components accumulated by the cultures, suggests that the biosynthetic pathway is most likely to be:-

CyaG → CyaGX → CyaGXG → CyaGXG-Acyl group

rather than:-

CyaG → CyaGG → CyaGG-acyl group → CyaGXG-acyl group.

4. ACKNOWLEDGMENTS

This work was supported by USDA-NRICGP Grant 94-37500-0512.

5. REFERENCES

Baker DC et al. (1994) Plant Cell Tiss. Org. Cult. 39, 79-91.
Dangles O et al. (1993) J. Am. Chem. Soc. 115, 3125-3132.
Dougall DK (1989) Plant Science 60, 259-262.
Dougall DK et al. (1997) Food Technology 51, 69-71.
Fujiwara H et al. (1997) Eur. J. Biochem. 249, 45-51.
Gläßgen WE, Seitz HU (1992) Planta 186, 582-585.
Gläßgen WE et al. (1992) Phytochemistry 31, 1593-1601.
Halaweish FT, Dougall DK (1990) Plant Science 71, 179-184.
Leznicki AJ, Bandurski RS (1988) Plant Physiol. 88, 1474-1480.
Yoshida K et al. (1991) Tetrahedron Letters 32, 5579-5580.

BIOSYNTHESIS OF ß-THUJAPLICIN WITH THE *CUPRESSUS LUSITANICA* SUSPENSION CELL CULTURE

K. Fujita, T. Yamaguchi and K. Sakai
Department of Forest Products, Faculty of Agriculture, Kyushu University
Fukuoka, Japan

Introduction

One of wood tropolones, ß-thujaplicin, exists in heartwood of some species in Cupressaceae. Biosynthesis pathway of ß-thujaplicin has been interested because it has unique conjugated seven-membered ring and there has been no experimental evidence which suggests its biosynthesis pathway. Therefore, first of all, we tried ^{14}C-geraniol feeding experiments in order to prove that geraniol is the precursor of ß-thujaplicin. Secondary, biosynthetic pathway of isoprene unit was studied by ^{13}C-glucose feeding experiments; especially, [U-^{13}C]glucose gave direct evidence distinguishing glycelaldehydephosphate (GAP)/pyruvate route from classical acetate/mevalonate pathway (Rohmer et al., 1993, Lichtenthaler et al., 1997). Moreover, labeled positions in ß-thujaplicin derived from [1-^{13}C]-, [2-^{13}C]-glucose gave the information to assume the transformation of carbon skeleton which is unique on ß-thujaplicin and it was confirmed by ^{13}C-geraniol feeding experiments.

Materials and methods

Culture cell line

The callus of *C. lusitanica* has been maintained and subcultured monthly for several years on solid Gamborg B5 medium containing 2% sucrose, 10μM naphtylacetic acid, 0.01μM 6-benzylaminopurine and 0.27% Gellan gum in the dark at 25°C (Gamborg et al., 1968).

Radioisotope tracer experiments

About 1g (fr.wt) of 4-week-old callus was transferred into 10ml of a modified liquid Gamborg B5 medium in a 50ml Erlenmeyer flask containing 1.5mg of Triton X100, ca 10mg of partially purified yeast extract and RI substrates (Itose, Sakai., 1997). After 5 days incubation at 25°C in the dark on a rotary shaker (70rpm), the mediums and cells were homogenized and extracted with EtOAc. ß-Thujaplicin was purified as BF_2 complex by normal phase and reversed phase HPLC (Endo M. et al., 1988), successively. All radioactivities were measured with a liquid scintillation counter. ß-Thujaplicin was

A. Altman et al. (eds.), Plant Biotechnology and In Vitro Biology in the 21st Century, 281–284.

quantified by UV absorbance on reversed phase HPLC.

Stable isotope tracer experiments

About 10g (fr.wt) of 4-week-old callus was transferred to a 100ml of modified Gamborg B5 medium in a 500ml Erlenmeyer flask containing elicitor and ^{13}C labeled substrates as shown in table 1 (Itose, Sakai, 1997). In case of feeding [10-^{13}C]geraniol, ^{13}C-free glucose was used to reduce the signal from natural abundant ^{13}C. After 3 to 5 days incubation at 25°C in the dark on a rotary shaker (70rpm), the medias and cells were homogenized and extracted with ether. ß-Thujaplicin in extractives were purified as BF_2- or Cu^{++}- complex by reversed phase HPLC. After removal of Cu by $K_3Fe(CN)_6$ treatment in case of ß-thujaplicin-Cu complex, ^{13}C-labeled positions were determined by ^{13}C-nmr with quantitative pulse sequence and couplings between intramolecular carbons were observed with a 2D-INADEQUATE pulse sequence.

Table 1 Composition of ^{13}C substrates and sugars for ^{13}C feeding experiments

^{13}C substrates (mg/L)		natural abundant glucose (mg/L)	mannitol (g/L)
[1-^{13}C]glucose	1000	0	19
[2-^{13}C]glucose	250	750	19
[U-^{13}C]glucose	250	750	19
[10-^{13}C]geraniol	150	[U-^{13}C]glucose 1000*	19

*added at 24h after ^{13}C-geraniol administration

Results and Discussion

Geraniol as a intermediate of ß-thujaplicin biosynthesis

The incorporation of radioactivity to ß-thujaplicin from [10-^{14}C]geraniol was more than [U-^{14}C]glucose (table 2). Incorporation of mevalonate and malonate was very poor as observed before (Yamaguchi T. et al., 1997). Therefore, it was concluded that geraniol was the intermediate of ß-thujaplicin biosynthesis.

Table 2 Incorporation of radioactivity into ß-thujaplicin from RI substrates

Substrates administrated	Radioactivity administrated (dpm)	Incorporation (%)
[U-^{14}C]glucose	1.05×10^7	0.041
[10-^{14}C]geraniol	0.40×10^7	0.093
[2-^{14}C]mevalonate	1.05×10^7	0.00046

GAP/pyruvate pathway and intramolecule rearrangement of carbon skeleton

Isoprenoids produced from [U-^{13}C]glucose through GAP/pyruvate pathway shows characteristic coupling spectrum observable by the 2D-INADEQUATE nmr method (Eisenrich et al., 1996, Schwender, et al., 1996). Because a set of three carbons from one glucose molecule are always incorporated into final products through the pathway, a long-range coupling between a pair of carbons and another carbon in the product from [U-^{13}C]glucose is observable at significant intensity by the 2D-INADEQUATE nmr method.

Table 3 Integration of Quantitative ^{13}C-NMR spectrum of ß-thujaplicin derived from [1-^{13}C]-,[2-^{13}C]-glucose and [10-^{13}C]geraniol.

Carbon position	[1-^{13}C] glucose	[2-^{13}C]glucose	[10-^{13}C]geraniol
1	1.06	2.17	1.08
2	1.06	0.73	0.96
3	8.24	1.00	1.04
4	1.17	2.23	1.03
5	8.18	1.13	1.06
6	1.00	2.21	0.99
7	8.24	0.90	1.22
8	1.03	1.97	0.95
9,10	10.21	2.16	1.94

In INADEQUATE experiments with developing time 120ms, the coupling signal between C-4 and C-9 (or 10), and among C-2, C-5 and C-6 of ß-thujaplicin was observed (fig. 1). These long-range coupling suggests that ß-thujaplicin is synthesized mostly via GAP/pyruvate but not via classical acetate/mevalonate pathway. The geraniol derived from [1-^{13}C]- and [2-^{13}C]-glucose via GAP/pyruvate pathway would be labeled at

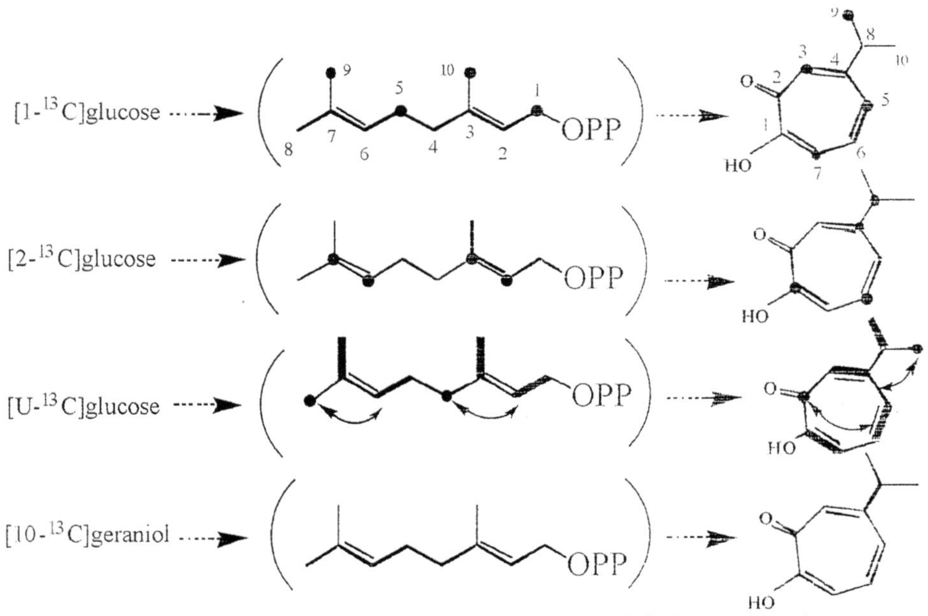

[1-^{13}C]glucose

[2-^{13}C]glucose

[U-^{13}C]glucose

[10-^{13}C]geraniol

Fig. 1 Labeled positions of ß-thujaplicin and geraniol through GAP/pyruvate pathway.
Dot: ^{13}C enriched. Bold line: short-range coupling. Arrow: long-range coupling.

positions 1, 5, 9 and 10 and at 1, 2, 5 and 6, respectively (table 3). Quantitative ^{13}C nmr measurements revealed that ß-thujaplicin derived from [1-^{13}C]- and [2-^{13}C]-glucose had high intensity of peaks at position of 3, 5, 7 and 9 and at 1, 4, 6 and 8, respectively (table 3). In case of [10-^{13}C]geraniol feeding, higher intensity at the 7-position was observed when natural abundant ^{13}C was diluted with [U-^{12}C]glucose. These labeled positions of the geraniol and ß-thujaplicin derived from [1-^{13}C]-, [2-^{13}C]-, [U-^{13}C]-glucose and [10-^{13}C]geraniol are summarized in fig. 1. Rearrangement of the isopropyl group of ß-thujaplicin is impossible, because the C-4 and C-8 were enriched by feeding [2-^{13}C]glucose and the long-range coupling between C-4 and C-9 (or 10) was observed in the [U-^{13}C]glucose feeding experiment. Therefore, the C-5, C-6, C-7, C-8 and C-9 of geraniol must correspond to the C-3, C-4, C-8, C-9 and C-10 of ß-thujaplicin, respectively. The C-2 of ß-thujaplicin, which was not labeled by [1-^{13}C]-or [2-^{13}C]-glucose feeding and had the long-range coupling with the C-5 in the [U-^{13}C]glucose feeding experiment, must come from the C-4 of geraniol. An adjacent pair (C-2 and C-3) in geraniol was separated by the carbon which was enriched by [1-^{13}C]glucose feeding. A pair (C-3 and C-10) of ^{13}Cs which does not have a long-range coupling with any other carbon in the [U-^{13}C]glucose feeding experiment still stays adjacently each other as the C-1 and C-7 of ß-thujaplicin. Hence, the methyl group at the 10-position in geraniol splits the bond between C-2- and C-3; therefore, the C-1, C-2, C-3 and C-10 of geraniol must correspond to C-5, C-6, C-10 and C-3 of ß-thujaplicin, respectively. These steps are summarized in fig. 2.

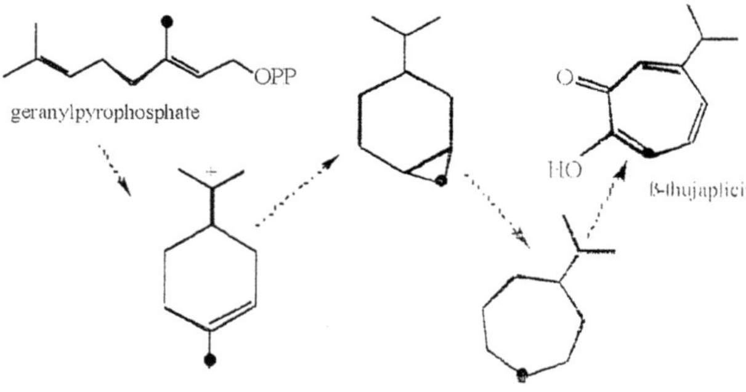

Fig. 2 Assumed carbon transformation during ß-thujaplicin biosynthesis from geranylpyrophosphate.
Dot: The carbon from 10-geranylpyrophosphate

References

Eisenrich, W. et al., (1996) Proc. Natl. Acad. Sci. USA, **93**, 6431-6436

Endo M. et al. (1988) J. Chromatography, **455**, 430-433.

Gamborg O. L. et al. (1968) Exp. Cell Res., **50**, 151-156.

Itose, R., Sakai, K. (1997) Plant Biotechnol., **14**, 163-167.

Lichtenthaler, H. K. et al. (1997) Physiol. Plant., **101**, 643-652.

Rohmer, M. et al. (1993) Biochem. J., **295**, 517-524.

Schwender, J. et al., (1996) Biochem. J., **316**, 73-80

Yamaguchi, T. et al. (1997) Mokuzai Gakkaishi, **43**, 696-698.

Biochemical and molecular bases of the carbon flux regulation between threonine and methionine in plants

B. Gakière, G. Curien, S. Ravanel, V. Verne, M. Droux, J. Yaxley, R. Douce and D. Job
Laboratoire mixte CNRS/Rhône-Poulenc (UMR041), Rhône-Poulenc Agrochimie, 14-20 rue Pierre Baizet 69263 Lyon cedex 9, France

1. Introduction

Methionine and threonine are two essential amino acids belonging to the aspartate-derived family of amino acids. In plants, they both originate from the same precursor, *O*-phosphohomoserine (OPH) (Fig. 1). This contrasts with the situation seen in bacteria where branching between the two pathways occurs upstream of OPH, at the level of homoserine (Saint-Girons *et al.*, 1998). Despite intensive studies (for reviews, see Giovanelli *et al.*, 1980; Leustek, 1996; Azevedo *et al.*, 1997; Hell, 1997; Ravanel, 1997; Ravanel *et al.*, 1998b), the specific mechanisms that are responsible for homeostatic regulation of methionine and threonine in plants are not fully understood.

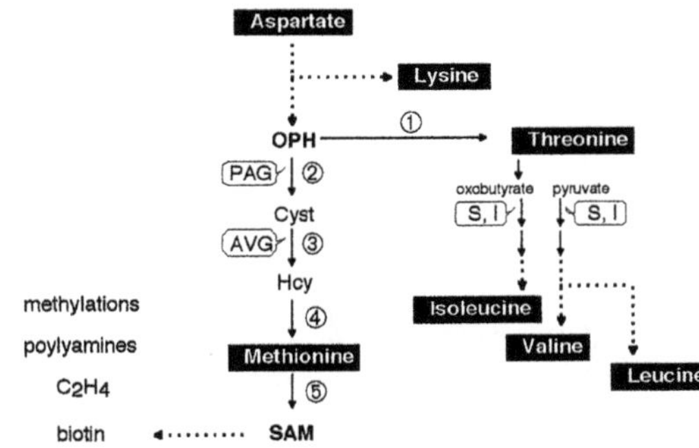

Figure 1. Biosynthetic pathways for the aspartate-derived amino acids in higher plants.
Enzymes: 1, threonine synthase; 2, cystathionine γ-synthase; 3, cystathionine β-lyase; 4, methionine synthase; 5, SAM synthetase.
Abbreviations: Cyst, cystathionine; Hcy, homocysteine; OPH, *O*-phosphohomoserine; SAM, *S*-adenosylmethionine.
Inhibitors: PAG, propargylglycine; AVG, aminoethoxyvinylglycine; S, sulphonylureas; I, imidazolinones.

This article is intended to review our present knowledge on the two committed enzymes for methionine and threonine syntheses in plants, namely cystathionine γ-synthase and threonine synthase, and to discuss possible mechanisms involved in regulation of the carbon flux between these two diverging branches.

A. Altman et al. (eds.), Plant Biotechnology and In Vitro Biology in the 21st Century, 285–292.

2. Threonine synthase

Plant threonine synthase, which is a pyridoxal 5'-phosphate (PLP)-dependent enzyme, catalyzes within plastids the last step of threonine formation (Madison, Thompson, 1976), converting *O*-phophohomoserine (OPH) to threonine and inorganic phosphate. Threonine is directly incorporated into proteins or further deaminated to yield oxobutyrate, which serves as a precursor for isoleucine biosynthesis (Fig. 1). In plant cells, OPH also serves as a precursor for methionine/S-adenosylmethionine (SAM) biosyntheses (Fig. 1). Thus, threonine synthase competes for OPH with cystathionine γ-synthase, the first committed enzyme in the methionine pathway (Giovanelli *et al.*, 1984). Of main interest is the considerable enhancement of the activity of the plant threonine synthase in the presence of SAM, the end-product of the methionine branch (Giovanelli *et al.*, 1984). This mechanism remains, however, largely unknown. It is worth noting that, contrary to the plant enzyme, bacterial threonine synthase, though catalysing the same reaction, is not subjected to kinetic modulation by SAM (Umbarger, 1978), presumably because in bacteria OPH is not a branch point metabolite. Indeed, in procaryotes, branching between threonine/isoleucine and methionine pathways occurs upstream of OPH, at the level of homoserine (Saint-Girons *et al.*, 1988).

To characterize further this unusual role for SAM, we have cloned the cDNA encoding the plastidial threonine synthase from *Arabidopsis thaliana*. This clone was further used to overproduce the mature form of the enzyme in *Escherichia coli* cells. The main features of these studies (Curien *et al.*, 1996, 1998) were: (i) the native and the recombinant *Arabidospsis* enzymes assume an homodimeric structure; (ii) the enzymes obey Michaelis-Menten kinetics whether they are assayed in the absence or presence of SAM; (iii) saturation isotherms of plant threonine synthase by SAM suggest the binding of at least two SAM molecules on each enzyme molecule, both of which affecting strongly the kinetic properties of the plant enzyme (Fig. 2).

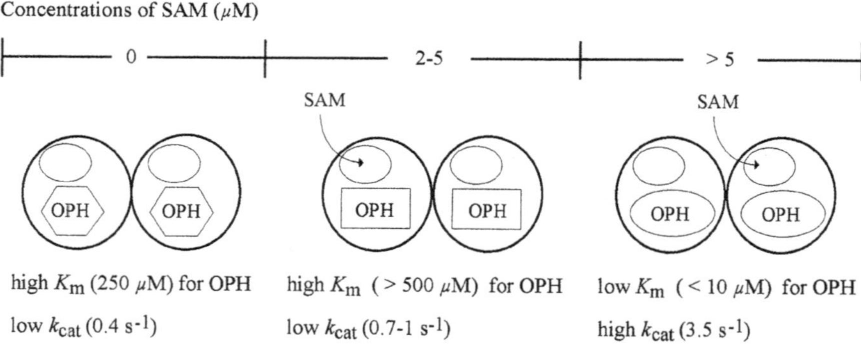

Figure 2. Kinetic properties of *Arabidopsis* threonine synthase in the absence or presence of SAM (adapted from Curien *et al.*, 1998).

It then appears that upon binding of SAM activator, plant threonine synthase acquires marked different properties allowing to decrease the K_m for the substrate (about 25-fold) and to increase the catalytic efficiency (8-fold) (Fig. 2). This might have physiological significance since the OPH concentration is below 100 μM in the chloroplast compartment (Datko *et al.*, 1984). From calculation of the specificity constants (k_{cat}/K_m), we can estimate that, in the absence of SAM, OPH would bind only very slowly to the enzyme, with a half-time value of approximately 10 s. On the other hand, the allosteric transition of plant threonine synthase induced by SAM binding allows the OPH substrate to diffuse at a much faster rate (200-fold) into the active site. In view of this, SAM may be considered as an essential activator of threonine synthase (Curien *et al.*, 1998).

Remarkably, mature *Arabidopsis* threonine synthase is characterized by a large N-terminal extension compared with corresponding bacterial enzymes. When the mature plant protein is truncated by 33 amino acids on its N-terminal part, the truncated derivative of the enzyme remains active but appeared to be poorly activated by SAM. This dramatic reduction of the stimulation by SAM therefore suggests a role of the N-terminal part of the plant threonine synthase in enzyme stimulation by SAM (Curien *et al.*, 1996).

3. Cystathionine γ-synthase

Cystathionine γ-synthase, which is a PLP-dependent enzyme, catalyzes the synthesis of cystathionine from cysteine and OPH in a γ-replacement reaction. Plants are unique by using OPH as the physiological α-aminobutyryl donor for cystathionine synthesis. In contrast, procaryotic cystathionine γ-synthases utilize either the succinyl or the acetyl derivatives of homoserine as substrates (Saint-Girons *et al.*, 1988). In plants, this enzyme is only present in the stromal space of plastids (Ravanel *et al.*, 1995a).

Genes and cDNAs for cystathionine γ-synthase are now available in several plant species. The deduced amino acid sequences exhibit several highly conserved regions (Kim, Leustek, 1996). Among them, a 23-residue block contains the active-site located lysine residue involved in Schiff base formation with the PLP cofactor (Ravanel *et al.*, 1998a). Interestingly, sequence comparisons also revealed that *Arabidopsis* cystathionine β-lyase (the second enzyme in methionine synthesis; see Fig. 1) exhibits significant homology with cystathionine γ-synthase from plant and bacterial sources and cystathionine γ-lyase from *S. cerevisiae* (Ravanel *et al.*, 1995b, 1996) (Fig. 3). All these enzymes are involved in the cysteine/methionine biosynthetic pathways and belong to the same class of PLP-dependent enzymes that catalyze γ-replacement reactions of C_4 amino acids and β- or γ-cleavage of cystathionine (Alexander *et al.*, 1994). This suggests that plant cystathionine γ-synthase and cystathionine β-lyase derived from a common ancestor, as proposed for the corresponding bacterial *metB* and *metC* genes (Belfaiza *et al.*, 1986).

Heterologous production of the *Arabidopsis* cystathionine γ-synthase as a functional mature enzyme (i.e., without transit peptide) was subsequently achieved in *E. coli* allowing characterization of its kinetic properties (Ravanel *et al.*, 1998a). The catalytic reaction proceeds through a ping-pong mechanism where OPH reacts first with the enzyme and cystathionine is the last product released. Despite a high catalytic constant (30 s^{-1}) the *Arabidopsis* enzyme discloses fairly high K_m values for OPH and cysteine

(2.5 and 0.46 mM, respectively) (Ravanel *et al.*, 1998a). This feature raises the question of the efficiency of this enzyme *in vivo* because the OPH and cysteine concentrations are below 100 μM in the chloroplast compartment (Datko *et al.*, 1984). In contrast, the K_m value of threonine synthase for OPH in the presence of its SAM activator is extremely low, being of the order of 5-10 μM (Curien *et al.*, 1996, 1998) (Fig. 2).

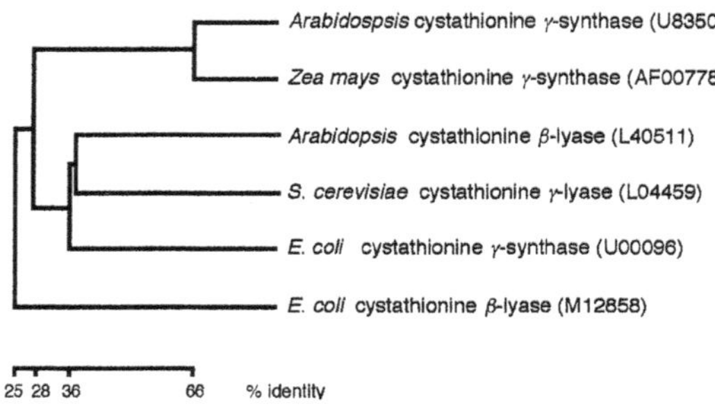

Figure 3. Amino acid sequence relationships between transsulfuration enzymes from various species. Multiple sequence alignment was created using progressive, pairwise alignment with the PileUp program from the Genetics Computer Group, Inc. The tree representation shows the clustering relationships used to create the alignment and indicates the percentage identity between each sequence. For the plant sequences, transit peptides were removed before performing the alignment. The GenBank accession numbers for each sequence are shown in brackets.

In the absence of cysteine, cystathionine γ-synthase from enteric bacteria efficiently catalyzes a γ-elimination reaction of *O*-succinylhomoserine yielding α-ketobutyrate, succinate and ammonia (Holbrook *et al.*, 1990). For the plant enzyme using OPH as substrate, this reaction is about 2,700-fold slower than the physiological γ-replacement reaction and 1,500-fold slower than the elimination reaction catalyzed by the *E. coli* enzyme (Ravanel *et al.*, 1998a). This difference arises from the inability of the plant cystathionine γ-synthase to accumulate a long-wavelength-absorbing species that is characteristic for the efficient elimination reaction catalyzed by the bacterial enzyme (Ravanel *et al.*, 1998a). It is worth noting that the K_m values for cysteine in the replacement reaction differ markedly between the bacterial and plant enzymes. While the former is about 50 μM (Holbrook *et al.*, 1990), the latter is one order of magnitude higher (Kreft *et al.*, 1994; Ravanel *et al.*, 1998a). Hence, a low rate of the elimination reaction prevents consumption of OPH through this futile process, especially in view of the low intracellular cysteine concentration in plants (Datko *et al.*, 1984).

4. Flux partitioning of OPH: kinetic considerations

Classically, it is admitted that the first enzymatic step in a pathway has the highest control coefficient. Accordingly, cystathionine γ-synthase and threonine synthase would occupy strategic positions for control. It is well established, however, that flux coordination between threonine and methionine branches in plants does not result from end-product retroinhibition of threonine synthase and/or cystathionine γ-synthase. Instead this coordination appears to result primarily in all plants so far tested *via* a strong stimulation of threonine synthase activity by SAM (Giovanelli *et al.*, 1984; Curien *et al.*, 1996, 1998).

Considering the kinetic parameters (k_{cat},'K_m) of cystathionine γ-synthase (Ravanel *et al.*, 1995, 1998a) and threonine synthase for OPH and also the experimentally-determined dependencies on SAM concentration of these kinetic parameters for the latter enzyme (Curien *et al.*, 1996, 1998), we may calculate, by computer simulation, what will be the expected range of variations in OPH concentration in response to fluctuations in SAM concentrations under conditions where the partition of the OPH flux between the two threonine and methionine branches is assumed to remain constant (i.e., this partitioning can be described by a proportionality constant, α, so that $v_{\text{threonine}} = \alpha \bullet v_{\text{cystathionine}}$). From these calculations, three conclusions can be drawn (Fig. 4).

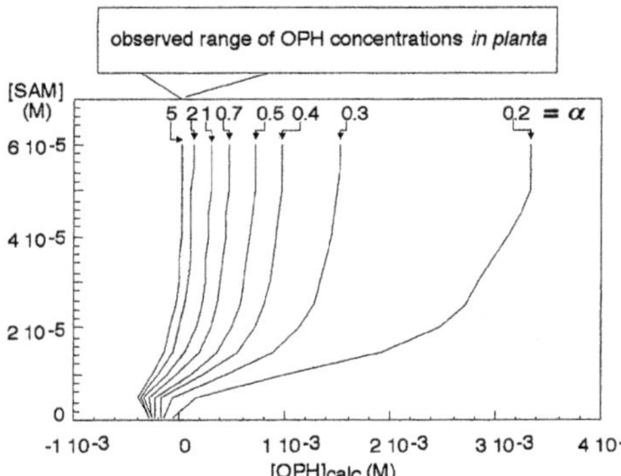

Figure 4. Computation of OPH concentrations (OPH_{calc}) from experimental values of the kinetic parameters (k_{cat}, K_m) of threonine synthase (TS) (Curien *et al.*, 1998) and cystathionine γ-synthase (CGS) (Ravanel *et al.*, 1998a) for the OPH substrate, and from the observed dependence of these kinetic parameters for threonine synthase on SAM concentration (Curien *et al.*, 1998).

The curves are calculated from the following model

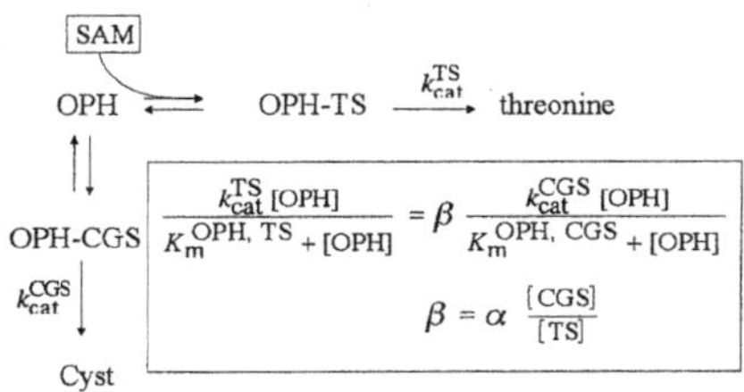

with $k_{cat}^{TS} = 0.4\ s^{-1}$ in the absence of SAM; $k_{cat}^{TS} = 3.3\ s^{-1}$ in the presence of 60 μM SAM; $K_m^{OPH,\,TS} = 270\ \mu M$ in the absence of SAM; $k_{cat}^{CGS} = 30\ s^{-1}$; $K_m^{OPH,\,CGS} = 2.5$ mM; and taking into account the k_{cat}^{TS} and $K_m^{OPH,\,TS}$ values measured at different SAM concentrations (Curien *et al.*, 1998; Ravanel *et al.*, 1998a). For the sake of simplicity, the model assumes equal concentrations of threonine synthase and cystathionine γ-synthase in the plastid compartment: if not this will change the α values used for the computer simulations.

First, since the OPH concentration in plant cells is extremely low (e.g. $< 100\ \mu M$), the α partitioning coefficient must necessarily be much greater than 1, meaning that most of the OPH is directed toward the threonine branch. Yet, the flux from OPH to methionine and SAM is expected to be higher than that to threonine, considering the general role of SAM in transmethylation reactions. In plants this problem is solved by the existence of an enzyme called *S*-adenosylhomocysteine (SAH) hydrolase that recycles efficiently SAH (as produced during transmethylation reactions involving the SAM-dependent methylases) to homocysteine (Kawalleck *et al.*, 1992; Tanaka *et al.*, 1996). We have cloned this enzyme from *Arabidopsis* (B. Gakière, J. Yaxley and D. Job, unpublished results). We also note that the occurrence of SAH hydrolase is required in all organisms for proper functioning of the SAM-dependent methylases, because SAH is a very potent inhibitor of the latter enzymes (Ravanel *et al.*, 1998b). In summary, owing to the existence of SAH hydrolase, a major part of homocysteine consumed through transmethylation reactions is saved thanks to a recycling phenomenon, thereby allowing the accumulation of sufficient methionine levels to sustain protein synthesis and metabolic activity of the plant cell despite a flux partitioning from OPH that favors strongly the synthesis of threonine at the expense of that of methionine.

Second, an examination of the computer curves shown in Figure 4 indicates that for the calculated OPH concentrations to be meaningfull (i.e., have positive values), the SAM concentration should at least be in the order of 10 μM. This finding is interesting because a SAM concentration of 10 μM corresponds precisely to the $[SAM]_{0.5}$ value (i.e., the concentration of SAM giving half-maximal allosteric and cooperative activation of threonine synthase) (Curien *et al.*, 1996, 1998). Thus, around this value, threonine synthase activity is extremely sensitive to fluctuations in SAM concentrations and consequently the flux partitioning of OPH between the threonine and methionine branches also will be highly sensitive to such fluctuations in SAM concentrations.

Third, Figure 4 shows that at constant [SAM], a decrease in the cystathionine γ-synthase concentration (e.g. in the presence of an inhibitor or by antisense expression) will entail a tremendous increase in the OPH concentration. Hence, the concentration ratio, [cystathionine γ-synthase]/[threonine synthase], is crucial for determining the OPH flux between methionine and threonine, and thus the amounts of these enzymes must be finely controled within the plastidial compartment .

5. Compartmentation of homocysteine metabolism

Expression of the regulatory potential of threonine synthase is dependent upon the existence of SAM within the plastidial compartment. Although it is well established that the conversion of aspartate to homocysteine is wholly confined to the chloroplast, there is good evidence that the final steps in the synthesis of methionine and SAM, as catalysed by methionine synthase and SAM synthetase, respectively, are only present in the cytosol. Also, SAH hydrolase is located in the cytosolic compartment (for a review, see Ravanel *et al.*, 1998b). Thus, in plants there exists an interaction of two different compartments in the regulation of the carbon flux between threonine and methionine. The molecular bases for such an interaction, which implies the existence of specific transporters of homocysteine, SAH and SAM are, however, completely unknown.

6. Flux partitioning of OPH: control of enzyme amounts

In yeast, biosynthesis of the sulfur amino acids is specifically controled *via* SAM-mediated negative transcriptional regulation (Thomas, Surdin-Kerjan, 1997). In enteric bacteria, methionine synthesis genes are also negatively regulated by SAM (Saint-Girons *et al.*, 1988). It has been suggested that control of methionine biosynthesis in plants also occurs at the level of gene expression and that the major site for such a regulation is probably cystathionine γ-synthase. In support of this contention, it was observed that (i) when *Lemna paucicostata* is grown in the presence of methionine the level of extractable cystathionine γ-synthase activity is reduced by 85% (Giovanelli *et al.*, 1985), (ii) culture conditions causing methionine starvation, e.g., in the presence of AVG (an inhibitor of cystathionine β-lyase, see Fig. 1) or in the presence of lysine plus threonine mixtures that inhibit an early step in OPH synthesis catalyzed by aspartate kinase (Galili *et al.*, 1995), were associated with a substantial increase of extractable cystathionine γ-synthase activity (Datko, Mudd, 1982; Thompson *et al.*, 1982a).

We have investigated the consequence of altering the level of cystathionine γ-synthase in plants by antisense repression of this enzyme in *Arabidopsis* (Gakière *et al.*, 1998; Ravanel *et al.*, 1998b). Transgenic plants exhibited two types of altered phenotypes, both of which were reversed by administration of exogenous methionine. The first, corresponding to homozygous plants, was lethal, underlying the importance of the methionine-biosynthetic enzymes as potential targets for herbicides. The second, corresponding to hemizygous plants, was less severe, being essentially characterized by a strong reduction in growth. Presumably, the expression level of antisense mRNA was higher in the homozygous than in the hemizygous plants. These observations were in agreement with previous inhibitor-based (PAG, an inhibitor of cystathinonine γ-synthase, see Fig. 1) studies showing that a basal level of cystathionine γ-synthase activity is essential for viability (Thompson *et al.*, 1982b). As expected, the accumulation of cystathionine γ-synthase was specifically depressed in the transformed hemizygous plants (to ca. 30% the level of wild type). Yet, the transgenic plants exhibited about a 3-fold increased level of cystathionine β-lyase, which could therefore compensate for the depressed cystathionine γ-synthase activity, and about a 2-fold increase of threonine synthase, presumably to compensate for a decrease in the availibility of the SAM activator and/or in response to an accumulation of OPH. Clearly, steps other than the first committed step in methionine synthesis are important to account for the homeostatic regulation of methionine and threonine in plants.

Finally, the OPH level was increased by at least 100-fold in the transgenic plants, although the methionine content of these plants was only slightly reduced (by ca. 20%) compared to that of the control plants. This feature can be fully accounted for by the kinetic properties of cystathionine γ-synthase (Ravanel *et al.*, 1995a, 1998a) and threonine synthase (Curien *et al.*, 1996, 1998) and the influence of enzyme amounts on the flux partitioning of OPH between the threonine and methionine branches (see Fig. 4).

7. Acknowledgments

This work has been supported by the CNRS and conducted in part under the *Bio Avenir* program funded by Rhône-Poulenc with the participation of the *Ministère de la Recherche et de l'Espace* and the *Ministère de l'Industrie et du Commerce Extérieur*. We thank Rhône-Poulenc Agriculture (UK) for a grant to one of us (B.G.). J.Y. was supported by the Goddard Sapin-Jaloustre Scholarship Trust.

8. References

Alexander, F.W., Sandmeier, E., Mehta, P.K. and Christen, P. 1994. *Eur. J. Biochem.* **219**, 953-960.

Azevedo, R.A., Arruda, P., Turner, W.L. and Lea, P.J. 1997. *Phytochemistry* **46**, 395-419.

Belfaiza, J., Parsot, C., Martel, A., Bouthier de la Tour, C., Margarita, D., Cohen, G.N. and Saint-Girons, I. 1986. *Proc. Natl. Acad. Sci. USA* **83**, 867-871.

Curien, G., Dumas, R., Ravanel, S. and Douce, R. 1996. *FEBS Lett.* **390**, 85-90.

Curien, G., Job, D., Douce, R. and Dumas, R. 1998. *Biochemistry*, in press.

Datko, A.H. and Mudd, S.H. 1982. *Plant Physiol.* **69**, 1070-1076.

Datko, A.H. and Mudd, S.H. 1984. *Plant Physiol.* **75**, 474-479.

Gakière, B., Ravanel, S., Droux, M., Douce, R. and Job, D. 1998. Submitted for publication.

Galili , G., Perl, A., Karchi, H. 1995. In: *Seed Development and Germination*, eds. Kigel, J. & Galili, G. (M. Dekker Inc., New York), pp. 791-810.

Giovanelli, J., Mudd, S.H. and Datko, A.H. 1980. In: *The Biochemistry of Plants* Vol. 5, ed. Miflin, B.J. (Academic Press, New York), pp. 468-487.

Giovanelli, J., Mudd, S.H. and Datko, A.H. 1985. *Plant Physiol.* **77**, 450-455.

Giovanelli, J., Veluthambi, K., Thompson, G.A., Mudd, S.H. and Datko, A.H. 1984. *Plant Physiol.* **76**, 285-292.

Hell, R. 1997. *Planta* **202**, 138-148.

Holbrook, E.L., Greene, R.C. and Krueger, J.H. 1990. *Biochemistry* **29**, 435-442.

Kawalleck, P., Plesch, G., Hahlbrock, K. and Somssich, I.E. 1992. *Proc. Natl. Acad. Sci. USA* **89**, 4713-4717.

Kim, J. and Leustek, T. 1996. *Plant Mol. Biol.* **36**, 1117-1124.

Kreft, B.D., Townsend, A., Pohlenz, H.-D. and Laber, B. 1994. *Plant Physiol.* **104**, 1215-1220.

Leustek, T. 1996. *Physiol. Plant.* **97**, 411-419.

Madison, J.T. and Thompson, J.F. 1976. *Biochem. Biophys. Res. Com.* **71**, 684-691.

Ravanel, S. 1997. *C. R. Acad. Sci. Paris* **320**, 497-504.

Ravanel, S., Droux, M. and Douce, R. 1995a. *Arch. Biochem. Biophys.* **316**, 572-584.

Ravanel, S., Gakière, B., Job, D. and Douce, R. 1998a. *Biochem. J.* **331**, 639-648.

Ravanel, S., Gakière, B., Job, D. and Douce, R. 1998b. *Proc. Natl. Acad. Sci. USA*, in press.

Ravanel, S., Job, D. and Douce, R. 1996. *Biochem. J.* **320**, 383-392.

Ravanel, S., Ruffet, M.-L. and Douce, R. 1995b. *Plant Mol. Biol.* **29**, 875-882.

Saint-Girons,I., Parsot, C., Zakin, M.M., Barzu, O. and Cohen, G.N. 1988. *CRC Crit. Rev. Biochem.* **23**, S1-S42.

Tanaka, H., Masuta, C., Kataoka, J., Kuwata, S., Koiwai, A. and Noma, M. 1996. *Plant Sci.* **113**, 167-174.

Thomas, D. and Surdin-Kerjan, Y. 1997. *Microbiol. Mol. Biol. Rev.* **61**, 503-532.

Thompson, G.A., Datko, A.H., Mudd, S.H. and Giovanelli, J. 1982a. *Plant Physiol.* **69**, 1077-1083.

Thompson, G.A., Datko, A.H. and Mudd, S.H. 1982b. *Plant Physiol.* **70**, 1347-1352.

Umbarger, H.E. 1978. *Ann. Rev. Biochem.* **47**, 533-606.

Expression of Soybean Vegetative Storage Proteins (VSPα) in Tobacco Leaves

D. Guenoune, R. Amir[1], S. Wolf[2], S. Wininger, H. Badani, B. Ben-Dor and S. Galili

Department of Field and Garden Crops, Agriculture Research Organization (Volcani Center), Bet-Dagan 50250, Israel

1- Department of Plant Physiology Migal Technological Center Kiryat Shmona 12100, Israel.

2- Department of Field and vegetable Crops, Faculty of Agriculture, Hebrew University, Rehovot 76100, Israel

Abstract

Vegetative storage proteins (VSPs) were originally identified as leaf proteins that accumulate to about 45% of the total soluble leaf proteins in depodded soybean plants (Staswick, 1990). In this work we expressed soybean VSPα in transgenic tobacco plants under the control of CaMV 35S promoter. Analysis by SDS polyacrylamide gels revealed a similar molecular weight for the soybean VSPα in transgenic and in wild type (WT) plants. This observation suggests that soybean VSPα in transgenic plants follows corrected processing. We infer that tobacco plants provide a suitable system for studying post-transcriptional and post translation regulation of VSP.

Introduction

Vegetative storage proteins (VSPs) were originally identified as leaf proteins that accumulate to about 45% of total soluble leaf protein in depodded soybean plants (Staswick, 1990). VSPs are abundant in young leaves and their level drops dramatically during the seed filling period. The expression of VSPs is regulated by several external stimuli including source-sink status, phloem blockage, nitrogen availability, wounding, water deficit, light, jasmonic acid and sucrose (for a review see Staswick, 1994; 1997). A gene encoding a VSP-specific thiol-protease has been isolated (Kalinski et al., 1997) suggesting that VSPs are also post-translationally regulated. Like seed storage proteins, VSPs are made in special cell types, stored in vacuoles, and later degraded for the nutritional needs of other organs. However, unlike seed storage proteins VSPs accumulate transiently and are degraded within a single life cycle. The VSPα and β of soybean are glycoproteins with molecular weight of 27 and 29 kD, respectively. They assemble as either hetro- or homo- dimers and have a signal peptide for

A. Altman et al. (eds.), Plant Biotechnology and In Vitro Biology in the 21st Century, 293–296.

entering into the ER. We expressed the soybean VSPα in transgenic tobacco plants under the control of CaMV 35S promoter for studying post-transcriptional factors involved in accumulation of VSPs in heterologous plants.

Material and methods

Construction of chimeric genes and plant transformation. The chimeric gene, containing the cauliflower mosaic virus (CaMV) 35S promoter, the translation enhancer Ω (Gallie et al., 1989), the *EcoR*I fragment of pKSH1 (coding sequence of VSPα)(Mason et al., 1988)), and the octopine synthase terminator (Grave et al., 1983), was cloned into the *Sma*I-*Sac*I sites of the binary vector pBINPLUS (van Engelen et al., 1995). The vector was introduced into *N. tabaccum* cv. Samsun NN by the leaf disc protocol (Horsch et al., 1985) (Fig. 1A).

For overexpression of the soybean VSPβ in *Escherichia. coli*, the coding sequence of the plasmid pKSH3 (Mason et al., 1988) was PCR amplified using POW DNA polymerase (Boehringer) to form a *Nhe*I site in frame in front of first codon and a *Xho*I site after the stop codon. The primers used were 5'-AGCTAGCTATGGTGAGCGTTCTTCG-3' and 5'-ACTCGAGCTACTCAATGTAGTAC-3'. The amplified fragment was cloned into pBluescript KS+ at the *Sma*I site, and PCR fragment was subcloned into the *Nhe*I and *Xho*I sites of pRSET B expression vector (Invitrogene) to form pHISB14.

Production of antibodies and western blotting. The pHISB14 was overexpressed in *E. coli* strain as previously described by Studier and Moffatt, 1987. The VSPβ was purified using His-bind TM Buffer kit (Novagin) according to the kit protocol. Proteins were extracted from 100 mg tobacco leaves with 0.5 ml PBS buffer (Sambrook et al., 1989) supplemented with protein inhibitors cocktail (Boehringer) followed by centrifugation at 15,000 x g for 30 min at 4° C. Equal amounts of proteins were separated on 10% SDS polyacrylamide gels (Laemmli et al., 1970). Proteins were then transferred to nitrocellulose membrane, stained by Ponsau-S, and reacted with anti-VSPβ serum using the ECL kit (Amersham) as recommended by the manufacturer.

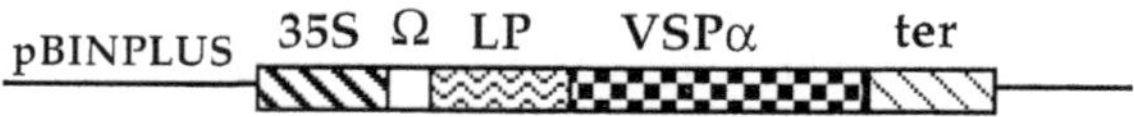

Fig 1. The chimeric gene utilized in this study: 35S, the CaMV 35S promoter; VSPα, the coding DNA sequence of the soybean vegetative storage protein α subunit; ter, the DNA sequence of the octopine synthase 3' terminator; LP, leader peptide.

Results and discussion

Rabbit antisera raised against the recombinant soybean VSPβ was found to detect the fusion protein in *E. coli* (data not shown) as well as the VSPα and VSPβ in crude protein extract from soybean leaves (Fig. 2B, lane S). Staswick (1994), showed that at high levels of nitrogen the anti soybean VSP cross reacted with one or two proteins both in moncot and dicot plants including tobacco. In our experimental conditions, however, no cross reactivity between the anti soybean VSPβ and proteins extracted from untransformed tobacco plants was observed (Fig. 2B, lane N). The soybean VSPα accumulated to relative high levels in leaves of all transgenic plants (Fig. 2 A).

A. Stain

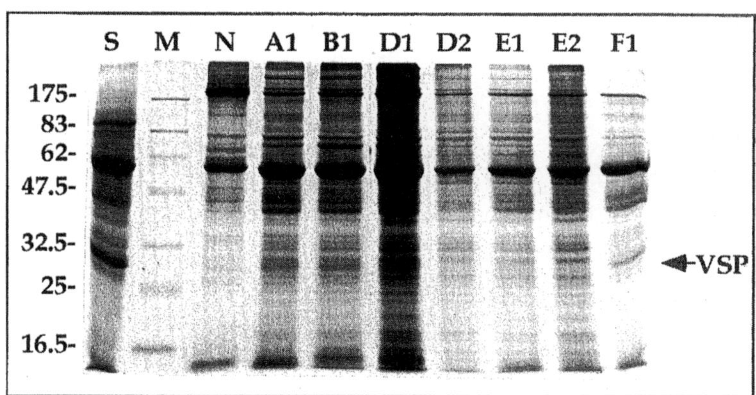

B. Western blot

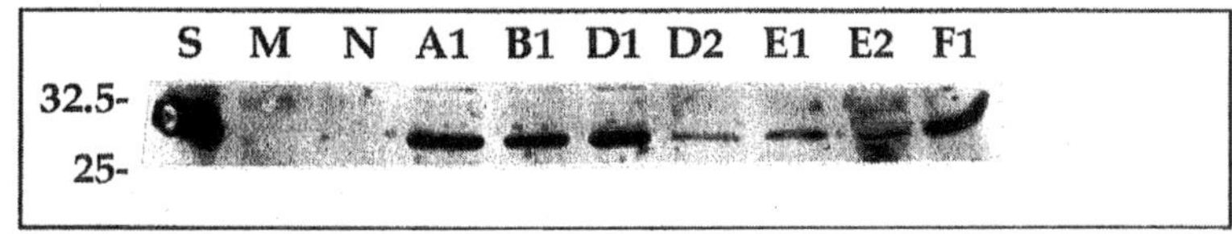

Fig 2. Coomassie Brilliant Blue staining (A) and western blot analysis (B) of soybean VSPα in several transgenic tobacco plants. In each, 20 μg of total soluble proteins extracted from the youngest leaves of 6-8 week-old plants were applied. S, soybean; N, untransformed tobacco; A1, B1, D1, D2, E1, E2 and F1 represent several transgenic tobacco lines; M, pre-stained molecular weight markers (their size in kD is indicated on the left).

The soybean VSPα in transgenic plants also reacted with antisera raised against the recombinant VSPβ protein Fig. 2B). In transgenic plants, the soybean VSPα migrated in SDS-PAGE identically to the wild type (WT) soybean VSPα in soybean. This observation suggests that the soybean VSPα in transgenic plants followed similar processing to that of the WT VSPα, including cleavage of the N-terminal leader peptide and glycosylation, (Staswick, 1994). The cleavage of the N-terminal also suggests that the soybean VSPα enters into the ER; however, we could not tell if it accumulates within the vacuoles. We conclude that tobacco plants provide a good system for studying post-transcriptional and post-translation regulation of VSP.

The significance of using the heterologous CaMV 35S promoter rather than the WT VSP is dual: (i) for overexpressing the soybean VSP in all vegetative tissues of tobacco, and (ii) for identifying post-transcriptional and post-translational factors that regulate VSP accumulation (for a review see Staswick, 1994; 1997).

Acknowledgments

The author wish to thank Dr. A. Sadka for providing the plasmids containg the soybean VSPs cDNA, pKSH1 and pKSH3.

References

Gallie DR et al (1989) Plant Cell 1, 301-311.

Grave HD et al (1983) J. Mol. Appl. Genet. 1,499-511.

Horsch RB et al (1985) Science 227, 1229-1231.

Kalinski A et al (1997) Plant Physiol. and Biochem. 35, 795-802.

Laemmli UK (1970) Nature 227, 680-685.

Mason HS et al (1988) Plant Mol Biol. 11, 845-856.

Sambrook J et al (1989) Molecular cloning: a laboratory manual, 2nd edn.. Cold Spring Harbor, New York.

Staswick PE (1990) Plant Cell 2, 1-6.

Staswick PE (1994) Annu. Rev. Plant Physiol. Plant Mol. Biol. 45, 303-322.

Staswick P (1997) J. of Experimental Botani 48, 2031-2036.

Studier WF, Moffat BA (1987) J. Mol. Biol. 189, 113-130.

van Engelen FA et al (1995) Transgenic Res. 4. 288-290.

Overexpression of L-phenylalanine ammonia-lyase and cinnamate 4-hydroxylase in tobacco cell suspension cultures

Paul Howles, Sameer A. Masoud, Jack W. Blount, Susanne Rasmussen, Chris Lamb, Richard A. Dixon

Plant Biology Division, Samuel Roberts Noble Foundation, PO Box 2180, Ardmore, Oklhoma 73402 (PAH, SAM, JWB, SR, RAD); Plant Biology Laboratory, Salk Institute for Biological Studies, 10010 North Torrey Pines Road, La Jolla, California 92037 (PAH, CL)

Cell suspension cultures were generated from transgenic tobacco plants expressing bean phenylalanine ammonia-lyase (PAL) or alfalfa cinnamate 4-hydroxylase (C4H) transgenes. PAL over-expressing cell lines exhibited high basal levels of PAL transcripts and enzymatic activity, but hardly exceeded wild-type cultures in the level of PAL activity obtained following elicitation. In contrast, expression of alfalfa C4H transcripts in unelicited cell suspension cultures did not result in a corresponding increase in C4H enzymatic activity. However, following elicitation, the level of C4H activity in over-expressing lines was significantly greater than in control lines.

It may be possible to increase production of natural products of commercial value by over-expression of genes encoding early biosynthetic pathway enzymes. Expression of bean phenylalanine ammonia-lyase (PAL) in transgenic tobacco resulted in leaf PAL activities up to 5 times greater than in wild-type plants, accompanied by increased levels of chlorogenic acid (CGA) (Howles et al. 1996). Over-expression of tryptophan decarboxylase in transgenic *Catharanthus roseus* led to accumulation of tryptamine, with no effects on the levels of indole alkaloids (Goddijn et al. 1995). In contrast, expression of hamster or *Hevea brasiliensis* 3-hydroxy-3-methylglutaryl coenzyme A reductase led to significant increases in the levels of downstream sterol metabolites in transgenic tobacco (Chappell et al. 1995; Schaller et al. 1995).

We have generated tobacco lines with altered expression of the second enzyme of the phenylpropanoid pathway, cinnamate 4-hydroxylase (C4H) (Sewalt et al. 1997). To evaluate whether over-expression of early phenylpropanoid pathway enzymes is maintained in cell suspension cultures, cultures were generated from leaf material of transgenic tobacco plants over-expressing PAL or C4H. We here describe the biochemical phenotypes of these cultures, and their response to elicitation.

Materials and Methods

A. Altman et al. (eds.), Plant Biotechnology and In Vitro Biology in the 21st Century, 297–301.

Production of Transgenic Cell Suspension Cultures: The alfalfa C4H cDNA sequence from the yeast expression vector W2A (Fahrendorf, Dixon 1993) was excised with *Bam*H1and *Sal*1, and cloned in place of the GUS gene in both sense and antisense orientations in the binary vector pBI121. *Agrobacterium tumefaciens* strain LBA4404 was transformed with the binary constructs, and transgenic tobacco (*Nicotiana tabacum* L. cv. Xanthi) plants produced by leaf disc transformation (Masoud et al. 1993), using kanamycin selection. Lines studied were #1, 7, and 32 (over-expressors containing the C4H sense transgene and exhibiting levels of C4H activity in leaves of 4-5 times control values), #2 (sense-suppressed, exhibiting levels of C4H activity in leaves of approximately 25% of control values), and 10 (empty vector control). Callus cultures were initiated from leaf discs on semi-solid BM medium supplemented with 10 μM NAA and 1 μM kinetin, and were subcultured every 2 to 3 weeks. Cell suspension cultures were established by transferring callus to SH liquid medium plus NAA and kinetin, and were subcultured every 10 to 14 days.

PAL over-expressing callus cultures were made from leaf tissue of two wild-type and two transgenic YE10-6T1 tobacco plants which consistently showed approximately 5 fold higher levels of extractable PAL activity than wild-type (Howles et al. 1996). The YE10-6T1 plants possess a bean *PAL2* transgene in the sense orientation with respect to a CaMV 35S promoter linked to a portion of native bean *PAL2* promoter (Elkind et al., 1990). Cell suspension cultures were initiated from these callus cultures as above.

Molecular and Biochemical Analyses: Analysis of PAL and C4H transcript levels by standard RNA blot procedures, of PAL enzymatic activity, and of soluble phenolic compounds was performed as described (Howles et al 1996). C4H was assayed by the method of Edwards and Kessmann (1992). NADPH cytochrome P450 reductase was measured in the same microsomal fractions as assayed for C4H activity using cytochrome C as substrate according to Sottocasa et al. (1967), but with the assay buffer changed to 100mM potassium phosphate, pH 7.0.

Results and Discussion

PAL transcripts and enzymatic activity in PAL2 transgenic cell suspension cultures: Newly initiated YE10-6T1 suspension cultures had significantly higher PAL activity (2-fold on average) than cultures derived from wild-type plants (Fig. 1A). When cultures were maintained for over two months there was a decrease in the basal PAL activity; this decrease was greater in wild-type than in YE10-6T1 cultures (data not shown). Yeast elicitor induced PAL activities in both wild-type and in some, but not all, YE10-6T1 suspension cultures (Fig. 1A). Maximum PAL activity in elicited YE10-6T1 cultures was generally, but not always, slightly more than in elicited wild-type cultures; one line that exhibited 10-fold higher basal PAL activity than the corresponding control line was totally unresponsive to elicitor (data not shown). The fold-increase in PAL activity was always greater in wild-type than in PAL over-expressing cultures. These results suggest that there is a limitation to the level of achievable PAL activity, probably because of feedback by pathway intermdiates (Mavandad et al 1990).

To confirm that the bean *PAL2* transgene is expressed in cell suspensions derived from YE10-6T1 plants, RNA gel blot analysis was performed using probes specific for both endogenous and transgene-encoded *PAL* transcripts (Howles et al. 1996). An elicitor-inducible transcript corresponding to the bean *PAL* transgene was detected in the YE10-6T1 cultures, regardless of age (Fig. 1B). The level of endogenous *PAL* transcripts was greater than that of the bean *PAL* transcripts in cultures less than two months old, but approximately equal in older cultures. Elicitation resulted in a massive increase in transcripts encoding the endogenous *PAL* genes (Fig. 1B).

C4H transcripts and enzymatic activity in C4H transgenic cell suspension cultures: A high level of alfalfa C4H transcripts was detected at zero time in the over-expressing and sense-suppressed lines #1, 2, 7 and 32 (Fig. 1C). In contrast, no C4H transcripts were detected in the empty vector control line at zero time, and the elicitor-induced increase in transcript levels in this line reflects induction of endogenous tobacco C4H transcripts. The high constitutive level of alfalfa C4H made it difficult to detect elicitor-induction of endogenous C4H transcripts in lines #1, 2 and 32. The level of C4H transcripts was similar in unelicited cell cultures of lines #1 (over-èxpressor) and 2 (suppressed), whereas, in leaves, transcript levels were higher in line #1 than in line #2 (data not shown).

In spite of the large difference in the level of C4H transcripts between unelicited C4H transgenic and empty vector control cell cultures, the level of C4H activity at zero time in lines #1, 2 and 10 was almost equal, with slightly lower activity in line #32 (Fig. 1D). However, following elicitation, C4H activity was induced to higher levels in C4H over-expressing lines #1 and 32 than in the control line #10, even though the 35S promoter controling expression of the C4H transgene is not elicitor-inducible (Oommen et al 1994). This pattern was reproducible in a second set of cell suspension cultures initiated from callus cultures that had been maintained by regular subculture for a further 12 months, although, as in the case of PAL, the overall enzymatic activities were lower. Therefore, in contrast to the situation for PAL, the high levels of C4H transcripts are either not translated, or are translated to inactive enzyme, in unelicited cells, and elicitation triggers translational/post-translational processes that result in active enzyme. Alternatively, the activity of C4H in the over-expressing lines may be limited by NADPH-cytochrome P450 reductase activity. This latter hypothesis is not supported by the fact that reductase activity was present at high basal levels (40-80 nmol/min/mg protein) and was induced 2-3 fold in the various lines (data not shown).

Phenylpropanoids in transgenic cell suspension cultures: Intact tobacco plants over-expressing PAL or C4H show quite large increases in the levels of phenolic compounds, particularly CGA (Howles et al., 1996; J.W. Blount et al., unpublished results). However, over-expression of PAL resulted in only an approximately 20% increase in the levels of

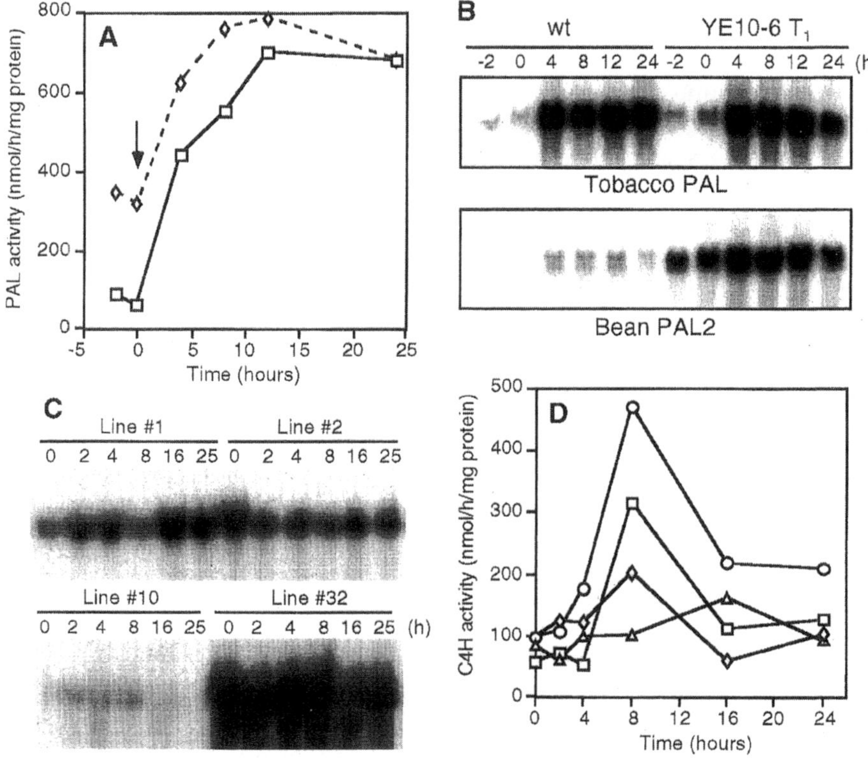

Fig. 1. A. Time course for the elicitation of PAL activity in newly initiated cell suspension cultures derived from wild-type (◊) and PAL over-expressing (YE10-6T1) (❍) lines. **B**. Levels of PAL transcripts in wild-type and PAL-overexpressing transgenic tobacco cell cultures. Total RNA isolated at the times shown relative to addition of yeast elicitor was subjected to northern blot analysis probing with labelled cDNAs recognizing tobacco PAL and bean PAL2, along with β-ATPase as loading control (data not shown). **C.** C4H transcripts levels determined by northern blot analysis at the times shown (h post-elicitation) in cell cultures of lines #1 and 32 (over-expressors), 2 (sense-suppressed) and 10 (empty vector control). The blot was probed with the full length alfalfa C4H cDNA. **D.** C4H activities in elicitor-treated transgenic tobacco cell suspension cultures. Cultures were op empty vector control (line #10) (◊), C4H over-expressor (lines #1 and #32) (O,) and C4H sense-suppressed (line #2) (Δ).

the major extractable phenolics in suspension cells (data not shown), and elicitation had little effect. Similarly, over-expression of C4H had little effect on soluble phenolic levels or profiles.

The reason for the different extents and metabolic outcomes of enzyme over-expression in intact plants and cell suspension cultures is not clear. Our data highlight the fact that endogenous feedback mechanisms and post-translational controls can impact the level to which an enzyme can be over-expressed in plant cells.

Acknowledgements

This work was supported by the Samuel Roberts Noble Foundation.

References

Chappell J et al (1995) Plant Physiol. 109, 1337-1343.
Edwards R, Kessmann H. (1992) In Molecular Plant Pathology- A Practical Approach, vol. 2 (ed. Gurr S J et al), IRL Press, Oxford, 45-62.
Elkind Y et al (1990). Proc. Natl. Acad. Sci. USA 87, 9057-9061.
Fahrendorf T, Dixon RA (1993). Arch. Biochem. Biophys. 305, 509-515.
Goddijn OJM et al (1995) Transgenic Res. 4, 315-323.
Howles PA et al (1996) Plant Physiol. 112, 1617-1624.
Lesot A et al (1990) Plant Cell Physiol. 31, 1177-1182.
Masoud SA et al (1993) Plant Mol. Biol. 21, 655-663.
Mavandad M et al (1990) Plant Physiol. 94, 671-680
Oommen A et al (1994) Plant Cell 6, 1789-1803.
Sewalt VJH et al (1997) Plant Physiol. 115, 41-50.
Sottocasa GL et al (1967) J. Cell Biol. 32, 415-438.

MANIPULATING ESSENTIAL AMINO ACID METABOLISM IN PLANTS

M. Jacobs, V. Frankard, M. Vauterin, E. Suharyanto and J. Tadesse
Laboratory of Plant Genetics, Vrije Universiteit Brussel, Institute of Molecular Biology, B1640 Sint-Genesius-Rode, Belgium

The nutritional quality of crop plants is determined by their content in essential amino acids provided as proteins in foods for humans or in feed for monogastric animals. Plant proteins are often deficient in some of the ten essential amino acids which are required in human and animal diet. In general cereals are deficient in lysine and legumes in the sulphur amino acids, methionine and cysteine.
Besides conventional genetic means, various strategies have been developed with the goal to improve the amino acid imbalance of crop-based diets. Modifying genes encoding storage proteins aiming at correcting their nutritional deficiency or expressing genes encoding nutritionally superior storage proteins were presented as possible strategies in order to improve protein quality, especially in seeds (J.E. Habben, B.A. Larkins, 1996).
An alternative way consists in producing plants containing elevated levels of specific amino acids by manipulating their metabolism. This can be achieved via alteration of key enzymes from those biosynthetic pathways by mutation and transformation. In the last years, we have focused on the aspartate derived amino acid biosynthetic pathway producing the four essential amino acids : lysine, threonine, isoleucine and methionine. Considerable research has been devoted to the regulation of this pathway with the aim of optimising amino acid content.

1. Biochemical regulation of lysine and threonine biosynthesis.

Higher plants synthesize lysine, threonine, isoleucine and methionine from aspartate via a branched pathway which is essentially regulated by endproduct feedback inhibition (see fig. 1). The first enzyme of the pathway, aspartate kinase (AK), is present as isoforms differentially feedback inhibited by either lysine or threonine. The other key enzyme of this pathway, dihydrodipicolinate synthase (DHDPS), the first enzyme of the lysine branch, is severely inhibited by lysine (K_i between 5 and 20 μM). Most of the biosynthetic enzymes of the pathway have been localized in the plastids.

2. Mutants overproducing free lysine or threonine.

In various species, mutants accumulating lysine or threonine in the soluble amino acid pool have been isolated by selecting for growth in the presence of the lysine amino acid

A. Altman et al. (eds.), Plant Biotechnology and In Vitro Biology in the 21st Century, 303–306.

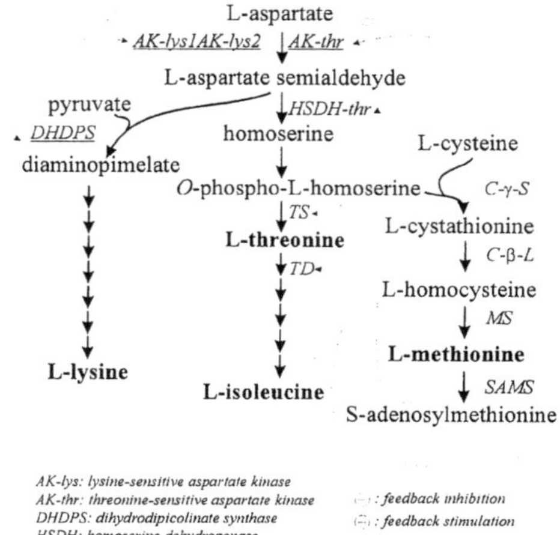

Figure 1: The aspartate biosynthetic family of amino acids, and major feedback controls

analog, S-2-(aminoethyl)-1-cysteine (AEC) or toxic combinations of lysine and threonine in the culture medium (M. Jacobs *et al.*, 1995). While threonine overproduction was detected in all tissues analysed and in particular in seeds, the increase of free lysine was only noticeable in leaves and calli. Moreover, an aberrant phenotype (reduction of foliar surface, absence of shoot elongation and sterility) was observed when high levels of free lysine accumulated (25% and more of the total free amino acid pool).

The biochemical characterization of these mutants has shown that we are dealing with a deregulation at the level of the two key enzymes, DHDPS in the lysine overproducer mutant and AK in the threonine overproducer. These mutated forms are both fully insensitive to feedback inhibition by lysine.

3. Cloning and characterization of the AK coding genes.

At least three forms of plant AK have been identified based on their differential feedback properties. The gene coding for a threonine-sensitive form (*ak-hsdh*) was isolated by heterologous hybridization in *A. thaliana*. The deduced amino acid sequence revealed the presence of a second region corresponding to the enzyme activity of homoserine dehydrogenase (HSDH), the first committed enzyme in the branch of the pathway that leads to threonine synthesis. We are thus dealing with a gene coding for a bifunctional protein with an AK activity at the NH_2-side and a HSDH activity at the COOH-terminus (M. Ghislain *et al.*, 1994). The expression of the *A. thaliana ak-hsdh/uidA* reporter gene encoding glucuronidase (GUS) was elevated in an array of young plant tissues containing actively growing cells (meristems, young leaves, cortical and vascular stem tissues, anthers, gynoecium, developing seeds). During the embryo development its expression appeared coordinated with the initiation and onset of storage protein synthesis (J.X.Zhu-Shimoni *et al.*, 1997).

As the major part of AK activity is sensitive to feedback inhibition by lysine, attempts have been done to clone the corresponding gene. Degenerate primers corresponding to conserved motives between lysine-sensitive bacterial AKs were used to clone two genes (*ak-lys1* and *ak-lys2*) encoding monofunctional AKs in *A. thaliana* (V. Frankard *et al.*, 1997; G. Tang *et al.*, 1997). The presence of two nuclear genes both encoding AK-lys proteins targeted to the chloroplast but with only 70% identity at deduced amino acid level raises questions about their respective roles in plant development.

Therefore, *A. thaliana* and *N. tabacum* have been transformed with the 5' upstream region of either *ak-lys* gene fused to the *uidA* gene. Staining for GUS activity shows that the genes have complementary expression patterns in vegetative tissues, floral organs and seeds. a*k-lys1* is expressed mainly in stems, which suggests its involvement in amino acid transport. This gene is also strongly expressed in developing seeds, and is thus probably involved in seed storage protein synthesis. a*k-lys2* is expressed in all other tissues, mainly young ones. It appears to have a more "house-keeping" function that *ak-lys1*. As revealed by northern analysis, it also reacts more to environmental conditions such as light regime and nitrogen supply. The two *ak-lys* genes encode thus distinct isozymes which are diffentially regulated.

The isolation and characterization of *A. thaliana* mutants displaying an AK-HSDH or an AK-lys isozyme less sensitive to respectively threonine and lysine feedback inhibiton (B. Heremans and M. Jacobs, 1995, 1997) will allow to identify at nucleotide level mutations leading to threonine accumulation. After incorporating the mutation into the corresponding c-DNA, this allele will be expressed in an appropriate construct to obtain threonine accumulation in specific tissues, and specially in seeds.

4. Cloning and characterization of the gene encoding DHDPS.

A mutant of *Nicotiana sylvestris* (RAEC-1) was shown to overproduce lysine due to a mutation in the *dhdps* gene which causes the DHDPS enzyme to be insensitive to the normal feedback inhibition by lysine. The *raec-1* mutation was identified as a substitution of two nucleotides changing asparagine into isoleucine in a conserved region of the protein (M. Ghislain *et al.*, 1995).

dhdps-encoding sequences from dicots were cloned by functional complementation in a bacterial DHDPS-deficient strain. This first clone allowed subsequently to isolate and sequence a full-length *Arabidopsis dhdps* cDNA (M. Vauterin and M. Jacobs, 1994). Constructs derived from the *Arabidopsis* cDNA allowed to generate through EMS *in vitro* mutagenesis clones encoding fully insensitive forms of the DHDPS protein (M. Vauterin, in press). Furthermore, the clones successfully isolated by functional complementation were all found to be insensitive which means that complementation selects for mutant *dhdps* clones encoding insensitive enzyme forms.

In a further step, the *Arabidopsis* promoter has been isolated and fused with the *uidA* gene. Expression of GUS was detected in meristems and vascular tissues of roots, in vascular tissues of stems and leaves and in the meristems of young shoots. In flowers, high expression was found in the carpels, pollen grain, young embryos, but not in endosperm of mature seeds. No lysine-induced repression of the *dhdps* gene could be detected.

5. Metabolic engineering for improved nutritional quality.

The manipulation of free amino acid pools of plants through expression of homologous, heterologous or mutated genes and the expression and stable accumulation of novel sink proteins rich or enriched in the respective amino acids must enable us to improve the nutritional quality of crop plants. This can be achieved by combining :

1) a balanced overproduction of free essential amino acids in the whole plant or targeted to sink organs.

2) a reduction of catabolic reactions to reduce amino acid degradation.
3) the exploitation of sink-source relationships to channel amino acids towards novel and nutritionally valuable storage proteins.
This balanced overproduction of amino acids in plants can be achieved by transferring genes coding for feedback insensitive key enzymes of bacterial and plant origin to plants.
Bacterial genes have first been used to transform several species like *Arabidopsis, Nicotiana tabacum,* barley, soybean and oilseed rape. In tobacco the expression of AK and DHDPS relatively insensitive to feedback inhibition caused important increases in the synthesis of both threonine and lysine (G. Galili, 1995). However, while threonine accumulated in seeds, variation in the level of free lysine was observed according to the species. This could be related to the activity of lysine-ketoglutarate reductase, the first lysine catabolic enzyme (H. Karchi *et al.,* 1994; S.C. Falco *et al.,* 1995). To prevent this lysine catabolism, an approach channeling the overproduced lysine to a nutritionally optimal sink protein can be envisaged.
In our laboratory, several chimeric constructs harbouring the *raec-1* mutated allele under the control of a constitutive or seed-specific promoter were transferred via *Agrobacterium* to *Nicotiana plumbaginifolia, Brassica napus* and biolistic to *Sorghum bicolor* (see accompanying paper). The expression of this plant gene also led to the synthesis of a feedback insensitive DHDPS, associated with high lysine content and eventually aberrant phenotypes as mentioned for the RAEC-1 mutant.
It appears that the expression of a single feedback insensitive enzyme as DHDPS exerts already a significant effect upon the flux through the aspartate pathway towards lysine. Coordinated alterations of rate-limiting enzymes will likely contribute to increase the flux even more towards such end products.

6. Acknowledgments

V. F. is recipient of a post-doc fellowship from the Fonds voor Wetenschappelijk Onderzoek - Vlaanderen (FWO). FWO also funds the research project G.0242.97.

7. References

Falco S.C. et al. (1995) Biotechnology 13, 577-82.
Frankard V. et al. (1992) Plant Physiol. 99, 1285-93.
Frankard V. et al. (1997) Plant Molec. Biol. 34 : 233-242.
Galili G. (1995). The Plant Cell 7, 899-906.
Ghislain M. et al. (1994) Plant Molec. Biol. 24, 835-851.
Ghislain M. et al. (1995) Plant J. 8, 101-111.
Habben J.E., Larkins B.A. (1996) Seed Development and Germination (Kigel O., Galili G. eds), pp. 791-809, Marcel Dekker Publisher, New York, U.S.A.
Heremans B., Jacobs M. (1995) J. Plant Physiol. 146, 249-257.
Heremans B., Jacobs M. (1997) Biochem. Genet. 35, 139-153.
Jacobs M. et al. (1995) Biochemistry of Plants : Intermediary Nitrogen Metabolism (Miflin B.J., Lea P.J. eds), pp. 161-195, Academic Press, New York USA
Karchi H. et al. (1994) Proc. Natl. Acad. Sci. USA 91 : 2577-2581.
Tang G. et al. (1997) Plant Molec. Biol. 34 :287-294.
Vauterin M., Jacobs M. (1994). Plant Molec. Biol. 25, 545-550.
Zhu-Shimoni J.X. et al. (1997). Plant Physiol. 113, 695-706.

A 28-kDa POD STORAGE PROTEIN OF *Phaseolus vulgaris*: CHARACTERIZATION, INDUCED ACCUMULATION, AND DEGRADATION

Takao Minamikawa, Pei-Ying Zhong and Daisuke Yamauchi
Department of Biological Sciences, Tokyo Metropolitan University, Minami-ohsawa 1-1, Hachioji, Tokyo, 192-0397 Japan

Introduction

During fruit development of French bean plants, a proportion of seed nitrogen is temporarily stored in pods as protein, amino acids and probably small peptides, which are then mobilized to support the growth of seeds (Endo et al., 1987). We observed a continued increase of endopeptidase activity throughout the development and subsequent senescence of French bean pods, and this endopeptidase activity is thought to participate in the protein mobilization in pods during fruit maturation (Tanaka et al., 1993). In the course of studies on endopeptidases in pods of French bean plants, we found that, when a large number of pods in early stages of fruit development were removed from a plant, relative levels of a specific 28-kDa protein were enhanced in the newly formed pods. The protein accumulated to high levels in the pods of plants that had been previously depodded, whereas a smaller amount was detected in the pods of non-treated plants. It was postulated that enhanced levels of the 28-kDa protein resulted from a reduction of plant sink size caused by removing pods in early stages of fruit development.

We isolated and characterized this 28-kDa protein from French bean pods and designate it 'pod storage protein' (PSP). Our current results have shown that PSP accumulated in stems, roots and other organs in addition to newly formed pods. Methyl jasmonate had no effect on the accumulation of PSP, but the accumulation was induced in leaves by wounding. In senescing pods, PSP was mainly degraded through 20-kDa and 17-kDa polypeptides as degradation intermediates.

Purification and Characterization of PSP

PSP was extracted from pods of the plants that had been depodded in early stages of fruit development. The crude extracts were fractionated by ammonium sulfate, and columns of DEAE-cellulose and Sephacryl S-200. The final gel filtration step provided an apparently homogeneous PSP preparation as examined by Coomassie blue staining after SDS-PAGE. However, analysis of the preparation by 2D-PAGE revealed three forms (A, B and C) of PSP with identical mobility (M_r 28,000) but different net charge. By the method of Hedrick and Smith (1986), the molecular mass of both forms A and C was estimated to be approximately 67 kDa, indicating that PSP is probably present as a dimer. Forms A and C shared at least 16 identical amino acid residues among the amino-terminal 20 residues. The nine amino-terminal amino acids of form B were identical with those of forms A and C, but Ala-1 of forms A and C was lacking in form B (Zhong et al., 1997).

A. Altman et al. (eds.), Plant Biotechnology and In Vitro Biology in the 21st Century, 307–310.

A cDNA expression library from French bean pods of stage V (Tanaka et al., 1991) was immnologically screened, and a clone carrying the longest cDNA insert was selected. This clone, designated λPSP1, included 765 bp of an open reading frame potentially encoding a polypeptide with 255 amino acid residues (calculated molecular mass 28,854 Da). The amino-terminal amino acid sequence of forms A and C of PSP corresponded to amino acid residues 37 to 52. The deduced amino acid sequence of the PSP cDNA has 71% and 65% identity to those of soybean vegetative storage proteins (VSP-α and VSP-β), respectively (Rhee, Staswick, 1992a; 1992b), and has 40% identity to acid phosphatase-1 of tomato (Erion et al., 1991). No putative glycosylation sites were contained in the deduced amino acid sequence of PSP cDNA, although VSPs (Wittenbach, 1983) and acid phosphatase-1 (Paul, Williamson, 1987) are glycoproteins. In fact, PSP was not stained with the periodic acid-Schiff's reagent, whereas VSPs were stained (Zhong et al., 1997).

Accumulation of PSP and its mRNA in pods

When a large number of pods in early stages of fruit development were removed from a plant of French bean, relative levels of PSP were enhanced in the newly formed pods. After SDS-PAGE/immunoblotting, amounts of PSP in the pod at various stages of fruit development was assessed by scanning the bands of the immunoreactive 28-kDa polypeptide with a densitometer (Fig. 1A). In pods of depodded plants, the amount of PSP per pod increased continously from stages II to V. In pods of non-treated control plants, the amount increased only from stages II to IV, after which the level was maintained until stage V. Thus the amount of PSP at stage V was doubled when plants were depodded. In pods of non-treated plants, the level of PSP mRNA increased gradually from stages II to III and then decreased. The level in pods of depodded plants at stage II was about two-fold higher than the control, and this level was maintained until

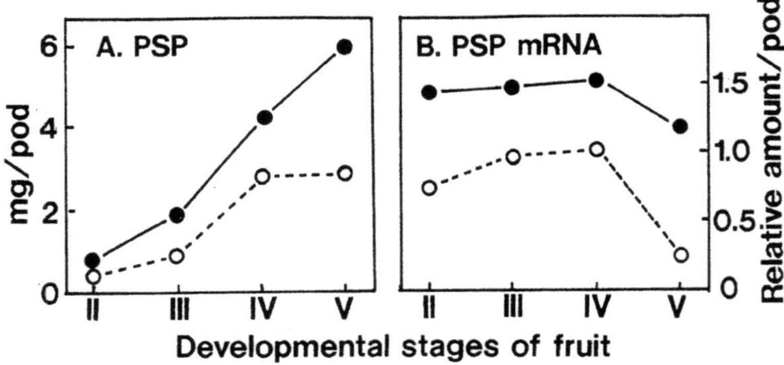

Figure 1. Effects of the removal of pods on the accumulation of PSP (A) and the levels of PSP mRNA (B) in pods. Pods were collected from the non-treated control plants (open circles) and depodded plants (closed circles). The amounts of PSP (A) and its mRNA (B) were assessed by scanning the immunoreactive bands and the autoradiographs, respectively, with a densitometer. II to V: developmental stages of fruit (Endo et al., 1987). The levels were averaged from three experiments.

stage V (Fig. 1B). The results suggest that the enhancement of PSP content in pods by pod removal is attributable to the increased level of PSP mRNA.

In contrast to the case of soybean VSP (Anderson et al., 1989), methyl jasmonate at 50 μg/ml, which was applied to pods had no significant effect to increase the accumulation of PSP. Sucrose, glucose or fructose at 2 to 5% added to a culture medium of pods at stage III showed little effect on the accumulation of PSP.

Accumulation of PSP and its mRNA in different organs

In non-treated plants, PSP accumulated in flowers in addition to pods but not in stems, leaves, roots or immature seeds. In these plants, PSP mRNA was detected in stems and flowers, but not in leaves or immature seeds. Pod removal induced the accumulation of PSP in stems and roots, but not in leaves or immature seeds (Fig. 2). In depodded plants, PSP mRNA was detectable in roots, and its level increased in stems, but no effect was observed in leaves and immature seeds. In plants carrying three to six leaves, PSP accumulated only in very young leaves. Relatively high levels of PSP were observed in young apical leaves of plants at flowering and podding stages. These results indicate that the sink removal induces or enhances the accumulation of PSP and its mRNA levels in an organ-specific manner. In a dark-grown seedling, the cotyledons and

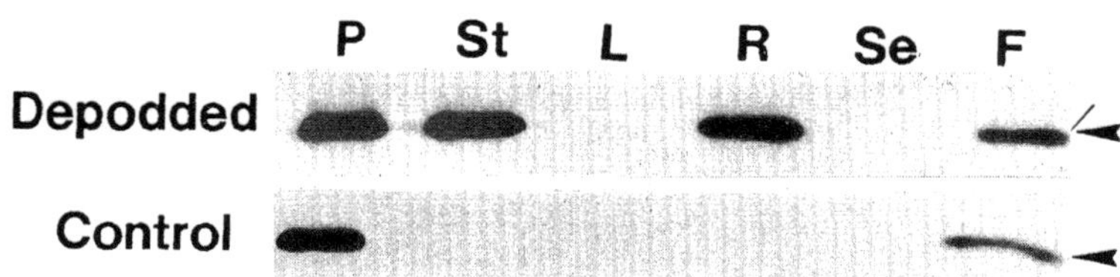

Figure 2. Organ specificity of the accumulation of PSP in the depodded and control plants. Crude extracts were prepared from pods (P), stems (St), leaves (L), roots (R) and whole seeds (Se), and analyzed by SDS-PAGE/immunoblotting. The arrowheads point to PSP.

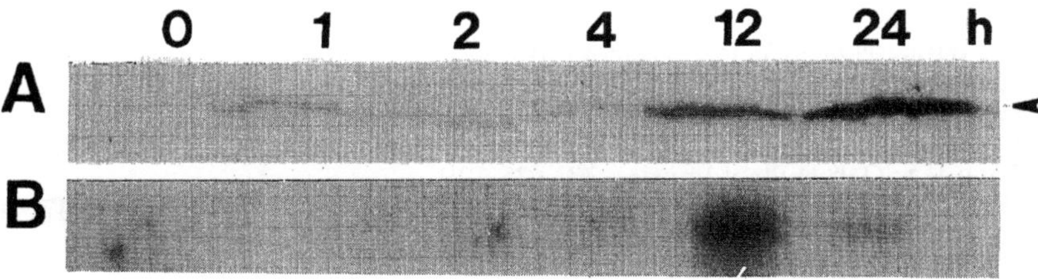

Figure 3. Changes with time in the amount of PSP (A) and levels of PSP mRNA (B) in wounded leaves. The 3rd leaves of young plants having six leaves were wounded by a razor blade. Protein and total RNA from the 3rd leaves 0 to 24 h after wounding were analyzed by immuno- and RNA-blotting, respectively. The arrowhead points to PSP.

embryonic axis correspond to source and sink organs, respectively. PSP was not detected in cotyledons during germination and early growth, but when cotyledons were detached from the embryonic axis and incubated, a 28-kDa polypeptide corresponding to PSP became detectable 9 to 11 days after the onset of imbibition. This suggests that the reduction of sink size by the removal of the axis induced the accumulation of PSP in the cotyledons.

Wounding enhanced the accumulation of PSP in leaves. When the 3rd leaf of a young plant having six leaves was wounded by a razor blade, the level of PSP in the wounded leaf increased gradually until 24 h of the treatment (Fig. 3A), and the level of PSP mRNA was detected at 4 h, became a maximum at 12 h and decreased at 24 h (Fig. 3B). In addition, in case that the 3rd leaf was wounded, the accumulation of PSP was induced systemically in the 2nd to 6th leaves but not in the 1st leaf.

Degradation of PSP in senescing pods

When pods were senescing, they were desiccated and the protein content in the pods decreased sharply. In later stages (38 to 45 DAF, days after flowering) of senescence of the pod, the amount of 28-kDa PSP, as estimated by SDS-PAGE/immunoblotting, was concurrently declined to less than one fifth of the content at stage V. In addition to 28-kDa PSP, however, two major polypeptides with molecular masses of 20 kDa and 17 kDa that were immunoreactive with the antibody against PSP, were detected in extracts from senescing pods. The amount of the 20-kDa polypeptide was higher than that of the 17-kDa polypeptide in both earlier (32 to 37 DAF) and later stages of senescence. The amino-terminal amino acid sequence of the 20-kDa polypeptide was identical to that of PSP, and that of the 17-kDa polypeptide corresponded to residues 127 to 132 of the amino acid sequence deduced from the nucleotide sequence of cDNA for PSP. Thus, in senescing pods, PSP may be degraded by limited proteolysis through 20-kDa and 17-kDa polypetides as earlier intermediates.

References

Anderson JM, Spilatro SR, Klauer SF, Franceschi VR (1989) Plant Sci. 62: 45-52.
Endo M, Minamikawa T, Yamauchi D, Mitsuhashi W (1987) J. Exp. Bot. 38: 1988-1995.
Erion JL, Ballo B, May L, Bussell J, Fox T, Thomas SR (1991) Plant Physiol. 97: 1462-1469.
Hedrick JL, Smith AJ (1986) Arch. Biochem. Biophys. 126: 155-164.
Paul EM, Williamson VM (1987) Plant Physiol. 84: 399-403.
Rhee Y, Staswick PE (1992a) Plant Physiol. 98: 792-793.
Rhee Y, Staswick PE (1992b) Plant Physiol. 98: 794-795.
Tanaka T, Minamikawa T, Yamauchi D, Ogushi Y (1993) Plant Physiol. 101: 421-428.
Wittenbach VA. (1983) Plant Physiol. 73: 125-129.
Zhong PY, Tanaka T, Yamauchi D, Minamikawa T (1997) Plant Physiol. 113: 479-485.

REGULATION OF LYSINE CATABOLISM IN PLANTS

D. Miron, G. Tang, H. Karchi, A. Schupper, S. Ben-Yaacob and G. Galili,

Department of Plant Sciences, The Weizmann Institute of Science, Rehovot 76100 Israel

Abstract

Seeds of many crop plants are considered as a poor nutritional value food source, due to very low levels of the essential amino acid lysine. Attempts to increase lysine content in seeds are hampered by the presence of an active catabolic process that degrades lysine via saccharopine into other metabolites. The first enzyme in this catabolic process is lysine ketoglutarate reductase (LKR). We have recently obtained evidence that strongly suggest that lysine finely controls its own level in seeds by maintaining LKR activity balanced using two separate complementary mechanisms. I) LKR activity is post-translationally stimulated by lysine via a Ca^{+2} dependent intracellular signaling cascade that ends in the phosphorylation of the enzyme, thus, preventing lysine to accumulate to high levels. II) subsequent binding of lysine to the active site of the enzyme may expose phosphate residues on the surface of LKR rendering them more accessible to dephosphorylation, thus allowing lysine to accumulate to a sufficient level needed for protein synthesis.

Introduction

The cellular level of the essential amino acid lysine is subject to tight regulation both in mammalians and plants. In both types of organisms, excess lysine is catabolized via saccharopine and α-amino adipic semialdehyde into α-amino adipic acid and glutamate (Markovitz *et al.*, 1984; Galili, 1995; Goncalves-Butruille *et al.*, 1996) . The first enzyme in the lysine catabolic pathway is lysine-ketoglutarate reductase (LKR), which condenses lysine and α-ketoglutarate into saccharopine, while the second enzyme, saccharopine dehydrogenase (SDH), converts saccharopine into a-amino adipic semialdehyde and glutamate (Galili, 1995) . Biochemical studies (Markovitz and Chuang, 1987; Goncalves-Butruille *et al.*, 1996) and gene cloning (Epelbaum *et al.*, 1997; Miron *et al.*, 1997; Tang *et al.*, 1997) have shown that in plants and in some mammalian species, LKR does not exist as a separate entity, but is linked to SDH, on a single bifunctional polypeptide. The molecular basis for this linkage has not been elucidated, but suggests that the structure of this bifunctional enzyme may also have an important regulatory role. We have previously demonstrated that injection of lysine to developing tobacco pods stimulated the activity of LKR in or tobacco seeds (Karchi *et al.*, 1994). A similar stimulation of this enzyme was also observed in transgenic tobacco seeds that overproduced lysine due to expression of a bacterial feedback-insensitive dihydrodipicolinate synthase (Karchi *et al.*, 1995) . This suggested that in plant cells, lysine may auto regulate its own catabolism *in vivo*. In addition, recent studies have shown that in tobacco seeds, the lysine-dependent stimulation of LKR activity was mediated by an intracellular signaling cascade requiring Ca^{2+} and protein phosphorylation (Karchi *et al.*, 1995) .

To further study the control of LKR and SDH activities in plants, we have purified the bifunctional LKR/SDH from soybean seeds to near homogeneity. Using the purified

A. Altman et al. (eds.), Plant Biotechnology and In Vitro Biology in the 21st Century, 311–314.

enzyme we tested whether these activities are modulated by direct phosphorylation and dephosphorylation, and whether this modulation is regulated by lysine. We found that *in vitro* dephosphorylation specifically inhibited the activity of LKR, but not of SDH, in a lysine-regulated manner (Miron et al., 1997).

Results

We have purified the LKR/SDH bifunctional enzyme from developing seeds. Using the purified enzyme, we have tested whether the LKR/SDH could be phosphorylated *in vitro,* in a reaction catalyzed by a recombinant casein kinase II (Lisitsky and Schuster 1995). This analysis showed that LKR/SDH is readily phosphorylated with this enzyme (data not shown).

To test whether *in vivo* phosphorylation may have an effect on LKR and SDH activities, we analyzed the response of these enzymes to dephosphorylation by alkaline phosphatase. As shown in Fig. 1A, the kinetics of LKR activity was progressively reduced with increasing concentrations of alkaline phosphatase, while addition of increasing concentrations of the phosphatase inhibitor β-Glycerol phosphate progressively decreased this inhibition (Fig. 1B).

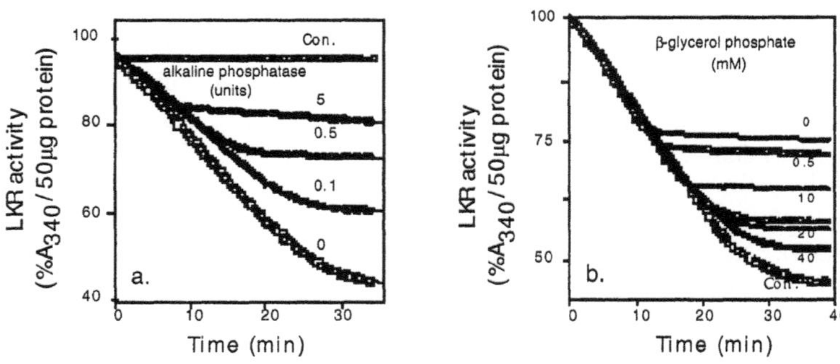

Fig. 1. Effect of dephosphorylation with alkaline phosphatase on LKR activity.
The kinetics of LKR activity was assayed spectrophotometrically by following the time-dependent reduction in A_{340} resulting from the oxidation of NADPH by LKR. (A) Assays were supplemented with 0-5 units/reaction of alkaline phosphatase as indicated in the panel. (B) Assays were supplemented with 5 units of alkaline phosphatase plus 0-40 mM of β-glycerol phosphate, as indicated in the panel.

We have also tested whether SDH, which is linked to LKR on a single bifunctional polypeptide, is also affected by alkaline phosphatase. we found that alkaline phosphatase had no effect on SDH activity (data not shown).

To examine whether lysine, being the substrate of LKR, may also modulate the response of this enzyme to dephosphorylation, purified LKR/SDH was pre-incubated with alkaline phosphatase in the absence or presence of lysine for periods of up to 60 min. This was followed by analyzing the kinetics of LKR activity for 7.5 min . As shown in Fig. 2, all pre-treatments with alkaline phosphatase in the absence of lysine did not significantly inhibit the activity of LKR, while pre-treatment with alkaline phosphatase for 20 min or more in the presence of lysine knocked out LKR activity.

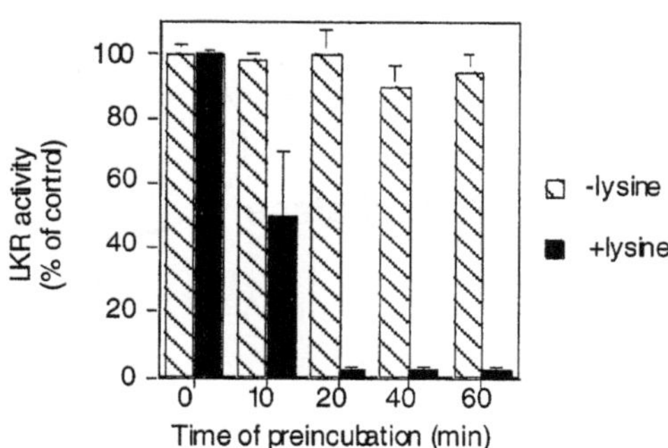

Fig. 2. **Effect of lysine on dephosphorylation of LKR by alkaline phosphatase.**

(A) LKR was pre-incubated at 30^{0}C for 10-60 min in reaction mixtures containing 2 units of alkaline phosphatase, with or without the substrate lysine. Following the pre-incubation, lysine was added to the reactions that did not contain it and the kinetics of LKR activity was assayed spectrophotometrically by following the time-dependent reduction in A_{340} resulting from the oxidation of NADPH by LKR. Values in each histogram represent the percentage of activity from a reaction that was pre-incubated for the same time period without alkaline phosphatase.

Discussion

Our results suggest that LKR represents a major regulatory step in lysine catabolism in plant seeds, and that its activity may be modulated by a counter-balanced action of protein kinases and phosphatases (Fig. 3). Moreover, these results also propose that the modulation of LKR activity is subject to a compound regulation by lysine. On one hand, excess cellular lysine may stimulate LKR activity, perhaps via a Ca^{+2}-mediated activation of a protein kinase that phosphorylates the LKR/SDH polypeptide (Karchi *et al.*, 1995). On the other hand, binding of lysine to the active site of LKR a route towards its catabolism, may bring about the exposure of concealed phosphate residue/s to the surface of the polypeptide, rendering them accessible to removal by protein phosphatases. This counter-balanced regulation, may enable the steady state LKR enzyme to be largely maintained as a phosphorylated active form under conditions of excess cellular lysine, while when lysine becomes limiting, LKR may be progressively inactivated by dephosphorylation. Such a process may enable an accurate control of free lysine levels in the cells, eliminating on one hand toxic effects of high lysine levels (Markovitz *et al.*, 1984; Falco *et al.*, 1995) , and on the other hand preventing its depletion from the free amino acid pool.

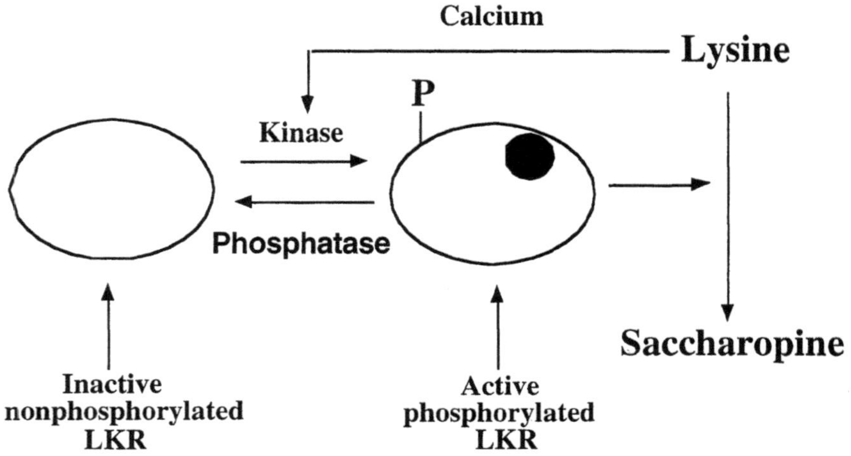

Fig. 3. Schematic diagram of our hypothesis for the lysine-dependent regulation of LKR activity in seeds. (P) phosphate residue. (Black Circle) active site

References

Epelbaum S et al (1997) Plant Mol. Biol. 35, 735-48.
Falco SC et al (1995) *Bio/Technol.* 13, 577-582.
Galili G (1995) Plant Cell 7, 899-906.
Goncalves-Butruille M et al (1996) Plant Physiol. 110, 765-771.
Karchi H et al (1994) Proc. Natl. Acad. Sci. USA 91, 2577-2581.
Karchi H et al (1995) Plant Cell 7, 1963-1970.
Lisitsky I and Schuster G (1995) Nucleic Acids Res. 23, 2506-2511.
Markovitz PJ et al (1984) J. Biol. Chem. 259, 11643-11646.
Markovitz PJ and Chuang DT (1987) J. Biol. Chem. 262, 9353-9358.
Miron D et al (1997) Plant J. 12, 1453-1458.
Tang G et al (1997) Plant Cell 9, 1-13.

Regulation of Higher Plant Leaf and Reserve Tissue Starch Synthesis

Jack Preiss, M.A. Ballicora, Y. Fu, J. Sheng and M. Wu, Dept. Biochemistry, Michigan State University, East Lansing, MI 48824 U.S.A.

Introduction. ADP-glucose is the glucosyl donor in the synthesis of starch and its synthesis is catalyzed by ADP-glucose pyrophosphorylase (reaction I, E.C. 2.7.7.27; ATP: α-D-glucose-1-phosphate adenylyltransferase). Reaction II, catalyzed by starch synthase (E.C. 2.4.1.21) ADP-glucose;1,4-α-D-glucan 4-α-glucosyltransferase, transfers the glucosyl residue from the sugar nucleotide to elongate the 1,4-α-D-glucan chain. Reaction III is catalyzed by branching enzyme (E.C. 2.4.1.18) 1,4-α-D-glucan 6-α-(1,4-α-glucano)-transferase and forms the branched chains found in amylopectin.

ATP + α-glucose-1-P <=> ADP-glucose + PPi (I)
ADP-glucose + α–1,4 glucan ==> α-1,4-glucosyl- α–1,4 glucan + ADP (II)
Elongated α-1,4-oligosaccharide chain ==> α-1,4-α-1,6 branched-glucan (amylopectin) (III)

Our report will be mainly concerned with studies describing the regulatory properties of the ADP-glucose pyrophosphorylase (ADPGlc PPase). Isolation of allosteric mutants of the ADPGlc PPase of Chlamydomonas (Ball et al., 1991) and of maize endosperm (Giroux et al., 1996), strongly suggest that in plant tissues the allosteric effects of 3-P-glycerate (3PGA) the activator, and inorganic phosphate, Pi, the inhibitor are important in regulation of starch synthesis. Indeed further evidence indicating that the ADPGlc PPase is a rate controlling enzyme for starch synthesis has been obtained by transforming several plant systems including potato tuber and tomato fruit and increasing starch about 30- to 60% in those plants (Stark et al., 1992).

The small subunit is the catalytic subunit and the large subunit modifies the regulatory properties of the higher plant ADPGlc PPase. The plant ADPGlc PPase is composed of two different subunits (Okita et al., 1990) and is a

A. Altman et al. (eds.), Plant Biotechnology and In Vitro Biology in the 21st Century, 315–319.

heterotetramer, $\alpha_2\beta_2$ (Preiss, 1998). Expression of the cDNA clones representing the potato tuber small and large subunits separately and together in *E. coli* (Iglesias et al., 1993; Ballicora et al., 1995) enabled one to determine if they had specific and separate functions. The potato tuber small subunit when expressed alone had high catalytic activity provided that the 3PGA concentration was increased to 20 mM (Ballicora et al., 1995). The 3PGA saturating concentration for the expressed transgenic or normal potato tuber heterotetrameric enzyme is 3 mM. It was found that the $A_{0.5}$ (concentration required for 50% of maximal activation) of the transgenic enzyme in ADPGlc synthesis is 0.10 mM while for the small subunit alone it is 2.4 mM. Thus, the small subunit by itself has about 24-fold lower apparent affinity for the activator. The small subunit expressed alone, is more sensitive to Pi inhibition than the transgenic heterotetrameric enzyme with an 8-fold lower Ki. The kinetics of 3PGA activation and the Pi inhibition were the main differences between the homotetrameric small subunit and the recombinant heterotetrameric ADPGlc PPase. These results are consistent with those obtained for the *Arabidopsis thaliana* mutant ADPGlc PPase lacking the large subunit (Lin et al., 1988). The mutant homotetrameric enzyme had lower affinity for the activator and higher sensitivity towards Pi inhibition than the heterotetrameric normal enzyme (Li, Preiss, 1992). The potato tuber large subunit expressed by itself has negligible activity. Thus the dominant function of the small subunit is catalysis while the major function of the large subunit is to increase the affinity of the small subunit for the activator, 3PGA and decrease the affinity for the inhibitor, Pi.

Modification of the Regulatory Sites of ADPGlc PPase by Site-directed mutagenesis. The Lys binding residues for 3PGA, the activator, has been determined for the spinach leaf ADPGlc PPase small (Morell et al., 1988) and large (Ball, Preiss, 1994) subunits and for the Anabaena ADPGlc PPase (Charng, et al., 1994; Sheng et al., 1996). The amino acid sequences surrounding these residues are seen in Table 1 along with the potato tuber ADPGlc PPase sequences. The residues in the *Anabaena* enzyme are Lys382 and Lys419. In the potato tuber the small subunit, they are Lys 404 and 441 and in the large subunit they are Lys 417 and 455.

Site-directed mutagenesis of Lys residue 419 of the *Anabaena* ADPGlc PPase to either Arg, Ala, Gln, or Glu yielded mutant enzymes with 25- to 150-fold higher, $A_{0.5}$ values (concentration of activator required for 50% of maximal activation), than that of wild-type enzyme (Charng, et al., 1994). In other words the mutants had much lower apparent affinities for the activator. Other kinetic constants such as affinity (Km) for substrates and the inhibitor, Pi, were not affected. Heat stability or catalytic efficiency of the enzyme were also not affected. Site II of the *Anabaena* enzyme corresponds to the lysyl residue, Lys440, of the spinach leaf small subunit (Morell et al., 1988). It corresponds to Lys468 and to Lys441 in the small subunits of rice seed and potato tuber ADPGlc PPase, respectively. Lys404 of the potato tuber large subunit corresponds to Site I of the *Anabaena* enzyme, Lys382. The amino acid sequence of the spinach leaf small subunit

Table 1. Amino Acid Sequences of the ADPGlc Activator Sites. The Lys residue known to bind the activator is in bold and the residue number is shown. The amino acids are designated in one letter code.

Source	Tissue	Site I	Site II
		419	382
Anabaena		IRRAIID**K**NAR	IVVVL**K**NAVITDGTII
Small subunit			
Solanum tuberosum	tuber	IKRAIID**K**NAR	IVTVI**K**DALIPSGIII
Spinacia oleracea	leaf	IKRAIID**K**NAR	IVTVI**K**DALIPSGIII
Large subunit			
Solanum tuberosum	tuber	IRKCIID**K**NAK	IIIILE**K**ATIRDGTVI
Spinacia oleracea	leaf	IKDAIID**K**NAR	ITVIF**K**NATIKDGVV

peptide containing the modified lysyl residue of site II is also highly conserved in the barley, wheat endosperm, maize, rice seed small subunits and the *Synechocystis* ADPGlc PPase subunit (see Preiss, Sivak, 1996). The amino acid sequence of site I of the spinach leaf large subunit is also highly conserved in the barley endosperm, maize, wheat leaf and wheat seed large subunit ADPGlc PPases (see Preiss, Sivak, 1996). Since the cDNA clones of the large and small subunits of the potato tuber ADPGlc PPase can be expressed in *E.coli*, this expression system has been used for site-directed mutagenesis experiments to characterize the allosteric function of the lysyl residues in the potato tuber ADPGlc PPase identified for the spinach enzyme.

Site directed mutagenesis of Lys 441 of the potato ADPGlc PPase small subunit to Glu and Ala results in mutant enzymes lower in their affinity, 30- to 83-fold, respectively, for 3PGA (Ballicora et al., 1996; Preiss et al., 1996; Ballicora et al., 1998). The conservative mutation to arginine resulted in only a two-fold increase in $A_{0.5}$, thus indicating that the positive charge of the cationic amino acid is important for the binding of the activator. Mutagenesis of Lysine residue 417 of the large subunit, the residue homologous to the *Anabaena* Lys residue 382 and to site I of the spinach leaf large subunit Lys residue was also done. When Lys417 was replaced by either Ala or Glu, the affinity for 3PGA decreased. However, the $A_{0.5}$ increase was only 3- to 13-fold and not as high as seen with the mutations of the small, 50 kDa subunit Lys441 residue. Mutation of small subunit Lys404 to Ala results in an enzyme that cannot be activated by 3PGA while mutation of large subunit Lys455 to Ala or Glu, resulted in only a 2- to 8-fold decrease in activator affinity. When both Lys residue 417 in the large (51 kDa) and Lys 441 in the small subunit were mutated, the decrease in affinity or increase in $A_{0.5}$, was additive. Thus, both subunit Lys residues do contribute to the binding of the activator. However, the regulatory sites I and II of the small subunit are more important sites as site-directed mutagenesis of the small subunit sites causes a greater lowering of the activator affinity than mutagenesis of the large subunit Lys residues. These studies indicate that the function of the large subunit via interaction with the small subunit increases the efficiency of the small subunit activator sites. Although the large subunit may have binding sites for the activator, the contribution of those sites are less compared to causing the activator sites of the small subunit to have higher affinity for 3PGA.

Inhibitor sites. In *Anabaena* as well as in all the higher plant ADPGlc PPases, there are five highly conserved arginine residues which are not conserved in the enteric bacterial ADPGlc PPases. The enteric ADPGlc PPases are not inhibited Pi but by 5-AMP. In *Anabaena* all the 5 conserved Arg residues, Arg 66, 105, 171, 294 and 385, were mutagenized to alanine (Sheng, Preiss, 1997). The Arg294Ala mutant resulted in an enzyme with more than 100-fold or 40-fold lower affinity for the inhibitor, phosphate, in the absence or presence of 3PGA. This mutation had no or lesser effects on the kinetic contants for the substrates or activator, 3PGA . Thus, Arg 294 of the *Anabaena* enzyme is involved in the binding of Pi. The activator, 3PGA and inhibitor, Pi obviously bind to different sites. The purified mutant Arg294Ala enzyme also had a specific activity, three-fold higher than the wild type enzyme suggesting that with disappearance of the inhibitor binding site there was also a conformational change causing the enzyme to have a higher catalytic efficiency. It would be of interest to use the gene of this mutant enzyme, P294A, for transformation of plants for the purpose of increasing starch content as has been been done with the E. coli mutant ADPGlc PPase gene (Stark et al., 1992).

References

Ball K, Preiss J (1994) J. Biol. Chem. 269, 24706-24711.
Ball S. et al (1991) Planta 185, 17-26.
Ballicora MA et al (1995) Plant Physiol. 109, 245-251.
Ballicora MA et al. (1998) Plant Physiol. In Press.
Ballicora MA et al. (1996) in NIAR/COE International Symposium Proceedings, Tsukuba, Japan, November 11-12, pp. 5-11.
Charng Y et al. (1994) J. Biol. Chem. 269, 24107-24113.
Giroux MJ et al. (1996) Proc. Nat. Acad. Sci., USA 93, 5824-5829.
Iglesias AA et al. (1993) J. Biol. Chem. 268, 1081-1086.
Li L, Preiss J. (1992) Carbohyd. Res. 227, 227-239.
Lin TP et al.(1988) Plant Physiol. 88, 1175-1181.
Morell M et al. (1988) J. Biol. Chem. 263, 633-637.
Okita TW et al. (1990) Plant Physiol. 93, 785-790.
Preiss J (1998) in Starch Chemistry and Technology, third edition, Whistler RL BeMiller

JN Paschall EF eds Academic Press Inc. San Diego, In Press.
Preiss J, Sivak MN (1996) in Photoassimilate distribution in plants and crops: Source-sink relationships pp. 63-96 (Zamski,E. and Schaffer, A.A., Eds.) Marcel Dekker Inc. New York.
Preiss J et al (1996) in Proc. Xth International Photosynthesis Congress Vol. 5, pp 47-52 (P. Mathis, Ed.) Kluwer Academic Publishers..
Sheng J, Preiss J. (1997) Biochem. 36, 13077-13084.
Sheng J et al (1996) Biochemistry 35, 3115-3121.
Stark DM et al (1992) Science 258, 287-292.

PLANT METABOLISM OF ORGANIC XENOBIOTICS. STATUS AND PROSPECTS OF THE 'GREEN LIVER' CONCEPT.

Heinrich Sandermann, Jr.
GSF - Forschungszentrum für Umwelt und Gesundheit GmbH
Institut für Biochemische Pflanzenpathologie
D - 85764 Oberschleißheim, Germany

Summary

The plant metabolism of xenobiotics resembles that of mammalian liver in terms of metabolite patterns, enzyme classes and protein as well as DNA sequences. Plants appear to contain isoenzymes with specificity for secondary plant substrates or xenobiotics. The derived 'green liver' concept is further supported by the following recent results:

O-, *N*-, and *S*-glucosyltransferases for xenobiotics occur in many higher and lower plant species. Cytochrome P450 monooxygenases for fatty acids and xenobiotics have been discovered in marine macroalgae. Glutathione-dependent formaldehyde dehydrogenase is identified as a progenitor of the plant alcohol dehydrogenase superfamily. Common indoor plants were active in the phytoremediation of air containing formaldehyde. Phytoremediation of soil and water also seems possible. However, new methods are required to overcome hydrophilic and lipophilic transport barriers. Several herbicides (glufosinate, glyphosate, isoproturon) were found to form metabolites that were identical in microbial, plant and animal systems.

The main prospects of the 'green liver' concept are seen in phytoremediation and in ecological genetics. A role of gene duplication and mutation, as well as genetic rearrangements is generally well established for the evolution of isoenzyme families. In addition, the existence of a gene pool in soil with potential accessibility to both microorganisms and plants is proposed because plant genes have recently been found to persist in soil over many months. A soil gene pool could contribute to processes such as soil adaptation to pesticides and horizontal gene transfer into microorganisms and plants.

1. Introduction

The 'green liver' concept was developed after plant cell suspension cultures had been employed as a new tool to study xenobiotic metabolism (Sandermann et al., 1977). The following observations were made:

A. Altman et al. (eds.), Plant Biotechnology and In Vitro Biology in the 21st Century, 321–328.

--- Non-polar xenobiotics (DDT, hexachlorobenzene, benzo[a]pyrene, diethylhexylphthalate) previously thought to be non-degradable by plants were converted to metabolites resembling those formed by liver enzymes. Later on, plant metabolic enzymes for benzo[a]pyrene (von der Trenck, Sandermann, 1980), diethylhexylphthalate (Krell, Sandermann, 1984), and DDT (Wetzel, Sandermann, 1994) were discovered.
--- An epoxide intermediate was for the first time trapped during a plant microsomal cytochrome P450 reaction. This provided an analogy to the mechanism of chemical carcinogenesis by liver cytochrome P450 species.
--- New plant glutathione S-transferases were discovered and provided an analogy to a main isoenzyme family of detoxification in mammalian liver.
--- Plant metabolism proceeded in the same sequence of phase I (transformation), phase II (conjugation) and phase III (excretion or internal compartmentation) that was known for liver systems.

The 'green liver' concept was subsequently extended to include the following enzyme classes of plant secondary metabolism: cytochromes P450, glutathione S-transferases, carboxylesterases, O- and N-glucosyltransferases and O- and N-malonyltransferases. Plant thus appeared to have families of closely related isoenzymes with specificity for either natural or xenobiotic substrates (Sandermann, 1992). The analogy to xenobiotic liver metabolism also included the inducibility of certain metabolic enzymes and the existence of processing pathways and associated compartmentation processes. A model for transport from the cytosol into the vacuole or into the apoplast was derived (Sandermann, 1994). In the literature, cDNA sequences of plant cytochromes P450 (Batard et al., 1998), glutathione S-transferases (Droog, 1997) and ABC-type transport proteins (Coleman et al., 1997) are being published at an accelerating pace. The status and prospects of the 'green liver' concept will be illustrated here by some recent results mainly from the author's laboratory.

2. Chemotaxonomy of glucosyltransferases

The 'green liver' concept had been largely derived from studies employing crop plant species. O-Glucosyltransferase activities for chlorinated phenols, for 2,2-bis-(4-chlorophenyl)-acetic acid (a DDT metabolite) and for indole-3-acetic acid have recently been determined in 59 different plant species and 4 plant cell suspension cultures. These enzymes are widely distributed in the plant kingdom and approached in some cases specific activities of 100 pkat/mg soluble protein (Pflugmacher, Sandermann, 1998a). A wide-spread occurrence and even higher activities were also found for N- and S-glucosyltransferases acting on chlorinated anilines and on 4-chlorothiophenol, respectively. These results as well as parallel studies on algal cytochromes P450 and on glutathione S-transferases for xenobiotics have shown that the 'green liver' concept can also be applied to lower plant species.

3. Cytochrome P450 monooxygenases for fatty acids and xenobiotics in marine macroalgae

Various marine macroalgae of the families *Chlorophyta*, *Chromophyta* and *Rhodophyta* have been studied as representatives of lower plant species (Pflugmacher, Sandermann, 1998b). Microsomes contained high oxidative activities for known cytochrome P450 substrates (fatty acids, 3- and 4-chlorobiphenyl, 2,3-dichlorobiphenyl, isoproturon; up to 16 pkat/mg protein). The presence of cytochrome P450 (~50 pmol/mg protein) in microsomes of the three algal families was demonstrated by CO-difference absorption spectra.

Intact algal tissue converted 3-chlorobiphenyl to the same monohydroxy-metabolite formed *in-vitro*. This conversion was five-fold stimulated after preincubation with phenobarbital, and was abolished upon pre-incubation with the known P450 inhibitor, 1-aminobenzotriazole. It was concluded that marine macroalgae could play a role as a metabolic sink for marine pollutants (Pflugmacher, Sandermann, 1998b).

4. Glutathione-dependent formaldehyde dehydrogenase: a progenitor of the alcohol dehydrogenase gene superfamily

The *fdh* gene encoding glutathione-dependent formaldehyde dehydrogenase (FDH) has been cloned from maize (Fliegmann, Sandermann, 1997). The authentic nature of the clone was shown by functional expression in *Escherichia coli* and by the perfect sequence alignment with tryptic peptides from maize FDH purified to homogeneity (U. Wippermann et al., submitted). The substrate specificity of the enzyme was unusual. Besides S-hydroxymethyl glutathione (the substrate derived from formaldehyde), pentanol-1, octanol-1 and ω-hydroxy fatty acids, were utilized (U. Wippermann et al., submitted). The latter substrates may be representative for certain medium- and long-chain lipid peroxidation products. The maize FDH sequence was different from the sequences of the multiple plant ADH isoenzymes utilizing ethanol as substrate, but was related to FDH sequences of animal sources. Phylogenetic analysis supported the convergent evolution of ethanol-consuming ADHs in animals and plants from formaldehyde-detoxifying ancestors. Similar results have been obtained for FDH from pea (Shafqat et al., 1996) and *Arabidopsis* (Martinez et al., 1996).

5. Phytoremediation of air

Metabolic studies with common indoor plants and plant cell suspension cultures (Giese et al. 1994; Sandermann et al., 1997a) suggest the metabolic scheme of **Figure 1** for formaldehyde. The latter is a suspected carcinogen whose indoor air concentration often exceeds the German limit value of 0.1 ppm.

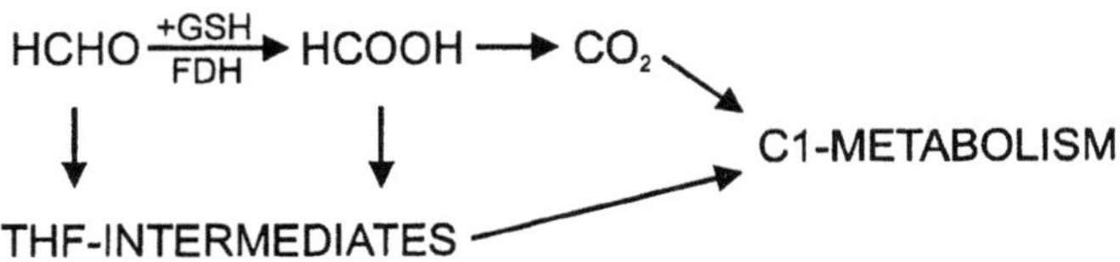

Figure 1. Plant metabolism of formaldehyde via the glutathione-dependent formaldehyde dehydrogenase (FDH) and tetrahydrofolate (THF) intermediates.

The major metabolic pathway is oxidation to formic acid by glutathione-dependent formaldehyde dehydrogenase (FDH), followed by normal C1 metabolism and assimilation into various natural products. Common room plants have been examined as room air filters for formaldehyde. In units (U) of mg formaldehyde turned over per kg fresh weight • h, the *in-vivo* rates of formaldehyde metabolism (about 0.1-1 U), and rates of stomatal uptake (about 4-5 U) were much lower than the maximal enzymatic FDH capacities of various leaf tissues (about 13-180 U). Therefore, stomatal conductance rather than enzyme activity seemed to be rate-limiting for *in-vivo* metabolism (Sandermann et al., 1997a). Stomatal conductance is strongly influenced by humidity, light supply and wind speed. The use of room plants as air filters therefore requires that plants be well watered, ventilated and lighted. Other gaseous organic air pollutants such as pentachlorphenol, vinylchloride or carbon monooxide remain to be studied.

Lipophilic aerosols or gaseous compounds may partition into the cuticle and may become inaccessible for cellular enzymes. This has for example been shown for the esterase acting on the lipophilic plasticizer chemical, diethylhexylphthalate. The high *in-vitro* cleavage rate was blocked when cuticles were added to the enzyme assay (Krell, Sandermann, 1985). No methods seem to exist at present to rapidly deplete plant cuticles of accumulated organic contaminants.

6. Phytoremediation of soil and water

The idea to employ plants to clean the environment is old. One early protagonist was K. Seidel who used aquatic plants and associated microbial communities to purify water heavily contaminated with phenols and other chemicals (Seidel, 1966). Certain plant species are about to be commercially used to clean water and soil contaminated with heavy metals (Watanabe, 1997). Procedures for organic pollutants in soil are only starting to be developed. The 'green liver' concept was in part derived from findings that plants contain metabolic enzymes even for highly persistent or highly polar chemicals (Komoßa et al., 1995). Plant cell suspension cultures appeared to have insignificant transport barriers because the velocity of the rate-limiting metabolic enzyme could predict the *in-vivo* metabolic rate within a factor of 10 (Sandermann, 1994). A major limitation of phytoremediation with intact plants seems to be in uptake and transport. For example, certain chemicals may adsorb to soil particles and not be bioavailable to plants. Adsorbed polar chemicals (e.g. glyphosate) may be made bioavailable by polyelectrolyte root exudates or salt solutions. Solubilized polar compounds may have low uptake rates that could be increased by overexpression of transport proteins in root plasma membranes. Soil-adsorbed non-polar chemicals (e.g. DDT) could be made bioavailable by lipophilic root exudates or by detergents. The degradative potential of rhizosphere microbial communities could generally be optimized.

7. Plant and microbial adaptation to xenobiotics

In 1982, the worldwide occurrence of herbicide resistant weeds was well documented (LeBaron, Gressel, 1982). Resistance development has since accelerated (Heap, 1997). As first shown for atrazine, a major reason for spontaneous resistance can be target site mutation (e.g. in the *psbA* gene). Certain weed species became resistant due to detoxifying enzyme activities (Gronwald, 1994), *Abutilon theophrasti* being a well-studied example. Resistant biotypes contained a glutathione S-transferase for atrazine. Such a transferase had previously been demonstrated in maize and sorghum.

Weed species such as *Lolium rigidum* and *Alopecurus myosuroides* became resistant to isoproturon and related herbicides due to a new detoxifying species of cytochrome P450. Such an enzyme had previously been demonstrated in wheat. The source of the new metabolic enzyme activities in weeds is not clear at present. Another ill-understood case of acquired metabolic capacity concerns the enhanced degradation of soil-applied pesticides even when they exert no selection pressure on soil micro-organisms (Racke, Coats, 1990).

8. Similarity of plant and microbial xenobiotic metabolism

The broad-spectrum herbicides, glufosinate and glyphosate, have been stated by their manufacturers to have no plant metabolism. However, it could be shown that both herbicides are metabolized in plant cell cultures and intact plant systems, the metabolite patterns being very similar to those known for microorganisms (**Figure 2**; Komoßa, et al., 1995). It has been pointed out (Gressel et al., 1996) that up-regulation of plant metabolic rates could be a way for weeds to develop resistance towards broad-spectrum herbicides. The latter have been classified as having a low risk of generating resistance in weeds. The first glyphosate-resistant weed species (*Lolium rigidum*) has recently been discovered (Heap, 1997). Studies of isoproturon metabolism have also revealed a considerable similarity between the metabolite patterns of soil microorganisms (Lehr et al., 1996) and of wheat cell suspension cultures and plants (W. Gläßgen et al., in preparation). The similarity of the metabolite patterns of Figure 2 may be due to the particular chemistry of the herbicide substrates and to an evolutionary relationship between microbial and plant enzymes. It should be noted that the conversions shown in Figure 2 are also known for animal systems.

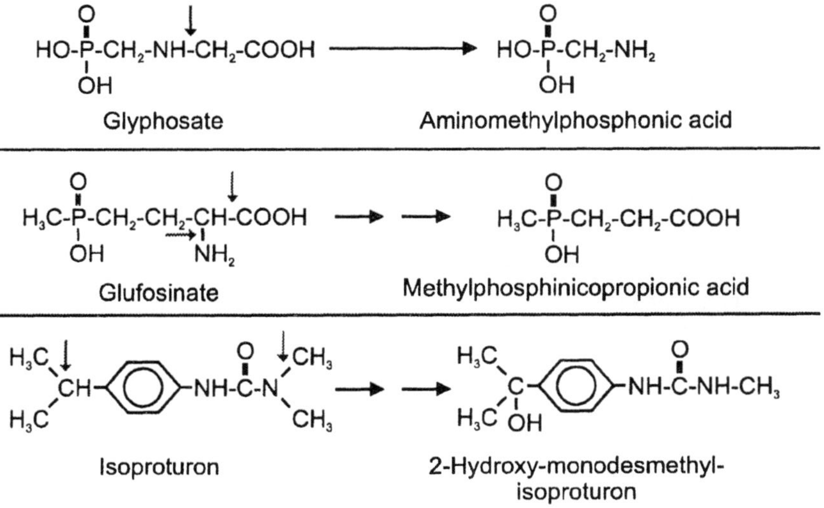

$HO\text{-}P(=O)(OH)\text{-}CH_2\text{-}NH\text{-}CH_2\text{-}COOH \longrightarrow HO\text{-}P(=O)(OH)\text{-}CH_2\text{-}NH_2$

Glyphosate → Aminomethylphosphonic acid

$H_3C\text{-}P(=O)(OH)\text{-}CH_2\text{-}CH_2\text{-}CH(NH_2)\text{-}COOH \longrightarrow \longrightarrow H_3C\text{-}P(=O)(OH)\text{-}CH_2\text{-}CH_2\text{-}COOH$

Glufosinate → Methylphosphinicopropionic acid

$(H_3C)_2CH\text{-}C_6H_4\text{-}NH\text{-}C(=O)\text{-}N(CH_3)_2 \longrightarrow \longrightarrow (H_3C)_2C(OH)\text{-}C_6H_4\text{-}NH\text{-}C(=O)\text{-}NH\text{-}CH_3$

Isoproturon → 2-Hydroxy-monodesmethyl-isoproturon

Figure 2. Common microbial, plant and animal metabolites of herbicides. The sites of oxidative attack are marked by arrows.

9. Ecological genetics

Genes for new metabolic isoenzymes generally arise by gene duplication and mutation or other gene rearrangements processes (Linhart, Grant, 1996). This can conceivably lead to new enzyme specificities, e.g. for atrazine or isoproturon. Most isoenzymes for xenobiotics have been isolated from plants sources not specially adapted to xenobiotics. The corresponding genes may therefore be due to random gene rearrangements. Recent results indicate that agricultural soils contain a long-lived pool of plant genes. For example, the *pat* - gene for resistance to glufosinate (Ernst et al., 1996) and several native plant genes (Sandermann et al., 1997b) have been detected by polymerase chain reaction over several months after the plant biomass had been ploughed into the soil. Soils are already known to contain an active gene pool from which antibiotic resistance genes can spread horizontally into numerous microorganisms (Davies, 1994). These genes are usually located on plasmids that can also serve as vehicle for degradative genes (Miller, 1998). The soil gene pool (**Figure 3**) is defined here as plasmid and chromosomal DNA present in, or released by, soil-associated microbial and plant cells.

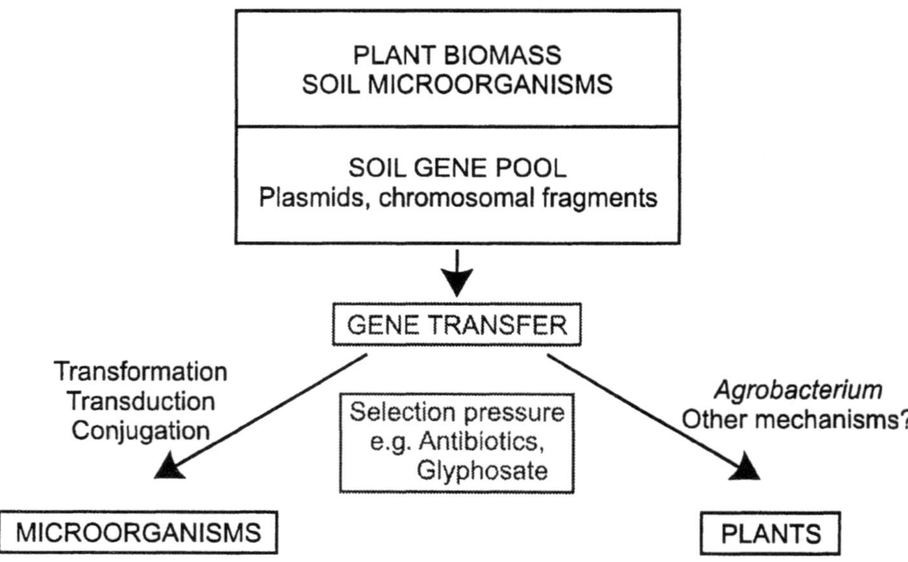

Figure 3. Hypothesis for the role of a soil gene pool in horizontal gene transfer to microorganisms and plants.

There are some examples for a transfer of chromosomal plant genes into soil microorganisms, and *Agrobacterium* can catalyze microbial gene transfer into plants via Ti-plasmids. More generally, examples for horizontal gene transfer for eucaryotes into microorganisms are known for evolutionary time intervals (Smith et al., 1992). Faster rates of horizontal gene transfer will depend on the presence of selection pressure, as well documented for antibiotic resistance genes (Davies, 1994; Miller, 1998). In this connection, it should be noted that glufosinate and glyphosate have a considerable potential to enhance horizontal gene transfer. These broad-spectrum herbicides exert a selection pressure not only on plants, but also on sensitive microorganisms such as *Rhizobium* spp. (Sandermann, 1997; Sandermann et al., 1997b).

10. Outlook

Many plant enzymes and genes for the metabolism of xenobiotics have been isolated and characterized since 1977. Research is likely to be extended to transport proteins and their genes. Main prospects of the 'green liver' concept are seen in phytoremediation and in ecological genetics. Plants by themselves or in association with microorganisms will be increasingly utilized to clean contaminated air, soil and water. The evolutionary origin of the plant genes for xenobiotics will likely be explored at the levels of plant-internal gene rearrangement processes and of the putative soil gene pool.

11. Acknowledgements

The excellent contributions of numerous research associates are gratefully acknowledged. This concerns in particular the group leaders of the Institute, Drs. D. Ernst, W. Heller, C. Langebartels, and A. Schäffner and their valuable improvements of the manuscript. Financial support came from BStMLU (München), Fondation Limagrain (Chappes, France) and Fonds der Chemischen Industrie (Frankfurt).

12. References

Batard Y et al. (1998) Plant J 14, 111-120.
Coleman JOD et al. (1997) Trends Plant Sci. 2, 144-151.
Davies J (1994) Science, 264, 375-382.
Droog F (1997) J Plant Growth Regul. 16, 95-107.
Ernst D et al. (1996) Plant Mol. Biol. Rep. 14, 143-148.
Fliegmann J, Sandermann H (1997) Plant Mol. Biol. 34, 843-854.
Giese M et al. (1994) Plant Physiol. 104, 1301-1309.
Gressel J et al. (1996) In: Engineering Plants for Commercial Products and Applications, (Collins GB, Shepherd RJ, eds.), Annals NY Acad. Sci. 792, 140-152.
Gronwald JW (1994) In: Herbicide Resistance in Plants. Biology and Biochemistry (Powles SB, Holtum JAM, eds.), pp. 27-60, CRC Press, USA.
Heap IM (1997) Pestic. Sci. 51, 235-243.
Komoßa D et al., (1995) In: Plant Contamination: Modeling and Simulation of Organic Chemical Processes. (Trapp S, McFarlane C, eds.), pp. 69-103, CRC Press, Boca Raton, USA.
Krell H-W, Sandermann H (1984) Eur. J. Biochem. 143, 57-62.
Krell H-W, Sandermann H (1985) Plant Sci. 40, 87-93.
LeBaron HM, Gressel J (eds.) (1982) Herbicide Resistance in Plants, John Wiley & Sons, New York, USA.
Lehr S et al. (1996) Intern. J. Environ. Anal. Chem. 65, 231-243.
Linhart YB, Grant MC (1996) Annu. Rev. Ecol. Syst. 27, 237-277.
Martinez MC et al. (1996) Eur. J. Biochem. 241, 849-857.
Miller RV (1998) Sci. Amer. (1), 47-51.
Pflugmacher S, Sandermann (1998a) Phytochemistry, in press.
Pflugmacher S, Sandermann H (1998b) Plant Physiol. 117, 123-128.
Racke KD, Coats JR (eds.) (1990) Enhanced Biodegradation of Pesticides in the Environment, American Chemical Society, ACS Symposium Series 426, Washington DC, USA.
Sandermann H (1992) Trends Biochem. Sci. 17, 82-84.
Sandermann H (1994) Pharmacogenetics 4, 225-241.
Sandermann H (1997) Spektrum der Wissenschaft (7) 38-41.
Sandermann H et al. (1977) In: Plant Tissue Culture and its Biotechnological Application, (Barz W et al., eds.) pp. 178-196, Springer, Berlin, Germany
Sandermann H et al. (1997a) In: Reinhaltung der Innenraumluft (FGU Berlin, ed.). Seminar Nr. 33, Berlin: UTECH '97, 77-88.
Sandermann H et al. (1997b) In: Zukunft der Gentechnik (Brand P, ed.), Birkhäuser, Basel, Switzerland, pp. 209-220.
Seidel K (1966) Naturwissenschaften 53, 289-297.
Shafqat J et al. (1996) Proc. Natl. Acad. Sci. USA, 93, 5595-5599.
Smith WW et al. (1992) Trends Biochem. Sci. 17, 489-493.
v d Trenck T, Sandermann H (1980) FEBS Letters 119, 227-231.
Watanabe ME (1997) Environ. Sci. Technol. 31, 182-186.
Wetzel A, Sandermann H (1994) Arch. Biochem. Biophys. 314, 323-328.

APPLICATION OF BIOREACTOR FOR THE PRODUCTION OF SAPONIN BY ADVENTITIOUS ROOTS CULTURES IN *PANAX GINSENG*

J.H. Seon[1], K.W. Yoo[1], Y.Y. Cui[1], M.H. Kim[2], S.J. Lee[2], S.H. Son[3], and K.Y. Paek[1]
[1]Research Center for the Development of Advanced Horticultural Technology, Chungbuk National Univ., Cheong-ju 361-763, [2] Hwan In Pharm.Co., Ansung-Kun, Kyunggi 456-800, and [3]Forest Research Institute, Suwon, Kyunggi 441-350, Korea

1. Introduction

Ginseng (*Panax ginseng* C.A. Meyer) is a herbaceous perennial plant cultivated in Korea, China, Japan and North America. Its roots have been used as traditional medicine for oriental people since ancient times. The crude root extracts possess wide range of saponins and sapogenins and are known to have anabolic, adaptogenic, antibiotic, minor hyperglycemic and anti-cancer activities (Brekman, Dardymov 1969). However, the commercial production of ginseng requires growth periods of four to six years and skillful techniques to manage the soil, climate and prevention of pests. Therefore, biotechnological approach as alternative method to produce saponin in vitro has been made. Since Butenko et al.(1968) first reported the conditions for tissue culture of ginseng root, ginseng cell and tissue for the production of the ginsenosides has been successfully achieved (Jhang et al. 1974, Furuya et al. 1983, Choi et al. 1990, 1995). Moreover, saponin production by cultures of transformed roots with *Agrobacterium rhizogenes* showed 2.4 times higher contents than native root (Yoshikawa, Furuya 1987). But high fluctuation of saponin content in cell cultures and problems related with human safety in transformed roots are big obstacles for commercialization. There has been very few reports on saponin production by adventitious or ordinary root culture in vitro. In this report, we demonstrated that the adventitious-root culture system by bioreactor is quite feasible for saponin production.

1. Materials and Methods

A. Altman et al. (eds.), Plant Biotechnology and In Vitro Biology in the 21st Century, 329–332.

2.1. Induction and maintenance of root culture

The callus was induced from four to six years aged ginseng roots and embryos native in Korea and maintained on the solidified B-5 medium supplemented with 1.0mg/L 2,4-D. For the induction of adventitious roots, 2-3 weeks aged calli were transferred on the B-5 solidified medium with 0.5 –2.0 mg/L NAA or 1.0 mg/L IBA. The roots emerged from callus after 6 weeks culture were collected and transferred into the various liquid media. They were maintained under dark on a rotary shaker (80 rpm) for further experiments.

2.2. Bioreactor culture

Three kinds of bioreactors – non stirred jar (20L), rotation drum (10L) and column with inner sieve (2L) were used. The membrane filtered air (0.1 vvm) was supplied through the air sparger into the medium. Several factors like ionic strengths, sucrose concentration and changes of culture medium which have influence on root growth and saponins production were investigated. The roots were harvested after 6~7 weeks for measurement of growth and gesenosides content. Crude saponin was extracted according to the method developed by 'Korea ginseng & Tobacco Research Institute'(1991). The content of ginsenosides were calculated by using HPLC (Waters, USA) on a NH_2 column. The amount of panaxadiol were calculated as the sum of Rb_1, Rd and Rc and the amount of panaxatriol were calculated as the sum of Re and Rg_1. The contents of sucrose, glucose and fructose in the culture medium were measured by HPLC attached with refractive index detector on a high performance carbohydrate column. Analysis of nutrients in the medium during culture period also carried out by using HPLC attached with suppressed conductivity instrument on Waters IC-Pak CM/D column for cations and Waters IC-Pak A AR column for anions.

3. Results

Figure 1 shows non-stirred jar bioreactor with medium supply and harvest system used for ginseng root culture. Roots were grown as clumps in a course of culture period and sunk to the bottom regardless of the high airflow rate. This system was quite effective for the removal of used medium or supply of fresh medium without an effort to subculture and contamination risks. Figure 2 represented nutrients consumption of culture medium during the culture period. Almost all of the ammonium were consumed within 2 weeks, while the nitrate were consumed relatively slowly. It meant that ammonium ion was preferentially used for root growth. The levels of Ca^{2+}, Mg^{+}, K^{+}, and PO_4^{-} ions were remained high until harvest. According to our results, it is suggested that the optimal salt strength or composition should be determined to obtain maximum roots growth.

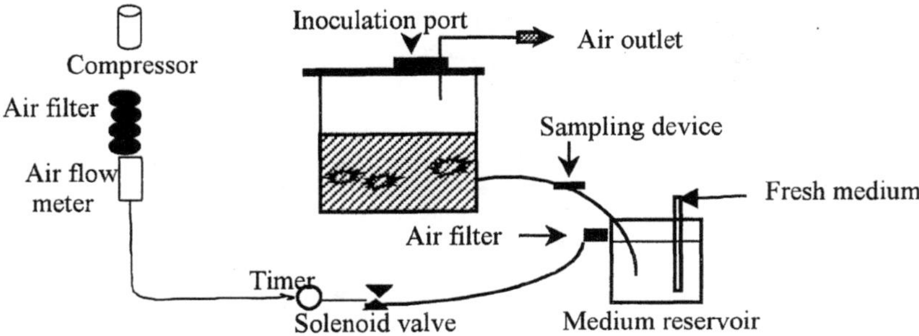

Fig1. A non-stirred jar bioreactor combined with a medium supply apparatus.

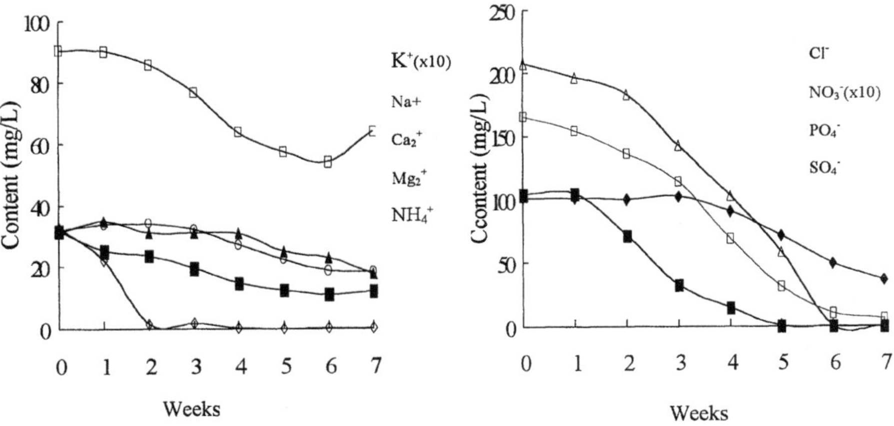

Fig 2. Changes of cations(left) and anions(right) in the medium ot root culture of *Panax ginseng* during the culture period..

In Fig. 3 is shown the relationship between root growth and saponin production as affected by sucrose concentration. Root growth was promoted by the increase of sucrose, while the maximal production of saponin was obtained at 5-7 %. Table 1 shows ginsenoside contents of callus, adventitious root and native roots. Although the total ginsenoside content of adventitious roots in vitro is no more than 1/3 of native roots, it showed that it has a higher possibility of economical production than native roots produced from field in terms of short culture period, labor saving and germ-free system.

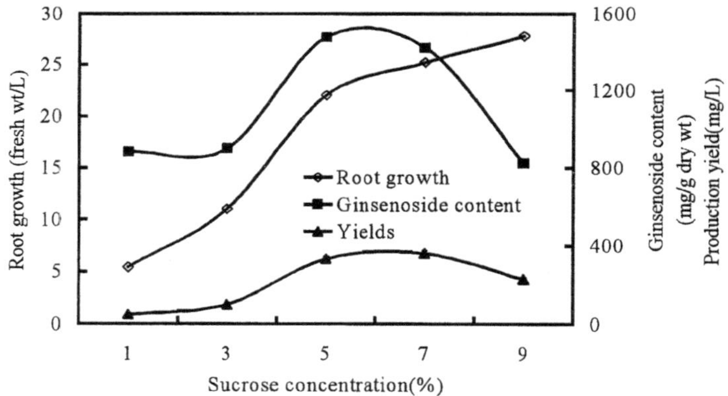

Fig. 3. The effect of sucrose concentrations on root growth and production of ginsenosides in *Panax ginseng* after 6 weeks culture.

Table 1. Gensenoside contents of callus, adventitious roots and native roots.

	Culture days	Panaxdiol (A)	Panaxtriol (B)	A/B	Total content
Callus	28	495.88	362.06	1.37	857.94
Adv. roots	49	799.79	677.11	1.18	1476.90
Native roots	(6 years)	3518.36	1266.64	2.77	4785.0

Panaxdiol : Rb_1 + Rd + Rc, Panaxtriol : Re + Rg_1 (mg/100g dry wt)

5. Reference

Brekman II , Dardymov IV (1969) Lloydia 32:46-51.

Butenko RG et al (1968) Bot. Zh 7:906-913.

Choi et al (1990) Korean J. Ginseng Sci. 14:107-111.

Choi et al (1995) Proc. Int. Ginseng Conf. Vancouver 1994, Canada.

Furuya Y et al (1983) Planta Med. 48:83-87.

Jhang JJ et al (1974) In vitro 9(4):253-259.

Yoshikawa T. and Furuya T. (1987) Plant Cell Rep. 6:449-453.

Stimulation of Bioactive Flavonoid Production in Suspension and Bioreactor-based Cell Cultures

Mary Ann Lila Smith and Marie-France Pépin,
University of Illinois, Urbana, IL 61801 USA

Why harvest bioflavonoids?

Complex bioflavonoid phytochemicals produced by certain plants, including the anthocyanin pigments, the proanthocyanidins, and other compounds in the same biosynthetic pathway, are now recognized as potent chemopreventive agents for disease prevention. Anticancer activity can in part be attributed to their role as *blocking agents* (Bomser et al., 1996; Kelloff et al., 1994) which can act to inhibit carcinogen uptake, activation of carcinogens, or carcinogen binding to DNA. Blocking agents can be inhibitory to enzymes that activate the cancer process, but they may also stimulate activity of enzymes that enhance DNA repair, or accomplish carcinogen detoxification. The bioflavonoids may act as *antiproliferative agents*, which can modulate signal transduction, inhibit polyamine metabolism and abnormal oncogene function, enhance intercellular communication, or restore tumor suppressor gene function. For example, an antiproliferative agent may inhibit stages of inflammation during carcinogenesis. The best known property of the bioflavonoids, however, is their well-recognized strong *antioxidant* activity in metabolic reactions; the ability to scavenge oxygen radicals and other reactive electrophiles, which have direct impact on aging, cancer incidence, and human health in general. Patterns of hydroxylation and glycosylation in anthocyanins appear to modulate their antioxidant capacity (Wang et al. 1997). Anthocyanin pigments demonstrated strong antioxidant capacity in a liposomal system recently (Tsuda et al. 1996), and in an oxygen radical absorbance capacity assay (Wang et al. 1997).

Rising consumer awareness about the health benefits of bioflavonoids in the diet has a strong and continuing impact on the food and supplement markets. Flavonoid pigments (anthocyanins) are valuable not only as nutraceuticals, but also as natural colorants for food and pharmaceutical products. Recently the size of the natural colorants market has increased to the point where it is nearly as large as the synthetic pigments sector; its growth rate has been twice that of synthetics. Finally, the antioxidant properties make them viable candidates for replacing synthetic antioxidants in food processing.

A. Altman et al. (eds.), Plant Biotechnology and In Vitro Biology in the 21st Century, 333–336.

The structures of most of these polyphenolic compounds are sufficiently complicated to preclude laboratory synthesis. Further, there is good evidence that associated groups of phytochemicals found in fruits, for example, work together *in situ*, and that human health benefits are only fully realized when anthocyanins, proanthocyanidins, and other flavonoids are associated in the food product, to allow additive or synergistic effects.

Why harvest bioactive flavonoids *in vitro*?

Although bioactive flavonoids have been discovered in a wide range of fruit, foliage, and vegetable crops, the active compounds comprise only a small percentage of the total content of the plant. For example, a typical anthocyanin-rich berry may contain approximately 85% water, ~13% carbohydrates (including the sugars), and ~3% fibers; only a fraction of a percentage of total content can be attributed to the bioflavonoids. Interference from associated, complex tissues can make efficient isolation of the active principles more challenging, and economic recovery costs high. Furthermore, the high pectin and sugar content found in some fruits interfere not only in the extraction/fractionation process, but also may create artifacts during bioactivity testing of the extracts.

Fortunately, cell cultures established from the same flavonoid-rich plants in nature can be induced to produce bioflavonoids *in vitro* (Decendit & Mérillon, 1996; Nawa et al. 1993; Pépin et al., 1995). The transition from field crop production (for fruit harvest and extraction) to harvest of uniform, aqueous cell cultures represents a tremendous reduction in space and production time, and independence from geographic restrictions. Plant cell culture is attractive from a commercial perspective, since it helps to bypass some of the difficulties associated with extraction of valuable phytochemicals from field crops, and lends itself to economies of scale (Dörnenburg and Knorr, 1997).

The rich flavonoid content can be proportionally higher in the cell culture than in the *in vivo* plant, and the former can be produced continuously on a year round basis. As a model system in our labs, germplasm from the genus *Vaccinium* (*V. pahalae*; oheloberry) has been stabilized for over three years in continuous callus culture (in darkness), can be scaled-up at high specific growth rate in liquid suspension cultures and to a 12L modified bioreactor, and routinely produces anthocyanins, proanthocyanidins, and other related flavonoids upon exposure to light. The cultures produce higher ratios of proanthocyanins/anthocyanidins than field or greenhouse specimens (Shibli et al., 1997; Smith et al., 1997). In an effort to investigate the potential for further scale-up, we have also stabilized a rapidly growing suspension cell line in the dark, which routinely produces 120 mg anthocyanins per liter in 12 days upon transfer to lighted conditions.

Cell culture extracts from this system can significantly inhibit the initiation stage of chemically-induced carcinogenesis (Madhavi et al., 1998; Smith et al., 1997). The cell culture source is free of pectin interference and contains less complex sugars, so that extraction, fractionation, and bioactivity testing is greatly facilitated. With less interference from tissues or other phytochemicals that could mask bioactive potential, the flavonoid extracts from ohelo cell cultures have demonstrated markedly higher antioxidant capacity in standard assays than parallel extracts from fruits (Table 1). Grape seed extracts, particularly rich in proanthocyanidins, are a well recognized natural antioxidant source, and predictably were the most potent in this assay, requiring a quite low Ec_{first} to induce the low half-life value, indicative of effective free radical quenching. However, the cell culture (ohelo) extracts were much more effective antioxidants than extracts from comparable fruits in nature, as they required a much lower concentration to achieve free radical quenching. Clearly, cell culture-derived bioflavonoids have significant advantages over parallel products extracted from field-grown or wild fruits.

Table 1. Free radical quenching of fruit and cell culture extracts in a galvinoxyl free radical assay.

Source of extract	EC_{first} (μg/ml)	$t_{1/2}$ (min)
elderberry	50	6.0
cranberry	50	29
grape	50	5.6
grape seed	0.5	14
ohelo cell culture	5.0	2.4

Ec_{first}: Effective concentration needed to quench the radical following first order kinetics

How can the microenvironment be used to trigger bioflavonoid synthesis *in vitro*?

Many of the same environmental factors that regulate plant secondary product formation in nature can be used to deliberately induce accumulation in tissue culture. We have concentrated on the systematic adjustment of *physical* microenvironmental conditions (including light quality) to control the synthesis of potentially valuable flavonoids by cultures, and have implemented a protocol for reliable monitoring of product accumulation by linking high-performance liquid chromatographic and machine-vision tools to streamline the analysis. For our *in vitro* work, anthocyanin pigments are used as convenient visible markers for bioflavonoid production; we are able to track the concurrent accumulation of proanthocyanidins and pigments in this way.

A large xenon lamp (2 kW solar simulator) with sufficient output to illuminate a 1.9 m^2 area with light energy equivalent to midsummer sunlight (approximating the solar spectrum, continuous throughout the UV and visible regions), and combinations of filters are used to adjust spectral quality and intensity. Cell cultures exposed for only 2 hours daily to 300 $\mu mol\ m^{-2}\ sec^{-1}$ (= 65 W m^{-2}, with 700 mW m^{-2} UV-B) provided by the xenon lamp elicited a higher anthocyanin production response (200 mg L^{-1}) than control cultures (120 mg L^{-1}) grown under the usual white continuous fluorescent culture room lighting (70 $\mu mol\ m^{-2} sec^{-1}$).

Oxygen availability to cells is an additional factor which exerts a strong influence on the biosynthesis of oxygenated organic compounds, since the enzymes involved in phenol production from cinnamic acid are oxygenases that use molecular oxygen as a substrate. As an example, we have increased gas transfer capacity and volumetric irradiance by increasing the surface-to-volume ratio of shake flask cultures from 0.3 to 0.6. This increased irradiance/oxygen availability resulted in a rapid increase in the anthocyanin content without impairing the biomass growth. Physical microenvironmental controls (e.g.

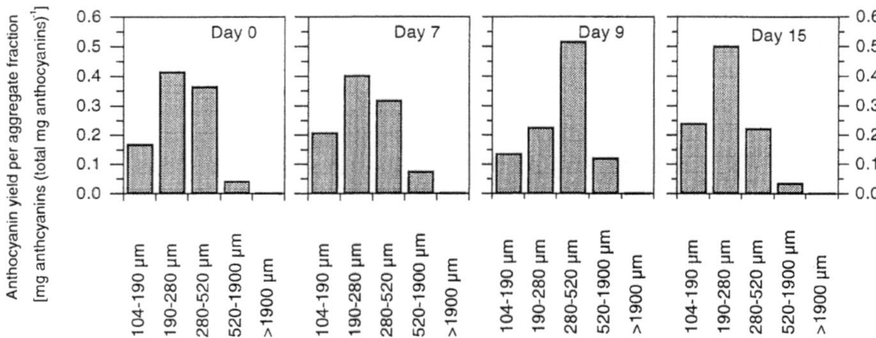

Figure 1 Progression in time of the proportions of total anthocyanin produced by aggregate size fractions of ohelo cells.

agitation rate and inoculum densities) are also effectively used in this model to limit aggregate size in culture, since flavonoid productivity is higher in small aggregate sizes in this system (Fig. 1).

Prospects

A well-defined predictable plant *in vitro* cell culture system for rapid production of bioflavonoid compounds is a valuable tool from a research standpoint (to investigate the key triggers for chemopreventive phytochemical synthesis), and as a model for production optimization. Directed manipulation of microenvironmental controls has permitted dramatic increases in yields and shifts in phytochemical profile, which promises to contribute an improved product for consumers.

References

Bomser J, Madhavi DL, Singletary K, Smith MAL (1996) Planta Medica 62: 212-216.

Decendit A, Mérillon JM (1996) Plant Cell Rep. 15: 762-765.

Dörnenburg H, Knorr D (1997) FoodTechnol. 51: 47-54.

Kelloff GC, Boone V, Steele J, Fay R, Lubet J, Crowell J, Sigman C (1994) J. Cell. Biochem. 520: 1-24.

Madhavi DL, Bomser J, Smith MAL, Singletary K (1998) Plant Sci. 131: 95-103.

Pépin M-F, Archambault J, Chavarie C, Cormier F (1995) Biotechnol. Bioeng. 47: 131-138.

Shibli R, Smith MAL, Kushad M (1997) Plant Growth Regul. 23: 201-205.

Smith MAL, Madhavi DL, Fang Y, Tomczak M (1997) J. Plant Physiol. 150: 462-466.

Tsuda T, Shiga K, Kawakishi S, Osawa T (1996) Biochem. Pharmacol. 52: 1033-1039.

Wang H, Cao G, Prior R 1997. Oxygen radical absorbing capacity of anthocyanins. J. Ag. Food Chem. 45: 304-309.

EXPRESSION OF ANTISENSE mRNA FOR ACID INVERTASES IN TRANSGENIC CARROT ALTERS PLANT DEVELOPMENT AND SUCROSE PARTITIONING

Guo Qing Tang and Arnd Sturm
Friedrich Miescher Institute, Maulbeerstrasse 66, CH-4058 Basel, Switzerland

1. Introduction

Invertase, which cleaves sucrose into hexoses, are present in most plant tissues in multiple forms. Acid invertases are ionically bound to the cell wall or accumulate as soluble proteins in the vacuole. Neutral and alkaline invertases are located in the cytoplasm.

Several functions have been proposed for invertases (Tymowska-Lalanne, Kreis, 1998). The specific roles of the different invertase isoforms are largely unknown and their elucidation is the goal of the present study. We previously characterized invertases from carrot (Sturm, 1996). In addition, we established an efficient protocol for the transformation of carrot by *Agrobacterium tumefaciens* (Hardegger, Sturm, 1998). These tools were used subsequently to study isoform functions by an antisense approach in transgenic carrot plants.

2. Results

The cDNAs for carrot cell wall invertase (cwi) or isoform I of vacuolar invertase (vi) were cloned in reverse orientation behind the 35S promoter of the binary vector pBI121 (Tang, Sturm, 1998). Transgenic calli with markedly reduced invertase protein levels were identified by western blotting. Suspension cultures were initiated and used for somatic embryogenesis.

The first differences between somatic embryos derived from control cells and antisense cells were visible at the cotyledon stage. In embryos expressing antisense mRNA for vi, the cotyledons were large and hypocotyls and roots stunted. In contrast, in somatic embryos expressing antisense mRNA for cwi, the cotyledon were not well separated.

Somatic embryos were transferred onto medium with 3% sucrose. Within 4-6 weeks, embryos of control cell lines developed into small plantlets with several leaves. In contrast, antisense embryos developed into poorly differentiated structures of green color (Figure 1).

An explanation for the aberrant development of the antisense plantlets grown on sucrose-containing medium may be an altered sucrose-to-hexose ratio in the tissues with reduced invertase activity. To circumvent this problem, embryos were grown on medium with a mixture of sucrose, glucose and fructose instead of only with sucrose. The malformation was clearly alleviated and the transgenic plantlets looked more or less normal.

When plantlets from hexose-containing media were transferred to soil, mature plants could be obtained. Control plants had well developed tap roots and 5 - 7 leaves. The dry

A. Altman et al. (eds.), Plant Biotechnology and In Vitro Biology in the 21st Century, 337–340.

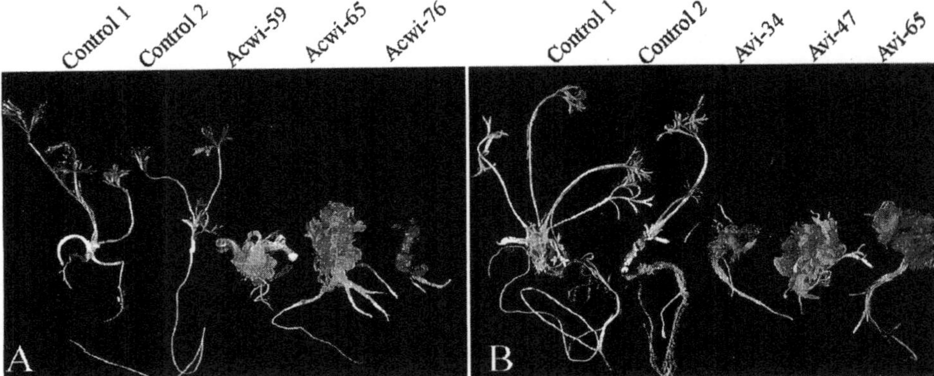

Fig. 1. Transgenic plantlets expressing antisense mRNA for cwi (A) or vi (B).

weight ratio of leaves to roots was 1:3. Plants expressing antisense mRNA for vi had smaller tap roots and more leaves (Figure 2B). The dry weight ratio of leaves-to-roots was 1.5:1. Plants expressing antisense mRNA for cwi had no tap roots and a bushy appearance (Figure 2A). Their dry weight ratio of leaves to roots was 14:1.

Soluble sugars were extracted from different organs of antisense plants and analyzed. The sugar content of antisense source leaves was up to 2 times higher than that of leaves of control plants (Figure 3A and B). Analysis of starch showed up to five times higher levels in source leaves of antisense plants than in source leaves of control plants (Figure 3C and D).

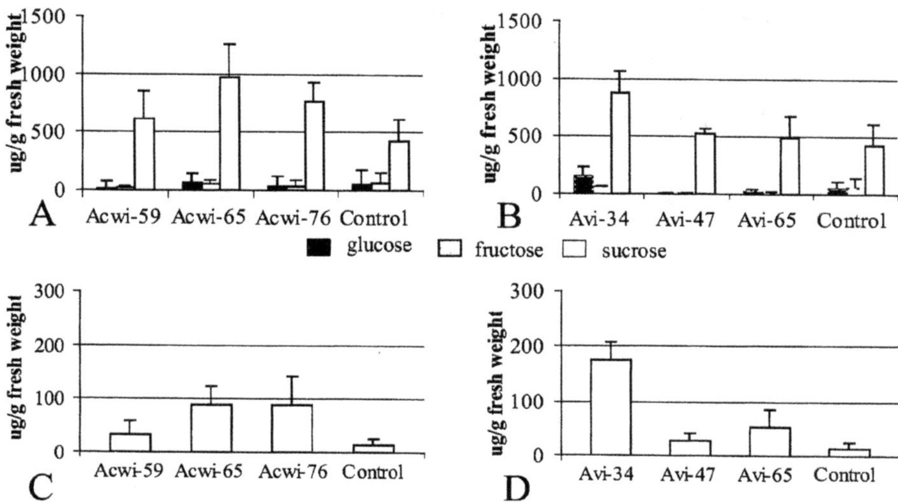

Fig. 3. Sugar (A, B) and starch (C, D) content of source leaves.

3. Discussion

An unexpected finding of our study is that the expression of antisense mRNA for vi or cwi in transgenic carrot led to phenotypical alterations as early as cotyledon stage somatic embryos. This is in line with recent reports on the importance of acid invertases in the early development of seeds of maize (Miller, Chourey, 1992) and bean (Weber et al., 1995).

When invertase antisense embryos were grown on sucrose-containing medium, they developed into stunted structures of green color. An explanation for the undifferentiated growth may be the sucrose-to-hexose level in these plantlets. This hypothesis is in line with the finding that sugars are not only substrates for growth but also regulate genes (Koch, 1996). Many of these genes code for enzymes of primary metabolism, but sugars also affect plant development (Jang et al., 1997). Especially glucose and fructose seem to be direct signal molecules and sucrose must be converted into hexoses before it can function as a signal (Jang, Shen, 1997). Furthermore, to function as a signal, the hexose concentration must be above a certain threshold level (Herbers et al., 1996). In the antisense plantlets, the reduced enzyme activity may have kept the hexose level below such a level. The assumed missignaling was compensated by addition of glucose and fructose, the products of the invertase reaction, and plants grown under these conditions developed normally.

The expression patterns of the acid invertase isoforms differ widely, suggesting isoform-specific functions. Because different physiological processes take place in source and sink, it has been suggested that the function of cwi is different in each organ (Sturm et al., 1995). The 35S promoter used in our studies is only poorly expressed in source leaves compared with tap roots (Hardegger, Sturm, 1998), Therefore, the expression of antisense mRNA for cwi may preferentially affect its root function. Here, cwi may cleave sucrose directly after phloem unloading, thereby creating the sucrose concentration gradient required for long-distance transport of photoassimilates. The phenotype of the transgenic plants obtained corresponds well with this hypothesis. The plants did not develop tap roots within the first 16 weeks and apparently used the surplus of assimilated carbon in the growth of additional leaves. Furthermore, the leaves accumulated elevated levels of sucrose and starch.

In young carrot seedlings, we previously found transcripts for isoforms I and II of vi in both leaves and roots. In older plants, their expression was restricted to roots, with isoform I highly expressed in primary roots and isoform II in developing tap roots (Sturm et al., 1995). Again, because the 35S promoter is only poorly expressed in carrot source leaves compared with tap roots, we expected that the expression of antisense mRNA for vi in developing plants would mainly affect the root functions of the enzyme. When sugars accumulate in the storage organ, vi may cleave sucrose to maintain the sucrose concentration gradient between leaves and roots (Sturm, 1996). The phenotype of the transgenic plants obtained are in good agreement with this prediction. Root development was slowed down and at a time when control plants already had large tap roots, those of the vi antisense plants were still very small. Because in these plants carbon assimilation was not impaired, the excess of sucrose was apparently invested in the growth of more leaves.

In conclusion, our data indicate that invertases have multiple functions. During early plant development, invertase levels seem to play an important role in plant development, most likely via control of sugar composition and metabolic fluxes. Later, when reduction of cwi or vi activity shift the leaf-to-root ratio towards the development of leaves, both isoenzymes appear to have important functions in sucrose partitioning.

4. Acknowledgements

We thank Marcel Lüscher for technical assistance with the HPLC analysis of sugars.

5. References

Hardegger M, Sturm A (1998) Mol. Breeding 4, 119-127.
Herbers K et al. (1996) Plant Cell 8, 793-803.
Jang JC et al. (1997) Plant Cell 9, 5-19.
Jang JC, Sheen J (1997) TIPS 2, 208-214.
Koch KE (1996) Annu. Rev. Plant Physiol. Plant Mol. Biol. 47, 509-540.
Miller ME, Chourey PS (1992) Plant Cell 4, 297-305.
Sturm A (1996) J. Exp. Bot. 47, 1187-1192.
Sturm A et al. (1995) Planta 195, 601-610.
Tang GQ, Sturm A (1998) submitted to Plant Cell.
Tymowska-Lalanne Z, Kreis M (1998) Adv. Bot. Research 28, 71-117.
Weber H et al (1995) Plant Cell 7, 1835-1846.

Isochorismate synthase cDNA isolation from cell cultures of *Catharanthus roseus* (L.)G.Don.

George J. Wullems[1], Léon van Tegelen[1], Marc Stalman[1], Anton F. Croes[1], Paolo R.H. Moreno[2*] and Robert Verpoorte[2].
[1]Department of Experimental Botany, Research group Molecular Plant Physiology. University of Nijmegen, The Netherlands.
[2]Leiden/Amsterdam Center for Drug Research, Division of Pharmacognosie, Gorleus Laboratories, Leiden University, The Netherlands
*Present adress: Dept. Quinica Fundamental. Instituto de Quinica, Universida de Sao Paolo, Sao Paolo, Brasil.

Introduction

Biosynthetic routes leading to plant secondary metabolism start from products of primary pathways. This makes the question, how primary and secondary metabolism are mutually regulated at the molecular level, an intriguing one. A major pathway in primary and secondary metabolism is the shikimate pathway (Herrmann, 1995). This pathway provides chorismate that is used not only for the synthesis of the aromatic amino acids, but also serves as precursor for a wide variety of aromatic substances (Herrmann, 1995; Weaver and Herrmann, 1997). In addition, chorismate is the starting point for a biosynthetic pathway leading to phylloquinones (vitamin K1) and anthraquinones (Poulsen, Verpoorte, 1991). The conversion of chorismate into isochorismate is the first committed step in this pathway. This step is catalysed by the enzyme isochorismate synthase (ICS) (Poulsen, Verpoorte, 1991). Since its substrate, chorismate, plays a dual role in the synthesis of compounds derived from this pathway and their distribution over the various sub-pathways, it is expected that this enzyme is tightly regulated. Model systems to study this regulation are elicited cell cultures of *Catharanthus roseus,* since they provide an example of the partitioning of chorismate. Elicitation of *C. roseus* with a fungal extract induces not only several enzymes of the indole alkaloid biosynthetic pathway (Pasquali *et al.*, 1992), but also ICS (Moreno *et al.*, 1994). Knowledge on the expression kinetics and on the biochemical characteristics of the enzymes that compete for chorismate (ICS, chorismate mutase, and anthranilate synthase) may help to understand the regulation of the distribution of this precursor over the various pathways. For chorismate mutase (Eberhard *et al.*, 1996) and for anthranilate synthase (Bohlmann *et al.*, 1995; Poulsen *et al.*, 1993) such information is already available, but not yet for ICS.

ICS plays an important role in bacterial and plant metabolism as precursor of *o*-succinylbenzoic acid, an intermediate in the biosynthesis of menaquinone (vitamin K2) (Weische, Leistner, 1985) and phylloquinone (vitamin K1) (Poulsen, Verpoorte, 1991). In

A. Altman et al. (eds.), Plant Biotechnology and In Vitro Biology in the 21st Century, 341–346.

micro-organisms, isochorismate is also a precursor for siderophores such as dihydrobenzoic acid (DHBA) (Young *et al.*, 1969), enterobactin (Walsh *et al.*, 1990), amonobactin (Barghouthi *et al.*, 1991) and salicylic acid (SA), (Serino *et al.*, 1995). Evidence from tobacco indicates that salicylic acid is derived from phenylalanine via benzoic acid (Yalpani *et al.*, 1993). It can however not be excluded that salicylic acid is also synthesised from isochorismate. Anyway, in plants isochorismate is a precursor for the biosynthesis of anthraquinones (Inoue *et al.*,1984; Sieweke, Leistner, 1992), naphtoquinones (Müller, Leistner, 1978), catalpalactone (Inouye *et al.*, 1975) and certain alkaloids in orchids (Leete, Bodem, 1976).

ICS was first extracted and partially purified from crude extracts of *Aerobacter aerogenes* (Young, Gibson, 1969). ICS activity has also been found in protein extracts of cell cutures from plants of the *Rubiaceae, Celastraceae* and *Apocynaceae* (Ledüc *et al.*, 1991;
Poulsen *et al.*, 1991, 1992). Genes encoding ICSs have been cloned from bacteria as *E. coli* (Ozenberger *et al.*, 1989), *Pseudomonas aeruginosa* (Serino *et al.*, 1995), *Aeromonas hydrophila* (Barghouthi *et al.*, 1991), *Haemophilus influenzae* (Fleischman *et al.*, 1995) and *Bacillus subtilis* (Rowland and Taber, 1996). Both *E. coli* and *B. subtilis* have two distinct ICS genes, one involved in siderophore production and the other in menaquinone synthesis (Daruwala *et al.*, 1996,1997; Rowland and Taber, 1996). The biochemical properties of the two ICS proteins from *E. coli* are different (Daruwala *et al.* 1997; Liu *et al.*, 1990). Sequence comparison has shown that the bacterial ICS proteins share homology with the chorismate utilising enzymes anthranilate synthase and para-aminobenzoate synthase. This suggests that they have a common evolutionary origin (Ozenberger *et al.*, 1989).

In recent years, a considerable amount of biochemical and molecular data have been obtained on the shikimate pathway in plants (Schmid, Amrhein, 1995; Weaver, Herrmann, 1997). Nevertheless, the research efforts on ICS from higher plants is sparse. Although the enzyme has been partially purified from *Galium mollugo* cell cultures (Ledüc *et al.*, 1991, 1997), purification of the ICS protein to homogeneity remained elusive, probably due to instability of the enzyme.

The main target of our research is to get information on the molecular regulation of ICS and its role in the regulation of the primary and secondary metabolism in higher plants. In this paper the purification of ICS to homogeneity and the cloning of the corresponding cDNA is reported.

Results and Discussion

For the study of the role of ICS in plant metabolism, the enzyme protein was first purified using a five step purification method (Van Tegelen *et al.*, submitted). With this method we were able for the first time to purify ICS to homogeneity from a plant source. The purification procedure yielded two isoforms of ICS (ICS I and ICS II). The ratio between the isoforms was the same in several independent purifications which indicated that the appearance of the two isoforms was not an artefact caused by the purification procedure. This conclusion was supported by re-injection experiments in which either of the two isoforms resulted in the occurrence of only the injected ICS in the chromatogram from the FPLC MonoQ column, indicating that the one isoform is not a breakdown product of the other. SDS-PAGE of the two isoforms of ICS revealed molecular weights of 65 kD and native gels 94 kD respectively. Both isoforms showed an identical pH dependency with a broad pH optimum between 7.0 and 9.0, and 50 % of the maximal activity at pH 6.6 and 10. With respect to the substrate dependency of the enzyme activity, both isoforms showed standard Michaelis-Menten kinetics for chorismate with K_m values of 558±5 μM and 319±41 μM.

In elicited *Rubia* cell cultures, ICS is also present in two isoforms, although in a different ratio. However, in *Galium mollugo*, only one isoform was found, but this might be due by the purification method used (Ledüc *et al.*, 1997). The bacteria *E. coli* and *B. subtillis* have also two isochorismate synthases, which are encode by two different genes (Daruwula *et al.*, 1996; Rowland and Taber, 1996). The functions of these ICS enzymes are different: one functions in the production of siderophores and the other in the synthesis of menaquinones. They also have different biochemical characteristics (Daruwula *et al.*, 1997). In contrast to the bacterial iso-enzymes, the *C. roseus* isoforms do not differ much in their biochemical properties.

In general, almost all enzymes in the shikimate pathway are present as two iso-enzymes, which are mostly regulated in a different manner (Weaver, Herrmann, 1997). For chorismate mutase however, three different isoforms of the protein originate from alternative splicing of two mRNA precursors (Görlach *et al.*, 1995). In general, only one isoform of the enzymes of the shikimate pathway is inducible.

The protein band containing ICS II was isolated from a native PAGE gel and digested with trypsin, which yielded about 50 peptides. Five peptides were isolated and sequenced. One of these peptides displayed high homology to bacterial ICS and a degenerated primer was designed against this peptide. Using PCR with this primer a cDNA encoding ICS was cloned. All peptides were present in the deduced amino acid sequence and also the homology with bacterial ICS showed that this cDNA encodes ICS. The protein encoded by this cDNA is 64 kD, which is in agreement with the molecular mass of the purified protein. If the transit peptide is taken into account, there is a discrepancy in molecular mass. However, since there is no obvious cleavage site found it is not possible the predict the exact molecular weight of the mature protein from the cDNA.

From the screening of the cDNAs only one class of cDNAs was obtained, all encoding the same protein. This indicates that both isoforms are encoded by the same cDNA and that differences among both forms are caused by post-transcriptional modifications. This conclusion is further supported by southern blot analysis, which strongly suggested the existence of one ICS gene.

The presumed plastidic import signal, encoded for by the ICS cDNA, indicates that the protein is located in the plastids. A plastidic import signal is present in almost all enzymes of the shikimate pathway and these are all located in the plastids (Weaver, Herrmann, 1997). One of the chorismate mutases, which lacks such a transit sequence, forms an exception. This enzyme most probably resides in the cytosol (Eberhard *et al.*, 1996). The fact that the complete chorismate producing pathway operates in the plastids, and the lack of evidence for the existence of a chorismate translocator that transports chorismate out of the plastids makes it plausible that also ICS is located in the plastids.

The predicted protein is about 30% identical (40% homologous) with isochorismate synthases from bacteria, with most homology in the C-terminal region. In addition, there is 30% homology (15% identity) between the ICS protein and the alpha-subunits of anthranilate synthases from *Arabidopsis thaliana* (Nigoyi, Fink, 1992) and *Ruta graveolens* (Bohlmann *et al.*, 1995). Also in this case the homology is most strong in the C-terminal part of the protein. This homology with anthranilate synthase is consistent with evidence that indicates a common reaction mechanism in which magnesium is involved in the activation of the active sites of both enzymes.

ICS activity of both isoforms was not inhibited by the presence of tyrosine, phenyalanine or tryptophan in the assay mixture. In this respect ICS and anthranilate synthase have a different allosteric feedback inhibition.

In *C. roseus* ICS activity is only measurable after elicitation (Moreno *et al.*, 1994). Chorismate is a common substrate and ICS has to compete for the available substrate with both anthranilate synthase and chorismate mutase, which synthesise precursors for the biosynthesis of the essential aromatic amino acids. The relatively low affinity of ICS for chorismate (558 μM and 319 μM for ICS I and II respectively) as compared to that of anthranilate synthase (67 μM) (Poulsen *et al.*, 1993) may prevent drainage of chorismate into isochorismate-derived products. For chorismate mutase no data are available from *C. roseus* to compare its affinity for chorismate with that of ICS.

In the secondary pathway to anthraquinones, strong indications exist that ICS activity is a limiting factor. This is shown with transgenic *Rubia peregrina* plants expressing a bacterial ICS gene (Lodhi *et al.*, 1996). These plants have a higher ICS activity than control plants and also the level of isochorismate derived anthraquinones is higher in the transgenic cultures. The observation that ICS activity in *C. roseus* cell cultures is only measurable after elicitation indicates that the flow of chorismate in the direction of isochorismate is controlled at the level of gene expression. The expression kinetic and regulation of gene expression of the chorismate utilising enzymes under various conditions will be target of further research to get more information on the regulation of metabolic trafficking and of the interaction between primary and secondary metabolism.

Acknowledgements

We wish to thank Dr. Johan Memelink for the gift of the *C. roseus* cDNA library. Bert Goossen and Karl Weisz are thanked for maintaining a steady supply of cultured cells.

References

Barghouthi, S., Payne, S.M., Arceaux, J.E.L. and Byers, B.R. (1991) *J. Bacteriol.* 173, 5121-5128.

Bohlmann, J., DeLuca, V., Eilert, U. and Martin, W. (1995) *Plant J.* 7, 491-502.

Daruwala, R., Kwon, O., Meganathan, R. and Hudspeth, M.E.S. (1996) *FEMS Microbiol. Lett.* 140, 159-163.

Daruwala, R., Bhattacharyya, D.K., Kwon, O. and Meganathan, R. (1997) *J.Bacteriol.* 179, 3133-3138. *Plant Mol. Biol.* 31, 917-922.

Fleischmann, R.D., Adams, M.D., White, O., Clayton, R.A., Kirkness E.F., Kerlavage, A.R., Bult, C.J., Tomb, J.-F., Dougherty, B.A., Merrick, J.M., McKenney, K., Sutton, G., Fitzhugh, W., Fields, C.A., Gocayne, J.D., Scott, J.D., Shirley, R., Liu, L.-I., Glodek, A., Kelley, J.M., Weidman, J.F., Phillips, C.A., Spriggs, T., Hedblom, E., Cotton, M.D., Utterback, T.R., Hanna, M.C., Nguyen, D.T., Saudek, D.M., Brandon, R.C., Fine, L.D., Fritchman, J.L., Fuhrmann J.L., Geoghagen N.S.M, Gnehm C.L., McDonald L.A., Small K.V., Fraser, C.M., Smith, H.O. and Venter, J.C. (1995) *Science* 269, 496-512.

Görlach, J., Raesecke, H.-R., Abel, G., Wehrli, R., Amrhein, N. and Schmid, J. (1995) *Plant J.* 8, 451-456.

Herrmann, K. (1995) *Plant Physiol.* 107, 7-12.

Inoue, K., Shiobara, Y., Nayeshiro, H., Inouye, H., Wilson, G. and Zenk, M.H. (1984) *Phytochemistry* 23, 307-311.

Ledüc, C., Ruhnau, P. and Leistner, E. (1991) *Plant Cell Rep.* 10, 334-337.

Ledüc, C., Birgel, I., Müller, R. and Leistner, E. (1997) *Planta* 202, 206-210.

Leete, E., Bodem, G.B. (1976) *J. Am. Chem. Soc.* 98, 6321-6329.

Leistner, E. (1975) *Planta Med. Supplement*, 214-224.

Liu, J., Quinn, N., Berchtold, G.A. and Walsh, C.T. (1990) *Biochemistry* 29, 1417-1425.

Lodhi, A.H., Bongaerts, R.J.M., Verpoorte, R., Coomber, S.A. and Charlwood, B.V. (1996) *Plant Cell Rep.* 16, 54-57.

Moreno, P.R.H., Van der Heijden, R. and Verpoorte, R. (1994) *Plant Cell Rep.* 14, 188-191.

Müller, W.U. and Leistner, E. (1978) Metabolic relation between naphthalene derivatives in *Juglans*. *Phytochemistry* 17, 1735-1738.

Nigoyi, K.K. and Fink, G.R. (1992). *Plant Cell* 4, 721-733.

Ozenberger, B.A., Brickman, T.J. and McIntosh, M.A. (1989) *J.Bacteriol.* 171, 775-783.

Pasquali, G., Goddijn, O.J.M., De Waal, A., Verpoorte, R., Schilperoort, R.A., Hoge, J.H.C. and Memelink, J. (1992) *Plant Mol. Biol.* 18, 1121-1131.

Poulsen, C. and Verpoorte, R. (1991) *Phytochemistry* 30, 377-386.

Poulsen, C., Van der Heijden, R. and Verpoorte, R. (1991). *Phytochemistry* 30, 2873-2878.

Poulsen, C. and Verpoorte, R. (1992) *Plant Physiol. Biochem.* 30, 105-113.

Poulsen., C., Bongaerts., R.J.M. and Verpoorte, R. (1993) *Eur. J. Biochem.* 212, 431-440.

Rowland, B.M. and Taber, H.W. (1996) *J.Bacteriol.* 178, 854-861.

Schmid, J. and Amrhein, N. (1995 *Phytochemistry* 39, 737-749.

Serino, L., Reimman, C., Baur, H., Beyeler, M., Visca, P. and Haas, D. (1995) *Mol. Gen. Genet.* 249, 217-228.

Sieweke, H.-J. and Leistner,E. (1992). *Phytochemistry* 31: 2329-2335.

Walsh, C.T., Liu, J., Rusnak, F. and Sakaitani, M. (1990). *Chem. Rev.* 90, 1105-1129.

Weaver, L.M. and Herrmann, K. (1997). *Trends Plant Sci.* 2, 346-351.
Weische, A. and Leistner, E. (1985). *Tetrahedron* 26, 1487-1490.
Yalpani,N., León,J., Lawton, M.A. and Raskin, I. (1993). *Plant Physiol.* 103: 314-321.
Young, I.G., Batterham, T.J. and Gibson, F. (1969) *Biochim. Biophys. Acta.* 177, 389-400.

THE BIOSYNTHETIC RELATIONSHIP BETWEEN LITTORINE AND HYOSCYAMINE IN *DATURA STRAMONIUM*

I. Zabetakis[1], C.W. Wong[1], R. Edwards[2] and D. O'Hagan[1]
[1]Department of Chemistry, University of Durham, Durham DH1 3LE, UK, and
[2]Department of Biology, University of Durham, Durham DH1 3LE, UK

1. Introduction

Tropane alkaloids, such as scopolamine and its precursor hyoscyamine, are biosynthesized in many solanaceous plants, notably in *Atropa*, *Duboisia*, *Hyoscyamus* and *Datura* species. These alkaloids are pharmacologically active (Brown, 1990). Since this time their biosynthesis has been the subject of intensive study and the details of the origin of the tropane ring system is still a topic of investigation in many laboratories (Robins et al., 1997).

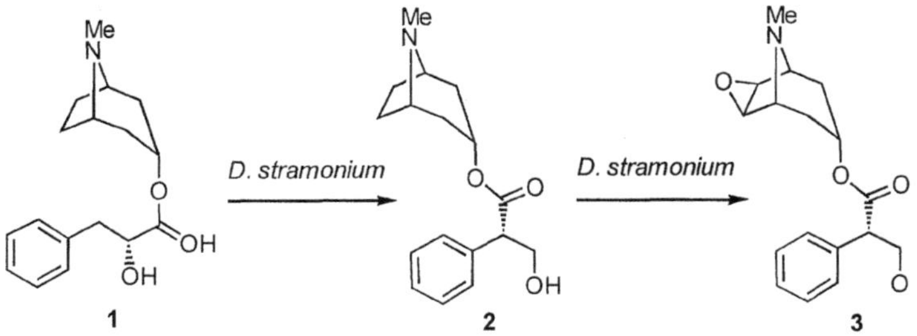

Also the origin of the branched aromatic tropic acid ester moiety of hyoscyamine **2** and scopolamine **3** is an intriguing problem and the details of this process are only coming to light. The conversion of littorine **1** to hyoscyamine **2** is a formal isomerisation and involves an intramolecular rearrangement of D-phenyllactate (itself derived from L-phenylalanine) to a tropate ester (Leete et al., 1975), a reaction which has intrigued chemists and biochemists for many years (Robinson, 1927). Recent isotopic labelling studies in *D. stramonium* have demonstrated that littorine **1**, the D-phenyllactate ester of tropine is the direct precursor to hyoscyamine, and not the free acid or a coenzyme-A ester (Robins et al., 1994). In the light of the importance of this observation we set out to reinforce the role of littorine **1** in hyoscyamine **2** biosynthesis by assessing the effect of added exogenous littorine **1** on endogenous hyoscyamine levels and *vice versa*.

A. Altman et al. (eds.), Plant Biotechnology and In Vitro Biology in the 21st Century, 347–350.

2. Materials and Methods

Plant culture studies: Root cultures of *Datura stramonium* were maintained in B50 medium as described previously (Robins et al., 1991). Alkaloids were extracted from the freeze-dried roots as described previously (Robins et al., 1994), using phenylacetic acid as an internal standard for quantitative GC analysis. Alkaloids were derivatised with MSTFA prior to the GC-FID analysis (Zabetakis et al., 1998). *Chemicals*: Hyoscyamine and available precursors were obtained from Aldrich. Racemic littorine was prepared from DL-phenyllactic acid and tropine as previously described and recrystallised prior to use (Robins et al., 1994). [2'-^{13}C]-Littorine was prepared from tropine and phenyl-[2-^{13}C]-lactic acid as previously described (Robins et al., 1995). For the inhibitor studies, clotrimazole was purchased from Sigma and was added to the cultures in 0.5 ml of MeOH.

3. Results and discussion

3.1. The effect of adding littorine **1** on hyoscyamine **2** levels in *D. stramonium*

The most recent biosynthesis experiments on the tropane alkaloids have been facilitated by the availability of root cultures of *Datura* species transformed with *Agrobacterium rhizogenes* (Robins *et al.* 1990, 1991). Transformed roots of *D. stramonium* were used throughout these studies. A series of experiments was conducted to determine if littorine **1**, when added to *D. stramonium* roots, stimulated the production of hyoscyamine **2**, consistent with the role of littorine as a precursor to hyoscyamine. When littorine was added at a final concentration of 0.1mM, quantitative GC analysis indicated an increase in hyoscyamine levels of 35% relative to control values. This suggested that 40% of the added racemic DL-littorine had been converted to hyoscyamine and therefore 80% of the D-enantiomer. Approximately 35% had remained unmetabolised which is presumably predominantly the L-enantiomer. A further 20% of the dose could be accounted for in the increased levels of phenyllactate and tropine and clearly these metabolites are derived after hydrolysis of the added littorine. When the added dose was reduced by half to a final concentration 0.05mM of DL-littorine, 32% was converted to hyoscyamine, 35% remained unmetabolised and 13% hydrolysed to tropine and phenyllactate.

In an additional study DL-[2'-^{13}C]-littorine was fed to *D. stramonium* cultures at a final concentration of 0.1mM and incubated for 10 days. The percentage of unchanged DL-[2'-^{13}C]-littorine and the amount converted to [3'-^{13}C]-hyoscyamine was determined by isotope incorporation values after GCMS analysis. In the event 32% of the supplied ^{13}C-littorine was converted to hyoscyamine and 37% remained unmetabolised (Zabetakis et al., 1998).

Hyoscyamine was similarly added to root cultures of *D. stramonium*; this led to increased levels of tropine and tropic acid which is consistent with hydrolase activity. However there was no apparent increase in littorine levels by GC (peak area) analysis at 0.1mM and 0.05mM levels. From these studies an efficient mass transfer (32-40%) of exogenously added DL-littorine to hyoscyamine is revealed consistent with its role as a precursor, however the phenomenon is *irreversible*.

3.2. The effect of clotrimazole on the alkaloid profile of *D. stramonium*

The mechanism of the isomerisation of **1** to **2** and the nature of the enzymes mediating the rearrangement are not clear, however there is some circumstantial evidence which suggests the action of a P-450 activity. P-450 enzymes have the ability to generate organic radical or carbocation intermediates, one of which is almost certainly required to prime this rearrangement (Akhtar and Wright, 1991). In view of this, we set out to test the involvement of a P-450 activity by studying the alkaloid profile after treatment of *D. stramonium* roots with the known P-450 inhibitor, clotrimazole (Bolwell et al., 1994). *D. stramonium* root cultures were grown in the presence of clotrimazole (0.1mM) and also with clotrimazole + littorine (0.1mM), the precursor to hyoscyamine. The alkaloid content was determined by quantative GC analysis after a 10 day incubation and the resultant data are summarised in Table 1. Notably, clortimazole had a 60% inhibitory effect on hyoscyamine production relative to littorine, suggesting that clotrimazole suppresses the conversion of littorine to hyoscyamine. When clotrimazole + littorine was fed to the roots, the stimulatory effect (~ 35-40%, see above) on hyoscyamine biosynthesis by littorine was neutralised, again consistent with the suppression of the conversion of littorine **1** to hyoscyamine **2** by clotrimazole. This series of experiments provides further evidence to support the involvement of a P-450 activity in tropic acid ester biosynthesis.

Table 1. The effect of incubating root cultures with littorine (Litt **1**), clotrimazole (Clotri) and Litt + Clotri. All values ($\mu mol.g^{-1}$ DW) represent the mean of duplicate incubations and ± the variation in the replicates from the mean.

	Control	Litt **1**	Clotri	Litt **1** + Clotri
Tropine	1.91 ± 0.15	3.75 ± 0.41	1.36 ± 0.05	2.33 ± 0.42
Tropic acid	0.94 ± 0.11	1.11 ± 0.14	0.63 ± 0.04	1.32 ± 0.12
PLA	7.14 ± 0.61	8.15 ± 0.40	7.51 ± 0.20	8.00 ± 0.30
Hyoscymine 2	**14.31 ± 0.55**	**16.53 ± 0.31**	**8.48 ± 0.12**	**7.70 ±0.41**
Littorine 1	**4.44 ± 0.51**	**10.35 ± 0.30**	**3.16 ± 0.14**	**7.56 ± 0.24**
Scopolamine **3**	1.80 ± 0.07	1.51 ± 0.12	1.71 ± 0.02	1.78 ± 0.14
FW (g)	3.78	3.90	2.21	2.31

3.3 Mechanistic Proposal

In our most recent studies (Wong and O'Hagan, 1998) we have incubated DL-phenyl-[2-^{18}O,^{2}H]-lactate with root cultures of *D. stramonium* to trace the fate of the C-2'

D. stramonium

D. stramonium

70-75% retention

oxygen-18

DL--phenyl-[2-^{18}O, ^{2}H]-lactate

1

2

phenyllactate oxygen atom of littorine **1** during the rearrangement of littorine to hyoscyamine. In the event 70-75% of the oxygen-18 was retained during the isomerisation of **1** to **2**. Our working hypothesis for the rearrangement of **1** to **2**, which is consistent with all of the data to date is shown in the scheme below and involves two enzymatic activities. The first is an iron-oxo enzyme (P-450) which mediates a one or a two electron oxidation to generate either a radical or a carbocation respectively at the C-3' position of littorine. After rearrangement, either the product radical is quenched by oxygen rebound in a classical manner and then collapses to aldehyde **4** or the carbocation collapses directly to aldehyde **4**. The high retention of ^{18}O found experimentally tends to favour direct collapse of the carbocation but there is some exchange of oxygen from aldehyde **4** with the medium, prior to reduction by a dehydrogenase. Further reinforcement of this hypothesis requires that the isomerase and dehydrogenase activities are demonstrated in cell free extracts, and this forms the focus of our ongoing investigations.

4. Acknowledgements

We thank Dr Richard J. Robins, CNRS, Nantes, France for supplying the root cultures, EPSRC for a studentship (CWW) and the Leverhulme Trust for funding.

5. References

Akhtar M et al (1991) Nat. Prod. Rep. 8, 527-551.
Bolwell et al. (1994) Phytochem. 37, 1491-1506.
Brown JH (1990) Goodman and Gilman's the pharmacological basis of therapeutics, pp 150-165, Pergamon Press, New York, United States.
Leete E et al (1975) J. Am. Chem. Soc. 97, 6826-6830.
Robins RJ et al. (1990) Planta 181, 414-422.
Robins RJ et al. (1991) Planta 183, 185-195.
Robins RJ et al. (1994) J. Chem. Soc. Perkin Trans. I, 615-619.
Robins RJ et al. (1995) J. Chem. Soc. Perkin Trans. I, 481-485.
Robins RJ et al. (1997) J.Am. Chem. Soc. 119, 10929-10934.
Robinson R (1927) Proc. University of Durham Philosophical Society, 8, 14-39.
Wong et al. (1998) Chem. Commun. 1045-1046.
Zabetakis I et al (1998) Plant Cell Rep. (in press)

CHARACTERIZATION OF A NOVEL FAMILY OF CALMODULIN-BINDING PLASMA MEMBRANE CHANNEL-LIKE PROTEINS

Tzahi Arazi and Hillel Fromm
Department of Plant Sciences, Weizmann Institute of Science, Rehovot, Israel.

Introduction

Calcium (Ca^{2+}) is a ubiquitous second messenger in all eukaryotes (reviewed by Clapham, 1995; Bush, 1995; McAinsh and Hetherington, 1998). Ca^{2+} signals are transduced primarily by Ca^{2+} modulated proteins such as calmodulin (CaM) (Van Eldik and Watterson, 1998). CaM has no catalytic activity of its own and its diverse functions are the result of its interactions with numerous downstream effectors such as protein kinases, phosphatases, ion channels, and cytoskeleton associated proteins (James et al., 1995). The role of CaM and CaM-related proteins in plants is being unraveled in recent years (Snedden and Fromm, 1998; Zielinski, 1998). It became apparent that in spite of the high similarity of CaM from plants with that in animals (close to 90% identity in amino acid sequence) the downstream targets of CaM in plants and animals are not all the same (Baum et al., 1996; Snedden and Fromm, 1998). In addition, plants possess a large repertoire of calmodulin-related proteins not present in other eukaryotes (Snedden and Fromm, 1998). Thus, plants use the Ca^{2+}/CaM messenger system in unique ways to accommodate their physiological requirements.

Another emerging group of second messengers in plants are cyclic nucleotides. Recent reports suggest their involvement in photomorphogenesis (Neuhaus et al., 1993; Bowler et al., 1994), ion transport across membranes (Kurosaki et al., 1994; Toshi, 1995; Gaymard et al., 1996; Volotovski et al., 1998) and cell cycle (Ehsan et al., 1998). However, little is known about the molecular components mediating cyclic nucleotide signal transduction in plants. One interesting example of cross-talk between Ca^{2+} and cyclic nucleotide signaling pathways in plants comes from studies of phytochrome-mediated photomorphogenesis. Ca^{2+}/CaM induces the expression of a number of genes associated with the development of photosynthetic complexes, whereas cGMP induces genes involved in anthocyanin biosynthesis (e.g. chalcone synthase). These signaling molecules operate in a reciprocally repressive manner; Ca^{2+}/CaM inhibits anthocyanin production, whereas cGMP inhibits the expression of the gene encoding the chloroplast chlorophyll A/B-binding protein (Bowler and Chua, 1994). Other evidence for the interaction between Ca^{2+}/CaM- and cyclic nucleotide-mediated signaling pathways in plants came from studies of cyclic nucleotide mediated Ca^{2+} efflux from inside-out plasma membrane vesicles of carrot cells (Kurosaki et al., 1994). In this study, cAMP activated Ca^{2+} transport whereas Ca^{2+}/CaM inactivated the transport. Consistent with these studies,

A. Altman et al. (eds.), Plant Biotechnology and In Vitro Biology in the 21st Century, 351–353.

recently it was shown that membrane permeable analogs of cAMP and cGMP induce an increase in cytosolic Ca^{2+} levels in tobacco protoplasts, suggesting the existence of plasma membrane Ca^{2+} channels that, directly or indirectly, are activated by cyclic nucleotides (Volotovski et al., 1998).

Here we present the characterization of novel tobacco Ca^{2+}/CaM-binding proteins with similarities to cyclic nucleotide gated cation and voltage activated K^+ channels.

Results

Using CaM-interaction screening, we have isolated cDNA encoding Ca^{2+}-dependent CaM -binding protein from tobacco (CBP4) that share sequence similarities with cyclic nucleotide-gated cation- and voltage-activated K^+-channels. Using sequence analysis, we have identified six putative transmembrane domains, a putative pore region, and a cyclic-nucleotide binding domain (Fig. 1).

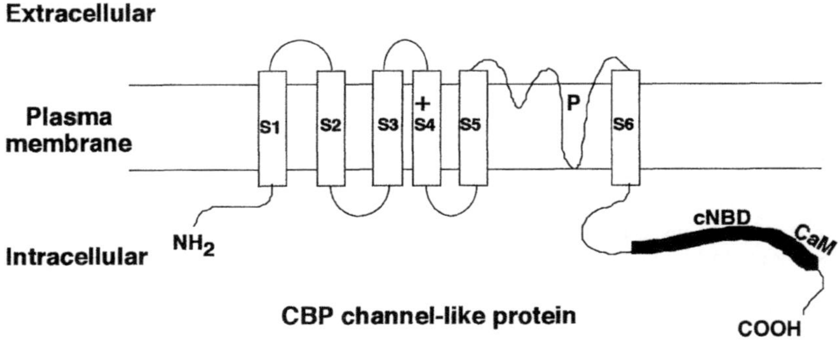

Figure 1. A structural model suggested for CBP4. The presumptive membrane-spanning and pore-lining regions are shown as S1-S6 and P, respectively. Putative cyclic nucleotide-binding domain (cNBD) and calmodulin-binding site (CaM) are indicated.

A full-length CBP4 protein expressed in Sf9 insect cells bound CaM in a Ca^{2+}-dependent manner, and a yeast interaction-trap assay mapped the CaM-binding activity to its C-terminal half. The binding of ^{35}S-CaM to CBP4 truncations further delimited the CaM-binding site to a 66-amino acid region. A 23-amino acid synthetic peptide from this region formed a complex with CaM in the presence of Ca^{2+} on a non-denaturing gel, indicating that this peptide represents the CBP4 CaM-binding domain (T. Arazi, unpublished) Measuring the fluorescence of dansyl-CaM interacting with this peptide revealed a dissociation constant of approximately 8 nM (T. Arazi, unpublished), which is within the physiological range of known CaM-protein interactions. Additional members of the CBP family were cloned in our laboratory from different organs and cell types of several plants, all of which have conserved cyclic nucleotide and CaM-binding domains (T. Arazi, unpublished). For biochemical and electrophysiological studies we expressed CBP4 in homologous transgenic tobacco plants system. In transgenic plants CBP4 is overexpressed and antibodies, prepared against the C-terminal of CBP4 were used to investigate its subcellular distribution. Membrane fractionation on sucrose gradients and two-phase partitioning suggests that tobacco CBP4 is associated with the plasma membrane (T. Arazi, unpublished). Taken together, the presence of conserved cyclic nucleotide and CaM binding domains in CBP4, and its localization to the plasma membrane of plant cells, suggest that this protein may function, as a component of cation non selective ion channels that play a fundamental role in Ca^{2+} and cyclic nucleotide signal transduction pathways in plants. Future studies will reveal the electrophysiological function of CBP4.

References

Baum, G., Lev-Yadun, S., Fridmann, Y., Arazi, T., Katsnelson, H., Zik, M., and Fromm, H. (1996) *EMBO J.* **15**, 2988-2996

Bowler, C., Neuhaus, G., Yamagata, H., and Chua, N.-H. (1994) *Cell* **77**, 73-81

Bush, D. S. (1995) *Annu. Rev. Plant Physiol. Plant Mol. Biol.* **46**, 95-122

Clapham, D. E. (1995) *Cell* **80**, 259-268

Devauchelle, G., Sentenac, H., and Thibaud, J.B. (1996) *J. Biol. Chem.* **271**, 2283-2287

Gaymard, F., Cerutti, M., Horeau, C., Lemaillet, G., Urbach, S., Revallec, M., Volotovski, I. D., Sokolovsky, S. G., Molchan, O. V., and Knight, M. R. (1998) *Plant Physiol.* **117**, 1023-1030

James, P., Vorherr, T., and Carafoli, E. (1995) *Trends Biochem. Sci.* **20**, 38-42

Kurosaki, F., Kaburaki, H., and Nishi, A. (1994) *FEBS Lett.* **340**, 193-196

McAinsh, M. R., and Hetherington, A. M. (1998) *Trends Plant Sci.* **3**, 32-36

Neuhaus, G., Bowler, C., Kern, R., and Chua, N.-H. (1993) *Cell* **73**, 937-952

Snedden, W. A., and Fromm, H. (1998) *Trends Plant Sci.* **3**, 299-304

Toshi, T. (1995) *J. Gen. Physiol.* **105**, 309-328

Van Eldik, L., and Watterson, M. D. (1998) *Calmodulin and Signal Transduction.* Academic Press, New York

Zielinski, R.N. (1998) *Annu. Rev. Plant Physiol. Plant Mol. Biol.* **49**, 697-725

IDENTIFICATION OF NOVEL CELL CYCLE GENES IN *ARABIDOPSIS THALIANA*

S. Burssens[1], L. De Veylder[1], M. Van Montagu[1] and D. Inzé[1,2]
[1] Laboratorium voor Genetica, Departement Genetica, Vlaams Interuniversitair Instituut voor Biotechnologie (VIB), Universiteit Gent, K.L. Ledeganckstraat 35, B-9000 Gent, Belgium; [2] Laboratoire Associé de l'Institut National de la Recherche Agronomique (France), Universiteit Gent, B-9000 Gent, Belgium

1. Introduction

The aim of our research is to understand the molecular basis of cell cycle control in higher plants, *in casu* in *Arabidopsis thaliana*, and how intrinsic developmental signals and environmental factors influence cell division and cell expansion. Progression through the eukaryotic cell cycle is regulated by protein kinase complexes, composed of a regulatory subunit, the cyclin, and a catalytic subunit, the cyclin-dependent kinase (CDK). The activity of CDK/cyclin complexes is regulated at multiple levels: (i) transcription of the CDKs and cyclins, (ii) association of the CDKs with specific cyclin partner(s), (iii) phosphorylation/dephosphorylation of the CDKs and cyclin subunits, (iv) interaction with other regulatory proteins such as SUC1/CKS1 homologues and CDK inhibitors (CKIs), and (v) cell cycle phase-dependent proteolytic destruction of components of the CDK complex.

Until now, two CDKs (CDC2a and CDC2b) (reviewed by Segers et al. 1997), 13 cyclins belonging to A-, B, or D- type (reviewed by Renaudin et al. 1996), and one CKI (Wang et al. 1997) have been described in *Arabidopsis thaliana*. To identify novel proteins interacting with the CDK, a two-hybrid screen was performed using a cDNA library of *A. thaliana*. We identified a putative substrate of CDC2aAt and Arath;*CYCD1;1* (De Veylder et al. 1997a), which had already been isolated (Soni et al. 1995). Also a novel D-type cyclin nominated Arath;*CYC;D4;1*, a *SUC1/CKS1* homologue nominated *CKS1At* (De Veylder et al. 1997b), and three CDK inihibitor (*CKI*) genes were among the positive clones. Recent molecular data and expression patterns of *CYCD4;1*, *CKS1At*, and the different *CKI* genes are discussed.

2. Results and Discussion

2.1. Arath;*CYCD4;1* encodes a novel D-type cyclin implicated in the initiation of lateral root formation

A. Altman et al. (eds.), Plant Biotechnology and In Vitro Biology in the 21st Century, 355–358.

D-type cyclins are believed to regulate the onset of cell division upon the perception of mitogenic stimuli and to play an important role at the entry of S phase. Previously, three different D-type cyclins have been identified in *A. thaliana* by their ability to rescue G1 cyclin-deficient mutant strains (Soni et al. 1995). By using CDC2aAt protein as a bait in a two-hybrid screen, another D-type cyclin, *CYCD4;1*, was isolated. Its sequence similarity with the other *A. thaliana* D-type cyclins is situated between 31.9% and 44.9%. In the cyclin core, CYCD4;1 shows 61%, 69%, and 66.6% similarity and, when the entire protein is considered, exhibits 37.2%, 44.4%, and 31% identity with CYCD1;1, CYCD2;1, and CYCD3;1, respectively.

CYCD2;1 expression has been shown to be induced by the addition of sucrose to a starved *A. thaliana* cell suspension culture, whereas the expression of *CYCD3;1* is up-regulated by cytokinins (Fuerst et al. 1996). By performing similar experiments, we could demonstrate that *CYCD4;1* expression is induced by sucrose. Moreover, the co-regulated expression of *CYCD4;1* and *CDC2aAt*, as observed in starved cell suspensions by simultaneous addition of cytokinins and auxins, suggests that a complex formation between these two partners is important for the resumption of cell division activity. Visualized on sections by *in situ* hybridization, *CYCD4;1* is expressed during vascular tissue development, embroygenesis, and lateral root primordia formation (LRP). *CycD4;1* expression can specifically be found in individual pericycle cells, just prior to the initial divisions that give rise to the actual LRP and also during the process of lateral root primordia formation. No *CycD4;1* transcripts were detected once the LRP was fully developed, suggesting that *CycD4;1* expression is required for initiation of lateral root organogenesis and not for maintaining the cell division activity.

2.2. CKS1At might play a role during both the endoreduplication and the mitotic cycle

More than one decade after their initial discovery in yeast and animals, the exact function of the *SUC1/CKS1* genes is still unclear. Based on crystallographic studies, the SUC1/CKS1 protein has been proposed to function as a docking factor for both positive and negative regulators of CDK complexes (Bourne et al. 1996). By the two-hybrid approach, we identified a SUC1/CKS homologue in *A. thaliana*, suggesting that, as part of the conserved toolbox of the eucaryotic cell cycle, essential functions are also attributed to this gene in regulating the plant cell cycle.

The *CKS1* gene of *A. thaliana* encodes a protein of 10.5 kDa and contains, in contrast to all other non-yeast homologues, a long polyglutamine-rich C-terminal domain. Deletion of the last 15 amino acids of CKS1At, including the polyglutamine stretch, shows that they are dispensable for the binding of both CDC2a and CDC2b. Presumably this part might be involved in the interaction with other proteins or may stabilize the CDK/CKS1At interaction. CKS1At is a functional homologue of the yeast SUC1, as shown by complementation analyses. Low to moderate expression levels of the CKS1At protein allowed a yeast temperature-sensitive *cdc2* mutant strain, altered at residues that are involved in the binding of SUC1/CKS1, to divide at the restrictive temperature. Additionally, CKS1At was able to complement a fission yeast SUC1-deleted strain. High expression levels of CKS1At inhibit cell division in both the mutant and wild-type strain, similar to what has been observed in yeast and *Xenopus* (Hayles et al. 1986; Patra, Dunphy 1996; Dunphy, Newport 1989). In contrast, overexpression of CKS1At mutated at the glutaminic acid residue (position 61) did not cause cell cycle arrest in yeast, confirming the essential importance of this amino acid for CDK binding in plants. By the two-hybrid approach and CKS1At-sepharose affinity selection, we could demonstrate that CKS1At binds to CDC2a and CDC2b *in vivo* and *in vitro*, suggesting a functional interaction with both CDKs, respectively. In an *A. thaliana* suspension culture, the CKS1At protein is only expressed in the exponential phase. To gain insight into possible functions of *CKS1At*, a comparitive expression analysis of *CKS1At* with other cell cycle genes, *CDC2aAt*, *CDC2bAt*, and *CYCB1;1* was performed in vegetative *A. thaliana* shoot apices (Hemerly et al., 1993; Ferreira et al., 1994; Segers et al., 1996). In endoreduplicating tissues, high and weak expression levels were observed for *CKS1At* and *CDC2aAt*, respectively, and strong expression of both genes was detected in actively dividing tissues. In contrast, *CDC2aAt* and *CycB1;1* transcripts were exclusively present in dividing cells. The observed coexpression of *CKS1At* with *CDC2aAt* and *CDC2bAt* supports the hypothesis that the *CKS1At* protein binds *in vivo* to both CDKs. However, the observed expression patterns indicate that CKS1At may operate in both mitotic and endoreduplication cycles whereas CDC2a, CDC2b, and CYCB1:1 only function during mitotic cell cycles. These data implicate that CKS1At may be functionally involved in the regulation of a not yet characterized CDK protein specialized in the entry and progression of S phase of the mitotic and endoreduplication cycles. By the generation of plants that overproduce CKS1At, we are currently investigating the role of CKS1At in the establishment of the spatial and temporal pattern of endoploidy during *A. thaliana* development.

2.3. The presence of a CKI family in plants

CKIs can inhibit CDK activity by their tight association with the CDK/cyclin complexes (reviewed by Martín-Castellanos, Moreno 1997). In mammals, two CKI families can be recognized based upon sequence similarity and mode of action: the INK4 and Kip1/Cip1 family. Among the genes yielded by the CDC2aAt two-hybrid screen, three genes were identified that show homology only to a 23-amino-acid region found in the human p27^{KIP} inhibitor. One of the clones turned out to be identical to the ICK1 (Wang et al. 1997). For two others, a flower cDNA library was screened to isolate the full lengths, nominated *FL39* and *FL66* and encoding proteins of 24 kDa and 23 kDa, respectively. When performing DNA gel blot analysis, an additional and weak band was observed for *FL66*, suggesting the presence of a fourth *FL66*-related gene. Only the FL39 protein contained a putative nuclear localization signal and a PEST-rich region was recognized, typically present in unstable proteins (Rogers et al. 1986). Sequence alignment of the three p27-related proteins revealed that they are mostly identical in the carboxyl-terminal domain. In the amino-terminal domain, short conserved motifs are present in all CKIs, often shared by only two of them. All inhibitors miss the LFG motif, which is essential for binding of the human p27^{KIP} for cyclin binding. This lack of similarity might be indicative for the divergence between plants and mammals during evolution in terms of specialization of cell cycle control.

By using the two-hybrid system, the binding specificity of the *A. thaliana* CKIs for CDC2aAt and CDC2bAt was tested. All CKIs associated exclusively with CDC2a, suggesting that they might operate at the G1/S and G2/M boundaries, where the CDC2a kinase activity is maximal. Most probably, the *A. thaliana* CKIs might influence the activity of the CDC2a complex at both transition points when, for example, the environmental conditions are inadequate. Similar to their mammalian counterparts, the *CKI* genes of *A. thaliana* might be also be involved in the regulation of the exit of the cell cycle linked to the onset of differentiation. At the moment, we are performing a detailed expression analysis of the different *A. thaliana CKI* genes to get a better understanding of their specific functions during the plant cell cycle.

3. Acknowledgments

This work was supported by grants from the Interuniversity Poles of Attraction Programme (Belgian State, Prime Minister's Office - Federal Office for Scientific, Technical and Cultural Affairs; P4/15), the European Commission BIOTECH program (ERBBIO4-CT96-O217), and the Fund for Scientific Research (Flanders) (G012196). L.D.V. was indebted to the Vlaams Instituut voor de Bevordering van het Wetenschappelijk-Technologisch Onderzoek in de Industrie for a predoctoral fellowship. D.I. is a Research Director of the Institut National de la Recherche Agronomique (France).

4. References

Bourne Y et al. (1996) Cell 84, 863-874.
De Veylder L et al. (1997a) J. Exp. Bot. 48, 2113-2114.
De Veylder L et al. (1997b) FEBS Lett. 412, 446-452.
Dunphy WG, Newport JW (1989) Cell 58, 181-191.
Ferreira PCG et al. (1994) Plant Cell 6, 1763-1774.
Fuerst RAUA et al. (1996) Plant Physiol. 112, 1023-1033.
Hayles J et al. (1986) EMBO J. 5, 3373-3379.
Hemerly AS et al. (1993) Plant Cell 5, 1711-1723.
Martín-Castellanos C, Moreno S (1997) Trends Cell Biol. 7, 95-98.
Patra D, Dunphy WG (1996) Genes Dev. 10, 1503-1515.
Renaudin J-P et al. (1996) Plant Mol. Biol. 32, 1003-1018.
Rogers S et al. (1986) Science 234, 364-368.
Segers G et al. (1996) Plant J. 10, 601-612.
Segers G et al. (1997) Plant Cell Proliferation and its Regulation in Growth and Development, pp 1-19, John Wiley & Sons, Chichester.
Soni R et al. (1995) Plant Cell 7, 85-103.
Wang H et al. (1997) Nature 386, 451-452.

Cell-to-Cell Movement of Tobacco Mosaic Virus

Vitaly Citovsky
Department of Biochemistry and Cell Biology, State University of New York, Stony Brook, NY 11794-5215.

1. Introduction

Following initial infection, many plant viruses spread from cell to cell through plant intercellular connections, the plasmodesmata. This cell-to-cell movement is mediated by virus-encoded movement proteins (MP) which act to increase plasmodesmal permeability and transport viral nucleic acids through the enlarged plasmodesmal channels (reviewed by Lucas, Gilbertson, 1994; Carrington et al., 1996; Ghoshroy et al., 1997).

The best studied MP is the 30 kDa protein of tobacco mosaic virus (TMV) (Deom et al., 1987). To date, MP has been suggested to possess four biological activities. (1) It is thought to bind TMV RNA, forming an extended MP-RNA complex that can penetrate the plasmodesmal channel (Citovsky et al., 1990; Citovsky et al., 1992). (2) It may interact with the cytoskeletal elements to facilitate transport of the MP-TMV RNA complexes from the cell cytoplasm to plasmodesmata (Heinlein et al., 1995; McLean et al., 1995). (3) MP also functions to increase the size exclusion limit of plasmodesmata (Wolf et al., 1989). (4) Finally, MP may interact with a cell wall-associated receptor which then phosphorylates the bound movement protein, inactivating its ability to gate plasmodesmata. Here, we examine these MP activities and integrate them into a model for the MP-mediated intercellular transport of TMV RNA.

2. MP-Nucleic Acid Complexes

In vitro studies showed that MP binds both RNA and single stranded (ss)DNA, but not double stranded (ds)DNA, cooperatively and without sequence specificity (Citovsky et al., 1990; Citovsky et al., 1992). This sequence non-specific binding explains the observations that coinfection with TMV can complement cell-to-cell movement of plant viruses which normally do not spread through plasmodesmata (reviewed by Carr, Kim, 1983; Atabekov, Taliansky, 1990). Electron microscopic observations revealed that MP binding unfolds the nucleic acid molecule creating an extended protein-RNA complex of 2.0-2.5 nm in diameter (Citovsky et al., 1992). Because the free folded TMV RNA has been estimated to be 10 nm in diameter (Gibbs, 1976), association with MP likely shapes it in a thinner transferable form capable of transport through plasmodesmal channels.

Cell-to-cell movement proteins of numerous other plant viruses were also found to bind nucleic acids (Citovsky et al., 1991; Osman et al., 1992; Schoumacher et al., 1992; Pascal et al., 1994; Thomas, Maule, 1995), suggesting that MP-nucleic acid complexes may represent cell-to-cell transport intermediates of many plant viruses.

A. Altman et al. (eds.), Plant Biotechnology and In Vitro Biology in the 21st Century, 359–363.

3. MP-Cytoskeleton Interaction

Recent data suggest that MP interacts with microtubules and, to a lesser extent, with actin microfilaments (Heinlein et al., 1995; McLean et al., 1995). This interaction was inferred from colocalization of MP with microtubules (Heinlein et al., 1995) as well as with actin filaments (McLean et al., 1995). MP association with actin and tubulin was also demonstrated *in vitro* (McLean et al., 1995). The biological role of MP-cytoskeleton interaction, however, remains unclear.

4. MP-Induced Increase in Plasmodesmal Permeability

MP-TMV RNA cell-to-cell transport complex must transverse plasmodesmata to enter the neighboring host cell. Although the estimated 2.0-2.5 nm diameter of this nucleoprotein complex (Citovsky et al., 1992) is relatively small, it still is incompatible with the 1.5 nm diameter of the intact plasmodesmal channel (Wolf et al., 1989). To allow movement, therefore, MP induces an increase in plasmodesmal permeability. The ability of MP to elevate plasmodesmal size exclusion limit was first detected by injection of fluorescently labeled dextrans into leaf mesophyll cells of transgenic tobacco plants expressing MP (Wolf et al., 1989). Unlike the wild-type tobacco mesophyll plasmodesmata which traffic only 0.75-1.0 kDa dextrans, the MP-expressing plants had a plasmodesmal size exclusion limit of almost 10 kDa (Wolf et al., 1989).

Direct microinjection of purified MP into the wild-type tobacco mesophyll cells resulted in a relatively fast (3-5 minutes) size exclusion limit increase up to 20 kDa, indicating that MP functions via an existing plasmodesmal transport machinery (Waigmann et al., 1994). Immunolocalization experiments demonstrated that microinjected MP itself moves between plant cells (Waigmann, Zambryski, 1995). Thus, the increased size exclusion limit of 10-20 kDa (Wolf et al., 1989; Waigmann et al., 1994) corresponds to a 5-9 nm diameter of the dilated channel, allowing traffic of 2.0-2.5 nm-wide MP-TMV RNA complexes (Citovsky et al., 1992).

5. MP Interaction with a Cell Wall-Associated Receptor

Presently, most of our knowledge about the process of TMV cell-to-cell movement derives from studies of its viral component, i.e. MP. Other than the cytoskeleton (Heinlein et al., 1995; McLean et al., 1995), no plant cell component that interacts with MP has been identified. Recently, we have isolated a plant cell receptor for MP. Using gel overlay assays, we detected a 38 kDa cell wall-associated protein, designated p38, that bound TMV MP (unpublished results). MP domains required for interaction with p38 were identified and compared to the known functional regions of MP. The first p38 binding domain, designated F1, was located at the N-terminus of MP; this protein region has been shown to be absolutely required for the movement function *in vivo* (Gafny et al., 1992; Lapidot et al., 1993). The second p38 binding domain, designated F2 and positioned in the C-terminal part of the protein, overlapped with the previously identified domain E required for the increase in plasmodesmal permeability and suggested to interact with a putative MP receptor (Waigmann et al., 1994). Potentially, these major binding domains are augmented by the context of the native protein.

Previously, we have shown that MP is phosphorylated by a cell wall-associated protein kinase at its C-terminal serine and threonine residues (Citovsky et al., 1993). p38 is also located in the cell wall. Thus, it may be able to phosphorylate its MP substrate directly on the PVDF membrane following binding. In these experiments, protein kinase activity was found consistently in a single band with the same electrophoretic mobility as p38, following incubation of the electrophoresed cell wall proteins with P30. No phosphorylation occurred when a MP mutant in which the phosphorylation site residues Ser-258, Thr-261, and Ser-265 are replaced with alanines (Citovsky et al., 1993) was used as substrate. However, this MP mutant retained its ability to bind p38, suggesting that the receptor and protein kinase activities of p38 are mutually independent and may represent two distinct protein functions.

Possible biological function of MP C-terminal phosphorylation was examined using negatively charged amino acid substitutions within the phosphorylation site, known to reveal the electrostatic effects of phosphorylation (Dean, Koshland, 1990). Our, microinjection experiments showed that substituting Ser-258, Thr-261, and Ser-265 with aspartate residues inactivated the MP ability to dilate plasmodesmata, suggesting the role of phosphorylation in down-regulation of MP activity.

The biological role of p38 probably is not limited to viral movement. Because viruses often adapt existing cellular machinery for their own needs, TMV likely employs an endogenous pathway for cell-to-cell transport of proteins and nucleic acids. Such plasmodesmal transport of RNA, for example, has been suggested to signal post-translational gene silencing (Palauqui et al., 1997). Thus, it is tempting to speculate that p38 has a natural cellular ligand involved in plasmodesmal traffic in healthy plants.

6. Possible Mechanisms for Plasmodesmal Transport

Transport through plasmodesmata most likely involves recognition of the transported molecule in the cell cytoplasm and its targeting to the plasmodesmal channel, followed by translocation (Figure 1). In the initially infected cell, MP is produced by translation of a subgenomic RNA produced during replication of the invading virus. Presumably, this protein then associates with a proportion of the viral RNA molecules, sequestering them from replication and mediating their transport into the neighboring uninfected host cells.

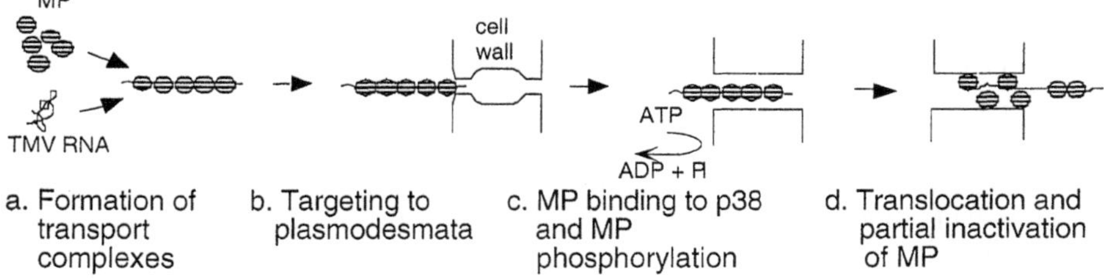

Figure 1. A model for plasmodesmal transport of MP-TMV RNA complexes.

Interaction of viral movement proteins with plasmodesmata may interfere with normal intercellular communication and, thus, be detrimental to the host plant. It is likely, therefore, that a mechanism exists to regulate the activity of MP and, possibly, cellular proteins capable of plasmodesmal transport (Lucas et al., 1995). Potentially, MP phosphorylation by p38 is involved in such regulation, minimizing MP interference with plasmodesmal permeability during viral infection. Thus, p38 may perform two distinct biological functions in the TMV cell-to-cell transport process. First, it may act as a specific receptor mediating MP targeting to the host cell wall and, probably, to plasmodesmata. This function may be essential for plasmodesmal transport; therefore, MP domains involved in p38 binding are required for its biological activity. Second, p38 probably acts as a protein kinase which phosphorylates MP that has already performed its function within plasmodesmata (Figure 1). This activity may play a role in down-regulation of MP but is not directly required for its function. Indeed, the phosphorylation domain D of MP is dispensable for the activity of this protein (Gafny et al., 1992).

7. Acknowledgment

Research in my lab is supported by grants from National Institutes of Health, U.S. Department of Agriculture, and U.S.-Israel Binational Research and Development Fund.

8. References

Atabekov JG, Taliansky ME (1990). Expression of a plant virus-coded transport function by different viral genomes. Adv. Virus Res. 38, 201-248.

Carr RJ, Kim KS (1983). Evidence that bean golden mosaic virus invaded non-phloem tissue in double infections with tobacco mosaic virus. J. Gen. Virol. 64, 2489-2492.

Carrington JC et al. (1996). Cell-to-cell and long-distance transport of viruses in plants. Plant Cell 8, 1669-1681.

Citovsky V et al. (1990). The P30 movement protein of tobacco mosaic virus is a single-strand nucleic acid binding protein. Cell 60, 637-647.

Citovsky V et al. (1991). Gene I, a potential movement locus of CaMV, encodes an RNA binding protein. Proc. Natl. Acad. Sci. USA 88, 2476-2480.

Citovsky V et al. (1993). Phosphorylation of tobacco mosaic virus cell-to-cell movement protein by a developmentally-regulated plant cell wall-associated protein kinase. Genes Dev. 7, 904-910.

Citovsky V et al. (1992). Visualization and characterization of tobacco mosaic virus movement protein binding to single-stranded nucleic acids. Plant Cell 4, 397-411.

Dean AM, Koshland DE (1990). Electrostatic and steric contributions to regulation at the active site of isocitrate dehydrogenase. Science 249, 1044-1046.

Deom CM et al. (1987). The 30-kilodalton gene product of tobacco mosaic virus potentiates virus movement. Science 327, 389-394.

Gafny R et al. (1992). Effects of terminal deletion mutations on function of the movement protein of tobacco mosaic virus. Virology 187, 499-507.

Ghoshroy S et al. (1997). Transport of proteins and nucleic acids through plasmodesmata. Annu. Rev. Plant Physiol. Plant Mol. Biol. 48, 27-49.

Gibbs AJ (1976). Viruses and plasmodesmata. In Intercellular Communication in Plants: Studies on Plasmodesmata. (ed. Gunning BES, Robards AW). Berlin, Springer-Verlag pp. 149-164.

Heinlein M et al. (1995). Interaction of tobamovirus movement proteins with the plant cytoskeleton. Science 270, 1983-1985.

Lapidot M et al. (1993). A dysfunctional movement protein of tobacco mosaic virus that partially modifies the plasmodesmata and limits virus spread in transgenic plants. Plant J. 4, 959-970.

Lucas WJ et al. (1995). Selective trafficking of KNOTTED1 homeodomain protein and its mRNA through plasmodesmata. Science 270, 1980-1983.

Lucas WJ, Gilbertson RL (1994). Plasmodesmata in relation to viral movement within leaf tissues. Annu. Rev. Phytopathol. 32, 387-411.

McLean BG et al. (1995). Tobacco mosaic virus movement protein associates with the cytoskeleton in tobacco cells. Plant Cell 7, 2101-2114.

Osman TAM et al. (1992). Cooperative binding of the red clover necrotic mosaic virus movement protein to single-stranded nucleic acids. J. Gen. Virol. 73, 223-227.

Palauqui JC et al. (1997). Systemic acquired silencing: transgene-specific post-translational silencing is transmitted by grafting from silenced stocks to non-silenced scions. EMBO J. 16, 4738-4745.
Pascal E et al. (1994). The geminivirus BR1 movement protein binds single-stranded DNA and localizes to the cell nucleus. Plant Cell 6, 995-1006.
Schoumacher F et al. (1992). Nucleic acid binding properties of the alfalfa mosaic virus movement protein produced in yeast. Virology 188, 896-899.
Thomas CL, Maule AJ (1995). Identification of the cauliflower mosaic virus movement protein RNA binding domain. Virology 206, 1145-1149.
Waigmann E et al. (1994). Direct functional assay for tobacco mosaic virus cell-to-cell movement protein and identification of a domain involved in increasing plasmodesmal permeability. Proc. Natl. Acad. Sci. USA 91, 1433-1437.
Waigmann E, Zambryski P (1995). Tobacco mosaic virus movement protein-mediated transport between trichome cells. Plant Cell 7, 2069-2079.
Wolf S et al. (1989). Movement protein of tobacco mosaic virus modifies plasmodesmatal size exclusion limit. Science 246, 377-379.

The role of PASTICCINO1, an FKBP-like protein, in plant development

Rachel J. Cowling, Paola Vittorioso, Jean-Denis Faure, Michel Caboche, Catherine Bellini
Laboratoire de Biologie Cellulaire, Institut National de Recherche Agronomique, route de St. Cyr, 78026 Versailles Cedex, FRANCE

Summary

The *pasticcino* (*pas*) mutants of *Arabidopsis thaliana* were identified due to their abnormal response to cytokinins. These mutants fall into 3 complementation groups: *pas1* (2 alleles), *pas2* (1 allele) and *pas3* (4 alleles). The mutants all have excessive and disorganised cell divisions leading to extra cell layers in the hypocotyl, abnormal apical meristems, and rosettes with fused vitreous leaves. This cell proliferation effect is exaggerated in the presence of cytokinins. The *PAS1* gene was cloned via a T-DNA insertion; it encodes for a protein with similarities to FK506-binding proteins (FKBPs), a class of immunophilins. Like other large FKBPs the PAS1 protein has several putative FKBP domains, a tetratricopeptide repeat domain and a calmodulin binding motif. PAS1 also has conserved nuclear localisation signals. Although the presence of immunophilins has been demonstrated in plants, their functions are unknown. *PAS1* is the first plant immunophilin-like gene to be disrupted and suggests a role for PAS1 in the control of plant cell division.

Screening for the *pas* mutants of *Arabidopsis thaliana* (Faure et al., 1998)

The *pas* mutants were selected from *Arabidopsis thaliana* populations which had been mutagenised by either EMS or T-DNA. The mutants were selected on the basis of phenotypic similarity to the *zea* mutants of *Nicotiana plumbaginifolia*, that is, for seedling hypertrophy when grown on medium containing cytokinins. The *zea* mutants are hypertrophic when grown in the presence of cytokinins due to a cell proliferation defect, which is specific to active cytokinins (Faure et al., 1994; Nogué et al., 1995) and in the case of *zea3* mutants is due to hypersensitivity to cytokinins (Martin et al., 1997). Alternatively, mutants were selected for altered hypocotyl length in the absence of cytokinins.

From these two screens and following genetic allelism tests three new complementation groups of nuclear recessive mutations were defined: *pas1* (2 alleles), *pas2* (1 allele) and *pas3* (4 alleles). On the basis of the differing root phenotypes (see below) it was shown that *pas3* is epistatic to *pas2*, which is epistatic to *pas1*.

The *pas* phenotype (Faure et al., 1998)

The earliest indication of *pas* phenotype is after the heart stage of embryogenesis. Mutant embryos fail to curve round within the seed because the cotyledons and hypocotyl

A. Altman et al. (eds.), Plant Biotechnology and In Vitro Biology in the 21st Century, 365–368.

do not form correctly. Upon germination the *pas* seedlings have shorter, broader hypocotyls than those of WT seedlings and the cotyledons remain undeveloped and unexpanded (Figure 1). Microscopic analysis of sections of seedlings shows that the cellular make-up of *pas* seedlings is disorganised compared to WT seedlings. Ectopic cell division leads to extra cell layers or parts of layers and there is a loss of cell adhesion in hypocotyls and cotyledons. This hyperproliferation is exaggerated when the mutants are grown on medium containing cytokinins.

The structure of the shoot apical meristem varies between individuals from being very small to being grossly enlarged, and the normally well-defined cell layers are not distinguishable. Adult plants have disorganised compact bushy rosette, with fused vitreous leaves. This latter phenotype is reminiscent of plants regenerated in tissue culture when auxin and cytokinin concentrations in the medium are unbalanced.

Although the *pas* mutants have similar aerial phenotypes the root phenotypes vary. The primary root of *pas1* is shorter than that of WT with rare lateral roots. The *pas3* roots are even shorter with occasional root branching. By contrast, *pas2* mutants are both longer and more branched than WT roots.

The PAS1 gene (Vittorioso et al., 1998)

Two alleles of the *pas1* mutant were identified; *pas1-1* is a T-DNA line and *pas1-2* is EMS-induced. The T-DNA was shown to segregate with the *pas1-1* phenotype. The DNA sequences flanking the T-DNA were isolated and a corresponding cDNA found. The T-DNA is inserted within a 4.2kb locus, specifying a 2.4kb mRNA, which encodes a 70kDa protein related to the FK506-binding protein (FKBP) class of immunophilins (see below). When the *pas1-2* mutant, which has a point mutation within the *PAS1* coding sequence, was transformed with the *PAS1* cDNA, plants with wild-type phenotype were recovered, showing that mutation in this gene is responsible for the *pas* phenotype.

The *PAS1* mRNA is detected in all plant tissues. In the *pas1-1* allele the T-DNA is inserted such that there is a translational fusion between PAS1 and β-glucuronidase (GUS), encoded by the T-DNA (Bouchez et al., 1993). Based on GUS activity staining of heterozygotes (with wild-type phenotype), the PAS1 protein is expressed in the shoot and root apical meristems. In the homozyogous mutants, the GUS activity is expressed throughout the hypocotyl and cotyledons, coinciding with the regions where cell division is deregulated.

PAS1 is an FK506-binding protein (FKBP)-like protein

The PAS1 protein has domains which, from sequence similarity, are involved in: nuclear localisation; FK506 binding and/or peptidylprolyl *cis-trans* isomerase (PPIase) activity; protein interaction via tetratricopeptide repeats (TPR); and calmodulin binding (Figure 2).

FKBPs are proteins which have similarity to FKBP12, a protein which binds the immunosuppressant drugs, FK506 and rapamycin (Siekierka et al., 1989; Kay, 1996). The presence of FKBPs in plants has been described (Breiman et al., 1992; Luan et al., 1994; 1996; Blecher et al., 1996; Vucich, Gasser, 1996; Kurek, Breiman, this volume; Kurek et al., this volume). Possible roles for FKBPs have been suggested: as molecular chaperones, in nuclear trafficking, in signal transduction, for example (Kay, 1996; Kurek, Breiman, this volume). However the precise function and mechanism of action of the large FKBPs in any organism is not known.

PAS1 is 71% similar to ROF1, an FKBP from arabidopsis and 54% similar to FKBP73 from wheat (Vittorioso et al., 1998; Vucich, Gasser, 1996; Blecher et al., 1996). PAS1 is approximately 50% similar to the mammalian FKBP59s, which are known to be components of steroid receptor complexes (Tai et al., 1992).

A common feature of FKBPs is their catalysis of protein folding at proline residues *in vitro*, the PPIase activity (Fischer et al., 1984). Although the *in vivo* significance of the PPIase activity of large FKBPs is not known, it is interesting to note that the nuclear-localised Pin1, a member of a different class of PPIases, has recently been shown to have direct effects on the control of mitosis in humans (Lu et al., 1996; Shen et al., 1998).

Current work

Our current work is to characterise the biochemical properties of the PAS1 protein and to discover the proteins which interact with PAS1 using the parallel approaches of immunoprecipitation and the yeast two-hybrid system. We are also studying the effect of the *pas* mutations on the expression of cell cycle genes.

Correct expression of the PAS1 protein is required to control precisely when and where plant cells are to divide, a process also influenced by cytokinins. Perhaps PAS1 acts by altering the conformation of nuclear-localised proteins either in a complex or in a receptor/signalling pathway. We speculate that the products of the *PAS2* and *PAS3* genes are involved in the same process.

Conclusion

PAS1 is the first plant immunophilin gene to be disrupted and studied genetically. The striking phenotype of *pas1* mutants suggests that PAS1 is important in the mechanism which controls the temporal and spatial regulation of plant cell division.

Acknowledgements

RJ Cowling is a post-doctoral fellow in the European Community Framework IV LATIN (Lateral Initiation) project.

References

Blecher O et al. (1996) Plant Mol. Biol. 32, 493-504
Bouchez D et al. (1993) C. R. Acad. Sci. Paris 316, 1188-1193
Breiman A et al. (1992) J. Biol. Chem. 267, 21293-21296
Faure J-D et al. (1994) Plant J. 5, 481-491
Faure J-D et al. (1998) Development 125, 909-918
Fischer G et al. (1984) Biomed. Biochim. Acta 43, 1011-1111
Kay JE (1996) Biochem. J. 314, 361-385
Lu KP et al. (1996) Nature 380, 544-547
Luan S et al. (1994) Proc. Natl. Acad. Sci USA 91, 984-988
Luan S et al. (1996) Proc. Natl. Acad. Sci USA 93, 6964-6969
Martin T et al. (1997) Plant Physiol. 114, 1177-1185
Nogué F et al. (1995) Plant Growth Reg. 17, 87-94
Shen M et al. (1998) Genes Dev. 12, 706-720
Siekierka JJ et al. (1989) Nature 341, 755-760
Tai et al. (1992) Science 256, 1315-1318
Vucich VA, Gasser CS (1996) Mol. Gen. Genet. 252, 510-517
Vittorioso P et al. (1998) Mol. Cell. Biol. 18, 3034-3043

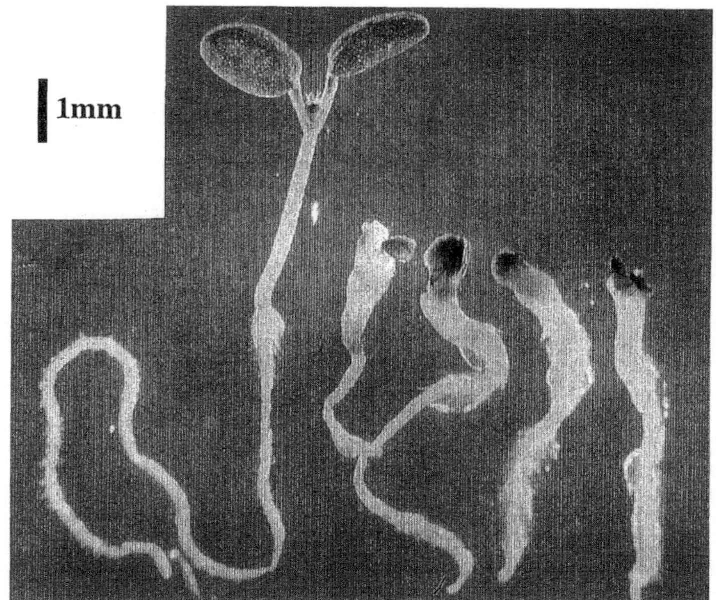

Figure 1 The *pas1* phenotype
Wild-type (left) and four *pas1-1* seedlings (right)

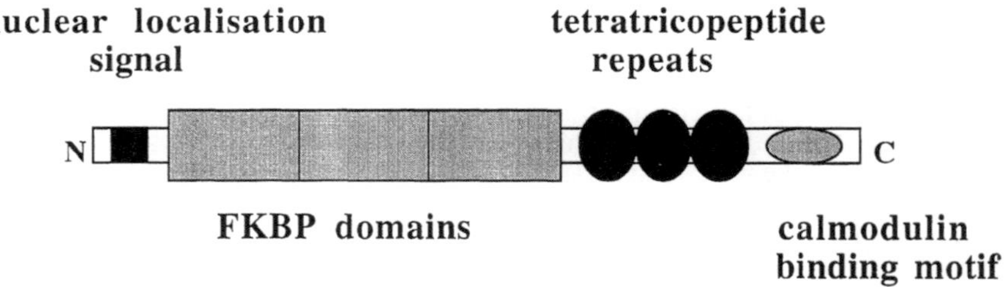

Figure 2 Schematic representation of the PAS1 protein

THE FORMATION OF THE PLANT VACUOLAR SYSTEM

Alexandre da Silva Conceição[1], Danièle Marty-Mazars[1], Natasha V. Raikhel[2], Francis Marty[1]. [1]Laboratoire de *phyto*Biologie Cellulaire, Université de Bourgogne, BP400, 21011 Dijon Cedex, France; [2]MSU-DOE Plant Research Laboratory, Michigan State University, East Lansing, MI 48824-1312, USA

The plant vacuole is a multifunctional subcellular compartment. In general, the vacuolar system is constituted of a single or few vacuoles. About 80% of the volume of mature plant cells is occupied by vacuoles. These organelles are limited by a single membrane, the tonoplast. According to the cell type and to particular functions, the vacuolar content can vary from a homogeneous consistency to a heterogeneous matrix containing crystalloid structures (Boller, Wiemken, 1986; Marty et al., 1980). From the biotechnological point of view, the endomembrane system of plants, particularly the ER and vacuoles, can be envisioned as the ultimate environment for stable protein accumulation. Several reports have successfully used seed storage vacuoles for the expression of heterologous proteins. We believe that the engineering of other plant tissues, and their vacuoles, to further develop transgenic plants as an economically feasible system for the large-scale production of recombinant proteins, will be possible in the future (da Silva Conceição, Raikhel, 1996).

As many other biological membranes, the tonoplast is constituted of a phospholipid bilayer in which different proteins, such as membrane transporters, are inserted. The passage of ions and other polar molecules is carried out by two types of transport proteins associated with the tonoplast. In one hand, channels form pores to facilitate the diffusion of molecules through the vacuolar membrane. This is the case of aquaporins, a family of tonoplastic intrinsic proteins (TIP) involved in the transport of water and the control of turgor pressure in plant cells. Aquaporins are the best characterized membrane markers of seed tonoplasts (Chrispeels, Agre, 1994; Maurel, 1997). Their homologues in vegetative cells have been studied in several plant species (Marty-Mazars et al., 1995; Barrieu et al., 1998). On the other hand, the active transport of molecules through the tonoplast is mediated by carriers and pumps. The vacuolar accumulation of ions and metabolites plays an important role in cell signaling, homeostasis, plant defense and detoxification (Leigh, Sanders, 1997).

The accumulation of soluble proteins in plant vacuoles makes use of the secretory pathway (Figure 1). Protein trafficking to the vacuole is mediated by transport vesicles and three different types of targeting signals have been identified so far in vacuolar proteins. These signals are recognized by a sorting machinery whose components are being characterized in plants (Bassham, Raikhel, 1997). In seeds and specialized vegetative tissues, storage proteins are mobilized as a source of amino acids. The processing enzymes needed for the maturation of storage proteins, and the acid hydrolases used for mobilization of reserves and plant defense, are other types of proteins accumulated in vacuoles (da Silva Conceição, Raikhel, 1996).

The plant vacuole has also an important role in plant morphogenesis because developmental and growth processes require turgor-driven cell expansion. Therefore,

A. Altman et al. (eds.), Plant Biotechnology and In Vitro Biology in the 21st Century, 369–372.

the identification of genes involved in vacuole development can be attained by genetic screenings based on morphogenetic phenotypes. This is the case of the *rhd3*, an Arabidopsis mutation which results in short and wavy root hairs with a volume less than one-third of the wild-type hairs. It has been suggested that the *RHD3* gene is required for vacuole enlargement during root hair cell expansion (Galway et al., 1997).

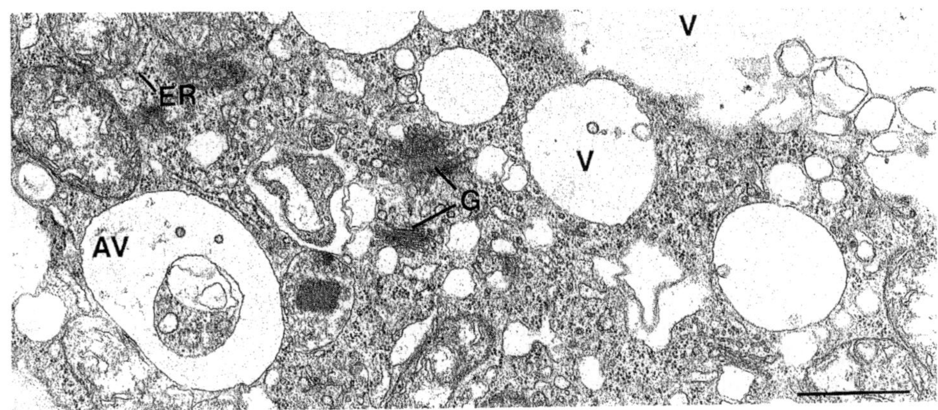

Figure 1: The vacuolar system of suspension-cultured cells from Sycamore (*Acer pseudoplatanus* L.). AV, autophagic vacuole; ER, endoplasmic reticulum; G, Golgi apparatus; V, vacuole. Bar: 1μm.

The Ontogenesis of Vacuoles

During plant development, the central vegetative vacuole (VV) carries out almost all functions of the vacuolar system. However, protein accumulation in seeds usually occurs inside specialized organelles, the protein storage vacuoles (PSVs). Early in seed formation, the initiation of protein deposition is made lining the tonoplast. At later stages of seed maturation, protein deposition proceeds through a fragmentation and/or budding process of PSVs to result in the formation of protein bodies (PBs). The physical properties related to changes in membrane molecular dynamics of the tonoplast during seed germination, have been monitored by Strzalka et al. (1995). These parameters indicated a more rigid nature of the membrane of dry seed PBs comparing to the tonoplast of 4-day-old seedlings. This is probably the consequence of a more restricted motional freedom of lipids. These results are consistent with the protective role of PB membranes for the encapsulation of storage protein crystalloids (Strzalka et al., 1995).

The ultrastructural analysis of vacuole biogenesis in meristematic cells of roots, indicates that vacuoles result from the autophagic activity of a system of provacuolar tubules. These structures accumulate vacuolar hydrolytic enzymes found in post-ER endomembrane compartments. Electron microscopic analysis showed that the autophagic process is cellular programmed through root development (Marty, 1978; Marty, 1997). Protein transport to the vacuole is mediated by Golgi-derived transport vesicles (Hara-Nishimura et al., 1998), suggesting that the Golgi-apparatus and the *trans*-Golgi network (TGN) are involved in the ontogenesis of the vacuolar system. However, other Golgi-independent transport pathways to vacuoles, via smooth regionally differentiated ER regions, have been described in plants (Levanony et al., 1992; Herman et al., 1994; Staehelin, 1997). Similarly, some workers suggest that storage protein deposition in pea cotyledons may occur in subdomains of the ER which eventually become vacuoles (Robinson, Hinz, 1997). Therefore, the origin of a provacuolar system needs to be further clarified by biochemical and molecular studies.

Traditionally the vacuolar system has been assumed to be the same compartment through all the cell types of a plant, in spite of the multiple functions of plant cell vacuoles. According to this view, the PSVs of maturing dicot seeds are thought to be the result from a fragmentation of the VV. Recently, different authors have demonstrated the existence of distinct vacuole populations in a single plant cell type. During the formation of PSVs, Hoh et al. (1995) have identified two vesicle populations. Vesicles of both populations possess a vacuolar proton-translocating inorganic pyrophasphatase (V-PPase) and a vacuolar-type H+-ATPase (V-ATPase). These proton pumps are responsible for the vacuole acidification that will power the vacuolar secondary transport systems in plants (Leigh, Sanders, 1997). Nevertheless, the two vesicle populations could be differentiated on the basis of aquaporin isoforms, suggesting that they correspond to two distinct vacuole types. Two types of vacuoles have also been characterized in vegetative tissues of root tips (Paris et al., 1996). In cotyledons, PSVs develop from a complex tubular-cisternal membrane system that seems to surround the pre-existing VVs. These results support the hypothesis that PSVs are formed *de novo* during cotyledon development (Hoh et al., 1995).

The central vegetative vacuole is reformed at different stages of plant development. In the case of seed germination and hydrolysis of storage proteins, the membranes of numerous PSVs are fused with each other to reform a single or few vacuoles. The incorporation of newly synthesized biochemical markers indicate that the membrane of PSVs are transformed into central vacuole membranes during the growth of seedlings (Maeshima et al., 1994). The disappearance of membrane proteins that are specific of PSVs, as for instance α-TIP, has also been reported. The mechanism of α-TIP removal appears to involve autophagic sequestering of membranes inside vacuoles (Melroy, Herman, 1991; Inoue et al., 1995).

Biochemical studies of reformation of the central vacuole throughout plant development have been hampered by the failure to isolate early vacuolar structures. To overcome this difficulty, evacuolated protoplasts have been used (Hörtensteiner et al., 1992). Furthermore, they represent a unique approach to study structural aspects of vacuole regeneration in synchronized cells (Newell et al., 1998).

Concluding Remarks

The plant vacuole has several major roles in development, growth and physiology of plants. Our current knowledge on vacuole formation is due to scientific approaches mainly based on microscopic and biochemical techniques. Several protein markers remain to be explored in detail to study the ontogenesis of vacuoles. This is the case of prevacuolar proteins as AtPEP12p (da Silva Conceição et al., 1997) and AtELP (Ahmed et al., 1997). However, new approaches as *in vitro* and free cell reconstitution systems, similar to those currently used to uncover the functioning mechanisms of the secretory pathway of yeast and mammalian cells, need to be introduced in plant cell biology.

REFERENCES

Ahmed, SU et al (1997) Plant Physiol. 114, 325-336.

Barrieu, F et al (1998) Planta 204, 335-344.

Bassham, DC and Raikhel, NV (1997) In: Advances in Botanical Research, The Plant Vacuole, Leigh, RA and Sanders, D ed., Academic Press, London, England, 43-58.

Boller, T and Wiemken, A (1986) In: Ann. Rev. Plant Physiol. 37, 137-164.

Chrispeels, MJ and Agre, P (1994) Trends Biochem. Sci. 19, 421-425.

da Silva Conceição, A et al (1997) Plant Cell 9, 571-582.

da Silva Conceição, A and Raikhel, NV (1996) In: Transgenic Plants, a Production System for Pharmaceutical and Industrial Proteins, Pen, J and Owen, M ed., John Wiley and Sons Ltd., West Sussex, England, 75-98.

Galway, ME et al (1997) Planta 201, 209-218.

Hara-Nishimura, I et al (1998) Plant Cell 10, 825-836.

Herman, EM et al (1994) Plant Physiol. 106, 1313-1324.

Hoh, B et al (1995) J. Cell Sci. 108, 299-310.

Hörstensteiner, S et al (1992) Planta 187, 113-121.

Inoue, K et al (1995) Plant J. 7, 235-243.

Leigh, RA and Sanders, D (1997) Advances in Botanical Research: The Plant Vacuole, Academic Press, London, England.

Levanony, H et al (1992) J. Cell Biol. 119, 1117-1128.

Maeshima, M et al (1994) Plant Physiol. 106, 61-69.

Marty, F. (1978) PNAS 75, 852-856.

Marty, F et al (1980) In: The Biochemistry of Plants, a Comprehensive Treatise, Stumpf, PK and Conn, EE ed., Academic Press, London, England, 1, 625-658.

Marty, F (1997) In: Advances in Botanical Research, The Plant Vacuole, Leigh, RA and Sanders, D ed., Academic Press, London, England, 1-42.

Marty-Mazars, D et al (1995) Eur. J. Cell Biol. 66, 106-118.

Maurel, C (1997) Ann. Rev. Plant Physiol. Plant Mol. Biol. 48, 399-429.

Melroy, DL and Herman, EM (1991) Planta 184, 113-122.

Newell, JM et al (1998) J. Exp. Bot. 49, 817-827.

Paris, N et al (1996) Cell 85, 563-572.

Robinson, DG and Hinz, G (1997) Protoplasma 197, 1-25.

Staehelin, LA (1997) Plant J. 11, 1151-1165.

Strzalka, K et al (1995) Biochim. Biophys. Acta 1239, 103-110.

SIGNALS AND THEIR TRANSDUCTION IN EARLY PLANT EMBRYOGENESIS.

Sacco de Vries, Agricultural University Wageningen, Department of Molecular Biology, Dreijenlaan 3, 6703 HA Wageningen, the Netherlands.

Abstract.
Several observations suggest that the development of somatic embryos from suspension cells in vitro depends on signals that derive from other suspension cells. If there is a correspondence between somatic and zygotic embryos, then the signals found in vitro must also be acting on zygotic embryos. Two examples of such signalling systems will be described. The first stems from the observation that carrot EP3 class IV endochitinases can rescue somatic embryos of the temperature-sensitive cell line ts11. Employing whole mount in situ hybridization it was found that a subset of the cells in embryogenic and non-embryogenic suspension cultures, including ts11, express *EP3* genes. No expression was found in somatic embryos. In carrot plants *EP3* genes are expressed in the inner integument cells of young fruits and in a specific subset of cells, located in the middle of the endosperm of mature seeds. No expression was found in zygotic embryos. These results suggest that the EP3 endochitinase has a "nursing" function during zygotic embryogenesis, and that this function can be mimicked by suspension cells during somatic embryogenesis. Signals aimed at the embryo must also be perceived, and as the second example of a signalling pathway involved in embryogenesis, the leucin-rich repeat containing Somatic Embryogenesis Receptor Kinase (SERK) will be discussed. During somatic embryogenesis, SERK expression is detected first in single cells and disappears at the early globular stage. During zygotic embryogenesis, SERK expression was detectable transiently in young zygotic embryos of up to 100 cells. These results demonstrate that competent cell formation and early somatic embryogenesis require a highly specific signal transduction chain also found during zygotic embryogenesis. Whether both examples of signal transduction chains are related is not known.

Introduction

Carrot cell cultures secrete many different proteins into the medium, a process that contributes to the conditioning of the medium. Different cell types contribute to the total pattern of proteins secreted into the culture medium (Van Engelen and De Vries 1993). Conditioned media are reported to have a promoting effect on the initiation of somatic embryogenesis (Hari 1980; Smith and Sung 1985). A causal relationship between secreted proteins and embryogenic potential exists thata has led to the identification of the extracellular protein 3 (EP3), identified as a chitinase. EP3 was originally purified as a protein capable of rescuing somatic embryos in the mutant carrot cell line ts11 at the non permissive temperature (De Jong *et al.* 1992). The acidic endochitinase EP3 was found to be a member of a small family of class IV chitinase

A. Altman et al. (eds.), Plant Biotechnology and In Vitro Biology in the 21st Century, 373–382.

genes (Kragh *et al.* 1996). Those highly homologous isoenzymes are encoded by at least 4 *EP3* genes. Two of these proteins, EP3-1 and EP3-3, were purified and shown to have subtly different effects on the formation of somatic embryos in newly initiated ts11 embryo cultures (Kragh et al 1996).

Since the effect of the chitinases was mimicked by *Rhizobium*-produced Nod factors (De Jong *et al.* 1993), it was proposed that the chitinases are involved in the generation of signal molecules essential for embryogenesis in ts11 (De Jong *et al.* 1993). The roots of leguminous plants are known to produce chitinases and during the interaction with *Rhizobium* these plant produced chitinases have been suggested to control the biological activity of Nod factors by cleaving and inactivating them. In this way, chitinases are proposed to have the potential to control plant morphogenesis and cell division (Staehelin *et al.* 1994).

In carrot, the formation of embryogenic cell cultures usually commences with the incubation of seed-derived seedling hypocotyl explants in auxin-containing medium (De Vries et al. 1988). The origin of embryogenic cells, that are usually present as clusters of small cytoplasm-rich cells (Komamine et al. 1990) is not clear and is thought to involve an auxin dependent transition stage occurring in single cells. It is generally assumed that the formation of plant embryos requires the activation of specific sets of genes (reviewed by Goldberg et al. 1994; Thomas et al. 1993) and many studies have employed differential screening techniques to identify such genes. Often, genes found to be expressed in early somatic embryos appeared to encode genes normally expressed late in zygotic embryogenesis or throughout plant development (reviewed by Zimmerman 1993). Others, such as *LTP* (Sterk et al. 1991) and *EMB-1* (Wurtele et al. 1993) are expressed at the corresponding, globular, stage in zygotic embryogenesis. Several genes have been reported (reviewed by Zimmerman 1993) that are putative markers for embryogenic cell clusters, but none have been described to date that are reliable markers for the preceding stage of competent cells. Several screens were carried out employing a series of carrot cell cultures with widely differing numbers of single competent cells as the starting material. One of the genes found encoded a receptor-like kinase that appears to mark competent and embryogenic cells and is also expressed in early somatic and zygotic embryos.

Results

Cell specific expression of the EP3 genes and localization of the encoded proteins in suspension cultures

To identify the suspension cells that express the *EP3* genes, whole mount in situ mRNA localization was employed on entire, immobilized suspension cultures. *EP3* mRNAs could only be detected in between 4-6% of the total number of cells in an embryogenic culture. The number of embryo-forming cells in a comparable culture does however not exceed 1 %,

(Toonen et al. 1994), suggesting there is no quantitative relation with *EP3* expressing cells. The highest concentration of *EP3* mRNAs, in both embryogenic as well as in non-embryogenic cultures, was found in single cells that were elongated and often strongly curved, coiled, elongated or rounded. Whole mount in situs on globular, heart or torpedo shaped somatic embryos never revealed *EP3* mRNAs. In order to determine when the first *EP3* expressing cells appeared during embryogenic cell formation, hypocotyl explants were treated with 2,4-D for a period of ten days, during which embryogenic cell formation occurs (Guzzo *et al.* 1994). In this system, cell division was reinitiated in cells of the vascular tissue, and generated a mass of rapidly proliferating cells. Only after ten days a very small number of cells at the periphery of the proliferating mass was found to contain *EP3* mRNA (Figure 1). This is three days later as the appearance of the first cells that are competent to form somatic embryos in this system (Schmidt *et al.* 1997). We conclude therefore that *EP3* gene expression is not directly correlated with embryogenic cell formation, nor with somatic embryo development.

In plants, *EP3* mRNAs could be found in developing and mature seeds. In early stages of seed development, approximately 5-6 DAP, *EP3* mRNA was detected in the inner integument cells, lining the surface of the embryo sac in which the zygote or the early embryo is located (Fig 2A). Up to at least 20 DAP, no expression of the EP3 genes was detected in the developing endosperm. In the endosperm of mature seeds, *EP3* mRNA was restricted to a narrow zone of endosperm cells starting at the cavity in which the embryo is located up to almost the opposite end of the endosperm (Fig 1B). Because EP3 chitinases are secreted proteins, we localized these enzymes during the development of seeds and in germinating seeds by tissue printing followed by EP3 antibody staining. The EP3 protein was uniformly spread in the integuments at 10 DAP but also in the developing endosperm at 20 DAP. This points to transport of the EP3 proteins from the maternal integument tissues towards the endosperm. However, we cannot completely rule out that a very low level of *EP3* gene expression that was below detection limits, occured in the peripheral endosperm cells. In mature seeds the protein was restricted to the inner cell layer of the cavity that surrounds the embryo and to a zone of cells that starts at this cavity and ends almost at the opposite side of the endosperm, which corresponds precisely with the expression pattern of the *EP3* genes as determined by in situ hybridization.

To summarize, the *EP3* genes are expressed in the maternal integument cells. The EP3 chitinases however, were detected in the developing endosperm, at a time that *EP3* chitinase gene products cannot be detected in the endosperm and as a consequence they must be largely of maternal origin. In mature seeds the integuments are completely degraded and then the EP3 proteins are produced in a subset of the endosperm cells.

Expression of the SERK gene during somatic and zygotic embryogenesis in carrot

Labeled probes for differential screening were obtained from RNA out of a <30 μm sieved subpopulation of cells from either embryogenic or non-embryogenic cell cultures. Employing these probes in a library screen of approximately 2000 plaques yielded 26 plaques that failed to

show any hybridization to either probe. These so-called cold plaques were purified and used in spot-dot Northern analysis. One clone showed low expression in all embryogenic cultures and in one non-embryogenic culture, but not in the others. The predicted amino-acid sequence shows homology with the structural features of plant and animal receptor protein kinases, and therefore the gene was named Somatic Embryogenesis Receptor Kinase (*SERK*). The SERK protein contains an N-terminal domain with five leucine-rich repeats (LRRs). Between the extracellular LRR domain of SERK and the membrane-spanning region there is a 32 amino acid region with 13 prolines, partly arranged in the sequence SPPPP, that is conserved in extensins, a class of universal plant cell wall proteins (Varner and Lin, 1989). The significance of this proline-rich box is not clear, it might act as a hinge region by providing flexibility to the extracellular part of the receptor, or act as a region for interaction with the cell wall. In extensins, usually all prolines in the SPPPP repeat are hydroxylated and are considered to be used as target for O-linked glycosylation. The proposed intracellular domain of the protein contains the 11 subdomains characteristic of the catalytic core of protein kinases, while sequence homology suggests a function as a serine / threonine kinase (Hanks et al. 1988). A bacterially expressed SERK fusion protein is indeed able to autophosphorylate on serines and threonines, confirming the predictions (Schmidt et al. 1997).
To determine directly whether *SERK* expressing cells indeed develop into somatic embryos, transformed carrot suspension cultures containing a *SERK* promoter-luciferase construct were analysed for luciferase expression in cell cultures sieved through a 50 μm mesh to enrich for single cells and small cell clusters. Development of the immobilized cells after recording the luciferase images was determined using automated cell tracking (Toonen et al. 1994). The origin of nine torpedo stage somatic embryos was determined this way. Of these, three developed from a single cell that showed luciferase activity at day 1, four developed from cell clusters consisting of 2-6 luciferase expressing cells while two embryos developed from single cells that failed to show a detectable level of luciferase activity at day 1. The somatic embryo shown in Figure 3 originated from a luciferase-expressing two-celled cluster. These results demonstrate that somatic embryos develop from single cells and small cell clusters expressing SERK. Somatic embryos up to the globular stage show luciferase expression, confirming the transient SERK gene expression pattern obtained with in situ mRNA hybridization (results not shown).

Discussion

We have shown that only a subset of cells present in an embryogenic carrot culture produce EP3 carrot chitinases. On the basis of the number, cell type and presence of *EP3* mRNA producing cells in embryogenic and non-embryogenic cultures, there was no correlation with the ability to produce somatic embryos and the presence of *EP3* mRNA. No *EP3* gene expression in somatic embryos was found. Because cells that produce EP3 do not develop into embryos themselves, it appears that EP3 chitinases, or products of their enzymatic activity, diffuse via the conditioned medium to cells that are able to help somatic embryos develop and in this way play a "nursing" role in the process of somatic embryogenesis.

In plants, *EP3* mRNA was found in the inner integuments, while the EP3 proteins are found in the endosperm. Later, cells in the centre of the endosperm do express the *EP3* genes and this is likely to be responsible for the presence of EP3 chitinases during imbition and germination. No *EP3* gene expression was found in zygotic embryos. It could therefore be expected that the chitinases play a similar “nursing” role in zygotic embryogenesis.

The EP3 chitinase proteins found in the endosperm of 20 DAP seeds are most likely produced by integument cells. This indicates a maternal contribution to the proteins that are present in the endosperm. Recent evidence for a role of maternal tissues in endosperm formation comes from the analysis of gametophytic mutations in Arabidopsis. In this species, *fie* (fertilization-independent endosperm) mutations are known that are female gametophytic and specifically affect endosperm formation (Ohad *et al.* 1996).

Regarding the biological function of the EP3 chitinases in plant embryogenesis, we propose that they are involved in reinitiating cell division in embryogenic cells and embryos as part of a “nursing cell” system, that is required for embryogenesis. This hypothesis is based on the following observations: 1.) The EP3 chitinases promote embryogenic cell formation and the number of somatic embryos as well as their progression in development when added to *ts11* cultures (De Jong *et al.* 1992). 2.) The expression of *EP3* genes in cells that do not develop into embryos in culture and an absence of expression in somatic embryos. 3.) The expression of *EP3* genes in maternal tissue and subsequent secretion of the encoded chitinase proteins, resulting in the presence of EP3 in the extracellular matrix of the endosperm surrounding globular-stage zygotic embryos. 4.) The absence of expression in zygotic embryos and expression in endosperm cells prior to and during germination. The role of EP3 could be direct and involve a structural property such as the chitin-binding domain of the protein. It appears more likely to envisage a function involving the catalytic properties of the enzyme through the release or modification of an N-acetylglucosamine containing signal molecule. Recently we obtained evidence that a particular class of soluble arabinogalactan proteins (AGPs) contains N-acetylglucosamine, and can be cleaved by endochitinases. We believe that this result will help to further unravel the role of chitin-based signalling molecules in plant embryogenesis.

The process of cellular reactivation and the subsequent formation of embryogenic cells in carrot explants has been described by Guzzo et al. (1994). That work showed that a particular elongated cell type appeared in culture, derived from small rapidly proliferating cytoplasmic cells that themselves derived from reactivated provascular cells. It was further shown cytologically, that some of the elongated cells underwent an asymmetrical division. After continued culture, small clusters of dividing cytoplasmic cells appeared that resemble the proembryogenic masses seen in established embryogenic suspension cultures (Guzzo et al. 1994, De Vries et al. 1988). To determine which single cells were competent to form embryos, markers are needed that distinguish precisely between competent and non-competent cells. The expression of the *SERK* gene described here was obtained by following the development of living SERK-expressing cells as visualized by luciferase expression under control of the SERK promoter. The predicted SERK protein sequence resembles a leucine-rich repeat (LRR) receptor kinase protein, a class of plant proteins that was originally described by Chang et al. (1992). While the avarage primary cell wall has a thickness of approximately 50 nm (Pruitt et al. 1993) and the maximum size of the entire extracellular domain can only be about 15 nm when present as an α helix, the extracellular ligand binding domain is likely to be completely embedded within the cell wall. The most likely type of ligand for SERK will therefore consist of a cell wall-diffusible small molecule such as a (glyco)peptide. Peptides effective in inducing plant responses, such as systemin (McGurl et al. 1992) and ENOD40 (Van de Sande et al. 1996) have been described.

Thus, it appears that while LRR containing receptor-like protein kinases play several roles in plant development, intercellular peptides are now being uncovered that are likely signal molecules that can activate developmental processes mediated through such receptors.

In the plant embryo sac and in the activated explant a situation may exist whereby an unknown inducer is present uniformly, while embryo formation awaits the presence of the *SERK* protein. Such a model may fit with the restricted expression pattern found for the *SERK* gene both *in vivo* and *in vitro*. It is also in line with the hypothesis that in plants inductive interactions mediated by diffusable signal molecules are an important regulatory mechanism (reviewed in Schmidt et al. 1994). The *SERK* gene described here may represent a significant part of a mechanism that is essential for the formation of plant cells destined to become embryo. Whether molecules resulting from the activity of the EP3 chitinase are possible ligands for SERK remains to be investigated.

Acknowledgements

Research in our group is supported by The Netherlands Organization for Scientific Research (A.v.H. and E.S.), and by the European Commision Biotechnology program PTP-Biotech (V.H.). I thank Arjon van Hengel, Ed Schmidt and Valérie Hecht for their help in preparing this Abstract.

Literature cited

Chang, C., Schaller, G.E., Patterson, S.E., Kwok, S.F., Meyerowitz, E.M. and Bleecker, A.B. (1992) The TMK1 gene from Arabidopsis codes for a protein with structural and biochemical characteristics of a receptor protein kinase. Cell 4, 1263-1271.

De Jong AJ, Cordewener J, Lo Schiavo F, Terzi M, Vandekerckhove J, Van Kammen A, De Vries SC (1992) A carrot somatic embryo mutant is rescued by chitinase. Plant Cell 4: 425-433

De Jong AJ, Heidstra R, Spaink HP, Hartog MV, Meijer EA, Hendriks T, Lo Schiavo F, Terzi M, Bisseling T, Van Kammen A, De Vries SC (1993) Rhizobium lipooligosaccharides rescue a carrot somatic embryo mutant. Plant Cell 5: 615-620

De Vries, S. C., Booij, H., Meyerink, P., Huisman, G., Wilde, H. D., Thomas, T. L. and Van Kammen, A. (1988a) Acquisition of embryogenic potential in carrot cell-suspension cultures. Planta 176, 196-204.

Goldberg, R. B., de Paiva, G. and Yadegari, R. (1994) Plant embryogenesis: zygote to seed. Science 266, 605-614.

Guzzo F, Baldan B, Mariani P, LoSchiavo F, Terzi M (1994) Studies on the origin of totipotent cells in explants of Daucus carota L. J. Exper. Botany 45: 1427-1432

Hanks, S. K., Quinn, A. M. and Hunter, T. (1988) The protein kinase family: conserved features and deduced phylogeny of the catalytic domains. Science 241, 42-52.

Hari V (1980) Effect of cell density changes and conditioned media on carrot cell embryogenesis. Z. Pflanzenphysiol. 96: 227-231

Komamine, A., Matsumoto, M., Tsukahara, M., Fujiwara, A., Kawahara, R., Ito, M., Nomura, K. and Fujimura, T. (1990) Mechanisms of somatic embryogenesis in cell cultures-physiology, biochemistry and molecular biology. In Progress in plant cellular and molecular biology (ed. H. J. J. Nijkamp, L. H. W. Van Der Plas, A. Van Aartrijk), pp. 307- 313. Kluwer Academic Publishers, Dordrecht.

Kragh KM, De Jong AJ, Hendriks T, Bucherna N, Højrup P, Mikkelsen JD, De Vries SC (1996) Characterization of chitinases able to rescue somatic embryos of the temperature-sensitive carrot variant ts11. Plant Molecular Biology 31: 631-645

McGurl, B., Pearce, G., Orozco-Cardenas, M. and Ryan, C.A. (1992) Structure, expression, and antisense inhibition of the systemin precursor gene. Science 257, 1570-1573.

Ohad N, Margossian L, Hsu Y, Williams C, Repetti P, Fischer RL (1996) A mutation that allows endosperm development without fertilization. Proc. Natl. Acad. Sci. 93: 5319-5324

Pruitt, R. E., Hulskamp, M., Kopczak, S. D., Ploense, S. E. and Schneitz, K. (1993) Molecular genetics of cell interactions in Arabidopsis.. Develop. 1993 suppl., pp. 77-84.

Schmidt EDL, Guzzo F, Toonen MAJ, De Vries SC (1997) A leucine-rich repeat containing receptor-like kinase marks somatic plant cells competent to form embryos. Development 124: 2049-2062

Smith JA, Sung ZR (1985) Increase in regeneration of plant cells by cross feeding with regenerating Daucus carota cells. In M Terzi, L Pitto, ZR Sung eds, Somatic Embryogenesis. IPRA, Rome, pp 133-137

Staehelin C, Schultze M, Kondorosi E, Mellor RB, Boller T, Kondorosi A (1994) Structural modifications in Rhizobium meliloti Nod factors influence their stability against hydrolysis by root chitinases. Plant Journal 5: 319-330

Sterk, P., Booij, H., Schellekens, G. A., Van Kammen, A. and De Vries, S. C. (1991) Cell-specific expression of the carrot EP2 lipid transfer protein gene. Plant Cell 3, 907-921.

Thomas, T. L. (1993) Gene expression during plant embryogenesis and germination: an overview. Plant Cell 5, 1401-1410.

Toonen, M. A. J., Hendriks, T., Schmidt, E. D. L., Verhoeven, H. A., Van Kammen, A and De Vries, S. C. (1994) Description of somatic-embryo-forming single cells in carrot suspension cultures employing video cell tracking. Planta 194, 565-572.

Van de Sande, K., Pawlowski, K., Czaja, I., Wieneke, K., Schell, J., Schmidt, J., Walden, R., Matvienko, M., Wellink, J., Van Kammen, A., Franssen, H. and Bisseling, T. (1996) Modification of phytohormone responseby a peptide encoded by ENOD40 of legumes and a nonlegume. Science 273,370-373.

Van Engelen FA, De Vries SC (1993) Secreted proteins in plant cell cultures. In KA Roubelakis-Angelakis, K Tran Thanh Van, eds, Markers of Plant Morphogenesis. Plenum Press, New York, pp 181-200

Varner, J. E. and Lin, L.-S. (1989) Plant cell wall architecture. Cell 56, 231-239.

Wilde, H.T., Nelson, W.S., Booij H., De Vries, S.C. and Thomas, T.L. (1988) Gene expression programs in embryogenic and non-embryogenic carrot cultures. Planta 176, 205-213.

Wurtele, E.S., Wang, H., Durgerian, S., Nikolau, B.J. and Ulrich, T.H. (1993) Characterization of a gene that is expressed early in somatic embryogenesis of Daucus carota. Plant Physiol. 102, 303-312.

Zimmerman, J. L. (1993) Somatic embryogenesis: a model for early development in higher plants. Plant Cell 5, 1411-1423.

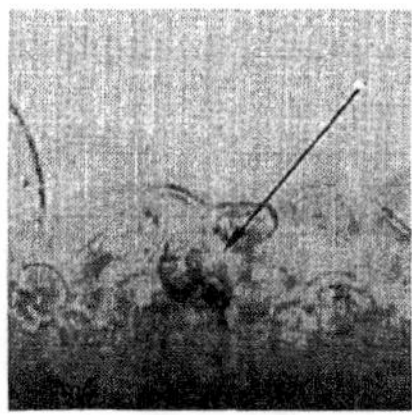

Figure 1. *EP3* gene expression in suspension cultures.
Plant material was analysed by whole mount in situ hybridization with an antisense *EP3* RNA probes. Light microscopy, coupled to Nomarski optics was used for visualization of the precipitate of the enzymatic detection using alkaline phosphatase. Cells present at the periphery of the proliferated cell mass on a section of a hypocotyl explant that was cultured in the presence of 2,4-D for 10 days are shown. The arrow points to a single *EP3* expressing cell.

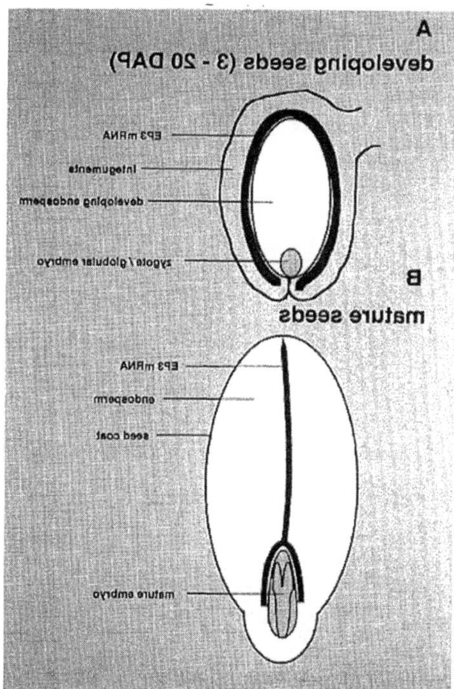

Figure 2. Summary of *EP3* gene expression in seeds.
I, integuments surrounding the developing embryo and endosperm; SC, seed coat; E, endosperm; EM, embryo. Bar: 50 µm.
A. Longitudinal section of a fruit, 7 DAP.
B. Longitudinal section of a mature seed.

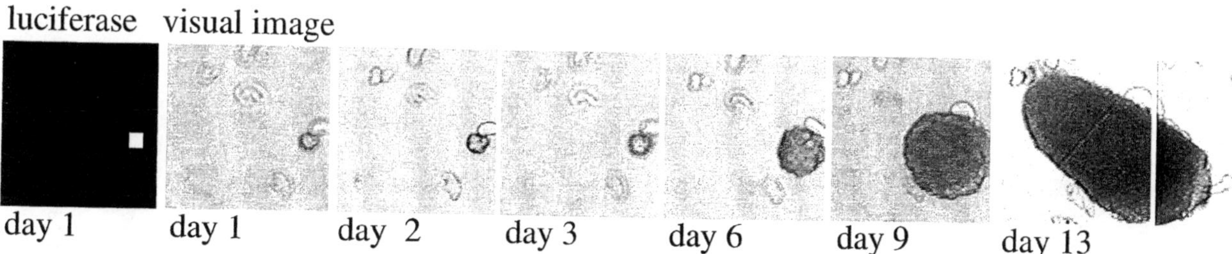

Figure 3. Luciferase expression under control of the *SERK* promoter. Luciferace activity of immobilised cells was recorded at day 1 with a CCD camera (left image). The single pixel in the left image measures 30 by 30 μm. Video cell-tracking of the cells was performed for a period of 13 days (light microscopic images). The light microscope images were sized to match the CCD image.

ATTAINMENT OF BILATERAL SYMMETRY IN MONOCOTS : INFLUENCE OF AUXIN POLAR TRANSPORT

Christiane Fischer and Gunther Neuhaus
University of Freiburg, Institute of Biology II, Cell Biology,
Schänzlestrasse 1, D-79104 Freiburg, Germany

INTRODUCTION

In recent decades, much attention has been devoted to study the mechanisms that enable plants to specify their overall body organization and tissue differentiation patterns. Genetic approaches using Arabidopsis have begun to reveal genes required for pattern formation, cell differentiation, and organ development (Mayer et al. 1991; Meinke, 1991; Goldberg et al. 1994). Some of these genes have been cloned and characterized (Shevell et al. 1994; Long et al. 1996; Lukowitz et al. 1996; Aida et al. 1997; Hardtke and Berleth, 1998). In monocots, many maize mutants that are blocked at different stages of embryogenesis have been identified (Clark and Sheridan, 1991). Most of the embryogenesis-related genes seem to be involved in basic cellular processes rather than in the regulation of specific developmental processes such as the networks of genes identified in *Drosophila* (Lukowitz et al.1996; Shevell et al.1994; Hardtke and Berleth, 1998).

A promising approach to achieve new insight into embryonic morphogenesis is the identification of signals and their molecular targets that direct pattern formation. The plant hormone auxin (indole-3-acetic acid) plays an important role in a wide variety of growth, developmental and physiological processes. Pharmacological studies on dicot somatic and zygotic embryos have revealed that auxin seems to be involved in early embryo development (Liu et al.1993; Schiavone and Cooke, 1987; Hadfi et al.1998). Further evidence for the determining role of auxin during early embryo development has been obtained from the analysis of the Arabidopsis *MONOPTEROS* mutant (*MP*). This mutant lacks the basal and central regions (root and hypocotyl) of the embryo body plan (Berleth and Jürgens, 1993). Embryos and leaves of adult plants of *MP* exhibit discontinuity in the vascular network (Berleth and Jürgens, 1993 ; Przemeck et al, 1996). Therefore the *MP* gene seems to have an early function in the establishment of vascular and body patterns in embryonic and postembryonic development. *MP* has been cloned recently and shown to encode a transcription factor likely to bind to auxin response elements (Hardtke and Berleth, 1998).

A. Altman et al. (eds.), Plant Biotechnology and In Vitro Biology in the 21st Century, 383–388.

Our aim is to evaluate the influence of auxin on the dramatic transformation of the embryo morphology that occur during the change from the globular to the transition stage of *Poaceae* embryos. This shift from radial to bilateral symmetry represents the initial delineation of two major embryonic structures: the embryonic axis (which contains the shoot and the root apical meristems) and the scutellum. Therefore this change is a prerequisite for the basic embryonic body formation. The approach chosen to address this question has been to manipulate *in vitro* the fate of zygotic embryos excised out of the kernel at different stages. Our aim is to gain new insights into embryo morphogenesis by perturbing the embryonic development with auxin and related substances. In particular, we have added exogenous auxins, auxin antagonists and auxin tranport inhibitors to the embryo culture medium in order to modulate the endogenous level, effect and distribution of auxin. *In vitro* treatments of wheat (*Triticum aestivum* L.) embryos lead to a wide range of morphogenetic alterations indicating that auxin plays a key role on embryonic pattern formation (Fischer and Neuhaus, 1996; Fischer *et al.* 1997).

RESULTS

Normal development of wheat embryos in vitro.

The globular proembryo is hemispherical in shape. The shift to bilateral symmetry starts with the growing of the scutellum in both the axial and the lateral directions, which results in a prominent structure on one side of the embryo proper. Almost at the same time, the swelling on the other side of the embryo proper indicates the initiation of the shoot apical meristem. The mature embryo has two primary organ structures: the embryonic axis, which contains the shoot and root meristems, and a scutellum (Figure 1, A). At this stage, the shoot apical meristem is entirely covered by the first leaf primordium, which is itself enclosed by the coleoptile. The coleoptilar ring primordium originates from the scutellum. The scutellum develops into an oval shield-like structure that is attached to one side of the embryonic axis (Figures 1, A).

Effect of exogenous auxins, antiauxins and auxin polar transport inhibitors on the in vitro development of wheat embryos.

Exposure of wheat globular embryos to different auxins, 2,4-D (2,4-dichlorophenoxyacetic acid), 2,4,5-T (2,4,5-trichlorophenoxyacetic acid), and IAA (indole-3-acetic acid) blocked the attainment of bilateral symmetry from radial symmetry resulting in a continuous uniform growth (Fischer and Neuhaus, 1996; Figure 1, B). Semi-thin sections of these embryos confirmed that they did not differentiate any visible protoderm, procambium, or meristems. The effect of 2,4,5-T and 2,4-D depended on their concentration and on the developmental stage of the isolated embryo. At 2 μM concentration of 2,4,5-T, none of embryos showed this radial growth phenotype whereas at 20 μM, 70 % of the cultured embryos developed this particular phenotype. All attempts to rescue these phenotypes were unsuccessful.

In contrast, PCIB (2-(p-chlorophenoxy)-2-methylpropionic acid), which is described as an auxin antagonist, has no visible direct effect on the wheat embryonic symmetry (Fischer and Neuhaus, 1996). Depending on the dose, PCIB affects cell division and cell shape leading in severe cases to an arrest of development. Therefore, we assume that auxin is involved in essential cellular processes during embryo development. No difference in the embryo response to PCIB due to the developmental stage of the isolated embryo has been observed.

Zygotic globular wheat embryos treated with TIBA (2,3,5-triiodobenzoic acid), an auxin polar transport inhibitor, developed an abnormal overall embryo symmetry (Fischer and Neuhaus, 1996; Figure 1, C). All embryos differentiated a shoot apical meristem as well as a scutellum, but the quality of their differentiation was altered compared with normal embryos. No root meristem was differentiated The position of the shoot apical meristem relative to the scutellum was anomalous. The shoot meristem, a broad abnormal structure, was shifted to the upper part of the embryo and not located as normal in the basal region near the suspensor. The scutellum did not develop laterally and axially but instead developed into a small cap-like structure above the meristem. A large area of cell elongation and cell division was observed in between the suspensor and the shoot apical meristem. The effect of TIBA depended on its concentration and on the developmental stage of the isolated embryos. At 1 μM of TIBA, 25 % of the cultured embryos had this phenotype, whereas at 10 μM of TIBA, 60 % of the embryos developed the phenotype. However, at 20 μM of TIBA, the compound had a toxic effect leading to a high rate of undifferentiated structures.

Although NPA (*N*-1-naphthylphthalamic acid) and quercetin (3,3',4',5,7-pentahydroxyflavone) belong to two different classes of auxin transport inhibitors, the phytotropins and the flavonoids, respectively, they induced the same specific abnormal phenotypes during embryo development (Fischer *et al.*1997; Figure 1, D and E). These abnormal embryos differentiated multiple meristems (i.e. multiple shoot and root meristems) and multiple organs (i.e. multiple coleoptiles and scutella). "Heart shaped Siamese embryos" always showed duplicated scutella, which had the same orientation, and were usually laterally attached (Figure1, D). A coleoptilar ring was differentiated from each of these scutella (Figure1, D). Each of these coleoptiles enclosed one to several shoot meristems. In "back to back Siamese embryos" up to five shoot meristems emerged in a ring around the initial embryo proper. When only two shoot meristems were differentiated, the orientation of the ectopic second shoot meristem was opposite to that of the first meristem so that their axes formed an angle of 180° (Figure 1, E). The additional ectopic shoot meristem(s) arose from a place where a scutellum would be expected. The scutella were also oriented oppositely (abaxial face against abaxial face) and were fused in their proximal part, leaving their distal tips free (Figure 1, E). The occurrence of multiple shoot meristem phenotypes depended on the concentration of the inhibitor and on the developmental stage of the isolated embryo.

Figure 1: Representative abnormal embryos induced by auxins and auxin polar transport inhibitors.

(A) Embryo isolated at the globular stage and cultured 5 days on medium free of auxin and auxin polar transport inhibitor. (B) Globular embryo cultured 13 days in the presence of 20 μM 2,4,5-T. (C) Globular embryo cultured 7 days in the presence of 10 μM TIBA. (D) Heart-shaped Siamese embryo induced in the presence of 1μM NPA after 6 days of culture. (E) Back-to-back Siamese embryo induced in the presence of 2 μM NPA after 12 days of culture.

cr: coleoptile or coleoptilar ring; ep: embryo-proper; lp: first leaf primordium; sc: scutellum; sm: shoot meristem.

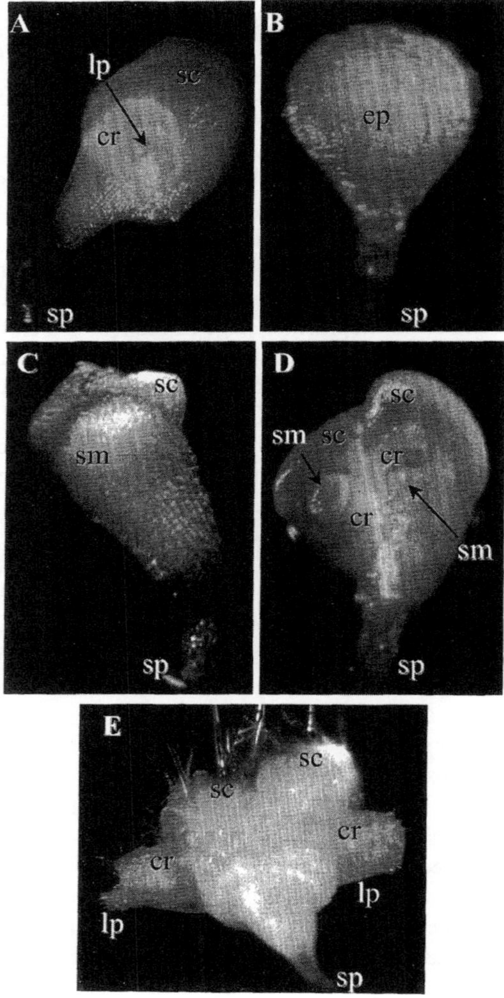

Approximately 35 % of the globular embryos cultured on a medium with 1 µM NPA differentiated multiple shoot meristems, whereas on a medium with 20 µM NPA, 85 % of the embryos developed multiple shoot meristems. The majority of embryos that had multiple shoot meristems differentiated only one primary embryonic root. However, multiple primary roots, especially duplicated primary roots, were also observed. The complete duplication of the axis as well as the scutellum can be defined as polyembryony. Multiple primary roots in combination with a normal shoot apical meristem and a normal scutellum were only differentiated by embryos isolated at the late globular-to-early transition stage and by embryos that were already bilaterally symmetrical at the onset of the treatment.

CONCLUSIONS

Our data support the hypothesis that auxin polar transport and auxin mediated processes play an important role in the differentiation of the embryonic axis and the scutellum. Based on this pharmacological study, a hypothetical model localizing auxin transport pathways and the site of auxin synthesis in the late globular wheat embryo is proposed (Fischer and Neuhaus, 1996; Fischer *et al.* 1997; Figure 2)

Figure 2 : Model localizing auxin transport pathways and the site of auxin synthesis in the globular embryo

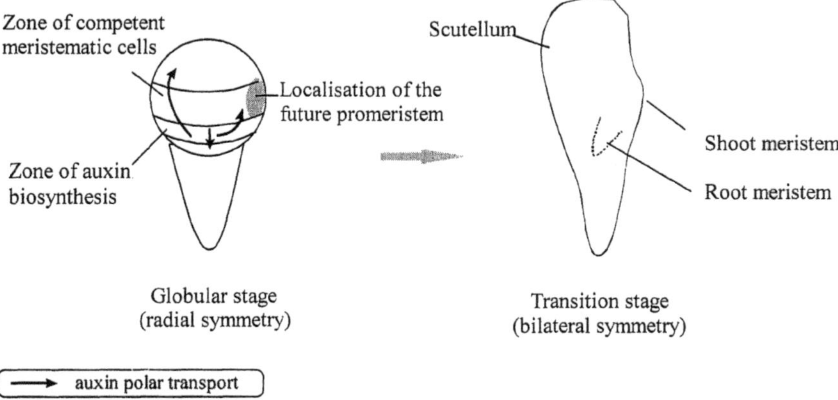

We suggest that the non-homogenous distribution of auxin within the embryo proper at the late globular and early transition stage might play a major role in the establishment of embryonic symmetry (Fischer and Neuhaus, 1996; Fischer et al. 1997). We assumed that auxin is synthesized in the lower part of the embryo proper and transported polarly in three main directions: along the longitudinal axis towards the area where the scutellum will differentiate, laterally towards the area where the promeristem will differentiate, and downwards towards the area where the embryonic root will differentiate.

Presently, we are evaluating the proposed model at the cellular and molecular level. Photoaffinity labelling and *in situ* hybridization experiments are currently being performed to determine the distribution of auxin in treated and untreated wheat embryos.

ACKNOWLEDGEMENTS

We thank Professor I. Potrykus (ETH-Zürich) in whose laboratory these experiments have been started. During this period of time, Ch. Fischer was supported by grant no. ALTF 287-1993 from the European Molecular Biology Organization (EMBO).

REFERENCES

Aida M et al. (1997) Plant Cell, 9, 841-857.
Berleth T, Jürgens G (1993) Development, 118, 575-587.
Clark JK, Sheridan WF (1991) Plant Cell, 3, 935-951.
Fischer C, Neuhaus G (1996) Plant Journal, 9, 659-669.
Fischer C et al. (1997) Plant Cell, 9, 1767-1780.
Goldberg RB et al. (1994) Science, 266, 605-614.
Hadfi K et al. (1998) Development, 125, 879-887.
Hardtke CS, Berleth T (1998) EMBO, 17, 1405-1411.
Liu C-M et al. (1993) Plant Cell, 5, 621-630.
Long JA et al. (1996) Nature, 379, 66-69.
Lukowitz W et al. (1996) Cell, 84, 61-71.
Mayer U et al. (1991) Nature, 353, 402-407.
Meinke DW (1991) Developmental Genetics, 12, 382-392.
Przemeck GK et al. (1996) Planta, 200, 229-237.
Shiavone FM, Cooke TJ (1987) Differentiation, 21, 53-62.
Shevell DE et al. (1994) Cell, 77, 1051-1062.

EXPRESSION OF CELL DIVISION CYCLE RELATED GENES IN *Beta vulgaris* L. STORAGE ORGANS AND CELLS IN SUSPENSION CULTURES

M. R. Fowler, M. J. Kirby, A. I. Atanassova, N. W. Scott,
A. Slater, and M. C. Elliott
The Norman Borlaug Institute for Plant Science Research,
De Montfort University, Scraptoft, Leicester LE7 9SU, UK

1. Introduction

The sucrose storage capacity of the sugar beet, *Beta vulgaris*, storage organ has been related to its anatomy. A transverse section of the storage organ reveals a pattern of concentric rings which is caused by the alternation of vascular zones with parenchymatous zones. Sucrose is transported to the storage organ via the phloem and moves laterally to the adjacent parenchymatous tissues, where it is stored in the cell vacuoles. The storage organ of sugar beet has 12-15 cambia, the divisions of the 6 inner rings provide the cells which make up some 75% of the storage root while the outer rings (9 and above) make almost no contribution to the bulk of the storage root. It has been found that the small parenchymatous cells close to the phloem accumulate sucrose to a higher concentration than that achieved by large (older) cells remote from the phloem. Hence, it is predicted that for optimum sucrose accumulation the storage organ should have a large number of parenchymatous zones, each of which offers a short diffusion path from the phloem to predominately small storage cells. In order to manipulate sugar beet development toward this optimum it is necessary to have an understanding of the processes that regulate cell division, cell expansion and differentiation. Towards this end we have begun to isolate and characterise cell division cycle related genes from sugar beet and we have developed a cell suspension culture system for analysis of their roles (Elliott et al. 1996).

2. Results and Discussion

The PCR has been used to amplify three *cdc2*-like protein kinase sequences and a cyclin sequence have been amplified from sugar beet mRNA or genomic DNA (Fowler et al. 1998b). The cdc2-like protein kinase genes have been designated *Bvcrk1*, *Bvcrk2* and *Bvcdc2*. They were originally obtained by using degenerate primers to the core region of *cdc2*-like kinases. *Bvcdc2* shows greater than 90% sequence identity with other plant *cdc2* sequences and has a perfectly conserved PSTAIRE region. The other two sequences have divergent PSTAIRE regions. *Bvcrk2* shows sequence homology to an *Arabidopsis* MHK gene and belongs to the MAK gene subgroup (Matsushime et al. 1990). The *Bvcrk1* gene codes for a novel protein kinase which shows structural similarities to members of the

A. Altman et al. (eds.), Plant Biotechnology and In Vitro Biology in the 21st Century, 389–392.

PITSLRE protein kinase group that are associated with apoptosis in animal cells. The gene has a consensus kinase core similar to that of *cdc2*-like genes, but in particular to the *CHED* gene (Lapidot-Lifson et al. 1992). Several genomic clones containing the complete coding region and the promoter region of the gene have been isolated. *Bvcrkl* has an extended 5' terminal region which has a glutamic acid rich domain. A cRACE (Maruyama et al. 1995) approach has been used to identify the start site of transcription. In this procedure single stranded cDNA copies of total mRNA were made and then circularised. Internal primers were then used to amplify across the ligated region. Several of the products have been sequenced and one 5' terminus has been regularly found. Analysis of the promoter region of the genomic clone indicates that this corresponds to a region very close to a consensus sequence for transcription initiation, a cap-site and a CAAT box. There is no TATA box in this region. However, further upstream is a second potential promoter site. This second site contains several potential binding sites for transcription factors. Motifs identified include a plant G-box, an auxin box, a site that corresponds to a mammalian Fos binding site and several consensus TATA boxes.

The cyclin obtained, Betvu;*CycA2*, is a partial sequence. It has been assigned to the A2 group of plant cyclins by classification on the basis of sequence similarity (Renaudin et al.1996).

The suspension culture system chosen for this study was developed in our laboratory to characterise the molecular and biochemical events involved in the transition from the G_0 to the S phase. The G_0 state has been extensively investigated in yeast and mammalian cells. G_0 cells, which have exited the normal cell division cycle, are generally considered to have a lower metabolic activity than cycling cells. However, G_0 cells can be present in organs exhibiting high metabolic activity, such as hepatocytes in the liver. Re-entry into the cell division cycle involves a defined pattern of gene expression. This is exemplified by serum deprived mammalian cells which can be induced to re-enter the cell division cycle by the addition of serum growth factors or mitogens. It has been shown that the G_0 to S phase transition is characterised by the expression of a cascade of genes (Fowler et al. 1998b). These have been categorised as immediate early or IE (including genes for transcription factors such as *c-fos*, *c-jun* and *c-myc*), delayed early or DE (CDK1, D-cyclins and ornithine decarboxylase) and late genes L (genes involved in DNA replication). It is not known whether plants exhibit this pattern of gene expression. In order to address this issue we have established several different *Beta vulgaris* cell suspension cultures as well characterised model systems. The cells are phytohormone autonomous, being routinely maintained in growth regulator-free medium, and are easily manipulated to enter into G_0 by nutrient depletion. This is usually done by extending the duration of the stationary phase of the batch culture cycle. The DNA of the quiescent cells remains intact but, in accordance with their G_0 status, there is no DNA replication. The total amount of protein synthesis in these cells is much lower than in dividing cells, protein half lives are reduced and protease activity is increased. It is however, possible to detect an increase in the synthesis of a small number of specific proteins in the G_0 cells. These quiescent cells remain competent for division and have been shown to express the *Beta vulgaris* homologue of *cdc2*, *Bvcdc2*, albeit at a low level as compared with dividing cells. On stimulation by subculture into fresh medium they re-enter the cell division cycle in a highly synchronised fashion. *Bvcdc2* expression is increased in this period. Division begins 6 days after subculture then cell numbers double within a 6 hour period. This synchrony has also been confirmed at a molecular level. Northern blot hybridisation analysis has been used to show that Histone H4 synthesis occurs specifically during the time period that corresponds to maximal DNA synthesis (i.e. the S phase) and the synthesis of sequences homologous to Arath;*CycB1*;1 a mitotic cyclin, occurs later than the histone and is maximal at the time of cell division (Fowler et al. 1998).

The expression patterns of *Bvcrk1* and *Bvcrk2* show interesting differences. *Bvcrk2* is expressed in G_0 cells and in dividing cells. However, there is no detectable expression of *Bvcrk1* in G_0 cells, but, on transfer of the cells to fresh medium the gene is expressed rapidly (within 30 minutes). This rapid induction is characteristic of so called IE genes in mammalian cells. Another important feature of the cell culture system is that it allows the early induction of gene expression to be uncoupled from cell division. This is facilitated by the hormone autonomous nature of the cultures and by the fact that the medium containing G_0 cells is nutrient depleted. G_0 cells will not proceed to a further round of cell division unless the nutrients are restored to the medium. This means that specific chemicals can be added to the medium containing G_0 cells and their ability to induce expression of specific genes can be assayed. Consequently, signals regulating gene expression can be identified. *Bvcrk1* expression in response to the addition of plant growth regulators and nutrients to medium containing G_0 cells was investigated by Northern blot hybridisation analysis. It was shown that *Bvcrk1* was induced by the addition of kinetin, IAA, sucrose and the polyamine putrescine and its precursor ornithine. A similar study has been carried out for the cyclin gene. Transcripts are not detectable in quiescent cells but, as with *Bvcrk1*, it is induced rapidly after subculture. In alfalfa the closest homologue of the sugar beet cyclin, Medsa;*CycA2*;1 is also induced very early during the G_0 to S phase transition (Meskiene et al. 1995) prior to the D-type cyclin Medsa;*CycD3*;1 (Dahl et al. 1995). The rapid induction of this sub-group of A-type cyclins is interesting as in mammalian cells D-type cyclins are normally considered to be the first cyclins to be induced after mitogenic stimulation. To add to the information from *cdc*-related genes and to understand the regulation of the G_0 to S transition in more detail we have begun to construct a map (Fig.1) of gene expression. This has been done using a number of techniques, in particular differential display and RAP-PCR (Fowler et al. 1998a), with a series of samples taken at various times early in the transition. We have obtained expression data for a number of the sequences and those showing the most interesting patterns have been characterised. In the main they have been shown to be the products of genes that code for signal transduction pathway proteins. We are investigating the expression patterns of a MAP kinase, a 14-3-3 protein, a cyclic nucleotide regulated ion channel protein, several other protein kinases and several chaperone proteins. The PCR data indicate that these genes are expressed early in the transition, usually within 6 hours. ODC, a well characterised DE gene in mammalian systems, is also expressed at this time, which coincides with a peak of ornithine decarboxylase activity (Fowler et al. 1996).

Expression of *Bvcrk1* in the storage organ has been studied using semi-quantitative PCR. Very little expression has been detected in leaf tissue or in very young roots, but high levels of expression are associated with storage organ development especially in regions of the organ in which there is cambial ring formation and cell division. This view is supported by *in situ* hybridisation studies where *Bvcrkl* expression is localised in the secondary cambia. Further analysis of a storage organ development sequence is being carried out using ISH to obtain a better understanding of the functions of *Bvcrkl* and Betvu;*CycA2*.

Figure 1. Map of gene expression during transition from G_0 to S in *Beta vulgaris* cells

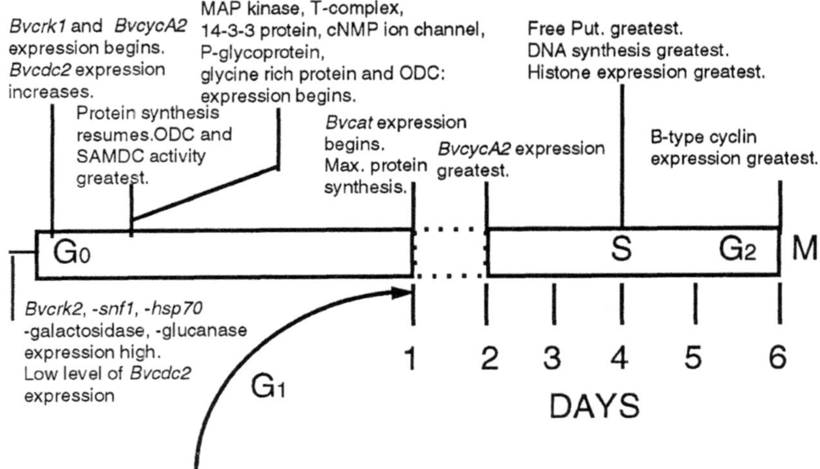

3. References

Dahl M et al. (1995) Plant Cell 7, 1847-1857.
Elliott MC et al. (1996) Russ. J. Plant Physiol. 43, 544-551.
Fowler MR et al. (1996) Physiol. Plant. 98, 439-446.
Fowler MR et al. (1998a) J. Exp. Bot. 49, 249-253.
Fowler MR et al. (1998b) Physiol. Plant. 102, 61-70.
Lapidot-Lifson Y et al. (1992) Proc. Natl. Acad. Sci. USA 89, 579-584.
Maruyama IN et al. (1995) Nucleic acids Res. 23, 3796-3797.
Matsushimi H et al. (1990) Mol. Cell. Biol. 10, 2261-2268.
Meskiene I et al. (1995) Plant Cell 7, 759-771.
Renaudin J-P et al. (1996) Plant Mol. Biol. 32, 1003-1018.

The plant endoplasmic reticulum and quality control of secretory proteins

Lorenzo Frigerio and Alessandro Vitale
Istituto Biosintesi Vegetali, Consiglio Nazionale delle Ricerche, via Bassini 15, 20133 Milano, Italy

1. Introduction.

The plant secretory pathway delivers proteins into the endoplasmic reticulum (ER), the Golgi complex, the vacuoles, the plasma membrane or the cell wall. The major biosynthetic trafficking route is from the ER to the Golgi complex and from there either to the vacuole and the tonoplast or to the cell wall and the plasma membrane. The ER can therefore be considered, both topologically and functionally, the principal compartment of the secretory pathway (Vitale et al., 1993). The plant ER is a complex and dynamic organelle, which appears to be composed of several morphologically and possibly functional subdomains (Staehelin, 1997). With regard to the synthesis and trafficking of secretory proteins, the ER is not only the site where proteins fold and assemble, but also the major checkpoint for these processes. In addition, the plant ER can act as a proper protein storage compartment. Here we discuss recent findings that shed light on the mechanisms of chaperone-assisted protein folding and quality control in the plant ER.

2. Chaperone-assisted protein folding and assembly

The plant ER contains a subset of resident proteins, collectively termed reticuloplasmins, which can act as molecular chaperones or folding catalysts. The best studied plant chaperone is the hsp70 - family member BiP (Pedrazzini and Vitale, 1996). BiP has been shown to bind to nascent polypeptides in an ATP-sensitive fashion. In vitro affinity studies with synthetic peptides suggest that in vivo BiP binds to hydrophobic sequences which are exposed in the nascent chain but successively become buried in the mature structure of the protein.Transient association of BiP with these sequences is likely to prevent incorrect inter- and intra-chain interactions and disordered aggregation (Blond-Elguindi et al., 1993). The first evidence for BiP association with nascent plant proteins comes from the study of rice prolamins. Prolamins are the major storage proteins in the seeds of monocot plants (Shewry et al., 1995): they assemble to form large protein bodies in the ER. BiP binds to nascent prolamin chains as they are translocated into the ER lumen (Li et al., 1993), but is also found at the periphery of the prolamin protein bodies: this suggests that BiP might play a role in protein body formation. Similarly, BiP has been found at the surface of ER protein bodies in maize (Zhang and Boston, 1992). Phaseolin is the best characterized plant storage protein shown to interact with BiP (Vitale et al., 1995). Phaseolin is a soluble, homotrimeric storage glycoprotein (Shewry et al., 1995) that, unlike prolamins, is transported through the Golgi complex to storage vacuoles, where it accumulates. During phaseolin synthesis in bean cotyledons, BiP

A. Altman et al. (eds.), Plant Biotechnology and In Vitro Biology in the 21st Century, 393–396.

associates with phaseolin monomers but is not found in association with trimers; this indicates that, once monomers are correctly folded, they still possess exposed chaperone binding sites, perhaps at positions involved in trimer formation (Vitale et al., 1995). In effect, a phaseolin mutant that is incapable of forming trimers is found in tight association with BiP (Pedrazzini et al., 1994), until it is eventually degraded (Pedrazzini et al., 1997). BiP also associates with newly synthesized β-conglycinin, the phaseolin homolog in soybean seeds, before trimer formation (Fontes et al., 1996). Recently, BiP has been found in association with a vacuolar type H+ ATPase, a multi-subunit protein: BiP could be co-immunoprecipitated with the entire ATPase complex using antibodies against a single peripheral subunit, suggesting that assembly of the multimer occurs in the ER and is chaperone-assisted (Li et al., 1998). In recent years, chaperone activity of other plant ER resident proteins has been investigated.For instance, homologs of calnexin and calreticulin (a soluble form of calnexin) have been cloned in plants (Denecke et al., 1995). Calnexin acts a chaperone in mammalian cells (Tatu and Helenius, 1997) and operates mostly on glycoproteins. Both calnexin and calreticulin are lectins that bind to monoglucosylated asparagine-linked oligosaccharide chains of glycoproteins: binding and release by the lectins is regulated by cycles of glucose trimming and re-glucosylation of the high mannose glycan chains of the substrate proteins: only upon correct folding are the glycoproteins released from the cycle. Two ER-located glucosidases are responsible for the removal of the glucose residues, whereas a glucosyltransferase is responsible for the re-addition of glucose to the glycan chains of those proteins which have not yet acquired their correct conformation, once again turning them into ligands for calnexin. In this model, the glucosyltransferase senses the folding state of the substrate protein, acting as the real chaperone, whereas calnexin (and/or calreticulin) is responsible for retention of the folding intermediates (Helenius, 1994). The first direct example of chaperone function for calnexin in plants was provided by the observation that calnexin (as well as BiP) associates with the vacuolar H+ ATPase (Li et al., 1998); Recently, the existence of a mechanism, possibily involving calnexin or calreticulin, similar to that described for mammalian cells, has been demonstrated in plants by Lupattelli et al. (1997). By studying the translocation of *in vitro* translated phaseolin into bean microsomes, the authors showed that the rate of phaseolin assembly into trimers is dependent on the glucosylation state of its oligosaccharide chains; inhibition of glucose removal led to an increase of the assembly rate *in vitro*, whereas a block at the level of partial glycan trimming inhibited assembly. This indicates that glucose trimming is the rate-limiting step in the formation of phaseolin trimers: this finding suggests that monomeric phaseolin interacts with calnexin or calreticulin (Lupattelli et al., 1997). Protein concentration in the ER is very high and there is growing evidence for the existence of large complexes of resident reticuloplasmins (Tatu and Helenius, 1997). Recently, BiP and calreticulin have been shown to form an abundant, high molecular weight complex in the plant ER: the complex is composed by more than 50% of cellular BiP, which is associated to a smaller proportion of calreticulin (Crofts, et al., 1998). Although the biological significance of such a complex remains to be elucidated it is suggested that it may represent a sort of reservoir for BiP, capable to modulate the availability of the chaperone in case of stress situations.

3. Quality control

Secretory proteins with irreversible folding defects, or incapable of assembly into oligomers are retained in the endoplasmic reticulum, by extended chaperone binding or by other unknown mechanisms, and are eventually targeted to degradation. This quality control system has been well characterized in mammalian cells (Hammond and Helenius, 1995). An ER-associated degradation pathway has been recently described in mammalian and yeast cells (Kopito, 1997): condemned proteins can be dislocated, in a chaperone-assisted process, through the translocation pore back to the cytosol (Wiertz et al., 1996; Plemper, et al., 1997); here they are degraded by the proteasome, in a process that can be ubiquitin-dependent (de Virgilio et al., 1998). An alternative degradation pathway, which involves targeting of defective proteins to the vacuole, has been identified in yeast (Hong et al., 1996). Evidence for the existence of a quality control system in the plant ER comes from the observation that a mutant of phaseolin, Δ363, which is incapable of assembly into trimers and remains monomeric, is found in prolonged association with BiP, and remains sequestered in the ER (or a closely related compartment) until it is eventually degraded, in a process that does not involve Golgi-mediated transport to the vacuole (Pedrazzini et al., 1997). A different phaseolin mutant, HiMet, that had been produced to increase its methionine content by the introduction of a 15-amino acid insert rich in methionine, has a remarkbly different fate: in transgenic tobacco seeds, HiMet phaseolin is quite unstable, but still capable of forming trimers (Hoffman et al., 1988), and is transported to the vacuole via the Golgi complex, like wild-type phaseolin. Interestingly, the addition of the C-terminal ER retention signal, HDEL, to HiMet phaseolin, is not sufficient to retain the protein in the ER and HiMet is still in large part transported to, and degraded in, the vacuole (Pueyo et al., 1995). Thus, the fact that HiMet forms trimers is probably sufficient for ER quality control to allow its exit from the ER: It is likely that trimerization masks chaperone-binding sites on phaseolin, whereas these sites are still exposed in the monomeric Δ363. Prolonged binding to BiP can thus be one of the mechanisms that avoid exit from the ER and transport to the Golgi complex of assembly defective phaseolin. The site of degradation of Δ363 is still undetermined, and the existence of an ER-to-cytosol dislocation pathway in plant has not been directly demonstrated. Recently, however, studies on the A subunit of the plant toxin ricin (a dimeric ribosome inactivating protein) have shown that the expression of ricin A chain into the plant ER leads to its Golgi-independent degradation, similarly to what occurs for Δ363: interestingly, degradation is paralleled by inactivation of the host cell ribosomes, indicating that a proportion of the A chain has reached the cytosol (Frigerio et al., 1998b). Co-expression of ricin A and B chain results in the formation of ricin dimers in the ER: dimer formation rescues the A chain from degradation and concomitantly reduces ribosome inactivation, further confirming that there is a link between the ER-associated retrotranslocation/degradation pathway and the presence of the A chain in the cytosol (Frigerio et al., 1998b). The ultimate site of degradation, however, still remains to be determined.What is the capacity of quality control? It is essential for cell viability that severely defective proteins do not reach their final compartment of residence, where they could negatively affect cell metabolism; quality control must thus be sufficiently flexible to cope with situations of stress. Indeed, when assembly-defective phaseolin is transiently expressed at very high levels in tobacco protoplasts, the whole of the protein is retained with high capacity by quality control and no saturation is observed (Frigerio et al., 1998a). In comparison, at similar levels of expression, wild-type phaseolin, that is targeted to the vacuole, saturates the vacuolar sorting mechanism and is in part secreted outside the cells (Frigerio et al., 1998a).

4. Conclusions

The existence of a mechanism for protein quality control in the endoplasmic reticulum of plant cells has only recently been established; the details of this mechanism still await elucidation: it will be particularly interesting to clarify the nature of the chaperone binding sites which determine recognition and retention of a misfolded protein by quality control; also it will be necessary to determine how the ER disposes of defective proteins, and whether plant cells possess unique ER-associated degradation pathways. The understanding of the details of this fundamental physiological process will be of great value for potential use of the plant secretory pathway as a site for the accumulation of biotechnologically important proteins.

5. References

Blond-Elguindi et al (1993) Cell 75, 717-728.
Crofts et al (1998) Plant Cell 10, 813-823.
de Virgilio M et al (1998) J. Biol. Chem. 273, 9734-9743.
Denecke J et al (1995) Plant Cell 7, 391-406.
Fontes EPB et al (1996) Brazilian J. Genet. 19, 305-312.
Frigerio L et al (1998a) Plant Cell 10 (6), in press.
Frigerio L et al (1998b) J. Biol. Chem. 273, 14194-14199.
Hammond C, Helenius A (1995) Curr. Opin. Cell Biol. 7, 523-529.
Helenius A (1994) Mol. Biol. Cell 5, 253-265.
Hoffman LM et al (1988) Plant Mol. Biol. 11, 717-729.
Hong E et al (1996) J. Cell Biol. 135, 623-633.
Kopito RR (1997) Cell 88, 427-430.
Li X et al. (1998) Plant Cell 10, 119-130.
Li X et al. (1993) Science 262, 1054-1056.
Lupattelli F et al (1997) Plant Cell 9, 597-609.
Pedrazzini E et al (1997) Plant Cell 9, 1869-1880.
Pedrazzini E et al (1994) Plant J. 5, 103-110.
Pedrazzini E, Vitale A (1996) Plant Physiol. Biochem. 34, 207-216.
Plemper RK et al (1997) Nature 388, 891-895.
Pueyo JJ et al (1995) Planta 196, 586-596.
Shewry PR et al (Plant Cell) 7, 945-956.
Staehelin LA (1997) Plant J. 11, 1151-1165.
Tatu U, Helenius A (1997) J. Cell Biol. 136, 555-565.
Vitale A et al (1995) Plant Physiol. 107, 1411-1418.
Vitale A et al (1993) J. Exp. Bot. 44, 1417-1444.
Wiertz EJHJ et al (1996) Nature 384, 432-438.
Zhang F, Boston RS (1992) Protoplasma 171, 142-152.

STUDIES ON SYMBIOTIC PROPERTIES OF PURINE AUXOTROPHS OF STRAIN *RMD* 201 OF *RHIZOBIUM MELILOTI*

Shruti Gupta[1], **S.P.S. Khanuja**[2], **Aqbal Singh**[3] **and Nam Prakash**[4],
[1]**Department of Botany, Daulat Ram College, University of Delhi, Delhi-110007,** [2]**Genetic Resources & Biotechnology, CIMAP, Lucknow-226015,** [3]**NRCPB, IARI, New Delhi-110012,** [4]**NRL, IARI, New Delhi-110012, India**

1. Introduction

The knowledge of the process of infection thread formation and establishment of functional nodules still remains obscure at molecular level. Yet studies on various rhizobial mutants clearly imply that certain deficiencies of bacteria can severely limit normal symbiotic process.

The symbiotic defect of *Rhizobium* purine auxotrophs has been observed in many laboratories using rhizobia with different hosts. In *R. meliloti* alfalfa symbiosis (Dickstein et al 1991; Kerppola, Khan 1988; Scherrer, Denarie 1971; Swamynathan, Singh 1992), purine auxotrophy appears specifically to affect the infection process (Djordjevic et al 1988; Kim et al 1988; Newman et al 1994; Noel et al 1988; Pain 1979; Pankhurst, Schwinghamer 1974). Although nodules induced by purine auxotrophs of *Rhizobium meliloti* do not fix nitrogen (Dickstein et al 1991; Kerppola, Khan 1988; Scherrer, Denarie 1971; Swamynathan, Singh 1992), it has been reported that they do have infection threads (Swamynathan, Singh 1992) and do not infect host cells (Dickstein et al 1991).

In the present investigation, the light and transmission electron microscopy studies of transverse sections of 30-day-old nodules elicited by wild type strain (*Rmd* 201) and four purine mutants showed that each mutant seemed to be blocked at different stages of nodule development. We have also investigated whether initial steps of the infection process are normal.

2. Materials and methods

Bacterial strains: The wild type strain *Rmd* 201and four purine auxotrophs, used here are streptomycin resistant derivatives of AK 631 (Swamynathan, Singh 1992).
Media and culture conditions: Complete medium (MSY) and minimal medium (RMM) for *Rhizobium* were used as described earlier (Swamynathan, Singh 1992).
Light and transmission electron microscopy: Freshly harvested nodules were dipped in Karnovsky's fixative for 3-5 hrs at room temperature. The nodules were rinsed with 0.1 M phosphate buffer twice and stored overnight at 4°C. Post-fixation was done with 2% OsO_4 in 1 M phosphate buffer for 10 min. After passing through the graded series of alcohol, nodules were embedded in Epon Araldite mixture (Mollenhaver 1964).

3. Results

Symbiotic efficiency of purine auxotrophs: The four purine auxotrophs considered for the present investigation *Rmd* 1101 (A_1), *Rmd* 1102 (A_2), *Rmd* 1105 (A_5) and *Rmd* 1106 (A_6) are symbiotically defective (Fix^-). They have exhibited blocks in functional nodule development at different steps during purine biosynthetic pathway (Swamynathan, Singh 1992).
Infection and nodule meristem initiation: All four purine auxotrophs induced nodules

A. Altman et al. (eds.), Plant Biotechnology and In Vitro Biology in the 21st Century, 397–400.

on their host, *Medicago sativa* but nodules showed altered phenotype and are ineffective (Swamynathan, Singh 1992). The prototrophic revertants of the purine auxotrophs elicited nodules similar in morphology to those of wild type strain *Rmd* 201.

Light microscopic studies of nodules elicited by purine auxotrophs: Transverse sections of nodules elicited by wild type *Rmd*201 showed all developmental zones of normally infected root nodules. Infection threads bordering the prefixation zone were observed in all mutants and appeared to be aborted.

TEM studies of ultrathin sections of nodules elicited by purine auxotrophs and wild type strain: Ultrastructural studies of transverse sections of nodules elicited by *Rmd* 201 revealed many bacteroids in the interzone region. There is branching of infection tubes (Fig. 1) and infection tubes proceed in cells through middle lamella in between cell walls. Infected cells adjoining empty cells showed starch granules near cell wall which were either electron dense or showed unstained texture. In nitrogen fixing zone some cells showed freshly released rhizobia but many cells have active bacteroids (Fig. 2). Young bacteroids showed peribacteroid membrane. When the bacteria had transformed into bacteroids, no PHB granules were observed. All organelles in cytoplasm were in active stage. Cytoplasm showed ribosomes, small lipid droplets and enlarged nucleus with irregular shapes. The analysis of ultrathin sections of nodules elicited by the four purine mutants are mentioned in table 1.

TABLE 1: Results of TEM studies of transverse sections of nodules elicited by purine mutants and wild type *Rmd* 201.

STRAIN	A1	A2	A5	A6	Wild type
Infection Tube					
Tube lysing	Yes(IC)	Yes(C)	Yes(IC)	Yes(IC)	Not
Tube matrix	Present	Present	Present	Absent	Present
Branching	++	-	-	++	+++
Initiation of thread	IC Space	IC Space	IC Not intact	IC Space	Middle lamella
Bacteria/Bacteroids					
Bacteroids	++ (lysing)	-	-	-	+++ (Not lysing)
Bacterial release	Yes, transforms to bacteroids but lysing	Yes	Lysed before release	Lysing	Released and not lysing

++ Good, +++ Very Good, IC-Intercellular, C-Cellular

4. Discussion

There are associations between nutritional requirement and effectiveness in nitrogen fixation. Purine auxotrophs used in the present study are not only infective but also form nodules. A normal mode of infection (Klein et al 1988 a, b) was observed initially. These mutants are capable of colonising nodules and bacteria can be recovered from nodules 30 days after bacterial inoculation. Nodules formed are small, spherical, white, more numerous and devoid of nitrogenase activity. Though infection threads were observed in all nodules elicited by auxotrophs but transformation of bacteria into effective bacteroids was lacking.

R. meliloti purine mutants used for the present study infect *Medicago* and colonise nodules even in the absence of external supplementation of AICAR or IMP. These results are contrary to the hypothesis proposed by Noel et al (1988), as this could have not been possible if signal for infection was missing. Light microscopic examination after 30 days revealed similar cellular structure as in wild type strain *Rmd*201. The nodules induced by purine mutants of *Rmd* 201 are round in shape and developmental zones are of normal size and shape. Peripheral vascular bundles are similar to true nodules, yet no bacteroid-filled cells were observed. Infection threads filled with bacteria were bordering the prefixation zone and appear to be aborted. It is possible that the bacteria were not released from infection threads or that plant cells containing just a few released bacteria were present but due to low number, could not be unambiguously identified by light microscopy.

In nodules infected with wild type bacteria, no amyloplasts were present in populated cells but starch granules were bordering the populated cells (Fig. 2) and its metabolites pass through plasmodesmata into populated cells. The host plant starch grains accumulated in prefixation zone and to some extent in intermediate zone of A_1 A_2 and A_5 mutants. They were absent in mutant A_6. Amyloplasts were seen in all mutants and were distorted in A_5. Mitochondria were seen in all mutants and wild type. The role of starch is to supply energy rich material for maintenance of nodule metabolism. In wild type, such starch was not found in N_2 fixing zone, and interzone indicating its full utilization. In mutants (Exo^+) A_1 and A_5, some starch was seen in prefixation zone indicating its utilization to some extent. In A_2 mutant (Exo^-), almost all starch remained unutilised. The differential degree of starch utilization in mutants is either due to their lysis as observed in A_2 or utilizing to some extent for repressing host immune response. Nodules elicited by purine mutants were not pink thereby lacking leghaemoglobin. Nitrogenase activity was not detected. Probably partial pressure of oxygen cannot be optimally maintained for nitrogenase to function and maintain its integrity leading to inffective nodules and starch accumulation in bordering cells. Studies of ultrathin tranverse sections of 30-day-old nodules formed by Pur^- mutants revealed blocks at different stages of nodule development. A_5 mutant is incapable of releasing the bacteria from infection thread (Fig. 5) whereas in A_2 mutant, lysis of infection tube occurs but released bacteria are unable to transform into bacteroids (Fig. 4). Transformation of bacteria to bacteroids occurs in A_1 and A_6 but in A_1 bacteriods show lysis (Fig. 3) unlike A_6 (Fig. 6), however, subsequent steps of nitrogen fixation pathway are lacking in both the mutants. Several possibilities concerning the manner in which purine auxotrophy can influence symbiosis have been considered earlier. Pankhurst, Schwinghamer (1974) suggested that adenine acts more as growth promoter since cytokinins which stimulate polyploid mitosis, an essential feature of nodule development, are synthesised from AMP (Morris, Powell 1987). But the fact that these purine mutants can form nodules, precludes this possibility. Probably they are utilising adenine provided by the host plant for their survival and for synthesizing cytokinin. One consistant observation is that purine auxotrophy somehow leads to development of ineffective nodules. One possibility is that ineffectiveness could be caused by lack of some metabolites for becterial metabolism. These defects could not be overcome by a plant supply because of lack of synthesis of suitable compounds by the host or because of too low rate of synthesis, or insufficient permeability of plant membrane envelope or bacterial membrane to these compounds. Another possibility is that metabolism of purines is directly linked to nitrogen fixation. Thus, all studies involving purine auxotrophs clearly indicate that effective nodules cannot be elicited in the absence of *de novo* purine biosynthesis.

5. References

1. Dickstein R et al (1991) Mol. Gen. Genet. 230, 423-432.
2. Djordjevic SP et al (1988) J. Bacteriol. 170, 1848-1857.
3. Kerppola TK, Khan ML (1988) J. Gen. Microbiol. 134, 913-919.
4. Kim CH et al (1988) Appl. Envior. Microbiol. 34, 423-427.
5. Klein S et al (1988a) Mol. Pl. Microbe Interact. 1, 94-100.
6. Klein S et al (1988b) J. Bacteriol. 170, 1003-1006.
7. Mollenhaver HH (1964) Stain Technol. 39, 111-114.
8. Morris RO, Powell GK (1987) Bioessays, 6, 23-28.
9. Newman JD et al (1994) J. Bacteriol. 176, 3286-3294.
10. Noel KD et al (1988) Arch. Microbiol. 149, 499-506.
11. Pain AN (1979) J. Appl. Bacteriol. 47, 53-64.
12. Pankhurst C E, Schwinghamer EA (1974) Arch. Microbiol. 100, 219-238.
13. Scherrer A, Denarie J (1971) Plant Soil Special, 39-45.
14. Swamynathan SK, Singh A (1992) J. Genet. 71, 11-21.

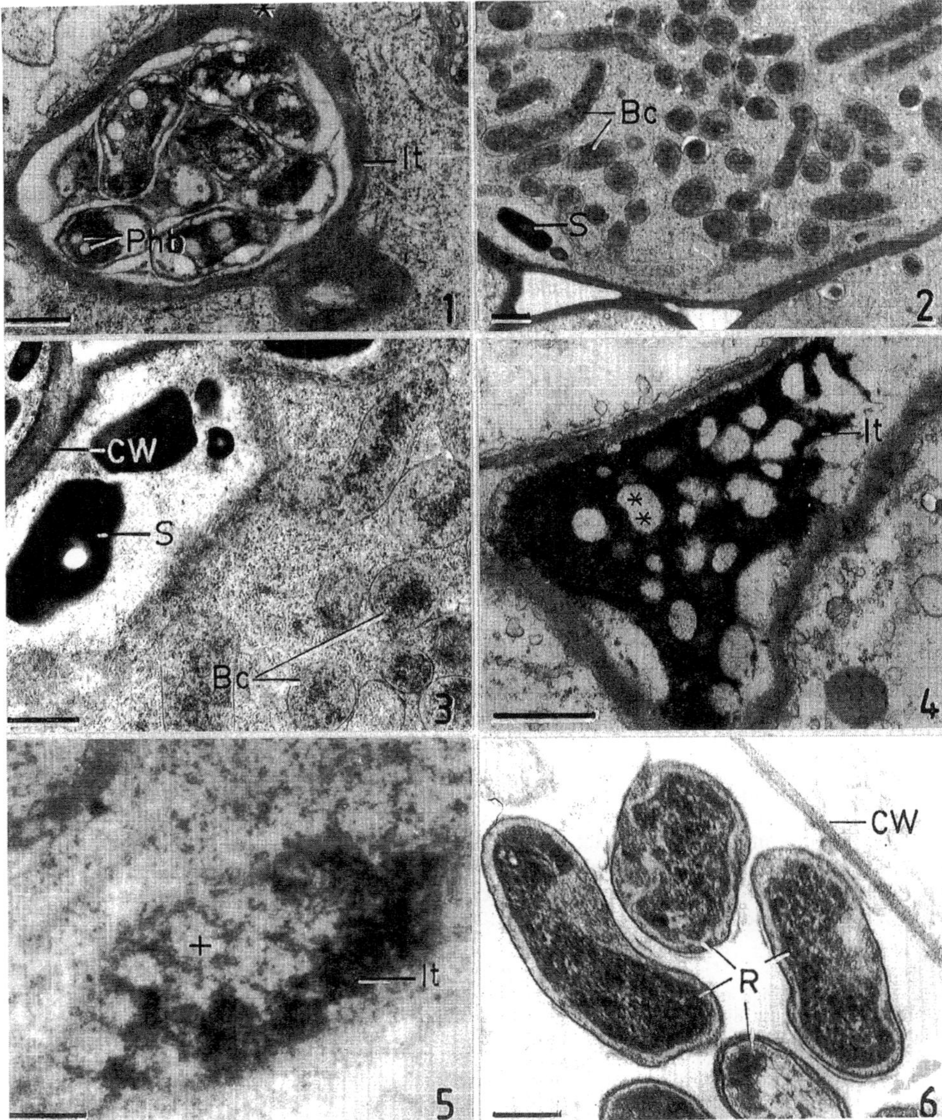

Fig. 1. Unreleased rhizobia in infection tube (WT), bar 1 μ. Fig. 2. Functional bacteroids in populated cells (WT), bar 1 μ. Fig. 3. A_1 mutant showing lysis of bacteroids, bar 1 μ. Fig. 4. Lysis of infection tube along with bacteria inside (A_2), bar 1 μ. Fig. 5. Aborting infection tube with lysed bacteria (A_5), bar 0.5 μ. Fig. 6. Partial transformation of bacteria into bacteroids (A_6), bar 0.25 μ. Infection thread (It), Cell wall material around infection tube (*), Polyhydroxy butyric acid granules (Phb), Cell wall (CW), Bacteroids (Bc), Starch grain (S), Tube matrix with lysed bacteria (**), Electron dense material (+), Rhizobia (R).

DEGRADATION OF POSTTRANSLATIONALLY UNSTABLE PROTEINS IN TRANSGENIC PLANTS UTILIZES MULTIPLE PATHWAYS TO THE VACUOLE

Eliot M. Herman[1], Maarten J. Chrispeels[2], Brian Larkins[3], and Gad Galili[4]

[1] Climate Stress Laboratory, USDA/ARS, Beltsville,MD, USA
[2] Biology Department, UCSD, San Diego, CA, USA
[3] Department of Plant Sciences, Univ. of Ariz., Tuscon, AZ, USA
[4] Department of Plant Biology, Weizmann Institute, Rehovot, Israel.

Abstract

Numerous examples of posttranslationally unstable foreign proteins have been identified in transgenic plants. Newly synthesized proteins exit the endoplasmic reticulum, either by progression through the endomembrane system or by assembly into ER-derived protein bodies, then converge in the seed protein storage vacuole (PSV) for degradation. The destabilized seed storage protein HiMet-phaseolin, produced by disrupting the primary sequence with an insertion, is posttranslationally unstable. Adding the ER-retention sequences K/HDEL indicates that the HiMet-phaseolin is stable in the ER/Golgi, but is degraded on progression to the PSV. Maize prolamin (zeins) proteins are assembled into protein bodies that are secreted into, and are transiently stable in the cytoplasm. Protein bodies assembled in seed cells are then sequestered in the PSV by apparently selective autophagy and degraded. The sequestration of materials observed within the PSV indicates that there are multiple mechanisms of sequestration that result in proteins resident in the cytoplasm, apoplast and endomembrane system transferred to the PSV for degradation. The functional role of the PSV in degrading cellular materials from cellular and extracellular locations has important implications for producing high levels of foreign proteins in plant cells.

Introduction

Modification of seeds through genetic engineering offers a direct approach to producing improved crops. Increased sulfur amino acid content of soybeans and increased lysine of cerealsare objectives that may be achieved with the expression of foreign genes in transgenic plants. Other possible uses of genetic engineering of crops include adding proteins to immunize against illness, adding proteins that can be accumulated and recovered for other uses, and producing seeds with altered characteristics that may improve their quality and utility as primary food substances and as products to be used in the processed food industry. There have been many experiments to produce transgenic seeds expressing foreign proteins in model plants (see Galili and Herman 1996 for review) and more recently in crops (see Shimoni et al 1997 for example). Although many of these experiments have been successful there are also many reports of foreign proteins although

A. Altman et al. (eds.), Plant Biotechnology and In Vitro Biology in the 21st Century, 401–406.

synthesized at high levels, are not accumulated due to posttranslational instability (Hoffman et al 1988; Ohtani et al 1991; Coleman et al 1996).

In this paper we report the results of studies using transgenic tobacco and soybean seeds as model systems. We demonstrate that the seed protein storage vacuole (PSV) is capable of degrading proteins that are initially accumulated in the endomembrane system, cytoplasm and extracellular space and are then sequestered in the PSV by secondary processes. Our results show that there are multiple pathways of protein sequestration in the PSV that apparently possess enzyme(s) capable of degrading some proteins while simultaneously accumulating endogenous storage proteins.

High Methionine (HiMet) phaseolin and selective degradation of a distorted protein

Hoffman et al (1988) published the results of an attempt to increase the sulfur amino acid content of a model legume storage protein, β-phaseolin. Phaseolin like other legume storage proteins is relatively deficient in sulfur-containing amino acids and introducing high-sulfur proteins into legumes, especially soybeans, would greatly increase its utility as animal feed. A β-phaseolin cDNA was modified by inserting a short oligonulceotide encoding an α-helical polypeptide into an α-helical domain of phaseolin. The mutant, termed HiMet-phaseolin, and a wild-type control were expressed in transgenic tobacco seeds under the control of the phaseolin seed-specific promoter. Both the wild-type and HiMet genes were expressed at similar levels, and both proteins were synthesized *in vivo* at high levels. Wild-type phaseolin was accumulated in the seed protein storage vacuole, the HiMet mutant was not accumulated and appeared to be degraded within several hours after its synthesis. An explanation of why HiMet is unstable emerged when the crystal structure of β-phaseolin was elucidated (Lawrence et al 1994). It appears that the modification used to produce HiMet lengthened an α-helical domain in a major ring structure and probably produced a large-scale distortion of the protein. Distorted HiMet apparently does form correct trimeric oligomers in the ER lumen, and it is also cotranslationally glycosylated (Hoffman et al 1988). Immunogold localization of HiMet and wild-type phaseolin in transgenic tobacco seeds showed that the wild type is accumulated within the PSVs, while the PSVs of the HiMet seeds exhibited no immunolabel. Immunogold labeling of HiMet was observed in small, dense vesicles that appear identical to the Golgi-derived secretion vesicles that carry storage proteins to the maturing PSV. The results of these experiments were interpeted to indicate that HiMet is postranslationally unstable in one or more compartments of the endomembrane system within a few hours of the protein's synthesis.

In order to determine the cellular site of HiMet phaseolin degradation, Pueyo et al (1995) added K/HDEL ER-retention sequences to the carboxy-terminus of the HiMet sequence. The addition of K/HDEL to HiMet resulted in increased stability and accumulation suggesting that HiMet degradation occurs ina post-ER compartment. Subcellular fractionation of maturing seeds expressing HiMet indicated that all of the protein is contained in a membrane-bound form. Localization of the HiMet-HDEL by immunogold EM showed most of the protein is contained within the ER lumen. HiMet-HDEL is apparently transport competent since it was also localized in the Golgi apparatus and secretion vesicles. However, HiMet was not accumulated in PSVs indicating that the protein is degraded after its deposition. The difference in stability of wild-type phaseolin and HiMet in tobacco PSVs indicates that the tobacco PSV possesses active proteases capable of selectively degrading the destabilized HiMet while not affecting the wild-type protein.

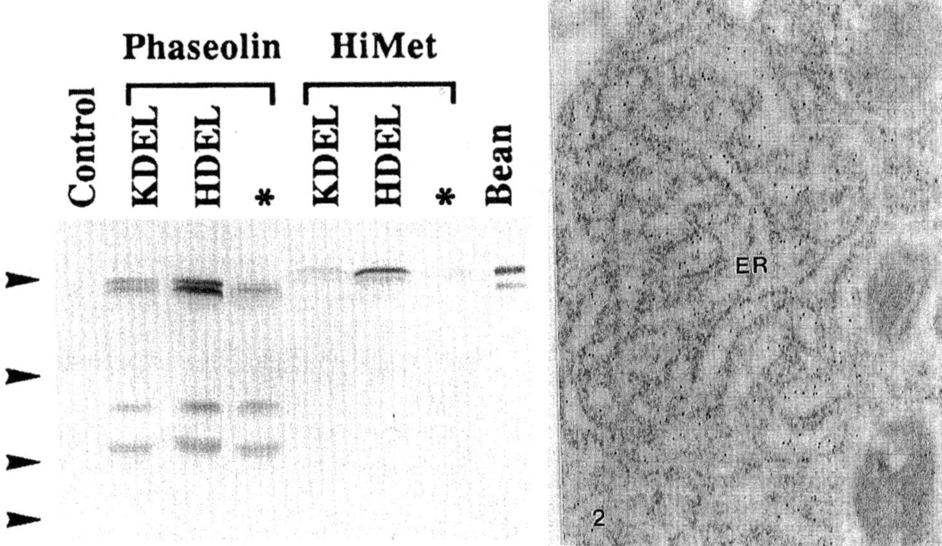

Figure 1. Immunoblot of HiMet and wild-type phaseolin with and without K/HDEL expression in transgenic tobacco seeds. The addition of ER retention sequences to HiMet results in enhanced stability of the protein by retarding it in the ER lumen.

Figure 2. Localization of HiMet-HDEL in the ER lumen in a maturing tobacco embryo cell.

Autophagy of cytosolic protein bodies

Cereal seed storage proteins, also known as prolamins are sequestered in cytosolic protein bodies that are directly assembled by, and derived from, the ER (See Galili and Herman 1997; Herman and Larkins 1999 for reviews). Prolamins are hydrophobic proteins that accrete in the ER lumen. The assembly of the protein accretions involves specific protein-protein interactions as well as the assistance of ER chaperone proteins (Li et al 1993). The resulting protein body that can include domains that sequester specific prolamins (Lending and Larkins 1989). Protein bodies may remain attached to the ER, as occurs in maize (Lending and Larkins 1989), or may become detached to produce a population of cytoplasmic protein bodies. The cytosolic protein bodies may either remain in the cytoplasm or become in stored in the vacuole, as occurs in wheat endosperm as the consequence of autophagy (Levanony et al 1992).

Zeins have been expressed in transgenic plants. Among the four types of zeins (α,β,γ,δ) tested, γ-zein and β-zein is accumulated at high levels when expressed alone in transgenic plants (Ohtani et al 1991; Hoffman et al 1987; Bagga et al 1995). The other zeins are either very unstable, or accumulated at low levels. Coexpression of α- and γ-zein intobacco seeds enhances the stability of α-zein (Coleman et al 1996). Similar experiments with the β and δ-zeins result in increased stability of the δ-zein (Bagga et al 1997). The cytoplasmic protein bodies appear to be stable, with a matrix of γ–zein that contains embedded locules of α-zein. The zein-containing protein bodies are also sequestered in the PSV (Coleman et al 1996, Bagga et al 1997). The protein bodies sequestered within PSV appear to be partially degraded and possess a lower concentration of immunolabeled γ-zein compared to the cytoplasmic-localized protein body. These results indicate that zein-containing protein bodies are stable as long as the organelle remains in the cytoplasm, but is vulnerable to the action of PSV-localized proteases once the protein body is sequestered within the PSV by autophagy.

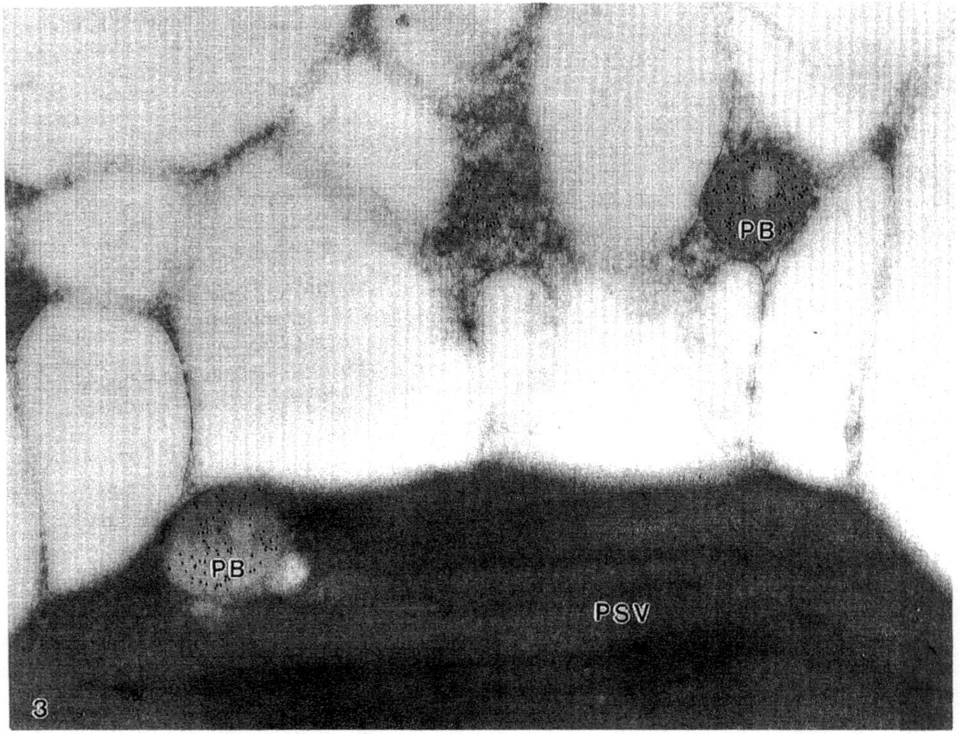

Figure 3. Immunogold assay of γ-zein localization in α/γ expressing tobacco seed endosperm. Two protein bodies (PB) are shown, one in the cytoplasm, and another sequestered within the protein storage vacuole (PSV) by autophagy.

Endocytosis and disposal in protein storage vacuoles

The suggestions that plant cells internalize extracellular material has a long and controversial history. Many plant scientists have asserted that the turgor pressure that inflates the plant cell against the cell wall constitutes a barrier to the uptake of large vesicles. Experimental evidence has indicated that the porosity of cell wall is limited to moderately small molecules of a few tens of kDa. However, there is ample cytological evidence that plant cells internalize extracellular materials. Plasma membrane elaborations consisting of vesicles or tubules termed plasmalemmasomes have been observed since EM has been applied to study plant cells. Similar structures, some of which are partially degraded, have been observed to be sequestered within the vacuole. The linkage between the extracellular cell surface material and material sequestered within the vacuole was established by Herman and Lamb (1992), who used EM immunocytochemistry to show that arabinogalactan proteins are localized on the cell surface, within plasmalemmasomes, in endosomes sequestered within the cytoplasm, and within the vacuole. Intravacuolar endosomes appeared to be disrupted and partially degraded. The presence of a cell surface protein within endosomes and vacuolar inclusions is strong evidence for a secondary route for disposal of material originating from the cell surface and apoplast. Plasmalemmasomes are also observed at the cell surface as well as internalized and sequestered in the vacuole in seed storage cells. This establishes that there is an active process that internalizes extracellular materials for vacuolar disposal. Although there is yet no reported instance where a protein expressed in a transgenic seed is transported to the cell surface and then returned back into the cell for disposal, the presence of this pathway in seed cells suggest that this may provide a third possible route for posttranslational instability of proteins in transgenic plants.

Vacuole-mediated degradation as a limiting factor in the accumulation of some proteins

The vacuole has long been recognized as a lytic compartment that is functionally equivalent to the lysosome of animal cells (Matile 1997 for review). Plant vacuoles differ from lysosomes storing as well as degrading proteins (Galili and Herman 1997; Herman and Larkins 1999 for reviews). Because specific proteases are expressed after germination to mediate mobilization of storage proteins (see Muntz 1996 for review), it has been assumed that maturing seeds accumulating storage proteins are protease deficient. Although this hypothesis is appealing from a mechanistic point of view, recent results from our laboratories suggest that maturing seeds do contain vacuolar proteases capable of degrading a wide variety of proteins. We demonstrated that destabilized and otherwise posttranslationally unstable storage proteins are degraded in the maturing PSV. Multiple cellular pathways converge to mediate the sequestration of proteins in the PSV. Unstable proteins can be stabilized by retaining the protein in a protease-deficient compartment such as the ER, as shown for HiMet (Pueyo et al 1995) and γ-gliadin (Napier et al 1997) or in the cytoplasm, as shown by the example of protein bodies containing α/γ zeins

(Coleman et al 1996). However, preventing translocation of proteins into the vacuole is hindered by a lack of information on how cells mediate protein and organelle sequesteration in the vacuole. An alternative approach may be to impair the proteolytic activity of the PSV, allowing it to accumulate not only the storage proteins but other labile proteins. Little is known about the protease inventory of maturing seed PSV. Identification and characterization of the PSV proteases offers the prospect of producing protease-deficient seeds. Disarming proteases has been used in several expression systems to improve the efficiency of protein accumulation. A similar approach in seeds could provide a basis to improve the ability to engineer changes in crops.

Acknowledgements

This work was supported in part by BARD (US-Israel Binational Agricultural Research and Development Fund) grant # US-2334-93 to EMH and GG.

References

Bagga S et al (1995) Plant Physiol. 107, 13-23
Bagga S et al (1997) Plant Cell 9, 1683-1696
Coleman CE (1996) Plant Cell 8, 2335-2345
Galili G, Herman EM (1997) Adv Bot Res 25,113-40
Herman E, Lamb CJ (1992) Plant Physiol. 98, 264-72
Herman EM, Larkins BA (1999) Plant Cell, in press
Hoffman LM et al (1987) EMBO J. 6. 3213-3221
Hoffman LM et al (1988) Plant Molec. Biol. 11, 717-729
Larkins BA , Hurkman WJ (1978) Plant Physiol. 62, 256-263
Lawrence MC et al (1994) J. Mol. Biol. 238, 748-776
Lending CR, Larkins BA (1989) Plant Cell 1, 1011-1023
Levanony H J.Cell Biol. 119, 1117-1128
Li X et al (1993) Science 262, 1054-1056
Matile Ph (1997) Adv. Bot. Res. 25, 87-112
Müntz K (1996) J. Exp. Bot. 298, 605-622.
Napier JA (1997) Planta 203, 488-494
Ohtani T et al (1991) Plant Mol. Biol. 16, 117-128
Pueyo JJ (1995) Planta 196, 586-596
Shimoni, Y et al (1997) J. Biol. Chem. 272, 15488-15495

The uptake and distribution of ^{14}C-NAA in cotyledons of cucumber (*Cucumis sativus* L.) *in vitro*

M. Klemš, M. Truksa, L. Havel, and S. Procházka
Mendel University of Agriculture and Forestry,
Dept. of Botany and Plant Physiology, Brno, Czech Republic

Introduction

1-naphtalene acetic acid (NAA) belongs to the group of auxin growth regulators and is widely used in *in vitro* cultures for the induction of morphogenesis or dedifferentiation. The morphogenic effects of NAA are closely related to its transport and accumulation as well as the interactions with other growth regulators (Davies et al., 1995; Coenen and Lomax, 1997). On the medium with NAA Van den Ende et al. (1984) induced the formation of flower buds on the tobacco thin tissue layers. Smulders et al. (1986) observed that the formation of flower buds depends on concentration of NAA in medium, its uptake and transport within the explant. The uptake of NAA into plant cell occurs mainly via free diffusion, whereas its export takes place mainly through membrane carriers (Delbarre et al., 1996). In the tobacco stem pith ^{14}C-NAA is taken up intensively already within the first hours of exposure (Abbas et al., 1995). The taken up NAA is rapidly metabolized. Smulders et al. (1990) reported that as much as 80 % of total ^{14}C-NAA taken up was metabolized already after 6 hours. Aranda et al. (1984) have identified the forms of NAA conjugated with glucose or aminoacids. The selective uptake, transport and metabolism of growth regulating substances depend strongly on the physiological stage and the degree of differentiation of the explant (Abbas et al., 1995).

We have studied the differences in uptake and distribution of ^{14}C-NAA in early (autotrophic) and fully developed (heterotrophic) cotyledons of cucumber kept on light and in dark with the aim to characterize the transport of NAA within explants. To study the transport processes of endogenous auxin ^{14}C-IAA was added to the same medium as for the NAA study. The distribution patterns of ^{14}C-activity after the uptake of labeled NAA and IAA were compared. The morphogenic response in relation to NAA exposure and uptake was observed.

Materials and methods

The sterile seedlings of cucumber (*Cucumis sativus* L. cv. Pálava) were grown on MS medium containing 5 g l^{-1} of sucrose. Cotyledons in two different developmental stages were used throughout the experiment. The early heterotrophic cotyledons (HC) were excised from seedlings 4 days after germination when the cotyledons are still covered by testa. Green autotrophic cotyledons (AC) were excised from the seedlings 7-8 days old. Cotyledons were cultured on MS medium supplemented with 0.18 mg.l^{-1} NAA and 1.5 mg.l^{-1} BAP.

^{14}C-NAA (Sigma, specific activity 199 Mbq.mmol^{-1} at 3.33 Mbq.l^{-1}) and ^{14}C-IAA (Izotope-Budapest, specific activity 457 Mbq.mmol^{-1} at 3.33 Mbq.l^{-1}) were

A. Altman et al. (eds.), Plant Biotechnology and In Vitro Biology in the 21st Century, 407–410.

used in the study. Cotyledons were kept on the media containing ^{14}C-NAA and ^{14}C-IAA for 2, 5, 8 and 20 hours either on light or in dark. The effect of 2,3,5-triiodebenzoic acid (TIBA) on the uptake and metabolism of NAA was studied by the 2 h preincubation of the cotyledons in 0.3% TIBA.

To study the morphological changes induced by growth regulators the cotyledons were exposed to the induction medium for 2 or 20 hours on light after which they were transferred to MS medium without growth regulators.

Distribution of ^{14}C-activity was determined after segmentation of cotyledons into basal, and apical part, the segments were dried at 105 °C to constant mass and combusted in the oxygen flow (Kala, Peška, 1980). The ^{14}C-activity was measured by liquid scintillation method on spectrometer Packard 2000 CA. Semiconductography (Tykva, 1995) was used as a reference method of the ^{14}C-activity determination.

To study the metabolism and degradation of ^{14}C-NAA and ^{14}C-IAA the methanolic extracts of the tissue were purified on C-18 cartridges (Sep-pak, Tessek), separation of growth regulators was performed by both TLC and HPLC.

Results

From the data illustrated in Fig.1 it is clear that the overall uptake of ^{14}C-NAA by AC is not affected by the light and increases with time during the period under study. The establishment of polar pattern of ^{14}C-activity distribution with higher values in the basal parts can be observed under both light and dark culture conditions already after 5 hours. The 2 h pre-treatment of cotyledons by TIBA reduces the amount of ^{14}C-NAA taken up and inhibits the establishment of polar distribution pattern.

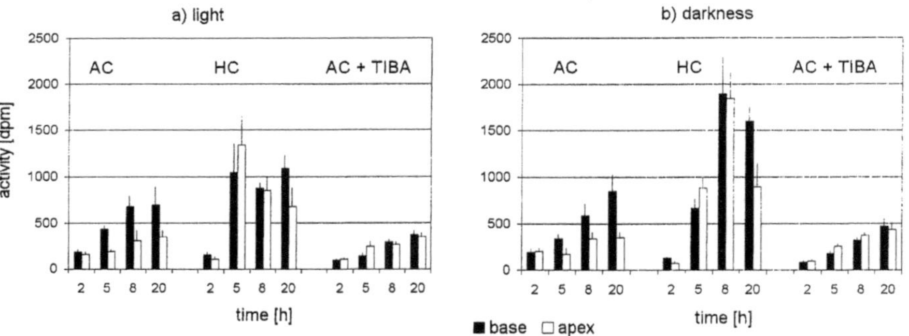

Fig. 1 ^{14}C-activity in basal (filled bars) and apical (empty bars) parts of cucumber cotyledons after exposure to ^{14}C-NAA for 2, 5, 8, and 20 hours in light (a) and darkness (b) (AC – developed autotrophic cotyledons, HC – early heterotrophic cotyledons, TIBA – 2,3,5-triiodobenzoic acid)

In the case of HC the overall uptake of ^{14}C-NAA is higher when compared to AC and the polar distribution of ^{14}C-activity similar to that of AC is established only after 20 hours. In the previous period the HC seem to undergo stages with ^{14}C-activity in apical and basal part either equal or higher in apex.

Autotrophic cotyledons take up approximately three times more ^{14}C-IAA on light than in the dark while higher ^{14}C-activity was always found in their basal parts (Fig. 2). Heterotrophic cotyledons take up comparable amount of ^{14}C-IAA under both light and dark conditions. In contrast to the autotrophic cotyledons the higher ^{14}C-activity was always found in the apical part.

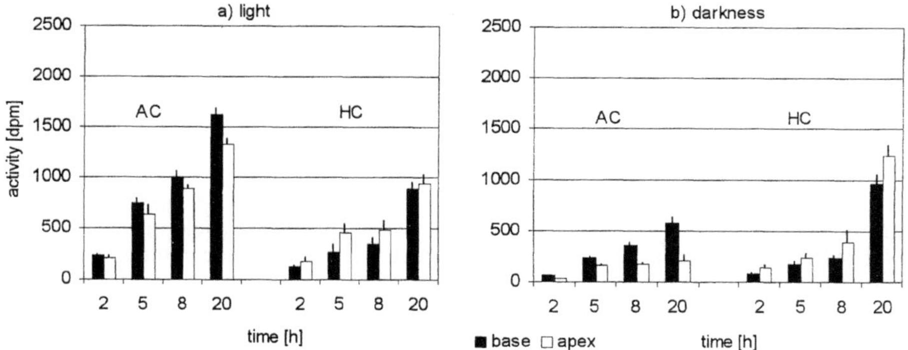

Fig. 2 ^{14}C-activity in basal (filled bars) and apical (empty bars) parts of cucumber cotyledons after exposure to ^{14}C-IAA for 2, 5, 8, and 20 hours in light (a) and darkness (b) (AC – developed autotrophic cotyledons, HC – early heterotrophic cotyledons)

The TLC analysis of free NAA and IAA and their metabolites revealed that the intensity of metabolism and degradation of IAA is substantially higher (Table 1). During the whole period under study the content of free IAA is around 4 % in autotrophic and 8 % in heterotrophic cotyledons. In the case of NAA the concentration of free acid is high in the beginning of culture, 80 % and 90 %, and the level drops after 20 h to 6 % and 16 % in autotrophic and heterotrophic cotyledons, respectively. The HPLC analysis of NAA metabolites of AC gave similar results. The peak of free NAA present in the analysis of samples after 2 h of culture diminishes after 20 h simultaneously with the appearance of a spectrum of unidentified NAA metabolites.

Table 1 Percentage of non-metabolized ^{14}C-NAA and ^{14}C-IAA in AC and HC during exposure to inductive medium (% of total ^{14}C-activity)

	^{14}C-NAA				^{14}C-IAA			
	2 h	**5 h**	**8 h**	**20 h**	**2 h**	**5 h**	**8 h**	**20 h**
AC	81.7	65.5	20.2	6.4	3.5	3.8	2.4	5.0
HC	91.7	75.8	72.5	16.4	10.0	8.3	8.1	7.5

Growth and developmental events in autotrophic cotyledons were related to the time of their exposure to the inductive medium (MS supplemented with NAA and BAP). The cotyledons induced for 2 h formed in the subsequent culture on medium without growth regulators roots on their basal parts. If the cotyledons were exposed to the inductive medium for 20 h small callus was formed instead of the roots on the basal parts.

Discussion

Due to the free diffusion as a prevailing mechanism of NAA uptake into cell (Delbarre et al., 1996) cotyledons within the 20 h of exposure take up NAA from medium intensively and accumulate it in the tissue. At the beginning of exposure the level of tracer is the same along the whole explant suggesting the equal uptake capability of all the tissue. However, in developed autotrophic cotyledons (AC), the clear pattern of higher accumulation of NAA and its metabolites in the basal part is established already after 5 h. That this redistribution of tracer activity is inhibited by TIBA pre-treatment points to the polar transport and the involvement of auxin specific cell efflux carriers (Morris, 1988). The transport inhibition reduces also the overall uptake capacity, leaving the level in both basal and apical part of TIBA pre-treated cotyledons equal and comparable to that of untreated apex.

The relatively larger amount of NAA taken up by early heterotrophic cotyledons (HC) could be explained by higher water uptake of developmentally younger explants and/or by the lack of functional efflux carriers causing accumulation of NAA in cells. The overall amount of NAA in HC tends to drop between 8 and 20 h of culture. Whether it is caused by the development of efflux carrier system and though the onset of polar auxin transport we can only speculate but this drop coincides with the redistribution of tracer in manner similar to more differentiated cotyledons. Simultaneously, from the 8 h on the metabolism of taken up NAA intensifies significantly.

The uptake of ^{14}C-IAA by cotyledons increases with time. In more developed AC the tracer redistribution pattern is similar to that of NAA with the larger portion of tracer in the basal part of cotyledons. The distribution pattern is reverted in younger HC even after 20 h, when distribution characteristic for AC is established in the case of NAA. However, the overall uptake of increases substantially between 8 and 20 h of exposure suggesting again the onset of functional auxin carrier system, in this case the auxin influx carrier that contributes substantially to the uptake of IAA into cell (Lomax, 1986; Delbarre, 1996). In contrast, the efflux carriers fail to produce the basipetal distribution pattern probably due to high level of accumulated NAA which occupies the carrier-mediated auxin efflux from the cell (Delbarre, 1996).

Acknowledgements

We would like to thank J. Balla and J. Flores for technical assistance. The work was supported by the grant of the Czech Ministry of Education (project No. VS96082) and the Grant Agency of the Czech Republic (project No. 96/K188).

References

Abbas E et al (1995) Structure and Function of Roots, pp 253–256, Kluwer Academic Publishers, Dordrecht, The Netherlands.

Aranda G et al (1984) Phytochemistry 23, 1221-1223.

Coenen C, Lomax T (1997) Trends Plant Sci. 2, 351-356.

Delbarre A et al (1996) Planta 198, 532-541.

Kala J, Peška J (1980) Use of nuclear methods and ionising radiation in genetics, and plant physiology (in Czech), pp 170-173, Agr. Univ., Brno, Czechoslovakia.

Smulders MJM et al (1986) Proceedings of international Symposium: Conjugated Plant Hormones, Structure, Metabolism, Function, pp 87-91, Gera, GDR.

Smulders MJM et al (1990) J. Plant Growth Regul. 9, 27-34.

Morris D (1988) Physiology and biochemistry of auxins in plants, pp 195-204, Academia, Praha, Czechoslovakia.

Lomax T (1986) Plant growth substances 1985, pp 209-213, Springer, Berlin, GDR.

Tykva R (1995) J. Radioanal. Nucl. Chem. 195, 327-334.

Van den Ende G et al (1984) Physiol. Plant. 61, 114-118.

NUCLEAR IMPORT OF THE CAPSID PROTEIN OF TOMATO YELLOW LEAF CURL VIRUS (TYLCV) IN PLANT CELLS.

T. Kunik[1], K. Palanichelvam[1], L. Mizrachy[1], H. Czosneck[2], V. Citovsky[3] and Y. Gafni[1]

[1]Volcani Center, Bet Dagan, Israel. [2]The Hebrew University of Jerusalem, Rehovot, Israel. [3] State University of New York, Stony Brook, NY, USA.

1. Introduction

Tomato yellow leaf curl virus (TYLCV) is a whitefly–transmitted monopartite geminivirus that causes a severe damage to tomato crops in many tropical and subtropical countries (Czosnek, *et al.*, 1988, 1990). The TYLCV found in Israel (TYLCVisr) has a monopartite-genome which consists of a single circular ssDNA molecule of 2,787 nucleotides encoding six open reading frames (ORFs): two on the virion (+) strand (including the capsid protein CP) and four on the complementary (-) strand (including the rep protein) (Navot *et al.*, 1991). The TYLCV replicates in the plant cell nucleus via a double-stranded DNA intermediate replicative forms, by a rolling circle mechanism (Laufs *et al.*, 1995; Saunders *et al.*, 1991; Stanley, 1995; Stenger *et al.*, 1991). The mechanism by which the viral DNA enters the plant cell nucleus during the early phase of infection is unknown. We suggest that the geminiviral capsid protein, which is the only known protein that serves for packaging the genomic DNA molecule into a viral particle, also transports this DNA into the host cell nucleus. For this, the CP must itself be karyophilic, possessing a specific nuclear localization signal (NLS) and most likely enters the nucleus by an active transport that needs the cytoplasmic factors karyopherin α/β, Ran/GTPase , NTF2 and GTP.

In order to investigate this possible role of the TYLCV CP in virus transport into the host cell nucleus, we monitored its specific nuclear accumulation in plant cells. In addition, we studied the CP interaction with the cytoplasm receptor karyopherin α and its ssDNA binding activity.

2. Procedure

To determine the intracellular localization of TYLCV CP, its DNA codoing sequence was fused in frame to the 3' end of the β-glucuronidase (GUS) gene. The resultant construct (pCP) was introduced into petunia protoplasts. Transient expression of the

A. Altman et al. (eds.), Plant Biotechnology and In Vitro Biology in the 21st Century, 411–415.

chimeric GUS-TYLCV CP gene was driven by a cauliflower mosaic virus 35S promoter. As a control, GUS alone was expressed under the same promoter. Following its expression, the GUS enzyme was localized histochemically. Whenever GUS alone was expressed in protoplasts, it remained in the cytoplasm (Fig. 1A). In contrast the GUS-TYLCV CP fusion protein was specifically localized to the cell nucleus (Fig. 1B). These results indicate that TYLCV CP is a karyophilic protein, most likely containing a functional NLS.

In order to identify the functional NLS sequence of TYLCV CP, we constructed a series of single and double deletion mutants in the gene coding for this protein. These mutants were fused to the GUS reporter gene and transiently expressed in petunia protoplasts. A construct containing only the first 38 amino acid residuse of CP was found exclusively in the nucleus, and its nuclear accumulation was similar to that of the full-length TYLCV CP (pCP) (Fig. 1C). These results indicate that the TYLCV CP NLS resides in the amino terminus of the protein

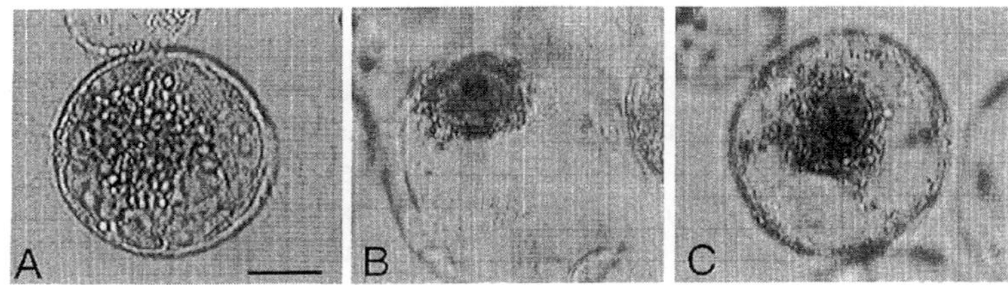

Figure 1. Nuclear import of TYLCV capsid protein (CP) in petunia protoplasts. Protoplasts were transfected by electroporation with 50μg plasmid DNA containing either GUS alone or the fused GUS-TYLCV CP. (A) GUS alone. (B) GUS-TYLCV CP. (C) GUS-TYLCV CPΔ39-260. Bar=10μm.

Next, we checked whether the TYLCV CP binds to the cytoplasmic factor karyopherin α that binds the NLS-protein and mediates its docking to the nuclear envelope. For that, a tomato cDNA library was screened for the gene encoding to karyopherin α. We used the homolog gene from arabidopsis as a probe. The gene for karyopherin α was cloned and we sequenced it (accession no. AF017252). Then, we used the two-hybrid system to check whether the karyopherin α binds to the capsid protein. When the capsid protein alone was introduced into the yeast cell it was insufficient for transcriptional activation of the *lacZ* gene. In contrast, when both the tomato karyopherin α and the TYLCV CP were introduced together, blue color appeared after 30 min to 12 hours indicating initiation of the *lacZ* gene transcription as a result of the interaction between those two proteins (results are not shown).

Gell mobility shift assay was used to study the ssDNA-binding activity of the purified TYLCV CP. As a probe for protein-DNA interactions we used a 422bp long PCR product, most of it from the intergenic region of the viral genome. This probe was radioactively end labeled and served as ssDNA probe (denatured) or as a dsDNA probe before incubation with the capsid protein (CP). ssDNA incubated with purified CP was strongly retarded (Fig. 2, 2) compared with the free ssDNA probe (Fig. 2, 1). Almost no retardation of the undenatured dsDNA probe could be detected under the same conditions (Fig. 2, 7) compared with the free dsDNA probe (Fig. 2, 6). Results obtained with DNA from non viral sources were similar to those described, indicating that the binding activity has no sequence specificity. Treatment of the reaction mixture with proteinase K abolished the binding (Fig. 2, 5), indicating that the retarded ssDNA probe indeed represented a protein-DNA complex. TYLCV CP association with the radioactively labeled ssDNA probe was blocked by the addition of 1,000-fold molar excess of the unlabeled probe (Fig. 2, 3). That TYLCV CP-ssDNA binding was sequence nonspecific was indicated by the finding that unlabeled ssDNA competitor derived from a nonviral source efficiently inhibited TYLCV CP association with the radioactively labeled viral ssDNA probe (Fig. 2, 4).

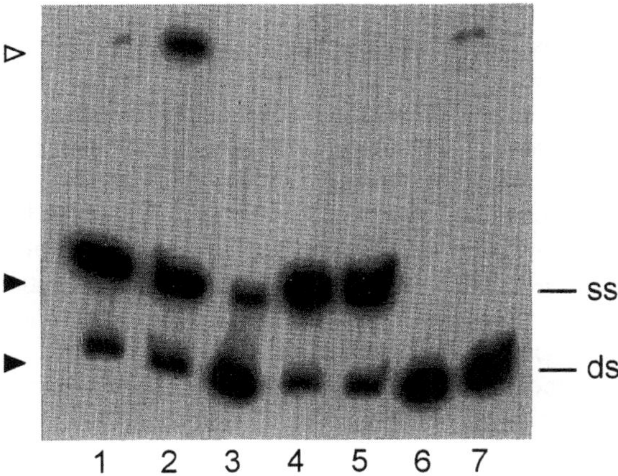

Figure 2. Interaction of TYLCV CP with ssDNA and dsDNA. Purified TYLCV CP (100 ng) was incubated with ssDNA or dsDNA probes and protein-DNA binding was analyzed by gel mobility shift~ assay as described in Methods. 1. ssDNA probe incubated in binding buffer alone 2. ssDNA probe incubated with TYLCV CP; 3. ssDNA probe incubated with TYLCV CP in the presence of a 1000-fold molar excess of unlabeled homologous ssDNA competitor; 4. ssDNA probe incubated with TYLCV CP in the presence of 1000-fold molar excess of unlabeled non-homologous ssDNA competitor; 5. ssDNA probe incubated with TYLCV CP and treated with proteinase K as described in Methods; 6. dsDNA probe incubated in binding buffer alone; 7. dsDNA probe incubated with TYLCV CP. Open arrowhead indicates the position of TYLCV-ssDNA complex; filled arrowheads indicate the positions of free ssDNA (top arrowhead) and free dsDNA (bottom arrowhead).

Material and Methods.

Protoplast isolation
Protoplasts were isolated from rapidly growing suspension cultures of Petunia hybrida cells (line 4544) as described by Kunik *et al.*, 1998. Protoplast transfecfion and GUS assays in protoplasts were done as described by Citovsky, 1994. GUS activity was detected 24-30 h after transfection, by histochemical staining.

Construction of GUS-TYLCV CP fusions
The *VI* gene of TYLCV, coding for CP, was fused in-frame to the *E. coli* GUS reporter gene in the expression vector pRTL-GUS/NlaΔBam developed by Carrington *et al.* (1991). Prior to subcloning *V1*, the Nla coding sequence was removed by digestion with Bglll and BarnHl restriction endonucleases.
Purification of TYLCV CP expressed in E. coil
Full length of TYLCV CP ORF was inserted to the expression vector pET28 (Novagen Inc., Madison, WI). As described by Palanichelvam *et al.*, 1998.The recombinant plasmid was introduced into E. coil strain BL21(DE3)pLysE. TYLCV CP was overproduced via the T7 RNA polymerase expression system (Studier *et al.*, 1990) and purified to near homogeneity as described in Citovsky *et al*, 1991.
Gel mobility shift assay
The virus- specific DNA probe included most of the intergenic region and a part of the *V1* gene of the TYLCV genome (Navot *et al.*, 1991) was end-labeled with [γ-^{32}P] ATP and T4 polynucleotide kinase (promega) as described by Palanichelvam *et al.*, 1998. Indicated amounts of TYLCCP were incubated for 30 min at 4^0C in 50 ul of binding buffer (10mM Tris/HCl, pH 8.0, 10% glycerol, 30mM NaCl) with 5 ng of double-stranded or heat-denatured, single-stranded probe DNA. After incubation, samples were electrophoresed on a 4% native polyacrylamide gel and then characterized by autoradiography, as described by Citovsky *at al.* (1990).
For proteinase K treatment, 1mg/ml of the enzyme was incubated for 30 min at 37^0C with the TYLCV CP-DNA mixture, prior to gel electrophoresis.

Conclusions

The TYLCV CP is a karyophilic protein containing a functional NLS in its N-terminus.
The TYLCV CP has ssDNA binding activity and this activity is sequence non-specific, strong and cooperative.
The TYLCV CP interacts with the cytoplasmic factor karyopherin α.

<u>References</u>

Carrington *et al* (1991). Plant Cell. 3:953-962.
Citovsky *et al* (1991) Proc. Natl. Acad. Sci. USA 88:2476-2480.
Citovsky *et al* (1990) Cell. 60:637-647.
Citovsky, V. (1994) Plant Molecular Biology Manual, pp 1-16, Kluwer Academic Publishers, Dordrecht, The Netherlands.
Czosnek *et al* (1988) Phythopathology 78:508-512.
Czosnek *et al* (1990) Phytopath. Mediterr. 29:1-6.
Kunik *et al* (1998) The Plant Jurnal 13:393-399.
Laufs *et al* (1995) Proc. Natl. Acad.Sci. USA 92:3879-3883.
Navot *et al* (1991) Virology, 185:151-161.
Palanichelvam *et al* (1998) J. Jen. Virol. June, 1988, in press.
Saunders *et al* (1991) Nucl. Acids. Res. 19:2325-2330.
Stanley, J. (1995) Virology, 206:707-712.
Stenger *et al* (1991) Proc. Natl. Acad. Sci. USA 88:8029-8033.
Studier *et al* (1990) Methods Enzimol. 185:60-89.

NOVEL WHEAT MOLECULAR CHAPERONES BELONGING TO THE PPIase-FKBP FAMILY: STUDIES ON THEIR REGULATION AND TISSUE SPECIFICITY

I. Kurek[1], N. Erel[1], C. Mayr[2], E. Herman[3], A. Breiman[1]
[1] Department of Plant Science, The George S. Wise Faculty of Life Sciences, Tel Aviv University, Israel; [2] Institut fur Biophysik und Physikalische Biochemie,Universitat Regensburg, Germany; [3] Climate Stress Laboratory, USDA/ARS Beltsville, MD, USA

INTRODUCTION

Protein folding *in vivo* is mediated by an array of proteins that act either as molecular chaperones or foldases or can have both functions. The molecular chaperones were originally defined as a group of unrelated classes of proteins that mediate the correct assembly of other proteins, but are not themselves components of the final functional structures (Ruddon, Bedows, 1997). They occur ubiquitously and many of them are classified as stress proteins, although they have essential functions under normal growth (Hartl, 1996; Buchner, 1996). When cells are exposed to elevated temperature or other environmental stresses, heat shock proteins belonging to several gene families are induced. Many heat shock proteins function as chaperones and play important roles in normal growth as well as in stress tolerance.

Recently, additional families of proteins the Peptidyl Prolyl cis trans Isomerases (PPIases) were classified as molecular chaperones. There are three structurally distinct classes of PPIases: the cyclophilins (CyP) which bind the immunosuppressive drug cyclosporin A (CsA) (Handscumacher et al., 1984), the FK506 binding proteins (FKBPs) which bind the macrolide drugs, FK506 and rapamycin (Harding et al., 1989; Siekierka et al., 1989), and the parvulin family (Dolinski, Heitman, 1997). Because of their drug binding activities, the PPIases have also been termed immunophilins. The binding of the drugs inhibit the PPIase activity (Schreiber, 1991), but the drug can serve as a molecular glue affecting the formation of novel protein complexes involved in signal transduction in the immune response pathway (Crabtree, Schreiber, 1996).

The two well characterized large cytosolic PPIases, the FKBP59 (also named FKBP52, Hsp56, p59, HBI) and CyP40 are components of the steroid hormone complex which includes the heat shock protein HSP90 (Sanchez, 1990; Peattie et al., 1992; Ratajczak et al., 1993). The FKBP59 and the CyP40 bind to the HSP90 via the tetratricopeptide (TPR) domain and form stable complexes with p23, p60 and the mammalian steroid receptor (Sanchez, 1990; Owens-Grillo et al., 1996).

Plant homologues of *FKBP59* were cloned and characterised: the Arabidopsis 62 kDa ROF1 (Vucich, Gasser, 1996), the wheat *wFKBP73* (Blecher et al., 1996), the *PASTICCINO1 (PAS1* FKBP70) (Vittorioso et al., 1998) and the maize FKBP66 (*mzFKBP-66*) (Hueros et al., 1998).

The large FKBPs from plants possess a similar structure with three or four FKBP12-like domains in the N-terminus portion of the protein, a tetratricopeptide repeat (TPR), which is thought to be involved in protein- protein interaction and a calmodulin binding domain. However their mRNA expression pattern is different. For example, the *ROF1* was expressed at low levels in all organs and induced by wounding and NaCl (Vucich, Gasser, 1996) whereas the *wFKBP73* was highly expressed in young tissues and not induced by stress (Blecher et al., 1996; Aviezer et al., 1998).

A. Altman et al. (eds.), Plant Biotechnology and In Vitro Biology in the 21st Century, 417–420.

Characterisation of the *pasticcino* mutants reveals for the first time the involvement of a member of the FKBP family in plant development. The inactivation of *PAS1* causes ectopic cell proliferation in cotyledons, extra layers of cells in the hypocotyl and abnormal apical meristem, a phenotype being correlated with cell division and cell elongation effect. It was also demonstrated that the mRNA steady- state level of the *pas*1 was increased in the presence of cytokinin. This is the first example of involvement of FKBP in the control of cell proliferation possibly connected to the cytokinin signal (Vittorioso et al., 1998).

Recently a novel heat stress induced PPIase, belonging to the FKBP family was isolated from a wheat cDNA library. The open reading frame encodes a heat induced 77 kDa protein, designated as the wheat FKBP77 (wFKBP77), which possesses three FKBP-12 like domains, a putative tetratricopeptide (TPR) motif and a calmodulin binding domain. The wFKBP77 exhibits 84% identity to the wheat FKBP73 isoform and 42% identity to the human FKBP59 (Kurek et al., 1998). Similar to the human FKBP59, the wheat FKBP77 as well as the wheat FKBP73 can bind to the heat shock protein HSP90, via the tetratricopeptide (TPR) domain in the presence of the p23, in wheat germ lysate (Reddy et al., 1998).

RESULTS AND DISCUSSION

We have previously showed that the *wFKBP77* could be detected after exposure to 37°C for 15 minutes, reaching its maximal level after 60 minutes, whereas the *wFKBP73* transcript level did not change during the heat stress (Kurek et al., 1998). Studies of the *wFKBP77* transcript abundance during plant development revealed that the *wFKBP77* and *wFKBP73* are highly expressed in young and reproductive tissues exposed for 2h at 37°C (Figure 1). In etiolated seedlings, the *wFKBP77* and *wFKBP73* are expressed in a similar pattern (Fig1. A). The transcript level of both wFKBPs was the highest in the two-day old shoots and decreased to 43% and 46% in the six-day old shoots. When seedlings were grown in light conditions (16 h light and 8 h dark), the RNA pattern dramatically changed. The decrease in the expression of *wFKBP77* was detected during seedling maturation in four and six-day old shoots being 15% and 8% respectively, as compared to two-day old shoots (Figure 1. B). The expression of the *wFKBP73* was reduced in the green shoots to a lesser extent, being 77% in the four-day old shoots and reaching a minimal expression of 14% in the six-day old shoots. In roots, the *wFKBP77* transcript level decreases gradually, whereas the *wFKBP73* transcripts was reduced drastically reaching 50% in the three-day old seedlings as compared to the two-day old seedlings (Figure 1. C). The *wFKBP77* is expressed in anthers and pistils at similar levels, whereas the *wFKBP73* is 8-fold higher in pistils as compared to anthers (Figure 1. D). In mature leaves, leaf sheaths and culms neither *wFKBP77* nor *wFKBP73* transcript could be detected (data not shown).

Since a difference in transcript level between green and etiolated shoots was found, 3 day-old dark grown seedlings were illuminated for 10 hours and exposed for 2 hours to heat stress and compared to light-grown seedlings and to etiolated seedlings, both exposed to the heat stress. It was found that the illumination treatment did not affect the *wFKBP77* transcript level as compared to the etiolated seedlings (Figure 2), indicating that the wFKBP77 is not directly regulated by light.

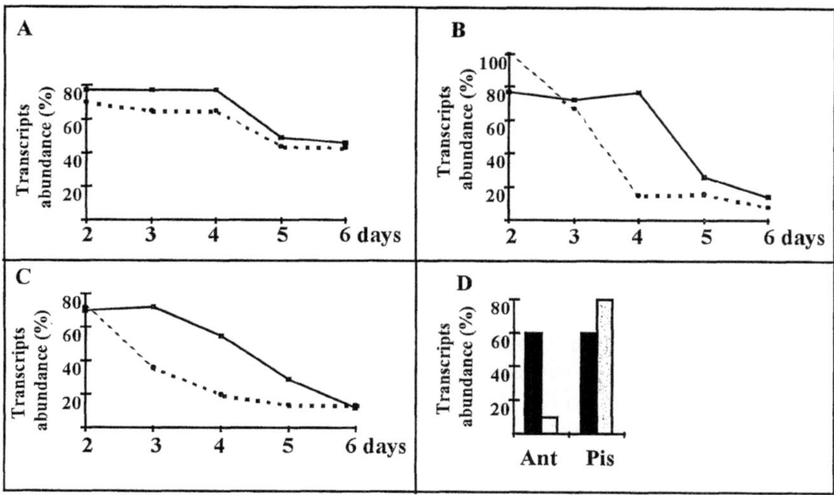

Figure 1. Transcripts abundance of *wFKBP77* (---, ■) and *wFKBP73* (–, □) in etiolated shoots (A), green shoots (B), roots (C) anther and pistils (D).

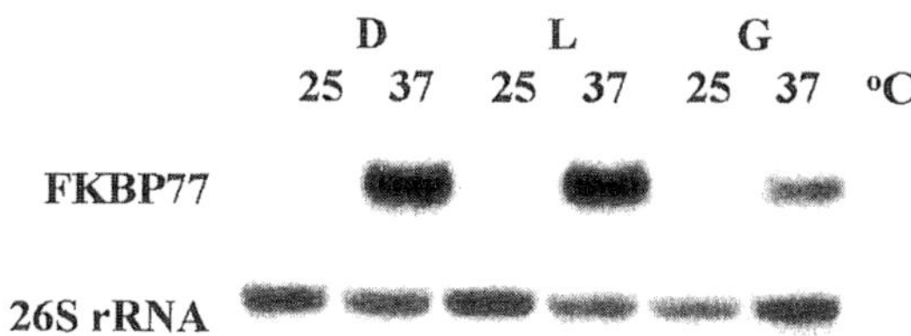

Figure 2. Northern blot analysis of 3 day-old dark grown seedlings (D), 3 day-old dark grown seedlings illuminated for 12 hours (L) and 3 day-old light grown seedlings (16 h light and 8 h dark cycles)(G).

We have characterized a novel heat induced wheat FKBP isoform that is developmentally regulated in a similar pattern to the wheat FKBP73 (Kurek et al., 1998). This observation defines the wFKBP77 as a heat stress dependent FKBP isoform, whereas the wFKBP73 can be defined as the cognate isoform since it is expressed under physiological conditions and is not induced by heat stress.

The presence of two wheat FKBPs is comparable with the situation of the HSPs. For example, developmental and heat stress dependent expression was reported for the HSP81 and HSP82 in maize where the highest expression was detected in young tassels. The HSP81 is only mildly heat inducible, whereas HSP82 shows strong inducibility (Marrs et al., 1993). In other cases, FKBPs were found to be increased after heat stress: the *VfFKBP15* which was induced only 4-5 fold after heat stress (Luan et al., 1996) and the mammalian HSP56 (also known as FKBP52) which has been characterized as a heat stress inducible protein (Sanchez, 1990; Tai et al.,1993). The presence of two isoforms in wheat, one expressed under normal growth and the other after heat stress indicates a more complex regulatory situation.

The higher abundance of *wFKBP77* and *wFKBP73* transcripts in young tissues and reproductive tissues can be compared to the expression of the human *FKBP51* which shows the highest levels in testis (Nair et al., 1997) and Arabidopsis cyclophilins which are highly expressed in floral tissues (Gasser et al., 1990) but different from the Arabidopsis *ROF1* and *VfFKBP15* which show similar levels of expression (Vucich, Gasser, 1996; Luan et al., 1996), or *Vf FKBP13* expressed preferentially in green tissues (Luan et al., 1994).

In order to obtain an insight in to the FKBPs role, transgenic rice and wheat plants overexpressing wFKBP77 and wFKBP73 have been produced. We hope to understand more about the function of wFKBP77 and wFKBP73 *in planta*, by analysing these transgenic rice and wheat plants in various stress conditions, known to increase the abundancy of malfoldeed proteins, such as heat stress.

REFERENCES

Aviezer K et al (1998) Plant Science. (in press).

Blecher O et al (1996) Plant Mol. Biol. 32, 493-504.

Buchner J (1996) FASEB J. 10, 10-19.

Crabtree GR, Schreiber SL (1996) TIBS 21, 418-422.

Dolinski K, Heitman E. (1997) Guidebook to Molecular Chaperones and Protein Folding Catalysis, Sambrook and Tooze, eds. Oxford: Oxford University Press, pp. 359-369.

Gasser CS et al (1990) Proc. Natl. Acad. Sci. 87, 9519-9523.

Harding MW et al (1989) Nature 341, 758-760.

Hartl FU (1996) Nature 381, 571-580.

Hueros G et al (1998) Planta 205, 121-131.

Kurek, I et al (1998) Plant Physiol. (in press).

Luan S et al (1994) Plant Cell 6, 885-892.

Luan S et al (1996) Proc. Natl. Acad. Sci. 93, 6964 -6969.

Marrs KA et al (1993) Develop. Genet. 14, 27-41.

Nair SC et al (1997) Mol. Cell Biol. 17, 594-603.

Owens-Grillo JK et al (1996) J Biol. Chem. 271, 13468-13475.

Peattie DA et al (1992) Proc. Natl. Acad. Sci. 89, 10974-10978.

Reddy RE et al (1998) Plant Physiol. (in press).

Ruddon RW, Bedows E (1997) J. Biol. Chem. 272, 3125-3128.

Sanchez ER. (1990) J Biol Chem 265, 22067-22070.

Schreiber SL (1991) Science 251, 283-287.

Siekierka JJ et al (1989) J. Biol. Chem. 265, 21011-21015.

Tai PK et al (1993) Biochemistry 32, 8842-8847.

Vittorioso P et al (1998) Mol. Cell Biol. 18, 3034-3043.

Vucich VA, Gasser CS (1996) Mol. Gen. Genet. 252, 510-517.

Regulation of Programmed Cell Death in Cultured Soybean Cells by a Cysteine Protease Inhibitor

A. Levine[1], B. Belinghi[1], M.Solomon[1], E.Menachem[1] and M.Delledonne[2]

[1]Hebrew University of Jerusalem, Dept. of Plant Sciences, Jerusalem 91904, Israel and [2]Istituto di Genetica, Universita' Cattolica S.C., Piacenza -Italy

Introduction

Regulation of cell death is among the most crucial tasks in the life of a cell. Recent data point to a complex regulatory system, composed of many signaling steps. Recent isolation of mutant arabidopsis and maize plants that undergo spontaneous cell death in unstimulated leaves implies that the cell death process is under genetic control (Dietrich et al., 1994; Greenberg et al., 1994). Many different stimuli have been shown to induce the process of programmed cell death (PCD) in a variety of plant species. The PCD response that occurs during the hypersensitive reaction of plants to avirulent pathogens is one of the most studied systems. When plants "sense" an infection by avirulent pathogens they begin to produce reactive oxygen species (ROS) that can trigger the cell death program in the reacting cells (Lamb and Dixon, 1997). Previous work has shown that the induction of PCD occurs only after the ROS have accumulated to a certain threshold value (Levine et al., 1994). Once however, the PCD has been activated, the initial stimulus can be removed without affecting the final outcome.

Our earlier studies have shown that the PCD process triggered by the oxidative stress can be efficiently blocked by synthetic protease inhibitors directed against serine and/or cysteine proteases (Levine et al., 1996). The broad activity of the synthetic inhibitors did not allow the precise determination of the protease responsible for PCD. Nevertheless, the strong inhibition of PCD by the inhibitors lead us to hypothesize that one or more proteolytic steps are involved in the PCD signal transduction pathway. Proteases and particularly cysteine proteases have been implicates in the animal and nematode apoptosis (Martin and Green, 1995). In these systems it was shown that a cascade of cysteine proteases with target specificity towards the amino acid aspartate at the N' term position control the animal apoptosis pathway. Recently, induction of cysteine proteases was also shown following programmed cell death stimuli in plants (Minami and Fukuda, 1995). In fact, the activation of cysteine proteases has been postulated as the most conserved event in the PCD of various organisms activated by many different stimulants.

A. Altman et al. (eds.), Plant Biotechnology and In Vitro Biology in the 21st Century, 421–424.

Results and Discussion

In order to test the involvement of proteases in PCD triggered by the oxidative burst, we first analyzed if the oxidative burst induced novel proteolytic activity. To this end suspension cultured soybean cells (cultivar William's *82*) were treated with 4 mM of hydrogen peroxide. The concentration was chosen on the basis of earlier studies which showed that this dosage induced a time dependent PCD process. After the oxidative stress treatment cells were harvested and analyzed for proteolytic activity in SDS-PAGE zymograms. Three induced bands were seen on the zymogram (98 kD, 64 kD, 50 kD) after the H_2O_2 treatment, while the activity of another 80 kD band that was very strong in uninduced cells decreased. These changes occurred very fast (20 min.) following the H_2O_2 treatment.

In order to analyze the target specificity of the induced proteases, the proteases were separated by anion exchange chromatography and each fraction tested with a set of short fluorogenic peptides. Such analysis revealed that the newly activated proteases had a strong preference for cleavage of peptides after Arginine (Z-Gly-Gly-Arg-AMC, boc-Gly-Lys-Arg-AMC, Z-phe-Arg-AMC), while the activity of proteases that exhibited preference for Phenylalanine or Tyrosine at this position (Z-Leu-Val-Tyr-AMC or Gly-Gly-Phe-AMC) was markedly reduced by the oxidative stress. No difference between the activity of induced and control fractions was seen with peptides that had Proline at this position. The proteolytic activity was fully active in the presence of EDTA, therefore, excluding the metalloproteases.

Given the immense importance of the proteases in animal apoptosis and the recent findings on the activation of protease genes in plant systems undergoing senescence and/or cell death, we were interested to see if the plant proteases have a regulatory role analogous to the animal systems. In order to evaluate if the activation of the proteases is essential for the execution of plant PCD, we decided to block the activity of specific classes of proteases through ectopic expression of specific plant protease inhibitor genes. When compared to using synthetic inhibitors this approach has the advantage of the high specificity inherent in the protein-protein interactions. Three well characterized endogenous inhibitors of soybean were chosen: Kunitz, CII and cystatin. Kunitz is a potent inhibitor of trypsin, CII is a Bowman-Birk inhibitor that acts against chimotrypsin and elastase and cystatin inhibits a number of cysteine proteases. All three genes were placed under the control of a constitutive CaMV 35S promoter. As control for transformation efficiency we included the b-glucuronidase gene, cloned behind the same promoter. This control also allowed us to evaluate the effects of the selected protease inhibitors on the soybean cell death against a background of similarly treated cultures,

The plasmids were inserted into *Agrobacterium tumefaciens* and the soybean cells were transformed via coinoculation. To estimate the transformation efficiency an aliquot of the GUS transformed cells was stained X-gluc. Judged from microscopic examination of X-gluc stained cells, this procedure resulted in >75% of transformed cells. After removal of agrobacteria, the cells were challenged with 5 mM of H_2O_2. Cell death was assayed 10 hours later by Evan's blue staining. The results presented in Figure 1A clearly show that the cystatin transformed cell cultures exhibited greatly reduced degree of dead cells. No protection from cell death was seen in the Kunitz transformed cells and only a limited inhibition of cell death was seen in the CII transformed cultures.

Few conclusions can be drawn from the data presented in Figure 1. First, the fact that the cell death of soybean cultures, which was stimulated by oxidative stress, could be inhibited by manipulation of intracellular metabolism argues that the H_2O_2 triggered cell death is not the direct result of cellular damage caused by peroxidation reactions but rather a result of an endogenous signaling pathway that terminated in cell death. In other words, H_2O_2 stimulated an active form of cell death, or PCD. Secondly, the inhibition of soybean PCD by the manipulation of intracellular proteolytic metabolism strongly implies that proteases play a regulatory role in the plant PCD process. The data also identify the class of cysteine proteases as the main type of proteases involved in the transduction of the oxidative stress. Most interesting is that the soybean response to oxidative stress could be controlled by ectopic expression of an endogenous gene.

In plants high levels of oxidative stress that result in necrotic lesions occur during pathogenesis response to avirulent pathogens as part of the hypersensitive reaction (HR). it has been shown that the oxidative burst plays a major role in PCD of pathogen challenged cells. We therefore tested if ectopic expression of cystatin can also modulate the degree of HR associated cell death. In soybean, HR is controlled by a well characterized gene-for-gene system that allows experimentation with isogenic lines of bacterial pathogens. In the specific case of soybean cultivar Williams82 HR occurs in response to infection by a strain of Pseudomonas syringae pv. glycinea that has the avrA avirulence gene, while the isogenic avrC harboring strain produces disease, with limited and delayed cell death. As shown in figure 1B cystatin expression greatly reduced the degree of cell death caused by the bacterial infection. Most notable is the effect caused by infection with the avirulent (avrA) strain, in which case cystatin overexpression almost completely blocked cell death. Interestingly, overexpression of cystatin also reduced the degree of cell death caused by the virulent strain.

In summary, our results show that in cultured soybean cells cysteine proteases play a key role in the programmed cell death response caused by avirulent pathogen or by direct oxidative stress. They also suggest an important role for endogenous cysteine protease inhibitors in the regulation of PCD.

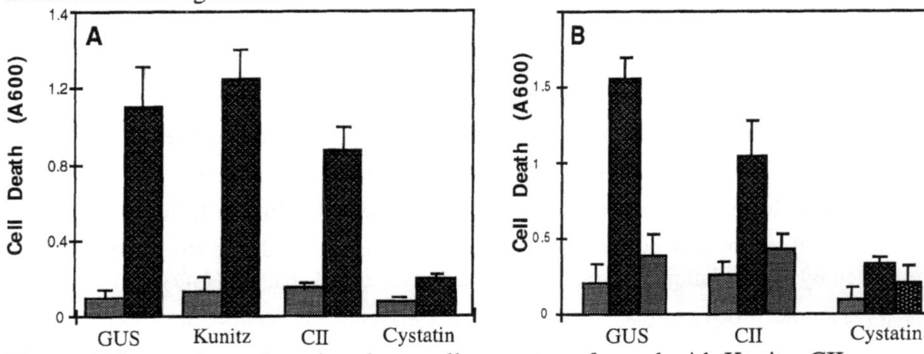

Figure 1. Suspension cultured soybean cells were transformed with Kunitz, CII, Cystatin or GUS as indicated. Cell death was assayed by Evan's blue (Levine et al., 1994). A: cells were left untreated (left bars) or challenged with H2O2 (right bars). B: cells were challenged with Pseudomonas syringae *avrA* (center bars) or *avrC* (right bars).

Materials and Methods

In-gel protease assay was done with 0.12% gelatin according to Heussen and Dowdle (1980). Anion exchange chromatography was done with UNO (BioRad) column using NaCl gradient elution. In-vitro protease activity was tested with peptides listed in the text. Fractions were incubated at 30C for 20 minutes and measured in a Biotek FL600 fluorometer. Inhibitor genes were prepared by PCR using primers to published DNA sequences. The genes were cloned into pBI121 (Clontech). Cell death was assayed as described in Levine et al (1994).

Acknowledgments
This work was supported in part by the Binational USA-Israel Science fund (BSF)

References

Dietrich RA et al (1994) Cell *77*, 565-77.
Greenberg JT et al (1994) Cell 77, 551-63.
Heussen and Dowdle (1980) Anal.Biochem. 102: 196-202
Lamb C and Dixon RA (1997) Plant Molec.Biol. 48, 251-275.
Levine A et al (1996) Current Biol. 6, 427-437.
Levine A et al (1994) Cell 79, 583-93.
Martin SJ and Green DR (1995) Cell 82, 349-52.
Minami A and Fukuda H (1995) Plant Cell Physiol. 36, 1599-606.

POLYADENYLATION AND DEGRADATION OF mRNA IN THE CHLOROPLAST

Irena Lisitsky and Gadi Schuster

Department of Biology, Technion - Israel Institute of Technology, Haifa 32000, Israel

The poly(A) tails found at the 3' ends of nearly all eukaryotic transcripts are added during RNA processing in the cell nucleus and are later shortened when RNA is transported to the cytoplasm (Jackson and Standart, 1990; Manley, 1988). In eukaryotic cells, the addition of multiple adenosine residues to the 3' end of nuclear RNA plays a key role in generating functional mRNAs and in regulating mRNA degradation. The presence or absence of poly(A) tail as well as its length can determine translational efficiency (Manley, 1995a; Wahle and Keller, 1996). In addition, deadenylation is the first and rate-limiting step in the breakdown of least some unstable eukaryotic mRNAs (Beelman and Parker, 1995). In bacteria, polyadenylation at the 3' end was shown to be involved in mRNA decay *in vivo* (Sarkar, 1997). The recent model for mRNA decay in *E. coli* suggests that a poly(A) tail targets mRNA for decay (Cohen, 1995; Manley, 1995b).

We have studied the post-transcriptional addition of poly(A)-rich sequences to mRNA in chloroplasts of higher plants (Schuster *et al.*, 1997). Several sites in the coding region and the mature end of spinach chloroplast *psbA* mRNA, which encodes the D1 protein of photosystem II, were detected as polyadenylated sites. The poly(A) moiety in the chloroplast, in contrast to that in eukaryotic nuclear encoded and bacterial RNAs, is not a ribohomopolymer of adenosine residues, but clusters of adenosines bounded mostly by guanosines and rarely by cytidines and uridines; it may be as long as several hundreds nucleotides (Lisitsky *et al.*, 1996). Further analysis of the initial steps of chloroplast *psbA* mRNA decay revealed specific endonuclease cleavage sites that perfectly matched the sites where poly(A)-rich sequences are added (Lisitsky *et al.*, 1996). Our results suggest a mechanism for the degradation of *psbA* mRNA in which endonucleolytic cleavages are followed by the addition of poly(A)-rich sequences to the upstream cleavage products, which target these RNAs for rapid decay.

To address the question whether the addition of a poly(A)-rich tail to RNA molecules is required for degradation by chloroplast exonuclease(s) the degradation assay based on lysed chloroplasts from spinach was used. In lysed chloroplasts, the addition of the polyadenylation inhibitor, cordycepin-triphosphate (3'-dATP), inhibited

A. Altman et al. (eds.), Plant Biotechnology and In Vitro Biology in the 21st Century, 425–428.

the degradation of *psbA* and *rbcL* mRNAs. Furthermore, degradation intermediates generated by endonucleolytic cleavages accumulated (Lisitsky *et al.*, 1997a). Similar results were obtained when yeast-tRNA was added to the mRNA degradation system as a non-specific exoribonuclease inhibitor (Klaff, 1995). Nevertheless, the stabilization mechanisms differ: while tRNA directly affects the exonuclease activity, 3'-dATP has an indirect effect by inhibiting polyadenylation. The results indicate that the addition of poly(A)-rich sequences to endonucleolytic cleavage products of chloroplast mRNA is required to target these RNAs for rapid exonucleolytic degradation (Lisitsky *et al.*, 1997a).

When *in vitro* transcribed RNAs were incubated with chloroplast-protein extract, competition between polyadenylated and non-polyadenylated RNA for degradation resulted in the rapid degradation of the polyadenylated molecules and stabilization of their non-polyadenylated counterparts (Lisitsky *et al.*, 1996). To elucidate the molecular mechanism governing this effect, we determined whether the chloroplast exoribonuclease 100RNP/PNPase (Hayes *et al.*, 1996) preferably degrades polyadenylated RNA. When separately incubated with each molecule, isolated 100RNP/PNPase degraded polyadenylated and non-polyadenylated RNAs at the same rate. However, when both molecules were mixed together, the polyadenylated RNA was degraded whereas the non-polyadenylated RNA was stabilized (Lisitsky *et al.*, 1997b). In RNA-binding experiments, 100RNP/PNPase bound the poly(A) sequence with much higher affinity than other RNA molecules, thereby defining the poly(A)-rich RNA as a preferential substrate for the enzyme. 100RNP/PNPase may therefore be involved in a mechanism in which post-transcriptional addition of poly(A)-rich sequence targets the chloroplast RNA for rapid exonucleolytic degradation (Lisitsky *et al.*, 1997b).

Table I: **Comparison of mRNA polyadenylation in bacteria, chloroplast and yeast (nuclear encoded mRNA)**

	E. coli	Chloroplast	Yeast
Length of poly(A) tail	14-60	17-270	55-60
Poly(A) tail content	Exclusively A	A - 75% G - 20%	Exclusively A
Polyadenylation sites	At accessible 3' ends	At accessible 3' ends	Downstream of consensus element
Function	Targeting mRNA for degradation	Targeting mRNA for exonucleolytic degradation	Formation of functional mRNA

In the last few years, our understanding of the chloroplast mRNA degradation pathway has already progressed. For certain mRNAs, such as *psbA*, the different steps of its specific decay have been disclosed (Schuster *et al.*, 1997). The succession of endonucleolytic-degradation events is reminiscent of the prokaryotic ancestor (Sarkar, 1997). This is supported by the observation that one of the enzymes involved has been identified as sharing structural and functional homology with its prokaryotic counterpart (Hayes *et al.*, 1996). The chloroplast, however, is in part regulated by the nucleus, as well as by external stimuli such as light. Moreover, the longevity of its mRNA better reflects the properties of plants than of bacteria (Gruissem and Schuster, 1993). Therefore, the chloroplast probably adopted an intermediate position by combining these different features. Following the biochemical pathway of plastid mRNA degradation, research will continue on the regulation of mRNA stability as part of the regulatory network which determines leaf development and adaptation to environmental conditions.

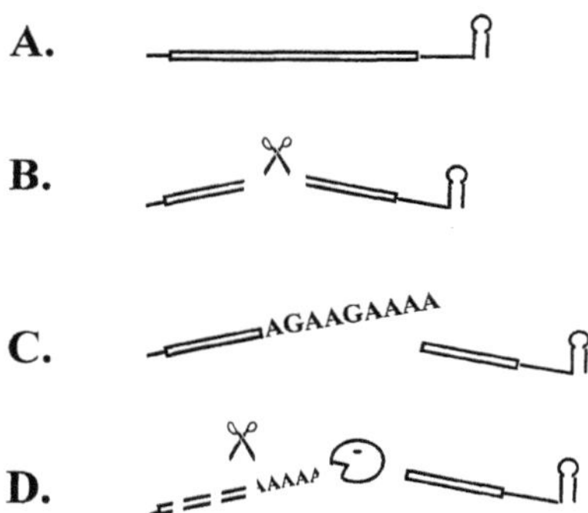

Figure 1. A model for the degradation pathway of mRNA in the chloroplast.
A. A schematic representation of the *psbA* mRNA molecule. The open box represents the amino-acid coding region, and stem-loop structures represent inverted repeats in the 3' untranslated region which potentially form stem-loop structure. B. The initial step in the mRNA degradation process is suggested to be endonucleolytic cleavage by an as yet unidentified endonuclease(s). The endonuclease is schematically symbolized by scissors. C. A poly(A)-rich tail, which can be up to several hundred nucleotides in length, is then added to the 3' end of the 5' endonucleolytic cleavage product. D. The polyadenylated RNA molecule is rapidly degraded by an exonuclease(s), possibly the 100RNP/PNPase.

ACKNOWLEDGMENTS

This work was supported by the Israel Science Foundation, administered by the Israel Academy of Science and Humanities, by a grant from the Niedersachsen Foundation for Science and Art and by a grant from the Israel-Japan Scientific Cooperation Research Foundation.

References

Beelman, C.A. and Parker, R. (1995) Cell, 81, 179-183.
Cohen, S.N. (1995) Cell, 80, 829-832.
Gruissem, W. and Schuster, G. (1993) In: Brawerman, G. and Belasco, J. (eds.), *Control of Messenger RNA Stability*. Academic Press, Orlando, FL, pp. 329-365.
Hayes, R., et al (1996) EMBO J., 15, 1132-1141.
Jackson, R.J. and Standart, N. (1990) Cell, 62, 15-24.
Klaff, P. (1995) Nucleic Acids Res, 23, 4885-4892.
Lisitsky, I, et al (1996) Proc. Natl. Acad. Sci. USA, 93, 13398-13403.
Lisitsky, I., et al (1997a) Plant J., 12, 1173-1178.
Lisitsky, I., et al (1997b) J. Biol. Chem., 272, 17648-17653.
Manley, J.L. (1988) Biochim. Biophys. Acta, 950, 1-12.
Manley, J.L. (1995a) Curr. Opin. Gen. Dev., 5, 222-228.
Manley, J.L. (1995b) Proc. Natl. Acad. Sci. USA, 93, 1800-1801.
Sarkar, N. (1997) Annu. Rev. Biochem., 66, 173-197.
Schuster, G., et al (1998) In: Bansal, K. (ed.) *Recent Advances in Plant Molecular Biology*. Oxford and IBH Publishers, New Delhi., Vol. in press.
Wahle, W. and Keller, W. (1996) Trends. Biochem. Sci., 21, 247-250.

FACTORS AFFECTING CELL DIVISON IN PLANT CELLS

T. Nagata, S. Ishida, S. Nagata, Y. Takahashi
Department of Biological Sciences, Graduate School of Science, University of Tokyo, 7-3-1 Hongo, Bunkyo-ku, Tokyo 113-0033, Japan

1. Introduction

Tobacco BY-2 cell line has unique characteristics (Nagata et al. 1992; Shibaoka 1993). Amongst, this cell line cultured at the optimal conditions showed very fast growth rate, as it showed a doubling time of 13-14 hrs and this cell line propagated 100-120 fold in a week. As a consequence, high synchrony of ca. 70% in terms of mitotic index (MI) starting from S phase is attained after the release of aphidicolin treatment, while much higher synchrony of ca. 95% MI starting from M phase can be obtained after the release of the sequential treatment of aphidicolin and propyzamide. This high synchrony is frequently being used for the study of various issues of plant cell cycle studies. We are further seeking how the cell cycle progression of this cell line could be stopped and re-started by other procedures than those described above. Then we found that cell cycle progression could be arrested and re-started at least by two treatments. One is phosphate starvation, while the other is auxin starvation. When stationary phase tobacco BY-2 cells were cultured in a medium without phosphate for 3 days, the cell cycle progression was arrested at G_0 phase. Subsequently, when phosphate was added to the medium, semi-synchronous cell division was induced. This cell division was preceded by DNA synthesis, indicating that the addition of phosphate regained the meristematic activity in the cells which were blocked at the G_0 phase of the cell cycle. Under this condition, a few phosphate up-regulated genes and a down-regulated gene were identified and characterization of these genes are currently being conducted (T. Sano, T. Nagata, unpublished results). Similar arrest of the cell cycle progression was also induced by the depletion of auxin from the culture medium and the re-addition of auxin to the auxin-starved tobacco BY-2 cells induced semi-synchronous cell division (Ishida et al. 1993). However, some significant difference was observed between these two procedures. Although the amount of phosphate that added to the culture medium was 2 mM and phosphate is involved in various metabolic pathways, 2,4-D, an auxin, was added only at the concentration of 0.90 μM. It is intriguing to elucidate how such a subtle change of auxin level affects the meristermatic activity upon plant cells. Thus, in this article we would like to describe our efforts to elucidate what kind of molecular

A. Altman et al. (eds.), Plant Biotechnology and In Vitro Biology in the 21st Century, 429–432.

changes could be induced by the addition of auxin to the auxin-starved tobacco BY-2 cells.

2. Procedure

Tobacco BY-2 cells derived from tobacco (*Nicotiana tabacum* L. cv. Bright Yellow 2) were cultured as described by Nagata et al. (1992). Auxin-starvation of tobacco BY-2 cells and the subsequent addition of 2,4-D to the auxin-starved BY-2 cells was conducted as described by Ishida et al. (1993). To monitor regain of meristermatic activity of tobacco BY-2 cells, MI was determined after DAPI staining under fluorescence microscopy. Simultaneously, DNA synthesis was assessed by the incorporation of BrdU into BY-2 cells, which was detected by staining with an antibody against BrdU under fluorescence microscopy.

In this study tobacco 2b13 cell line (Nakajima et al. 1979), which was derived from tobacco BY-2 cells and is growing without the addition of auxin for ca. 20 years, was also used.

3. Results and Discussion

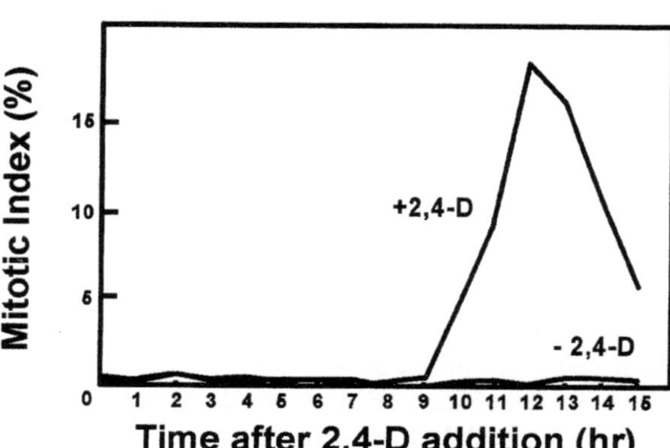

First, we examined whether DNA synthesis is involved upon the induction of semi-synchronous cell division by the addition of 2,4-D to the auxin-starved tobacco BY-2 cells. Immunostaining with the antibody against BrdU revealed that the incorporation of BrdU into nuclei was detected from 4 hrs to 10 hrs. Thus, it was concluded that the addition of 2,4-D facilitated the progression of the cell cycle of auxin-starved tobacco BY-2 cells from G_0 phase to S phase. Subsequently cell division was observed from 9 hrs, reaching a maximum at 12 hrs after the addition of 2,4-D, while no cell division was observed without the addition of 2,4-D (Figure). Although the regain of meristematic activity by the addition of phosphate to phosphate-starved tobacco BY-2 cells was similar to that described above, there are several differences between the two. Meristematic activity was recovered by the addition of phosphate to the phosphate-starved cells until the 7th day of culture, while that by the addition of auxin was only possible until the 4th day of culture.

Search for differentially expressing genes regarding 2,4-D in the cDNA library by the differential screening method allowed us to isolate one cDNA clone, which was named *arcA*. Though *arcA* was induced specifically by auxin, its expression was ubiquitously observed in the whole plant body. Auxin-responsive regions of *arcA* promoter was confined to some specific regions in the promoter of *arcA* (Ishida et al.

1996). *arcA* was not found in the search for phosphate-induced genes upon addition of phosphate to phosphate-starved tobaco BY-2 cells (T. Sano, T. Nagata, unpublished results). The predicted gene product of *arcA*, which has a molecular mass of 35,825 kDa, belongs to WD-40 repeat proteins, in which the most typical one is β-subunit of G proteins (Ishida et al. 1993). However, accumulated information in the data base revealed that arcA is classified into a group of receptor for activated C-Kinase (RACK) 1 among the WD40 repeat proteins. RACK1 was identified biochemically in animal cells as the anchoring protein for activated Protein Kinase C (PKC), which was transferred from cytoplasm to cell membranes upon the induction by growth hormones (Mochly-Rosen 1995). RACK1-like proteins were widely found in eukaryotic cells; human, mouse, chicken, *Neurospora*, *Saccharomyces*, *Chlamydomonas* and tobacco. Protein structures of RACK1s are extremely well preserved, because amino acid sequences of these proteins among vertebrates from human to avian was 100% identical. Recently, other plant homologs were found in *Brassica napus* and *Medicago sativa*. In the latter case, *Msgbl* was isolated as a gene that was expressed in the early stage of nodule formation of *M. sativa* by the infection of *Rhizobium meliloti* (McMhann et al. 1997). Thus, it may be reasonable to speculate that arcA and its plant homologs could play a role for the receptors for activated PKC in plants. However, there has not been any experimental evidence for the presence of PKC in plants.

Thus, we tried to find cDNAs, whose gene products could have protein-protein interaction with arcA. When we have carried out the screening of such genes in tobacco cDNA libraries by the yeast two hybrid systems using *arcA* as a bait, we have isolated several cDNAs. Among them, a cDNA clone T218 was found to be a homolog of β-subunit of α-dendrotoxin-sensitive K^+-channel, which was shown to play a pivotal role in regulating K^+-channel in animal cells (Scott et al. 1994). Thus far, we confirmed biochemically the binding of arcA with the gene product of cDNA clone T218, since the fusion protein of maltose-binding protein (MBP) and arcA, that was trapped to the amylose column, bound specifically the gene product of T218. Introduction of point mutations into either of these partners cancelled this specific binding. However, the functional significance of this β-subunit of K^+-channel in plants has not been clarified yet.

To circumvent this difficulty in using plant cells, we also tried to elucidate the function of *arcA* using *S. cerevisiae*, as we have identified an *arcA* homolog in *S. cerevisiae*. The gene disruption of *arcA* homolog in *S. serevisiae* seemed to show some defects in meiosis. Further, though we are also examining the interaction of *arcA* with PKC in *S. serevisiae*, we have not got conclusive results yet, but we may find a clue to identify the function of *arcA* in plants in this manner.

Finally, I would like to come back again to factors that induce cell division in the auxin-starved tobacco BY-2 cells. Very recently when we replaced auxin with cell exudates from tobacco 2b13 cells, we could see the induction of cell division by the cell exudates. As described in the Procedure, tobacco 2b13 is a so-called habituated cell line for auxin. Furthermore, as it was confirmed that the cell line does not produce much amount of indole-3-acetic acid (Nakajima et al. 1979), a possibility that this factor could be auxin is excluded. Thus, this result indicated that cell exudates from tobacco 2b13 could be replaced with the effect of auxin and we are currently trying to identify what kinds of components in the cell exudates is responsible for inducing cell division in the auxin-starved cells. Though the molecular mechanism of habituation has not been clarified yet since its discovery, our experiment could give a clue to answer this long lasting question.

In conclusion, cell cycle progression of tobacco BY-2 cells was blocked by the depletion of auxin from the medium and the addition of auxin to these auxin-starved cells induced semi-synchronous cell division. Upon addition of auxin, was induced an auxin-regulated gene *arcA*, whose mode of expression suggests the presence of novel signal-transduction pathways regarding auxin, which would have a close link to the induction of cell division in these cells.

4. References

Ishida S et al. (1993) **Proc. Natl. Acad. Sci. USA** 90, 11152-11156.
Ishida S et al. (1996) **Plant Cell Physiol.** 37, 439-448.
McKhann HI et al. (1997) **Plant Mol. Biol.** 34, 771-780.
Mochly-Rosen D (1995) **Science** 268, 247-251.
Nagata T et al. (1992) **Int. Rev. Cytol.** 132, 1-30.
Nakajima H et al. (1979) **Plant Cell Physiol.** 29, 1489-1499.
Scott VES et al. (1994) **Proc. Natl. Acad. Sci. USA** 91, 1637-1641.
Shibaoka H (1993) **J. Plant Res.** Special Issue 3, 3-15.

CELLULAR INTEGRATION OF AUXIN- AND LIGHT SIGNALS DURING WHEAT SOMATIC EMBRYOGENESIS

A Nato[1], C Fresneau[2], J De Buyser[1], D Lavergne[1], G Ducreux[1] and Y Henry[3]
1: Laboratoire M.V.E., 2: Laboratoire de Bioénergétique Membranaire, 3: Laboratoire B.D.P. - Université Paris XI, 91405 Orsay, France.

Introduction

Plants respond to stimuli such as light and phytohormones involved in the control of cell division, differentiation, nutrient transport and accumulation. Research is being conducted to identify signaling elements that couple the stimuli to physiological responses. We hypothesize that signaling cascades which convert extracellular stimuli to intracellular metabolic changes are highly conserved through evolution. In this work, we analysed molecules known to play an important role in animal signaling. The visual transduction cascade in retinal cells constitutes our model for the study of putative elements which may be involved in auxin and light desensitization within plant cells and tissues (Fig. 1).

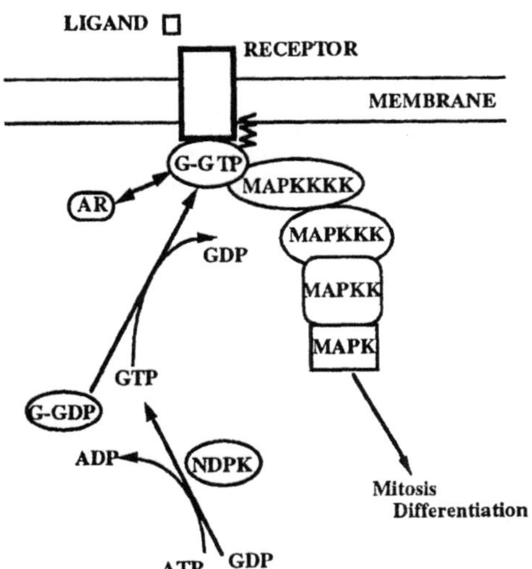

Fig. 1: Model for signaling pathways involved in plant development. AR = arrestin, G = G-protein, NDPK = nucleoside diphosphate kinase, MAPK = Mitogen-Activated Protein Kinase,

A. Altman et al. (eds.), Plant Biotechnology and In Vitro Biology in the 21st Century, 433–436.

Arrestin, a new class of protein observed in a wide range of animal tissues, is involved in the regulation of light signaling through G-protein interaction (Wilden et al. 1986). Recently, arrestin proteins were detected both in *Euglena gracilis* (Razaghi et al. 1997) and in the plant kingdom (Mirshahi et al. 1991) particularly during wheat somatic embryogenesis (Nato et al. 1995, 1997). Studies have shown that in higher plants, G-proteins are involved in hormonal and light signal transduction (Warpeha et al. 1991; Novikova et al. 1997) and in plant embryogenesis processes (Fabry, Beyser 1996; Zwaal et al. 1996). We also focused on NDP kinase (NDPK) expression and activity due to its role in controlling the supply of trinucleotide during embryogenesis (Yano et al. 1995) and to the striking homology of the NDPK amino acid sequence from spinach with both the human Nm23 and the *Drosophila melanogaster* Awd NDPKs.
The involvement of NDPK, G proteins and arrestin proteins was tested through temporal expression studies during wheat somatic embryogenesis induction from immature zygotic embryos by 2,4-D and during successive subcultures under light or darkness. The present work shows that under auxin treatments, molecules immunoreactive with antibodies to G-protein, arrestin and NDPK proteins as well as [α-^{32}P]GTP-binding proteins, were expressed during early wheat somatic embryogenesis. Photoregulation of these proteins was also observed during successive subcultures of somatic tissues.

Material and Methods

Hexaploid wheat Chinese Spring (CS) cultivar was used due to its high performance in somatic tissue culture (De Buyser et al. 1992). Immature embryo culture was performed according to the procedure described by Henry et al. (1994). The monoclonal antibodies raised against amino acid sequences of the bovine arrestin, the 143-2 anti-NDPK monoclonal antibody (a gift from F. Traincard, Institut Pasteur, Paris) and the polyclonal antibodies raised against the synthetic oligopeptide GAGESGKSTIVKQM representing a highly conserved region of mammalian α subunit of heterotrimeric G-proteins have been previously described (Nato et al. 1997). The wheat proteins were extracted, electrophoresed and immunoblotted as described in Nato et al. (1997). NDP kinase activity was measured by a classical PK-LDH coupled assay with ATP as the donor nucleotide and 8-Br-IDP as the acceptor nucleotide. For affinity labelling using [α-^{32}P]-GTP, the membranes were incubated for 1 hour in 50 mM Tris-HCl buffer (pH 7.5) containing 200 mM Na Cl, 5 mM $MgCl_2$, 1 mM EGTA and 0.3% Tween 20. They were then incubated for 1.5 h in 125 mCi (370 MBq/ml in water) [α–^{32}P]-GTP from Amersham and 20 mM cold ATP or GTP. Five washings of 30 min were performed in the incubation buffer containing 20 mM cold ATP or GTP.

Results

Influence of 2,4-D on somatic embryogenesis initiation

Antisera raised against a peptide sequence homologous to the highly conserved region of the α subunit of heterotrimeric G-proteins cross reacted with 58, 38, 32 and 25 kDa proteins from the soluble extracts of 7 days *in vitro*-cultured immature embryos. The [α-^{32}P]-GTP-binding assays labelled polypeptides of relative molecular mass of 48, 36, 32, 29, 23 and 22 kDa, which are close to the reported molecular mass of G-proteins. For the control sample (immature embryos) at time 0, four [α-^{32}P]GTP-binding proteins were present at 36, 32, 29 and 23 kDa but hardly perceptible. These results suggest that the *in vitro* induction of somatic tissue culture on MS medium containing 2,4-D triggers off a significant overexpression of soluble GTP-binding proteins whatever the embryogenicity of the material grown under darkness or under light.

The immature embryos cultivated for 7 days on medium containing 2,4-D overexpressed a 16 kDa NDPK band. The embryogenic process was characterized by the appearance of an additional NDPK form expressed as a 32 kDa band. *In vitro* culture of wheat immature zygotic embryos significantly stimulated NDPK activity, whereas somatic embryogenesis induction by the auxin 2,4-D had no significant effect.

The immature wheat embryo soluble extract contained a low amount of a 40 kDa arrestin form. The *in vitro* culture conditions in a 2,4-D containing medium, induced the overexpression of this 40 kDa protein as early as 7 days. After 21 days of *in vitro* culture the full expression of somatic embryogenesis was associated with the appearance of an additional 29 kDa arrestin form.

Influence of light during somatic tissue subculture

The G-protein analysis of embryogenic cultures under darkness revealed that the soluble extract contained two weak bands at 58 and 32 kDa. Both were overexpressed after subculture either in light or darkness. Bands at 18 and 22 kDa appeared. Soluble GTP-binding protein patterns were similar for the embryogenic and nonembryogenic tissue cultures grown either in darkness or light. On the other hand, when proteins from the microsomal fraction were labelled with [α-^{32}P]GTP, a strong specific GTP binding at 20 kDa was observed for the embryogenic cultures. This form is dramatically increased under light.

The 16 kDa classical NDPK was expressed at the same level either in light or in total darkness. In contrast, an additional 32 kDa polypeptide immunoreactive to NDPK antibody appeared as early as 5 days after subculture under light.

The 40 kDa plant arrestin form was detected whatever the culture conditions. The soluble 29 kDa form expressed in embryogenic cultures was under the control of light. Indeed, in a 12 h photoperiod, this arrestin form was expressed 5 days after subculture.

Expression in membrane proteins under 2,4-D or light conditions

Microsomal protein immunoblotting assays revealed a significant increase both in the expression of the 58 kDa Gα protein and of the 40 kDa arrestin protein, either in darkness or in light conditions. Light also promoted the appearance of the membrane-bound 32 kDa NDPK, the membrane-bound 29 kDa arrestin and a 35 kDa G-protein.

Discussion

In vitro culture conditions are responsible for important changes in signal transduction mechanisms. *In vitro*-cultured wheat immature embryos responded to the auxin 2,4-D and to light signals by modifying expression of G-proteins (as described in tobacco BY-2 cells by Ishida et al. 1993), NDPK and arrestin proteins. Our present work revealed that the cell proliferation process both under auxin and light conditions was characterized by the appearance of new immunoreactive forms of G-proteins and by the expression of an additional NDPK form (32 kDa). No significant effect of the 2,4-D treatment was detected on NDPK activity, in agreement with recent results from Novikova et al. (1997). Recently, Schmidt et al. (1997) have suggested that a putative specific signal transduction chain is expressed during early embryogenesis in carrot tissue culture. Roles for Gα

proteins in the control of cell proliferation have been identified in mammals (Denis-Henriot et al. 1996). Affinity probing with [α-^{32}P]-GTP showed pronounced GTP binding to polypeptides with relative molecular masses of 22 to 48 kDa. Bominaar et al. (1993) suggest that receptor-stimulated NDPK contributes to the mediation of hormone action by providing GTP which activates GTP-binding proteins. The expression of the 40 kDa plant arrestin-like form is highly stimulated in the presence of auxin, suggesting a possible function in the regulation of the hormonal transduction pathway. The cell differentiation process during somatic embryogenesis was characterized by the appearance of a photoregulated 29 kDa arrestin form observed in both the soluble and membrane fractions.

The specific role of G-proteins, NDPK and arrestins remains to be determined, but their expression in plant light-grown tissues suggests a possible function in the regulation of cell photomorphogenesis. The widespread tissue distribution of the signaling elements described here and their expression patterns modulated by light and hormones support the view that they may be involved in roles through G-protein-modulated signaling pathways. The stimuli-directed expression of these signaling molecules is an interesting feature and may be a useful tool for future research. Another emerging theme of plant signal transduction through G-protein-coupled receptor is the study of the desensitization process, *i.e.* the possible rapid loss of cellular sensitivity following presentation of a stimulus.

References

Bominaar AA, Molijn AC, Pestel M, Veron M, Van Haastert PJM (1993) Embo J 12: 2275-2279

De Buyser J, Marcotte JL, Henry Y (1992) Euphytica 63: 265-270

Denis-Henriot D, de Mazancourt P, Goldsmith PK, Giudicelli Y (1996) Cell Signal 8: 225-234

Fabry R, Beyser K (1996) Protoplasma 190: 79-87

Henry Y, Marcotte JL, De Buyser J (1994) Theor Appl Genet 89: 344-350

Ishida S, Takahashi Y, Nagata T (1993) Proc Natl Acad Sci USA 90: 11152-11156

Mirshahi M, Nato A, Razaghi A, Mirshahi A, Faure JP (1991) C R Acad Sc Paris 312: 441-448

Nato A, Mirshahi A, Cavalcante-Alvez JM, Lavergne D, Ducreux G, Mirshahi M, Faure J-P, De Buyser J, Tichtinsky G, Kreis M, Henry Y (1995) Current Issues in Plant Molecular and Cellular Biology, pp 519-524, Kluwer Academic Publishers, Dordrecht, The Netherlands

Nato A, Mirshahi A, Tichtinsky G, Mirshahi M, Faure J-P, Lavergne D, De Buyser J, Jean C, Ducreux G, Henry Y (1997) Plant Physiol 113: 801-807

Novikova G, Moshkov I, Smith AR, Hall MA (1997) Planta 201: 1-8

Razaghi A, Bonaly J, Chacun H, Faure JP, Mirshahi M, Barque JP (1997) BBRC 233: 601-605

Schmidt EDL, Guzzo F, Toonen MAJ, De Vries SC (1997) Development 124: 2049-2062

Warpeha KMF, Hamm HE, Rasenick MM, Kaufman LS (1991) Proc Natl Acad Sci USA 88: 8925-8929

Wilden U, Hall SW, Kühn K (1986) Proc Natl Acad Sci USA 83: 1174-1178

Yano A, Umeda M, Uchimiya H (1995) Plant Mol Biol 27: 1053-1058

Zwaal RR, Ahringer J, van Luenen HGAM, Rushforth A, Anderson P, Plasterk RHA (1996) Cell 86: 619-629

A MULTI-PARAMETER APPROACH FOR THE STUDY OF *IN VITRO* PHOTOSYNTHESIS

A. Rival*, K. Triques*, T. Beulé*, A. Nato**, D. Lavergne**, J. Santamaria***, J.-L. Verdeil*, V. Hocher*, C. Oropeza*** & S. Hamon*.

* CIRAD-CP/ORSTOM. GeneTrop Laboratory, ORSTOM, P.O. Box # 5045, 34032 Montpellier Cedex 01, France. e-mail: rival@mpl.orstom.fr;
** LMVE. Université de Paris Sud. Bat. 360, 91405 Orsay Cedex, France;
*** CICY. Biotechnology, Apdo 87, 97310, Cordemex. Merida, Mexico.

1. Introduction

During transfer to *ex vitro* conditions, the physiological status of *in vitro* grown plantlets is an important factor determining success rates (Van Huylenbroeck and Debergh, 1996). Apart from the control of water loss (Santamaria *et al*, 1994), the *in vitro* acquisition of photosynthetic ability is of paramount importance. The present paper describes a multi-parametric approach, combining both *in planta* and *in vitro* analysis for the study of photosynthesis in coconut (*Cocos nucifera* L.) and oil palm (*Elaeis guineensis* Jacq.) *in vitro* grown plantlets.

2. Material and methods

2.1. Plant material. The genetic origin of plant material and the protocols for *in vitro* culture have already been described for oil palm (Rival *et al.*, 1997) and coconut (Assy-Bah *et al.* 1989; Triques *et al.*, 1997a).

2.2. Analysis of photosynthetic parameters. The various methods employed for the estimation of chlorophyll fluorescence, the measurement of CO_2 exchanges and transpiration rates and the estimation of chlorophyll concentrations have been previously described (Rival *et al.*, 1997, Triques *et al.*, 1997 *a,b*).

3. Results

3.1. Chlorophyll fluorescence. $Ø_P^{MAX}$ was very low in dark-grown plantlets, then increased during the greening of leaves as early as 2 weeks after cultivation under PAR (Table 1). Values for $Ø_P$ and $Ø_P^{MAX}$ were not significantly different in *in vitro* grown plantlets (after 4 weeks under PAR) and in the acclimatised coconut palm. In oil palm (data not shown), *in vivo* chlorophyll fluorescence measurements indicated that the maximum photochemical activity of PhotosystemII (PSII) was very low in proliferating embryos and strongly increased in later development stages, finally reaching an activity very close to that measured in acclimatised plants. The quantum yield of photosynthetic electron transport followed the same trend except that a marked depression of electron transport activity was observed in the rooted plantlets.

A. Altman et al. (eds.), Plant Biotechnology and In Vitro Biology in the 21st Century, 437–440.

Table 1 - Changes in the maximal quantum yield ($Ø_P^{MAX}$) of PSII photochemistry (in dark-adapted leaves) and the actual quantum yield ($Ø_P$) of PSII photochemistry (in light-adapted leaves) in coconut leaves sampled at various stages of *in vitro* development. *Reported values are the means of 3 independent measurements ± SD. Means followed by the same letter are not significantly different according to the Newman and Keul's test.*

Culture stage	$Ø_P^{MAX}$ (Fm-F0)/Fm	$Ø_P$ (F'm-Fs)/F'm
dark grown plantlets	0.29 ± 0.02*a*	ND
1 week PAR	0.58 ± 0.02*a*	0.33 ± 0.03*a*
2 weeks PAR	0.71 ± 0.05*c*	0.41 ± 0.09*b*
4 weeks PAR	0.72 ± 0.04*c*	0.45 ± 0.03*b*
autotrophic plant	0.76 ± 0.01*c*	0.50 ± 0.03*b*

3.2. Net photosynthesis and transpiration. Net photosynthesis rates were measured through CO_2 exchange in leaves from *in vitro* grown plantlets (Table 2). During the greening of leaves, the net CO_2 exchange increased. The photosynthesis rate in vitroplants was then half of that measured in the autotrophic coconut palm.

Table 2 - Net photosynthesis estimated through CO_2 exchanges and transpiration rates in leaves from *in vitro* grown coconut plantlets.

Culture stage	Photosynthetic rate $\mu mol_{CO2}.m^{-2}\,s^{-1}$	Transpiration rate $mmol_{H2O}.m^{-2}\,s^{-1}$
dark grown plantlet	-0.500*a*	0.040*a*
1 week PAR	0.271*ab*	0.340*a*
2 weeks PAR	0.940*b*	1.135*b*
4 weeks PAR	1.144*b*	1.140*b*
autotrophic plant (reference)	2.430*c*	1.137*b*

3.3. RubisCO and PEPC capacities and PEPC:RubisCO ratio. Carboxylase (PEPC and RubisCO) capacities were measured in mature zygotic embryos, etiolated leaves and greening leaves (Table 3). During the *in vitro* culture process, the PEPC capacity of leaves drastically decreased. In contrast, the RubisCO capacity increased throughout the *in vitro* culture period. Consequently, the PEPC:RubisCO ratio dropped from 89.17 in the mature embryo down to 0.03 in ready-to-acclimatise plantlets (6 weeks under PAR), a ratio similar to that found in the autotrophic coconut palm. Furthermore, we have demonstrated (with a different batch of plantlets) that *in vitro* grown coconut plantlets during acclimatisation showed a faster decrease in their PEPC:RubisCO ratio than the seedlings, suggesting that an earlier transition from a heterotrophic to an autotrophic mode of carbon fixation takes place in the *in vitro*-cultured material (Triques *et al.*, 1997b). Just before acclimatisation, the RubisCO activity in vitroplants was lower than that in seedlings of the same age. Nevertheless, after acclimatisation, RubisCO activities were comparable in both *in vitro* and *in planta* germinated material.

3.4. Quantification of RubisCO. RubisCO was quantified in coconut mature embryos, dark grown plantlets and in greening leaves. RubisCO content increased from 0 $mg.g^{-1}_{TSP}$ in dark-grown leaves to 172.8 $mg.g^{-1}_{TSP}$ in leaves after 4 weeks under PAR. The RubisCO content was found to be 217.6 $mg.g^{-1}_{TSP}$ in the autotrophic coconut palm. In oil palm, the relative amount of RubisCO increased during somatic embryo development (from 32 in proliferating embryos to 388 $mg.g^{-1}_{TSP}$ in 2nd-cycle-shootlets), then it decreased during the rooting treatment (264 $mg.g^{-1}_{TSP}$).

Table 3 - Changes in total soluble protein contents and RubisCO and PEPC specific capacities during the *in vitro* development of coconut zygotic embryos.

Culture stage	TSP content $mg.g^{-1}_{FW}$	PEPC capacity $\mu mol_{CO2}.h^{-1}.mg^{-1}_{TSP}$	RubisCO capacity $\mu mol_{CO2}.h^{-1}.mg^{-1}_{TSP}$	PEPC:RubisCO ratio
mature embryo	14.40 ± 1.71^{b}	17.83 ± 0.51^{a}	0.20 ± 0.00^{a}	89.17 ± 2.56^{a}
dark grown plantlet	0.94 ± 0.00^{a}	40.30 ± 11.31^{a}	0.95 ± 0.21^{ab}	42.14 ± 2.49^{a}
1 week PAR	3.07 ± 0.49^{a}	4.17 ± 2.55^{b}	2.12 ± 1.02^{bc}	2.31 ± 1.44^{b}
2 weeks PAR	5.18 ± 0.25^{a}	3.14 ± 0.79^{b}	5.41 ± 1.19^{d}	0.61 ± 0.24^{b}
4 weeks PAR	5.13 ± 1.09^{a}	1.38 ± 1.00^{b}	3.83 ± 0.41^{c}	0.36 ± 0.26^{b}
6 weeks PAR*	20.56 ± 3.38^{c}	0.08 ± 0.03^{b}	2.83 ± 0.34^{bc}	0.03 ± 0.01^{b}
autotrophic plant	14.26 ± 5.09^{b}	0.33 ± 0.36^{b}	6.60 ± 1.56^{d}	0.04 ± 0.04^{b}

4. Discussion

Our study suggests that there is a high level of PSII activity in the vitroplant. Our data are consistent with the $Ø_P^{MAX}$ values measured in several species cultivated *in vitro*, such as tobacco (0.82) or potato (0.73) (Pospisilova *et al.*, 1993). $Ø_P$ is a reliable index of quantum yield of PSII photochemistry in illuminated leaves (Genty *et al.*, 1989) and reflects a fully functional linear electron transport chain in *in vitro* grown plantlets.

Both photosynthesis and chlorophyll fluorescence were found to increase concomitantly during the *in vitro* culture process, suggesting an increase in CO_2 assimilation. The existence of a correlation between $Ø_P$ and CO_2 fixation measurements under non-photorespiratory conditions has been previously reported (Genty *et al.*, 1989; Krause and Weis, 1991). Nevertheless, the photosynthesis rate measured in *in vitro* grown plantlets remained half as much as that of the autotrophic palm.

At the early stages of *in vitro* culture (*i.e.* 1 week under PAR) the PEPC:RubisCO ratio was very high, due to a high PEPC capacity. High PEPC capacities (6.6 $\mu mol\ CO_2.\ H^{-1}.mg^{-1}_{TSP}$) were also measured in shoot-forming cotyledons of *Pinus radiata* (Kumar *et al.*, 1988) and in young somatic embryos of oil palm (5.2 $\mu mol\ CO_2.\ h^{-1}.mg^{-1}_{TSP}$) (Rival *et al.*, 1997). The occurrence of a transient preferential CO_2 fixation through PEPC when C3 plants or isolated cells are cultivated *in vitro* has been described in several species (Nato *et al.*, 1981; Hdider and Desjardins, 1994). The PEPC:RubisCO ratio decreased in vitroplants down to 0.03, a value similar to the one measured in autotrophic coconut palm. Similar patterns were observed in *in vitro* grown plantlets of oil palm (Rival *et al.*, 1996), in which a depletion of the PEPC:RubisCO ratio (down to 0.06) was noted during the *in vitro* development of somatic embryos. During *in vitro* growth and development, coconut plantlets showed a transition from a heterotrophic to an autotrophic (RubisCO-mediated) mode of carbon fixation. Indeed, a marked decrease in PEPC, concomitant with substantial increase in RubisCO capacity, was observed. In our case, the RubisCO capacity and content were lower than in the adult autotrophic coconut palm and it could explain the low rates of CO_2 assimilation found in *in vitro* grown plantlets. The high level of sucrose present in the culture medium affects the RubisCO capacity.

5. Conclusion

In vitro-cultured coconut and oil palm plantlets displayed an early initiation of a photosynthetic metabolism. Concomitant changes in several parameters ($Ø_P^{MAX}$, $Ø_P$, CO_2 fixation, PEPC:RubisCO ratio and transpiration rates) were measured. However, a lower rate of net photosynthesis was recorded in *in vitro* grown plantlets as compared with the acclimatised palm. This could be explained by a lower RubisCO content and activity, together with a lower chlorophyll content compared to the acclimatised palm. With respect to Grout's classification (1988), it can be assumed that both oil palm and coconut belongs to the class of plants in which in vitro-grown leaves can contribute to autotrophy and then play an active part in acclimatisation. This is clearly indicated by the dramatic decrease in the PEPC/RubisCO activity ratio and the increase in the photochemical activity of PSII and photosynthetic electron transport observed. This work now needs to be complemented on both palm species by the monitoring of the parameters studied here as a whole during the subsequent stage of acclimatisation of plantlets, with reference to seedlings.

Acknowledgements. *Our work was conducted under a joint research programme between ORSTOM (Institut Français de Recherche Scientifique pour le Développement en Coopération) and CIRAD (Centre de Coopération Internationale en Recherche Agronomique pour le Développement). The present study was partly supported by the Commission of European Communities (contract number: ERBTS3*CT940298). Thanks are due to Adriana Quiroz and Gaby Fuentes from CICY for their participation in the project.*

References

Assy-Bah B., Durand-Gasselin T., Engelmann F., and Pannetier C.,1989. Oléagineux, 44 (11): 515 - 523.

Genty B., Briantais J.M., and Baker N.R., 1989.. Biochim. Biophys. Acta., 990: 87 - 92.

Grout B.W.W.,1988. Acta Hort. *230,* 129-135 (1988).

Hdider C. and Desjardins Y., 1994.. Can. J. Plant Sci., 74: 827 - 831.

Kozai T., 1991. in: Y.P.S. Bajaj (Ed.) Biotechnology in Agriculture and Forestry, Springer Verlag, Berlin, Heidelberg, vol. 17: 313-343.

Krause G.H., and Weis E., 1991. Ann. Rev. Plant Physiol. Plant Mol. Biol., 42: 313 - 349.

Kumar P.P., Bender L. and Thorpe T., 1988. Plant Physiol., 87: 675 - 679.

Lavergne D., Nato A., Dupuis J.M., Pean M., and Chagvardieff P., 1992. Physiol. Plant., 84: 292 - 300.

Lichtentaller H.K., Buschmann C., Döll M., Fietz H.J., Bach T., Kozel U., Meier D., and Rahmsdorf U., 1981. Photosynth. Res. 2: 115 - 141.

Nato A., Mathieu Y., and Brangeon J., 1981 Physiol. Plant., 53: 335 - 341.

Pospisilova J., Catsky J., Synkova H., Machackova I., and Solarova J., 1993. Photosynthetica, 29(1): 1 - 12.

Rival A., Beulé T., Nato A., and Lavergne D., 1996. Plantations, Research, Development, 3(6): 418 - 428.

Rival A., Beulé T., Lavergne D., Nato A., Havaux M., and Puard M., 1997. J. Plant Physiol., 150(5): 520 - 527.

Santamaria J.M., Davies W.J., and Atkinson C. 1994. J. Exp. Bot. 44, 99-107.

Triques K., Rival A., Beulé T., Puard M., Roy J., Nato A., Lavergne D., Havaux M., Verdeil J.L., Sangare A., and Hamon S., 1997a.. Plant Sci. 127 : 39-51.

Triques K., Rival A., Beulé T., Dussert S., Hocher V., Verdeil J.L., and Hamon S., 1997b. Plant Cell Tiss. Org. Cult. 49: 227-231.V

Van Huylenbroeck J.M., and Debergh P.C., 1996. Plant Tiss. Cult. Biotech., 2(3): 136 - 141.

Plant D-type (CycD) cyclins and the regulation of the plant cell cycle

D. A. Sorrell[1], C. Riou-Khamlichi[1], N. Chaubet-Gigot[2], B. Combettes[2], C. Gigot[2] and J. A. H. Murray[1]
[1]Institute of Biotechnology, University of Cambridge, Tennis Court Road, Cambridge CB2 1QT, UK, and [2]Institut de Biologie Moléculaire des Plantes, Centre National de la Recherche Scientifique, Université Louis Pasteur, 67084 Strasbourg Cedex, France

1. Introduction

Progression through the eukaryotic cell cycle is driven by the sequential activation and destruction of kinase complexes, consisting of a serine/threonine cyclin-dependent kinase (cdk) bound to a regulatory cyclin. The cdk is dependent on binding the cyclin for activity, and the cyclin determines the substrate specificity and subcellular localisation of the complex. In animals multiple cdks and cyclins are involved in the control of cell cycle progression. Cyclins are classified as being mitotic (A or B-type) or G1 (D or E-type) cyclins based on sequence criteria and the region of the cell cycle they regulate.

External signals impinge on the mammalian cell cycle at a principal control point in late G1, called the Restriction (R) point. Depending on the nature of these signals, cells either commit to a further round of cell division or exit the cell cycle, entering either a state of quiescence or a pathway of terminal differentiation. Progression through the R point requires D-type cyclins whose transcription is absolutely dependent on the presence of serum growth factors. Transcript levels decline rapidly upon growth factor withdrawal, and reappear rapidly upon re-addition. Cyclin D-dependent kinases drive cells through the R point by phosphorylating the retinoblastoma (Rb) protein in mid- to late G1, causing the release of E2F transcription factors which then activate genes required for S-phase.

A, B and D-cyclin homologues have been isolated from plants, but little is known about their cell cycle function. In plants, a G1 control point also appears to be important to the regulation of the commitment to further divisions or differentiation. In order to understand how this regulation occurs at the molecular level, we and others have focused on plant D-type cyclins. Here we briefly review work on the isolation of plant D-type cyclins, and summarise our recent data on the expression of plant D-type cyclins in model Arabidopsis and tobacco cell culture systems.

2. Plant D-type (CycD) cyclin isolation and sequence relationships

Plant D-type cyclin (CycD) cDNAs were first isolated from Arabidopsis (Soni et al., 1995) and alfalfa (Dahl et al., 1995) by their ability to functionally complement yeast strains defective in G1 (CLN) cyclin activity. Subsequently CycDs have been isolated from *Antirrhinum* (Gaudin et al., submitted), *Helianthus* (D. Freeman, J.A.H. Murray, unpublished), tobacco (Sorrell et al., submitted) and *Chenopodium* (Fountain et al., unpublished) by screening cDNA libraries with CycD cDNAs as probes. These cyclins form three distinct groups designated CycD1, CycD2 and CycD3 (Figure 1; Renaudin et

A. Altman et al. (eds.), Plant Biotechnology and In Vitro Biology in the 21st Century, 441–444.

al., 1996; Murray et al., 1998). Interestingly, two distinct CycD3s have been isolated from both *Antirrhinum* and tobacco. These CycD3 cyclins are differentially expressed in *Antirrhinum* meristems (Gaudin et al., unpublished) and tobacco cell cultures (Sorrell et al., unpublished; see below), suggesting they are not functionally redundant.

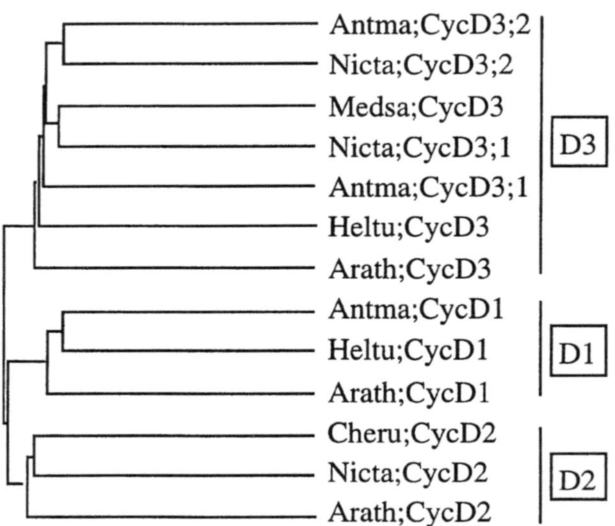

Figure 1: Plant D-type cyclins form three distinct groups, CycD1, CycD2 and CycD3. The putative amino acid sequences of all CycD cyclins available were compared using the GCG program PILEUP. The names are in accordance with official nomenclature (Renaudin et al., 1996). The final suffix indicates the order that cyclins within the same group from a single species were isolated, and not a structural relationship. Antma, *Antirrrhinum majus*; Arath, *Arabidopsis thaliana*; Cheru, *Chenopodium rubrum*; Heltu, *Helianthus tuberosus*; Medsa, *Medicago sativa* (alfafa); Nicta, *Nicotiana tabacum* (tobacco).

CycD cyclins all contain an N-terminal LxCxE Rb protein binding motif (in single-letter code; x being any amino acid), the distinguishing feature of animal D-type cyclins. This motif enables D-cyclin-associated kinases to bind with and phosphorylate Rb proteins. Like animal D-cyclins, CycDs lack the 'destruction box' sequence required for the degradation of mitotic cyclins. Instead both plant and animal D-cyclins, generally contain PEST sequences, regions rich in proline, glutamate, serine and threonine, thought to be involved in the rapid degradation of some proteins.

3. Expression of plant D cyclins

3.1 Expression during the cell cycle

Mammalian D-cyclin transcript levels remain constant during the cell cycle in contrast to the periodicity exhibited by cyclin A, B and E transcripts. Previous analyses of CycD2 and

CycD3 transcripts in Arabidopsis cell suspensions or liquid cultured callus material, synchronised with chemical cell cycle inhibitors, suggested that Arabidopsis CycD cyclins also remain at a constant level during the cell cycle (Soni et al., 1995, Fuerst et al., 1996).

We have now examined tobacco CycD cyclin transcript levels during the cell cycle in synchronised tobacco BY-2 cells (Sorrell et al., submitted). These cells can be synchronised to a high level using the S-phase inhibitor aphidicolin (Nagata et al., 1992). We found that CycD3;2 transcript levels remained constant during the cell cycle, consistent with the expression of Arabidopsis and mammalian D-type cyclins. In contrast, tobacco CycD2 and CycD3;1 transcripts accumulated in mitosis, a pattern of expression not normally associated with D-cyclins, perhaps suggesting a novel function for plant D-type cyclins in mitotic cells.

3.2 Expression in quiescent cells re-entering the cell cycle

D-cyclin transcripts accumulate rapidly when quiescent (G0) mammalian cells are stimulated to re-enter the cell cycle with serum growth factors. We examined CycD transcripts in sucrose depleted Arabidopsis suspension cultured cells stimulated to re-enter the cell cycle by sucrose re-addition. Upon re-addition CycD2 transcripts accumulated rapidly (within 30 min), while CycD3 was induced later in G1, just prior to the S-phase marker gene histone H4. These results are consistent with our previous studies using the liquid cultured callus material (Soni et al., 1995; Fuerst et al., 1996).

We investigated the timing of tobacco CycD3;2 expression in stationary BY-2 cells stimulated to re-enter the cell cycle by dilution into fresh medium. Similar to the Arabidopsis CycD3, we found that tobacco CycD3;2 was induced in late G1, approximately one hour before the S-phase marker gene histone H4.

The alfalfa CycD3 gene is also induced prior to the onset of S-phase in quiescent cells re-entering the cell cycle (Dahl et al., 1995). Taken together, these data suggest that plant CycD cyclins have a role in G1 phase progression similar to mammalian D-cyclins, and that CycD3 may be involved in the G1-to-S-phase transit.

3.3 Expression in response to plant growth regulators and sucrose

Mammalian D-type cyclins integrate growth signals with the cell cycle. Using the liquid cultured callus material, we have previously provided evidence for the specific induction of CycD2 and CycD3 by sucrose and cytokinin respectively (Soni et al., 1995).

Further studies have been undertaken using the Arabidopsis cell suspension system. Cytokinin re-addition to cells deprived of exogenous plant growth regulators (PGRs; auxin and cytokinin) resulted in the rapid induction of CycD3, a response most sensitive to zeatin (C. Riou-Khamlichi, J. A. H. Murray, in preparation). This induction was not blocked by cycloheximide at concentrations known to inhibit *de novo* protein synthesis and cell cycle progression, suggesting that it is in direct response to a cytokinin triggered signal cascade and not to cell cycle position. CycD3 was also induced by other growth promoting PGRs, but to a lesser extent compared with cytokinin. The induction of CycD3 by PGRs was dependent on the presence of sucrose. In contrast, CycD2 transcript levels were unaffected by PGRs. CycD2 transcripts disappeared in cells deprived of sucrose. When sucrose was re-added CycD2 was rapidly induced (see section 4.2), in both the presence or absence of PGRs. This induction was not blocked by cycloheximide, suggesting that it is a direct response to sucrose.

These data strengthen the hypothesis that CycD2 and CycD3 are differentially regulated by carbon source availability and PGRs, and are consistent with CycDs having a role in transducing signals into the cell cycle similar to mammalian D-cyclins. Based on these observations and the timing of expression in quiescent cells re-entering the cell cycle we present a model for G1/S phase progression in Arabidopsis (Figure 2).

4. Conclusions

Mammalian D-cyclins regulate the progress of cells through the G1 phase of the cell cycle in response to extracellular growth signals through the interaction with Rb. Plant D-cyclins (CycDs) with sequence features and expression patterns reminiscent of mammalian D-type cyclins have been isolated, suggesting that CycDs may also function as integrators of signals into the cell cycle via interactions with plant Rbs. In support of this hypothesis, we have also recently shown that Arabidopsis CycDs bind to both plant and human Rb proteins via the LxCxE motif (Huntley et al., 1998). Further studies both *in vitro* and *in planta* will determine the exact relationship between CycDs and the commitment of cells to further proliferation or differentiation.

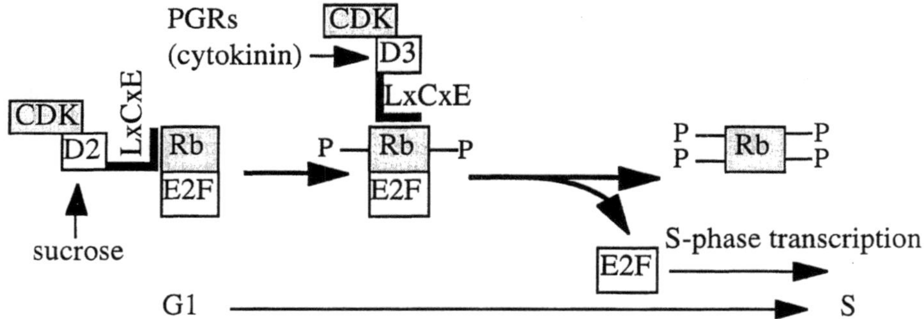

Figure 2: Simplified model for G1/S progression in Arabidopsis.
We propose that G1-S progression is driven by CycD-dependent kinase complexes. CycD2 is induced by sucrose upstream of CycD3, which is induced later in G1 by cytokinin and possibly other plant growth regulators. We speculate that the CycDs drive cells into S-phase by binding to Rb proteins via the LxCxE motif and causing Rb phosphorylation, we propose this results in the release of as yet unidentified plant E2F transcription factors which activate genes required for S-phase.

5. References

Dahl M et al (1995) Plant Cell 7,1847-1857
Fuerst RAUL et al (1996) Plant Physiol. 112, 1023-1033
Huntley R (1998) Plant Mol. Biol. 37, 155-169
Murray JAH et al (1998) in Plant Cell Division (Francis D, Dudits D and Inzé D, eds.), pp. 99-127, Portland Press, London
Nagata T et al (1992) Int. Rev. Cytol. 132, 1-30
Renaudin JP et al (1996) Plant Mol. Biol. 32, 1003-1018
Soni R et al (1995) Plant Cell 7, 85-103

Genetic and Biochemical analysis of arabidopsis SPY

T. Thornton[1], L. Kreppel[2], G. Hart[2] and N. Olszewski[1]
[1] Department of Plant Biology, University of Minnesota
[2] Department of Biological Chemistry, Johns Hopkins University, School of Medicine

Abstract. The arabidopsis SPY protein is a negative regulator of gibberellin signal transduction and based on protein sequence similarity it is hypothesized to be a O-GlcNAc transferase (OGT). Proteins from *spy* mutants were found to exhibit allele specific alterations in the pattern of GlcNAc modification. Insect cell expressed SPY is GlcNAc modified and preliminary activity assays indicated that this protein has OGT activity.

1. Introduction

The *spy* mutants of arabidopsis were identified in a genetic screen for gibberellin (GA) response mutants (Jacobsen, Olszewski, 1993). Genetic analysis suggests that SPY functions as a negative regulator of GA signal transduction. Mutations at the *SPY* locus partially suppress the phenotypes of GA deficiency and also largely suppress the phenotypes of *gai*, a semi-dominant GA-insensitive dwarf (Jacobsen et al., 1996).

The *SPY* gene encodes a predicted protein with significant similarity to O-GlcNAc transferases (OGTs) (Kreppel et al., 1997; Lubas et al., 1997). SPY, like the OGTs, has N-terminal tetratricopeptide repeats (TPR) followed by the putative catalytic domain. OGTs add single N-acetylglucosamine in an O-linkage to serine or threonine residues of many proteins (Hart, 1997; Snow, Hart, 1998). O-GlcNAcylation is dynamic and is hypothesized to be a regulatory modification analogous to protein phosphorylation (Hart, 1997; Snow, Hart, 1998). The TPR domains are thought to interact with additional proteins that control the specificity and/or modify the activity of the OGT enzyme (Kreppel et al., 1997). OGTs have also been demonstrated to be both phosphorylated and O-GlcNAcylated and these posttranslational modifications are believed to regulate their activity (Kreppel et al., 1997). Here we report the results of several experiments designed to test the hypothesis that SPY is a functional OGT.

2. Materials and Methods

Above ground plant material was harvested and immediately frozen in liquid nitrogen and ground to a fine paste in 1 ml/g 2X SDS-PAGE sample buffer (Laemmli, 1970). The paste was transferred to a tube and boiled for 5 min. and insoluble material

A. Altman et al. (eds.), Plant Biotechnology and In Vitro Biology in the 21st Century, 445–448.

was pelleted by centrifugation. Proteins were separated by SDS-PAGE and blotted onto a PVDF membrane. Galactosyltransferase on-the-blot assays was performed as described in Heese-Peck et al. (1995). PNGase treatment of blots was performed essentially as described by Weitzhandler et al. (1993). Fluorography was performed as described in Heese-Peck et al. (1995).

The *SPY* cDNA was cloned into pBlueBacHis (Invitrogen) and SPY protein was produced in *Sf9* cells. Affinity purification of the SPY protein was performed using Talon Metal Affinity Resin (Clontech) essentially as described by the manufacturer. *E. coli* expression of gp40 and all western blots were performed essentially as described in Heese-Peck et al. (1998).

Affinity purified SPY was desalted as described by Haltiwanger et al. (1992). The OGT activity assay was modified from Haltiwanger et al. (1990) and Lubas et al. (1995). Protein blots were incubated for 1h at room temperature in a 200 μl total reaction containing 50 mM Tris pH 7.4, 12.5 mM $MnCl_2$, 1 mg/ml BSA, 25 μM NaF, 245 μM UDP-GlcNAc, and 100 μl desalted SPY or 100 μl desalted affinity column flowthrough. Blots were washed 2X for 10 min. with TBS and 3X for 10 min. in GalT labeling buffer. GalT labeling and PNGase treatment were then performed as described above.

3. Results and Discussion

Total protein from wildtype and *spy* mutants bearing terminal GlcNAc residues were labeled with ^{3}H-galactose using GalT. PNGase treatment was then used to remove most N-linked sugars and labeled proteins were visualized by fluorography.

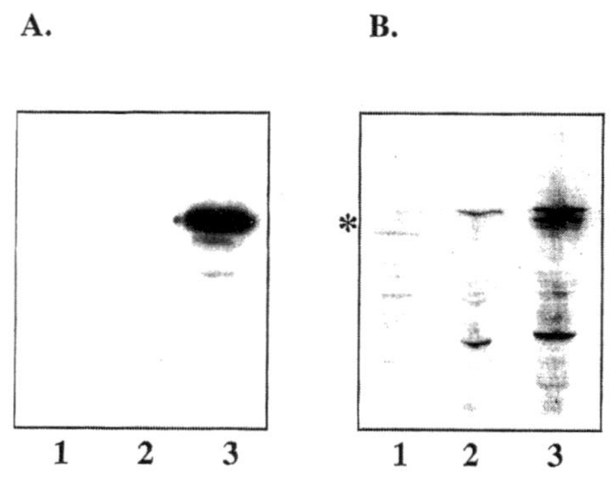

Figure 2

Intense labeling of a 45 kDa protein, indicated by the asterisk, is apparent in the wildtype proteins (lane 1) but not those from *spy-4* (lane 2), an allele causing a severe phenotype and greatly reduced amounts of *SPY* mRNA. This observation is consistent with reductions in SPY activity causing a reduction in OGT activity. The lack of major effects on protein modification can be explained in two ways. It is possible that many of the modifications are N-linkages that are refractory to PNGase (Tretter et al., 1991). Alternatively, the labeling may be due to a second OGT (Hartweck, Olszewski, unpublished).

The *spy-2* mutation causes a 22 amino acid deletion affecting TPR 8 and TPR 9 (Jacobsen et al., 1996). This allele strongly suppresses the phenotypes of the dominant *gai* mutation. A genomic *SPY* transgene complements the *spy-2* mutation and transgenic *spy-2 gai* plants resemble *gai*. Figure 1B shows the pattern of terminal GlcNAc modified proteins from *gai* (lane 1), *spy-2 gai* (lane 2), *spy-2 gai* complemented with a genomic *SPY* transgene (lane 3), and wildtype (lane 4). A 65 kDa protein with enhanced labeling, indicated by the asterisk, is detected in *spy-2 gai* (lane 2) but it is not apparent in the *gai* (lane 1) or wildtype (lane 4). This GlcNAc modified protein is also detected in *spy-2 gai* plants containing the genomic *SPY* transgene (lane 3). Using wheat germ agglutinin blotting, an additional 65 kDa GlcNAc modified protein was also detected in *spy-2* but not in wildtype or *spy-4* (data not shown). The additional GlcNAcylated protein in the *spy-2* mutant suggests that the TPR domains play a role in directing SPY to its substrates. Since the *spy-2* and *-4* mutations have similar effects on GA signal transduction but affect the modification of the 45 and 65 kDa proteins differently, it is unlikely that the modification of either protein is involved in GA signal transduction.

Because the GalT assay can detect a range of structures containing terminal GlcNAc residues, further analysis is needed to determine if, as is the case for proteins modified by animal OGT, the 65 and 45 kDa proteins have simple O-GlcNAc modifications.

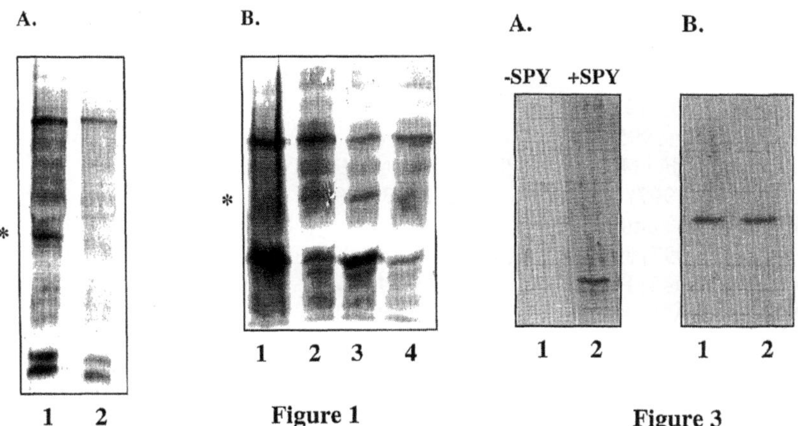

Figure 1 **Figure 3**

Histidine-tagged SPY protein was produced in insect cells and assayed for GlcNAc modifications and enzymatic activity. Figure 2A is a western blot probed with anti-SPY antibodies. The uninfected cells (lane 1) and infected cells producing a control protein (lane 2) have no reacting proteins, while the cells producing SPY contain a reacting protein, indicated by the asterisk, of the expected size. GalT labeling and PNGase treatment of a duplicate blot shown in figure 2B, detects a GlcNAc modified protein that has the same mobility as the recombinant SPY protein (lane 3) and is not present in the uninfected cells (lane 1) and control protein cells (lane 2). Thus, like other OGTs (Kreppel et al., 1997), SPY may be O-GlcNAc modified.

The nuclear pore protein from tobacco, gp40, is known to have terminal O-GlcNAc modifications (Heese-Peck, Raikhel, 1998). Therefore, we used total proteins from *E. coli* expressing gp40 as a substrate to assay SPY's OGT activity. Figure 3A shows the results from the reactions containing affinity purified SPY (lane 2) or affinity column flowthrough (lane 1). GalT labeling was used to visualize incorporated GlcNAc. GlcNAc modified proteins are only detected in the sample incubated with SPY suggesting that SPY is an OGT. However, additional experiments to rule out the possibility that the insect OGT has co-purified with SPY are needed. The substrate of SPY in these reactions is unclear because the proteins detected by GalT are not the same size as the protein detected using anti-gp40 antibodies (Figure 3B). However, no GlcNAc modified proteins were detected when the assay was performed using extracts that did not contain gp40 (data not shown).

4. Acknowledgments

We thank C. Asleson and J. Berman for the helicase recombinant baculovirus and A. Heese-Peck and N.V. Raikhel for the gp40 clone and antibodies. This work was supported by an NSF Grant (MCB 96-04126) to N.O.

5. References

Jacobsen SE et al. (1996) Proc. Natl. Acad. Sci. 93, 9292-9296
Jacobsen SE, Olszewski NE (1993) Plant Cell 5, 887-896
Haltiwanger RS et al. (1992) J. Biol. Chem. 267, 9005-9013
Haltiwanger RS et al. (1990) J. Biol. Chem. 265, 2563-2568
Heese-Peck A et al. (1995) Plant Cell 7, 1459-1471
Heese-Peck A, Raikhel NV (1998) Plant Cell 10, 599-612
Kreppel LK et al. (1997) J. Biol. Chem. 272, 9308-9315
Laemmli EK (1970) Nature 227, 680-685
Lubas WA et al. (1997) J. Biol. Chem. 272, 9316-9324
Lubas WA et al. (1995) Biochemistry 34, 1686-1694
Snow DM, Hart GW (1998) Int. Rev. Cytol. 181, 43-74
Tretter V et al. (1991) Eur. J. Biochem. 199, 647-652
Weitzhandler M et al. (1993) J. Biol. Chem. 268, 5121-5130

GENETIC ENGINEERING OF DISEASE RESISTANCE IN APPLE FRUIT CULTIVARS AND ROOTSTOCKS

Herb S. Aldwinckle[1], John L. Norelli[1], Jyothi Prakash Bolar[1], Kisung Ko[2] Gary E. Harman[2], and Susan K. Brown[2]

Departments of Plant Pathology[1] and Horticultural Sciences[2]
Cornell University, NYS Agricultural Experiment Station
Geneva, NY 14456 USA

Abstract

Apple fruit cultivars and rootstocks are propagated vegetatively, and are characterized by extreme genotype preference by nurseries, growers, and consumers. Furthermore, apple is generally heterozygous, shows severe inbreeding depression, and has an excessive generation time. Thus genetic engineering, *a priori*, is preferred to sexual breeding for improving established fruit cultivars and rootstocks. Efficient *Agrobacterium tumefaciens*-mediated transformation using binary vectors and kanamycin resistance selection has now been well established. Prime targets for genetic engineering are disease and insect resistance, which would address the perceived environmental and public health concerns of substantial use of chemical pesticides in apple production. Promising results have been obtained in terms of increasing the resistance to the most important fungal (apple scab) and bacterial (fire blight) diseases of apple worldwide, using transgenes for chitinolytic enzymes and for lytic proteins, respectively. Thus far, positive results have been obtained with single transgenes. Based on *in vitro* synergism against the pathogens, it is likely that plants containing two chitinolytic or lytic protein transgenes will have greatly enhanced resistance to fungi and bacteria, respectively. Such transgenic plants have now been obtained and are being evaluated for disease resistance.

1. Introduction

Only a few apple cultivars are responsible for a large proportion of total production in most growing regions. These cultivars are valued by consumers, supermarkets, and growers for their appearance, quality, flavor, storability, and orchard characteristics. They are sold in most developed countries by names, which cannot legally be used for other cultivars. To retain an apple cultivar's desirable characteristics and to introduce disease resistance genes by conventional breeding methods is virtually impossible because of apple's heterozygosity, long generation time, and self-incompatibility, which make the necessary backcross programs too lengthy and costly. Genetic engineering would bypass these hurdles by introducing resistance genes directly into

A. Altman et al. (eds.), Plant Biotechnology and In Vitro Biology in the 21st Century, 449–451.

the present desirable commercial cultivars and producing resistant clones with the same qualities and names.

Fire blight, caused by *Erwinia amylovora,* is the most important bacterial disease of apple worldwide, causing serious losses in many growing regions. It is now only partially controlled by orchard sprays of antibiotics and other less efficient methods. Most commercial cultivars and rootstocks are susceptible or very susceptible to the disease. Apple scab, caused by the fungus, *Venturia inaequalis,* is the most prevalent disease of apple worldwide, the most expensive to control, and results in the greatest orchard use chemical fungicides. All commercial cultivars are susceptible to scab. These diseases were targeted by genetic engineering for environmental, health, and economic reasons.

2. Materials and Methods

Native (attacin E) and synthetic (SB-37, Shiva 1) genes derived from the saturniid moth, *Hyalophora cecropia,* as well as lysozyme genes from hen eggwhite and T4 bacteriophagewere transferred into apple cultivars, Royal Gala and Galaxy, and rootstocks, M.7 and M.26, using *Agrobacterium*-mediated transformation of wounded leaf pieces (Norelli et al., 1994; Norelli et al, 1996; Ko et al, this proceedings). Apple cultivars, Royal Gala and Marshall McIntosh were similarly transformed with endochitinase(*ThEn-42*) and N-acetylglucosaminidase (*ThNag-70*) genes from the biocontrol fungus, *Trichoderma harzianum* (Wong, personal communication; Bolar et al., this proceedings). Transformations with multiple gene constructs were also made.

Transformed regenerants were selected by expression of the *nptII* selectable marker gene on genotype-specific media containing kanamycin, and cefotaxime, and subcultured on paromomycin media. Transformation was confirmed sequentialy by NPTII ELISA, PCR with suitable primers, and Southern analysis. Gene expression was quantified by enzymatic assay or ELISA.

Transgenic apple plants transformed for resistance to fire blight were evaluated for resistance by direct inoculation of their actively growing shoot tips with inoculum of a virulent strain of *E. amylovora*. The length of the resultant necrotic lesion was expressed as the proportion of total shoot length. Plants transformed for resistance to apple scab were evaluated for resistance by atomizing a suspension of conidia of mixed virulent isolates of 5 races of *V. inaequalis* on to the actively growing apical leaves. Plants were incubated in a chamber with 100% RH at 20°C for 48 hours and then kept in a greenhouse until synptoms developed fully. Number of lesions per leaf, proportion of leaf surface infected, and number of conidia rinsed from each leaf were recorded.

3. Results and Discussion

Several lines of apple cultivars Royal Gala and Galaxy, and M.7 and M.26 rootstocks transgenic for attacin E and SB-37 showed increased resistance to fire blight compared with non-transgenic parent cultivars. Some lines with T4 lysozyme gene also had increased resistance. Variance among and within transgenic lines, and among experiments was noted.

Resistance to apple scab was clearly increased over parent cultivars by transformation with the *T. harzianum* endochitinase gene. Resistance was correlated with endochitinase expression. An effect on vigor of growth of transgenic lines with greatest expression of this gene was observed.

Results indicate that resistance in apple cultivars to fire blight and scab, and in rootstocks to fire blight can be increased by introduction of heterologous genes for antibacterial and antifungal proteins, respectively.

The challenges are to achieve consistent high levels of resistance with no undesirable side effects and with minimal levels of heterogene expression, and to properly evaluate the transgenic cultivars and rootstocks for orchard performance.

4. Acknowledgments

The work reported here was supported in part by the New York Apple Research and Development Program, funded by New York apple growers; the New York Apple Research Association, funded by New York apple processors; the New York State Center for Advanced Technology's Biotechnology Program; and USDA-CSREES.

5. References

Norelli JL et al (1994) Euphytica 77:123-128
Norelli JL et al (1996) HortScience 31:1026-1027.

BARLEY TRYPSIN INHIBITOR CME CONFERS INSECT RESISTANCE TO WHEAT

Fredy Altpeter[1+], Isabel Diaz[2], Heather McAuslane[3], Kamel Gaddour[2], Pilar Carbonero[2] and Indra K. Vasil[1]*

*[1]Laboratory of Plant Cell and Molecular Biology, 1143 Fifield Hall, University of Florida, Gainesville, FL 32611-0690, USA., *Corresponding author (e-mail: ikv@gnv.ifas.ufl.edu). [2]Laboratorio de Bioquímica y Biología Molecular. Dpto. de Biotecnología-UPM. ETSI Agrónomos. 28040 Madrid. Spain. [3] Department of Entomology and Nematology, P.O. Box 110620, Gainesville FL 32611-0620. [+]Present address: IPK Gatersleben, Gene Transfer Group, Corrensstr. 3, 06466 Gatersleben, Germany*

1. Introduction

Proteinase inhibitors have been proposed to be involved in the defence response against herbivorous pests (Hoffmann et al. 1992). The efficacy of a specific inhibitor depends on the structural compatibility of its reactive site with the substrate-binding site of the targeted proteinase. For example, trypsin-inhibitors include either an arginyl or lysyl residue, which is recognized by trypsin-like enzymes. Barley trypsin inhibitor CMe (BTI-CMe), an abundant protein in barley endosperm, is one of the best characterized members of the cereal multigene family of trypsin/α-amylase inhibitors (Carbonero and Gracia-Olmedo, 1998) and was first purified from barley flour as a protein of 14 KDa that was specifically active *in vitro* against trypsin. Recently BTI-CMe was shown to inhibit specifically the trypsin-like proteases of the gut extracts of the fall armyworm, *Spodoptera frugiperda* (Lepidoptera: Noctuidae) (Alfonso et al. 1997), whereas the BTI-CMe inhibitor is rapidly degraded in the digestive tract of mammals (unpublished data, Carbonero et al.).

We have transformed wheat (*Triticum aestivum* L.) with the *Itr1* gene encoding the BTI-CMe (Altpeter et al. 1998), in order to evaluate its potential for improvement of resistance against a major storage pest in many developing countries: The Angoumois Grain Moth (*Sitotroga cerealella,* Lepidoptera: Gelechiidae).

2. Results and Discussion

Each of the 30 wheat lines expressing the selectable *bar* gene (overall transformation efficiency of 1.5 %) was normal and fertile, probably as a result of the short period in tissue culture (Altpeter et al. 1996). Integration of the *Itr1* gene was confirmed by PCR. The Southern analysis suggested a transgene copy number of 2 - 5 in most of the lines. Each of the five lines (12, 25, 30, 51, 69) selected for a detailed study stably transmitted the transgene in a Mendelian fashion up to the third seed generation, consistent with

A. Altman et al. (eds.), Plant Biotechnology and In Vitro Biology in the 21st Century, 453–456.

integration at a single locus. The transgenic proteinase inhibitor was properly expressed and processed as shown by Northern- and Western-blot. The expression level of BTI-CMe protein in transgenic wheat seed was up to 1.1 % of the total extracted protein, whereas expression in nontransformed wheat seeds was undetectable (tab. 1). The functional integrity of BTI-CMe was confirmed with a trypsin inhibitor activity assay (fig. 1).

Table 1. BTI-CMe protein in homozygous seeds (R_3) of transgenic wheat lines, nontransformed control seeds (NC) and barley endosperm (BE) in % of total extracted seed-protein.

Wheat lines	2	12	25	30	51	69	NC	BE
BTI-CMe protein (%)	0.2	0.7	0.6	1.1	0.7	0.5	0.0	0.9

Data represent avaerage of three independent assays expressed as % of the total extracted seed-protein (Bio-Rad kit).

We found a significant reduction in the survival of the Angoumois Grain Moth (*Sitotroga cerealella,* Lepidoptera: Gelechiidae), reared on transgenic wheat seeds expressing the trypsin inhibitor BTI-CMe, compared to the nontransformed control (tab. 2). Trypsin inhibitor expression of approximately 1 % of the total soluble protein in tobacco was also sufficient to significantly reduce survival of a lepidopteran pest (Hilder et al. 1987).

Table 2. Survival rates (%) of neonate larvae of *S. cerealella* reared on transgenic seeds of lines 12, 25, 30, 51, 69 or nontransformed wheat seeds (NC).

Wheat lines	12	25	30	51	69	NC
Survival rate[A] (%)	61^{BC}	69^{BC}	66^{BC}	75^{AB}	64^{BC}	86^{A}

[A] Least squares means followed by a different letter differed significantly at p = 0.05.

BTI-CMe inhibitor had also a significant effect on the weight gain of the insect, but no effect on the developmental period of the insect. Cross-sectioning of seeds 25 days after infestation with neonate larvae of *S. cerealella* revealed that only the very early developmental stage of the larva is inhibited and once the larvae have passed this stage, they can overcome the inhibition (tab. 3). Hoffmann et al. (1992) also found that the early developmental stage of a lepidopteran pest was most susceptible to mortality after feeding on transgenic tobacco expressing a trypsin inhibitor. Jongsma et al. (1995) reported that some insect species are able to induce gut proteinase activity insensitive to inhibition by a specific trypsin inhibitor and become resistant to this proteinase inhibitor.

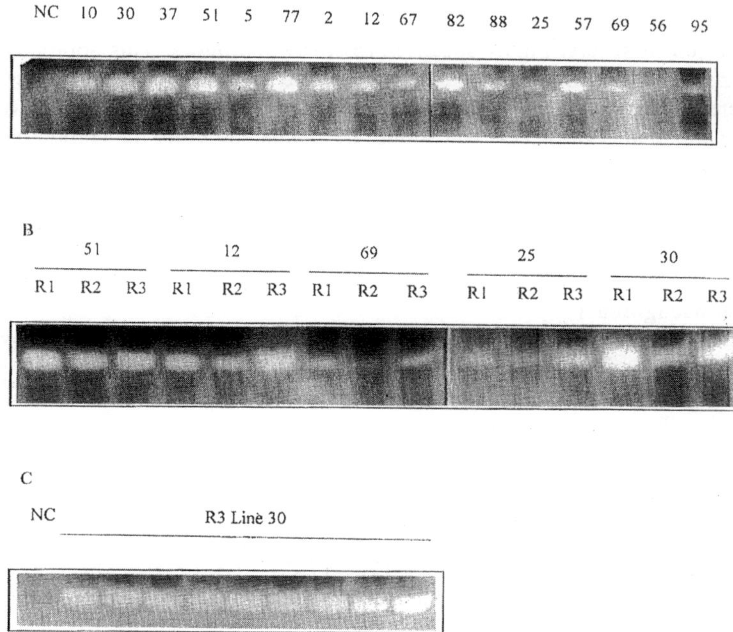

Figure 1. Trypsin inhibitor activity was visualized following incubation in N-Acetyl-$_{DL}$-Phenylalanine-β-Naphthyl Ester (APNE) as a decolorized band after separation of the proteins in non-denaturing PAGE and incubation of the gels in a trypsin solution.
A: Protein extracts from R_1 seeds of lines 2, 5, 10, 12, 25, 30, 37, 51, 56, 57, 67, 69, 77, 82, 88, 95 and from nontransformed wheat seed (NC). B: Protein extracts from R_1, R_2 and R_3 seeds of lines 12, 25, 30, 51 and 69. C: Protein extracts from independent homozygous R_3 seeds of line 30 and from nontransformed wheat seed (NC).

Table 3. Examination of seeds for entrance holes and cross-sections of seeds for *Sitotroga cerealella* reared on transgenic seeds expressing the CMe trypsin inhibitor (12, 30) or the nontransformed control seeds.

	seeds with entrance holes	early instar larvae or small cavities	late instar larvae or pupae
control	80 %	0 %	80 %
line 12	80 %	23 %	57 %
line 30	83 %	23 %	60 %

The effectiveness of the protection and the range of target insects may be extended by engineering highly specific inhibitors by phage display (Roberts et al. 1992) and coexpression of several highly specific proteinase inhibitors. This approach should also minimize the risk of insects developing resistance to the inhibitor. However, it has to be considered that some proteinase inhibitors are anti-nutritional for humans or animals (Gumbmann et al. 1986). This fact should be evaluated before the commercial release of a transgenic crop.

In summary, we have shown that expression of high levels of a proteinase inhibitor in transgenic wheat seeds increased its resistance against an important storage pest.

3. Acknowledgments

We thank Dr. David Weaver for advice and experimental material (*S. cerealella*), Dr. Phil Bruckner for field planting of wheat, Dr. John Capinera for the grasshopper colony, Debbie Boyd, Jason Squitier and J. Garcia for assistance and Linda McCandless for maintaining the *S. cerealella* colony.

4. References

Alfonso J, Ortego F, Sanchez-Monge R, Garcia-Casado G, Pujol I, Castanera P and Salcedo G (1997) J Chem Ecol 23, 1729-1741

Altpeter F, Vasil V, Srivastava V, Stoeger E and Vasil IK (1996) Plant Cell Rep 16, 12-17

Altpeter F, Diaz I, McAuslane H, Gaddour K, Carbonero P and Vasil IK (1998) Mol Breed (submitted)

Carbonero P and Garcia-Olmedo F (1998). Cereal proteins, Chapman and Hall, London

Gumbmann MR, Spangler WL, Dugan GM and Rackis JJ (1986) Nutrional and Toxicological Significance of Enzyme Inhibitors in Foods, Plenum, New York

Hilder V A, Gatehouse AMR, Sheerman SE, Barker RF and Boulter D (1987) Nature 330, 160-163

Hoffmann MP, Zalom FG, Wilson LT, Smilanick JM, Malyj LD, Kiser J, Hilder VA and Barnes WM (1992) J Econ Entom 85, 2516-2522

Jongsma MA, Bakker P L, Peters J, Bosch D and Stiekema WJ (1995) Proc Natl Acad Sci USA 92, 8041-8045

Roberts BL, Markland W, Ley AC, Kent RB, White DW, Guterman SK and Ladner RC (1992) Proc Natl Acad Sci USA 89, 2429-2433

SALT TOLERANCE AND OXIDATIVE STRESS AS STUDIED BY THE REGULATION OF PHOSPHOLIPID HYDROPEROXIDE GLUTATHIONE PEROXIDASE IN SALT-SENSITIVE AND SALT-TOLERANT CITRUS CELLS

O. Avsian-Kretchmer, Y. Gueta-Dahan, Y. Eshdat and G. Ben-Hayyim
Institute of Horticulture, ARO, The Volcani Center, Bet Dagan, Israel

Introduction

In recent years there has been an increasing interest in genes and proteins that are involved in plant resistance to salinity and drought. Plants exposed to salt stress must undergo changes in their metabolism for adaptation to the altered environment. They have to cope with a change in water potential, a high level of toxic ions such as Na^+ and Cl^-, and the interference of salt with the uptake of essential nutrients. Recently, oxidative stress was demonstrated as an additional factor resulting from salt stress (Gossett *et al.* 1996; Gueta-Dahan *et al.* 1997; Hernandez *et al.* 1993, 1994, 1995).

Genes and proteins which are induced by NaCl were isolated in a search for biochemical functions associated with salt tolerance. In our laboratory, we had previously isolated a salt-associated citrus protein and its corresponding gene, *csa* (Ben-Hayyim *et al.* 1993; Holland *et al.* 1993). Sequence analysis of *csa* and its encoded protein revealed a significant homology to the animal phospholipid hydroperoxide glutathione peroxidase (PHGPX) (Holland *et al.* 1993).

In a previous report we showed that the salt-induced increase in the protein level of PHGPX was similar in salt-sensitive and salt-tolerant cells when determined at the end of the growth cycle, and under conditions designed to achieve 50% inhibition of growth (Gueta-Dahan *et al.* 1997). This report describes early responses of the citrus PHGPX and its encoding gene, *csa* at the mRNA transcript and protein levels. Our results suggest that NaCl-induced *csa* expression is mediated by reactive oxygen species.

Experimental procedures

Plant material and growth conditions. Shamouti orange (*Citrus sinensis* L. Osbeck) ovular callus cells (L5) and an adapted salt-tolerant cell line derived from it (R10) were grown as previously described (Ben-Hayyim, Kochba 1982). Cells, at the end of their growth cycle were transferred to a fresh medium, subjected to the various treatments, harvested, frozen immediately in liquid N_2 and kept at -80°C.

Extraction of RNA and Northern blot analysis. Total RNA was extracted from

A. Altman et al. (eds.), Plant Biotechnology and In Vitro Biology in the 21st Century, 457–460.

100-200 mg of frozen citrus cells using 1 ml of the TriPure™ isolation reagent (Boehringer Mannheim). Glyoxal/DMSO denatured RNA samples were electrophoresed and transferred overnight to a positively charged nylon membrane (Sambrook *et al.* 1989). cDNA clones labeled with Klenow enzyme by the random prime method were added to prehybridized membrane and hybridizations were carried out overnight at 60°C. Membranes were rinsed with high stringency wash buffer, and were exposed to X-ray film (Biomax, Kodak). Autoradiograms were scanned and quantified with the software TINA 2.07d.

Extraction of protein and Western blot analysis. Cells (100-200 mg) were ground in a minimal volume of extraction buffer and samples containing equal amounts of protein were subjected to denatured PAGE and immunoblot analysis using polyclonal antibodies raised against the purified citrus PHGPX protein, as previously described, (Ben-Hayyim *et al.* 1993).

Results

Salt-sensitive and salt-tolerant citrus cells were exposed to the same external NaCl concentrations. As shown in Figure 1a, the *csa* mRNA transcript level in the salt-sensitive cells was induced within 7 h after exposure to salt, when no such induction could be detected in the salt-tolerant cells. A longer exposure to salt showed that the increased *csa* mRNA transcript level in the salt-sensitive cells was transient, exhibiting a maximum at 7 h, and then decreasing to a lower level between 24 and 72 h. In salt-tolerant cells, a significant increase in *csa* mRNA transcript level was observed only after 16 h of exposure, and the level had risen about tenfold at the end of the growth cycle. Changes in the level of PHGPX protein were correlated with its mRNA transcript level with respect to the earlier increase observed in the salt-sensitive cells, but the PHGPX induction was not of a transient nature (Fig. 1b).

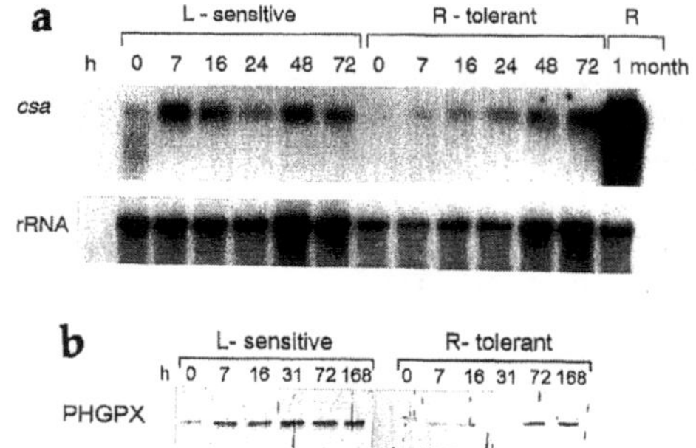

Fig. 1. Salt-induced response of csa mRNA transcript level (a) and PHGPX protein amount (b).

We exposed the salt-sensitive and the salt-tolerant cells to t-butylhydroperoxide (tBH, 5 mM), a peroxide which has been demonstrated to serve as a substrate for PHGPX (Faltin *et al.* 1998). As seen in Figure 2, the *csa* mRNA steady-state level increased significantly after 2 h of exposure and remained high for 7 h. Unlike previous results related to salt treatment, the

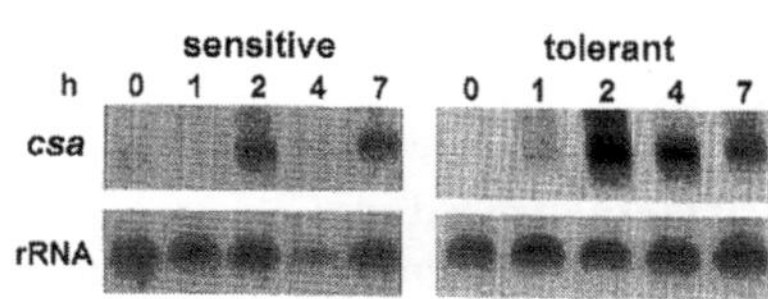

Fig. 2. Effect of tBH on csa mRNA transcript level in citrus cells.

induction of *csa* transcript was similar in both types of cells and occurred earlier than the induction observed in cells exposed to NaCl.

Overnight preincubation of cells with the antioxidants kampherol (0.1 mM) or dithiothreitol (DTT, 10 mM), eliminated the observed salt-induced *csa* expression in the salt-sensitive cells (S) after 7 h (Figure 3). It should be noted that DTT caused an increase in *csa* expression in the absence of salt, in the salt-sensitive and in the salt-tolerance cells (T).

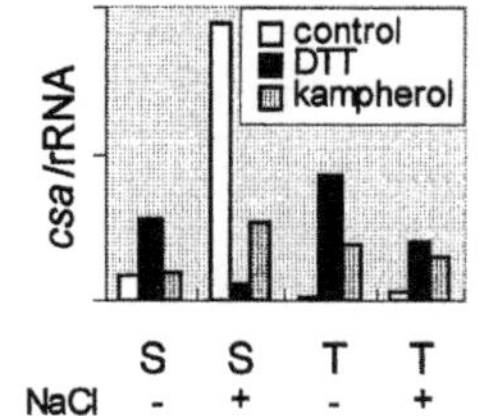

Fig. 3. Effect of antioxidants on csa expression.

Discussion

Glutathione peroxidases are a family of isozymes which catalyze the reduction of H_2O_2, and organic and lipid hydroperoxides by reduced glutathione, and thus help to protect cells against oxidative damage (Flohe, Gunzler 1984). In view of the fact that many stresses, e.g., drought, cold, biotic and, probably, salt stress induce the formation of reactive oxygen species (ROS) (Gueta-Dahan *et al.* 1997), the induction of PHGPX may be important for ROS scavenging.

The short-term exposure to salt elicited different patterns of response from the two types of cells.

In a model that we have suggested elsewhere, PHGPX was induced under conditions where salt stress resulted in excess H_2O_2, which was not efficiently scavenged and reacted with lipids to form hydroperoxides, the substrate of the enzyme (Gueta-Dahan *et al.* 1997). This situation was obtained when salt increased superoxide dismutase (SOD) activity and concomitantly reduced ascorbate peroxidase (APX) activity. However, in paraquat-treated cells, where both enzymatic activities were elevated, no increase in the level of PHGPX was observed. It should be noted that the constitutive level of APX in the salt-sensitive cells was much lower than that observed in the salt-tolerant ones, thus rendering them less efficient in scavenging H_2O_2 and more prone to production of hydroperoxides (Gueta-Dahan *et al.* 1997).

Further support for the idea that *csa* mRNA induction is related to ROS was given by the present finding that tBH, a substrate of PHGPX, induced its mRNA independently of salt sensitivity (Figure 2). This induction was specific to *csa* and did not occur with another salt-induced gene, C3 (Naot et al 1994). Moreover, it cannot be due to a general oxidative stress, since paraquat and H_2O_2, which does not serve as a substrate, failed to induce the *csa* mRNA or its protein product (Gueta-Dahan *et al.* 1997, and data not shown). These observations and the fact that the induction of *csa* by tBH was already significant after 2 h, earlier than that observed with salt (Figure 2), suggest that *csa* is directly induced by the substrate of its encoded enzyme PHGPX. Our results are in agreement with results obtained with the yeast *Hansenula mrakii*, in which a direct induction of glutathione peroxidase by its substrate was demonstrated (Inoue *et al.* 1990). Activity of this enzyme was detected in the soluble fraction of cells grown in the presence of linoleic acid hydroperoxide, but not in its absence. The presence of the substrate in the growth medium did not inhibit growth, whereas growth of another yeast, *Saccharomyces cerevisiae*, which lacks this enzyme, was totally inhibited under the same conditions (Inoue *et al.* 1990). It has also been reported that in this strain of yeast hydrogen peroxide failed to induce glutathione peroxidase, while ROS, in the form of hydroxyl and superoxide radicals did so (Tran *et al.,* 993). We suggest that the addition of antioxidants enables a better scavenging of NaCl-induced ROS, and as a consequence, hydroperoxides are not produced and csa expression is not required (Figure 3).

In conclusion, the pattern of salt-induced expression differed between salt-sensitive and salt-tolerant cells, but was similar upon exposure to tBH. On the basis of these results we propose that *csa* expression is directly induced by the substrate of its encoded protein product, the enzyme PHGPX. The induction by salt is indirect and occurs mainly via an oxidative stress which leads to the formation of hydroperoxides. Our results suggest that one aspect of salt tolerance is a better capacity to deal with salt-induced ROS production and, hence, a reduction in the formation of damaging hydroperoxides.

References

Beeor-Tzahar T et al (1995) FEBS Lett. 366, 151-155.
Ben-Hayyim G, Kochba J (1982) Plant Sci. Lett. 27, 87-94.
Ben-Hayyim G et al (1993) Plant Sci. 88, 129-140.
Eshdat Y et al (1997) *Physiol. Plant.* 100, 234-240.
Faltin Z et al (1998) Physiol. Plant. (in press).
Flohe L, Gunzler WA (1984) Methods Enzymol. 105, 114-121.
Gossett DR et al (1996) Plant Physiol. 112, 803-909.
Gueta-Dahan Y et al (1997) Planta, 203, 460-469.
Hernandez JA et al (1993) Physiol Plant. 89, 103-110.
Hernandez JA et al (1994) New Phytol. 126, 37-44.
Hernandez JA et al (1995) Plant Sci. 105, 151-167.
Holland D et al (1993) Plant Mol. Biol. 21, 923-927.
Inoue Y et al (1990) Agric. Biol. Chem. 54, 3289-3293
Naot D et al (1995b) Gene, 161, 171-173.
Sambrook J et al (1989) Molecular Cloning: a Laboratory Manual. Cold Spring Harbor, NY: Cold Spring Harbor Laboratory Press.
Tran LT et al (1993). Biochem. Biophys. Acta, 1164, 166-172.

Callus from the resurrection plant *Craterostigma plantagineum* is a versatile tool to study ABA-mediated gene expression

D. Bartels
Botanisches Institut der Universität Bonn, D-53115 Bonn
and Max-Planck-Institut für Züchtungsforschung, D-50829 Köln, Germany

Desiccation is lethal to most higher plants only the small group of the so-called resurrection plants have evolved mechanisms to tolerate extreme dehydration. Such plants can tolerate up to a 98% loss of relative water content and recover within several hours of rehydration. This extreme desiccation tolerance is the basis for using *C. plantagineum* as an axperimental system to investigate the molecular basis of desiccation tolerance (Ingram and Bartels 1996). Many desiccation-inducible genes encoding structural proteins, enzymes or regulatory proteins have been isolated and characterized. Callus from this resurrection plant is not normally desiccation tolerant but it acquires desiccation tolerance after it has been treated with exogenous ABA (Bartels et al. 1990) Thus it has an absolute requirement for exogenously applied ABA to withstand drying.

Experimental protocol for obtaining

desiccation tolerant callus of C. plantagineum

Grow callus on MS medium

↓

Transfer callus to MS medium containing 5 mg l^{-1} ABA for four days

↓

Place callus on dry filter paper in the airstream of a ventilating hood for at least 16 hours

↓

Transfer callus to MS medium

↓

Monitor regrowth

The callus system is particularly suited to study the involvement of ABA in gene induction and desiccation tolerance. The ABA treatment induces the same set of transcripts which are expressed upon drying in the whole plant. A good correlation was observed between gene expression and desiccation tolerance in callus for many genes

A. Altman et al. (eds.), Plant Biotechnology and In Vitro Biology in the 21st Century, 461–463.

tested. For instance genes with homologies to late embryogenesis abundant genes (LEA) in seeds are only expressed in ABA-treated and ABA-treated dried callus but not in dried only callus (Piatkowski et al. 1990) Several experimental approaches have been carried out to analyse the ABA-mediated gene expression. The strategies will be discussed in the following text.

ABA analogs and gene expression

The effect of different ABA analogs on callus viability and the expression of a set of desiccation - induced transcripts was analysed (Chandler et al. 1997). For this callus was treated with (S)-(+)-ABA and (R)-(-)-ABA and two isomer derivatives of each. Both stereoisomers alone induced transcript expression and viability after dehydration. Two derivatives of (R)-(-)-ABA gave hardly any transcript expression and no viability. One (S)-(+)-ABA derivative induced transcripts and viability, however the other analog derivative led to the induction of transcripts but not to viability, suggesting that one analog must induce all genes necessary for viability whereas the other analog induces only a subset of genes, and some factors crucial for survival after desiccation are missing. This differential effect of the two analogs on gene expression and callus viability will be used as a tool in a subtractive screen to isolate genes essential for callus viability.

Isolation of mutants using activation tagging in *C. plantagineum* callus genetic mutations would be an ideal tool to identify genes important in desiccation tolerance. However, *C.plantagineum* is not suitable for genetic approaches because its genome is polyploid. The only possibility for a mutational approach is to obtain dominant mutations. The observation that wild-type callus of *C. plantagineum* does not survive desiccation without ABA treatment offers an excellent way to select for dominant mutations that activate the ABA and/or drought signalling pathways and thus confer desiccation tolerance to callus. For this objective leaf discs of *C. plantagineum* were transformed with a suitable T-DNA vector. Transgenic callus was selected capable of surviving a severe dehydration treatment in the absence of exogenously supplied ABA. One transgenic callus was selected which was able to withstand the dehydration process without ABA incubation (Furini et al. 1997). Analysis of RNA transcripts in this callus showed that genes are constitutively expressed in the callus which in the non-transformed tissues can only be induced by ABA treatment or desiccation. This indicates the activation of a gene downstream of ABA. The tagged gene (cDT-1) and corresponding cDNA clones were isolated. cDT-1 is constitutively expressed in the transgenic callus line and by ABA treatment in the untransformed callus or by drying in leaves. The cDT-1 gene is present in multiple copies in the *Craterostigma* genome. DNA sequence analysis of cDNA and genomic clones revealed some unusual features of the cDT-1 gene such as the absence of a plausible open reading frame and the presence of a stretch of adenosines. The mechanism of cDT-1 action is not yet understood and under study.

Callus as a source for ABA-related DNA binding proteins.

Callus can be grown in large amounts and it is very suitable to reproducibly isolate protoplasts. Transient expression assays can be performed with these protoplasts and it is thus possible to define promoter sequences relevant for ABA induction. These assays were used to reveal 5' sequences which confer ABA responsiveness in the promoter of the Lea-like dehydration and ABA inducible gene CDeT27-45 (Michel et al. 1993). A procedure was devised to isolate sufficient amounts of nuclear proteins from ABA-treated and untreated calli. The proteins were used in gel shift assays and it was shown that nuclear proteins from ABA treated callus bind specific sequence elements in the promoter region by the CDeT27-45 gene. By electrophoretic mobility shift assays ABA independent and ABA dependent binding activities could be distinguished. The ABA inducible binding acitvity is correlated with mRNA and protein accumulation and acquisition of desiccation tolerance (Nelson et al. 1993). Subsequently a protein purification procedure was established which allowed to purify the binding protein.

Conclusions

The induction of desiccation tolerance through ABA in callus presents a unique experimental system to dissect the ABA mediated signalling pathway in *C. plantagineum.* In combination with an efficient transformation protocol it is a powerful potential system to isolate mutants. A subtractive gene screening procedure may lead to the identification of essential genes for desiccation tolerance utilizing ABA analogs. It further allows to study activities of promoter sequences in the homologous system from which subsequently DNA binding proteins can be isolated.

References

Bartels D et al (1990) Planta 181, 27-34.
Chandler J et al (1997) Physiologia Plantarum 99, 465-469.
Furini A et al (1997) The Embo Journal 16, No. 12, 3599-3608
Ingram J., Bartels D. (1996). Annu. Rev. Plant Physiol. Plant Mol. Biol. 47, 377-403.
Michel D et al (1993) The Plant Journal 4 (1), 29-40
Nelson D et al (1994) The Plant Journal 5 (4), 451-458
Piatkowski et al (1990) Plant Physiology 94, 1682-1688.

Expression of fungal chitinolytic enzymes in transgenic apples confers high levels of resistance to scab.

Jyothi Prakash Bolar (1), John L. Norelli (1), Gary E. Harman (2), Susan K. Brown (2) and Herb S. Aldwinckle (1).

(1) Department of Plant Pathology, (2) Department of Horticultural Sciences, Cornell University, New York State Agricultural Experiment Station, Geneva, NY 14456.

Abstract

The overall goal of this research is to develop scab resistance in apples by transforming them with genes coding for chitinolytic enzymes from the biocontrol fungus *Trichoderma harzianum*. *Agrobacterium*- mediated transformation was used to transfer an endochitinase (ThEn-42) gene into Marshall McIntosh (MMc) apples. Varying levels of expression of ThEn-42 were found in different lines. Transgenic MMc lines expressing ThEn-42 were more resistant to apple scab than the parental lines as indicated with both grafted and own-rooted plants. There was a negative correlation between level of ThEn-42 expression and plant vigor. However, plants with intermediate levels of expression were substantially less susceptible than nontransgenics and some plants with intermediate levels of resistance had good vigor. Recently, we cloned an exochitinase gene encoding N-acetyl-ß-D-glucosaminidase (ThNag-70) from *T. harzianum*, and obtained ThNag-70 transgenic-MMc lines. A very high efficiency of transformation was obtained using the *Agrobacterium* strain EHA105-PCH32. Previous reports have demonstrated a strong in vitro synergistic activity between ThEn-42 and ThNag-70 against fungal pathogens. Both ThEn-42 and ThNag-70 were cloned into a single construct, and transgenic lines expressing both these enzymes were obtained. Presently the ThNag-70 transgenic lines and the lines expressing both ThEn-42 and ThNag-70 are being tested for scab resistance. We hypothesize that plants with "pyramided" genes, that express more than one synergistic protein, will be resistant to diseases even at relatively low levels of enzyme production and also have good vigor.

1. Introduction

Apple scab, caused by the fungus *Venturia inaequalis* (Cooke) G. Wint, is the most widespread disease in apple orchards around the world, and accounts for substantial pesticide usage on apples. Many commercially important cultivars, including McIntosh, are very susceptible to apple scab (Merwin et al. 1994). Uncontrolled scab can lead to total crop loss, defoliated trees, increased susceptibility to winter cold injury, and decreased bloom and crop in subsequent years (Anderson, 1956). The need for disease-resistant varieties to reduce the use of chemical fungicides has been well documented. The reasons for developing transgenic forms of our preferred commercial

A. Altman et al. (eds.), Plant Biotechnology and In Vitro Biology in the 21st Century, 465–468.

varieties by gene therapy, rather than developing completely new resistant varieties by breeding, are several. Most significant is the ability to preserve the desirable varietal characters and to add disease resistance.

Chitinolytic enzymes, which break down chitin, an important component of the cell walls of most fungi but not of plants or vertebrates, have been isolated from the biological control fungus, *Trichoderma harzianum* (Harman et al. 1993). These enzymes not only are highly active against fungal pathogens, but that they are also synergistic with each other and with different antifungal compounds (Lorito et. al. 1996). Among the more strongly synergistic combinations are an endochitinase (ThEn-42), which randomly cleaves chitin polymers and an N-acetyl-ß-D-glucosaminidase (ThNag-70) , which hydrolyzes nonreducing N-acetyl-ß-glucosamine residues from terminal ends (Tronsmo, Harman, 1993; Lorito et. al. 1996). Sequence information of ThEn-42 (Hayes et al. 1994) and ThNag-70 (Peterbauer et al. 1996) have been reported. 'Pyramiding' these genes in a single construct and transforming apples should be an effective strategy in increasing fungal resistance. The goal of this research was to obtain transgenic Marshall McIntosh (MMc) apple lines expressing ThEn-42, ThNag-70, and both enzymes, in order to evaluate resistance of these transgenics to *V. inaequalis*.

2. Materials and Methods

2.1 Plasmid Binary Vectors

Plasmids p35S-ThEn42 and pBIN19ESR (obtained from C.K. Hayes and K. Wong respectively, Cornell University, Geneva, NY) contain the genomic and cDNA respectively, encoding ThEn-42 from *T. harzianum* strain P1. Both the cDNA and genomic versions of *ThNag-70* were cloned using a PCR based strategy; their sequences were confirmed; and plasmids pBI121-cNag (cDNA) and pBI121-gNag (genomic) were constructed. pBIN(endo+cNag) was constructed, containing both the cDNA's of *ThEn-42* and *ThNag-70* driven by separate CAMV-35S promoters in the same construct. All five plasmids were transferred into *Agrobacterium tumefaciens* strain EHA105. In addition, plasmids pBI121-cNag, pBI121-gNag and pBIN(endo+cNag) were transferred into *A. tumefaciens* strain EHA105-PCH32.

In vitro grown leaves of MMc were inoculated with disarmed *A. tumefaciens* strains carrying the genes encoding the chitinolytic enzymes. Procedures for inoculation, selection and regeneration of transgenics have been previously described (Bolar et al. 1995, 1998a). The explants were regenerated on a modified N6 medium in the presence of 25 mg L^{-1} kanamycin, 300 mg L^{-1} cefotaxime, and the medium was solidified with agar (5.2 g L^{-1}) and Gelrite (0.6 g L^{-1}) (Bolar et al. 1995, 1998a).

2.2 Identification of transgenic plants

The regenerating shoots were transferred to jars containing proliferation medium amended with Paromomycin 100 mg L^{-1} (Bolar et al. 1995, 1998a). Shoots that grew on this medium were considered putative transformants, and activity of the *nptII* (marker gene) was confirmed by a standard NPTII ELISA. Transformation was confirmed by PCR analysis with primers specific for various fragments of the T-DNA, and by Southern analysis. An activity assay was developed to quantify the expression of the transgenic enzyme. Activity of ThEn-42 was detected using a fluorescent substrate, 4-methylumbelliferyl-ß-D-N,N',N"-triacetylchitotrioside and that of ThNag-70 was detected using the substrate 4-methylumbelliferyl N-acetyl-ß-D-glucosaminide. Western analysis was conducted using antibodies specific to the transgenic enzymes.

2.3 Propagation of transgenic plants and disease screening

Preliminary experiments were conducted to optimize the inoculum concentration of *V. inaequalis* and establish parameters for quantifying disease severity. A technique to own-root transgenic apple lines was developed (Bolar et al. 1998b) that increased the screening efficiency. Endochitinase transgenic apple lines were grown as both grafted plants on seedling rootstocks and as own-rooted plants. Propagated plants were inoculated in the greenhouse with conidial suspension of *V. inaequalis* and incubated at 20°C and 100% RH to favor disease development. Their degree of infection was quantified at intervals as numbers of sporulating lesions, percent leaf area infected, and number of conidia recovered per leaf, and compared with infection on control plants of non-transformed MMc and Liberty (resistant). The experiment was repeated twice.

3. Results and discussion

We have successfully transferred the genes coding for chitinolytic enzymes into MMc apples. The recovered lines produced differing levels of the transgenic enzyme ranging from low, moderate to high levels of expression. Transgenics produced from the cDNA constructs of *ThEn-42* (pBIN19ESR) or *ThNag-70* (pBI121-cNag) had a higher level of enzyme expression than lines produced from genomic constructs (p35S-ThEn42 or pBI121-gNag). These results may suggest that the plants were not efficient in splicing heterologous fungal introns.

There was a very high correlation between level of expression of ThEn-42 and disease resistance. Lines expressing moderate to high levels of ThEn-42 were significantly resistant to scab compared to the non-transformed control. In very high expression of ThEn-42 protection probably resulted from direct effect due to the degradation of the chitin of growing hyphae, as reported in chitinase-transgenic canola plants (Benhamou et al. 1993). Further cytological data are needed to test this hypothesis. Interestingly, moderate producers of ThEn-42 also had very high level of disease resistance, probably resulted from a direct effect of the transgenic enzyme, and also an indirect effect due to the release of chitin oligomers. Garcia et al (1997) reported that resistance level against *Rhizoctonia solani* by the ThEn-42 expression in transgenic tobacco was independent of enzyme level, and suggested that the transgenic protein released chitinolytic elicitors of plant defense from the pathogen cell walls. However, we also found that growth and vigor of the lines with very high expression of endochitinase was substantially reduced relative to the nontransformed control or lines with lower expression. At intermediate levels of expression, however, resistance still was seen but reduction in plant vigor was variable and plants that had near-normal levels of vigor were obtained.

Both the cDNA and genomic constructs carrying *ThNag-70* resulted in a higher efficiency of transformation when the *A. tumefaciens* strain EHA105-PCH32 was used than when strain EHA105 was used. EHA105-PCH32 had an extra copy of *vir*G (vir inducer protein), *vir*E1 and *vir*E2 (single strand binding protein) (Hamilton et al. 1997). Different lines produced a range of levels of expression of ThNag-70. Activity of both ThEn-42 and ThNag-70 were detected in lines transformed with pBIN(endo+cNag).

4. Future work

We have propagated *ThEn-42* transgenic lines that have shown increased resistance in the greenhouse for field trails. In the field these transgenic lines will be evaluated for disease resistance and horticultural qualities. Presently, ThNag-70 transgenic lines and the lines expressing both ThEn-42 and ThNag-70 are being propagated for disease evaluation. Based on the results of these tests, we hope to select resistant lines that would significantly reduce the use of chemical fungicides and result in economic, environmental, and health benefits.

5. Acknowledgments

We thank Christopher K. Hayes and Kwai-Weng Wong for providing plasmids, p35S-ThEn42 and pBIN19ESR, used in this study. This research was funded in part by the New York State Apple Research and Development Program and the New York State Center for Advanced Technology Biotechnology Program.

6. References

Anderson HW (1956) Diseases of Fruit Crops, McGraw-Hill, New York, USA.
Benhamou N et al (1993) Plant J. 4, 295-305.
Bolar JP et al (1995) HortSc. 30, 875.
Bolar JP et al (1998a) Plant Cell Tiss. Org. Cult. (in press).
Bolar JP et al (1998b) HortSc. (in press).
Garcia I et al (1997) Phytopathology 87, S32.
Hamilton et al (1997) Gene 200, 107-116.
Harman GE et al (1993) Phytopathology 83, 313-318.
Hayes CK et al (1994) Gene 138,143-148.
Lorito M et al (1996) Mol. Plant-Microbe Interact. 9, 206-213.
Merwin IA et al (1994) Plant Dis. 78, 4-10.
Peterbauer C et al (1996) Curr. Genet. 30, 325-331.
Tronsmo A, Harman GE (1993) Anal. Biochem. 208, 74-79.

ISOLATION AND EXPRESSION OF AN NADPH-DEPENDENT REDUCTASE GENE ABLE TO DETOXIFY THE *EUTYPA LATA* TOXIN IN GRAPEVINE.

S. Dalmayrac, S. Colrat, P. Guillén, M. Guis, G. Martínez-Reina, C. Deswarte, M. Bouzayen, J. P. Roustan, J. Fallot, J. C. Pech and A. Latché.

Ecole Nationale Supérieure Agronomique de Toulouse, UA INRA, Avenue de l'Agrobiopole BP 107, Auzeville-Tolosan, 31326 Castanet Tolosan Cedex, France.

1. Introduction

Eutypa dieback, caused by the ascomycete fungus *Eutypa lata* (Pers. : Fr.) Tul. and C. Tul. is currently the most serious disease of the grapevine. It affects a great number of vineyards throughout the world and is becoming the primary constraint on vineyard longevity and can also indirectly affect the quality of wine (Moller and Kasimatis, 1981; Munkvold et al. 1994). To date no active chemicals are available for the destruction of the parasitic fungus in the plant. Eutypa dieback is only managed by practices such as removing and destroying infected plants, delaying pruning, and protecting pruning wounds (Fallot et al. 1997). However, these practices are labour intensive, expensive and poorly efficient.

The ascospores of the pathogen infect and colonize the xylem tissue through pruning wounds, and then propagate into the cambium and phloem in the vine trunk and arms (Duthie et al. 1991). After an incubation period of 3 years or more, a canker forms around the infected wound and the symptoms appear on the herbaceous part of the plant, leading to dwarfed and withered new growth, necrosis at the margins of the leaves, dryness of the inflorescences, and finally death of one or more branches.

It has been suggested that the symptoms are due to a toxin, 4-hydroxy-3-(3-methyl-3-butene-1-ynyl) benzaldehyde, named eutypine, synthesised by the mycelium present in the trunk and transported by the sap (Tey-Rulh et al. 1991). Several facts suggest that eutypine is involved in the development of eutypa dieback symptoms (Tey-Rulh et al. 1991, Deswarte et al. 1994). It has been shown that eutypine behaves as a weak acid, exhibits a marked lipophilic character, crosses the plasmalemma by passive diffusion and accumulates in cell cytoplasm (Deswarte et al. 1996a). Furthermore, eutypine uncouples mitochondrial oxidative phosphorylation by increasing proton leakage through a cyclic protonophore mechanism (Deswarte et al. 1996b).

Recently, we have shown that eutypine is metabolised in grapevine cells into a compound identified as 4-hydroxy-3-(3-methyl-3-butene-1-ynyl) benzylic alcohol, named eutypinol (Fallot et al. 1997). This compound was found to be non-toxic for grapevine and to have no protonophoric activity. Since it has been found that the level of sensitivity to eutypa dieback is dependent on the genotype, a relationship between the tolerance of some grape varieties to the disease and their capacity to inactivate the toxin was established (Deswarte, 1994).

The progress made in understanding eutypine's mechanism of action has opened up new prospects regarding the development of an efficient strategy for stopping this disease based on the research of a toxin tolerance gene that could lead to the eutypine detoxification. We report here the isolation of a cDNA encoding a protein exhibiting an eutypine reductase activity from *Vigna radiata*. Transgenic grapevine cells and plantlets, over-expressing this gene, were obtained and the tolerance to eutypine was evaluated.

A. Altman et al. (eds.), Plant Biotechnology and In Vitro Biology in the 21st Century, 469–472.

The outcome of these findings is discussed in terms of disease prevention through genetic engineering approaches.

2. Material and Methods

2.1. Plant material

Seeds of *Vigna radiata* L. R. Wilcz were purchased from Cereal Wander Nutrition Annonay, France. Callus and cell suspension cultures of *Vitis vinifera* cv. Gamay were grown as described by Ambid et al. (1983). The embryogenic cultures of rootstock 110 Richter (*V. rupestris x V. Berlandieri*) were provided by A. Bouquet (INRA, Montpellier, France) and grown as described by Le Gall et al. (1994).

2.2. Cloning of cDNA VR-ERE from Vigna radiata and expression in E. coli

The purification procedure of the protein VR-ERE has been described by Colrat (1998). Based on the amino acid sequence of the peptide, a cDNA was cloned by iPCR (Guillén, unpublished data). The amino acid sequence corresponding to the nucleotide sequence of the cDNA was predicted using the TRANSL program of the PC Gene software package (Department of Medical Biochemistry, University of Geneva, Switzerland). The sequence of the oligonucleotide sequence and the deduced peptidic sequence have been described (Patents n°FR9700962). To study the expression of the VR-ERE gene in *E. coli*, the coding region of the VR-ERE cDNA was cloned into the *Eco*RI site of the expression vector, pT7-7, as described by Tabor and Richardson (1985).

2.3. Transformation of grapevine cells cultured in vitro

Transformation was performed using 10-day-old callus cultures of *Vitis vinifera* cv. Gamay and 110 Richter using *Agrobacterium tumefaciens,* C58, containing the binary vector pGA643 (An et al. 1988) with a Nos/NPTII chimeric gene and the VR-ERE cDNA under the transcriptional control of the CaMV 35S promoter. Calli were submerged for 5 min in an overnight culture of *A. tumefaciens* (10^6 to 10^8 cells mL^{-1}), dry blotted, then co-cultivated on growth medium in the dark for 48 h at 28°C. Finally, transformed cells were selected on growth medium containing 100 mg/mL kanamycin and 500 mg/ml carbenicillin. The embryogenic calli were cultured on the embryogenic medium and the regenerated plantlets were cultured in conditions previously described (Le Gall et al. 1994).

3. Results and Discussion

3.1. Characterization of VR-ERE cDNA clone from Vigna radiata

After screening of different plant species for their ability to metabolise eutypine, a 36 kDa enzyme, named VR-ERE (*V. radiata* - Eutypine Reducing Enzyme), was purified from *Vigna radiata*. Degenerated oligonucleotides were designed, based on the amino acid sequence of the peptide. A full-length clone of 1254 bp was subsequently obtained by iPCR. It contained an open reading frame of 975 nucleotides starting at ATG and ending with GTT, 995 bp upstream from the poly $(A)^+$ tail. The deduced amino-acid sequence corresponds to a 325 amino-acid polypeptide with an expected molecular mass of 35.9 kDa.

The VR-ERE cDNA encoded a protein that showed high homology (86.8% identity) with the drought-inducible CPRD 14 gene isolated from *V. unguiculata* whose physiological function is unknown (Iuchi et al. 1996) and the cDNA of *Eucalyptus gunnii* encoding for a cinnamyl alcohol dehydrogenase (71.7 %, Goffner et al. 1998). The VR-ERE protein exhibited a low homology with a cinnamoyl-CoA reductase from *E. gunnii* (Lacombe et al. 1997) and dihydroflavonol reductases.

3.2. Biochemical characteristics of recombinant VR-ERE

The functional expression of VR-ERE cDNA was confirmed in *E. coli*. The recombinant protein exhibited a K_m of 12.5 μM and 8.4 μM towards eutypine and NADPH respectively, and an optimum pH of 6.5. It was verified that the reaction product was eutypinol. These results suggest that VR-ERE is an NADPH-dependent oxidoreductase.

The biochemical analysis of the VR-ERE protein revealed that the enzyme reduced a broad range of aromatic and aliphatic aldehydes. However, it showed low affinity towards the cinnamyl aldehyde precursors of lignin such as coniferyl aldehyde (K_m: 83μM), sinapyl aldehyde (K_m: 117μM) and cinnamaldehyde (K_m: 128μM). It also failed to reduce the major dihydroflavonols found in *V. radiata*. These data indicate that the VR-ERE protein cannot be considered as CAD or DFR, although well-conserved regions are shared by all three proteins.

3.3. Expression of the VR-ERE gene in V. vinifera cells

The capacity of VR-ERE to confer resistance to the toxin eutypine was evaluated by over-expressing VR-ERE cDNA in *V. vinifera* cv. Gamay. The construct was transferred into grapevine cells via *Agrobacterium*-mediated transformation (Fig. 1).

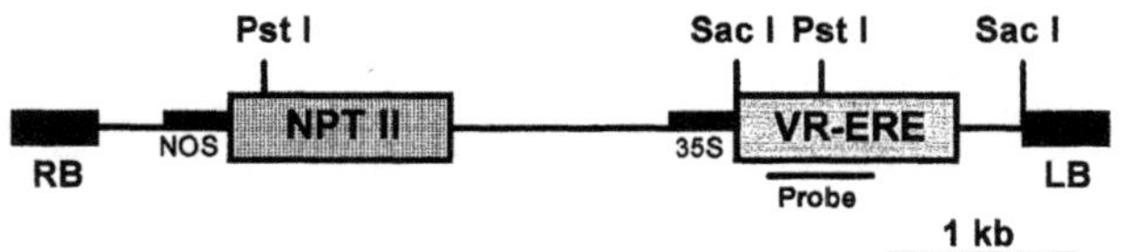

Figure 1. Construction of the pGA-VR-ERE plasmid binary vector.

Southern blot analysis using a fragment of VR-ERE as a probe confirmed the integration of the VR-ERE cDNA into the genome of the three selected transgenic callus lines. No hybridisation was observed with DNA extracted from wild-type grapevine calli under the same conditions. Northern blot and Western blot analysis confirmed gene expression in the transgenic cells. The three lines displayed eutypine-reducing activity ranging from 7 to 12 fold higher than that of the wild type calli.

In order to demonstrate that the over-expression of VR-ERE protein can improve *V. vinifera* cell resistance to eutypine, the wild-type and transgenic calli were grown under several concentrations of eutypine ranging from 0 to 1000 μM (Fig. 2). Transgenic callus growth was not affected by 500 μM eutypine while wild-type *V. vinifera* cell growth was completely inhibited by the same eutypine concentration. These data incate that the VR-ERE gene constitutes an attractive candidate to confer eutypine tolerance to grapevine tissues.

3.4. Production of transgenic 110 Richter plants expressing the VR-ERE gene

In order to obtain trangenic plants, 110 Richter somatic embryogenic cells were transformed using the pGA-VR-ERE construction previously described (Fig. 1). The putative transformed cells were cultured in the regeneration medium and several plantlets were obtained. The presence of the transgene was demonstrated by Southern blot and its expression verified by Western blot. The ability of the trangenic leaf tissues to metabolise eutypine *in vivo* was found to be increased by more than 50 % compared to that of the wild-type tissues. The first observations on the development of the transgenic grapevine vitroplants cultured in the presence of eutypine suggest that the transgenic 110 Richter plants were more tolerant to eutypine than the wild-type plants.

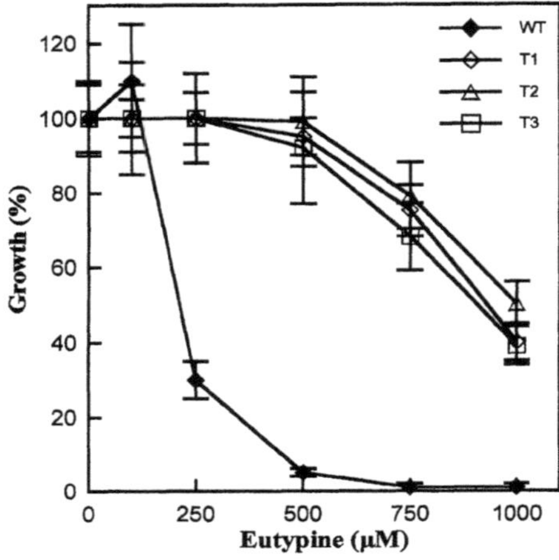

Figure 2. Influence of eutypine on the growth of VR-ERE transgenic lines (T1, T2, T3), and wild-type (WT) callus of *Vitis vinifera*, cv. Gamay. The calli were cultured in the presence of a range of eutypine concentrations. Callus weight was determined after 4 weeks in culture. Each point represents the mean of 40 replicates.

4. Conclusion

The data presented in this paper constitute an important contribution to the understanding of the interaction between *Eutypa lata* and grapevine. Furthermore, the characterization of the VR-ERE gene contributes to increasing understanding of the detoxification system for fungal toxins. The VR-ERE gene is an attractive candidate to confer resistance to eutypine. The behaviour of the transgenic plants in the presence of *Eutypa lata* fungus may to confirm the role of eutypine in symptom development and to study the efficiency of the VR-ERE gene in the fight againts eutypa dieback.

5. Acknowledgements

This work was supported in part by EEC program N° 8001-CT91.205, the French 'Ministère de l'Enseignement Supérieur et de la Recherche' and Midi-Pyrénées Region. P. Guillèn P. and G.R. Martinez were recipient of a EEC fellowship. M. Guis and S. Colrat hold grants from the 'Ministère de l'Enseignement Supérieur et de la Recherche'. The 110 Richter embryogenic cultures were kindly provided by A. Bouquet, INRA, Montpellier, France. The authors would like to thank P. Winterton for rewieving English manuscript and H. Mondiès for this skilful technical assistance.

6. References

Ambid C et al (1983) Physiol. Vég. 21, 82-92
An G et al (1988) Plant Molecular Biology. Kluwer Acad. Pub. The Netherlands
Colrat S (1998) PhD, INP-ENSAT, Toulouse, Fance, pp 1-105
Deswarte C (1994) PhD, INP-ENSAT, Toulouse, France, pp 1-103
Deswarte C et al (1996a) J. Plant Physiol. 334, 200-205
Deswarte C et al (1996b) Arch. Biochem. Biophys. 149, 336-342
Duthie JA et al (1991) Phytopathol. 81, 1183
Goffner D et al (1998) Plant Molec. Biol. 36, 755-765
Iuchi S et al (1996) Plant Cell Physiol. 37, 1073-1082
Lacombe et al (1997) Plant J. 11, 429-441
Le Gall O et al (1994) Plant Science 102, 161-170
Moller WJ et al (1981) Plant Disease Reporter 65, 429-431
Munkvold GP et al (1994) Phytopath. 84, 186-192
Tabor S et al (1985) Proc. Natl. Acad. Sci. USA 82, 1047-1078
Tey-Rulh P et al (1991) Phytochem. 30, 471-473

INDUCED DEFENSE MECHANISMS IN PLANT-FUNGUS INTERACTIONS: DIFFERENCES BETWEEN CELLS IN CULTURE AND LEAF TISSUE

S. Dorey, M. Kopp, B. Fritig & S. Kauffmann
Institut de Biologie Moléculaire des Plantes du C.N.R.S.
Université Louis Pasteur
12, rue du Général Zimmer, 67084 Strasbourg, France

ABSTRACT

Cultured plant cells are model systems of reduced complexity to decipher the mechanisms of plant defense responses to pathogens. Their use led to major advances, particularly in the detection and characterization of early defense events, occurring within minutes. The connection of these with later events, occurring within hours, remains, however, to be uncovered. In the present study, we compared the responses of tobacco cells in culture to those of leaf tissues after treatment with 3 fungal elicitors.

1. INTRODUCTION

Plants can cope with invading pathogens by inducing a powerful disease resistance mechanism called the hypersensitive reaction (HR) which results in plant cell death at the site of pathogen penetration. Efficient defense also implies that plant cells undergo a wide range of metabolic alterations (stimulation of secondary metabolic pathways, accumulation of a broad range of defense proteins with antimicrobial activity) which occur within hours (referred to as late events) and which are produced in a spatio-temporally regulated manner (Kombrink, Somssich, 1995). These alterations are downstream of a cascade of events triggered very rapidly, within minutes, after the plant has perceived the pathogen or a microbial component called elicitor. Recognized early events include ion fluxes accross the plasmamembrane and an oxidative burst (Hammond-Kosack, Jones 1996). While late events can be studied in whole plants infected with pathogens or treated with elicitors, such *in planta* systems disclose experimental drawbacks to investigate early events. Model systems of reduced complexity consisting of cultured plant cells treated with pathogen-derived signals are of easier use in this respect. One major challenge is to establish models which give access to the connections between early and late events. It opens also the question whether responses observed in cell cultures are expressed in a strictly similar manner as when *in planta* systems are considered. To bring out some new elements, we compared tobacco cell cultures and tobacco leaf tissues for some of their responses to 3 *Phytophthora megasperma* elicitors,

A. Altman et al. (eds.), Plant Biotechnology and In Vitro Biology in the 21st Century, 473–476.

i.e., 2 elicitins called α- and ß-megaspermin and a 32 kD glycoprotein.

2. MATERIALS AND METHODS

The glycoprotein, α- and ß-megaspermin were purified from the culture medium of *P. megasperma* H20 (Baillieul et al., 1995). Elicitors were infiltrated into leaves of *Nicotiana tabacum* cv. Samsun NN to cover areas of about 2 cm^2, and applied to tobacco cell cultures (*N. tabacum* cv. Bright Yellow) maintained in the dark at 25°C in a modified Murashig-Skoog medium. Cell death was monitored on leaf tissue (Dorey et al., 1997), and cells using 400 µL aliquots incubated with 0,05% Evans blue and then washed extensively. The dye was solubilized in 50% methanol with 1% SDS for 30 min at 50°C and quantified by the absorbance at 600 nm. H_2O_2 was measured by chemiluminescence using 100 µL of cells to which peroxidase (60 units) and luminol (25 µL) were added.

3. RESULTS AND DISCUSSION

3.1 Induction of cell death

When infiltrated into tobacco leaves at 50 nM the 3 elicitors are potent inducers of HR cell death visualized as tissue necrosis (Baillieul et al., 1996). Table 1 shows that increasing elicitor concentrations still caused tissue necrosis remaining strictly localized to the infiltrated tissue. Lowering the amounts to 5 nM resulted in partial necrosis. Apparently, the procedure created a slight gradient of elicitor concentration decreasing from the site of infiltration to the external limit of liquid spread. With a 0.5 nM concentration of each elicitor the tissue remained symptomless. Hence, the 2 elicitins and the glycoprotein showed the same biological activities when infiltrated into tobacco leaves, i.e., induction of HR cell death with a similar threshold level and induction of typical defense responses (Kauffmann et al., 1993; Keller et al., 1996).

nM	α-megaspermin	ß-megaspermin	glycoprotein
0.5	-	-	-
5	+/-	+/-	+/-
50	+	+	+
250	+	+	+
500	+	+	+

Table 1: Necrosis of tobacco leaf tissues treated with 3 *P. megasperma* elicitors. -: no necrosis; +/-: limited necrosis; +: necrosis of the infiltrated tissue. The 3 elicitors were shown to be similarly active when applied to different *N. tabacum* cultivars.

Application of the elicitors to cell cultures revealed a very different pattern in respect to cell death induction. The glycoprotein and α-megaspermin did not cause significant cell death when applied at up to 250 nM, while 5 nM ß-megaspermin did (figure 1). The slight differences observed in the absorbance values when either the glycoprotein or α-megaspermin were used reflected variations in the assay rather than real cell death.

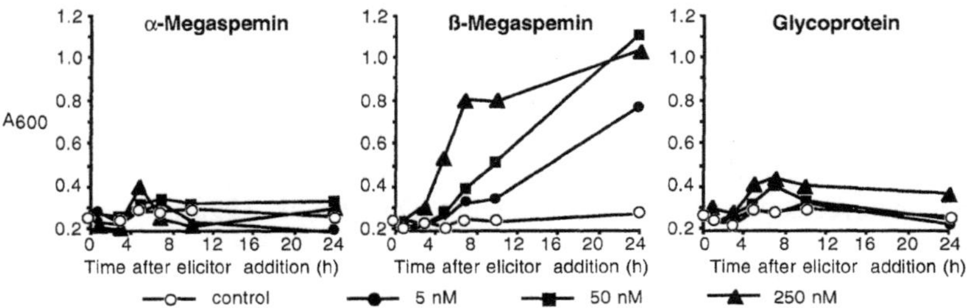

Figure 1: Cell death of cultured tobacco cells treated with 3 *P. megasperma* elicitors.

Increasing ß-megaspermin amounts increased cell death. In our conditions, an absorbance of 1.2, which is reached at 24 h when 50 or 250 nM ß-megaspermin was applied, corresponded to death of about 35% of cells. It was rather unexpected that α-megaspermin did not cause death of cultured cells, since α- and ß-megaspermin display more than 85% amino acid sequence identity. One possible explanation for the lack of cell death inducing activity of α-megaspermin, and also of the glycoprotein, could be a lack of perception by plant cells in culture.

3.2 Perception of the 3 P. megasperma elicitors by tobacco cell suspensions

A highly sensitive perception system, based on the alkalinization of the growth medium of cell cultures, has been established (Boller, 1995). Alkalinization is a consequence of ion fluxes through the plasmamembrane, and occur very early. All 3 elicitors induced alkalinization, but with different kinetics and/or amplitudes: transient with the glycoprotein, more durable with the 2 elicitins, ß-megaspermin being the most active (figure 2). Hence, the absence of cell death inducing activity of α-megaspermin and of the glycoprotein was not caused by a lack of perception. Furthemore, no qualitative correlation was apparent between cell death induction and medium alkalinization.

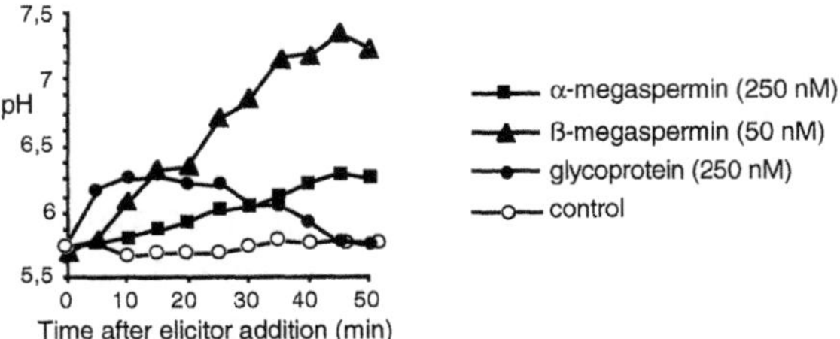

Figure 2: Medium alkalinization of cultured cells treated with 3 *P. megasperma* elicitors.

3.3 Induction of the oxidative burst

A hallmark of the HR is the occurrence of an oxidative burst rapidly after pathogen or elicitor perception by the plants cells (Hammond-Kosack, Jones, 1996). In our cell suspension system, treatment with the 2 elicitins caused an oxidative burst measured as an accumulation of H_2O_2 in the extracellular medium (figure 3). Furthermore, there was no correlation between the oxidative burst inducing activity and the cell death inducing activity of the 2 closely related elicitins. Quite unexpected was the fact that the glycoprotein failed to induce such a burst in culture whereas a large increase in H_2O_2 had been previously observed in leaf tissue infiltrated with 50 nM glycoprotein, though in this case it occurred several hours after treatment (Dorey et al., 1998).

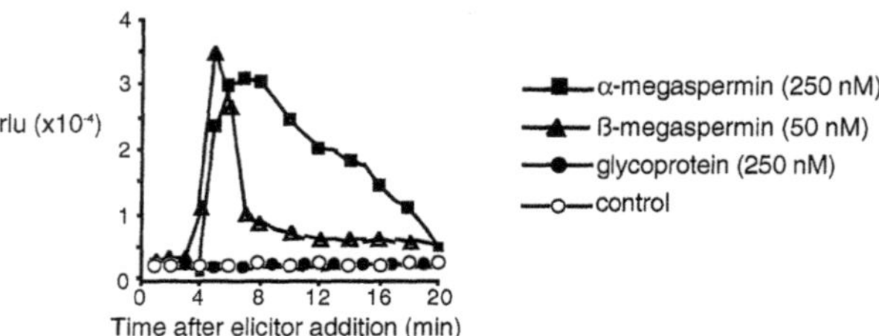

Figure 3: Oxidative burst in cell cultures treated with 3 *P. megasperma* elicitors.

In conclusion, the results indicate that while leaf tissues exhibit responses typical of the HR after treatment with any of the elicitors, cultured cells did not always develop HR cell death even when they had perceived the elicitor as shown by growth medium alkalinization. Furthermore, the rapid oxidative burst (within 10 min) after elicitor treatment of cultured cells was not sufficient to cause HR cell death. Though cultured cells are powerful tools to uncover many aspects of plant defense responses, they lack important features, i.e. tissue organization, which may be important for crosstalks between signalling pathways and, indeed, in intercellular signalling.

REFERENCES

Baillieul F et al. (1995) Plant J. 8: 551-560.
Baillieul F et al. (1996) Mol. Plant-Microbe Interact. 9: 214-216.
Boller T (1995) Annu. Rev. Plant Physiol. Plant Mol. Biol. 46: 189-214.
Dorey S et al. (1997) Mol. Plant-Microbe Interact. 10: 646-655.
Dorey S et al. (1998) Mol. Plant-Microbe Interact. 11: in press.
Hammond-Kosack KE, Jones JDG (1996) Plant Cell 8: 1773-1791.
Kauffmann S et al. (1993) Developments in Plant Pathology : Mechanisms of Plant Defense Responses, pp 140-143, Kluwer Acad. Publishers, Dordrecht, The Netherlands.
Keller H et al (1996) Mol. Plant-Microbe Interact. 9: 696-703.
Kombrink E, Somssich IE (1995) Adv. Bot. Res. 21: 1-34.

SELECTION OF HERBICIDE RESISTANT CELL LINES AND SUBSEQUENT IDENTIFICATION OF HERBICIDE TARGETS

J. Freitag[1], F. Schmidt[2], H.-D. Pohlenz[3], R. Walden[4]and R. Höfgen[1]
[1]MPI of Molecular Plant Physiology, Golm; [2]AgrEvo, Frankfurt; [3]Schering Agrochemicals / AgrEvo, Berlin and [4]MPI of Breeding Research, Cologne, Germany

1. Introduction

Plants can built up various mechanisms of resistance against herbicides. The frequency of generation or selection of resistance is rather low. Herbicide resistance can be caused by detoxification processes like destruction, modification, conjugation and sequestration of active components of the herbicides. Other possibilities of occurrence of resistance mechanisms are an altered uptake of the toxin or mutations of the target sites. Target amplifications were also described as mutations leading to herbicide resistance either due to, gene amplification or increased expression of RNA. We tried to combine approaches of molecular biology and tissue culture to device a method leading to the same effect, i.e. overexpression of the target, by applying genetic engineering of target plants. It was our main interest to generate herbicide resistant cell lines for a subsequent cloning of target genes coding for enzymes inhibited by toxic compounds. The determination and cloning of the molecular targets of these substances will allow overexpression and purification of the target protein for implementation into automated screening procedures. Furthermore, this will allow the biochemical characterization of the target protein to support approaches for improvement of the inhibitor structure or rational drug design strategies. Additionally, mutated or overexpressed target genes can be used in plant breeding to confer resistance to crop plants.
We adopted the T-DNA activation tagging technique (Hayashi et al., 1992) for the cloning of target genes coding for enzymes inhibited by herbicides. The *Agrobacterium* mediated gene transfer delivers the T-DNA randomly into the plant genome. The overexpression of the tagged gene is caused by the stable integration of a strong enhancer tetramer derived from the 35S CaMV promoter into the plant genome. This dominant enhancer mutation allows a positive selection for resistance to a phytotoxic compound. Tobacco protoplasts (cv SR1) were co-cultivated with *Agrobacterium* harbouring the vector pPCVICEn4HPT (Hayashi et al., 1992) allowing generation of a large number of transformants. Such a large number of successful transformations is indispensable to reach a saturation of the plant genome with the enhancer tag. This approach should lead

A. Altman et al. (eds.), Plant Biotechnology and In Vitro Biology in the 21st Century, 477–479.

to cell lines or plants resistant to herbicides by overproduction of the corresponding target enzyme due to an enhancement of the gene expression or an enhanced expression of other mechanisms like an overproduction of destroying or modifying enzymes.

2. Results and discussion

We have been selected several cell lines in tissue culture to be resistant to different phytotoxic compounds. Here, we will present data for glyphosate as one of the phytotoxic compounds we used. Glyphosate is the active component of the herbicide Roundup inhibiting the enolpyrovylshikimate-3-phosphate synthase (EPSPS) (Mousdale,Coggins, 1991). The enzyme EPSPS converts shikimate to enolpyrovylshikimate-3-phosphate in the aromatic amino acids pathway finally leading to the biosynthesis of phenylalanin, tyrosin and tryptophan. Resistance to glyphosate was already described and was caused by an increased transcription of the EPSPS gene or by overproduction of EPSPS due to an amplification of the encoding gene (gene copy number) (Jones et al., 1996). Therefore we choose glyphosate for the validation of our experimental strategy.

The selective conditions for protoplasts treated with glyphosate were established in 'kill curve' experiments. Efficient selection of protoplasts was observed at a concentration of $5x10^{-4}$M glyphosate in the cultivation media. For the performed tagging experiment we used a 10^{-3}M concentration of glyphosate in the media to minimise the incidencies of spontaneously occurring resistances in tissue culture. To diminish the risk of false positives the transformed protoplasts were additionally cultivated under selective conditions for T-DNA integration. A plant selective marker gene conferring hygromycin resistance (HPT) was transfered together with the enhancer tetramer. Hygromycin selection was performed at a concentration of 20 mg/l in the media.

In the glyphosate tagging experiment more than $3x10^7$ protoplasts were transformed and led to formation of fourteen independent calli. We regenerated plants from all fourteen calli applying 2mg/l zeatine and 2mg/l BAP for shoot induction.

Integration of the T-DNA was determined by Southern blot analysis with callus and plant DNA. The genomic DNA was digested with the restriction enzyme EcoRI that does not cut within the T-DNA. The Southern blot experiments were performed using the enhancer and the HPT gene as probes. To investigate for the occurrence of chimerics we isolated several plants from one callus. Southern blot analysis of these plants revealed similar hybridisation signals indicating that these plants were isogenic. A fragment of a *N. tabacum* EPSP synthase (*Nt*-EPSPS) of 1.6 kb size amplified by PCR was used as a probe in a further Southern experiment in order to analyse for band shifts caused by T-DNA integrations compared to a wild type plant.

Gene expression was tested in Northern blot experiments where additionally HPT and *Nt*-EPSPS were used as probes. For an increase of EPSPS mRNA we found only a weak evidence in the performed Northern experiments. The hygromycin resistance gene was expressed in all mutants.

To test for glyphosate resistance of the regenerated transgenic plants greenhouse spraying experiments were performed. All transgenic lines being positive in Southern and Northern experiments were propagated after self-pollination of the T_0 generation. Thus, we were

able to investigate a large number of individuals (T_1) of each transgenic genotype. This segregating populations were subjected to spraying experiments. Seedlings (T_1) segregated into the wild type genotype were excluded by hygromycin pre-selection. Moreover segregating analysis exhibited additionally that only some of the isolated genotypes segregated in a mendelian manner. Five weeks old plants were tested in spraying experiments applying several concentrations of the herbicide 'Roundup ultra'. Two transgenic lines showed tolerance, but no complete resistance to an otherwise toxic condition of 360g AS/ha Roundup.
For the cloning of genomic DNA the plasmid rescue approach was used (Fritze, Walden, 1995). This involved cleaving of the genomic DNA with restriction enzymes that do not cut within the T-DNA. The T-DNA carried an *E.coli* origin of replication and an ampicilin resistance gene for selection in *E.coli.* After digestion of the callus DNA with EcoRI we religated the genomic DNA and transformed these circular fragments into competent cells. We obtained twelve rescued T-DNA including clones deriving from four independent calli. One of these clones was analysed further because it derived from one of the two tolerant genotypes identified in the spraying experiment.
The insert of flanking plant DNA in that clone was approximately 6 kb in size. We rescued a 2.8 kb genomic fragment neighbouring the enhancer and a 3.2 kb fragment flanking the left T-DNA border. The rescued clone was subjected to sequencing.

3. References

Fritze K, Walden R. (1995) Methods in Molecular Biology 44, *Agrobacterium* protocols pp 281-294, Humana Press Inc., Totowa, NJ, USA
Hayashi H et al. (1992) Science 258, 1350-1353
Jones JD et al. (1996) Plant Cell Reports 15, 431-436
Mousdale DM, Coggins JR (1991) Target Sites for Herbicide Action, pp 29-56, Plenum Press, New York, USA

4. Acknowledgements

I wish to thank Renate Lührs, Astrid Blau, Yvonne Maas, Antje Fischer, Josef Bergstein and Lothar Willmitzer for their scientific support, advice, suggestions and for all the help.

HERBICIDE TOLERANCE IN CROPS : A COMMERCIAL REALITY

G. FREYSSINET[1] and D. J. COLE[2]
[1]Rhône-Poulenc, 25 Quai Paul Doumer, 92408 Courbevoie Cedex, France
[2]Rhône-Poulenc Agriculture, Fyfield Road, Ongar, Essex, CM5 OHW, U.K.

1. Introduction

Weed infestation of crops causes extensive loss in the production of our food. Estimations range from 10 to 30% of crop yield annually. Despite a multitude of herbicides being available to farmers, there are considerable limitations in their effectiveness. With this in mind, engineering herbicide tolerance in crops has been seen as a worthy agricultural and commercial goal. In this short review we will present the reasons for such a goal and the strategies and technologies for gene transfer to confer herbicide tolerance in crops. We will finish by a market evaluation of what is now and will be available by the next century.

2. Why develop herbicide tolerant crops?

They are at least four reasons to develop such material :

- Agriculture still faces unsolved problems with the control of weeds in crops.

- Herbicides conventionally are of two types, those which are selective in certain crops and those which are 'total' herbicides. Selective herbicides are limited to specific crops and cannot always discriminate between closely related weed and crop plants. 'Total' herbicides offer good spectrum of weed control but they cannot be used in crops. It thus would be useful to be able to use good herbicides across a wider range of crops without risking crop damage.

- Our extensive knowledge of the biochemical mode of action of herbicides and their degradation can underpin strategies for engineering new selectivities.

- For an Agrochemical company this new approach offers the opportunity to expand the market share of a given herbicide, to extend its market or even to develop new herbicides which will have not been developed because of lack of selectivity.

Taking all these reasons into account it appeared obvious that the first application of genetic engineering in plants would target tolerance to herbicides.

A. Altman et al. (eds.), Plant Biotechnology and In Vitro Biology in the 21st Century, 481–485.

3. The strategies used for the introduction of herbicide tolerance

Two main strategies have been used based on what is seen in nature for herbicide tolerance.

- The first strategy consists in either an overexpression of a sensitive target site or the expression of a target which is atypically insensitive. In this second case, the target is isolated either from an organism naturally resistant to the herbicide or by making and selecting mutated tolerant targets in the laboratory.

- The second strategy uses an enzyme which renders the herbicide non-phytotoxic. Such enzymes are present either in plants which are tolerant to the herbicide through a detoxification process or in soil bacteria which are responsible for the degradation of the herbicide in the environment.

4. The technologies utilised for the transfer of the trait into the crops

Three different strategies have been used when we look at crops which have reached the market.

- The first approach was used in the late 1970s by Canadian researchers. They crossed a resistant mustard tolerant to atrazine with either *Brassica napus* or *B. rapa*. After a conventional back-crossing programme they obtained Canola tolerant to atrazine. This Canola was suited to a specific market in a region of Canada with heavy infestation of brassica weeds which contaminates the harvested seeds, increasing the erucic acid content of Canola oil.

- The second approach was first achieved in maize using *in vitro* selection against imidazolinone herbicides. Fertile plants were regenerated and extensive back-crossing was needed to recover elite maize lines with good agronomic performance. Such maize has been on the market since 1992.

- The third approach uses genetic engineering techniques. Genes for tolerance are derived from plants or micro-organisms. They are put under the control of regulatory sequences allowing expression of the gene in tissues normally targeted by the herbicide. In certain cases targeting of the gene product to the chloroplast is needed. The two main approaches of gene introduction, *Agrobacterium* and biolistics, have been utilised. The choice of technique used depends on what is feasible with the plant species into which the gene construct is introduced.

5. Which herbicide tolerances have been created so far?

Table 1 summarises available data. In the case of target modification the target was known before and, in certain cases, mutated targets were already available. For detoxification, generally the type of enzyme needed was also known. For most of the cases the genes for detoxification have been isolated from soil treated with the herbicide, or from effluent from the industrial plant where the herbicide is produced. In the case of bialafos, the gene is derived from the bacteria normally used for the production of the herbicide.

Table 1. Some genes used for the introduction of herbicide tolerance into crops

A. Modified target

Herbicide	Gene product	Gene origin
Asulam	dihydropteroate synthase	bacteria
Atrazine	D1 protein	plant
Dalapon	dehalogenase	bacteria
Diflufenican	phytoene desaturase	bacteria
Glyphosate / Sulfosate	enolpyruvyl-shikimic acid phosphate synthase	bacteria or plant
Imidazolinone / Sulfonylurea	acetolactate synthase	bacteria or plant
Quizalofop / Sethoxydim	acetyl CoA carboxylase	plant

B. Detoxification

Herbicide	Gene product	Gene origin
2,4-D	2,4-D monoxygenase	bacteria
Glyphosate	oxido-reductase	bacteria
Glufosinate / Bialaphos	*N*-acetyl transferase	bacteria
Bromoxynil / Ioxynil	nitrilase	bacteria
Thiazopyr	esterase	animal
Chlorotoluron	cytochrome P450	animal
Alachlor	glutathione transferase	plant

Depending on the herbicide the gene has been transferred to several different plant species. Such systems can be used not only to produce crops tolerant to the herbicide, they are also useful systems for the laboratory selection of transformed plant cells. For instance, Rhône-Poulenc's nitrilase / Oxynil system (see below) can be used for selection not only *in vitro* to select the transformed tissue but also *in vivo*. This allows rapid selection of plants having received the gene during the back-crossing process.

6. Crops tolerant to Oxynils

In 1984 RP started a programme, in collaboration with Calgene (California), to isolate a gene conferring tolerance to the Oxynils. Oxynils are a family of contact herbicides acting on photosystem II. They are quickly degraded in the soil and in many monocotyledons. A gene coding for a specific nitrilase has been isolated from a bacterium, *Klebsiella ozaenae,* isolated from soil treated with the herbicide. This nitrilase converts the Oxynils into an acidic residue which is non-phytotoxic. This gene has been put under the control of constitutive or light-inducible promoters and introduced into several dicot plant species such as tobacco, tomato, cotton, canola, carrot, clover and egg-plant. In all these plant species it confers tolerance to more than ten times the agronomic application rate. Cotton tolerant to Oxynils has been produced by Calgene and was among the first transgenic crops introduced into the market in the U.S., in 1995. Canola with the same trait should reach the Canadian market before the end of the century, and maybe the European market if the appropriate authorisations are obtained.

7. The market

The first crops tolerant to herbicides reached the market in 1995. Since that time they have expanded considerably. Table 2 gives an estimate of the usage in 1997 and the estimation for 1998.

Table 2 : Market for herbicide tolerant crops

Country	Crop	Herbicide	'97(Mha)	'98 (Mha)
Argentina	Soybean	Glyphosate	1.400	3.000
Canada	Canola	Imi / Sul	0.825	1.135
		Glufosinate	0.900	0.825
		Glyphosate	0.500	1.240
Mexico	Cotton	Glyphosate	0.004	0.050
U.S.A.	Cotton	Bromoxynil	0.100	0.600
		Glyphosate	0.330	1.800
	Maize	Glyphosate	0	0.310
		Imi	1.500	1.600
		Sethoxydim	0.200	0.250
	Soybean	Glyphosate	3.600	10.300
		Sul	0.800	0.900

Imi = Imidazolinone ; Sul = Sulfonylurea

The market for such crops looks set in most developed countries; Europe lags behind for regulatory and public perception reasons. Unless this situation changes, the European farmers will be penalised. The first maize tolerant to glufosinate and resistant to insects has been planted on about 2, 000 hectares in 1998. The increase in area from 1996 to 1997 was more than 300 %. These increases will continue and in Canada, Canola tolerant to herbicides will account for more than 50% of the planted area and a similar situation will exist for soybean in the US. In certain cases 100% of the area will be planted with herbicide tolerant cultivars. In the future more transgenic crop species will be available and other herbicide tolerances will be marketed.

8. Conclusions

This short review demonstrates that herbicide tolerant crops are a market reality. The first market introduction of a genetically modified crop was cotton tolerant to bromoxynil in 1995 in the USA. Since then several other systems have appeared and now there is a market presence in Canada, USA, Mexico and Argentina. In Europe the first maize tolerant to glufosinate and resistant to insects has been introduced. The area is low but should expand next year.

These results demonstrate that the technology is feasible and that the farmers are willing to accept it if it brings a benefit for them. This benefit can be a mixture of : low cost of treatment, better weed control, easy management of the crop, flexibility of use, new tools to manage herbicide-resistant weeds. In most countries the public acceptance is good.

The impact on the market is a reduction of herbicide used (low application rates with more environmentally friendly molecules). It is also a reduction in the cost of the herbicide chemicals. The Agrochemical and seed companies share the benefits with the farmers. In certain cases, mainly for glyphosate tolerance (Round-Up Ready) there is a technology fee on the seeds, but, in this case, the herbicide cost is low.

There are still a few questions pending such as : What are the risks of gene transfer to wild relatives? How to manage rotations ? This will become more critical as several crops will carry the same tolerance. Due to pollen transfer, multiple herbicide tolerance may develop, what will be the impact ?

Herbicide tolerance will not solve all the problems of weed control. For the time being it seems to be a beneficial technology for better crop management. A careful monitoring of the market and weed control is needed to ensure that this technology can play a useful role for many years in the future.

Identification of regulatory and metabolic genes associated with leaf senescence

T. Hajouj and S. Gepstein
Department of Biology, Technion-Israel Institute of Technology, Haifa 32000, Israel

Introduction

Senescence is characterized by a controlled series of biochemical and physiological events comprising the final stage of development. There is a dramatic shift in metabolic pathways and reactions during the onset of senescence (Nooden, 1988a). The metabolic transition is controlled by a genetically programmed sequence requiring new and specific gene expression (Smart, 1994). The progression of a leaf through the senescence program is visibly marked by the loss of chlorophyll and consequent yellowing as a result of the disassembly of the chloroplast (Woolhouse, 1984). Leaf senescence involves degradation of proteins, nucleic acids and membranes, and the subsequent transport of the nutrients resulting from this degradation to other organs, such as developing seeds, leaves or storage organs (Smart, 1994; Thomas, Stoddart, 1980). A senescence "genetic switch" is triggered and modifies gene expression by causing switching off nuclear and plastid genes encoding for photosynthetic and other functional proteins (Senescence down regulated genes) and switching on genes that are associated with the catabolic biochemical processes (SAG- Senescence associated genes).

Major plant hormones have been implicated as being involved in the senescence process, but only cytokinins and ethylene have been shown definitively to have a role in the regulation of senescence (Smart, 1994). The phytohormone cytokinin is generally regarded as an inhibitor of senescence (Gan, Amasino, 1997; Gan, Amasino, 1995; Van Staden et al., 1988), while ethylene is a potent promoter of leaf senescence which can accelerate many of physiological changes normally associated with leaf senescence (Mattoo, Aharoni, 1988). Expression of several genes associated with ethylene biosynthesis was found to increase during fruit ripening and leaf senescence (Davies, Grierson, 1989; John et al., 1995).

Among these genes, ACC synthase and ACC oxidase were found to increase their expression during leaf senescence. Direct evidence for the role of ethylene in leaf senescence was obtained from experiments with transgenic tomato plants which were ethylene-deficient due to antisense inhibition of the ACC oxidase

A. Altman et al. (eds.), Plant Biotechnology and In Vitro Biology in the 21st Century, 487–490.

gene (John et al, 1995). The ethylene deficiency resulted in a delay of leaf senescence by about 10-14 days (Hamilton et al., 1990; John et al., 1995; Picton et al., 1993).

The regulatory mechanisms underlying leaf senescence are largely unknown due to lack of information regarding the identity of the senescence- associated regulatory genes (SARG). The present report summarizes our attempt in identifying senescence associated genes (SAG) as well as the regulatory ones (SARG), that are exclusively expressed during leaf senescence.

Materials and methods

Plant material. Seeds of bean (*Phaseolus vulgaris* L. cv. Kinghorn) were grown in the greenhouse under optimal growth conditions of natural light at 24 °C. Primary leaves were excised in different ages and were used for RNA isolation as described by Puissant (1990).

Differential display. Differential display of mRNA was performed as described by Liang and Pardee (1992). 100 ng of poly $(A)^+$ mRNA from young (15 days after planting) and senescent (45 days after planting) were reverse-transcribed using various anchor primers.

Sequence analysis. DNA sequence was determined and analysis was carried out using the BLAST network services.

Results and discussion

For the differential display assays, mRNA of mature (fully expanded) and senescing leaves were isolated, reverse transcribed, PCR amplified and separated on polyacrylamide sequencing gel. Two cDNA bands that exclusively appeared in senescing leaf extracts were excised, labeled and used as probes for Northern blot analysis. The probes hybridized to transcripts of about 3.2 and 1.6 kb and both showed characteristic but different pattern of temporal expression. The 1.6 kb transcripts accumulated at low level in mature leaves and increased with the progression of leaf senescence (Fig. 1).

Sequence analysis and database search of the cDNA fragment showed similarity (90% identity) to S-adenosylmethionine synthase (SAM synthase).

SAM synthase catalyzes the reaction involved in the synthesis of S-adenosylmethionine, a precursor of the phytohormone ethylene. It has been proven that ethylene plays a key role in the control of leaf senescence and fruit ripening (Gepstein, Thimann, 1980). Indeed, the synthesis of the immediate precursor of the ethylene, ACC increases significantly during the onset of leaf senescence. The two enzymes participated in the ethylene biosynthetic pathway, the ACC synthase and ACC oxidase show the characteristic temporal pattern of senescence associated genes (SAGs), and increase during the onset of senescence (Buchanan-Wollaston 1997).

Since SAM is the substrate for ACC and ethylene synthesis, a continuos supply of SAM and high activity of SAM synthase during leaf senescence is required. These results, indicate that in addition to ACC synthase and ACC oxidase, the temporal expression of the SAM synthase gene resemble the characteristic behaviour of senescence associated gene. However, gene expression of SAM synthase is not restricted to the late stages of senescence, and also occurs in early stages of leaf development. SAM is also the precursor for polyamines, which have been shown to act as anti-senescence agents in early stages of development (Adams, Yang, 1977).

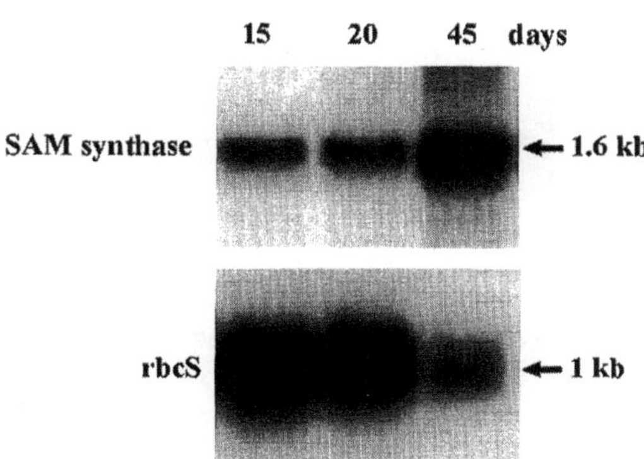

Figure 1. Expression of the SAM synthase mRNA during leaf development. Northern blot carrying RNA isolated from leaves at 3 different developmental stages was hybridized with ^{32}P-labeled fragment obtained by differential display. Rubisco small subunit (rbcS) was used as a probe representing photosynthetic genes. 10 μg of total RNA was loaded in each lane.

Moreover, SAM is known as the donor for methyl groups required for DNA methylation, a process involved in organ differentiation. Since SAM participates in these two major processes during leaf development, it is not surprising that gene expression of SAM synthase takes place during early stages of leaf development as well as during leaf senescence.

The cDNA with an apparent size of 3.2 kb, represent a gene product that is exclusively exist during the onset of leaf senescence. In contrast to the SAM synthase gene, mRNA of this gene could not be detected before the onset of senescence and its initial rise appeared before any morphological or biochemical senescence symptoms could be detected (Fig. 2).

Full-length cDNA was isolated following screening of cDNA library of senescing leaf and its nucleotide sequence shows homologies to characteristic domains of genes encode for protein kinase receptor. Since kinases known as components of regulatory mechanisms in other developmental of processes, it is most likely that this novel gene is a senescence associated regulatory gene (SARG).

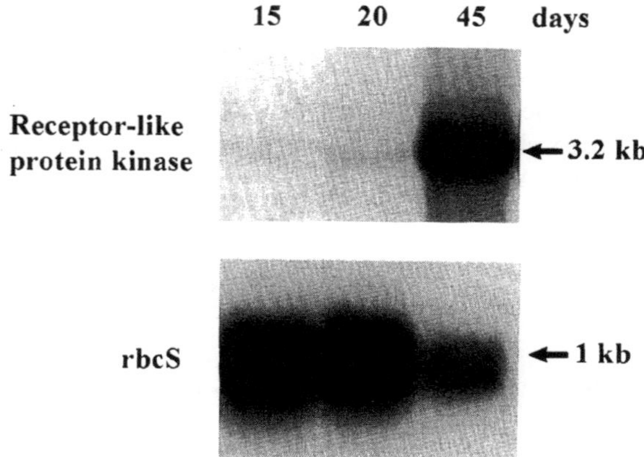

Figure 2. Expression of the Receptor-like protein kinase mRNA during leaf development. Northern blot carrying RNA isolated from leaves at 3 different stages of development was hybridized with ^{32}P-labeled fragment obtained by differential display. Rubisco small subunit (rbcS) was used as a probe representing photosynthetic genes. 10 μg of total RNA was loaded in each lane.

References

Adams DO, Yang SF (1977) Plant Physiology 60, 892-896.

Buchanan-Wollaston V (1997) Journal of Experimental Botany 48,181-199.

Davies KM, Grierson D (1989) Planta 179, 73-80.

Gan S, Amasino RM (1997) Plant Physiol. 113, 313-319.

Gan S, Amasino RM (1995) Science 270, 1966-1967.

Gepstein S, Thimann KV (1981) Plant Physiology 68, 349-354.

Hamilton et al. (1990) Nature 346, 284-287.

John et al. (1995) The Plant Journal 7(3), 483-490.

Liang P, Pardee AB (1992) Science 257, 967-971.

Mattoo KM, Aharoni N (1988) Senescence and Aging in Plants, PP 41-279, Academic Press, San Diego.

Nooden LD (1988a) Senescence and Aging in Plants, PP 391-439, Academic Press, San Diego, CA.

Picton et al. (1993) Plant Journal 3, 469-481.

Puissant C, Houdebine LM (1990) Biotechniques 8, 148-149.

Smart CM (1994) New Phytology 126, 419-448.

Thomas H, Stoddart JL (1980) Annual Review of Plant Physiology 31, 83-111.

Van Staden J et al. (1988) Senescence and Aging in Plants, pp 281- 328, Academic Press, San Diego.

Woolhouse HW (1984) Cell Aging and Cell Death, pp 123-153, Cambridge: Cambridge University Press.

RESISTANCE TO GLYPHOSATE IN A POPULATION OF *LOLIUM RIGIDUM*

T.R. Hawkes[1], D.F. Lorraine-Collwill[2], P.H. Williams[1], S.A.J. Warner[1], P.B. Sutton[1], S.B. Powles[3,4] and C. Preston[2,3]

[1]Dept. of Invention, Zeneca Agrochemicals, Jealott's Hill, Bracknell, Berks. RG42 6EY, United Kingdom

[2]Dept. of Crop Protection, Waite Campus, University of Adelaide, PMB1 Glen Osmond SA 5064, Australia

[3]Cooperative Research Centre for Weed Mangagement Systems, Waite Campus, University of Adelaide, PMB1 Glen Osmond SA 5064, Australia

[4]Western Weeds Initiative, Faculty of Agriculture, University of Western Australia, Nedlands WA 6907, Australia

1. Introduction

Lolium rigidum is a grass weed which is widely established throughout the cropping regions of Southern Australia. It has displayed a marked propensity to evolve resistance to herbicides and, indeed, there now exist some populations which have become resistant to most of the major herbicide chemistries used in Australia (Preston et al 1996; Powles et al 1997). Until recently glyphosate has seemed relatively immune to this problem. However, it now seems that at least two populations of *lolium rigidum* have evolved resistance in response to the repeated use of glyphosate over many years (Pratley et al 1996; Powles et al 1998). In the case of the population described by Powles et al, glyphosate failed to control a population of *lolium rigidum* in an orchard in Orange, NSW, Australia following 15 years of previously successful use. Resistance was confirmed in pot dose experiments and survivors of a treatment of 450 g ai ha^{-1} were further intercrossed to yield a population which consistently exhibited 7-11 fold resistance when compared to the survival rates of a standard susceptible population. Here we describe the results of a study aimed to elucidate the mechanism of this recently evolved resistance.

2. Materials and Methods

Pot dose experiments were as described by Powles et al (1998). For enzyme assays, stem and leaf material from seedling plants (up to 5 leaf) was frozen in liquid nitrogen, ground to a powder with polyvinylpolypyrrolidone and resuspended in chilled 90 mM Hepes-KOH buffer (pH 7.50) containing 0.2 mM ammonium molybdate, 9 mM potassium fluoride, 5 mM dithiothreitol (DTT), 10 mM ascorbate, 0.7 mM 4-(2-aminoethyl)-benzenesulfonyl fluoride (ABSF) and 10% v/v glycerol. Following centrifugation to remove debris, the supernatant was exchanged down Sephadex G25 into the same buffer but lacking ABSF. Assays for 5-enolpyruvylshikimate-3-phosphate (EPSP) synthase were as described by Boerboom et al (1990) except that shikimate-3-Pi was at 0.75 mM, phosphoenolpyruvate

A. Altman et al. (eds.), Plant Biotechnology and In Vitro Biology in the 21st Century, 491–494.

was at 0.5 mM and 20 μl of the above extract was included in a final assay volume of 35 μl. For assays of 3-deoxy-D-arabinoheptulosonic acid (DAHP) synthase extracts (above) were first exchanged into 10 mM Hepes-KOH buffer (pH 7.3) containing 1 mM DTT, loaded onto a 1 ml MonoQ anion exchange column equilibrated with the same buffer, washed with 0.1 M NaCl and the active fraction eluted in 0.2M NaCl. Assays of the plastidic form of the enzyme were then carried out as described by Siehl (1997). Northern blotting of the polyA mRNA fraction from resistant and susceptible plants using a partial cDNA of the lolium EPSP synthase gene as probe was carried out as described by Sambrook et al (1989). Refluxed methanol extracts of ground up plant material were cleared by centrifugation and analysed for shikimate by HPLC using a Lichrosorb-NH_2 column fitted with a UV detector and eluted isocratically with acetonitrile; water; phosphoric acid (95:4:1).

3. Results and Discussion

Glyphosate acts by inhibiting 5-enolpyruvylshikimate-3 phospahte (EPSP) synthase a plastidic enzyme involved in the biosysnthesis of chorismate, a common intermediate required for the biosynthesis of tyrosine, tryptophan and phenylalanine as well as other essential components of the plant (Steinrucken, Amrhein 1984). An early effect of treatment with glyphosate is the rapid accumulation of shikimate (derived from shikimate-3 phosphate) in affected tissues and it is thought that this unregulated diversion of carbon and the subsequent swelling and bursting of plastids is the major factor responsible for the herbicidal effects of glyphosate (Mollenhauer et al 1987; Siehl 1997). We therefore carried out an experiment to examine whether shikimate accumulates to any lesser extent in treated resistant plants. Seedlings of resistant and susceptible plants were treated with 112.5 g ha^{-1} of glyphosate and then, at intervals up to 7 days, sets of 12 whole plants were pooled and analysed for shikimate. Two days after treament with glyphosate the shikimate content of susceptible plants had increased by 3.7 fold as compared to only a 1.7 fold increase in resistant plants. Seven days after treament, the shikimate content of resisistant plants had declined to near its basal level whereas the shikimate content of susceptible plants remained elevated (by ~ 2X). The decreased accumulation of shikimate in resistant plants could, in principle, result from 1) a reduction in the sensitivity of EPSP synthase to inhibition by glyphosate 2) an increased level of EPSP synthase 3) a decrease in the flux through the steps of the pathway which precede EPSP synthase 4) reduced uptake and/or transport of glyphosate to the target site or 5) an increase in the rate of metabolic inactivation of glyphosate.

The EPSP synthase extracted from resistant plants appeared no less sensitive to inhibition by glyphosate than did that from susceptible plants (IC_{50} values of 1.4 +/- 0.25 and 1.2 +/- 0.25 μM, respectively). Resistant plants did not contain measurably more EPSP synthase activity than did susceptible plants (4.2 +/- 0.9 as compared to 3.7 +/- 0.9 nmol EPSP formed min^{-1} mg^{-1} protein, averaged from 6 extracts). Neither, on the basis of Northern blotting, did resistant plants contain more mRNA for EPSP synthase. An apparent ~ 20%

increase in extractable activity of EPSP synthase was observed in both resistant and susceptible biotypes 17 h after application of 125 g ha^{-1} ofglyphosate.

DAHP synthase which is located in the plastid catalyses the first step in the pathway from erythrose-4-phosphate to chorismate. Feedback regulatation of the pathway is thought to be exerted *via* inhibition of DAHP synthase by arogenate (Siehl, 1997). Based on 3 extracts the DAHP synthase activity in resistant plants (7.5 +/- 2 nmol min^{-1} mg^{-1}) was certainly no lower and possibly somewhat higher than that (5.6 +/- 2.2 nmol min^{-1} mg^{-1}) in susceptible plants. Similar to EPSP synthase, pretreatment of plants with glyphosate appeared to cause a slight increase in the extractable activity of DAHP synthase.

A possible basis for resistance would be reduced uptake or translocation of glyphosate. [^{14}C] glyphosate was spotted onto a point just below the midpoint of the larger leaf of 2-leaf stage plants. 24 h after treatment the leaf was washed to remove herbicide that had not been absorbed. Leaf and stem tissue *at* the site of application, *above* the site of application (the top of the leaf) and *below* the site of application (the rest of the plant) was then combusted and the ^{14}CO2 released counted. Both resistant and suceptible plants absorbed more than 75% of the applied herbicide. In both cases, this was almost fully translocated with < 4% remaining at the site of application, ~35% moving below and ~ 25% above. Not all of the applied radioactivity was recovered and it is presumed that the missing ~ 30% was in the roots or lost *via* the roots to soil.

Preliminary investigations of glyphosate metabolism indicated that glyphosate is not rapidly metabolised in either the resistant or susceptible biotype.

The likely basis for the observed resistance could not easily be construed on the basis of the data above. Additional clues were sought through a further study of cross resistance. Powles et al (1998) showed that the glyphosate-resistant biotype displayed no detectable cross-resistance to a wide range of other herbicide chemistries. There was, however, some cross-tolerance to diclofop methyl. We have since tested further and have also observed cross-resistance, to a structural analogue of glyphosate, 2-hydroxy-3-(1,2,4-triazol-1-yl) propyl phosphonate (TP). There was no detectable cross resistance to more distant structural analogues of glyphosate such as phosphinothricin. TP is an experimental herbicide with a mode of action (inhibition of hisitidine biosynthesis) which is quite unrelated to that of glyphosate but which, also acts at a target site (imidazole glycerol phosphate dehydratase) located within the plastid (Hawkes et al 1993). Given this commonality we speculate that, at least in part, resistance has arisen from a mutation leading to reduced import of glyphosate to its site of action within the plastid. Glyphosate and TP bear a structural resemblance to phosphate esters shuttled between the stroma and the cytoplasm and it is possible that import of phosphonate herbicides is dependent upon the same carriers as involved in the shuttles.

4. Conclusions

No measurable differences were observed between resistant and susceptible biotypes in respect of i) the target enzyme (EPSP synthase), ii) DAHP synthase, the first enzyme of the target (shikimate) pathway, iii) absorption of glyphosate or iv) translocation. The resistant plants exhibited a degree of cross resistance to TP, a herbicide which although structurally similar to glyphosate has an unrelated mode of action. On the basis of these observations we speculate that some change in the import of glyphosate to its site of action within the plastid is involved in the mechanism of resistance. However, it may be misleading to try and attribute the relatively moderate level of resistance observed to any single cause. As a widely distributed outcrossing weed, rigid rye grass is undoubtedly genetically diverse and at least a proportion of the resistance may arise from the sum effect of many small differences relative to suceptible biotypes. The question of whether or not multiple mechanisms are involved should become clearer once the inheritance of resistance is examined.

5. References

Boerboom, CM et al (1990) Weed Science, 38, 463-467.
Hawkes, TR et al (1993) in Proc. Brighton Crop. Conf., 2 , 739-744
Mollenhauer, C et al (1987) Pest. Biochem. Physiol., 29, 55-65
Pratley J et al (1996) in 'Proceedings of the 11'th Annual Conference of the Grasslands Society of NSW' (J. Virgona and D. Michalk, eds) (The Grasslands Society of NSW, Inc., Wagga Wagga, Australia) p 122
Powles, SB et al (1998) Weed Sci. (in press)
Preston C et al (1996) in 'Herbicide Resistance: Biology and Biochemistry' (T.M. Brown, ed) American Chemical Society, Washington, DC, pp 117-129
Sambrook, J et al (1989) Molecular cloning - a laboratory manual. Cold Spring Harbour press, 2nd edition.
Siehl, DL (1997) in 'Herbicide Activity: Toxicology, Biochemistry and Molecular Biology' (R.M. Roe et al ed.) IOS press, pp 37-66.
Steinrucken, HC, Amrhein, N (1984) EMBO J., 143, 351-357

Molecular identification of herbicide targets applying transgenic approaches in tissue culture

R. Hoefgen
Max-Planck-Institute of Molecular Plant Physiology, D-14476 Golm / Potsdam, Germany

1. Introduction

Molecular approaches and techniques are an integral part of plant molecular biology, plant physiology and plant breeding. In agriculture the application of transgenic plants providing new or improved traits will become an increasingly important part of breeding programmes and will support strategies of sustainable agriculture and integrated pest management. However, in agrochemical research the increasing knowledge of molecular biology is not used to such an extent to substantially support lead finding and herbicide development. On the other hand the still dominating approaches based on chemical synthesis programmes produce steadily increasing numbers of compounds but, unfortunately, not a likewise increase of hits leading to pesticides, ultimately. Thus, to our understanding molecular and biochemical methods and approaches have to be developed or implemented, not to replace but to support lead finding strategies in agrochemical research programmes and marketing strategies of pesticidal compounds, especially herbicides (Foster, Hoefgen, 1993, Abell, 1996).

Two strategic approaches will be presented as examples for experimental designs based on molecular techniques which might support agrochemical research. The first example will be target validation using antisense inhibition of endogenous genes which should be a prerequisite of biochemical approaches for herbicide discovery such as high throughput screening (HTS) or rational design approaches based on either enzyme activity studies or protein crystallisation (Hoefgen et al, 1993,1995). Antisense validation can provide a valuable hint whether a certain enzyme might be suited as a potential target or should better be neglected in further synthesis programmes or screening procedures (Siehl, 1992). The second example will describe the target identification based on the application of the T-DNA tagging approach (Hayashi et al, 1992) to determine and clone the unknown molecular target of a given phytotoxic compound or herbicide. This is of special interest in cases where conventional biochemical methods do not readily lead to the determination of a target enzyme. Moreover, this knowledge and the obtention of the respective target gene and it's molecular and biochemical analysis will accelerate strategies for synthesising improved or optimised compounds starting from known

A. Altman et al. (eds.), Plant Biotechnology and In Vitro Biology in the 21st Century, 495–498.

substances as lead structures. Furthermore, the detailed knowledge of a target enzyme will support the registration of compounds and sustain marketing strategies.

2. Results and Discussion

Target validation

Antisense inhibition of previously cloned homologous genes is well established to investigate the role of enzymes encoded by these genes in a biochemical pathway and to analyse the physiological effects of their inhibition (Hoefgen, Willmitzer, 1992). The desired genes can be obtained by conventional methods of molecular biology or by taking advantage of the ever increasing data bases generated through the diverse genome projects. The objectives of these approaches were first, to show lethality or severe growth retardation upon inhibition of the putative target, second to mimic *in vivo* the effect of a herbicide inhibiting the corresponding enzyme (phenocopies) and third, to determine effective threshold levels of remaining enzyme activity. To determine the herbicidal potential of a presumed target we followed to lines of experimental set-ups: constitutive and inducible control of the antisense gene. In order to obtain also transgenic plant lines with a lethal phenotype under constitutive expression of the antisense gene supplementation of the tissue culture media during transformation, regeneration and propagation is required if ever possible. Antisense inhibition, unlike other systems such as mutational approaches or insertion mutagenesis regularly yields a gradient of individually inhibited plants (transgenic lines) which can be exploited to determine threshold levels.
Antisense inhibition of the chlorophyll biosynthetic gene glutamate-1-semialdehyde aminotransferase (GSA) and of two amino acid biosynthetic genes of the essential amino acids isoleucine (threonine deaminase, TD) and isoleucine, leucine and valine (acetolactate synthase, ALS) were performed using the constitutive 35 S promoter. GSA antisense was done in tobacco and TD and ALS antisense was done in potato. GSA antisense (Hoefgen et al, 1993) resulted in readily recognisable ‚albino' phenotypes, ranging from pale green to almost white in the homozygous state, when offspring was analysed. All plants could be propagated in tissue culture 2MS medium, however plants with less than 25% chlorophyll content when compared to wild type died under greenhouse conditions when subjected to normal or high light illumination. The symptoms observed resembled the effects of bleaching herbicides and photodynamic herbicides. The threshold level determined indicated that a reduction of 75% of the wild type enzyme activity is sufficient to produce a lethal phenotype, thus suggesting GSA to be a valid target for biochemical design of herbicides. Antisense inhibition of TD (Hoefgen et al, 1994), however, yielded slight changes in phenotype such as a temporary growth retardation, a partial loss of apical dominance and changes in tuber and flower morphology. Especially interesting is a complete loss of fertility due to pollen sterility. TD deaminase activity is extremely low in potato leaves and tubers but stress inducible (e.g. by drought, ABA, jasmonic acid, Hildmann et al., 1992). Though antisense TD plants always showed extremely low enzyme activities no exact determination of the remaining enzyme activity could be executed. Due to the strong inducibility of TD we would not recommend it as a target. ALS, a known target of several highly efficient

herbicide classes such as sulfonylureas or imidazolinones, proved also in the antisense system it's validity as a herbicide target. ALS antisense plants with only 50 % reduction in enzyme activity already showed slight phenotypic effects as growth retardation, an inhibition of about 80% resulted in severely stunted, chlorotic and necrotic plants, which hardly survived under greenhouse conditions and no plants with less than 15% remaining activity could be detected, even with amino acid supplementation in tissue culture (Hoefgen et al, 1995). Thus, antisense inhibition could be successfully established as a pre-screening system for evaluating the validity of potential target enzymes. In a second series of experiments we analysed inducible antisense expression systems, which should correspond more closely to the application of a herbicide to a before healthy plant. However, these approaches partially failed in our hands due to the lack of suitable inducible promoters which should ideally be tightly controlled (chemically or environmentally), also during tissue culture procedures, strongly inducible and easily applicable as well under tissue culture as under greenhouse conditions. We tested, among others, a tetracycline inducible system which was strongly inducible but failed in the other two features, a potato proteinase inhibitor II promoter inducible by ABA or jasmonic acid, which was not tightly controlled as these substances are endogenous to the plant and a heat inducible promoter which was tight and inducible but induction led to pleiotropic effects making screening more difficult (Hoefgen, unpublished).

Target identification

Agrobacterium – mediated random integration of T-DNA tagging vectors harbouring a strong enhancer next to the right border of the T-DNA into protoplasts of *Nicotiana tabacum* and their regeneration to microcalli and, eventually, transgenic plants (Hayashi et al, 1992) was adopted to select for those cell lines resistant to otherwise toxic substances, due to the overexpression of the target enzyme. A set of 6 different herbicides was tested of which the results for glyphosate are presented in this publication and in the accompanying paper by Jens Freitag et al. (this issue). After establishing a 'kill curve' determining the selective conditions for glyphosate under tissue culture conditions if applied to protoplasts of tobacco cv SR1, protoplasts were cocultivated with *Agrobacteria* harbouring the vector pPCVICEn4HPT and subjected to selective conditions of 20 mg/l hygromycine to select for T-DNA integration and 10^{-3}M glyphosate to select for independent cell lines resistant to glyphosate of which 14 were initially obtained. Integration of the T-DNA was determined by Southern blot analysis and expression of the hygromycine gene was checked in Northern blot hybridisations as well from callus or tissue culture material as from greenhouse grown plants. All lines proved to be positive in this stage. Genomic DNA neighbouring the integrated T-DNA was cloned by plasmid rescue as the T-DNA contained an *E. coli* origin of replication and an ampicillin resistance gene for selection in *E. coli* and is currently subjected to sequencing. Stem cuttings of original transformants (T_0) or seedlings after first selfing (T_1) were subjected to spraying experiments in the greenhouse applying 900 mgAS/l of glyphosate which has been shown to effectively inhibit growth of untransformed controls. Two transgenic lines showed tolerance, but no complete resistance to otherwise toxic conditions of glyphosate. Northern blots hybridised to an EPSPS cloned by PCR from tobacco revealed no strong overexpression of the EPSPS gene. Thus, at the present state, it is still ambiguous whether we succeeded in tagging the known target, EPSPS, or whether we induced another type of resistance mechanism. The molecular basis of the observed resistance to glyphosate on protoplast and whole plant level will be investigated.

A central feature of both approaches is the cloning of target genes and the analysis of *in vivo* effects of either tissue culture lines or transgenic plants. In the case of target validation cloning of a putative target gene is the prerequisite of this approach which will finally provide information whether a certain target enzyme might be suitable for further research. This information will be gained prior to extensive chemical synthesis and biochemical efforts and should be used to direct these programmes at the very beginning. On the other hand, T-DNA activation tagging suggested for target identification will result in identifying previously unknown targets of phytotoxic compounds and to will allow to develop or identify new resistance mechanisms to known or newly developed herbicides either by overexpressing the target enzyme or by inducing pre-existing resistance mechanisms in the plant. As the responsible genes can be ‚easily‘ cloned, these resistance mechanisms can be transferred to other crop plants as well and can be introduced into the plant material of crop breeding programmes.

3. Literature

Abell LM (1996) Weed Science 44, 734-742

Foster SG, Hoefgen R (1993) Opportunities for molecular biology in crop production, pp. 75-81, British Crop Protection Council Monograph, Cambridge, UK

Hayashietal (1992) Science

Hildmann et al (1992) Plant Cell 4, 1157-1170

Höfgen R and L. Willmitzer (1992) Plant Science 87, 45 - 54

Höfgen R et al (1994) Proc Natl Acad Sci USA 91, 1726-1730

Höfgen R et al (1995) Plant Physiol. 107, 469-477

Höfgen R et al (1995) Pesticide Science 43, 175 – 177

Siehl, DL (1992) Biosynthesis and Molecular Regulation of Amino Acids in Plants, pp 146-162 American Society of Plant Physiologists Rockville USA

4. Acknowledgement

The author is in debt to Dr. Hans-Dieter Pohlenz, Schering Agrochemicals-Berlin, now Schering Pharmaceuticals-Berlin, and Prof. Dr. Lothar Willmitzer, IGF-Berlin, now MPI of Molecular Plant Physiology - Golm for long discussions generating the above mentioned projects and support throughout their performance. Furthermore I would like to thank Drs. Bernd Laber, Wolfgang Streber, Rick Walden and Frank Schmidt as well as Uli Kutschka, Iris Schuettke, Ann-Kristin Klonus and Astrid Blau. Especially grateful I am to Jens Freitag who devoted his PhD work to establishing the ‚target identification system‘.

TRANSFORMATION OF CROPS TO HERBICIDE-RESISTANCE AND THEIR USE AGAINST PARASITIC WEEDS

D. M. Joel[1], D. Aviv[2], T. Surov[2,] V. Portnoy[1], T. Goldman-Guez[1] and J. Gressel[2],
[1] Agricultural Research Organization, Newe-Ya'ar 30095, and [2]The Weizmann Institute of Science, Rehovot 76100, Israel

1. Introduction

Two of the most devastating pests in agricultural fields, mainly in Africa, the middle east and southern Europe, are the broomrapes (*Orobanche* spp.) and the witchweeds (*Striga* spp.). These flowering plants parasitize roots of many crops, exploiting photosynthate and water and significantly reducing and damaging the yields (Joel et al. 1995).

The broomrapes are non-photosynthetic parasites that devastate legumes and cause heavy damage to vegetables, sunflower and fodder crops. The witchweeds affect the grains of sub-Saharan Africa. Similarly, parasitic plants like the dodders (*Cuscuta* spp.) and mistletoes that attack shoots and branches of host plants, cause considerable damage to agriculture and forestry (Parker and Riches, 1993).

While effective control methods are available for most weeds, there are so far only limited ways to control parasitic weeds (Parker and Riches, 1993). The impact of parasitic weeds on the world economy is therefore tremendous.

The development of genetically engineered herbicide-resistant crops recently allowed us to control the parasitic weeds *Orobanche* on these crops with the help of the respective herbicides (Joel et al. 1995). This was based on the hypothesis that transgenic crops with resistance due to a modification of the target site of a herbicide would allow the systemic translocation of the unmetabolized herbicide to the attached parasite (Gressel 1992). Crops with metabolic resistance would be useful only if the herbicide is directly applied to the parasite underground, because the crop would degrades the foliar-applied herbicide before it gets to the parasite through host tissues (Gressel et al., 1994). Indeed we were able to achieve nearly complete control of broomrape with model transgenic crops bearing genes for target-site resistance to three different herbicides (Joel et al 1995). World griculture could gain control over parasitic weeds if effort is put in transfer of the available target-site resistances into the genomes of local varieties of susceptible hosts of broomrapes. This has been a major aim in our project. We also extended our efforts to herbicide treatments of crop seeds before sowing. This may reduce the environmental damage caused by the herbicides, and reduce the amounts of the chemicals needed for parasitic weed control.

A. Altman et al. (eds.), Plant Biotechnology and In Vitro Biology in the 21st Century, 499–502.

2. Material and Methods

2.1. Engineering asulam resistant potato: Plasmid pJIT118 containing the bacterial *sul* gene, which encodes a modified dihydropteroate synthase (DHPS) insensitive to inhibition by asulam was kindly provided by Drs. G. Freyssinet and R. deRose of Rhone-Poulenc. pJIT118 contains a KpnI fragment that is composed of the *sul* gene preceded by a chloroplast transit peptide and a double 35S promoter and followed by CaMV poly A site (Guerineau et al. 1990). The KpnI fragment was transferred into pGA492, and introduced into *Agrobacterium tumefaciens* strain EHA101 by electroporation. Potato tubers were cut into slices and incubated in *A. tumefaciens* EHA101-sul for 30 min. The slices were then transferred for co-cultivation on regeneration plates containing MS medium with zeatin riboside and indole-3-acetyl-L-aspartic acid, and transferred onto selection agar plates containing a sulfonamide (sulfadiazine or asulam). Two to four weeks later, regenerated shoots were transferred for rooting on plates containing sulfonamide.
Tubers of putative transformant lines were germinated in pots. The emerging shoots were cut and transferred to flasks containing an aqueous dilution of asulam. The flasks were kept in the greenhouse for 2-3 weeks for rooting.
Young asulam-resistant potato plants of different clones were transplanted into pails filled with soil that was artificially inoculated with seeds of *Orobanche aegyptiaca* . Different doses of asulam were sprayed on the plant foliage, and potato shoot survival was recorded. As controls we used treated non-transgenic plants as well as untreated transgenic plants, with and without broomrape.

2.2. Seed treatments: Transgenic rapeseed, engineered with the aroA gene that encodes a modified enolphosphate-shikimate phosphate (EPSP) synthase gene (Della-Cioppa et al., 1987) conferring target-site resistance to glyphosate that were used for the seed treatments were kindly provided by Drs. G. Kishore and S. Padgette of Monsanto. Seeds of transgenic rapeseed were imbibed in different solutions of glyphosate, for various durations. The seeds were dried and then sown in broomrape-infested soil. Infection by broomrape on the rape plants was followed, as also the growth of the plants. In a similar manner, seeds of the weeds *Phalaris minor, Avena sterilis, Amaranthus retroflexus, Solanum nigrum* were soaked in†glyphosate under†similar conditions. They were then sown as above.

3. Results and discussion

3.1. In several transformation experiments we selected five independent asulam resistant potato lines. Seeds that were obtained from these lines and germinated on asulam containing medium formed normal roots, compared to seeds of control non-transformed potato that

developed only very short roots, indicating that the trait is dominant and is transmitted through a meiotic cycle. In a similar manner, cut stems of transformed lines formed roots in the presence of 650 μM asulam, while non transgenic plants did not.
Transplants of asulam-resistant transformants were grown in *Orobanche*-inoculated soil in pails. Plants were sprayed with various concentrations of asulam at different times after shoot emergence. *Orobanche* that developed on untreated potato plants caused their rapid death. On the other hand, broomrape was completely suppressed on asulam-treated potato transformant , while preserving normal potato viability. Differences in the survival rates and in broomrape infectivity after asulam treatments were nevertheless found between transgenic potato clones. Some clones were more vigorous than others, and they also differed in their response to broomrape infestation with and without asulam treatments. The early treatment with asulam at 15 days after transplanting (about the time when broomrape begin to attach to potato roots) did not give reasonable control of broomrape. Only later treatments provided protection against the parasite. The rates of asulam used (16 to 32 kg.ha-1) are much higher than those typically used in agriculture (1 to 7.5 kg.ha-1,), indicating the robustness of the transformants to cope with high rates of asulam. We must now ascertain the lowest effective concentrations for broomrape eradication in the field. Asulam is not widely used in agriculture, as only few crops are resistant to it. Thus, the potential increase in its use could add to the diversity of herbicides used, relieving the pressure on heavily used herbicides.

3.2. We hypothesized that loading the herbicide during imbibition is more effective than later, because at this stage herbicides can be carried by water entering the seed, rather than diffuse into fully expanded cells. And indeed, whereas seeds that were treated with the herbicide after imbibition were not "immune" against broomrape, those treated during imbibition showed positive results.
Broomrape developed only on rapeseed plants that emerged from seeds that were treated with the herbicide concentrations lower than 0.7M. Treatments with 0.7M glyphosate completely prevented parasitism by broomrape at crop plant maturity and allowed the plants to reach maximum height as in the non-infected control (Fig. 2). The average height of the crop plants was reduced by herbicide concentrations that were higher than 0.7M. We can therefore see that under the experimental conditions, treating seeds with the herbicide is effective for broomrape control without causing damage to the host plant. Further experiments are needed to develop a reliable method that will exploit these promising preliminary results.

3.3. Having succeeded with seed treatments we also examined an additional hypothesis: we assumed that treating seed stocks of herbicide-resistant crops will allow the control of weed seeds that contaminate the stock. In order to examine this hypothesis we optimized the seed treatments for broomrape control and then tested the influence of treatments with these optimal concentrations on seeds of various troublesome weeds. And indeed, soaking seeds of Phalaris minor, Avena sterilis, Amaranthus retroflexus, and Solanum nigrum in glyphosate under similar conditions resulted in failure of the emerging weed seedlings to establish and survive at 0.7M. The results recently appeared in a patent application. Treatments such as these may therefore selectively control contaminating weed seeds in seed stocks of herbicide-resistant crops.

5. Conclusions

It was possible to achieve nearly complete control of broomrape and to prevent its damage using transgenic herbicide-resistant crops. As presumed, applying the respective herbicides to target-site herbicide-resistant host plants alleviated parasitism. This can be done either by directly treating mature host plants, or by treatment of crop seeds prior to their exposure to the parasite. Another advantage of herbicide resistance conferred to agricultural crops is the ability to control weed seeds that contaminate the seed stocks of these crops. World agriculture could gain control over parasitic and non-parasitic weeds if effort is put in the transfer of the available herbicide resistances into local varieties of the many crops that are susceptible to these pests, and extend the methods here described to the control of similar pests.

6. Acknowledgments

We thank G. Kishore and S. Padgette (Monsanto) for the glyphosate-resistant oilseed rape, and G. Freyssinett and B. Pelissier (Rhone Poulenc) for the plasmid containing the *sul* gene. This work was supported by the United States Agency for International Development, Trilateral Egypt-USA-Israel Program grant HNE-0158-G-00-3060-00.

7. References

Gressel J et al (1994) Biology and Management of Orobanche, pp. 406-418, Royal Tropical Institute, Amsterdam, .

Guerineau F et al (1990) Plant Mol. Biol. 15, 127-136.

Joel DM et al (1995) Nature 374, 220-221.

Parker C and Riches C (1993) Parasitic Weeds of the World: Biology and Control, CAB International, London .

Surov T et al (1998) Theor. Appl. Genetics 96, 132-137.

DIFFERENT RESPONSES BETWEEN ANTHOCYANIN-PRODUCING AND NON-PRODUCING CELL CULTURES OF *GLEHNIA LITTORALIS* TO STRESS

Y. Kitamura, M. Ohata, T. Ikenaga and M. Watanabe
School of Pharmaceutical Sciences, Nagasaki University, Bunkyo-machi 1-14, Nagasaki 852, Japan.

ABSTRACT. Using anthocyanin-producing (Violet) and non-producing (White) cell lines of *G. littoralis,* effects of various stresses on cell growth, anthocyanin content, furanocoumarin induction, and PAL activity were determined. Yeast extract (YE) and free radical generators (H_2O_2, AAPH and X-ray) were used as potent stress inducers. Cell growth and anthocyanin content was not effected with X-ray or YE treatment in both cell lines, whereas not only the decrease of cell growth but also bleaching of anthocyanin in Violet cell cultures was observed with H_2O_2 and AAPH. AAPH did not cause furanocoumarin induction either in Violet or White cell cultures. X-ray irradiation, H_2O_2 and YE treatments induced bergapten formation in White cell cultures, whereas no stress compound was found at all in Violet cell cultures by any treatment. Increase of PAL activity in Violet cell cultures correlated with furanocoumarin production. The reasons why Violet cells did not produce furanocoumarin are discussed.

1. Introduction

An individual plant can produce more than one class of defense compound against herbivores, pathogens and environmental stresses; some are constitutively accumulated in plant cells, but others are induced in response to stimuli. However, the interaction between compounds in such cases has not yet been clarified. In fact, it is uncertain whether a plant cell produce different classes of defense compounds at the same time or not, or whether different classes of defense compounds work cooperatively or independently.

As an approach to these questions, we introduce here as simple model systems two cell lines of *Glehnia littoralis* (Umbelliferae) ; a cell line which accumulates anthocyanins constitutively (Miura et al, 1998) and a second line which does not produce pigment but

A. Altman et al. (eds.), Plant Biotechnology and In Vitro Biology in the 21st Century, 503–506.

is induced to form furanocoumarins by elicitor treatment (Kitamura et al, 1998). Between these cell lines, we determine different responses to various free radical generators in addition to yeast extract. AAPH, hydrogen peroxide and X-ray are selected as typical generators of organic radicals, inorganic radicals and both radicals, respectively.

2. Material and Methods

2.1. Plant material and culture methods

Callus cultures were originally induced from a petiole of a *G. littoralis* seedling on MS basal agar medium supplemented with 2,4-D (1 mg /L), kinetin (0.01 mg /L) and 3 % sucrose. They were subcultured on the same medium at 25℃ in the dark at one-month intervals. After one year of subculture, a reddish-purple spot appeared in the white calli. The pigmented parts were isolated mechanically and subcultured under the same culture conditions. The most intensively coloured parts of the calli were repeatedly selected by visual discrimination check more than 7 times. Finally we established a stable, very dark purple callus line together with non-pigment-producing callus line.

2.2. Stress treatment

Cell suspension cultures established from the two callus lines were subcultured every 21 days; 10-day-old cultures were treated with potent stress inducer and then incubated for an additional day. Cells and medium were separated and analyzed.

2.3. Potent stress inducers

Yeast extract and free radical generators, H_2O_2 and 2,2'-azobis(2-amidinopropane) dihydrochloride (AAPH)(A. B. Ohlsson et al., 1995) were used as chemicals. X-ray was irradiated at 0.44 Gy/min (150 kV, 5 mA) as another stress inducer, using soft X ray-generator (Softex M-150WE).

2.4. Analysis

Furanocoumarin extracted with EtOAc from cultured medium was analyzed by HPLC. A Wakosil-II5C 18RS column was used and eluted with acetonitrile-H_2O (35:65) at a constant flow rate (1.5 ml/min). The eluent was monitored at 254 nm. Anthocyanin in Violet cells was extracted with MeOH containing 0.1 % HCl overnight at 4℃. The absorbance of the supernatant containing anthocyanins was measured at 535 nm. The anthocyanin content was calculated using log ε 4.47 for cyanidin 3-monoglucoside. PAL activity was determined by a radiochemical methods using (U-^{14}C)L-phenylalanine as a substrate. Crude enzyme preparation and enzyme reaction was carried out according to the previous method (Y. Kitamura et al., 1998).

3. Results

Effects of free radical generators (H_2O_2, AAPH and X-ray) as well as yeast extract (YE) on cell growth, anthocyanin content, furanocoumarin induction as stress compound, and PAL activity in anthocyanin-producing and non-producing cell cultures were determined.

X-ray irradiation (up to10 Gy) as well as YE treatment (up to 10 g/L) did not effect on cell growth or anthocyanin content, but the cell growth decreased in a dose dependent manner up to 10 mM with H_2O_2 and AAPH in both cell cultures. Bleaching of anthocyanin and inhibition of PAL activity in Violet cell cultures was also observed after treatments of AAPH and H_2O_2. Violet cell cultures more survived than White cell cultures under AAPH and H_2O_2 treatments.

AAPH did not cause furanocoumarin induction either in Violet or White cell cultures. X-ray irradiation and H_2O_2 and YE treatments induced bergapten formation in White cell cultures, whereas no stress compound was found at all in Violet cell cultures by any treatment. Increase of PAL activity correlated with furanocoumarin production in White cell cultures (Fig. 1).

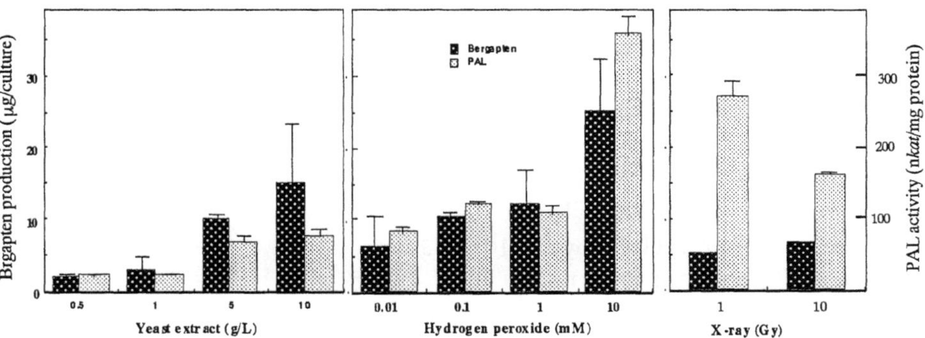

Fig. 1. Induction of bergapten synthesis and PAL activity after various stress treatments in non-anthocyanin-producing cell cultures of *G. littoralis*.

4. Discussion

We successfully induced furanocoumarin biosynthesis by stress treatment in non-anthocyanin-producing cell cultures. Among free radical generators, X-ray and hydrogen peroxide worked as stress compound inducers, but organic radical generator AAPH did not. Free radicals are thought to participate in signaling mechanisms and a transient accumulation of H_2O_2 and O_2^- in plant cells has been detected after stress treatments (M. C. Mehdy, 1994; L. Legendre et al, 1993). Even exogenous H_2O_2 induced transcription of the gene encoding glutathione S-transferase (A. Levin et al, 1994) and activated glyceollin synthesis (N. Degousee et al., 1994) in soybean cells. These results suggest that inorganic radical but not organic radical must participate this reaction.

Two hypotheses are proposed as the reasons why *G. littoralis* suspension cells do not produce furanocoumarin in anthocyanin-producing cells.

(1) Since anthocyanin has quite strong potency as a radical scavenger *in vitro*, it seems possible that anthocyanin captures radicals formed directly and indirectly, and terminates further physical and biological responses, such as the induction of furanocoumarin biosynthesis. Since anthocyanins are accumulated in vacuoles and radicals are generated around the cell surface (M. C. Mehdy, 1994), it is uncertain how a reaction between them might occur *in vivo*, though.

(2) Cells expressing one metabolic pathway may be inhibited in the expression of another, competing or functionally-related, pathway. Both the anthocyanin and furanocoumarin pathways branch off from phenylpropanoid metabolism in *G. littoralis*. Anthocyanin-producing cells of *G. littoralis* suspension cultures may metabolically, though not morphologically, differentiate into specialized cells which can not express another metabolic pathway.

5. References

Degousee N et al (1994) Plant Physiol. 104, 945-952.
Legendre L et al (1993) Plant Physiol. 102, 233-240.
Levin A et al (1994) Cell 79, 583-593.
Kitamura Y et al (1998) Phytochemistry 47 (in press).
Mehdy MC (1994) Plant Physiol. 105, 473-497.
Miura H et al (1998) Phytochemistry 47 (in press).
Ohlsson AB et al (1995) Free Rad. Biol. Med. 19, 319-327.

Galaxy Lines Transgenic for Attacin E AND T4 lysozyme Genes Have Increased Resistance to Fire Blight

K. Ko(1), J.L. Norelli(2), S.K. Brown(1), H.S. Aldwinckle(2), and K. Düring(3)

Dept. of (1)Horticultural Sciences, (2)Plant Pathology, Cornell Univ., Geneva, NY 14456, USA and (3)Federal Center for Breeding Research on Cultivated Plants, D-06484 Quedlinburg, Germany

Abstract

Genes encoding lytic proteins, attacin E (att E) and T4 lysozyme (T4 L), were cloned into 6 different plasmid binary vectors for use in *Agrobacterium*-mediated transformation of apple. All 6 constructs containing att E, T4 L, or both were successfully transferred to Galaxy by *Agrobacterium*-mediated transformation. Apple transgenics have been recovered with all six constructs. NPTII ELISA and transgene PCR have confirmed transgene transformation with *att* E and *T4 L*. Transformation rate varied among different plasmid constructs. The transformation rate of pCa2AMVAtt plasmid vector was lower than with other plasmids. All *att* E transgenic lines expressed att E. However, the protein expression level was different among transgenic lines. pCa2AMVSPAtt transgenic lines had low expression of *att* E compared to other *att* E transgenic lines. In pCa2AMVSPAtt transgenic lines, att E was detected in total apple protein but not in intercellular fluid. Inoculation of transgenic lines indicated that some lines have increased resistance to fire blight. Fusion att E was produced using the pRSET *E. coli* expression vector and purified. The purified fusion att E and T4 L were used successfully to obtain polyclonal antibodies for protein analysis.

1. Introduction

Fire blight is a devastating bacterial disease caused by *Erwinia amylovora* on apple (*Malus* X *domestica* Borkh.). The disease was first reported in North America in 1780, and originated there (Aldwinckle and Beer, 1979). During this century, it has been dispersed from North America to New Zealand, Europe, the Middle East, and other countries. Fire blight is poorly controlled because of sporadic infection and very limited chemical options. Many apple scion varieties, such as Gala, Idared, and Jonathan, and rootstocks, such as M.9 and M.26 are highly susceptible.

Introduction of antibacterial peptides and protein genes into plants could be an effective way of enhancing resistance to phytopathogenic bacteria (Jaynes et al., 1993; Düring et al., 1993; Norelli et al., 1994; Huang et al., 1996; Huang and McBeath, 1997). Att E is an antibacterial protein, synthesized by pupae of the giant silk moth, *H. cecropia*, in response to bacterial infection (Hultmark et al., 1983). *att* E has been transferred to tobacco (*Nicotiana tabacum* var. *Xanthi*) (Destéfano-Beltrán, 1991), M.7 apple rootstock (Norelli et al., 1994), and anthurium (*Anthurium andraeanum*) (Chen and Kuehne, 1996) to increase resistance to bacteria. In transgenic M.7 apple rootstock

A. Altman et al. (eds.), Plant Biotechnology and In Vitro Biology in the 21st Century, 507–511.

lines, the insertion and transcription of the *att* E have been confirmed and fire blight resistance has been enhanced (Norelli et al., 1994).

Another antimicrobial compound is the bacteriophage T4 lysozyme, which is a muramidase that cleaves the glycosidic bond between an N-acetlymuraminc acid (MurNAc) and N-acetylglucosamine (Glc-NAc) of peptidoglycan of gram positive and gram negative bacteria. Transgenic potato plants expressing the *T4 L* gene have been reported to have increased resistance to *Erwinia carotovora atioseptica*, causing potato soft rot (Düring et al., 1993).

The effectiveness of these antimicrobial proteins in transgenic plants could be affected by the expression level and the targeted location (Destéfano-Beltrán, 1991; Düring, 1996). The expression level of transgenes, ß-glucuronidase (*uid*A) (Datla et al., 1993), ß-phaseolin (Oliver et al., 1993), and luciferase (Gallie and Kobayashi, 1994) was reported to be increased because of elevated translation of the transcripts by AMV (untranslated reader sequence of Alfalfa Mosaic Virus RNA4) in transgenic plants. Foreign proteins, by the addition of signal sequences of patatin, glucose oxidase, or the pathogenesis-related protein 1b (SP), have been intercellularly secreted in transgenic plants (Iturriaga et al., 1989; Denecke et al., 1990; Wu et al., 1995). The secretion of antibacterial proteins to the intercellular spaces at the site of invading phytopathogenic bacteria has been believed to be an essential to prevent bacteria colonization in plant tissue (Düring et al., 1993; Düring, 1996). Düring (1996) reported that intercellularly secreted foreign T4 L in transgenic potato plants provided resistance against *E. carotovora* even at a low T4 L expression level.

Combination of these different gene regulatory elements and both antimicrobial proteins in the plant binary vectors could allow us to understand the effect of the transgene expression level and location of the expressed proteins on disease resistance of transgenics.

Therefore, the main goal of this research is to genetically transform Galaxy with genes for lytic proteins known to inhibit bacteria and to determine their effect on fire blight resistance. New plasmid binary vectors have been constructed 1) to investigate the AMV effect on the lytic protein expression, 2) to study the SP effect on lytic protein location , and 3) to allow for combined expression of both *att* E and *T4 L* in transgenic Galaxy plants.

2. Materials and methods

2.1. Plant Binary vectors

Genes encoding lytic proteins, att E and T4 L, were cloned into 6 different plasmid binary vectors for use in *Agrobacterium*-mediated transformation of Galaxy apple (Ko et al., 1997). The plasmids were: pBINCa2Att (enhanced CaMV 35S promoter (Ca2)/*att* E/NOS terminator in pBIN19), pCa2AMVAtt (Ca2/AMV/*att* E/NOS terminator in pBI121), pCa2AMVSPAtt (Ca2/AMV/SP/*att* E/NOS terminator in pBI121), pWIAtt (potato proteinase inhibitor II promoter (WI)/*att* E/WI terminator in pBI121), pWIAttCa2AMVT4 (WI/*att* E/WI terminator/Ca2/AMV/*T4 L* /NOS terminator in pBI121), and pCa2AMVT4 (Ca2/AMV/*T4L*/NOS terminator in pBI121). These plasmid vectors were transferred to *A. tumefaciens* (EHA105).

2.2. Gene transfer, Plant regeneration, and Selection

These 6 different plasmid binary vectors have been transferred to Galaxy apple using *Agrobacterium*-mediated transformation and transformants were regenerated on LSG medium containing 100 mg/L of kanamycin as described in a previous study (Norelli et al., 1996). Transformation of Galaxy apple with binary vectors was confirmed by PCR and NPTII ELISA analysis as previously described (Ko et al., 1998). Western analysis was used to study gene expression in transgenics.

2.3. Rooting culture and Disease evaluation

Transgenic Galaxy lines were obtained, rooted, acclimated to soil, and grown in a growth chamber at 26°C, 80 RH (Norelli et al., 1996; Bolar et al., 1998). When apple plants were vigorously growing, shoot tips were inoculated by injection of 7.5 x 10^7 cfu/ml of *E. amylovora* (Ea273) (Norelli et al., 1994). Length of necrotic lesion was measured at intervals around 4 weeks after inoculation, and expressed as % of total shoot length that blighted.

2.4. Production of att E using expression vector

Att E coding region was inserted using PCR based cloning into the pRSET *E. coli* expression vector (Invitrogen Corp., San Diego, CA) for att E protein purification. Western analysis was used to confirm fusion *att* E expression. Purified fusion att E and T4 L (kindly donated from Dr. B.W. Matthews, Univ. of Oregon) was injected into a rabbit to produce polyclonal antibody.

3. Results and Discussion

All 6 constructs containing *att* E, *T4 L* or both were transferred to Galaxy by *Agrobacterium*-mediated transformation and have been regenerated and propagated. NPTII ELISA and transgene PCR have confirmed transgene transformation. pCa2AMVAtt transformation rate was lower than other plasmid transformation rates. Expression of *att* E in *att* E transgenic lines was confirmed and quantified by western analysis. All *att* E transgenics produced att E protein. However, the expression level differed among the transgenic lines. Generally, pCa2AMVSPAtt transgenic lines have lower expression of *att* E than other *att* E transgenic lines. In pCa2AMVSPAtt transgenic lines, att E was detected in total apple protein while no att E was detected in intercellular fluid. This failure of att E detection in intercellular space could be explained by the low rate of secretion of total expressed protein, decreased efficiency of translation, or protein degradation in intercellular space (Lippincott-Schwartz et al., 1988; Denecke et al., 1990; Mills et al., 1994). Att E protein in hemolymph of *H. cecropia* pupae was partially degraded in nontransgenic apple intercellular fluid. Therefore, this failure of att E detection in intercellular space could be due to degradation of att E protein.

T4 L expression was detected in total protein of a *T4 L* transgenic potato line (Düring et al., 1993) but was virtually undetectable in that of *T4 L* transgenic Galaxy apple lines using western blot. *T4 L* gene sequence has its own signal peptide to secrete T4 L into intercellular space of apple cells. It is considered that the failure to detect the T4 L could be due to T4 L degradation in intercellular space or other post translational events as for att E in this study. Also, translational inefficiency and transcriptional gene silencing of transgene in transgenic apple are plausible explanations for the failure of T4 L detection (Hightower et al., 1994; Ko et al., 1998).

Disease resistance of different construct transgenic lines has been tested. Some transgenic lines had significant disease resistance compared to nontransgenic Galaxy. Resistance level varied among transgenic lines from the same plasmid. This could be due to variable expression level of att E. However, no correlation between *att* E expression level and disease resistance was observed.

Fusion att E was produced using the pRSET *E. coli* expression vector and purified. Collecting time for maximum expression of fusion att E after induction of the pRSET vector was optimized. The purified fusion att E and T4 L were used successfully to obtain polyclonal antibodies for protein analysis. Purification of att E from *E. coli* expressed fusion att E was conducted. However, the rate of recovery of att E from fusion att E was too low for efficient production. Att E purified from hemolymph of *H. cecropia* pupae and T4 L will be used for in vitro assays of antimicrobial activity. Current work is focusing on characterizing the Galaxy transgenics and evaluating the effect of the lytic proteins on disease resistance.

4. Future Work

The synergism of att E and T4 L will be investigated through disease evaluation of transgenics and in vitro test. For this study in transgenics, att E and T4 L transgenic lines are being established and propagated and western and northern analysis to detect T4 L protein and mRNA in transgenics, respectively, are being developed. Also, the effect of AMV and SP on *att* E expression and location and fire blight disease resistance in transgenics will be studied. In addition, the antimicrobial activity of att E and T4 L will be tested in vitro to determine the synergism between these lytic proteins.

5. Acknowledgments

We thank Dr. B.W. Matthews for providing T4 L used to produce T4 L polyclonal antibody in this study. This work was supported by grants from the New York Apple Research and Development Program, and from the Cornell Center for Advanced Technology in Biotechnology which is sponsored by the New York State Science and Technology Foundation and industrial partners.

6. References

Aldwinckle and Beer (1979) Horticultural Reviews 1, 423-474.
Bolar et al. (1998) Hortscience (In press).
Chen and Kuehne (1996) J.Amer.Soc.Hort.Sci. 121, 47-51
Datla et al. (1993) Plant Sci. 94, 139-149.
Denecke et al. (1990) Plant Cell 2, 51-59.
Destéfano-Beltrán (1991) Ph.D thesis, Ohio.
Düring et al. (1993) Plant Journal 3, 587-598.
Düring (1996) Molecular Breeding 2, 297-305.
Gallie and Kobayashi (1994) Gene 142, 159-165.
Hightower et al. (1994) Plant Cell Rep. 13, 295-299.
Huang et al. (1996) Science in China Series C Life Sciences 39, 652-661.
Huang and McBeath (1997) Phytopathology 87, S45
Hultmark et al. (1983) EMBO 2, 571-576.
Iturriaga et al. (1989) Plant Cell 1, 381-390.
Jaynes et al. (1993) Plant Sci. 89, 43-53.
Ko et al. (1997) Phytopathology 87, S53.
Ko et al. (1998) J.Amer.Soc.Hort.Sci. 123, 11-18.
Lippincott-Schwartz et al. (1988) Cell 54, 209-220.
Mills et al. (1994) Plant Sci. 104, 17-22.
Norelli et al.(1994) Euphytica 77, 123-128.
Norelli et al. (1996) Hortscience 31, 1026-1027.
Oliver et al. (1993) J. Experimental Botany 44, 1239-1244.
Wu et al. (1995) Plant Cell 7,1357-1368.

THE HEAT INDUCED INACTIVATION OF THE HERBICIDE RESISTANCE GENE *PAT* ISOLATED FROM *STREPTOMYCES VIRIDOCHROMOGENES* DEPENDS ON ITS DNA SEQUENCE

S. Köhne, K. Neumann, I. Broer; Universität Rostock, Fachbereich Biologie, Zellphysiologie, Doberanerstr. 143, D-18051 Rostock, Germany

Abstract

After 10 days of cultivation at 37°C, the herbicide resistance encoded by the chimaric pat41 gene (coding region from Streptomyces viridochromogenes fused to the 828bp CaMV35S promoter) was strongly reduced in 100% of the 27 independently transformed Nicotiana tabacum SRI lines analyzed. This reversible reduction occurred in sterile and unsterile culture in the first and second generation and even when the overnight temperature was reduced to 24°C. Neither the enzyme activity, the protein nor the pat41 specific RNA could be detected in the heat treated plants, regardless of the number of copies and the hemi -or homozygous state of the transgene. In contrast to this, the expression of the synthetic patS coding region fused to the 534 bp CaMV35S promoter and coding for essentially the same protein, was stable in heat treated plants. The exchange of the GC rich coding region of the pat41 gene by the AT rich synthetic DNA fragment carrying the patS coding region led to the stabilization of the specific RNA steady state level. However, the presence of the transgene-encoded protein at 37°C could only be achieved by the usage of the specific 5' and 3' untranslated regions present in the synthetic patS gene.

Introduction

It has already be demonstrated that changes in environmental conditions are able to influence gene and even transgene activity in plants. One of the first examples for the reduction of transgene activity by environmental factor published occurred during the deliberate release of transgenic *Petunia hybrida* plants carrying the maize *A*1 gene (Linn et al., 1990). Following a period of high light intensity and elevated temperatures up to 36 °C about 60 % of the transgenic plants grown on the field had partially or even completely lost the transgene-encoded color. Nevertheless, heat does not seem to be the main environmental factor triggering the A1 inactivation. Although transgenic plants cultivated in the greenhouse where exposed to temperatures higher than 48 °C, transgene inactivation was detectable for only 5% of the plants (Meyer at al., 1992). A direct correlation between heat-treatment and transgene inactivation could be demonstrated in transgenic *Nicotiana tabacum* plants. After ten days of incubation at 37 °C the plants lost there enzymatic activity encoded by the luciferase gene (*luc*) from *Photinus pyralis*. This reaction was reversible and could be observed in up to 40 % of the independent transgenic lines analyzed. The temperature sensitivity or stability of the *luc* gene expression seemed to depend on the DNA sequences of the integration locus. Analysis of the expression of the *npt*II gene located on the same T-DNA showed that this gene seemed to undergo the same regulatory phenomenon (Broer , 1996; Neumann et al., 1997).

The heat-induced inactivation of transgenes may be of importance, when transgenes are used in agriculture. Already 1992 Walter et al. could show that in transgenic suspension culture of Medicago sativa, the *pat* gene encoded resistance against the commercially available herbicides BASTA™ and Liberty™ both containing the active ingredient Phosphinothricin (L-Pt) was suppressed by a heat treatment. Pt is produced as phosphinotricyl-alanyl-alanin (Ptt) by *Streptomyces hygroscopicus* (Kondo et al., 1973) and

A. Altman et al. (eds.), Plant Biotechnology and In Vitro Biology in the 21st Century, 513–516.

S.viridochromogenes (Bayer et al., 1972). Pt is a structural analogue of glutamic acid, which competitively inhibits bacterial and plant glutamine synthetases (Bayer et al., 1972; Lea et al., 1984). Resistance genes have been isolated from both producer strains (Thompson et al., 1987; Strauch et al., 1988) and transferred to several plant species (De Block et al., 1988). Both genes code for a phosphinothricin-*N*-acetyl-transferase which inactivates Pt. After a heat treatment of 37°C for 10 days, the phosphinothricin-*N*-acetyl-transferase activity decreased in 95% of the cells resulting in Pt-sensitivity. In the control culture only 12% of the cells lost the Pt-resistant phenotype.

Because suspension cultures are artificial systems, the heat-induced loss of *pat* gene expression was analyzed in a plant system (Neumann et al.1997, Köhne et al., 1998). To enable expression in transgenic plants, the coding region of the natural *pat* gene from *Streptomyces viridochromogenes* (Strauch et al., 1988) has been modified by the addition of 5' and 3' DNA sequences and fused to the 823 bp CaMV35S promoter and nopaline synthase termination signal sequences resulting in the *pat*41 gene (Dröge et al., 1992). Plants carrying this *pat*41 gene were named SRI/16.41. At 24°C the *pat*41 gene expression leads to resistance to more then 50 g L-Pt·L^{-1} (Dröge et al., 1992). In the absence of the herbicide these plants showed no visible heat damage or changes of the growth rate, leaf size and morphology when cultivated at 37 °C. Already after the application of an aqueous solution containing 4 g L-Pt·L^{-1}, 100 % of the heat treated plants displayed symptoms of L-Pt-sensitivity like bleaching of leaves. This behavior was independent of the culture medium (sterile / unsterile conditions, hydroponic culture and soil). Furthermore we could show that neither the copy number nor the hemi- or homozygous state of the transgene did influence the reaction to a 37 °C treatment (Köhne et al. 1998).

Materials and Methods

see Köhne et al., 1998

Results

The heat induced inactvation of the pat gene expression was influence by the coding region and the 5' and 3' UTRs.

SRI/16.41 plants were able to acetylate 50% of the ^{14}C labeled L-Pt present in the PAT-assay (Fig. 1, lane 1). After the heat-treatment no PAT-activity could be detected (Fig. 1, lane 2). The same reaction was observed on the protein and RNA level. Small amounts of PAT-protein and *pat*41 RNA were present in plants cultivated at 24 °C (Fig.1 lane 1). After the incubation at 37 °C no *pat* encoded protein or RNA were detectable by western or northern blot analysis (Fig. 1, lane 2).

Using the *pat* DNA sequence (Wohlleben et al., 1988) as a template, a synthetic *pat* gene coding for the same protein, herein named *pat*S, has been constructed. The sequence of the *pat*S was adapted to the plant codon usage (Eckes at al., 1989). As a result, the GC-content of the *pat*S coding region, was reduced to 49 % compared to 69 % of the *pat*41 coding region The synthetic DNA fragment was fused to the 534 bp CaMV promoter and the camv termination signal sequence. Plants carrying this *pat*S gene were named SRI/OCA. This transgenic plants were able to acetylate 100% of the ^{14}C labeled L-Pt present in the PAT-assay (Fig.1 lane 3). This activity was not reduced during a five day heat-treatment (Fig. 1, lane 4). The same phenomenon held true for the *pat*S encoded protein and RNA levels. At 24 °C PAT-protein and *pat*S RNA could be detected in higher amounts in the SRI/OCA plants compared to the SRI/16.41 plants (Fig .1, lane 3). After the 37 °C incubation even a noticeable increase of the amounts of PAT-protein and *pat*S RNA was verifiable (Fig. 1, lane 4).

Through the construction of different T-DNAs we were able to show that neither the T-DNA regions surrounding the transgene nor the orientation of the gene on the T-DNA influenced the transgene expression during a 37 °C incubation (Köhne et al., 1998). The exchange of the *pat*41 3' region did not influence the heat induced instability of transgene expression (Köhne et al., 1998). The influence of the different coding regions on the stability of transgene expression was analyzed by the exchange of the *pat*41coding region with a DNA fragment carrying the *pat*S coding region leading to the *pat*43 gene (Köhne et al., 1998). Plants carrying the gene were named SRI/16.43. At 24 °C enzyme activity and amounts of PAT protein and *pat* RNA in these plants were similar to those observed in the SRI/16.41 lines (Fig. 1, lane 5). Surprisingly, at 37

°C, the SRI/16.43 plants displayed no decrease in the amount of *pat*43 encoded RNA like it was observed in the SRI/16.41 plants. In contrast, even a slightly increased *pat*43 RNA steady state level could be verified. Nevertheless, no PAT protein was detectable (Fig.1, lane 6).

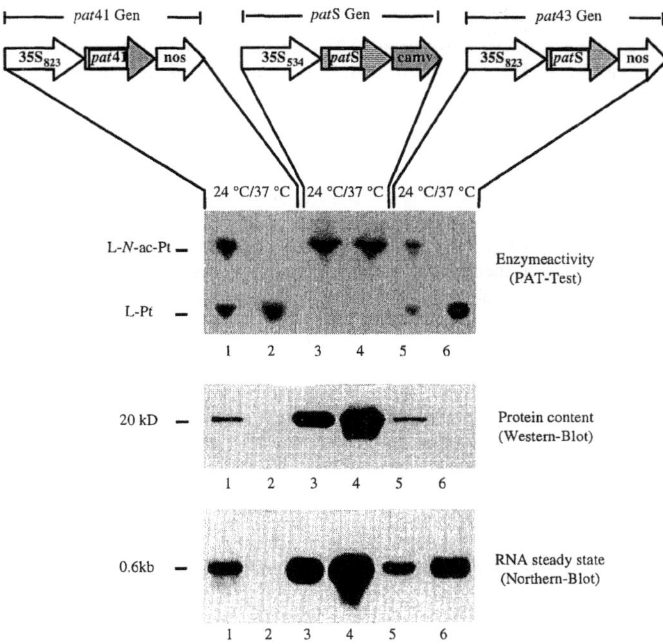

Fig. 1: Molecular analysis of independent tobacco lines transformed with different *pat* genes.
PAT-Test: PAT-activity in different tobacco lines cultivated at 24 °C or 37 °C. L-Pt, L- Phosphinothricin; L-*N*-ac-Pt, L-*N*-acetyl-phosphinothricin; **Western-Blot:** Total protein from plants analyzed in the PAT-test was incubarted with a polyclonal PAT-antibody (Vinnemeier et al., 1995); **Northern-Blot:** Total RNA isolated from the same plants used for the western analysis. A ^{32}P labeled *patS* or *pat*41-fragment were used as hybridization probe; **CaMV$_{823}$**, 823 bp promoter of the *cauliflower mosaic* virus 35S-RNA gene (Baulcombe et al., 1986); ***pat*41**, 5' - 3' modified phosphinothricin-*N*-acetyltransferase gene of *Streptomyces viridochromogenes* (Dröge et al., 1992); **nos**, transcriptional termination signal of the nopaline synthase gene from *Agrobacterium tumefaciens* (Dhaese et al., 1993); **CaMV$_{534}$**, 534 bp promoter of the *cauliflower mosaic* virus 35S-RNA gene lacking 300 bp in the 5' region (Eckes et al., 1989); ***pat*S**, the coding region of the synthetic phosphinothricin-*N*-acetyltransferase gene (Eckes et al., 1989); **camv**, transcriptional termination signal of the *cauliflower mosaic* virus 35S-RNA gene (Eckes et al., 1989).

Discussion

The heat induced inactivation of different transgenes seems to be caused by various mechanisms. In the case of the *luc* gene, the identical DNA sequence can be subject to transgene inactivation during the heat treatment or not, according to Neumann et al. (1997) the decision seems to depend on the insertion locus. In addition, the steady state level of the *luc* gene encoded RNA is even increased in all plants during the heat treatment, while the amount of protein is strongly reduced only in the heat sensitive plants. In contrast to this, the *pat*41 gene expression is heat sensitive in all 27 plants analyzed, no matter where the transgene is located, and the protein and RNA level is reduced. It therefore was obvious to assume, that there are two different levels of transgene inactivation present in the plant. The first level is the RNA steady state level, possibly influenced by the transcription rate or RNA stability in the heat treated plant. This theory was supported by the mode of expression observed for the *pat*43 gene. Since the reaction did not depend on the promoter sequence used, it might be assumed that changes in the RNA stability are responsible for the decrease of *pat*41 RNA. According to our results, these changes might depend on the DNA sequence of the coding region, either on the GC-contend or on specific elements possibly present in the *pat*41 coding region. The second level is the presence of the transgene encoded protein, either influenced by protein stability or translation. Since the protein encoded by the *pat*S and *pat*43 genes are identical, we have to conclude that the instability is due to a specific inhibition of the translation of *pat* RNA, possibly depending on the 5' or 3' untranslated regions fused to the coding region.

The elements described in this paper might be useful for the stabilization of transgene expression during periods of mild heat stress in the field.

References

Baulcombe DC et al (1986) Nature 321, 446 -229
Bayer E et al (1972) Helvetica Chimica Acta 55, 224-239
Broer I (1996) Field Crop Res. 45, 19-25
De Block M et al (1987) EMBO J 6, 2513-2518
Dhaese P et al. (1983) EMBO J. 2, 419-429
Dröge W et al (1992) PLANTA 187, 142-151
Eckes P et al (1989) J. Cell Biochem. supplement 13D, 334
Köhne S (1998) J. Plant Physiol. accepted
Kondo Y et al (1973) Sci. Rep. Meiji Seika 13, 34-41
Lea PJ et al (1984) Phytochem. 23, 1-6
Linn F et al (1990) Mol. Gen. Genet. 222, 329-336
Meyer P et al (1992) Mol.Gen.Genet. 231, 345-352
Neumann K et al (1997) Plant Physiol. 115, 939-947
Strauch E et al (1988) Gene 63, 65-74
Thompson CJ et al (1987) EMBO J. 6, 2519-2523
Vinnemeier J et al (1995) Zeitschrift für Naturforschung 0939-5075/95/1100-0796
Walter C et al (1992) Mol. Gen. Genet. 235, 189-196
Wohlleben W et al (1988) Gene 70, 25-37

PHYSIOLOGICAL MARKERS FOR SALINITY TOLERANCE IN PLANTS

A. Läuchli
University of California, Department of Land, Air and Water Resources, Davis, CA 95616-8627, USA

1. Introduction

Salinity is an environmental constraint that limits plant growth and productivity through several effects, i.e., osmotic effects, ion toxicity, interaction with nutrient ion acquisition, and oxidative stress. In arid and semiarid climate regions, where irrigation agriculture is prominent and salinity is a serious threat to crop production, it is highly desirable to produce by genetic manipulation new varieties of crops with enhanced tolerance to salinity and high yield potential (Epstein et al 1980). Efforts through plant breeding have thus far been relatively disappointing, but the available methodology of producing transgenic lines through molecular approaches (Potrykus, 1991) offers new opportunities for developing salt tolerant varieties of crop species (Bartels and Nelson 1994). Nevertheless, salt tolerance is a multigenic trait (see Cushman et al 1990) and it is still debated, which physiological or biochemical processes should be targeted for identification and transfer of the genes that control the expression of the targeted processes and could therefore serve as markers for specific components of salt tolerance.

2. Osmotically-Induced Proline Accumulation in Roots

Maintenance of root growth in saline soils is critical for plant health (Kafkafi and Bernstein 1996). Saline soils are characterized by low water potentials and high salt concentrations in the soil solution. Low soil water potentials have osmotic effects on plants, causing growth reduction particularly over the short term (Munns et al 1995). Protection against the osmotic effects on roots can be achieved by osmotic adjustment in the growing region through accumulation of compatible osmolytes in the cells (Voetberg and Sharp 1991). Proline accumulation in root tips may contribute to osmotic adjustment and sustained root growth under salinity (Colmer et al 1996, Rodriguez et al 1997). Enhanced proline accumulation in root tips is proposed as a physiological marker for the osmotic component of salinity stress, and specifically for maintenance of osmotic homeostasis in growing roots.

A. Altman et al. (eds.), Plant Biotechnology and In Vitro Biology in the 21st Century, 517–520.

Accumulation of proline in root tips of maize was first recognized by Voetberg and Sharp (1991) to contribute to osmotic adjustment and root growth under drought stress. We obtained similar results in salt-stressed sorghum but only in the presence of elevated Ca supply where root growth was maintained (Colmer et al (1996). Proline concentration under these conditions increased by about 50-fold, and only in the tip 10 mm region but not in the older part of the root. Similar results were more recently obtained by Rodriguez et al (1997) using salt-stressed maize seedlings. It is assumed that proline contributes to osmoregulation in plants (see Delauney and Verma 1993) by accumulating in the cytoplasm as a compatible osmolyte. Thus, it is a component in the osmotic adjustment of growing root cells to salt stress. Proline may also protect proteins and enzymes from the inhibitory effects of high Na concentrations. More research is needed to unravel the role of proline in salt tolerance of higher plants. In this regard an interesting hypothesis was recently advanced by Zhu et al (1997). They showed that the salt-hypersensitive Arabidopsis mutant sos1 accumulates much more proline under salt stress than wild-type Arabidopsis (Liu and Zhu, 1997b). Furthermore, root growth of this mutant was only hypersensitive to osmotic stress caused by mannitol treatment at low stress levels, whereas at high osmotic stress no further reduction in root growth was apparent. The latter effect was possibly related to accumulation of proline (Zhu et al (1997). Thus, proline accumulation may be critical in osmotic adjustment to high stress levels.

3. Maintenance of K/Na-Selectivity in Roots at High Sodium Concentrations

In many crops salinity interferes with the uptake and transport of K, thus disrupting K/Na-selectivity and ion homeostasis (Niu et al 1995, Lauchli 1998). The maintenance of adequate net uptake of K by plants at high Na concentrations is important, since the physiological functions of K in plants cannot be substituted by Na, except for the osmotic role of Na in the vacuoles (Flowers and Lauchli 1983). Sustained K/Na-selectivity is proposed as a physiological marker for the ionic component of salt stress, providing ion homeostasis in growing roots.

That maintenance of K/Na-selectivity is important for root growth under salinity has been demonstrated by several researchers in recent years. In a number of crop species K/Na-selectivity can be maintained by supplemental Ca supply (Lauchli 1990, 1998). The Arabidopsis mutant sos1 (Wu et al 1996) is an outstanding example that salt hypersensitivity is associated with defective high-affinity K uptake, leading to a loss of K/Na-selectivity. Such a loss appears to be caused by several plasma membrane-associated events (Fig. 1). Na displaces Ca from Ca-binding sites at the plasma membrane, allowing an increase in K efflux ((Cramer et al 1985, Yermiyahu et al 1994), possibly through outward rectifying K channels. High Na can also block high-affinity K transporters (Gassman et al 1996), the result of which is decreased K influx and increased Na influx, and it also causes membrane depolarization followed by increased K efflux and hence further reduction of net K uptake (Lauchli and Schubert 1989). High

Na may also increase Na influx by transport through nonselective cation channels. Supplemental Ca partly mitigates these Na/K-interactions (Zhong and Lauchli 1994, Liu and Zhu 1997a), probably through an intracellular Ca sensor (Liu and Zhu, 1998), and protects salt-sensitive plants from the loss of K/Na-selectivity at high salinity.

4. References

Bartels D, Nelson D (1994) Plant, Cell Env. 17, 659-667.
Colmer TD et al (1996) Physiol. Plant. 97, 421-424.
Cramer GR et al (1985) Plant Physiol. 79, 207-211.
Cushman JC et al (1990) Environmental Injury to Plants, pp. 173-203, Academic Press, San Diego CA.
Delauney AJ, Verma DPS (1993) Plant J. 4, 215-223.
Epstein E et al (1980) Science 210, 399-404.
Gassmann W et al (1996) Plant J. 10, 869-882.
Kafkafi U, Bernstein N (1996) Plant Roots, the Hidden Half, 2nd ed, pp. 435-451. M. Dekker, New York.
Lauchli A (1990) Calcium in Plant Growth and Development, pp. 26-35, Amer. Soc. Plant Physiol. Symp. Series Vol. 4, Rockville MD.
Lauchli A (1998) Frontiers in Potassium Nutrition, Amer. Soc. Agron., Madison WI, in press.
Lauchli A, Schubert S (1989) Environmental Stress in Plants, pp. 131-138, NATO ASI Series Vol. G19. Springer, Berlin.
Liu J, Zhu JK (1997a) Proc. Natl. Acad. Sci. USA 94, 14960-14964.
Liu J, Zhu JK (1997b) Plant Physiol. 114, 591-596.
Liu J, Zhu Jk (1998) Science 280, 1943-1945.
Munns R et al (1995) Aust J Plant Physiol. 22, 561-569.
Niu X et al (1995) Plant Physiol. 109, 735-742.
Potrykus I (1991) Annu. Rev. Plant Physiol. 42, 205-225.
Rodriguez HG et al (1997) Plant Physiol. 113, 881-893.
Voetberg GS, Sharp RE (1991) Plant Physiol. 96, 1125-1130.
Wu SJ et al (1996) Plant Cell 8, 617-627.
Yermiyahu U et al (1994) J. Membrane Biol. 138, 55-63.
Zhong H, Lauchli A (1994) Planta 194, 34-41.
Zhu JK et al (1997) Crit. Rev. Plant Sci. 16, 253-277.

Legend

Fig. 1. Plasma membrane of root cells: primary events at high Na leading to loss of K/Na-selectivity

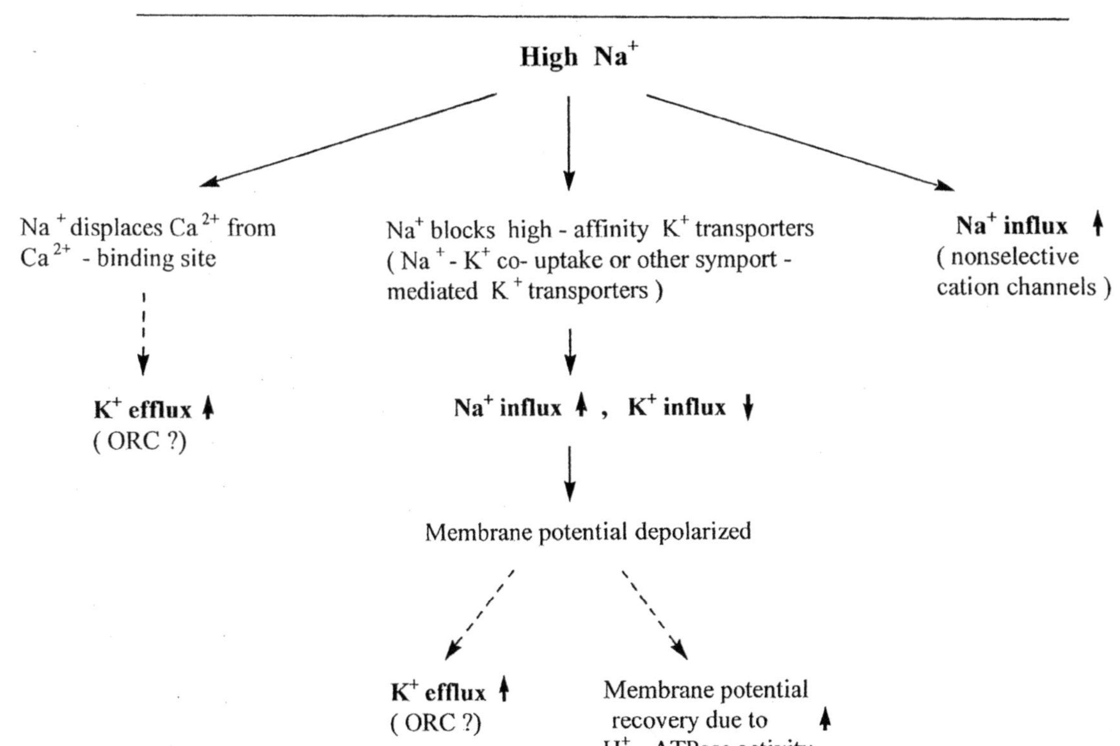
Plasma Membrane of Root Cells : Primary Events at high Na+ leading to Loss of K / Na Selectivity
High Na+
Na+ displaces Ca2+ from Ca2+ - binding site
K+ efflux ↑ (ORC ?)
Na+ blocks high - affinity K+ transporters (Na+ - K+ co- uptake or other symport - mediated K+ transporters)
Na+ influx ↑ , K+ influx ↓
Membrane potential depolarized
K+ efflux ↑ (ORC ?)
Membrane potential recovery due to H+ - ATPase activity ↑
Na+ influx ↑ (nonselective cation channels)
(ORC = outward rectifying K+ channel)

EXPRESSION OF ROTAVIRUS PROTEINS IN TRANSGENIC PLANTS

Tsafrir S. Mor, Liz Richter, and Hugh S. Mason
Boyce Thompson Institute for Plant Research,
Tower Rd. Ithaca, NY 14853, USA

1. Introduction

The concept of transgenic plants as a production and delivery system for subunit vaccines was introuduced in 1992 in a report on the expression of the hepatitis B surface antigen in tobacco(Mason et al. 1992). Subsequently, additional viral and bacterial antigens, including the capsid protein of Norwalk virus and the B subunit of the heat-labile toxin of *Escherichia coli*, were successfully expressed in transgenic plants (Haq et al. 1995; Mason et al. 1996). The immunogenicity of these antigens was verified in animal models and more recently in human clinical trials (Mason et al. 1998; Tacket et al. 1998; Thanavala et al. 1995).

Because most human pathogens initiate infection through mucosal surfaces, the mucosal immune system is in fact the body's first line of defence and also the most effective; its sensitisation at one location very often stimulates responses at distal mucosal surfaces and also ignites systemic immune responses (Mestecky et al. 1997). It is generally recognised that particulate antigens elicit stronger mucosal immune responses than soluble antigens, which may repress the immune response by inducing immuno-tolerance (Garside, Mowat 1997). Particulate antigens may also be more resistant to the proteolytic and acidic environment of the stomach.

Capsid proteins of many types of viruses can assemble into virus like particles (VLPs). These VLPs, devoid of the viral genetic material, often resemble the native virions in their morphology, antigenic properties and stability. VLPs are therefore predicted to make excellent mucosal vaccines which combine safety and effectiveness (Estes et al. 1997). Our group has previously reported the expression of Norwalk virus (responsible for gastro-enteritis outbreaks) and hepatitis B virus VLPs in transgenic plants. These homo-oligomeric VLPs are similar to the respective native viruses in their morphology and in antigenic properties (Mason et al. 1996; Mason et al. 1992; Thanavala et al. 1995). Encouraged by these results, we have now turned to the more complex rotavirus.

Rotavirus is the most important etiologic agent of acute infantile gastro-enteritis, causing serious dehydrating diarrhoea in humans and other mammals. In underdeveloped countries poor sanitation and a want of health-care facilities and personnel combine to aggravate the consequences of rotaviral infection with a death toll of about 900,000 each year (Pérez-Schael et al. 1997). Although rotavirus causes few deaths in industrialised countries, virtually all children under 5 are infected with the virus with more than 1% requiring hospitalisation for severe disease. It is estimated that in the United States alone,

A. Altman et al. (eds.), Plant Biotechnology and In Vitro Biology in the 21st Century, 521–524.

direct medical costs mount to more than 560 million dollars per year (Tucker et al. 1998). A live human-rhesus reassortant vaccine is being reviewed for licensing in the United States (Pérez-Schael et al. 1997). However, at about $60 for the required three doses, the vaccine is prohibitively expensive in developing countries. Production of subunit oral vaccines in transgenic plants in is an attractive cost-effective alternative to the more traditional technology and would allow its global.

2. Materials and Methods

DNA fragments containing the coding sequences of the VP2, VP4, VP6 and VP7 of rotavirus SA11 were amplified from plasmids containing their cDNA sequences (kind gifts of Dr. M. Estes) to introduce an NcoI site and a SacI site at their 5' and 3' termini respectively. These fragments were cloned into plant expression vectors under the control of the cauliflower mosaic virus (CaMV) 35S promoter. The sequence of the VP2 gene was redesigned as described in the next section and the synthetic VP2 gene (sVP2) was constructed by the assembly PCR method (Stemmer et al. 1995). The sequence of the sVP2 gene was verified by automated sequencing and in one of the clones obtained the single inadvertent mutation (R142G) was corrected by site-directed base modification (QuikChange™ Site-Directed Mutagenesis Kit of Stratagene). Co-expression plasmids were created to have expression cassettes of the VP6 and sVP2 genes in tandem. Potato (*Solanum tuberosum* L. cv. FL1607) and tomato (*Lycopersicon esculentum* cv. TA234TM2R) plants were transformed by co-cultivation with *Agrobacterium tumefaciens* LBA4404 (Filatti et al. 1987; Wenzler et al. 1989). Transformants were selected on kanamycin and screened for expression of rotavirus genes. Total RNA was extracted from plant tissues, fractionated by denaturing electrophoresis, transferred to nylon membranes and hybridised as described in (Mason et al. 1992). Crude soluble proteins were prepared by homogenizing leaves in buffer (10 mM Tris pH 7.4, 0.1 mM EDTA) and collecting the supernatant after centrifugation at 3000 g in a minifuge. Rotavirus proteins were assayed by ELISA using rotavirus specific antibodies (generously contributed by Dr. M. Estes (Crawford et al. 1994). Protein content was determined by a modified Bradford assay (Zor, Selinger 1996).

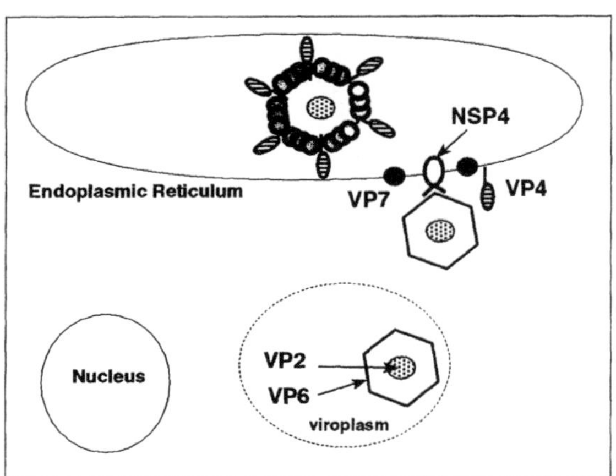

Figure 1. Assembly of rotavirus

3. Results and Discussion

The virion of rotavirus (family *Reoviridae*) is a large (100 nm) icosahedral structure arranged in three concentric layers (Estes 1996). The core consists of the 11 segments of the dsRNA genome, two minor structural proteins involved in viral replication and VP2, which forms the inner shell. The middle protein layer consists of the VP6 protein and the outer layer contains the glycoprotein VP7 and the spike protein VP4. In an elaborate assembly process (Fig. 1) (Estes 1996), VP2 and VP6 form a double layered particle in the cytoplasm; the double layered particle buds into the lumen of the endoplasmic reticulum, transiently acquiring a membrane envelope in the process; then the outer layer, which consists of VP7 and VP4, is added and the particle looses the envelope. The virus toxin, NSP4, is thought to mediate this process possibly by serving as a receptor for VP6 and by releasing calcium ions from the endoplasmic reticulum (Ball et al. 1996).

In order to obtain triple-layered rotavirus VLPs, the four capsid proteins have to be co-expressed, they have to accumulate to high enough levels and they have to assemble. Although co-expression of four different genes in transgenic plants was demonstrated before, the case of rotavirus is more complicated by the need to synthesise the components in two different compartments and then bring them together for the final assembly steps. Furthermore, the viral proteins may prove to have a deleterious impact on cellular physiology. Nonetheless, substantial progress has been made toward the goal of creating a plant-based rotavirus VLP.

We have transformed potato plants with the four individual genes encoding the capsid proteins of rotavirus SA11 under the constitutive CaMV 35S promoter. Of the four, only the VP6 expression cassette was able to drive the accumulation of the VP6 mRNA and its protein product in the transgenic plants to a high level (Richter et al., unpublished). Expression from the other genes was not nearly as good. While many factors can contribute to the low level of expression of the other genes, including instability of individual subunits, a closer examination of the sequences of the genes encoding VP2, VP4 and VP7 revealed many features that could reduce the transcription level, transcript stability or translatability *in planta*. These features include low GC content, unfavourable codon usage, RNA destabilising sequences, cryptic polyadenylation signals, adverse intron recognition sequences, premature termination signals and presence of potential methylation sites.

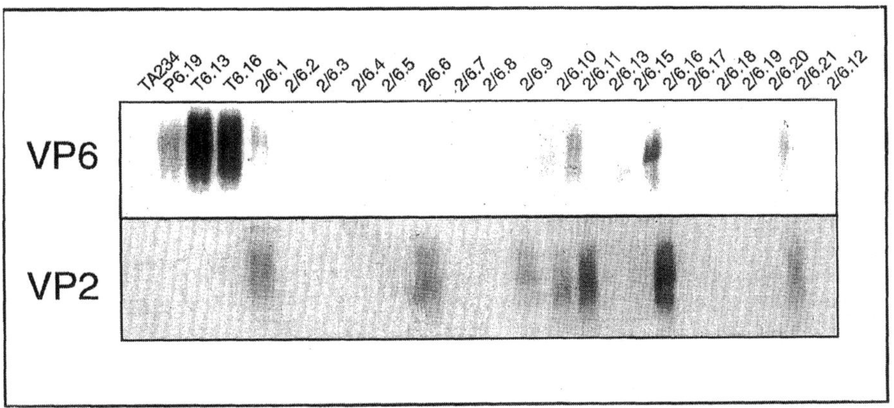

Figure 2. Northern blot analysis of total RNA hybridized with VP6 (upper panel) or sVP2 (lower panel) probes. Membrane was stripped between probings.

In order to optimise the expression of the VP2 gene, we have designed a synthetic gene in which the problematic features inherent to the VP2 gene sequence were eliminated without changing the predicted amino-acid sequence (Richter et al. unpublished). The synthetic version of the 2.7 kb long VP2 gene was constructed and was cloned into a plant expression vector under the control of the CaMV 35S promoter. Similarly, co-expression plasmids were created in which the VP6 and VP2 expression cassettes were cloned in tandem. These constructs were used in *A tumefaciens* mediated transformation to transform potato plants and tomato plants. Kanamycin resistant transformants were screened by total RNA hybridisation to assay the levels of sVP2 and VP6 transcripts. Accumulation of the VP6 protein was assayed by ELISA. Several lines that accumulate transcripts of both genes were selected and are currently under further analysis (Fig. 2).

Although some reports suggest that mucosal administration of rotavirus VLPs that consist of only VP2 and VP6 elicits protective immunity (O'Neal et al. 1997), it might still be beneficial to produce the complete triple-layered VLPs in plants. The larger, more native-like structure may be more stable and both VP4 and VP7 contain important epitopes for the induction of neutralising antibodies. We are currently redesigning the rotavirus genes encoding VP4 and VP7 to optimise their expression in plants. Although NSP4 is required for the assembly of rotavirus virion in infected cells, triple layered VLPs were formed in its absence in insect cells by the baculovirus expression system (Crawford et al. 1994). The importance of NSP4 in the transgenic plant system presented here is yet to be determined.

4. Acknowledgements

This research is supported by the Thrasher Research Fund (Award #02904-2) and by a grant from the World Health Organisation (#V27/181/80). T.S. Mor is supported by a Postdoctoral Award (No. FI-251-97) from The United States – Israel Binational Agricultural Research and Development Fund (BARD). We thank Joan Sesing Lenz for her plant tissue culture expertise.

5. References

Ball JM et al (1996) Science 272, 101-104.
Crawford SE et al (1994) J. Virology 68, 5945-5952.
Estes MK. (1996) Fields Virology, pp. 1625-1655, Lippincott-Raven, Philadelphia.
Estes MK et al (1997) Adv. Exp. Med. Biol. 412, 387-395.
Filatti JJ et al (1987) Bio/Technology 5, 726-730.
Garside P, Mowat AM (1997) Crit. Rev. Immunol. 17, 119-137.
Haq TA et al (1995) Science 268, 714-716.
Mason HS et al (1996) Proc. Natl. Acad. Sci. USA 93, 5335-5340.
Mason HS et al (1998) Vaccine in press.
Mason HS et al (1992) Proc. Natl. Acad. Sci. USA 89, 11745-11749.
Mestecky J et al (1997) Behring. Inst. Mitt. 98, 33-43.
O'Neal CM et al (1997) J. Virol. 71, 8707-8717.
Pérez-Schael I et al (1997) N. Engl. J. Med. 337, 1181-1187.
Stemmer WPC et al (1995) Gene 164, 49-53.
Tacket CO et al (1998) Nat. Med. 4, 607-609.
Thanavala Y et al (1995) Proc.Natl. Acad. Sci. USA 92, 3358-3361.
Tucker AW et al (1998) JAMA 279, 1371-1376.
Wenzler H et al (1989) Plant Sci. 63, 79-86.
Zor T, Selinger Z (1996) Anal. Biochem. 236, 302-308.

MANIPULATING OXIDATIVE STRESS RESPONSES USING TRANSGENIC PLANTS: SUCCESSES AND DANGERS

P. Mullineaux and G. Creissen
John Innes Centre, Norwich Research Park, Colney, Norwich, NR4 7UH, United Kingdom

1. Introduction. Oxidative stress and antioxidant systems in plants

A bi-product of oxygen metabolism in all aerobes is the production of active oxygen species (AOS) such as the superoxide anion ($O_2^{\cdot-}$), hydrogen peroxide (H_2O_2) and the hydroxyl radical ($OH^{\cdot}$). If the production of AOS exceeds the capacity of enzymic and non-enzymic antioxidant systems to scavenge these species, or if these protective systems are compromised, then oxidative stress occurs. This manifests itself as oxidative damage to membranes and macromolecules such as DNA and proteins. The reader is referred to Foyer, Mullineaux (1994) and Scandalios (1997) for more detailed considerations of the processes briefly described below.

Plants are prone to oxidative stress as a consequence of oxygenic photosynthesis occurring in the vicinity of highly energised pigment beds and very active photosynthetic electron transport chains. This susceptibility is in addition to metabolic processes of oxygen, such as respiration, which are common to all aerobic organisms. Many environmental stresses, such as drought, chilling in high light, saline conditions, exposure to gaseous pollutants, ultraviolet irradiation and attack by plant pathogens have an oxidative stress component.Therefore, the role of the AOS scavenging and antioxidant systems of plants are considered to be crucial in determining their ability to cope with and adapt to the fluctuating environment in the field.

The ascorbate-glutathione (AG) cycle is regarded as the main way in which AOS are reduced to water. This cycle in the chloroplast is shown in Figure 1. The central feature is the series of linked oxidation and reduction reactions around ascorbate and glutathione, the net affect of which is to reduce H_2O_2 to water (Foyer, Halliwell 1976). The reduction of superoxide anion to H_2O_2 catalysed by superoxide dismutase is included as well as the reduction of monodehydroascorbate free radical to ascorbate catalysed by monodehydroascorbate free radical reductase (Fig.1). Recently, the realisation that at least one class of phospholipid hydroperoxide glutathione peroxidase (PHGPX) is localised in the chloroplast (Mullineaux et al 1998) provides a theoretical link in plants between the antioxidant function of α-tocopherol (vitamin E), lipid peroxidation and the cycling of oxidised and reduced glutathione (Fig.1).

Since, the ascorbate-glutathione cycle was proposed for the chloroplast, variations on this theme have been suggested for several subcellular compartments. Extraplastidial AG cycles could also include the alternative enzymes which scavenge AOS such as the catalases and the

A. Altman et al. (eds.), *Plant Biotechnology and In Vitro Biology in the 21st Century*, 525–532.

glutathione peroxidase activity of some glutathione-S-transferases (GSTs; Willekens et al 1994, Bartling et al 1993).

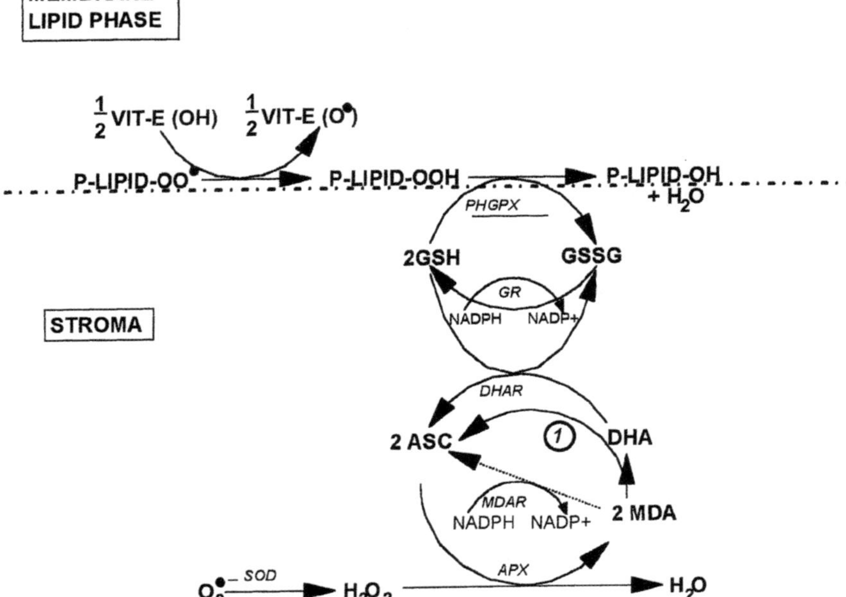

Figure 1. The scavenging of active oxygen species in the chloroplast in both lipid membrane phase and the stroma, linked to redox cycles for ascorbate and glutathione and the oxidation of α-tocopherol (vitamin E). The displayed scheme is combined from the ascorbate-glutathione cycle (Foyer and Halliwell 1976) with modifications described in Scandalios (1997). The reactions involving vitamin E are from Scandalios (1997). The presence of the linking PHGPX-catalysed reaction (underlined) are combined from Mullineaux et al (1998) and Beeor-Tzahar et al (1995). Abbreviations are as follows: P-LIPID-OO , phospholipid peroxy radical; P-LIPID-OOH, phospholipid peroxide; P-LIPID-OH, phospholipid alcohol; VIT-E(OH), α-tocopherol (vitamin E); VIT-E (O), α-chromanoxyl radical; PHGPX, phospholipid hydroperoxide-dependent glutathione peroxidase; GSH, reduced glutathione; GSSG, glutathione disulphide; GR, glutathione reductase; DHAR, dehydroascorbate reductase; ASC, ascorbic acid; DHA; dehydroascorbate; MDA, mondehydroascorbate free radical; MDAR, monodehydroascorbate free radical reductase; APX, ascorbate reductase; SOD, superoxide dismutase; $O_2^{\cdot-}$, superoxide anion; H_2O_2; hydrogen peroxide. Reaction 1 is the non-enzyme catalysed spontaneous dismutation of 2 MDA molecules to 1 ASC and 1 DHA respectively.

2. Overexpression of enzymes of the ascorbate-glutathione cycle in transgenic plants

In various plant species subjected to environmental stresses, the levels both the enzyme activities of the AG cycle and AOS scavenging and/or their corresponding mRNAs often rise

(Foyer, Mullineaux 1994; Inzé,Van Montagu 1995). This has provided the incentive for several laboratories to generate transgenic plants which overexpress single components of the protective systems with a view towards understanding the roles of specific enzymes in particular subcellular compartments and improving tolerance to oxidative stress which therefore, could improve tolerance to a broad spectrum of environmental stresses.
In this context, the most frequently studied transgenic plants are those overexpressing one of the three isoforms of SOD in either the chloroplastic, mitochondrial or cytosolic compartments (Table 1). In addition, glutathione reductase (GR), ascorbate peroxidase (APX) and glutathione peroxidase (GPX) have been overexpressed, principally in transgenic tobacco (Table 2).

2.1. Analysis of GR transgenic tobacco: An example of genetic manipulation for stress tolerance

In principle, the insertion of a single gene into a plant should provide a direct test of the importance of that gene product in the stress tolerance mechanisms of that species. However, the outcome of such experiments may not be so easy to interpret, as illustrated by the experience in our laboratory in manipulating GR expression in the chloroplast, chloroplast plus mitochondrion and the cytosol (Broadbent et al 1995; Creissen et al 1995). T2 progeny, arising from self-crosses of a total of 12 primary transgenic parents were studied for their tolerance to methyl viologen and ozone. These lines were selected because they gave up to an average of 4.5-fold elevation in GR activity in total leaf extracts. Transgene-recessive siblings for each transgenic line were used as controls.

Tolerance to methyl viologen was only detected in two chloroplast GR- and one cytosol GR-overexpressor. None of the co-expressors of high levels of GR combined in both chloroplasts and mitochondria (cotargeted GR; Creissen et al 1995) showed tolerance to this oxidative stress-inducing herbicide. Conversely, two out of 4 of the co-targeted GR transgenic lines displayed tolerance to a single high dose of ozone (200nl l^{-1} for 8h) while none of the other classes of transgenic plants displayed any tolerance to this stress. This included all of the lines designated as showing reduced sensitivity to methyl viologen. Thus, overexpression of GR was no guarantee of improved tolerance (or decreased sensitivity) to oxidative stress. Nevertheless, in those lines which showed some decreased sensitivity there was a segregation of this phenotype with the transgene, implying in these lines that the transgene was responsible, at least in part, for the improved performance of these plants. A further curious observation was that the same lines did not give tolerance to both stress treatments, implying that while both tolerances were mediated by enhanced GR activity, these were achieved by different mechanisms.

2.2 Survey of genetic manipulation of plants for increased oxidative stress tolerance

From the combined published work of several laboratories shown in Tables 1 and 2, it is possible to make several comments :

- Overexpression of the same type of enzyme in the same transgenic species does not universally lead to the detection of plants with altered responses to oxidative stress. This may be because of the many different variations on the way stress exposure experiments can be conducted. These could include variations in the timing and dose of stress exposures, the ages of the plant, the way in which the response to the stress is monitored. Furthermore, different genotypes (varieties) of a particular species may been used.

- The types of control plant vary. These can include non-transformed wild-type lines, tissue culture regenerants, control transformants but without the specific transgene in question, and transgene recessive siblings from a segregating population of progeny. If tissue culture with its attendant stresses imposes some selection on plant regenerants, then this could be an important consideration.

- Most work has been done on progeny from transformed plants. However this is not immediately experimentally possible for some species (eg. in poplar, alfalfa and potato). Therefore, somaclonal variation, leading to epigenetic effects can only be ruled out where many independent transformants show the same trends and control transgenic material does nothing (Foyer et al 1995).

- The best transgene expressing lines do not necessarily give rise to plants which are tolerant to oxidative stress.

- Many papers report the generation of considerable numbers of transgenic lines, but only very small numbers are investigated at the physiological level and are subjected to stress. This is presumably because of space limitations and the logistics of carrying out large scale whole plant physiology experiments in controlled conditions. In some cases the criterion for selection has been the best transgene expressors. However, in other cases the selection criteria are not apparent.

-Methyl viologen treatment has proved very popular in screening transgenic lines for evidence of increased tolerance to oxidative stress. However, plants showing reduced sensitivity to methyl viologen do not necessarily show any increased tolerance ot other stresses.

- Success in generating oxidative stress tolerance in one species with a particular transgene does not mean that the same construct will work in other plant species.

2.3. What does all this mean?

From the considerations above, two factors stand out prominently. First, the differing ways stress experiments are conducted may account for the variation in data from laboratory to laboratory. Second, other factors are at work which could promote or negate the effect of the transgene. These factors could possibly arise as a consequence of plant tissue culture and transformation and could perhaps be epigenetic effects which influence other aspects of whole plant responses to oxidative stress. This notion is supported by work which shows that tolerance to oxidative stress can also be enhanced by cycling plants through a tissue culture règime and applying some appropriate selection. In this way, paraquat-tolerant tobacco (also more tolerant to SO_2) and salt-tolerant alfalfa have been generated (Tanaka et al 1988; Safarnejad et al 1996). In both of these examples the activities of enzymes of the AG cycle

Enzyme Activity	Plant Species	Subcellular Location	Stress Applied	Tolerance shown?	No. of transgenic lines generated	No. of transgenic lines examined	Reference
Cu/Zn SOD	tobacco	chloroplast	MV, chilling + high light	NO	15	1	1
Cu/Zn SOD	tomato	chloroplast	MV, chilling + high light	NO	?	2	1
Mn-SOD	tobacco	chloroplast	MV (light +dark)	YES	15	2	2
Mn-SOD	tobacco	mitochond.	MV (dark)	YES	15	2	2
Cu/Zn SOD	tobacco	chloroplast	ozone	NO	2	2	3
MnSOD	alfalfa	chloroplast	freezing	YES	33	2	4
MnSOD	alfalfa	mitochond.	freezing	NO	40	2	4
Cu/Zn SOD	potato	chloroplast	MV	YES	40	2	5
Cu/Zn SOD	potato	cytosol	MV	YES	40	2-3	5
Cu/Zn SOD	tobacco	chloroplast	MV, chilling + high light	YES	>20	1	6,7
Mn-SOD	tobacco	chloroplast	ozone	YES	2	2	8
Mn-SOD	tobacco	chloroplast	MV	YES	N/A	2	9
Cu/Zn SOD	tobacco	cytosol	ozone	YES	?	6	10
Fe-SOD	tobacco	chloroplast	MV, salt, chilling	YES NO NO	?	>1	11
Mn-SOD	cotton	chloroplast	MV, chilling + high light	NO	>6	3	12

Table 1. Refereed papers, listed in chronological order, reporting overexpression of superoxide dismutase (SOD) isoforms in various species of transgenic plants. Oxidative stress tolerance is scored as YES if at least one plant was reported to score positive. The numbers refer to the total number of independent transgenic lines generated. The number of lines examined means those independent

transgenic lines which were assayed for oxidative stress tolerance and for which data were presented in the report. Abbreviations are mitochond., mitochondrion; MV, methyl viologen (paraquat); ?, where the number of lines generated cannot be readily discerned. The references are as follows:
1, Tepperman, Dunsmuir (1990); 2, Bowler et al (1991); 3, Pitcher et al (1991); 4, McKersie et al (1993); 5, Perl et al, (1993); 6, Sen Gupta et al (1993a); 7, Sen Gupta et al (1993b); 8, Van Camp et al (1994); 9, Slooten et al (1995); 10, Pitcher and Zilinskas (1996); 11, Van Camp et al (1996); 12, Payton et al (1997).

Enzyme Activity	Plant Species	Subcellular Location	Stress Applied	Tolerance shown?	No. of transgenic lines generated	No. of transgenic lines examined	Reference
GR(*)	tobacco	cytosol	MV	NO	>3	1	1
GR(*)	tobacco	cytosol	MV	YES	26	1	2
GR(*)	tobacco	chloroplast	MV, SO_2	YES	?	1	3
GR	tobacco	mitochond. and/or chloroplast	MV, ozone	YES	9	9	4,5
GR	tobacco	cytosol	MV	YES	3	3	4
GR(*)	poplar	chloroplast	chilling+ high light	YES	5	2	6
GR(*)	poplar	cytosol	MV	NO	5	1	6
cAPX	tobacco	cytosol	MV	NO	>3	3	7
cAPX	tobacco	chloroplast	ozone	NO	5	up to 5	8
GST/GPX	tobacco	cytosol	chilling, salt	YES	5	up to 5	9

Table 2. Refereed papers, listed in chronological order, reporting overexpression of glutathione reductase (GR), cytosolic ascorbate peroxidase (cAPX) and glutathione-S-transferase/glutathione peroxidase (GST/GPX) in transgenic tobacco or poplar. GR(*) means the enzyme is from *Escherichia coli*, the rest of the enzymes were encoded by plant-derived sequences. Oxidative stress tolerance is scored as YES if at least one plant was reported to scored positive. The numbers refer to the total number of independent transgenic lines generated. The number of lines examined means those independent transgenic lines which were assayed for oxidative stress tolerance and for which data were presented in the report. Abbreviations are mitochond., mitochondrion; MV, methyl viologen (paraquat); ?, where the number of lines generated cannot be readily discerned. The references are as follows:1, Foyer et al (1991); 2, Aono et al (1991); 3, Aona et al (1993); 4, Broadbent et al (1995); 5, Creissen et al (1995); 6, Foyer et al (1995); 7, Saji et al (1996); 8, Torsethaugen et al (1997); 9, Roxas et al (1997).

(Fig.1) were elevated. Thus, the assertion that oxidative stress tolerance can be enhanced by overexpressing a single component of the antioxidant defence systems has to be treated with caution. Epigenetic effects may partly determine how a transgene encoding an AG cycle enzyme influences whole plant responses to oxidative stress.

An example of these considerations is the apparent selection of alfalfa plants transgenic for chloroplast-targeted Mn-SOD (McKersie et al 1993; Table 1), some lines of which showed enhanced capacity for regrowth after freezing. That these alfalfa lines show enhanced tolerance to oxidative stress is not in question, vegetatively propagated progeny from this material showed good levels of tolerance to stress in field trials (McKersie et al 1996). However, seed-propagated material showed no correlation between field performance indicators and enhanced SOD as assayed for those lines in the glasshouse (B. McKersie, S.R. Bowley, personal communication). It may be possible that the stress tolerance may have nothing to do with the expression of the Mn-SOD transgene.

2.4. Successes with transgenic plants and oxidative stress

The original rationale for using a transgenic plant,viz., that of achieving a single gene change in an otherwise isogenic background, remains a valid approach. A good example is the recent work of Foyer and co-workers (reviewed in Noctor et al 1998) in which the foliar levels of glutathione have been elevated by over-expressing a transgene encoding γ-glutamylcysteine synthetase (γ-ECS) in the cytosol of poplar. Expressing this gene had a clear biochemical effect and this did not give rise to any oxidative stress tolerance in the few lines that were tested. Therefore, some of the assumptions of the role of glutathione in the cytosol may have to be questioned in the light of these data. In this context, the value of transgenic plants is manifest.

Other observations on the functioning and regulation of antioxidant defences have come from transgenic plant studies. For example, enhanced activities and transcripts of enzymes of the AG cycle have been observed in transgenic tobacco plants overexpressing SOD in their chloroplasts (Sen Gupta et al 1993b; Van Camp et al 1996). Similarily, the co-targeting of pea GR to both chloroplasts and mitochondria was a surprising observation in GR-overexpressing transgenic tobacco (Creissen et al 1995) and may have contributed to reduced ozone sensitivity in some of these lines (Broadbent et al 1995). Knowledge about the way in which amino acid metabolism allocates precursors to glutathione biosynthesis has been greatly advanced using transgenic poplar with elevated activities of the enzymes of glutathione synthesis (Noctor et al 1998).

2.5. Conclusions and prospects

Transgenic plants have proved to be a very useful tool in helping us to begin to understand plants' responses to oxidative stress, although this might not have been for the expected reasons! The use of transgenic plants will remain for the foreseeable future, but the transgenes employed may be different. For example, stacking of multiple transgenes to attempt to stimulate whole sections of the AOS scavenging system or the use of genes regulating sections of antioxidant metabolism might become an option.

3. References

Aono M et al (1991) Plant Cell Physiol. 32, 691-697.
Aono M et al (1993) Plant Cell Physiol. 34, 129-135.
Bartling D et al (1993) Eur. J. Biochem. 216, 579-586.
Beeor-Tzahar T et al (1995) FEBS Lett. 366, 151-155.
Bowler C et al (1991) EMBO J. 10, 1723-1732.
Broadbent P et al (1995) Plant J. 8, 247-255.
Creissen G et al (1995) Plant J. 8, 167-175.
Foyer C, Halliwell B (1976) Planta 133, 21-25.
Foyer C et al (1991) Plant Physiol. 97, 863-872.
Foyer CH, Mullineaux PM (1994) Causes of Photooxidative Stress and Amelioration of Defense Systems in Plants, CRC Press, Boca Raton, Florida, USA.
Foyer C et al (1995) Plant Physiol. 109, 1047-1057.
Inzé D, Van Montagu M (1995) Curr. Opin. Biotech. 6, 153-158.
McKersie B et al (1993) Plant Physiol. 103, 1155-1163.
McKersie B et al (1996) Plant Physiol. 111, 1177-1181.
Mullineaux PM et al (1998) Plant J. 13, 375-379.
Noctor G et al (1998) J. Exp. Bot. 49, 623-647.
Payton P (1997) Photosynth. Res. 52, 233-244.
Perl A et al (1993) Theor. Appl. Gen. 85, 568-576.
Pitcher LH et al (1991) Plant Physiol. 97, 452-455.
Pitcher LH, Zilinskas BA (1996) Plant Physiol. 110, 583-588.
Roxas P et al (1997) Nature Biotech. 15, 988-991.
Sen Gupta A et al (1993a) Proc. Natl. Acad. Sci. USA 90, 1629-1633.
Sen Gupta A et al (1993b) Plant Physiol. 103, 1067-1073.
Safarnejad A et al (1996) Euphytica 92, 55-61.
Saji H et al (1996) Environ. Sci. 9, 241-248.
Scandalios JG (1997) Oxidative Stress and the Molecular Biology of Antioxidant Defenses, monograph 34, Cold Spring Harbor Laboratory Press, New York, USA.
Slooten L et al (1995) Plant Physiol. 107, 737-750.
Tanaka K et al (1988) Plant Cell Physiol. 29, 743-746.
Tepperman JM, Dunsmuir P (1990) Plant Mol. Biol. 14, 501-511.
Torsethaugen G et al (1997) Plant Physiol. 114, 529-537.
Van Camp W et al (1994) Bio/Technology 12, 165-168.
Van Camp W et al (1996) Plant Physiol. 112, 1703-1714.
Willekens H et al (1994) Proc. Natl. Acad. Sci. USA 91, 10450-10454.

TRANSFORMATION OF TOBACCO AND ASPEN PLANTS WITH THE *ITA* LOCUS OF AN INCQ PLASMID CONFERS RESISTANCE TO *AGROBACTERIUM TUMEFACIENS* INFECTION

M. Ovadis[1], L. Chernin[1], T. Tzfira[1], V. Canaan[1], A. Aharoni[1], D. Sakar[2], A. Altman[1] and A. Vainstein[1]

[1]The Otto Warburg Center for Biotechnology in Agriculture and the Kennedy-Leigh Centre for Horticultural Research, The Hebrew University of Jerusalem, Rehovot 76-100 Israel; [2]Faculty of Agriculture, Dicle University, Diyarbakir, Turkey

1. Introduction

Agrobacterium tumefaciens is well known as a tool for the genetic engineering of plant species. However, *A. tumefaciens* is also a highly pathological bacterium and the cause of 'crown gall' disease in a wide range of plant species (DeCleen and DeLey 1976), including many economically important ones (e.g. grapevine, tomato, poplar, rose and chrysanthemum). The disease, characterized by the production of tumors on plant-wound sites, is a result of the plant's stable genetic transformation by phytopathogenic *A. tumefaciens* strains, which carry large, tumor-inducing (Ti) plasmids. During the infection process, the *Agrobacterium* transfers a defined DNA fragment (the T-DNA)—located on the Ti plasmid—to the plant cells (Zambryski 1992). Following the T-DNA's integration, several oncogenic (*onc*) genes encoded by it are highly expressed, altering the plant's hormonal balance (Gaudin et al. 1994). This eventually leads to extensive and uncontrolled plant cell division and to the formation of a tumor.

When present in the bacterial cells, several R-plasmids (e.g. IncW and IncQ plasmids) have been found to suppress the tumorigenicity of various types of *A. tumefaciens* to many plant species, including kalanchoe, tomato, pea and sunflower (Loper and Kado, 1979; Farrand et al. 1981; Chernin et al. 1984; Avdienko et al. 1987). Production of IAA and cytokinins by *A. tumefaciens* derivatives, carrying both the Ti and IncQ or IncW plasmids, significantly decreased relative to the parent *A. tumefaciens* strain (Chernin et al. 1984; Avdienko et al. 1987; Miklashevichs et al. 1988). Tumorigenicity of these *A. tumefaciens* derivatives can be restored *ex planta* by applying exogenous IAA to infected plants (Chernin et al. 1984; Avdienko et al. 1987) or *in planta* by infecting tobacco lines that are transgenic for the *ipt* gene from plasmid pTiC58 and hence produce higher levels of trans-zeatine riboside (Chernin et al. 1991b).

The *ita* (inhibition of tumorigenicity of *Agrobacterium*) locus of the RSF1010 (IncQ) plasmid has been found to specifically reduce the tumorigenicity of *Agrobacterium* strains virulent to kalanchoe and tomato (Chernin et al. 1991a). To further investigate the antitumorigenic effect of the *ita* locus we expressed it in two unrelated model species, tobacco and aspen. Selected transgenic clones exhibited decreased sensitivity to

A. Altman et al. (eds.), Plant Biotechnology and In Vitro Biology in the 21st Century, 533–536.

virulent *Agrobacterium* infection, revealing the agricultural potential of the *ita* locus for introducing *Agrobacterium* tolerance in transgenic plants.

2. Material and Methods

Genetic transformation of tobacco and aspen plants. An EcoRV-B fragment of the RSF1010 plasmid was cloned into a binary plasmid under the control of a 35S constitutive promoter. Transgenic tobacco and aspen plants were generated and selected under hygromycin (*hptII*) using the previously described leaf disc (Horsch et al. 1988) and stem explant (Tzfira et al. 1997) transformation methods, respectively.
Bacterial culture and preparation. *A. tumefaciens* strains 1D1 (pTi1D1, octopine type), C58 (pTiC58, nopaline type) and A281 (pTiBo542, succinaminopine type) were grown overnight in potato-dextrose or LB media. Bacteria were prepared by either centrifugation and resuspension in sterile water (for tobacco inoculation), or dilution to an OD_{550} of 0.1-0.2 in fresh LB medium (for aspen inoculation).
Infection of transgenic plants. *Ex vitro* tobacco and *in vitro* aspen transgenic plants were screened for their susceptibility to several wild-type *Agrobacterium* strains. Greenhouse-grown tobacco plants were infected by puncturing their stems between the first and second internodes with a sterile needle (21G) and injecting 10-20 μl of bacterial suspension. After 3 weeks, the tumors were scored according to size, dissected from the plants and weighed. Stem segments of *in vitro*-grown aspen plants were dip-inoculated for 30 min in an overnight bacterial suspension and cultured on half-strength MS-based medium (Murashige and Skoog 1962). After 2 days of co-cultivation, the segments were washed and replanted in fresh medium supplemented with 300 mg/l carbenicillin. After 3 weeks, the number of stem segments forming galls and the gall's fresh weight were determined.

3. Results

Non-transformed tobacco plants were highly susceptible to wild-type *A. tumefaciens* strain 1D1, as reflected by the frequency of gall formation and gall size. A screening of 20 independent tobacco *ita*-transformants revealed various responses to 1D1 infection. In three different transformants, the disease symptoms (frequency of gall formation and gall size) were significantly lower as compared with control (non-transformed and *hptII*-transformed) plants (Fig. 1A). In lines T6 and T13, the average gall size was reduced by ca. 90% relative to the size of tumors formed by the infected control plants.

Similar observations were made in an *ita*-transgenic aspen plant (line C5). Controls (non-transformed and *hptII*-transgenic) were screened for their susceptibility to different *Agrobacterium* strains. Although no gall formation was observed when these plants were inoculated with strain 1D1, a high infection rate and large gall production on both cut surfaces of the stem segment were observed with strains C58 or A281 (Fig. 1C). Dip-inoculation of line C5 plants resulted in a very low rate of gall formation (e.g. Fig. 1B, a), with reductions of 95% in gall fresh weight and 80% in gall diameter as compared to control plants (Fig. 1B, b and c).

Several more studies were performed with *ita*-transgenic plants in order to further elucidate the possible mechanism of disease suppression in *ita*-transgenes. Leaf discs of *ita*-transgenic tobacco plants were inoculated with a disarmed *A. tumefaciens* strain, carrying a binary vector harboring the *uidA* (GUS) reporter gene and the *nptII* selection gene. Frequency of transient GUS expression was similar in both non-transformed and *ita*-transgenic tobacco leaf discs. Furthermore, the yield of transgenic plants regenerated from leaf disks was similar in both control and *ita*-transgenic plants. Hence neither the T-DNA's ability to enter the nucleus and express itself nor the stable integration and expression of the T-DNA in transgenic cells was affected. It is therefore possible that the *ita* locus plays a more specific role in reducing crown-gall formation in transgenic plants.

The combined action of both auxin and cytokinin genes (Gaudin et al. 1994), leading to altered hormonal balance in transgenic cells, is required for gall formation (Akiyoshi et al. 1983). Thus, we hypothesized that the *ita* locus could have an effect on the plant's biosynthesis and/or sensitivity to hormones. To test this, we evaluated the sensitivity of aspen stem segments to the addition of exogenous cytokinin *in vitro*. Indeed, *ita*-transgenic C5 plants cultured on TDZ-containing WMP medium (Tzfira et al. 1997) developed ca. five times more adventitious shoot buds than control plants. Moreover, root formation was drastically suppressed in C5 plants.

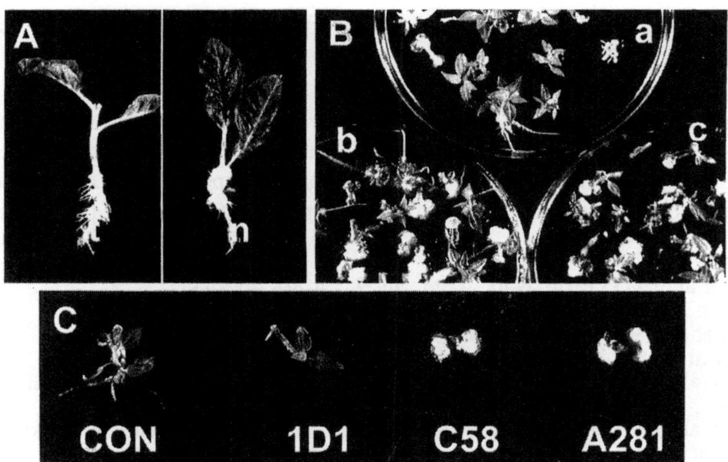

Fig 1. Suppression of crown gall disease in *ita*-transgenes. (A) Reduced crown gall size in *ita*-transgenic tobacco plant (t) infected with wild type *A. tumefaciens* strain 1D1. n, non-transformed plant. (B) *In vitro ita*-transgenic aspen plants(a) exhibiting a high tolerance to wild-type *A. tumefaciens* strain C58. Gall production was clearly visible in non-transformed (b) and *hptII*-transgenic (c) aspen plants. (C) Response of aspen stem segments to infection with different wild-type strains of *A. tumefaciens in vitro*.

4. Discussion

In this study we present first-time evidence of the antitumorigenic effect of an *ita* fragment from IncQ in transgenic plants. Transgenic plants exhibited a reduced rate of gall formation as well as reduced gall size and weight. Previous studies have shown that the presence of antitumorigenic IncW or IncQ plasmids in *Agrobacterium* does not affect either the stability of the pTi-DNA or specific plasmid functions such as the ability to code for opine catabolism (Farrand et al. 1981; Chernin et al. 1984; Avdienko et al. 1987). Here we show that *ita*-transgenic tobacco plants are not affected in their susceptibility to stable genetic transformation by disarmed *A. tumef.* *ita*-Transgenic aspen plants exhibited a higher sensitivity to exogenous cytokinin, as inferred from much higher shoot production in the presence of TDZ relative to control plants. In view of reports on reduced auxin and cytokinin production by *A. tumefaciens* strains carrying the RSF1010 plasmid (Avdienko et al. 1987; Miklashevichs et al. 1988), we suggest that the inserted RSF1010 (*ita*) region somehow affects phytohormone production and/or sensitivity in transformed plants. Moreover, Beneddra et al. (1996) reported a correlation between sensitivity to cytokinin and susceptibility to crown gall, and a similar correlation has been reported for *Arabidopsis* (Tourneur et al. 1985) and tobacco (Lincoln et al. 1992) mutants.

Further study is required to better understand the mechanism of the *ita* locus's effect, i.e. to identify its product and its relation with hormonal balance/sensitivity. Nevertheless, these primary results clearly demonstrate the potential of the *ita* locus from the IncQ plasmid RSF1010 for introducing resistance to crown gall disease in transgenic plants.

5. Acknowledgments

This work was supported in part by a grant from the Horowitz Foundation, Yissum R&D Company of the Hebrew University of Jerusalem, and by the Association of Israeli Flower Growers. We are thankful to Mrs. I. Kushner and Dr. T. Denisova (Moscow, Russia) for their participation in constructing a binary vector carrying the *ita* locus.

6. References

Akiyoshi et al (1983) Proc. Natl. Acad. Sci. USA 80, 407-411.
Avdienko et al (1987) Agric. Biol. 11, 91-97.
Beneddra et al (1996) Phytopathology 86, 225-231.
Chernin et al (1984) Mol. Gen. Genet. 195, 195-199.
Chernin et al (1991a) In: Pathology and Molecular Biology of Crown Gall (Tempe J., ed.). CNRS, Inst. des Sciences Vegetales, Gif sur Yvette, France.
Chernin et al (1991b) In: Plant Biotechnology and Molecular Biology (Skryabin K.G., ed.). Abstracts of the Symp., p. 46. Pushchino, Moscow Region.
DeCleen M and DeLey J (1976) Bot. Rev. 42, 389-466.
Farrand et al (1981) Mol. Gen. Genet. 181, 44-51.
Gaudin et al (1994) Plant Physiol. Biochem. 32, 11-29.
Horsch et al (1988) Plant Molecular Biology Manual A5, 1-9.
Lincoln et al (1992) Plant Physiol. 98, 979-983.
Loper J and Kado C (1979) J. Bacteriol. 139, 591-596.
Miklashevichs et al (1988) Geneticka 24, 250-258.
Murashige T and Skoog F (1962) Physiol. Plant. 15, 473-497.
Tourneur et al (1985) In: Plant Genetics (Freeling M., ed.) pp. 791-797. Alan R. Liss Inc., NY.
Tzfira et al (1997) Plant Mol. Biol. Rep. 15, 219-235.
Zambryski P (1992) Ann. Rev. Plant Physiol. Plant Mol. Biol. 43, 465-490.

GROWTH CHARACTERISTIC CHANGES DURING SALINITY ADAPTATION OF THE WILD TOMATO SPECIES *L. pennellii*

A. Rus, S. Rios, E. Olmos and M.C. Bolarin
Dpto. Riego y Salinidad, CEBAS-CSIC, Apdo 4195, 30080-Murcia, Spain

Summary

Calluses of the wild salt-tolerant tomato species *L. pennellii* have been used to determine the changes induced by salinity on the growth characteristics as well as on the ionic contents along the first four subcultures, with the aim of determining the adaptation capacity of the cells to salinity. The calluses grown in saline medium showed similar RGRs to those in control medium up to the 3rd subculture; in the 4th, the callus RGR increased significantly with salinity, indicating the high adaptation capacity of the salt tolerant species to salinity. The salt-adaptation of *L. pennellii* resulted not only in an increased growth rate but also in increased cell volume. The enhanced growth of salt adapted cells was related to the capacity to maintain the water content and to the increased saline ion uptake.

Introduction

Despite numerous attempts to enhance the salt-tolerance of the cultivated tomato using the tolerance of the related wild species of *Lycopersicon*, practical results have not been as satisfactory as expected. One of the major problems is the difficulty of screening thousands of plants without having available and rapid traits. It would be very interesting to be able to evaluate the salt tolerance by *in vitro* culture because of the relatively little space, the lower time required for the selection, and finally the controlled environment (Blits et al., 1993). However, it is necessary to take into account the high adaptation capacity of the cells to salinity, a fact which makes it impossible to distinguish between adapted and salt-tolerant cells (Tal, 1993). Most studies on the adaptation capacity of the cells to salinity have been carried out in glycophytes. In the present work, calluses from leaf of the wild salt-tolerant tomato species, *L. pennellii*, have been used to determine the changes induced by salinity on the growth characteristics as well as on the ionic contents along the first four subcultures.

Material and Methods

A salt-tolerant accession of the wild species *L. pennellii* Corr. D'Arcy (PE-47) was used. Callus initiation and culture was carried out as previously described (Cano et al.,

A. Altman et al. (eds.), Plant Biotechnology and In Vitro Biology in the 21st Century, 537–540.

1996). Briefly, explants of leaves were transferred to plastic Petri dishes containing 25 ml sterile growth medium. This medium was suitable for callus initiation and culture. After initiation, calluses were subcultured every 28 days on fresh medium either without NaCl (control) or containing 175 mol m^{-3} NaCl. Callus cultures were maintained until the fourth subculture. At least 60 replicates (15 Petri dishes with 4 calluses each) were used for each treatment. In each subculture, calluses were weighed under sterile conditions every week (days 7, 14, 21 and 28), and the relative growth rate (RGR) was calculated on the basis of fresh weight.

At the end of each subculture (day 28), cell volume was estimated. Quantitative and morphometric analysis was realized using an automatic image analyse equipment. Cell length and width were measured, and cell volume was estimated by using simple equations for cylindrical and ellipsoid solids in accordance with Iraki et al. (1989).

On day 28, the calluses were lyophylized and weighed (dry weight). Water content was calculated as (FW-DW)/DW ratio. On dry material, Na^+ and Cl^- contents were determined according to Perez-Alfocea et al. (1994).

Results

Growth of *L. pennellii* callus, measured by the RGR during the growth cycle, changed with the subculture in control as in saline media (Fig. 1). In control medium, RGR tended to decrease with the subculture, especially between the 1st and 2nd subculture. This could be due to the high osmotic pressure of the basal medium used in *in vitro* culture (approximately 0.4 MPa), which may be inducing an osmotic stress (Dracup, 1991).

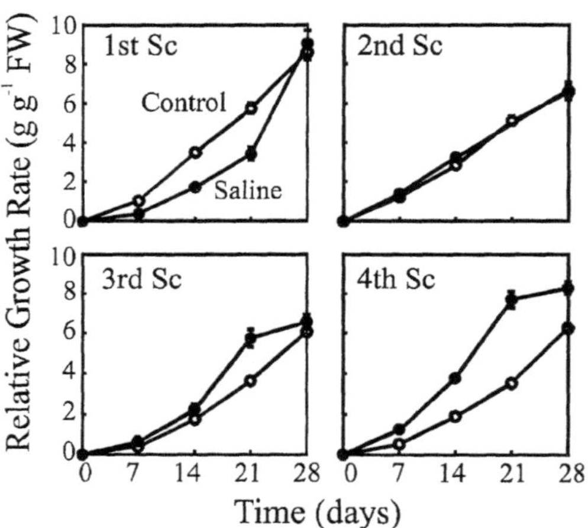

Fig.1. Growth curves of *L. pennellii* calluses grown on control and saline (175 mol m^{-3} NaCl) media along the first four subcultures. Bars on data point are ± S.E.

If we compare the RGR of the calluses grown in control and saline media, it is observed that callus growth did not decreased with salinity in any of the first four subcultures at the end of the growth cycle, which is in agreement with the high salt tolerance reported for this wild species both at the whole plant and cell levels (Perez-Alfocea et al., 1994).

In the last subculture, callus RGR was even greater in saline medium than in the control one. It has been observed that cells are capable of adapting to high NaCl (Dracup, 1991). The adaptation capacity of callus cultures to salt stress could depend on the tolerance level of the genotype and the salt level used. According to our results, we can conclude that the salt-tolerant wild species is able to adapt very rapidly to a salt medium (after only 3 passages).

The cell volume changes at the end (day 28) of the 1st, 2nd and 4th subculture are shown (Fig. 2). In the 1st subculture, cells grown in control medium exhibited variability, showing two major groups, a feature which did not occur in saline medium. As the subculture advanced, the salt-cultured cells acquired a higher and more spherical volume; thus in the 4th subculture, the cell volume was 4 times higher in salt-adapted than in control cells. The reduction of cell volume in response to salt stress has been reported in salt-adapted cells of glycophytes (Binzel et al., 1985; Hasegawa et al., 1986; Iraki et al., 1989). However, the adaptation to salt in the halophyte *Atriplex nummularia* did not result in a reduced rate of cell expansion or a reduced maximal cell size (Casas et al. 1991) in spite of the fact that the high salt level to which the halophytic cells adapted (342 mol m^{-3} NaCl) induced a growth reduction. In this study, the cell volume not only was maintained but also was enhanced, a fact which could be due to the lower salt level used (175 mol m^{-3} NaCl).

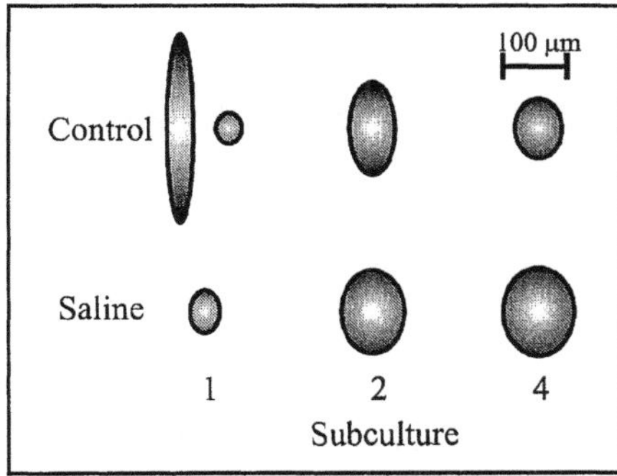

Fig 2. Cell volume changes along the subcultures of *L. pennellii* cells proceeding from calluses grown on control and saline (175 mol m^{-3} NaCl) media at the end of the growth cycle (day 28). Cell volume was estimated on the basis of length and width of the cells.

L. pennellii was able to adapt more rapidly to the salt medium than to the basal medium, according to the callus growth and cell volume changes. This could be due to the availability of saline ions to adjust osmotically; thus, the enhanced growth of salt-adapted cells in the 4th subculture was related to the increased Na^+ and Cl^- uptake (Table 1). The capacity of inclusion of saline ions has also been described in the wild species at the whole plant level (Perez-Alfocea et al., 1993).

Table 1. Water (g g^{-1} DW) and ion saline (μmol g^{-1} DW) content changes in *L pennellii* calluses grown on control and saline media (175 mol m^{-3} NaCl) at the end of the growth cycle (day 28). In each column, mean values having the same letter are not statistically different ($p < 0.05$).

	Subculture	Water	Na^+	Cl^-
Control	1	21.9 a	173 c	270 c
	2	20.0 b	235 c	263 c
	3	17.6 c	203 c	261 c
	4	16.5 c	138 c	247 c
Saline	1	20.9 ab	2895 b	2425 b
	2	17.2 c	2957 b	2821 b
	3	16.3 c	2759 b	2689 b
	4	16.8 c	4139 a	3757 a

The water content was reduced by salinity only in the 2nd subculture, whereas in the other subcultures, there was no significant differences between the water contents of the calluses grown in control and saline media (Table 1). The high salt tolerance of *L. pennellii* seems to be related to the capacity to maintain the water content, as it occurred in salt-adapted cells of tobacco, which showed an enhanced capacity to retain water on exposure to NaCl during the process of adaptation (Hasegawa et al., 1986).

In conclusion, the wild species was able to adapt quickly to salt stress. Moreover the adaptation of *L. pennellii* to 175 mM NaCl resulted in increased growth rate and cell size. These results, as well as the higher capability to retain water, contribute to diluting the high accumulation of saline ions in salt-adapted cells of the wild tomato species, and are indicative of an halophytic behaviour in a saline environment.

Acknowledgments

The authors are grateful to "Ministerio de Educación y Ciencia" for the "Acción Integrada" HB96-0083.

References

Binzel ML et al (1985) Plant Physiol. 79, 118-125.
Blits KC et al (1993) Journal of Exp. Bot. 44, 681-684.
Cano EA et al (1996) Plant Cell Reports 15, 791-794.
Casas AM et al (1991) Plant Cell Reports 10, 81-84.
Dracup M (1991) Aust. J. Plant Physiol. 18, 1-15.
Hasegawa PM et al (1986) HortScience 21, 1317-1324.
Iraki NM et al (1989) Plant Physiol. 91, 39-47.
Perez-Alfocea F et al (1993) Physiol. Plant. 87, 493-498.
Perez-Alfocea F et al (1994) J. Hort. Sci. 69, 639-644.
Tal M et al (1978) Z. Planzenphysiol. 86, 231-240.
Tal M (1993) Acta Hort. 336,69-78.

Biochemical characterization of the Pto protein kinase conferring speck disease resistance in tomato

Guido Sessa, Mark D'Ascenzo and Gregory B. Martin

Department of Agronomy, Lilly Hall of Life Sciences,
Purdue University, West Lafayette, IN 47907-1150, USA.

Summary

In tomato plants, the Pto kinase confers resistance to the bacterial speck pathogen, *Pseudomonas syringae* pv. *tomato*, that expresses the avirulence gene *avrPto*. The resistant phenotype is the final result of a phosphorylation cascade triggered by the specific recognition between the AvrPto signal molecule and the Pto kinase, which is thought to function as an intracellular receptor. As a first step in the investigation of the molecular mechanisms leading to Pto activation, we characterized *in vitro* the biochemical properties of Pto autophosphorylation and phosphorylation of its putative effector Pti1. Pto autophosphorylated at multiple sites, as determined by phosphopeptide mapping. In addition, Pto autophosphorylation occurred *via* an intramolecular mechanism, as the specific activity of the kinase was not affected by its molar concentration in the assay. Kinetic analysis of Pti1 phosphorylation by Pto revealed that this reaction proceeds with a K_m of 4.1 μM and V_{max} of 0.55 nmole/min/mg. Finally, Pto phosphorylated Pti1 at one major and at several minor sites.

Introduction

In many plant-pathogen interactions the activation of the defense response is mediated by a specific recognition involving the product of an avirulence (*avr*) gene in the pathogen and the corresponding resistance (*R*) gene in the plant (Flor, 1973).

Proteins encoded by *R* genes are postulated to function as receptors that bind the cognate *avr* gene product and activate defense responses. An excellent paradigm for this type of mechanism is offered by the *Pto*-mediated disease resistance in tomato (Fig. 1). The *Pto* gene encodes a serine/threonine protein kinase and confers resistance specifically to the bacterial pathogen *Pseudomonas syringae* pv. tomato expressing the avirulence gene *avrPto* (Martin et al., 1993). A direct interaction occurs between the Pto product and the product of the *P. syringae* pv. tomato *avrPto* gene when tested in the yeast two-hybrid system (Tang et al., 1996; Scofield et al., 1996).

The recognition between the tomato Pto kinase and the bacterial AvrPto protein initiates a signal transduction pathway that involves downstream effectors and ultimately leads to disease resistance. The *Prf* gene is required for Pto-mediated resistance and encodes a protein with a leucine zipper, a nucleotide binding site, and

A. Altman et al. (eds.), Plant Biotechnology and In Vitro Biology in the 21st Century, 541–544.

leucine-rich repeats (Salmeron et al., 1996). However, the role of Prf in the Pto pathway remains unclear. Other putative effectors were identified by their specific interaction with the Pto kinase in the yeast two hybrid system. Among them are Pti1, a serine/threonine protein kinase that is specifically phosphorylated *in vitro* by Pto and is involved in the hypersensitive response (Zhou et al., 1995), and Pti4, Pti5 and Pti6, putative transcription factors which are similar to the tobacco ethylene-responsive element binding proteins (Zhou et al., 1997).

To start investigating the role of Pto kinase activity in the molecular mechanisms leading to Pto activation, we characterized *in vitro* the biochemical properties of Pto autophosphorylation and phosphorylation of its putative effector Pti1.

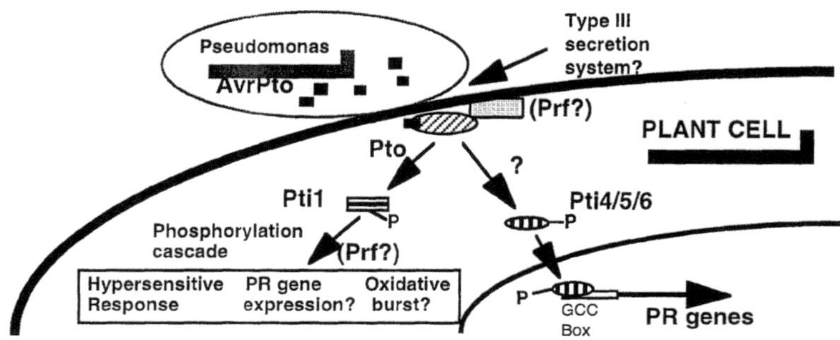

Fig. 1. **A model for signal recognition and transduction leading to speck disease resistance in tomato.**

Results and Discussion

Pto autophosphorylation occurs through an intramolecular mechanism

The Pto protein is an active serine/threonine kinase when tested *in vitro* (Loh et al., 1995). In addition, its interaction within the plant cell with the AvrPto protein defines Pto as an intracellular receptor or as part of a receptor complex (Tang et al., 1997; Scofield et al., 1997). To further investigate the mechanism of Pto activation, we tested whether Pto autophosphorylation is intramolecular (first order with respect to enzyme concentration) or intermolecular (second order with respect to enzyme concentration). Pto was expressed in bacteria as a maltose binding protein fusion (MBP-Pto), and the effect of its molar concentration on the autophosphorylation reaction was studied. The rate of autophosphorylation was linear with respect to enzyme concentration (Fig. 2A), and the phosphate incorporation per molecule was constant when MBP-Pto concentration varied by 60-fold (Fig 2B). These data indicate that MBP-Pto autophosphorylation is an intramolecular mechanism.

An intramolecular mechanism for Pto autophosphorylation was further supported by the evidence that an active glutathione *S*-transferase-Pto fusion protein failed to phosphorylate an inactive MBP-Pto molecule (Sessa et al., 1998).

These results make unlikely a Pto activation similar to that documented in mammals for extracellular receptors with tyrosine kinase activity (Heldin, 1995). Ligand binding to the extracellular domain of these receptors induces dimerization which in turn activates intermolecular authophosphorylation, mediating interactions between the activated receptor and cytoplasmic proteins. Alternative modes of Pto activation by AvrPto include induction of conformational changes that expose sites, not previously available, to autophosphorylation or phosphorylation by an additional kinase. Pto autophosphorylation activity of Pto may also represent a prerequisite for the interaction with AvrPto. Interestingly, a potential site for Pto autophosphorylation, threonine 204, is required for Pto-AvrPto interaction in the yeast two-hybrid system and for recognition specificity in a tobacco leaf transient assay (Frederick et al., 1998).

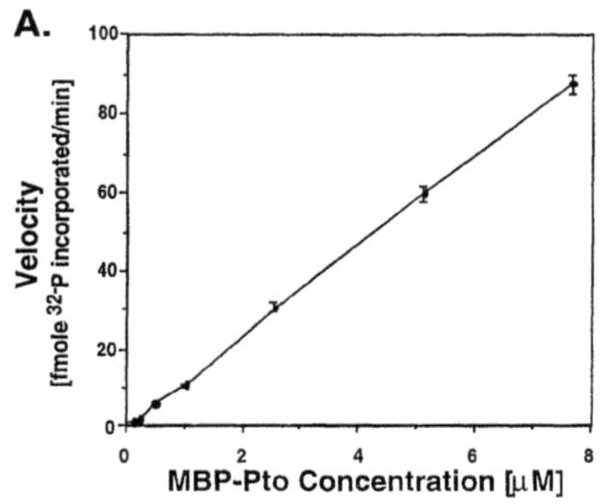

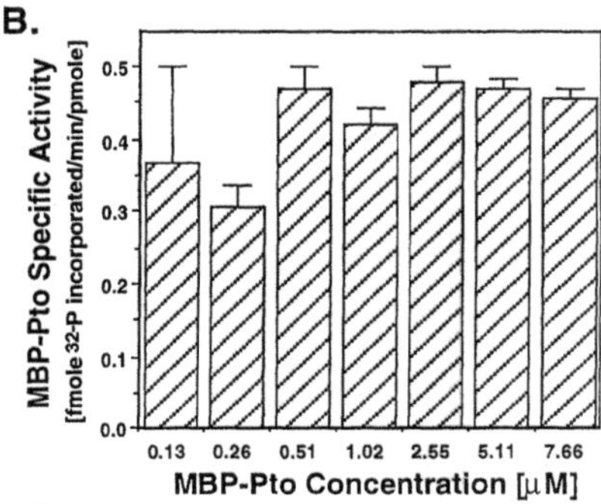

Fig. 2. **Effect of enzyme concentration on the autophosphorylation of MBP-Pto fusion protein.** Autophosphorylation activity of MBP-Pto was tested in an *in vitro* kinase assay at enzyme concentrations ranging from 0.72 to 17.9 μM. *A*, Plot of phosphate incorporation rate versus MBP-Pto concentration in the assay. *B*, Specific activity of MBP-Pto at different enzyme concentrations.

Kinetic analysis of Pti1 phosphorylation by Pto

The Pti1 kinase was previously shown to be specifically phosphorylated *in vitro* by the Pto kinase (Zhou et al., 1995). To study the kinetics of this reaction, the initial velocity of the phosphorylation of a kinase deficient mutant GST-Pti1(K96N) by Pto-MBP was analyzed at different substrate concentrations (Sessa et al., 1998). Non-linear least squares fitting analysis of the data estimated K_m and $V_{\max}$ values for the GST-Pti1(K96N) substrate as 4.1±0.6 μM and 0.55±0.03 nmole/min/mg, respectively.

Tryptic phosphopeptide mapping of Pto and Pti1

To investigate Pto autophosphorylation sites in more detail, MBP-Pto fusion protein was autophosphorylated *in vitro* and digested with trypsin as described (Sessa et al., 1998). The tryptic digest was resolved horizontally by two dimensional electrophoresis and chromatography. Digestion of autophosphorylated MBP-Pto generated one major and at least four minor phosphopeptides (Fig. 3A). The ratio between the intensity of the major spot and minor spots was about 5:1. The variable intensity of the labeling of different peptides might be due to the degree of enzyme affinity for different sites and to the number of phosphorylation sites in each peptide.

To examine the Pti1 sites phosphorylated by Pto, the kinase deficient mutant GST-Pti1(K96N) was phosphorylated by an MBP-Pto fusion protein and then digested by trypsin. A phosphopeptide map of the Pti1 digestion products revealed the presence of one major and several minor phosphorylated peptides. The ratio between the intensity of the major spot and minor spots was about 50:1.

The determination of phosphopeptide maps for Pto autophosphorylation and Pti1 phosphorylation by Pto represents a first step toward identification of phosphorylated residues in these proteins and examination of their role in resistance.

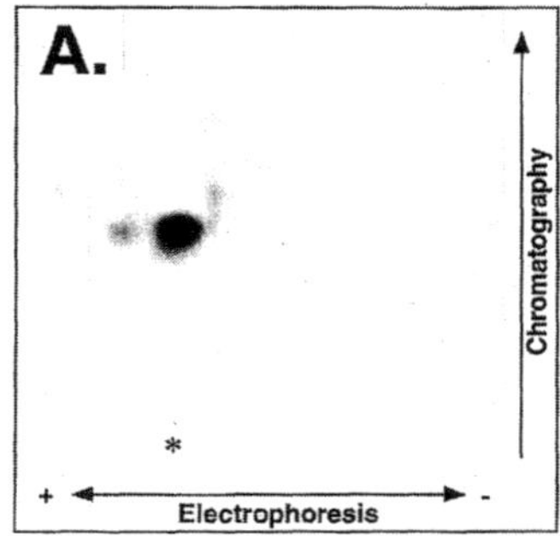

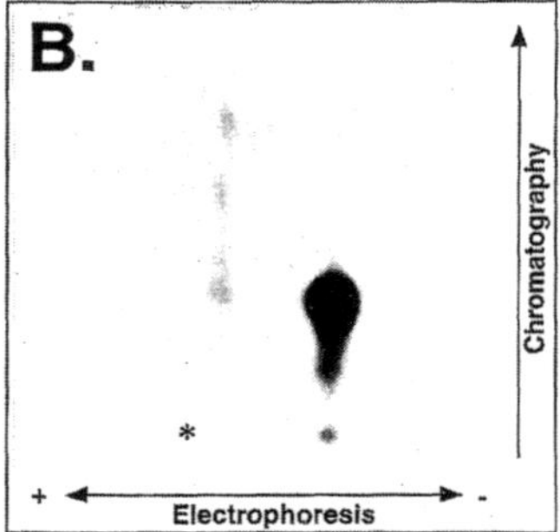

Fig. 3. Tryptic phosphopeptide mapping of phosphorylated Pto and Pti1. Autoradiograms of tryptic phosphopeptide maps of *A*, Autophosphorylated MBP-Pto; *B*, Kinase-deficient GST-Pti1(K96N) phosphorylated by MBP-Pto. Proteins were phosphorylated *in vitro*, digested by trypsin and loaded onto cellulose TLC plates. Peptides were then fractionated horizontally by thin layer electrophoresis at pH 4.7 and vertically by chromatography. The origin is indicated by an asterisk in each plate.

Acknowledgments

This research was supported, in part, by Postdoctoral Award No. FI-248-97 from BARD, the United States-Israel Binational Agricultural Research and Development Fund (to G.S.), National Science Foundation grant MCB-93-03359 (to G.B.M.), and a David and Lucile Packard Foundation Fellowship (to G.B.M.).

References

Flor, H. H. (1971) Annu. Rev. Phytopathol. 9, 275-296.
Frederick R D et al (1998) Mol. Cell, in press.
Heldin, C-H (1995) Cell 80, 213-223.
Loh Y-T, Martin, G B (1995) Plant Physiol. 108, 1735-1739.
Martin G B et al (1993) Science 262, 1432-1436.
Salmeron J M et al (1996) Cell 86, 123-133.
Sessa G et al (1998) J. Biol. Chem., in press.
Scofield S R et al (1996) Science 274, 2063-2065.
Tang X et al (1996) Science 274, 2060-2063.
Zhou J et al (1995) Cell 83, 925-935.
Zhou J et al (1997) EMBO J. 16, 3207-3218.

IN VITRO ELIMINATION OF PRUNUS NECROTIC RINGSPOT VIRUS IN A PLUM CULTIVAR

S. Spiegel[1], Y.Tam[1], A. Rosner[1], M. Brison[2], B. Helliot[2], A. Pierronnet[3], J.P. Eyquard[3], M.T. de Boucaud[2]
[1]Dept. of Virology, The Volcani Center, Bet-Dagan, Israel. [2]Lab. de Physiologie Cellulaire Vegetale, Univ. de Bordeaux I, Talence Cedex, and [3]INRA, Unite de Recherche sur les Especes Fruitieres et la Vigne, Villenave d'Ornon cedex, France

Production of vegetatively propagated plant material is based on building up stocks from source plants. Viruses entering a propagation system at the breeding or germ plasm level are transmitted vertically down the propagation chain . The ultimate outcome of this process is virus-infected planting material. Elimination of viruses at the beginning of the propagation chain will generally result in virus-free regenerated plants which can serve as a source for propagation. The most effective method currently used for elimination of viruses from plant material is a combination of thermotherapy followed by shoot-tip culture. Chemotherapy has also been applied for virus elimination (Walkey, 1991). Following exposure to a high temperature or chemotherapeutant for several weeks, shoot-tips are excised from treated plants and either regenerated in vitro on a suitable culture medium into a new plant or grafted onto a virus-free plant. The virological status of the resulting plants is assesed several times over a given period of time during which virus-infected plants are discarded. The success of the three methods mentioned above, applied traditionally to developed plants grown in pots depends, among others, on the regenerability of small apices (1 mm or less), heat tolerance of treated plants and phytotoxicity of chemotherapeutants. Several woody species (e.g. *Prunus*) are very sensitive to heat and their exposure to elevated temperatures result in poor survival. Recently, cryopreservation was evaluated for it's potential to eliminate plum pox virus from a *Prunus* rootstock (Brison et al., 1997). This method involves chilling excised shoot-tips to -196°C in a protective solution and regrowing them in vitro on a suitable medium. Applying a virus elimination procedure to numerous apices in several hours rather than weeks may shorten the process considerably and is worthwhile evaluating for various host - virus systems.

Therapy procedures, increasingly applied to in vitro grown plant materials allow controlled modifications of media components and environmental factors which improve survival and regeneration of virus free apices (Stein et al., 1991; Spiegel et al., 1995).

In this study, in vitro grown shoots of plum (cv. Ogden) infected with prunus necrotic ringspot virus (PNRSV) were subjected to the following virus elimination treatments:

A. Altman et al. (eds.), Plant Biotechnology and In Vitro Biology in the 21st Century, 545–547.

(a) Shoot-tip culture - excision of shoot-tips (0.5-5 mm long) from shoots cultures grown at 23°C and regeneration in vitro.
(b) Thermotherapy - alternating high (38°C for 16 hrs, light) and low (28°C for 8 hrs, dark) temperatures for twenty eight days.
(c) Chemotherapy - shoot cultures grown for twenty eight days at 23°C on a culture medium with virazole (20 mg/1L medium).
(d) Combination thermo-/chemotherapy - temperature regime (as in b) applied to shoot cultures grown on medium with virazole (as in c) for twenty eight days.
Survival of shoots was monitored visually. Following treatments (b), (c) and (d), apices (about 5 mm long), were excised from the main shoot-tips (ms) and axillary buds (ab), subcultured and regenerated. Virus status in regenrated shoots, maintained clonaly, was determined at time intervals by ELISA and by RT-PCR. PNRSV-positive shoots (as derermined by ELISA) were discarded. Virus-free shoots (ELISA and PCR-negative) have been subcultured every month and following a dormancy period will be retested. Results in Table 1show that regenerated shoots from 1 mm long apices were infected with PNRSV. This finding suggests that PNRSV is present in the meristems and/or first leaf primordia and therefore shoot-tip culture is not a suitable method. The same applies for virazole in the medium. Alternating high and low temperatures and the combination of virazole and heat treatment are efficient and result in more than 50% of virus free main shoots and more than 80% healthy axillary buds which developed during the treatment. Additional thermotherapy and combined thermo-/chemotherapy experiments are underway. The possibility to use cryopreservation procedures for PNRSV elimination is also being studied.

Acknowledgments
This research was supported in part under Grant No. C16-059, U.S. Israel CDR program, USAID and Arc en Ciel - Keshet 1997, France - Israel Program.

References

Brison M et al (1997) Plant Science 123, 189-196.
Spiegel S et al (1995) Acta Hortic. 386, 419-420.
Stein A et al (1991) Ann. appl. Biol. 119, 265-271.
Walkey DGA (1991) Production of virus-free plants, pp 270-292 in: Applied Plant Virology. John Wiley and Sons, New York.

Table 1. Effect of virus elimination procedures on survival, and infection with PNRSV in plum cv. Ogden shoot cultures grown in vitro.

Treatment	No. of treated	No. of regenerated infected/survivors
shoot-tip 0.5-5 mm excision	91	91/91
Alternate high/low temp.	57	15/36ms 3/21ab
Virazole & Alternate high/low temp	24	0/14ms 0/ 4ab

*main shoot (ms); **axillary buds (ab)

RESISTANCE TO TOMATO SPOTTED WILT VIRUS IN TRANSGENIC TOMATO VARIETIES

P. Stoeva[1], D. Hristova[2], P. Donkova[1], M. Petrova[1], N. Gorinova[1], M. Yankulova[1], D. Inze[3], W. van Kamp[3], A. Atanassov[1]*
[1]Institute of Genetic Engineering, 2232 Kostinbrod-2, Bulgaria; [2]Institute of Plant Protection, 2232 Kostinbrod-2, Bulgaria, [3]VIB, RUG, B-9000, Gent, Belgium

1. Introduction

Tomato spotted wilt virus (TSWV) is a thrips transmitted plant virus which infects more than 650 plant species belonging to different botanical families (Peters et al.1990, Goldbach, Peters 1994). Tomato appears to be one of the most vulnerable to TSWV economically important plant species due to its prolonged cultivation period in the greenhouse and the monoculture field cultivation. TSWV is among the most important viral diseases affecting tomato and causing significant economic losses. The infected plants are stunted and with reduced yield, the fruits are blemished with necrotic and chlorotic ringspots, making them unmarketable. Sources of genetic resistance have been reported to exist in genus Lycopersicon and tomato cultivars descending from such accessions have been described to be resistant to TSWV and not isolate specific (Stevens et al., 1994; Boitieaux, Giordano 1993). However in the field, in plants carrying a resistance gene from L. peruvianum, the virus still accumulates resulting in the development of disease symptoms on the fruit (Ultzen et al. 1995; Rosello et al. 1996). Efficient engineered pathogen derived TSWV resistance in tomato is reported by Ultzen et al. (1995) and Kim et al.(1994).
The objective of this research is to study the resistance to TSWV in transgenic tomato genotypes incorporating two different chimaeric genes: the pathogen-derived nucleoprotein (N) gene from the Bulgarian tobacco isolate of TSWV L3 (Maiss et al. 1998) and the mitochondrial MnSOD gene from *N. plumbaginifilia* (Bowler et al. 1991).

2.1. Materials and Methods

2.1.1. Plant material

In *Agrobacterium*-mediated transformation experiments (unpublished results) were included *in vitro* plants from the Bulgarian tomato field varieties Helios and Miliana. and the selection lines 4d and 6d developed in the Institute of Genetics, Sofia. Primary transgenic plants (R_0) were micropropagated and transferred to soil and grown under greenhouse conditions. R_1-R_3 progeny lines were produced by selfpollination of selected individual plants.

2.1.2 Protein analysis in transgenic tomato plants

The N - protein analysis was carried as described (Stoeva et al. 1998). Tests were considered positive when the absorbance value at OD_{405nm} was at least twice that of the healthy control. Transgenic plants with extract optical density readings lower or equal to this value were considered as nonexpressors, those with higher values were classified as expressors.

A. Altman et al. (eds.), Plant Biotechnology and In Vitro Biology in the 21st Century, 549–552.

2.1.3. Tospovirus strain and inoculation of transgenic and control plants.
Bulgarian TSWV-isolate (1D-94) from greenhouse infected tomatoes was maintained by grafting on tomato plants. The inoculum was prepared and applied as described (Stoeva et al. 1998). The mechanical inoculation was carried out on R_0 plants after adapting the plants to soil at 2 - 6 leaf stage, and on $R_1 - R_2$ seedlings at the stage of 2 well developed leaves. A second inoculation was carried out 2 weeks later. Only plants remaining symptomless following two challenge inoculations over a period of 30 days were classified as resistant. Spread of infection was monitored at three-day intervals. The transgenic plants exhibiting symptoms after the second mechanical inoculation were counted as susceptible.

2.1.4. Isozyme analysis
Isozymes were extracted from leaves in two volumes (v/w) of the extraction buffer: 0.05M Tris-HCl, pH 6.8, 0.2% 2-mercaptoethanol and 12 % glycerol. The isozymes were separated by discontinuous polyacrylamide gel (10%) electrophoresis (PAGE). The electrode buffer system used was Tris-glycine, pH 8.3. The enzyme-activity stains for superoxide dismutase (SOD) were prepared according to Beauchamp, Fridovich (1971).

2.1.5. DNA analysis
Total DNA from leaves was isolated according to Fulton et al. (1995). Ten μg of DNA were digested with *Dra*I restriction enzyme (there is no *Dra*I restriction site in T-DNA inserts) (Boehringer Mannheim) and fractionated by electrophoresis on 0.8% agarose gels. The fractionated DNA was blotted and hybridized to a DIG-labeled NPT II gene according to the manufacturer's protocols of the Gene Screen Plus membranes (DuPont) and of the Non-radioactive Labeling and Detection of Nucleic Acids system (Boehringer Mannheim). The PCR analysis for the NPT II gene was performed according to Wang et al. (1993)

2.1.6. Analysis of transgenic plants
All putative transformants rooting on selective media were tested for the expression of the NPT II selectable marker gene (McDonnell et al. 1987) or by PCR. All NPT II positive, N gene carrying primary transformants, originating from the different cultivars were analyzed by ELISA for the presence and the level of accumulation of the of viral N protein. R_0 plants carrying either of the chimaeric genes were evaluated for resistance to TSWV by mechanical inoculation under controlled conditions. Thirty seedlings per $R_1 - R_2$ progeny line carrying the N-gene, were screened for the translational product of the N transgene and then were evaluated for resistance to TSWV. A second assay of nucleoprotein level was carried out in the selected TSWV resistant $R_1 - R_2$ plants. Thirty seedlings per $R_1 - R_3$ progeny lines carrying the MnSOD gene were evaluated for resistance to TSWV by mechanical inoculation under controlled conditions. Three assays of the nucleoprotein level were carried out in the selected TSWV resistant $R_1 - R_3$ plants 30, 60 and 90 days post inoculation. Selected plants from $R_1 - R_3$ progeny lines were analyzed for the SOD enzyme. Selected plants were analyzed by PCR and DNA gel blot analysis for the integration of the NPT II selectable marker gene.

2.2. Results and discussion

On the basis of the developed genotype independent protocol for genetic transformation of tomato, the N gene of TSWV and MnSOD gene from *N. plumbaginifolia* were introduced in different tomato varieties and lines. NPT II positive primary transgenic plants were obtained with both constructs. Thirteen transformants carrying the MnSOD gene and 14 carrying the N-gene were screened for resistance to the TSWV isolate 1D-94 upon mechanical inoculation. Among the R_0 plants carrying the MnSOD 53.85% were resistant, while the N-gene protected primary transformants were 71.43% (Table 1).

Table 1. Resistance to TSWV in transgenic tomato plants

Transgenic plants and progeny lines	% TSWV protected plants
Carrying the MnSOD	
R_0	53,85
R_1 - Helios/1	77,34
R_2 - Helios/1	78.34-100
R_3 - Helios/1-1	100
R_1 - Helios/2	65,21
R_2 - Helios/2	37,50-100
R_1 - Helios/4	57,14
R_2 - Helios/4	46,15-100
Carrying the N gene	
R_0	71,43
R_1 - Helios/3	73,33
R_2 - Helios/3	40,00-75.00
R_1 - line4d/8	62,5
R_2 - line4d/8	34,09-36.36
R_1 - Miliana/7	83,33
R_2 - Miliana/7	25.00-92,31
R_1 - line 6d/9	57,14
R_2 - line 6d/9	30,00-55.17
Control	0

The established resistance to TSWV in the R_1 progenies of transgenic plants carrying the MnSOD gene varied from 57.14 to 77.34 % suggesting the incorporation of one to several transgene inserts. In R_2 the percentage of the resistant plant was in the range 37.5 - 100% depending on the progenitor transformant. In R_2 were selected 100% TSWV protected progenies and the inheritance of the full resistance was proven in R_3 generation (Table 1). The ELISA of individual symptomless transgenic plants 30 - 60 days postinoculation showed that some of the selected plants performed as tolerant to the virus - the OD_{405nm} readings were positive. All plants remained simptomless during their development. The third immunoprotein analysis after 90 days showed that all plants were virus free. The integration of the selectable marker gene was proven by Southern blot analysis carried out for the NPT II PCR product of individual resistant transgenic plants from R_2 Helios/2 and R_3 Helios/3 (Table 1). The Southern blot analysis of *Dra*I digested DNA from the completely protected line R_3 Helios/1 was performed with the NPT II was homozygous for the transgene gene. The analysis revealed that the progeny and had three inserts. PAGE analysis was performed on individual resistant R_1 - R_3 plants originating from different primary transformants. In resistant R_1 progeny plants originating from transgenic plant R_0 Helios-1 an additional isozyme band was detected in comparison to the control. The study of the isozyme pattern of the resistant transgenic plants from R_2 Helios/2 and R_3 Helios/1 lines did not confirm the established polymorphism. In comparison to the control a decrease in the intensity of some of the low mobility SOD bands was detected in the screened transgenic plants. These results indicate that the resistance to TSWV cannot be correlated to the accumulation of the MnSOD enzyme encoded by the transgene. The SOD overexpression in the chloroplasts has been correlated to enhanced stress tolerance to ozone, low temperature, to metylviologen and photoinhibition (Bowler et al. 1991; Slooten et al. 1992; Van Camp et al. 1994; McKersie et al. 1993; Sen Gupta et al. 1993), indicating that a similar antioxidant response operates against different stress factors. Further studies are in progress to establish the role of the level of MnSOD (respectively the level of superoxide radicals) and the established resistance to TSWV.

The established resistance to TSWV in the R_1 progenies of transgenic plants carrying the N gene varied from 57.14 to 83.33 % suggesting the incorporation of different number of transgene inserts. In R_2 the percentage of the resistant plant was in the range 25 - 93.31% depending on the progenitor transformant (Table 1). Both expressor and nonexpressor tomato transformants were TSWV protected. There was no viral replication in the resistant transgenic plants and the OD_{405nm} readings taken before and post inoculation were similar. The integration of the selectable marker gene was proven by Southern blot analysis for the NPT II gene PCR product of individual

resistant transgenic plants from different R_2 and R_3 progeny lines. Further studies are in progress to select from all tomato genotypes completely protected lines. The obtained results demonstrated that transgenic tomato plants carrying the N-gene from tobacco isolate L3 are protected to the heterologous tomato isolate. Our results confirm the efficiency of the N-gene mediated protection establish for tomato (Ultzen et al. 1995, Kim et al. 1994) and other plant species like tobacco, lettice, peanut(De Haan et al. 1992; Gielen et al. 1991; Li et al. 1997; MacKenzie, Ellis 1992; Pang et al. 1992; 1993; 1996).
).

2.3. Conclusion

1. Transgenic tomato plants, carrying the mitochondrial MnSOD gene *from N. plumbaginifolia* (Bowler et al. 1991), are immune or tolerant to the Bulgarian greenhouse tomato isolate TSWV -1D upon mechanical inoculation. (Maiss et al. 1991)
2. Transgenic expressor and nonexpressor tomato plants are carrying the TSWV N-gene from the Bulgarian tobacco isolate L3 were completely protected to the heterologous greenhouse tomato isolate 1D-94.

3. References

1. Beauchamp, Fridovich (1971) Ann Biochem 44:276-278
2. Boitieaux, Giordano (1993) Euphytica 71, 151-154
3. Bowler et al. (1991) EMBO J 10:1723-1732
4. De Haan et al. (1992) Bio/Technology 10, 1133-1137
5. Fulton et. al. (1995) Plant Mol Biol Rep 13, 3, 207-209
6. Gielen et al. (1991) Bio/Technology 9, 1363-1367
7. Goldbach, Peters (1994) Seminars in Virology 5, 113-120
8. Kim et al. (1994) Plant disease 78, 615-621
9. Li et al.(1997) Transgenic Research, 6, 297-305.
10. MacKenzie D., Ellis, P (1992). Mol. Plant-Microbe Interact. 5, 34-40
11. Maiss et al. (1991) J Gen Virol 72, 461-464
12. McDonnell et al. (1987) Plant Mol Biol Rep 4, 380-386
13. McKersie et al. (1993) Plant Physiol 103:1155-1163
14. Pang et al (1993) Bio/Technology 11, 819-824.
15. Pang et al. (1992 Phytopathology 82, 1223-1229.
16. Pang et al. (1996) Plant Journal 9, 899-909.
17. Pang et al. Phytopathology 84, 234-249.
18. Peters et al. (1990) Virus-Thrips-Plant Interactions of TSWV. Proc USDA Workshop, Beltsville, MD., pp. 1-14. Springfield: National Technical Information Service
19. Rosello et al. (1996) Scientia Horticultirae,6-7:117-150
20. Sen Gupta et al. (1993) Plant Physiol 103:1067-1073
21. Slooten et al. (1995) Plant Physiol. 107:737-750
22. Stevens et al. (1994) Euphytica 80:79-84
23. Stoeva et al. (1998) Molecular Breeding 4:155-164
24. Ultzen et al. (1995) Euphytica 85:159-168.
25. Van Camp et al. (1994) Biotechnology 12:165-168
26. Wang Het al. (1993) Nucleic Acid Research, v.21, 17:4153-4154

Arabidopsis thaliana in Culture: A Powerful Tool to Decipher the Mode of Action/Target Sites of Herbicides with Antimetabolite Activity

M.V. Subramanian, E.W. Walters, O. Lyzwanski and D.L. Siehl
Novartis Crop Protection, 975, California Avenue, Palo Alto, California 94304, USA

Abstract

Arabidopsis thaliana in culture is a powerful tool to determine the mode of action of herbicides with antimetabolite activity. The culture media can be manipulated to identify various biosynthetic pathways blocked by probe compounds. For example, inhibition of Arabidopsis in culture caused by four standards, viz., asulam, glyphosate, sulcotrione, and pyrithiobac-sodium (PTB) were specifically reversed by p-aminobenzoate (PABA), aromatic amino acids, homogentisic acid and branched chain amino acids, respectively. These standards are known inhibitors of 7,8-dihydropteroate synthase (DPT synthase, in the folic acid biosynthesis pathway), 5-enolpyruvylshikimate-3-phosphate synthase (EPSP synthase, in the shikimate pathway), p-hydroxyphenylpyruvate dioxygenase (HPPD, in the plastoquinone biosynthesis pathway) and acetolactate synthase (ALS, in the pathway for branched chain amino acids), respectively. This technique was used to investigate two compounds with previously unknown modes of action. Inhibition of Arabidopsis growth by hydantocidin (Hy), and 6-methylanthranilate (MA), was specifically reversed by adenosine-5'-monophosphate (AMP, for Hy), and anthranilate or tryptophan (for MA). Hy was thus suspected to block purine biosynthesis, while MA was proposed to block the biosynthesis of tryptophan. Follow-up studies revealed that Hy and MA are proherbicides. The herbicidal forms were identified as hydantocidin-5'-phosphate (HP) and 4-methyltryptophan (4MT), respectively. Target sites for HP and 4MT were found to be adenylosuccinate synthetase (ADSS) in the purine biosynthesis pathway and anthranilate synthase (AS) in the tryptophan biosynthesis pathway. Observations made with Hy were also confirmed with a known inhibitor of ADSS, hadacidin (Ha). ADSS was inhibited competitively by both Ha and HP, but with respect to aspartate and inosine-5'-monophosphate (IMP), respectively. HP was found to be about two orders of magnitude more potent an inhibitor than Ha. Further, HP was co-crystallized with ADSS complexed to its substrates. A detailed analysis of the crystal structure should help in the design of new inhibitors as possible herbicides.

A. Altman et al. (eds.), Plant Biotechnology and In Vitro Biology in the 21st Century, 553–556.

Introduction

The study of mode of action of herbicides received a major thrust in the mid eighties with several agrochemical industries initiating research on their proprietary herbicides and lead molecules. It was felt that identifying the target sites and mechanism of action of herbicides should help in the discovery of novel compounds with new modes of action by: (a) Establishing structure-activity parameters to facilitate analog synthesis and lead optimization (Kleschick et al. 1992). (b) Initiating target site screens of chemical libraries to discover new inhibitors (Siehl et al. 1997) and (c) Intuitively recruiting new related target sites and establishing new screens (Halgand et al. 1998). In addition, understanding the biochemistry of herbicides and lead molecules should help in predicting toxic effects as well as the potential for development of resistance among weed populations (Subramanian et al. 1996).

Several experimental models including bacteria (LaRossa & Schloss 1984), pea roots (Ray 1984), cell culture (Subramanian et al. 1989) and *Arabidopsis* (Siehl et al. 1996) have been used for deciphering the mode of action of herbicides. *Arabidopsis*, in particular, is a powerful tool to decipher the mode of action of antimetabolites. There are several advantages in using *Arabidopsis* to study the mode of action of probe compounds: (a) It is very sensitive; hence potency of a probe compounds can be determined fairly accurately even if the target site is unknown (b) *Arabidopsis* growth media can be manipulated with specific components, just like bacterial media (Siehl et al. 1996) (c) There is virtually no contamination problem while handling *Arabidopsis* and the sterile seeds can be stored for more than four weeks at 4°C (d) Most biosynthetic pathways of the plant kingdom are elaborated in *Arabidopsis*, which thus accurately represents pathways in weeds. One exception here is in the first committed enzyme in fatty acid biosynthesis, acetyl CoA carboxylase. While most grasses have a multifunctional enzyme in the chloroplast which is sensitive to aryloxyphenoxypropionate and cyclohexanedione herbicides (grass herbicides), broad-leaved plants (Sasaki et al. 1995) including *Arabidopsis* have a plastidic multicomponent enzyme that is insensitive to grass herbicides (e) Tests with *Arabidopsis* can be conducted in microtiter format with several replicates (Siehl et al. 1996) (f) Finally, test results can be scored visually, in 5-7 days.

Materials and Methods

Arabidopsis thaliana was grown in 24-well microtiter plates as described previously (Siehl et al. 1996). Probe compounds and antidotes were dissolved in DMSO, or methanol or water, and added as appropriate, directly to the microtiter wells. Final concentration of DMSO did not exceed 2% v/v. Herbicide standards (Table 1) and probe compounds, MA, Hy, HP and Ha (Table 2) were synthesized by chemists at Novartis Crop Protection. Enzyme assays for inhibition of HPPD (Secor 1994), ALS (Subramanian et al. 1991), AS (Siehl et al. 1997) and ADSS (Walters et al. 1997) by sulcotione, PTB, MT and HP respectively, were conducted as described in literature.

Results and Discussion

Reversion of growth inhibition of *Arabidopsis* caused by four standards *viz.*, asulam, glyphosate, sulcotrione and PTB, whose target sites are known, were antidoted by appropriate metabolites. The results are shown in Table 1 along with the K_i or I_{50} values for the herbicides on the respective target sites.

Table 1

Herbicide	Target site / I_{50}[1]	Effect on *Arabidopsis*	Antidoting agent(s)/ Effect on growth inhibition
Glyphosate	EPSP synthase / 30 μM[2]	100% growth inhibition at 800 μM	Complete reversion by 0.5 mM tyr, trp and phe
Asulam	DPT synthase / 40 μM[3]	100% growth inhibition at 40 μM	Complete reversion by 0.2 mM p-aminobenzoate
Sulcotrione	HPPD / 0.8 μM[4]	Completely bleached plants at 0.8 μM	Complete reversion by 0.5 mM homogentisic acid
Pyrithiobac-sodium	ALS / 0.03 μM[4]	100% growth inhibition at 0.8 μM	Complete reversion by 0.3 mM val, leu, ileu

[1]Concentration of the herbicide required for 50% inhibition of the enzyme. [2]Padgette et al. (1991) and [3]Veerasekaran et al. (1981). [4]Present work based on HPPD from maize and ALS from cotton.

Inhibition of *Arabidopsis* caused by glyphosate was reversed only in the presence of all three aromatic amino acids. PABA, though a shikimate pathway derived product, was not required for reversion. Inhibition of *Arabidopsis* caused by asulam was reversed only by PABA. Bleaching of *Arabidopsis* caused by sulcotrione was reversed specifically by homogentisic acid and not by tyrosine. All three branched chain amino acids were required for complete reversion of *Arabidopsis* inhibition caused by PTB. These results are consistent with the inhibition of specific target sites in various pathways by the standards.

The technique of antidoting *Arabidopsis* growth inhibition was used to study the mode of action of two compounds with previously unknown modes of action, *viz.*, MA and Hy. *Arabidopsis* growth inhibition caused by MA was reversed completely by anthranilate or tryptophan (Table 2). Thus, it was proposed that MA was converted to 4MT, which blocked the biosynthesis of tryptophan by inhibiting AS. Subsequently, it was found that MA was readily converted to 4MT by *Arabidopsis* seedlings (data not shown). Further, it was shown that AS was inhibited by DL-4MT ($I_{50} = 2$ μM) as well as tryptophan ($I_{50} = 5$ μM), but not by MA (Siehl et al. 1997).

Arabidopsis growth inhibition caused by Hy was fully reversed by yeast extract (Siehl et al. 1996). Further analysis revealed that growth inhibition was reversed by AMP, adenine or adenosine but not by IMP and guanosine-5'-monophosphate (GMP, Siehl et al. 1996). The same pattern of reversion was observed with Ha, a known inhibitor of ADSS, suggesting that Hy may also inhibit ADSS. However, ADSS and adenylosuccinate lyase, the enzymes involved in the biosynthesis of AMP from IMP, were not inhibited by Hy (Siehl et al. 1996). Based on the fact that all intermediates in the purine biosynthetic pathway have a 5'-phosphate group on the ribose moiety, it was hypothesized that Hy is a proherbicide and that the actual inhibitor could be HP. Consistent with this hypothesis, a synthetic sample of HP inhibited ADSS with an I_{50} of 0.2 μM (Siehl et al. 1996). HP also showed the same reversion pattern with respect to inhibition of *Arabidopsis* (Table 2). HP was found to be a competitive inhibitor of ADSS with respect to IMP (K_i = 0.05 μM, Walters et al. 1997). In contrast, hadacidin is competitive with respect to aspartate and is about two orders of magnitude less potent that HP. Recently, in collaboration with Dr. R. Honzatko of Iowa State University, HP was co-crystallized with ADSS complexed with all of its substrates (Poland et al. 1996). A detailed analysis of the crystal structure should help design new analogs of HP as possible herbicides.

Table 2

Probe Compound	Effect on *Arabidopsis*	Antidoting agent(s)/ Effect on growth inhibition	Identified Target site / I_{50}
MA	100% growth inhibition at 50 μM	Complete reversion by 0.2 mM anthranilate or trp	AS / 2 μM (for 4MT)
Hy	100% growth inhibition at 1.2 μM	Complete reversion by 0. 5 mM AMP, adenosine or adenine	ADSS / No inhibition (0.2 μM for HP)
HP	100% growth inhibition at 13 μM	Same as above	ADSS / 0.2 μM
Hadacidin	100% growth inhibition at 70 μM	Same as above	ADSS / 5 μM

References

Halgand F et al. (1998) Biochemistry 37, 4773-4781.
Kleschick WK et al. (1992) ACS Symposium Series 504, 10-16.
LaRossa RA and Schloss JV (1984) J. Biol. Chem. 259, 8753-8757.
Poland BW et al. (1996) Biochemistry 35, 15753-15759.
Padgette et al. (1991) J. Biol. Chem. 266, 22364-22369.
Ray TB (1984) Plant Physiol. 75, 827-831.
Sasaki Y et al. (1995) Plant Physiol. 108, 445-449.
Secor J (1994) Plant Physiol. 106, 1429-1433.
Siehl DL et al. (1996) Plant Physiol. 110, 753-758.
Siehl DL et. al. (1997) Weed Science 45, 628-633.
Subramanian MV et. al. (1989) Prospects for Amino acid Biosynthesis Inhibitors in Crop Protection and Pharmaceutical Chemistry. BCPC Monograph No. 42, 97-100.
Subramanian MV et al. (1991) Plant Physiol. 96, 310-313.
Subramanian MV et al. (1996) Proceedings of the Second International Weed Control Congress Vol II, 447-454.
Veerasekaran P et al. (1981) Pesticide Science 12, 330-338.
Walters EW et al. (1997) Plant Physiol. 114, 549-555.

PRODUCTION OF POTATO GENOTYPES WITH RESISTANCE TO POTATO VIRUS Y (PVY) BY BIOTECHNOLOGICAL METHODS

R. Thieme[1], T. Gavrilenko[2], T. Thieme[3] and U. Heimbach[4]

[1]Federal Centre for Breeding Research on Cultivated Plants, Institutes for Breeding and Breeding Methods of Crop Plants, D-18190 Groß Lüsewitz, Germany
[2]N. I. Vavilov Institute of Plant Industry 190000 St. Petersburg, Russia
[3]BTL Bio-Test Labor GmbH Sagerheide, D-18184 Sagerheide, Germany
[4]Federal Biological Research Centre for Agriculture and Forestry, D-38104 Braunschweig, Germany

1. Introduction

In potato breeding crossing programmes are performed for the incorporation of desirable traits of related potato species into the genepool of a given potato variety. For example, derivatives from *Solanum demissum* have been utilised in 77 % of all foreign gene introgressions carried out in the European cultivars because of its resistance to late blight and its positive contribution to agronomic performance (Hermsen 1988, 1994). Another promising source for increasing the level of resistance to pathogens in commercial cultivars is to use exotic sources such as wild *Solanum* species. However, some of these candidates cannot be crossed with the cultivated potato. Thus, approaches like bridging species, embryo rescue techniques or somatic hybridization were used to overcome sexual barriers. *S. brevidens* has been successfully used as a source of resistance to *Erwinia* soft rot, potato leaf roll virus, PVX and PVY *via* backcrossing of fertile interspecific somatic hybrids with *S. tuberosum* (Austin *et al.* 1988, Pehu *et al.* 1990).

The goals of the present study are (1) to screen wild potato species for their resistance to PVY, (2) to study the resistance mechanisms, and (3) to produce hybrid clones with a high level of PVY resistance by use of biotechnological methods.

2. Materials and Methods

2.1. Materials

Wild species were selected according to the information on their resistance to PVY in the literature. The potato species, breeding clones and cultivars were obtained from N. I. Vavilov Institute of Plant Industry (**VIR**), St. Petersburg, Russia; Genbank External Branch „North" Groß Lüsewitz (**GLKS**), Germany; Scottish Crop Institute Pentlandfield (**SCIP**) and Institutes for Breeding and Breeding Methods of Crop Plants, Groß Lüsewitz (**BAZ**) (Table 1). The dihaploid breeding clones T14, T17 and T67 are distinguished by desirable agronomical characteristics, i.e., high starch content, high yield, nematode resistance as well as wart and cooking quality. The potato cultivars and the dihaploid clone PDH40 were used as virus susceptible controls.

2.2. Methods

2.2.1. Protoplast fusion, identification and characterization of the somatic hybrids

Protoplast isolation, fusion and culture were as described by Thieme *et al.* (1997).

A. Altman et al. (eds.), Plant Biotechnology and In Vitro Biology in the 21st Century, 557–560.

The somatic hybrids were identified by isoelectric focussing of isozymes and by microsatellite analysis. The PCR reactions were carried out as described by Thieme, Hackauf (1998).

Ploidy levels of the somatic hybrids were estimated by flow cytometry. Somatic hybrids with tetraploid, hexaploid or octaploid level were transferred into the greenhouse to compare the morphological traits of parental clones and hybrids according to standard catalogues and to carry out backcrossing experiments.

Table 1: Accessions used in the present study

Species	Accessions	Resistance to PVY	References
S. etuberosum	VIR k-91411	+	Valkonen *et al.* 1992
S. tarnii	GLKS 96. 203.1		
S. cardiophyllum	GLKS 93. 6/26/2	-	Rothacker 1961
S. berthaultii	GLKS 96. 181/3	+	Rothacker 1961, Hawkes 1994
S. pinnatisectum	GLKS 71/1. 71/6	+	Rothacker 1961, Hawkes 1994
S. tbr PDH 40	CSIP	-	Gibson *et al.* 1988
S. tbr T 14	BAZ GL 6.063 108-86N		
S. tbr T 17	BAZ GL 6.086 006-86		
S. tbr T 67	BAZ GL 6.090 004-86N		
S. tbr cv. 'King Edward'	GLKS 35	-	Anonymous 1998a
S. tbr cv. 'Bintje'	GLKS 81	-	Anonymous 1998b

S. tbr = Solanum tuberosum

GISH was performed on hybrid mitotic metaphase chromosomes with total DNA of *S. etuberosum* as a probe directly labelled by nick translation with FITC-12-dUTP according to Jacobsen *et al.* (1995).

2.2.2. Assessment of PVY resistance

Mechanical inoculation of potted and *in vitro* plants was carried out using plant sap of tobacco plants infected with PVY^{O} or PVY^{N}. Virus transmission was carried out using apterous aphids (*Myzus nicotianae*) transferred to the plants with an acquisition time of 15 min being allowed. After separate inoculation and aphid transfers potted and *in vitro* plants were cultured for three weeks at 25°C. Virus transmission was assessed for thirty clonal repetitions per genotype by ELISA.

3. Results and Discussion

3.1. Screening of potato genotypes in relation to PVY resistance

Virus transmissions were not observed in three of the wild species using both mechanical and vector inoculation (Table 2). In contrast to the literature *S. berthaultii* seems to be susceptible to PVY. In contrast, *S. etuberosum* displayed the expected virus resistance (Table 2). *S. etuberosum* is a diploid (2n=2x=24), nontuberous species with virus and frost resistance (Valkonen *et al.* 1992, Novy, Helgeson 1994) which, however, cannot be crossed with the cultivated potato. Thus, *S. etuberosum* and the breeding clone T67 were selected for protoplast fusion.

3.2. Protoplast fusion, culture and hybrid identification
After fusion the first 178 regenerated shoots from 2567 calli were analyzed and a fusion success of 90% was achieved (Table 3). The clone T67 was characterised by a high combining ability and regeneration capacity in different fusion experiments for the production of intraspecific somatic hybrids.
In isozyme and microsatellite marker analysis the interspecific hybrids displayed the additive banding patterns of the parents.

Table 2: Frequence of PVY transfer (%) to *Solanum* spp. by mechanical inoculation and vector transfer to plants under greenhouse and *in vitro* conditions (n ≥ 30)

Species	**Potted**				***In vitro***			
	Mechanical		**Vector**		**Mechanical**		**Vector**	
	PVY^O	**PVY^N**	**PVY^O**	**PVY^N**	**PVY^O**	**PVY^N**	**PVY^O**	**PVY^N**
S. etuberosum	0	0	0	0	0	0	0	0
S. tarnii					0	0	0	0
S. cardiophyllum	0	0	0	0	0	0	0	0
S. berthaultii	80	93	7	3	36	80	3	3
S. pinnatisectum					93	96	40	0
S. tbr PDH 40	46	39	0	0	88	68	14	0
S. tbr T 14	36	35	0	36	80	85	19	0
S. tbr T 17			0	3	16	100	10	0
S. tbr T 67	47	33	3	14	35	97	11	9
S. tbr cv.'King Edward'	43	19	36	41	100	85	9	7
S. tbr cv. 'Bintje'	72	55	33	0	33	72		

3.3. Ploidy level and morphological characters of the somatic hybrids
Sixty % of the somatic hybrids were tetraploid (2n=4x=48, Table 3). Because of multifusion or polyploidization events 30 % of the hybrids were hexaploid (2n=6x=72) or octaploid (2n=8x=96). Fiftyfive tetraploid, seven hexaploid and five octaploid somatic hybrids were transferred into the greenhouse. A phenotypic variation was observed among these hybrids. Plant habit, flower and leaf characters were intermediate compared to those of the parents. Other traits like anthocyanin pigmentation were inherited as dominant characters traits of parental clones and the interspecific somatic hybrids.

Table 3: Regenerated plants and somatic hybrids of T67 (+) *S. etuberosum*

No. of calli	**No. of shoots**	**No. of shoots tested**	**No. of identified hybrids (fusion success)**	**Ploidy level**			
				4C	**6C**	**8C**	**6-8C**
2567	178	101	91 (90%)	60	14	15	2

3.4. Genomic *in situ* hybridization (GISH) analysis
Besides the tetraploid hybrids with the expected 24 chromosomes of each *S. tbr* and *S.*

etuberosum aneuploid hybrids were observed which had lost two potato chromosomes.

Table 4: Virus transmission (%) of 30 tetraploid somatic hybrids after mechanical PVY inoculation of *in vitro* plants (n = 30), *S. etuberosum* = 0%, clone T67 = 100%

	Virus transmission			
	0	**3-20**	**23-40**	**47-77**
No. of hybrids	8	12	7	3

3.5. Assessment of PVY resistance of the somatic hybrids

First results obtained with mechanical inoculation of thirty somatic hybrids of T67 (+) *S. etuberosum* reveal a large variation in PVY transmission (Table 4). Eight hybrids showed no virus transmission according to ELISA. These hybrids are currently tested by use of different inoculation methods under greenhouse and field conditions.

4. Conclusions

- A four-step test method of PVY transmission was used which includes mechanical inoculation and transfer of virus by aphids to potted and *in vitro* plants.
- PVY resistant and susceptible genotypes of *Solanum* spp. could be selected.
- Interspecific hybrids between a dihaploid potato breeding clone and the wild species *S. etuberosum* were produced by protoplast fusion.
- Using GISH interspecific hybrids with the expected chromosome composition and aneuploid genotypes were identified.
- Somatic hybrids showed variability in morphological characters as well as virus transmission.
- Future experiments will be performed to study the influence of *S. etuberosum* on the behaviour and the biology of different vectors for understanding the resistance mechanism.

5. References

Anonymous (1998a) Recommended list of potatoes including details of other varieties available. NIAB Potato Variety Handbook, pp. 33, Plumridge Linton, Cambridge.

Anonymous (1998b) 73. Rassenlyst Landbouw., p 180, CPRO-DLO Wageningen.

Austin SE *et al.* (1988) Phytopathology 78, 1216-1220.

Gibson RW *et al.* (1988) Theor. Appl. Genet. 76, 113-117.

Hawkes J. (1994) Potato Genetics, pp 515-538, CAB Intern. Wallingford, UK.

Hermsen JHTH (1988) The use of plant genetic ressources, pp 68-87, Cambridge University Press, UK.

Hermsen JGTH (1994) Potato Genetics, pp 515-538, CAB Intern. Wallingford, UK.

Jacobsen E et al. (1995) Heredity 74, 250-257.

Novy RG, Helgeson JP (1994) Theor. Appl. Genet. 89, 775-782.

Pehu E et al. (1990) Plant. Sci. 69, 95-101.

Rothacker (1961) Die Kartoffel Band 1, pp 353-558, Landwirtschaftsverlag, Berlin.

Thieme R et al. (1997) Euphytica 97, 189-200.

Thieme R, Hackauf B (1998) Vortr. Pflanzenzüchtg. 42, 149-151.

Valkonen J et al. (1992) Ann. Appl. Biol. 120, 301-313.

PLANT TOLERANCE TO WATER AND SALT STRESS: THE EXPRESSION PATTERN OF A WATER STRESS RESPONSIVE PROTEIN (BspA) IN TRANSGENIC ASPEN PLANTS

W. X. Wang, T. Tzfira, N. Levin, O. Shoseyov and A. Altman
The Kennedy-Leigh Centre for Horticultural Research and The Otto Warburg Center for Biotechnology in Agriculture, The Hebrew University of Jerusalem, Rehovot 76-100, Israel.

1. Introduction

Abiotic stress conditions, such as drought, salt, cold and heat, are major environmental factors affecting plant growth and productivity. Plants have developed several strategies to adapt or to tolerate stress conditions, one of which is the accumulation of compatible solutes helping to maintain an osmotic balance with environment (Rudulier et al 1984). Another important strategy is the expression of specific stress-responsive proteins which may function in plant tolerance (Dure 1993; Close 1997). A hydrophilic boiling stable protein (designated as BspA) has been identified by us as a major water stress and ABA-induced protein in aspen (*Populus tremula* L.) plants (Pelah et al 1995, 1997a). A differential expression pattern of BspA in poplar genotypes differing in their water stress tolerance, and a positive correlation between BspA accumulation and reduced ion leakage upon water stress was previously reported (Pelah et al 1997b), yet direct evidences for its precise mode of action are still lacking.

In order to further study the correlation between BspA expression and plant tolerance to water and salt stress we produced transgenic aspen plants over expressing a full-length *bspA* cDNA fragment. A large number of transgenic aspen plants, resulting from independent transformation events, were rescued and screened using PCR and northern blot analyses. A variant levels of *bspA* expression and BspA accumulation, as well as cellular response and growth performance under water and salt stress, were observed among several selected clones.

2. Materials and Methods

In vitro drought stress. *In vitro* transgenic and non-transformed (NT) plants were wilted under ambient conditions until 20% loss of their initial fresh weight. Stressed plants were then kept in closed petri-dishes for additional 3 h. Fresh plants served as control and kept in humid petri-dishes. All plants were subjected for either western blotting (according to Pelah et al 1995) or electrolyte leakage measurement. For osmotic potential measurement, the plants were first rehydrated in deionized water in vials for additional 24 h.

A. Altman et al. (eds.), Plant Biotechnology and In Vitro Biology in the 21st Century, 561–565.

In vitro salt stress. *In vitro* plants were cultured hydroponically, either in 150 mM NaCl or in deionized water, for 24 h. Treated plants were used either for western blotting or for direct osmotic potential measurement.

Ex vitro drought stress. Acclimatized potted plants were withheld irrigation until wilting, followed by rewatering for 48 h. The treatment was repeated twice, and afterwards plants were subjected to measurements of leaf abscission, stem elongation, water potential, osmotic potential and western blotting.

Osmotic potential measurements. About 150 mg fresh weight leaf tissue samples were collected in an Eppendrof tube and immediately frozen in liquid nitrogen. The sample were then thawed at room temperature for few minutes and centrifuged for 5 additional minutes. This procedure was repeated 2-3 times until tissue sap appeared. The osmotic potential of the collected tissue sap was measured in a WESCOR 5500 vapor pressure osmometer (WESCOR, INC).

Electrolyte leakage measurements. Shoots of individual stressed and non-stressed plants were immersed in 10 ml deionized water and the electrical conductivity of the solution was measured at 10, 20, 30, 60, 120 minutes. The samples (deionized water plus the shoot tissue) were then autoclaved for 4 min and the total conductivity of the solution was measured. The electrolyte leakage (E) at each time point (t) was expressed as % of total conductivity as: $E\%=[(Tt/Ta)-(Ct/Ca)]\times 100$ (Tt: conductivity at time t of treated plant; Ta: conductivity of autoclaved treated plants; Ct: conductivity at time t of control plants; Ca: conductivity of autoclaved control plants.

3. Results and Discussion

Osmotic adjustment of transgenic plants. Both drought stressed transgenic and NT *in vitro* plants reached 20% water loss at about the same time, exhibited a similar severe leaf wilting and similar recovery, after 24-h of rehydration. However, a decrease of leaf osmotic potential was observed in *bspA*-transgenic aspen, as compare with NT plants (Table 1), resulting in a decrease of about 400% in lines G3 and L3. *bspA*-transgenic plants exhibited a decrease in their osmotic potential also when subjected to salt stress, resulting in a about 220% decrease in line L3. A decrease in osmotic potential was also observed in *ex vitro bspA*-transgenic plants subjected to drought stress (Table 1), thus indicating better osmotic adjustment as compare with NT plants.

Table 1. Percentage decrease of leaf osmotic potential.

	In vitro culture plants		*Ex vitro* plants
Line	Drought stress	Salt stress	Drought stress
NT	4.9	14.6	38.7
E1	11.0	ND	41.5
G3	23.2	20.8	ND
H3	6.3	26.8	ND
L3	22.6	32.0	47.8

ND, not determined

Electrolyte leakage. Electrolyte leakage is frequently used to assess plant membrane damage by desiccation. Drought-stressed *bspA*-transgenic plants exhibited varied degree of electrolyte leakage as compare with NT plants. Some selected lines, i.e. line G3, exhibited extremely low level of electrolyte leakage upon water stress (Fig. 1). *In vitro* plants of line G3 also exhibited a better osmotic adjustment (Table 1), yet no similar pattern was observed in other transgenic lines (e.g. line E1). Since membrane damage, measured by electrolyte leakage, and osmotic adjustment, measure by osmotic potential, are different parameters representing two different mechanisms of plant stress tolerance, additional studies are required to clarified the role of *bspA* gene in reducing electrolyte leakage.

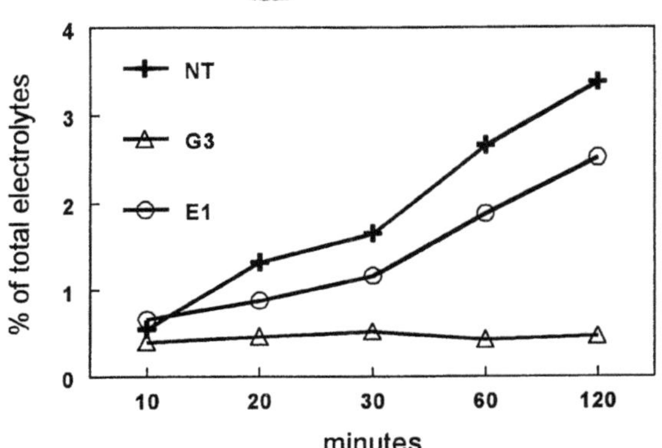

Fig. 1. Electrolyte leakage in *bspA*-transgenic and NT plants. NT, non-transformed plants; G3 and E1: two independent *bspA*-transgenic lines.

Growth characteristics of potted transgenic plants. *Ex vitro bspA*-transgenic plants and NT plants exhibited a similar severe stress following withholding irrigation. Shoot water potential of *ex vitro*-stressed-plants was -2.5, -2.4 and -2.0 Mpa in *bspA*-transgenic lines E1 and L3 and NT, respectively, as compare with about -1.1 Mpa of non-stressed *bspA*-transgenic and NT plants. However, the time to second wilting was delayed: 15.7 and 18.0 days in *bspA*-transgenic lines E1 and L3, respectively, as compared with NT plants of 14.3 days (Table 2). Stem elongation was inhibited in both *bspA*-transgenic and NT stressed plants, however, this inhibition was less severe in *bspA*-transgenic lines (Table 2). About 40% of the leaves abscised from the lower part of *bspA*-transgenic plants during the stress period, as compared with 36% in NT plants (Table 2). Leaf senescence, shifting nutrients from old leaves is considered one of the strategies of plant survival under stress conditions, thus maintaining the growth of younger leaves. Taking together with osmotic adjustment in *bspA*-transgenic lines (Table 2), the improved growth performance of *bspA*-transgenic plants may be due to the combination of both nutrient transport and osmotic adjustment.

BspA expression. An elevated BspA accumulation levels were observed in non-stressed *bspA*-transgenic lines L3, E1, G3 and H3 as compared with NT plants (Fig. 2). Although the *bspA* cDNA was driven by the 35S CaMV constitutive promoter, some transgenic lines exhibited a decrease in BspA level under stress condition (e.g. E1, Fig. 2A; G3, Fig.

2B and H3, Fig. 2C). Although some *bspA*-transgenic lines had an improved tolerance to water and salt stress as evidence by growth performance and osmotic adjustment, a positive correlation between stress tolerance and the level of BspA expression has been found in only one transgenic line (L3), while in other lines variable BspA expression patterns were observed.

Table 2. Growth performance of transgenic plants under drought stress.

Line	Stem Elongation			Leaf Abscission	Days to 2nd Wilting
	Non stressed (cm)	Stressed (cm)	Decrease (% of control)	Stressed (% of control)	Stressed (days)
NT	13.8 ±1.2	5.1 ±0.7	63.1	36.3 ±3.9	14.3 ±0.6
E1	11.8 ±0.7	5.4 ±1.0	44.3	39.2 ±2.2	15.7 ±0.9
L3	13.2 ±1.1	6.0 ±0.4	54.4	40.8 ±3.5	18.0 ±0.6

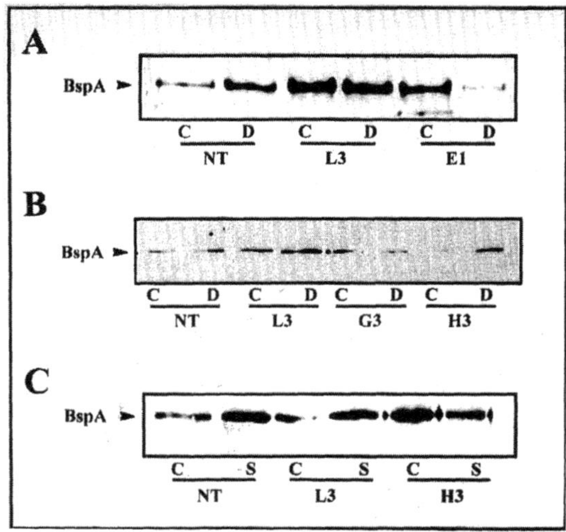

Fig 2. Western blot analysis of BspA expression in *bspA*- transgenic and NT aspen plants. (A) Drought stress of *ex vitro* plants. (B) drought stress of *in vitro* plants. (C) salt stress of *in vitro* plants. NT: non-transformed plants; L3, E1, G3, H3; *bspA*-transgenic lines. C: non stress; D: drought stress; S: salt stress.

Recently, transgenic plants expressing stress relevant genes, e.g. *colA* coding for choline oxidase (Hayashi et al 1997), *hav1* a LEA family protein (Xu et al 1996) and the *mtlD* coding for mannitol-1-phosphate dehydrogenase (Tarczynski et al 1993), exhibited an enhanced stress tolerance. The presented BspA expression pattern and plant stress tolerance, additional data of experiments in progress, indicate a highly positive effect of BspA on stress tolerance. To determine a true correlation between BspA expression and plant tolerance, an evaluation of the level of BspA should also be included.

4. References

Close TJ (1997) Physiol. Plant. 100: 291-296.

Dure L (1993) In: Plant Response to Cellular Dehydration During Environmental Stress. pp. 91-103. (Close T.J. and Bray E.A. eds.) American Society of Plant Physiologists, MA.

Hayashi H et al. (1997) Plant J. 12: 133-142.

Pelah D et al. (1995) Tree Physiol. 15: 673-678.

Pelah D et al. (1997a) J. Plant Physiol. 151: 96-100.

Pelah D et al. (1997b) Physiol. Plant. 99: 153-159.

Rudulier DL (1984) Science 224: 1064-1068.

Tarczynski MC et al. (1993) Science 259: 508-510.

Xu D et al. (1996) Plant Physiol. 110: 249-257.

"EXPRESSION OF RECOMBINANT ANTIBODIES IN PLANTS AND CELL SUSPENSION CULTURES"

Rainer Fischer*, Jürgen Drossard, Stephan Hellwig, Neil Emans and Stefan Schillberg

*Institut für Biologie I (Botanik und Molekulargenetik), RWTH Aachen, Worringerweg 1, D-52074 Aachen, Germany, *corresponding author: Phone: +49 241 806631, Fax: +49 241 871062, E-mail: fischer @ bio1.rwth-aachen.de.*

INTRODUCTION

Monoclonal antibodies (mAbs)[7] are extremely useful tools in medicine, biology and biotechnology. Combining hybridoma and recombinant DNA technology gave access not only to full-size molecules but also to various recombinant antibody fragments (rAbs) and fusion proteins[19], broadening the range of possible applications. Recent improvements in heterologous gene expression systems[5,9,10,11,15] and development of phage display technologies[19] make it possible to design and express rAbs against almost any antigen and to fine tune these molecules for improved performance.

Although the production of mAbs and rAbs is a standard technique, the ideal expression system has yet to be developed. Currently, bacteria[15] and mammalian cell cultures[1] are the most established systems while yeast and baculovirus-infected insect cells play a minor role[16]. Due to the delicate nature of mammalian cells, cultivation can be difficult and requires expensive equipment and media. In downstream processing, care must be taken to remove oncogenic sequences or viral contamination for use *in vivo*. The use of animals as a source of mAbs is becoming limited by legal and ethical restrictions. Bacteria do not produce glycosylated full-size antibodies, contaminating endotoxins are difficult to remove and recombinant proteins often form inclusion bodies, making labour- and cost-intensive *in vitro* refolding necessary.

Based on the recent demand for bulk quantities of functional, active recombinant proteins, alternatives to expression in microbes and animal cells are required. Transgenic plants[12] offer alternative approaches for mass production of recombinant proteins and antibodies[6,8,18],with their enormous levels of biomass[13]. Plants are easy to grow and can be cultivated as calli, cell suspensions or hairy roots using certified "Good Laboratory Practice" (cGLP) and "Good Manufacturing Practice" (cGMP)[9]. Heterologous proteins accumulate to high levels in plant cells and plant-derived antibodies are indistinguishable from those produced by hybridomas[6,18]. Protein synthesis, secretion, folding and post-translational modifications are similar in plant and animal cells. Co-purification of blood-borne pathogens and oncogenic sequences is avoided during downstream processing of recombinant proteins from plants giving a homogenous, safer product.

Here, we demonstrate that transgenic *Nicotiana tabacum* plants and suspension cultures can be used for production and purification of recombinant antibodies. Expression of full-size heavy and light chain cDNAs permitted assembly of rAbs, which were exported to the intercellular space[18]. Suspension cell lines were used to analyze media and physico-chemical parameters for rAb expression. Furthermore, rapid, efficient purification strategies were developed to preserve antibody activity. Upscaling and downstream processing was successful up to a level of 5 kg of plant material in shake flasks or 30-L fermentation.

A. Altman et al. (eds.), Plant Biotechnology and In Vitro Biology in the 21st Century, 567–570.

RESULTS

Design of expression constructs in pSS

Full size cDNAs of the IgG_{2b}-heavy and kappa light chain were cloned from the hybridoma secreting TMV-neutralising mAb24. Modified cDNAs were integrated in tandem array into the plant expression vector pSS. Fab fragments were constructed by introducing a TAG stop codon downstream of the first cysteine within the hinge region of the IgG_{2b} heavy chain. For $F(ab')_2$ fragments, the stop codon was inserted after the second hinge region cysteine. ScFvs and bispecific antibodies were prepared by splice overlap extension PCR. For ER retention, a C-terminal KDEL-sequence was added.

Characterization of transgenic Tobacco plant lines

All expression constructs were introduced into *N. tabacum* c.v. Petite Havana SR1 using *Agrobacterium*-mediated transformation and regenerants were kanamycin selected. Individual transformed plants were transferred to soil and used for further analysis. All transgenic plants obtained appeared phenotypically normal compared to control Tobacco plants.

Transgenic plants were tested for expression of functional rAbs by ELISA. Up to 15 μg antibody per gram leaf material was present in transgenic T_0-plant lines secreting the full size antibody. Yields were increased in double transgenic T_2-plants. $F(ab')_2$ and biscFvs were expressed at comparable molar levels. ScFvs and biscFvs expression levels could be increased >10 fold by the addition of a KDEL ER retention motif. Cytosolic scFv and biscFv expression was detectable but low.

➢ Table 1: Expression levels (μg rAb/g leaf tissue) of rAbs in different compartments of T_0 transgenic Tobacco plants. n.d.= not determined.

rAb	apoplast	ER	cytosol
full size	15 (154 T_2)	n.d.	-
$F(ab')_2$	10.5	n.d.	-
Fab	0.5	n.d.	-
scFv	0.55	20	0.002
biscFv	7.85	79	0.001

Establishment of transgenic SR1 and BY-2 suspension cell lines

Leaf discs from high producer transgenic plants were cultivated in petri dishes on selective MS agar medium containing 75 mg/l kanamycin and phytohormones to facilitate callus induction and callus was sub-cultured for liquid cell suspensions. The fast growing Tobacco cell line Bright Yellow-2 (BY-2) was used for establishing transgenic suspension cell cultures by cocultivation with Agrobacteria carrying Fab, scFv or biscFv cDNAs. The size of the rAb had a clear effect on rAb localisation: Full-size antibodies (150 kDa) were almost exclusively apoplastic, Fabs (50 kDa) were evenly distributed between supernatant and cell extract and scFvs (30 kDa) were found predominantly in the supernatant. Addition of a C-terminal KDEL-tag resulted in ER retention. Due to the weak activity of the enhanced 35S promoter in the BY-2 line, Fab, scFv and biscFv expression levels were 3-5 fold lower than in an SR1 suspension line.

Table 2: Expression levels (µg rAb/g fresh weight) and localisation of rAbs in suspension cultures obtained from transgenic plants ([1]) or transgenic BY-2 cells ([2]).

rAb	cell extract	super-natant	standard expression	amino acid addition
full size ([1])	> 95%	< 5%	15	145
Fab ([2])	50%	50%	0.55	1.6
scFv ([2])	< 25%	> 75%	0.15	0.35
biscFv-KDEL ([2])	>99%	<1%	1.8	n.d.

Downstream processing of rAbs

Proteins for therapeutic applications must be highly purified before use to minimise or even eliminate adverse clinical reactions against contaminants. For diagnostic tests, contaminating biomolecules, that may interfere with the specificity and sensitivity of the rAb, have to be removed to guarantee product conformity and reliability. In order to achieve this, we established an efficient purification protocol for full size TMV-specific rAbs.

Isolation of recombinant proteins from transgenic plant cells faces unique problems with release of proteases and oxidizing agents (phenols, tannins) from subcellular compartments after cell disruption. Furthermore, all cell debris must be removed from the crude extract and a large volume has to be processed during the initial downstream processing step. Our protocol combines enzymatic cell lysis, cross-flow filtration for clarification and fast-flow affinity chromatography for rapid processing. Full-size rAbs were purified from plant cell extracts on protein-A and protein-G affinity columns in a similar manner to purification from mammalian sources. Incorporation of cross-flow filtration significantly simplified clarification of the crude cell extract. Initial affinity chromatography on Protein-A gave efficient removal of contaminants and approximately 100-fold concentration of the recombinant protein. ELISA assays showed that 80% of rAb in the crude extract could be recovered from up to 5kg fresh cell weight. Therapeutic applications[9] of Tobacco-derived rAbs may require additional purification steps to ensure product quality and conformity and to remove contaminating noxious plant products.

Specificity and affinity of rAb24 purified from transgenic plants

ELISA was used to determine the plant rAb24 specificity and affinity for TMV. It confirmed that expressed heavy and light chain polypeptides assembled into functional antibodies and bound to TMV-*v*, TMV-*d* (dahlemense) and TMV-HR (Holmes' Ribgrass) but not other plant viruses. Titration ELISA showed that the specificity and affinity (5×10^9 M^{-1}) of the murine mAb24 and plant expressed rAb24 were identical, indicating the maintenance of the neotope specificity and affinity for TMV. Similar data were obtained for the $F(ab')_2$, Fab, biscFv and scFv-fragments.

Currently, investigations are underway to gain an indepth understanding of the structural and functional properties of affinity-purified TMV-specific rAbs using comparative mass spectroscopy and NMR. Preliminary results indicate that the N-glycosylation pattern in the plant-derived rAb24 is similar to that in mammalian cells, although there may be differences in terminal sugar residues.

Future potential of plant suspension cell lines for production of rAbs

Currently, plant cell cultures are not widely commercially used to produce recombinant products, largely because the development of biopharming and large-scale cultivation is still in its infancy and because their potential has been underestimated[16]. Plant suspension cultures can be scaled up to the 10,000 litre range[13] for production of recombinant proteins under cGLP- and cGMP-conditions. Our goal is to increase recombinant protein yields by using very high cell density cultures with improved productivity per cell.

Based on the successful expression of TMV-specific full-size rAb forms belonging to the IgG_{2b} subclass, improvements under investigation are: a) expression of other murine isotypes (IgG_{2a}, and IgG_1), b) expression of mouse-human chimeric and human antibodies, c) expression of diabodies and scFv fusion proteins, d) expression of rAbs with diagnostic or therapeutic interest, e) enhancement of expression by modifying plant expression vector regulatory elements, f) use of other plant species for faster biomass production and reduced levels of noxious compounds, g) pilot-scale production by using fermentation technologies (e.g. stirred tank reactor or airlift-fermenter) and h) choice of media and addition of selective supplements.

Our Tobacco systems show that transgenic plants and cell suspension cultures are an untapped source for large scale production of rAbs. The feasibility of plant derived rAbs and fragments for disease therapy and diagnosis are being investigated in several laboratories because of the potential to generate large quantities of correctly folded, specific, high quality rAbs at low cost. To enhance the potential of plant-derived antibodies, it will be important to switch from laboratory- to pilot-scale fermentations and rAb production in seed, combined with state of the art purification. The latest developments in downstream processing (perfusion chromatography, expanded bed technology and engineered affinity ligands), will enable improved handling of large sample volumes, minimise processing time and avoid proteolytic and oxidative degradation of recombinant proteins. This combinatorial approach together with improvements in plant gene expression and the choice of expression hosts, will provide breakthroughs in the development of large scale production of inexpensive diagnostic and therapeutic antibodies by biopharming.

REFERENCES

1. Bebbington CR et al (1992) Bio/Technology 10. 169-175.
2. Blobel G, Dobberstein B (1975) J. Cell Biol. 67. 835-851.
3. Denecke J et al (1991) Plant Cell 3. 1025-1035 .
4. Düring K et al (1990) Plant Mol. Biol. 15. 281-293.
5. Ford CF et al (1991) Protein expression and purification 2. 95-107.
6. Hiatt A et al (1989) Nature 342. 76-78.
7. Köhler G, Milstein C (1975) Nature 256. 52-53.
8. Ma JK-C et al (1995) Science 268. 716-719.
9. Manohar V, Hoffman T (1992) TIBTECH 10. 305-309.
10. Nygren PA et al (1994) TIBTECH 12. 184-188.
11. Plückthun A (1991) Curr. Opin. in Biotechnol. 2. 238-246.
12. Potrykus I (1991) Ann. Rev. Plant Physiol. Plant Mol. Biol. 42. 205-225.
13. Scragg AH (1992) Curr. Opin. in Biotechnol. 3. 105-109.
14. Siden E et al (1981) Proc. Natl. Acad. Sci. USA 78. 1823-1828.
15. Skerra A (1993) Curr. Opin. in Immunol. 5. 256-262.
16. Taticek RA et al (1994) Curr. Opin. in Biotechnol. 5. 165-174
17. Van Regenmortel MHV (1986) Plant Viruses, Vol 2. 79-104. Plenum Press, New York.
18. Voss A et al (1995) Molecular Breeding 1. 39-50.
19. Winter G, Milstein C (1991) Nature 349. 293-299.

BIOTECHNOLOGICAL APPROACHES TO MAKING VACCINE IN PLANTS

J. Kapusta[1], A. Modelska[2a], M. Figlerowicz[2b] J. Podkowinski[1], T. Pniewski[1] O. Lisowa[1], P. Berbezy[1], D. Helias[1], J. Biesiadka[1], , I. Femiak[1], M. Letellier[1], V. Yusibov[4], A. Plucienniczak[3], H. Koprowski[4] A.B. Legocki[1]
[1]Institute of Bioorganic Chemistry, Noskowskiego 12/14, 61-704 Poznań, Poland, [2a]University of Medical Sciences, Department of Endocrinology, Przybyszewskiego 49, 60-355 Poznan, Poland, [2b]University of Medical Sciences, Department of Pediatric, Szpitalna 27/33, 60-572 Poznan, Poland, [3]Institute of Biotechnology & Antibiotics, Staroscinska 5, Warsaw, Poland, [4]Thomas Jefferson University, 1020 Locust Street, Philadelphia, PA 19107, USA

Introduction

From the early 1990s the recombinant DNA technology led researchers to explore plants as a novel source for antigen expression. The concept of production edible vaccines in plants has been recognized to be attractive for its safety, simplicity and low farming cost (Mason et al 1992). The latter might be especially valid for developing countries where classical ways of vaccination often encounter various limitations. Plants are potentially useful for delivering vaccines as it is widely known that plants themselves produce a variety of medically valuable compounds. There are two well established systems for antigen expression in plant.

1. Recombinant plant virus engineered to express antigen

Alfalfa mosaic virus (AIMV) provides an example of a plant virus recombined to harboring antigen protein able to replicate and spread in plants. Coat protein (CP) of AIMV was engineered as a carrier antigenic peptides from rabies virus and HIV (Yusibov et al. 1997). Mice fed with AIMV recombined virus infected spinach leaves or immunized intraperitoneally with leaf extracts can be protected against challenge infection of lethal dose rabies virus (Modelska et al. 1997).

2. Direct expression of antigen in transgenic plants.

The HBsAg was expressed for the first time in tobacco (Mason et al. 1992). It was shown that purified tobacco derived HBsAg elicit qualitatively similar immunological response as yeast produced commercial anti-HBV vaccine (Tahanavala et al 1995).
Other antigens produced in plants retain competence to stimulate of animals (Hag et al 1995, McGarvey et al. 1995, Mason et al. 1996) as well as human immune system (Tacket

A. Altman et al. (eds.), Plant Biotechnology and In Vitro Biology in the 21st Century, 571–574.

et al. 1998) after oral administration. Transgenic plants as alternative source of antigen might be considered as potentially stable system in which antigens can assemble into immunogenic molecule. Moreover, transgenes integrated into plant genome are capable of being transferred to next generations.

Legume plants as carriers of edible vaccines

Legume plants are commonly known as a rich source of proteins. We consider this economically and ecologically important group of plants as an appropriate target of edible vaccines.

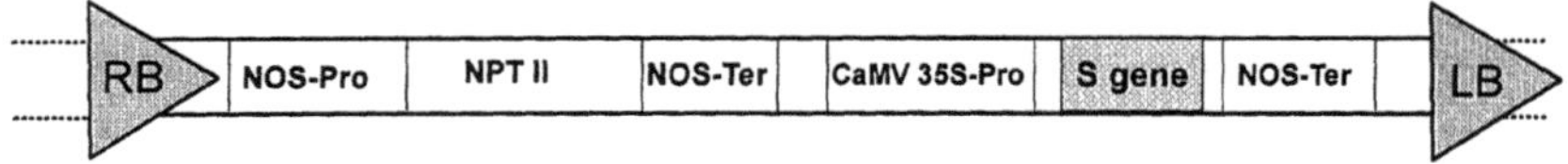

Fig. 1 Scheme of *Agrobacterium tumefaciens* binary plasmid carrying NPTII selection gene and S gene sequence of hepatitis B virus.

We have taken the advantage of our previous results concerning technology transformation of lupin callus (Kapusta et al.1996, Kapusta et al 1998) and the most recent data from regeneration and transformation of lupin meristem (Letellier et al unpublished). Lupin with its seed storage protein content up to 40% of dry weight may serve as suitable pasture plant to antigen expression for veterinary vaccines.

The goal of this study is to develop transgenic plants-based vaccine against classic swine fever virus (CSFV) and hepatitis B virus (HBV).

Classic swine fever virus

CSFV is an etiologic agent of classical swine fever (CSF) also termed hog cholera. CSF is a highly infectious disease with high mortality rates. Many huge pig outbreaks have been reported in the past. The most recent outbreak in 1997 caused death of millions of pigs in Holland and Germany. CSFV, the etiologic agent of this disease is a single-stranded RNA virus. The genome of CSFV varied from 9 - 13 kb.
Based on the known sequence of large envelope polyprotein three plasmids were designed and constructed: pKPE0, pKPE1 and pKPE2. This plasmids carry corresponding antigens E0, E1 and E2. Tobacco has been transformed by *A. tumefaciens* harboring constructs encoding E0 and E1 proteins. Regenerated plants were verified with PCR as transgenic (data not shown).

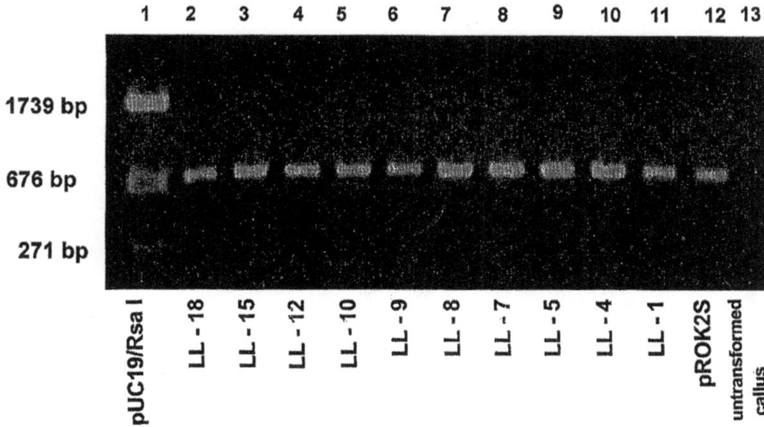

Fig. 2 PCR analysis of yellow lupin callus. Lanes: 1, molecular size marker; 2-11, gene S carrying transformed lupin cell lines; 12, pROK2S - positive control; 13, untransformed callus - negative control.

Hepatitis B virus

Plasmid pROK2S (Fig. 2) has been constructed by cloning the coding sequence of HBsAg into pROK2 plasmid under 35S promoter. Yellow lupin (*Lupinus luteus* L.) seeds were infected with *Agrobacterium* strain carrying pROK2S. Callus derived on a selection medium was analyzed with PCR, which showed its transgenic character (Fig. 2). The ABBOTT kit designed to enzyme immunoassay detection of HBV surface antigen in serum or plasma has successfully been used to assay HBsAg in plant tissue protein extracts (data not shown). The best callus line LL-10 producing c.a. 150 ng HBsAg/g fresh tissue was used in oral immunization experiment with mice (Fig. 3). Mice, BALB/c, male 6-8 weeks old were fed with callus harboring HBsAg. They received 5g of fresh tissue, and this dose was repeated a month later. The control group was fed with transgenic callus without HBV antigen. Serum was collected two weeks after each immunization. ELISA was used for the detection of anti-HBsAg antibodies. Plates were coated with HBsAg (1mg/ml) and the secondary peroxidase conjugated anti-mouse IgG and IgA antibodies were used in the assay.

The results from the experiments with yellow lupin callus show for the first time that oral administration of transgenic lupin tissue carrying HBsAg is able to induce production of anti-HBsAg IgA and IgG immunoglobulin after first immunization and then more pronounce after busting. The clear tendency of antibodies elevating after oral immunization of mice fed with transgenic lupin callus appears to be caused by HBsAg antigen present in food.

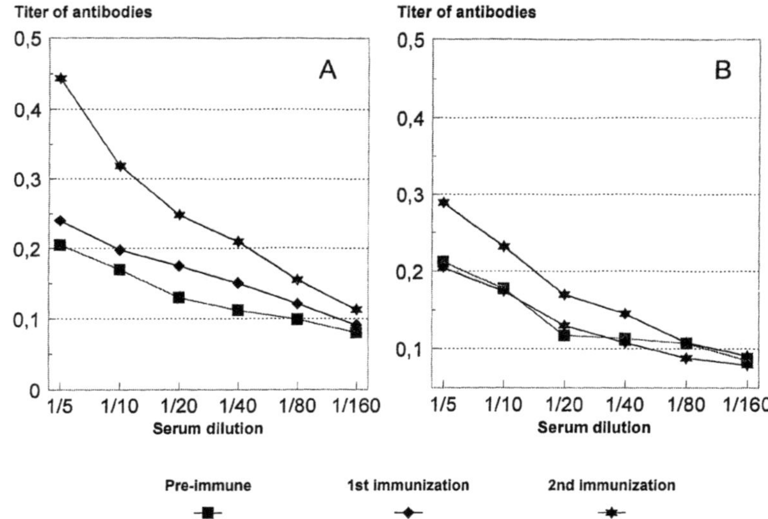

Fig. 3 Evaluation of anti-HBsAg antibodies (IgA & IgG) titer in serum of mice fed with transgenic lupin callus carrying HBsAg (A) or control callus without HBV antigen (B). Mice were primed with 5g/mouse lupin callus on the 1st day (1st immunization) and busted on the 30th day (2nd immunization). Experimental groups represent 5 male 6-8 weeks old BALB/c mice for (A)and 4 animals for (B) treatment. Each data represent an average of 5 (A) or 4 (B) individual assessments.

Acknowledgments

Research partially supported by the State Committee for Scientific Research. Research project No 6 P04B 010 09

Literature

Hag TA et al (1995) Science 268, 714-716.

Kapusta J et al (1996) Book of Abst of Eigth Intr. Lupin Con., Asillomar USA, May 11 - 15, 1996.

Kapusta J et al (1998) International Conference Nucleic Acid and their Constituents: Chemical Evolution Underlying Biological Evolution, Poznan, May 10-13, 1998 p. 109

Mason HS et al (1992) Proc. Natl. Acad. Sci. USA 89, 11745-11749.

Mason HS et al (1996) Proc. Natl. Acad. Sci. USA 95, 5335-5340.

McGarvey PB et al (1995) BioTechnology 13, 1484-1487.

Modelska A et al. (1997) Proc. Natl. Acad. Sci. USA 95, 2481-2485.

Tacket CO et al (1998) Nature Medicine 4, 607-609.

Thanavala Y et al (1995) Proc. Natl. Acad. Sci. USA 92, 3358-3361

Yusibov V et al. (1997) Proc. Natl. Acad. Sci. USA 94, 5784-5788.

GENETIC ENGINEERING OF SEED OIL FATTY ACID COMPOSITION

Deborah S. Knutzon[1], George M. Chan[1], Pradip Mukerji[2], Jennifer M. Thurmond[2], Sunita Chaudhary[2], and Yung-Sheng Huang[2]

[1]Calgene LLC, Davis, CA, USA
[2]Ross Products Division, Abbott Laboratories, Columbus, OH, USA

1. Introduction

Polyunsaturated fatty acids (PUFAs) are important components of infant as well as adult nutrition due to their roles as both structural elements of cell membranes and as precursors to eicosanoids, a family of potent biological effectors such as prostaglandins and prostacyclins. PUFAs have been reported to be of clinical significance in fetal growth and development, cardiovascular disease, skin disease and certain inflammatory conditions (Das et al 1988). There are two main families of PUFAs, n-6 and n-3, in which the n-designation refers to the position of the last double bond relative to the methyl end of the molecule. The n-6 and n-3 PUFAs are derived from linoleic acid (LA; 18:2n-6) or alpha-linolenic acid (18:3n-3), respectively, via an alternating series of desaturation and two-carbon elongation steps. The same enzymes are believed to catalyze reactions in both pathways although this has not been shown for most systems. In the n-6 pathway, γ-linolenic acid (GLA; 18:3n-6) is formed from LA by the action of a Δ6-desaturase. GLA is elongated to produce dihomo-γ-linolenic acid (DGLA; 20:3n-6), which then undergoes Δ5-desaturation to produce arachidonic acid (ARA; 20:4n-6). A similar series of reactions in the n-3 pathway results in the production of eicosapentaenoic acid (EPA; 20:5n-3).

Current sources of PUFAs include fish oils, microorganisms such as *Mortierella, Porphyridium,* and *Thraustochytrium,* and plant oils such as evening primrose and borage (Ratledge 1993). None of these sources are economically well-suited for use in human nutrition however. The ability to engineer the fatty acid biosynthetic pathway of an oilseed crop to produce these valuable PUFAs would be of considerable interest. Commercial oilseeds do not carry out desaturation beyond LA or ALA, therefore the introduction of Δ6 and/or Δ5 desaturase activity would be needed to achieve synthesis of GLA or ARA. Clones encoding Δ12 and Δ15 desaturases involved in the production of LA and ALA, respectively, have been identified from a number of sources (Okuley et al 1994), and Δ6 desaturases have been cloned from cyanobacteria (Reddy et al 1993), borage (Sayanova et al 1997) and *Caenorhabditis elegans* (Napier

A. Altman et al. (eds.), Plant Biotechnology and In Vitro Biology in the 21st Century, 575–578.

et al 1998). All of these membrane-bound enzymes share some common features. The most striking of these is the presence of three histidine-rich regions (His-boxes) that are conserved both in sequence and in spacing relative to one another. These histidine-rich motifs may contribute to iron binding in the active site (Okuley et al 1994). There have been no prior reports of a cloned Δ5 desaturase, the enzyme catalyzing the final step in synthesis of ARA.

2. Procedure

Total RNA was isolated from a three-day old culture of *Mortierella alpina* (ATCC 32221) grown in Difco PD medium. First-strand *M. alpina* cDNA was used as a template in PCR reactions using degenerate oligonucleotide primers based on regions of amino acid sequence homology between known desaturase sequences. The forward primer contained all possible codons corresponding to the peptide sequence HHTYTN and the reverse primer contained all possible codons corresponding to the peptide sequence HHLFP. PCR-derived DNA fragments were used as probes to isolate full-length cDNA clones from the library. In addition, random sequencing of clones from the library was undertaken to identify cDNAs with homology to desaturases.

Verification of substrate specificity of desaturase clones was obtained by expression in recombinant *Saccharomyces cerevisiae* and transgenic canola seeds. For yeast expression, the cDNA clones were expressed from the GalI promoter of pYES2 (Invitrogen). Cultures were grown in minimal media containing 2% glucose and lacking uracil. Various fatty acids at a concentration of 25 uM were added to the cultures to serve as substrates for the desaturases. After galactose induction and growth for 48 hours, the cells were pelleted and the extracted lipids were analyzed by GC to determine the extent conversion of various substrates to products.

For expression in canola, the cDNAs were inserted into a napin expression cassette (Kridl et al 1991) and transferred into a binary vector (McBride, Summerfelt 1990) for *Agrobacterium*-mediated transformation (Radke et al 1987) of a low-linolenic variety of *Brassica napus*. Regenerated plants were grown in the greenhouse and pools of 20 segregating T2 seeds from each independent transformed line were analyzed by GC for fatty acid composition.

3. Results

PCR amplification of *M. alpina* cDNA using primers corresponding to conserved histidine-rich regions of known desaturases resulted in isolation of a cDNA clone, Ma29, that showed significant homology to the known Δ6-desaturases from cyanobacteria. Two additional cDNA clones were identified by random sequencing of the *M. alpina* library; Ma648 displayed significant homology to known plant microsomal Δ12 desaturases, and Ma524 showed the most homology to the borage Δ6 desaturase and a sunflower protein of unknown function (Sperling et al 1995). Figure 1 shows the relationship of these desaturases based on amino acid similarity. Ma524 and Ma29 share two characteristics that distinguish them from Ma648: the presence of an N-terminal domain with homology to cytochrome b5, and a change of amino acid sequence from HXXHH to QXXHH at the third "his-box".

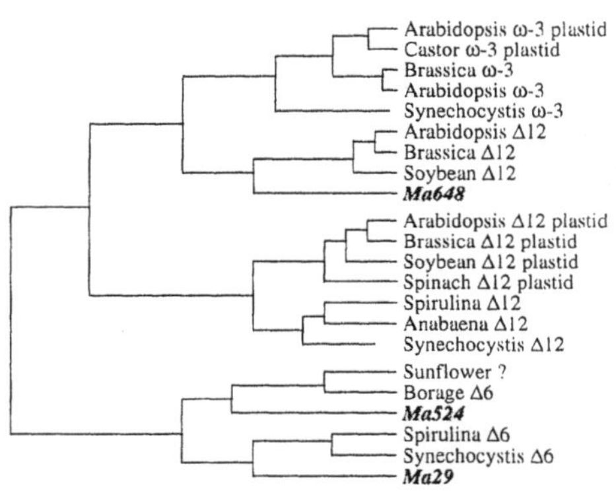

Figure 1. Dendogram of desaturase amino acid sequences from different sources.

The substrate specificities of the three Mortierella alpina desaturases were determined by expression in yeast and transgenic canola. Expression of Ma29 grown in the presence of DGLA results in conversion of approximately 14% of the incorporated DGLA to ARA, indicating Δ5-desaturase activity. Canola seeds do not contain DGLA, but expression of the cDNA in transgenic seeds results in the production of Δ5-desaturation products of oleic acid and LA, the most prominent fatty acids in the seed (Figure 2; Knutzon et al 1998).

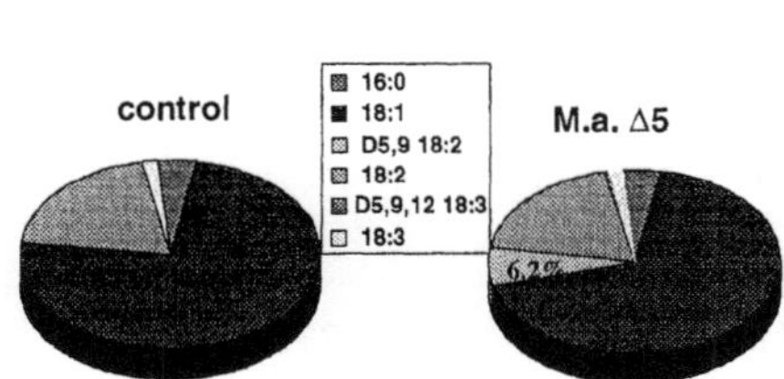

Figure 2. Fatty acid composition of T2 seed pool from transgenic canola expressing M. alpina D5 desaturase, Ma29.

Expression of Ma648 in yeast resulted in good conversion of endogenous oleic acid to LA, indicative of Δ12-desaturase activity. In canola seeds, expression of this cDNA leads to an increase of LA from approximately 23% to >45% in T2 seed pools. Ma524 was identified as a Δ6 desaturase based on its ability to convert exogenously supplied LA to GLA in yeast. Expression of Ma524 in canola leads to >15% GLA in transgenic seeds, but as with the Δ5 desaturase, desaturation of oleic acid is observed, resulting in accumulation of 9Δ6,9 18:2. Simultaneous expression of both the Δ6 and Δ12 desaturases (driven by having both cDNAs on the same T-DNA) leads to the accumulation of >30% GLA and a dramatic decrease in the Δ6,9 18:2 fatty acid in the seeds. Figure 3 shows the fatty acid composition of a T2 seed pool from one such transformant.

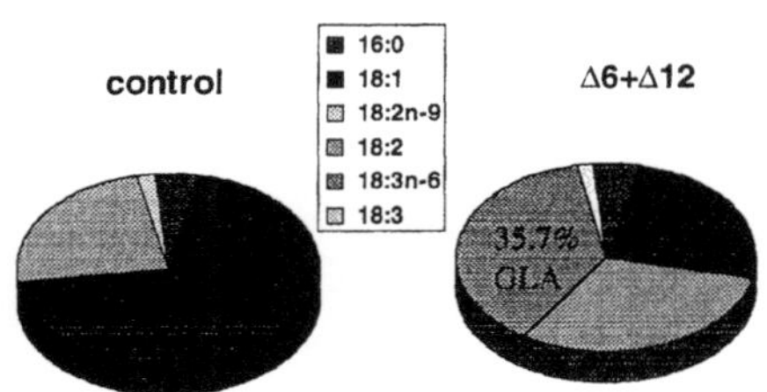

Figure 3 Expression of M. alpina D6 and D12 desaturases in transgenic canola

4. Conclusion

We have isolated cDNA clones encoding Δ5, Δ6, and Δ12 desaturases from the filamentous fungus, *M. alpina*, and demonstrated their substrate specificities by expression in yeast and canola seeds. These desaturases should prove useful in the production of PUFAs in transgenic plants and/or microbes.

5. References

Das UN et al (1998) Nutrition 4, 337-341.
Knutzon DS et al (1998) J. Biol. Chem. In press
Kridl JC et al (1991) Seed Sci. Res. 1, 209-219.
McBride KE, Summerfelt KR (1990) Plant Mol. Biol. 14, 269-276.
Napier JA et al (1998) Biochem. J. 330, 611-614.
Okuley J et al (1994) Plant Cell 6, 147-158.
Radke SE et al (1987) Theor. Appl. Genet. 75, 685-694.
Ratledge C (1993) Trends Biotechnol.11, 278-284.
Reddy AS et al (1993) Plant Mol. Biol. 22, 293-300.
Sayanova O et al (1997) Proc. Natl. Acad. Sci. USA 94, 4211-4216.
Sperling P et al (1995) Eur. J. Biochem. 232, 798-805.

ENGINEERING CARBOHYDRATE METABOLISM IN TRANSGENIC PLANTS

J. Kossmann[1], J. Lloyd[1], J. Riesmeier[2] and L. Willmitzer[1]
[1]Max-Planck-Institut für molekulare Pflanzenphysiologie, Karl-Liebknecht-Str. 20, 14476 Golm, and [2]Planttec Biotechnologie GmbH Forschung und Entwicklung, Hermannswerder 14, 14473 Potsdam, Germany

1. Introduction

Genetic engineeringenables the manipulation of key steps of carbohydrate metabolism leadingto drastic alterations of sink-source relations, and to profound modifications of starch structure in transgenic potato plants. Examples are discussed with special emphasis on crop productivity, tuber development, and starch biosynthesis.

The term `sink-source relations′ describes the complex interaction of different plant organs for an optimal allocation of water, mineral nutrients and assimilates. We will focus on the partitioning of carbohydrates, which is best described by the harvest-index, the ratio between the dry matter accumulated in the harvestable organs and the total dry matter of a plant. The total plant dry weight produced per hectare has not changed drastically in the recent past, whereas the harvest-index has almost doubled, indicating that the yield increases are mainly due to an optimisation of sink-source relationships rather than to increased photosynthetic efficiency (Gifford et al. 1984). A very drastic picture emerges when a wild-type potato species, *Solanum demissum*, is compared to modern high-yielding varieties. The harvest-index has increased from 0.07 to 0.81 (Inoue, Tanaka; 1978).

Starch is found as a storage carbohydrate in seeds, tubers and roots, or is synthesized as a transient carbon pool in photosynthetically active tissues in a diverse range of plant species. Wild type starch is normally composed of essentially linear (0.1% branchpoints) α-1,4-glucans (amylose), and via α-1,6-glycosidic bonds branched (4-5% branchpoints) α-1,4-glucans (amylopectin). For industrial applications the major source is maize, but to a lesser extent wheat and potato starch are also used. For most industrial processes (55%) starch hydrolysates are used. Around 30% of the starch is used in its native form and 15% in a chemically modified form. These chemical modifications alter the physicochemical properties of the different starches in order to optimize the material for use in food and non food applications.

The techniques enabling the efficient genetic transformation of potato plants have made it possible to very specifically manipulate certain enzymatic steps important for carbohydrate metabolism. This can be achieved either through reducing the expression of

A. Altman et al. (eds.), Plant Biotechnology and In Vitro Biology in the 21st Century, 579–582.

endogenous genes, or the ectopic expression of genes from heterologous sources. If it were possible to manipulate key steps of starch biosynthesis, which are of major importance in determining certain properties of the starch synthesized, genetic engineering of plants might serve as a tool either to replace some of the chemical modifications or to generate novel starches for novel applications. The results of genetic modifications of carbohydrate metabolism on potato tuber development, tuber yield, dry matter content, and starch properties will be discussed.

2. The influence of modified sucrose synthesis in photosynthetic tissues on tuber yield

In photosynthetic active tissues light energy is captured and converted into chemical energy in the form of ATP and NADPH in the light driven photosynthetic processes. In the dark reactions, ATP and NADPH are used to assimilate and reduce CO_2, which takes place in the Calvin cycle. The products of the cycle are subsequently converted into sucrose, which can be transported into heterotrophic tissues. If, during the course of a day, sucrose is accumulating in the leaves, possibly due to an imbalance of supply and demand for carbohydrates, the surplus of carbohydrate can also be converted into transitory starch, which is degraded and converted into sucrose during the night.
It has been a matter of long debate whether the photosynthetic activity of plants is limiting to crop yield. In the last years transgenic potato plants have been developed, where the photosynthetic capacity was either downregulated directly or indirectly by manipulating the expression of Calvin cycle enzymes or other essential steps of photosynthetic carbon metabolism through antisense RNA technology.
One example are plants, where the expression of the plastidic fructose-1,6-bisphosphatase (cp-FBPase) was reduced (Koßmann et al., 1994). The cp-FBPase is one of the key regulatory enzymes of the Calvin cycle. Transgenic plants were selected which still retained 12%, 14% and 36% of wild-type levels of cp-FBPase activity. When photosynthesis was measured in a saturated CO_2-atmosphere (5%) and under high light intensity (1000 mmol quanta $m^{-2}s^{-1}$) all of the selected lines showed a decrease in photosynthetic capacity. The line which still retained 36% of the enzyme activity reached about 80% of the wild-type levels of photosynthesic activity. This reduction in maximal photosynthetic activity, however, did not lead to a decrease in crop productivity, since no tuber yield penalty was observed. On the other hand, plants with only 12% of the wild-type FBPase activity produced only 25% of tubers on a freshweight basis. Similar results have been obtained with tobacco plants with reduced ribulose-1,5-bisphosphate carboxylase levels, where even higher reductions in photosynthetic rates had only very little effects on the relative growth rate, even under high nitrogen supply (Stitt, Schulze; 1994). Tobacco plants do not have a significant storage sink organ, therefore one could argue that photosynthesis in tobacco is always sink-limited.

The examples discussed here demonstrate the extreme plasticity of whole plant carbon metabolism. The output (tuber yield) remains unchanged over a wide range of reduction of and does not seem to be limited by the photosynthetic capacity of potato plants.

3. Attempts to modify the starch content in potato tubers

Starch is the main storage carbohydrate in potato tubers and constitutes up to 80% of the dry matter of the organ. The main route directing carbon from sucrose into starch is believed to be as follows: Glucose-1-phosphate is activated by the enzyme ADP-glucose pyrophosphorylase leading to the formation of ADP-glucose, the substrate of the different isoforms of starch synthase. ADP-glucose pyrophosphorylase is the key regulatory enzyme of starch biosynthesis, since it is allosterically regulated by the levels of 3-phosphoglycerate and inorganic phosphate.

The important role of ADP-glucose pyrophosphorylase was demonstrated when plants were generated expressing very low levels of the enzyme (Müller-Röber et al., 1992). Tubers derived from these plants contain only minute amounts of starch, but high levels of soluble sugars, mainly sucrose and glucose.

In a complementary approach an increase of the starch content of potato tubers was achieved (Stark et al., 1992). Here an ADP-glucose pyrophosphorylase with modified allosteric properties was overexpressed in transgenic plants. However, contrasting reports have been made for other varieties of potato (Sweetlove et al., 1996). In this study the increased flux of carbon into starch was compensated for by an increase in starch degradation.

Other steps in the pathway from sucrose to starch were also subjected to genetic manipulation. One of the most prominent examples is the introduction of a yeast invertase into the apoplastic space of tubers (Sonnewald et al., 1997). This manipulation gave rise to plants forming fewer tubers with increased size, leading to a narrower tuber size distribution range. However, a reduction of the starch content also was measured. The mechanism leading to the reduction of the starch content is still under investigation.

4. Modification of starch structure in transgenic plants

To date only a few aspects of a rather complex pathway are understood. In plants the general pathway of starch synthesis comprises three enzymatic steps (Smith et al., 1997):

(1) the conversion of glucose-1-phosphate to ADP-glucose by the ADP-glucose pyrophosphorylase, the key regulatory enzyme of the pathway

(2) the transfer of glucose from ADP-glucose to an α-1,4-glucan by different isoforms of starch synthase (granule-bound or soluble)

(3) the introduction of branchpoints by different isoforms of branching enzyme.

Nearly all of these enzymes have been purified and biochemically characterized. However, *in vitro* only the production of soluble glucans has been demonstrated, the synthesis of starch granules using these enzymes has never been achieved. This indicates

that other enzymes are also involved in determining starch structure. The starch phosphorylases, glucanotransferases such as disproportionating enzyme, glucosyltransferases like the T-enzyme, and different hydrolases like the α- and β-amylases are candidates for further investigation. For the debranching enzymes (isoamylase or pullulanase) it is known that they are important in determining the final structure of starch. Mutants lacking isoamylase synthesize a more highly branched, water soluble polysaccharide called phytoglycogen (Ball et al., 1996).
Recently a novel cDNA encoding for the protein, which is responsible for the phosphorylation of potato starch, was described (Lorberth et al., 1998). Amylopectin can contain, depending on the plant organ where it is manufactured, different levels of phosphate monoesters. In potato tubers the phosphate content is exceptionally high as compared with other plant storage organs. Even though many starch biosynthetic enzymes are known, it has never been possible to explain how the phosphate monoesters are incorporated into starch. In order to address this question proteins bound to potato starch granules were isolated and corresponding cDNAs were cloned. One of the resulting clones, designated R1, encodes a protein of 160 kDa. If the expression of this gene is inhibited in transgenic potato plants, the phosphate monoester content of the starch synthesized in the transgenic lines is reduced down to 10% as compared with wild-type plants. This indicates that the R1 protein is responsible for the phosphorylation of starch. This is also supported by the fact that the expression of the protein in *Escherichia coli* leads to elevated phosphate contents of the glycogen synthesized. The existance of the R1 protein is not confined to Solanaceous species, but corresponding sequences are also found in Arabidopsis and rice, but not in bacteria, mammals and yeasts, indicating that it is a general but unique component of starch biosynthesis.

5. References

Ball et al. (1996) Cell 86, 349-352
Gifford RM et al. (1984) Science 225, 801-808
Inoue H, Tanaka A (1978) J Sc Soil Man Japan 49, 321-327
Koßmann J et al.(1994) Plant J. 6, 637-650
Lorberth et al. (1998) Nature Biotech. 16, 473-477
Müller-Röber B et al.(1992) EMBO J. 11, 1229-1238
Smith et al. (1997) Ann. Rev. Plant Physiol. Plant Mol. Biol. 48, 67-87
Sonnewald et al. (1997) Nature Biotech. 15, 794-797
Stark DM et al. (1992) Science 258, 287-292
Stitt M, Schulze D (1994) Plant Cell Environ. 17, 465-487
Sweetlove et al. (1996) Biochem. J. 320, 493-498

GENETIC MANIPULATION FOR THE PRODUCTION OF HIGH LYSINE CORN

B.A. Larkins, X. Wang, Y. Sun, P.A. Hughes and A.M. Clore
Department of Plant Sciences, University of Arizona, Tucson, AZ 85721

1. Introduction

Cereals provide 50% of the dietary protein for humans and can comprise 70% of the protein intake for people in developing countries (Deutscher, 1978). It is expected that the demand for cereal grains will increase dramatically in the future, as a consequence of the expanding human population, which could double by the year 2030 (Mann, 1997). Unfortunately, most cereals do not provide a nutritionally balanced source of protein (FAO/WHO, 1985). The abundant proteins they contain, the storage proteins or prolamins, are devoid of several amino acids essential for monogastric animals. The most limiting of these is lysine (Nelson, 1969). There are several naturally occurring mutants in cereals, such as *opaque*2 (*o*2) and *floury*2 (*fl*2) in maize (Nelson, 1969), that decrease prolamin (zein) synthesis and increase the lysine content of the endosperm, but the pleiotropic effect by which these mutations increase the synthesis of lysine-containing proteins is not understood.

We initiated research to determine the origin and genetic regulation of lysine-rich proteins in maize endosperm. After identifying proteins whose synthesis is increased in *o*2 mutants, we determined the concentration of elongation factor-1α (EF-1α or EF-1A) is significantly elevated (Habben et al., 1995). Indeed, we found the concentration of EF-1A is highly correlated with the total lysine content of the endosperm ($r^2 = 0.9$), even though the protein itself accounts for only 1-2% of the lysine (Sun et al., 1997). The nature of the other lysine-rich proteins with which EF-1A is allied is unknown.

EF-1A is a multifunctional protein, and the basis for the increase of its synthesis in *o*2 mutants could result from one or more of its roles. Since the amount of EF-1A cannot explain the correlation with lysine content, there must be a stoichiometric relationship between the concentration of EF-1A and other lysine-rich proteins. To gain insight into the nature of these proteins, we have localized EF-1A in endosperm cells, and we have begun to purify and characterize isoforms of EF-1A.

A. Altman et al. (eds.), Plant Biotechnology and In Vitro Biology in the 21st Century, 583–590.

2. Materials and Methods

Indirect immunofluorescence and laser scanning confocal microscopy were used to visualize the cytoskeleton in unembedded sections of maize endosperm tissue. Sections 30 to 50 μm thick of fresh kernels were cut with a microtome. The sections were immediately treated with 100 μM m-maleimidobenzoyl N-hydroxysuccinimide estser (MBS) before a 1-hr fixation in 4% formaldehyde. Both the MBS stabilization and fixation steps took place in cytoskeletal stabilizing buffer (Abe et al., 1991). The sections were permeabilized with cellulase, followed by a 30-min incubation in 0.1 % TritonX-100 in PBS. Before incubation with antisera, the sections were treated with a blocking solution (Clore et al., 1996). Sections were analyzed with a Bio-Rad MRC 600 confocal laser scanning microscope with rhodamine or fluorescein filter sets.

For EF-1A purification, 100 g of 16 day after pollination (DAP) maize kernels was ground in 200 ml of buffer A (20 mM KPi, pH 7.2, containing 25% glycerol, 10 mM 2-mercaptoethanol, 0.1 mM PMSF and 1 mM iodoacetic acid) using a Polytron. The homogenate was centrifuged at 7000 rpm with a GSA rotor, filtered through 2 layers of Miracloth, and adjusted to 40, 60 and 80% NH_4SO_4. The pellet precipitated by 60% to 80% ammonium sulfate was collected, resuspended in buffer B (50 mM Tris-HCl, pH 8.0, containing 10% glycerol, 10 mM 2-ME and 0.1 mM PMSF) and loaded on an SP Sepharose column that was pre-equilibrated in buffer B. After washing, the column was eluted with a linear gradient of 0 to 1.0 NaCl in buffer B. The EF-1A purified by ion-exchange chromatography was dialyzed against buffer C (25 mM triethylamine, pH 10.5, containing 10% glycerol, 10 mM 2-ME and 0.1 mM PMSF) overnight and loaded onto a PBE 118 column equilibrated in the same buffer. After washing in buffer C, the "free (F)" form EF-1A was eluted with Pharmalyte 10.5-8 containing 10% glycerol, 10 mM 2-ME and 0.1 mM PMSF at pH 8.0. The "sticky (S)" form of EF-1A was subsequently eluted from the Pharmalyte column with a gradient of 0 to 1.0 M NaCl. A third, protein body-associated (P) form of EF-1A was purified following sucrose gradient fractionation of endosperm homogenate (Habben et al., 1993). After aspirating the protein bodies from the gradient, EF-1AP was solubilized using 1% deoxycholate; the protein was then purified by SP-Sepharose chromatography.

3.1 EF-1A is complexed in a cytoskeletal network surrounding protein bodies in developing maize endosperm

Based on the association of EF-1A with the ER and cytoskeleton (Merrick, 1992), and the *in vitro* association of polysomes and protein bodies with the actin cytoskeleton (Abe et al., 1991), we hypothesized that protein bodies, EF-1A, and the cytoskeleton are associated *in vivo*. To test this hypothesis, we examined the spatial relationships of microtubules, microfilaments, EF-1A, and protein bodies at early and late stages of endosperm development (Clore et al., 1996). During the early stage of development, the endosperm cells had not yet accumulated starch or storage proteins, while the older cells were actively synthesizing these products. These older cells were found to have an unusual cytoplasm; they did not have a central vacuole but contained numerous plastids and protein bodies. The protein bodies occur in the RER, as 1 μm spherical structures. Despite the fact that these cells were densely packed with storage products, we could elucidate the relative distributions of actin, microtubules, EF-1A, and protein bodies using laser scanning confocal microscopy.

We demonstrated that maize endosperm cells have an extensive cytoskeleton that changes with the onset of storage product deposition (Clore et al., 1996). In young cells, the cytoskeleton and EF-1A are usually in association with nuclei and/or in the cell cortex, while in older cells, they are associated with developing protein bodies. For example, when young endosperm cells were labeled using an actin antibody, followed by a secondary antibody conjugated to a fluorophore, signal was found both as meshworks around nuclei and as cables extending from the nuclei (Fig. 1A). In contrast, when the same antibodies were used to label the actin in older cells (Fig. 1B), actin filaments were found both in a transversely-oriented array in the cell cortex and also throughout the cytoplasm between the starch grains. In addition, clusters of actin were visualized between the starch grains (Fig. 1B). By performing dual immunofluorescence to simultaneously label actin and γ-zein (which resides on the periphery of protein bodies), we demonstrated that these actin clusters closely surround protein bodies (not shown). This finding corroborated those of previous studies which described protein bodies isolated from endosperm homogenates as being associated with actin (Abe et al., 1991; Habben et al., 1993).

Microtubules were labeled using α-tubulin monoclonal antibodies. In young cells (Fig. 1C), microtubules were found to emanate from the nucleus and in a roughly parallel array adjacent to the cell wall. Microtubules in older endosperm cells are distributed in two arrays: one consists of roughly parallel, cortical microtubules beneath the cell wall (not shown), while a second, multidirectional network permeates the cytoplasm between the starch grains (Fig. 1D). Dual immunofluorescence revealed these microtubules to be closely juxtaposed with protein bodies (Fig. 1D). The cortical array is commonly described in interphase plant cells, but the multi-directional cytoplasmic array is more unusual, and its function is poorly characterized (Baskin et al., 1992; Huang, Russell, 1994; Webb, Gunning, 1994).

In young cells not yet undergoing storage product deposition, EF-1A is found in and around the nucleus and in membranous structures emanating from the nucleus (Fig. 1E). In the older cells, this protein is found between starch grains surrounding spherical structures 1 μm in diameter (Fig. 1F). These structures were determined to be protein bodies (Clore et al., 1996). Simultaneous labeling of EF-1A and actin demonstrated that the two are colocalized around protein bodies, and pretreatment of the endosperm with cytochalasin D, which disrupts filamentous actin, caused EF-1A to become dispersed throughout the cytoplasm (not shown). These findings suggest that EF-1A exists in a complex with actin and may bundle actin filaments around the ER. Indeed, we have shown that maize EF-1A can bundle yeast actin *in vitro* (Sun et al., 1997).

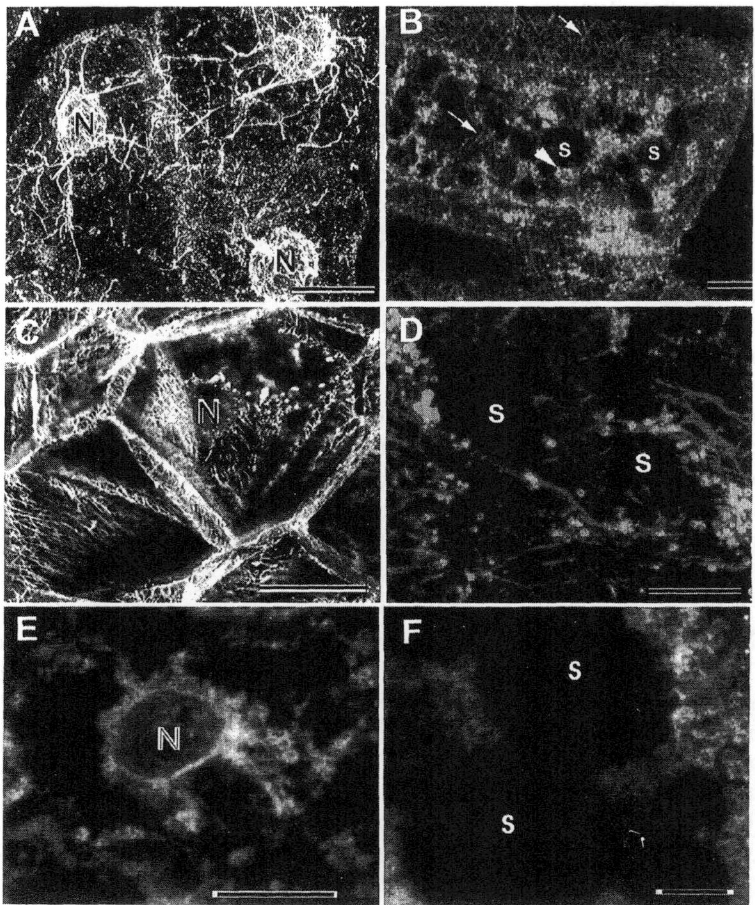

Fig. l. Actin (A and B), microtubules (C and D) and EF-1A (E and F) in young maize endosperm cells prior to storage product (starch and storage protein) deposition (A, C, and E), and in older endosperm cells after the onset of storage product deposition (B, D, and F). Visualization of these components using indirect immunofluorescence and confocal microscopy revealed changes in their distributions during development. (A) Actin in young cells. At this stage, actin is found in "baskets" around the nuclei (N) and in strands emanating from the nuclei toward the cell wall. Bar = 15 μm. (B) Actin in an older cell. Actin is now found in filaments (arrows) both adjacent to the cell wall and in the cytoplasm between the starch grains (S). Actin is also found in clusters (e.g. arrowhead) between the starch grains. Bar = 20 μm. (C) Microtubules in young cells. Microtubules emanate in multiple directions from the surface of the nuclei (N) and are also found in parallel arrays in the cell cortex (beneath a wall parallel to the plane of section in the cell to the lower left). Bar = 15 μm. (D) Microtubules (gray filamentous structures) and protein bodies (light gray circles) in an older cell. In such cells, the microtubules exist in a parallel array in the cell cortex (not shown), as well as in a multidirectional array between the starch grains (S). Protein bodies are found adjacent to these microtubules. Bar = 10 μm. (E) EF-1A in a young endosperm cell. EF-1A is found primarily around the nucleus (N) and in membranous structures near the nucleus. A lower level of signal is also found inside the nucleus. Bar = 15 μm. (F) EF-1A in an older endosperm cell. This protein is now found around spherical structures 1 μm in diameter which were determined to be protein bodies (Clore et al., 1996). S = starch grains. Bar = 5 μm. (Figure reprinted from Clore, Larkins, 1998.)

3.2 Biochemical analysis of EF-1A isoforms in maize endosperm

We have identified three distinct forms of EF-1A in maize endosperm: free (F), sticky (S) and protein body bound (P). Each of these may reflect a different physiological function of the protein. We have recovered EF-1AF and EF-1AS in nearly homogeneous preparations. Both proteins are precipitated with 60 to 80% ammonium sulfate and adsorbed on an SP Sepharose column at pH 8.0. Following elution, EF-1AF and EF-1AS are separated with a chromatofocusing column (PBE 118) (Figs. 2 and 3). EF-1AF is eluted with Polybuffer 10.5-8, and EF-1AS is then recovered with a linear gradient (0-1.0 M) of NaCl. The two proteins have nearly identical molecular weights and strongly react with polyclonal antibodies against maize EF-1A.

We have begun to characterize the biochemical properties of EF-1AF and EF-1AS. Dr. Wendy Boss' laboratory showed that EF-1A purified from carrot suspension cultures is able to activate phosphatidylinositol-4 kinase (PI-4K) activity (Yang et al., 1993), and that this activity is positively correlated with the phosphorylation of EF-1A by a calcium dependent protein kinase (CDPK). In collaboration with Dr. Boss, we analyzed the ability of EF-1AF and EF-1AS to activate PI-4K. The results showed EF-1AF is three-times more active than EF-1AS in stimulating PI-4K activity (Fig. 4). When we tested the ability of these isoforms of eEF1 to be phosphorylated by CDPK, we found EF-1AF can be phosphorylated, while EF-1AS cannot (Fig. 5). Thus, it appears these proteins are biochemically distinct.

Recently, we succeeded in purifying a third isoform, EF-1AP, that is associated with the RER surrounding protein bodies. This isoform is biochemically distinct from the S and F proteins, being much more hydrophobic. We do not know the basis of the biochemical differences between the three proteins, but experiments are in progress to compare peptides maps based on two dimensional HPLC-MS. This analysis will reveal the basis for differences in primary amino acid sequences as well as post-translational modifications. We intend to compare the biological activities of these different forms of EF-1A, based on their phosphorylation by CDPK, activation of PI-4K, aminoacyl-tRNA

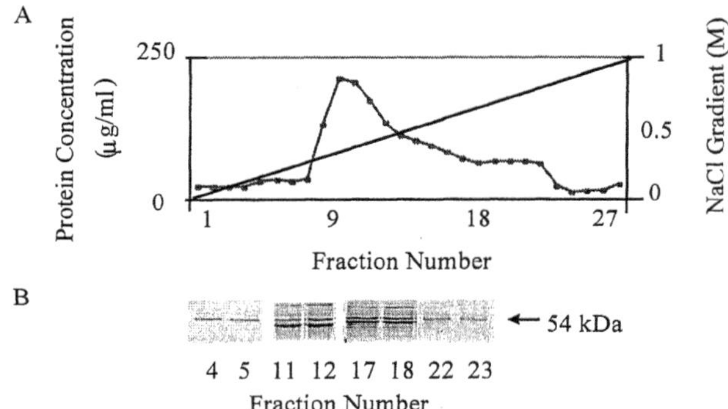

Fig. 2. Separation of EF-1AS from EF-1AF by PBE 118 chromatography. (A) EF-1AF was eluted with Polybuffer 10.5-8, and EF-1AS was subsequently recovered using a 0 to 1.0 M NaCl gradient. Proteins were monitored with a Bio-Rad Protein Assay reagent. (B) Aliquots of 20 µl from each fraction were separated by SDS-PAGE and stained with Coomassie-blue.

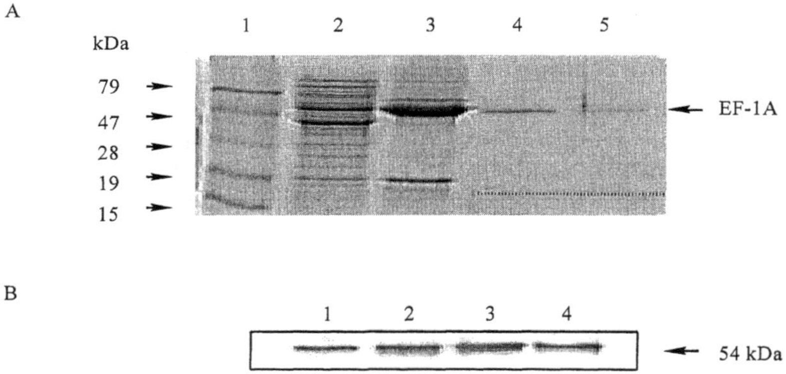

Fig. 3. SDS-Page and immunoblotting of EF-1A during purification. (A) EF-1A from 60-80% ammonium sulfate pellet (Lane 2) was enriched by ion-exchange SP Sepharose chromatography (Lane 3) and separated into EF-1AF (Lane 4) and EF-1AS (Lane 5). Lane1, prestained protein markers. Proteins from each purification step were separated by SDS-PAGE and stained with Coomassie-blue. (B) Immunoblotted EF-1A from the from the 80% ammonium sulfate pellet (Lane 1), SP Sepharose column (Lane 2), and EF-1AF (Lane 3) and EF-1AS (Lane 4) from the PBE 118 column. The immunoblot was stained with polyclonal antibodies against a recombinant maize EF-1A.

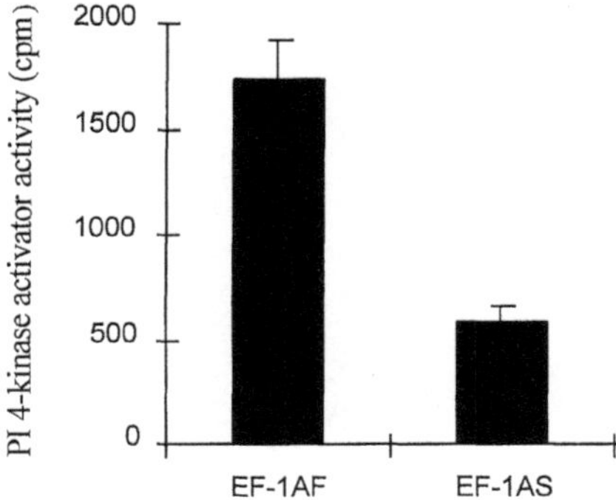

Fig. 4. Activation of PI-kinase by EF-1AF and EF-1AS. The activity of PI-kinase (assayed by ^{32}P incorporation into PI) was measured in counts per minute (cpm), and the activator activity was calculated by subtracting the control (without addition of EF-1A). The error bars indicate the standard deviation from three assays.

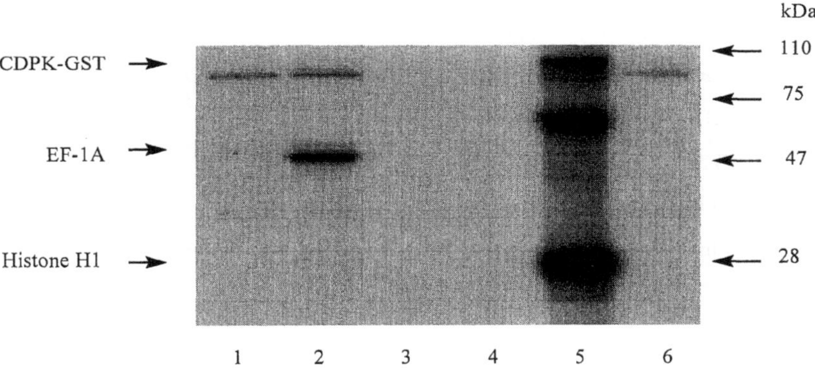

Fig. 5. Autoradiogram showing differential phosphorylation of EF-1AF and EF-1AS after treatment with CDPK. Lane 1, EF-1AS treated with CDPK; Lane 2, EF-1AF treated with CDPK; Lane 3, EF-1AS without CDPK; Lane 4, EF-1AF without CDPK; Lane 5, a positive control showing histone H1 treated with CDPK; and Lane 6, CDPK by itself.

binding to ribosomes, and actin and microtubule bundling. By determining the relative level and biological activities of each of the isoforms and their physiological properties, we hope to gain insight into the process or structure modified in *o2* mutants that underlies the pleiotropic increase in lysine-rich proteins.

4. References

Abe S et al (1991) Protoplasma 165, 139-149.

Baskin TI et al (1992) Planta 187, 405-413.

Clore AM et al (1996) Plant Cell 8, 2003-2014.

Clore AM, Larkins BA (1998) Plant Physiol. (in press).

Deutscher D (1978) Nutritional Improvement of Food and Feed Grains, pp. 281-300, Plenum, New York.

FAO/WHO (1985) Energy and Protein Requirements, WHO Tech. Rept. Ser. No. 724, World Health Organization, Geneva.

Habben JE et al (1993) Plant Mol. Biol. 23, 825-838.

Habben JE et al (1995) Proc. Natl. Acad. Sci. USA 92, 8640-8644.

Huang B-Q, Russell SD (1994) Planta 194, 200-214.

Mann C (1997) Science 277, 1038-1043.

Merrick WC (1992) Micro. Rev. 56, 291-315.

Nelson OE (1969) Adv. Agron. 21, 171-194.

Sun Y et al (1997) Plant Physiol. 115, 1101-1107.

Webb MC, Gunning BE (1994) Sex Plant Reprod. 7, 153-163.

Yang W et al (1993) J. Biol. Chem. 268, 392-398.

ETHYLENE, POLYAMINES AND FRUIT RIPENING

ROSHNI MEHTA[1], AVTAR HANDA[2], AUTAR MATTOO[1,3]

[1]USDA Vegetable Laboratory, Building 010A, Agricultural Research Center-W, Beltsville, MD 20705, USA, and [2]Department of Horticulture, Purdue University, W. Lafayette, IN 47907, USA. [3]Address correspondence to this author

1. Introduction

Ethylene is a gaseous plant hormone which influences many aspects of plant growth, development and senescence. In addition, ethylene is produced by fungi and bacteria, and potentially influences host-pathogen and host-pest interactions (1,2). Although some progress has been made in identifying the receptors for ethylene action and in elucidating ethylene signal transduction pathway in higher plants (3,4), not much is known about how microbial pathogens or phytophagous insects recognize ethylene or how its action gets transmitted. Ethylene likely influences some or more of the metabolic processes via interactions with other plant growth regulators and hormones that regulate plant growth and development (5). Historically, five principal classes of phytohormones are recognized: auxins, gibberellins, cytokinins, abscisic acid, ethylene. To these, newly established ones should be added, brassinosteroids (6) and methyljasmonate (7). In addition, other growth regulators such as polyamines (8) and salicylic acid (9) likely interact with these hormones to specifically regulate plant metabolism or plant responses to biotic and abiotic stresses. Auxins, gibberellins, cytokinins, brassinosteroids, and polyamines are generally considered as promotors of growth and development while methyljasmonate, abscisic acid and ethylene promote senescence and cell death. The shift of a cell from growth and development to senescence is a commitment determined by not only the relative levels of these two sets of growth regulators, but also by the sensitivity of that particular cell to perceive different hormones, individually or in a certain combination.

2. Cross-talk between polyamines and ethylene biosyntheses pathways

We are interested in understanding the key molecular mechanisms involved in hormonal regulation of plant metabolism. Of particular interest are the factors that predispose a fruit or a leaf cell to senesce, and means by which these factors could be neutralized in a timely manner. Ethylene production in the fruit is induced after growth and cell expansion are completed, which results in the promotion of the senescence process, a predominantly catabolic one. Polyamines (putrescine, spermidine, spermine), on the other hand, accumulate and function during cell division and growth of the plant, and act as anti-senescence growth factors (8). A temporal relationship has been observed

A. Altman et al. (eds.), Plant Biotechnology and In Vitro Biology in the 21st Century, 591–595.

between polyamines and ethylene during plant development - this has led to suggestions that changes in the levels of polyamines and ethylene influence specific physiological processes in plants (10). Both share a common intermediate, S-adenosylmethionine (SAM), and a common byproduct, methylthioadenosine (11-13). Interestingly, *in vitro* studies have shown that polyamines inhibit ethylene biosynthesis in a variety of fruit and vegetative tissues, while ethylene suppresses the accumulation of polyamines (14,15). It has been shown that polyamines inhibit ethylene biosynthesis by suppressing the induction of ACC synthase, which catalyzes the formation of 1-aminocyclopropane-1-carboxylic acid (ACC) from SAM, and ACC oxidase, which converts ACC into ethylene. The inhibition of the ethylene biosynthesis enzymes generates a feed back causing SAM to accumulate which is then channeled into polyamine biosynthesis (16,17). Biosynthesis of polyamines initiates with the synthesis of diamines, putrescine and cadaverine, from arginine or lysine, respectively (8). Aminopropyl group from decarboxylated SAM, formed from SAM by SAM decarboxylase, is donated to form spermidine, and, in turn, spermidine reacts with another aminopropyl group from decarboxylated SAM to form spermine. Ethylene-mediated suppression of polyamine biosynthesis seems to occur by an inhibition of SAM decarboxylase, a key enzyme in the polyamine biosynthesis. Thus, these results have led to the hypothesis that there is a cross-talk between these two biosynthetic pathways, viz., ethylene biosynthesis and polyamines biosynthesis (18). If this were true, then the implications of these findings are that one can interfere with the ripening process by increasing the production of polyamines during the senescence phase and empower the plant cell to reverse parts of the senescence process, maintain or overaccumulate desirable nutrients, and prolong the shelf-life of the produce. The exact nature and extent of interrelationships between ethylene and polyamines at both the physiological and biosynthetical levels still remain to be resolved. Further, although polyamines have been implicated in many physiological processes in plants, a definite role of these metabolites in plant metabolism, growth and development has not yet been demonstrated.

Kushad and Dumbroff (19) suggested several approaches to probe the proposed relationships between polyamines and ethylene: one, to monitor the levels of polyamines and ethylene, the corresponding key enzyme activities, and the metabolic flux of radiolabeled metabolites; second, to develop and use appropriate genetic mutants that show delayed senescence or a wide variation in their levels of individual polyamines; and third, to produce transgenic plants transformed with antisense gene constructs for the key genes involved, ACC synthase, spermidine synthase or SAM decarboxylase fused to regulatable promoters. Another approach of choice to address the role of polyamines in fruit ripening is to introduce and express a reconstructed polyamine-producing gene in the fruit following initiation of ripening so that polyamines accumulate at a stage when they are normally at very low levels (18,20).

3. Accumulation of polyamines by regulated expression of yeast SAM decarboxylase in ripe tomato fruit

To test if tomato metabolism could be modifed by introducing at the initiation of normal senescence a key gene necessary for production of polyamines, we genetically transformed tomato plants with a construct of S-adenosylmethionine decarboxylase, the enzyme that catalyzes the first commited step in polyamine biosynthesis, fused to the regulatable E-8 promoter to enable accumulation of polyamines in the fruit only upon ripening. The E-8 promoter is developmentally (ripening) regulated and by ethylene (21).

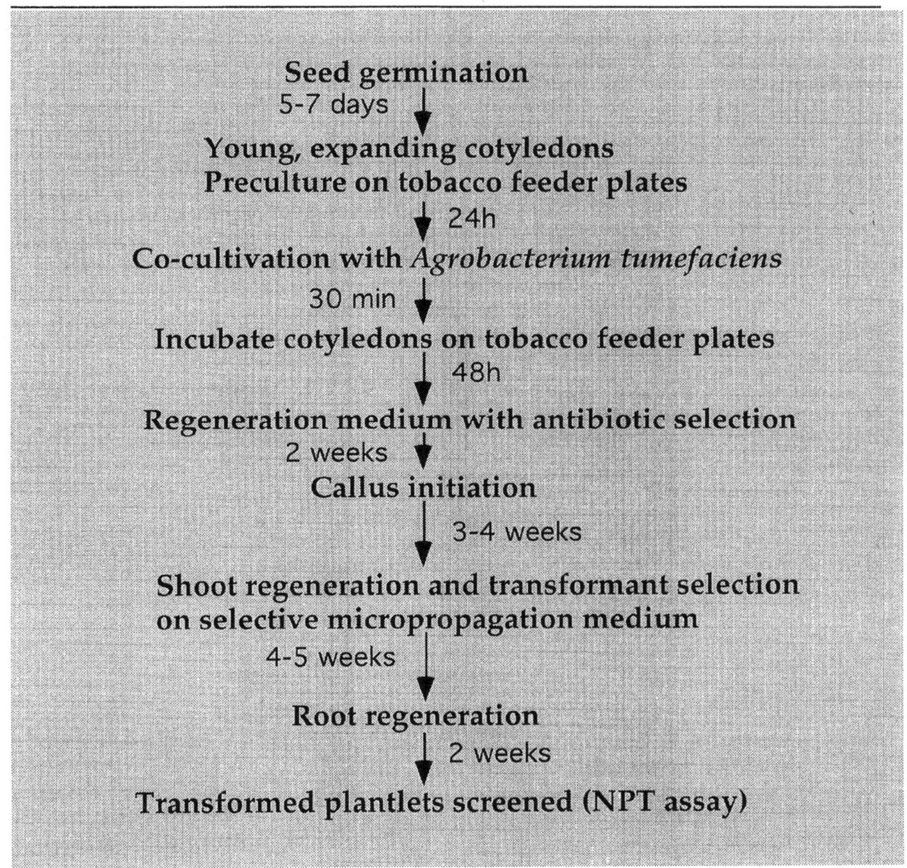

Figure 1
Protocol for transformation and regeneration of transgenic tomato plants

Analysis of such fruit should reveal if continued accumulation of polyamines interferes with the expression of ethylene biosynthesis genes and the ethylene receptor gene family, and thereby result in modulating ripening process. Such transgenic fruit can also be used as a model to study the role of polyamines in the ripening/senescence process. For instance, accumulation of polyamines may change the redox of the cells due to their property of being anti-oxidative in nature (22) and by their ability to stabilize cellular membranes polyamines may be conducive for tilting the balance towards anabolic from catabolic metabolism (23).

Tomato transformation has become a routine in several laboratories. A summary is given in Figure 1. Putative transformants were selected using kanamycin - antibiotic marker selection, and the tissue grown into seedlings. A number of R_0 transformed lines showed growth aberrations and in some cases seeds did not germinate, when grown under greenhouse conditions. The number of seeds per fruit in some of the transformed lines varied from 2 to 78 as compared to an average of 89 in the non-transformed wild type. The fruit from the several transformed lines was analyzed for stable integration of the gene, polyamine content, soluble solids, color development, and shelf-life. Gene integration was checked by Southern blotting of polymerase chain reaction (PCR)-amplified tomato genomic DNA. Eeach transformant integrated one or more copies of the SAMdc gene.

Confirmation that the transgenic lines had integrated the heterologous SAM decarboxylase gene in a functional manner came from assaying the levels of polyamines, putrescine (Put), spermidine (Spd) and spermine (Spm) in the fruit of these lines as well as in the SAMDC15 azygous line and non-transformed wild type. The data indicated that indeed, in the red ripe fruit of some lines, polyamines did accumulate - lines 1 and 16 preferentially accumulated Put whereas line 2 predominantly accumulated Spd, Spm being almost undetectable in lines 1, 6 and 10.

Finally, we tested the ability of these fruit to produce ethylene. Ethylene production in three transgenic lines was one-half or one-third of the control, in two other lines it was 2 to 3 times higher than the control, while another line produced as much ethylene as the control, wild-type fruit. These various lines provide us valuable tools to determine any linkages between the ability of the fruit to accumulate polyamines and their levels of ethylene production and if ethylene perception is affected. In order to draw concrete conclusions, these plants need to be taken to the next generation, segregation analysis done and more detailed biochemical and physiological analyses performed on azygous, hemizygous and homozygous lines.

4. References

1. Mattoo, A.K. and J. Suttle (1991) *The Plant Hormone Ethylene*. CRC Press. Boca Raton, Florida, 337 pp.
2. Abeles, F. B., P.W. Morgan, and M. E. Saltveit Jr (1992) *Ethylene in Plant Biology*, (San Diego, California: Academic Press), p.414
3. Smalle, J., and Van Der Straeten, D. (1997) Ethylene and vegetative development. *Physiol. Plant.* **100**, 593-605.
4. Brown, K.M. (1997) Ethylene and abscission. *Physiol. Plant.* **100**, 567-576.
5. Suttle, J.C. (1991) Ethylene interactions with other endogenous growth substances, in *The Plant Hormone Ethylene*, A.K. Mattoo and J.C. Suttle, eds., (Boca Raton, Florida: CRC Press, Inc.), pp. 115-131.
6. Clouse, S.D., and Sasse, J.M. (1998) Brassinosteroids: Essential regulators of plant growth and development. *Annu. Rev. Plant Physiol. Plant Mol. Biol.* **49**, 427-451.
7. Creelman, R.A., Tierney, M.L., and Mullet, J.E. (1992) Jasmonic acid/methyl jasmonate accumulate in wounded soybean hypocotyls and modulate wound gene expression. *Proc. Natl. Acad. Sci. U.S.A.* **89**, 4938-4941.
8. Slocum, R.D., and Flores, H.E. (1991) Eds. *Biochemistry and Physiology of*

Polyamines, CRC Press, Boca Raton, Florida, p 264.
9. Leslie, C.A., and Romani, R.J. (1986) Salicylic acid: A new inhibitor of ethylene biosynthesis. *Plant Cell Reports* **5**, 144-146.
10. Mattoo, A.K., and White, B. (991) Regulation of ethylene biosynthesis, in*The Plant Hormone Ethylene*, A.K. Mattoo and J.C. Suttle, eds., (Boca Raton, Florida: CRC Press, Inc.), pp. 21-42.
11. Fluhr, R., and Mattoo, A.K. (1996) Ethylene: biosynthesis and perception. *Crit. Rev. Plant Sci.* , B.V. Conger, ed., CRC Press, Inc., Boca Raton, Florida, **15**: 479-523.
12. Imaseki, H. (1991) The biochemistry of ethylene biosynthesis, in *The Plant Hormone Ethylene*, A.K. Mattoo and J.C. Suttle, eds., (Boca Raton, Florida: CRC Press, Inc.), pp. 1-20.
13. Kende, H. (1993) Ethylene biosynthesis. *Annu. Rev. Plant Physiol. Plant Mol. Biol.* **44**: 283-307.
14. Apelbaum, A., Burgoon, A.C., Anderson, J.D., Lieberman, M., Ben-Arie, R., and Mattoo, A.K. (1981) Polyamines inhibit biosynthesis of ethylene in higher plant tissue and protoplasts. *Plant Physiol.* **68**, 453-456.
15. Suttle, J.C. (1981) Effect of polyamines on ethylene production. *Phytochemistry* **20**, 1477-1480.
16. Even-Chen, Z., Mattoo, A.K., and Goren, R. (1982) Inhibition of ethylene biosynthesis by aminoethoxyvinylglycine and by polyamines shunts label from 3,4-(^{14}C)methionine into spermidine in aged orange peel discs. *Plant Physiol.* **69**: 385-388.
17. Roberts, D.R., Walker, M.A., Thompson, J.E., and Dumbroff, E.B. (1984) The effects of inhibitors of polyamine and ethylene biosynthesis on senescence, ethylene production and polyamine levels in cut carnation flowers. *Plant Cell Physiol.* **25**: 315-322.
18. Mehta, R.A., Handa, A., and Mattoo, A.K. (1997) Interactions of ethylene and polyamines in regulating fruit ripening, in A.K. Kanellis et al. (eds.), *Biology and Biotechnology of the Plant Hormone Ethylene*, Kluwer Academic Publ., The Netherlands, pp. 321-326.
19. Kushad, M.M., and Dumbroff, E.B. (1991) Metabolic and physiological relationships between the polyamine and ethylene biosynthetic pathways, in R.D. Slocum and H.E. Flores, (eds.) *Biochemistry and Physiology of Polyamines*, CRC Press, Boca Raton, Florida, pp. 77-92.
20. Mehta, R.A. and A.K. Mattoo (1995) Gene expression and protein dynamics during tomato fruit ripening, in *Postharvest Physiology, Pathology and Technologies for Horticultural Commodities: Recent Advances* (A. Ait-Oubahou and M. El-Otmani, eds.), (Institut Agronomique et Veterinaire Hassan II, Agadir, Morocco, pp. 343-352.
21. Deikman, J., Kline, R., and Fischer, R.L. (1992) Organization of ripening and ethylene regulatory regions in a fruit-specific promoter from tomato (*Lycopersicon esculentum*). *Plant Physiol.* **100**, 2013-2017.
22. Drolet, G., Dumbroff, E.B., Legge, R.L., and Thompson, J.E. (1986) Radical scavenging properties of polyamines. *Phytochemistry* **25**, 367-371.
23. Ben-Arie, R., Lurie, S., and Mattoo, A.K. (1982) Temperature-dependent inhibitory effects of calcium and spermine on ethylene biosynthesis in apple discs correlate with changes in microsomal membrane microviscosity. *Plant Science Lett.* , 239-247.

HOW GENES PAINT FLOWERS

Joseph Mol* , Erich Grotewold** and Ronald Koes*
*Dept. of Genetics, Vrije Universiteit, De Boelelaan 1087, 1081HV Amsterdam, The Netherlands, **Dept. of Plant Biology, 218 Rightmire Hall, Ohio State University, Columbus, Ohio, USA.

Introduction

The biosynthesis of flower pigments is perfectly suited to study the spatio-temporal control of gene expression and the compartmentalization of metabolites. Because of the commercial value of flowers, their pigmentation has been a subject of applied research for nearly four centuries. Until recently, it was only possible to develop new varieties by traditional breeding techniques (i.e. continuous crossing and selection) and, to a lesser extent, by mutation breeding. Today, however, transgenic approaches are also being introduced. The impressive results of traditional breeding are obvious to anyone who visits flower exhibitions and auctions. These varieties with their range of colours and patterns, constitute an enormous mutant collection for scientific study.
We will briefly summarize our current understanding of the molecular mechanisms underlying flower colouration (a more exhaustive account in given in Mol, 1998).

Factors involved in the determination of flower colour

The colouration of flowers and fruits is due to the accumulation of flavonoids (including anthocyanins), carotenoids and betalains. Anthocyanins are the major flower pigments in higher plants and have been studied extensively during the past decades (Elooma, Holton 1994; Koes et al. 1994; Holton, Cornish 1995; Mol et al. 1996; Mol et al. 1998). Although the accumulation of anthocyanins is a prerequisite for pigmentation, a range of additional factors such as co-pigmentation, vacuolar pH and cell shape, determine flower hue.

Anthocyanin synthesis and storage

The biochemistry and enzymology of the anthocyanin pathway is well understood and virtually all the genes that encode the biosynthesis enzymes have been isolated (Holton, Cornish 1995). In many cases mutations in such genes result in the accumulation of pathway intermediates giving new flower or seed colours. However, no plant species displays all the possible flower colour. In contrast to *Petunia hybrida* (petunia), species such as *Rosa hybrida* (rose) and chrysanthemum do not synthesize the purple delphinidin derivatives because they lack flavonoid 3'5'-hydroxylase (F3'5'H) activity (Elomaa, Holton 1994). Similarly, orange pelargonidin-type anthocyanins are not found in petunia, because the petunia dihydroflavonol 4-reductase (DFR) enzyme does not accept the dihydrokaempferol precursor as a substrate (Meyer et al. 1987).

A. Altman et al. (eds.), Plant Biotechnology and In Vitro Biology in the 21st Century, 597–600.

The anthocyanin pathway enzymes, with the exception of the cytochrome P450 mono-oxygenases (F3'H, F3'5'H), are recovered in cytoplasmic extracts whereas the anthocyanins are stored in the vacuole. Recently, the first clues have been obtained on the mechanism that transports anthocyanins into the vacuole. In maize, mutations of the *bronze-2* gene (*bz2*) result in kernels that do not contain anthocyanins. A similar mutation has been identified in petunia (an9; Alfenito et al. 1998). The BZ2 and AN9 proteins share homology to glutathione S-transferases and are able to conjugate a glutathione moiety to artificial substrates (Marrs 1995, 1996). It is conceivable that the addition of glutathione to anthocyanins tags them for transport into the vacuole. The details of this reaction are still unclear (Marrs 1996).

Co-pigmentation
The colour of anthocyanins is influenced by the presence of metal ions and copigments such as flavonols and flavones. Co-pigmentation involves the formation of stacked complexes that cause a blue shift in the anthocyanin absorption spectrum. Flavonols are yellow or colourless flavonoids and, like anthocyanins, are derived from dihydroflavonols. In petunia, the *Fl* locus encodes flavonol synthase (FLS), a 2-oxoglutarate-dependent oxygenase. The enzyme influences flower colour in different ways: by co-pigmentation and by substrate competition. Competition between FLS and DFR for common dihydroflavonol substrates can shift the flavonol:anthocyanin ratio (Holton et al. 1993). Flavonols impart an ivory colour to 'white' flowers but insects recognize flavonol-producing flowers much better by virtue of their strong UV absorption.

Vacuolar pH
Flowers of many plant species, when senescing, display a blue shift in colour. This is accompanied by an increase of the vacuolar (Yoshida et al. 1995). Both environmental and genetic factors appear to control this feature. In petunia, seven loci have been defined (*ph1-ph7*) that, when mutated, cause blueing of the flower and increase the pH of petal extracts. One of these loci, *ph6*, was isolated by transposon tagging using the maize element *Ac* (Chuck et al. 1993). The recent cloning of *anthocyanin1* (*an1*), which encodes a bHLH regulator of anthocyanin biosynthesis, revealed that *an1* and *ph6* are alleles of the same locus (C. Spelt, unpublished). *An2* and *an11*, encoding a MYB and WD40 regulator of anthocyanin biosynthesis respectively, also display a pH effect when mutated (F. Quattrocchio, unpublished). This indicates that an1, an2 and an11 downstream genes should contribute to pH control and raises the question as to whether a genetic relationship exists between regulatory *an* and *ph* loci. Epistatis studies suggest that *ph4* and *an1* operate in the same pH pathway; *ph3* appears to participate in a second pathway (F. Quattrocchio, unpublished).

Cell shape
The shape of the cells that accumulate anthocyanin pigments influences their optical properties and thereby the colour that one perceives. Cells of the inner epidermis of wild-type *Antirrhinum majus* petals are conical which confers greater light absorption and a velvet sheen. The fainter colour of *mixta* mutants results from flattening of these epidermal cell (Noda et al. 1994). The homology of *mixta* with genes encoding *Myb*-domain proteins suggests that the MIXTA protein controls cell shape by regulating gene expression. Mutation of a homologous gene in petunia (*mybPh1*) results in a similar phenotype (van Houwelingen et al. 1998; L. Mur, unpublished). In petunia *shrivelled up* (*shp*) mutants, a large fraction of the cells in the petal epidermis have a collapsed 'flat tire' appearance, that results in a drastic change of the flower colour (van Houwelingen et al.

1998). Neither the *mixta* nor *shp* downstream target genes have been identified, and therefore the molecular mechanism underlying these cell-shape changes remains to be resolved.

Transcriptional regulation and control of pigmentation pattern

Cell-specific accumulation of anthocyanins produces pigmentation patterns. The expression of anthocyanin biosynthesis genes is largely regulated at the transcriptional level and thus, pigmentation patterns reflect the cell-specific expression patterns of the regulatory genes.

Mutants define anthocyanin-specific regulators

Two classes of mutants in which anthocyanin synthesis is abolished have been identified in numerous plant species. The first affects flavonoid biosynthesis enzymes whereas the second influences regulatory loci. Molecular analysis of anthocyanin regulatory loci in maize showed that they represent two gene families encoding MYB-domain (*c1*) and bHLH (*r*) transcription factors resp. Both families comprise multiple paralogous genes that encode similar proteins with different expression patterns . The *c1* and *r* families activate the entire set of anthocyanin biosynthesis genes. Thus, their expression patterns could directly determine the pigmentation pattern in maize. In support of this, ectopic expression of *r* and *c1* genes in maize is sufficient to induce anthocyanin synthesis in otherwise unpigmented maize tissues (Ludwig et al. 1990).

More recent analyses of anthocyanin biosynthesis in dicot flowers revealed large variations on this theme. First, ectopic expression of *r* and *c1* in dicots increases anthocyanin pigmentation in some but not all tissues (Lloyd et al. 1992; Quattrocchio et al. 1993,1998). Second, the regulatory anthocyanin loci identified in *Antirrhinum majus* and petunia control the anthocyanin-specific genes, but they differ in their control over the early genes that are common to the synthesis of multiple classes of flavonoids. Nevertheless, they all encode similar proteins. The *an2* gene from petunia encodes a MYB-domain protein with functional and structural similarity to C1. Transient expression assays, in which these MYB and bHLH-type regulators were swapped between maize and petunia, indicate that the species-specific regulation resides in the promoters of the downstream anthocyanin biosynthesis genes, rather than in the regulatory proteins themselves (Quattrocchio et al. 1998). There are various evolutionary mechanisms that might account for this variation in the control of common genes (Koes et al. 1994; Quattrocchio et al. 1993).

The petunia *an11* locus encodes a cytosolic protein with five WD40 repeats which are thought to mediate protein-protein interactions. Transcription of the *an11* locus is ubiquitous and does not require the *an1* and *an2* loci, indicating that AN11 is a more general and upstream regulator. This is supported by the observation that *an11* mutants are rescued by transient expression of *an2* and by its conservation in fungi and animals (de Vetten et al. 1997).

Negative regulators of anthocyanin biosynthesis have been identified in maize (*intensifier1*) and *Arabidopsis*. (*banyuls* ; *icx1*). *In1* encodes a bHLH protein with similarity to R; *banyuls* and *icx1* have not been cloned yet (for more details consult Weisshaar, Jenkins 1998; Mol et al. 1998).

Specificity of regulatory proteins

The anthocyanin branch of flavonoid biosynthesis is regulated by the co-action of a MYB and bHLH transcription factor. Some of the early 'common' flavonoid genes have redundant regulatory circuits. For example, in Antirrhinum *pal, chi* and *f3h* seem to be regulated by two redundant *myb* genes (Moyano et al. 1996), whereas in maize *chs, chi* and *dfr* genes are triggered by a MYB/bHLH couple (anthocyanin biosynthesis), but also by a single *myb* gene product termed P that does not seem to require a bHLH co-factor (Phlobaphene biosynthesis; Grotewold et al. 1994). This redundancy serves two purposes: first, competition for common binding sites combined with different transcriptional activatory potentials of redundant *myb* genes may provide gearing and temporal control of target gene expression. These features could be fine-tuned by phosphorylation/dephosphorylation and by association with other proteins. Secondly, the presence of high-affinity and low-affinity binding sites in a single promoter could ensure the activation of distinct branches of phenylpropanoid biosynthesis (Grotewold et al. 1994, Sainz et al. 1997).

Towards a blue rose

Fundamental research in the field of flower pigmentation also has applied aspects. Production of anthocyanins in vegetative tissues by forced expression of *r* and *c1* genes is feasible in plant species with the maize-type R/C1 control. In all other cases, only a colour enhancement may be seen in anthocyanin-producing cell types (Lloyd et al. 1992; Quattrocchio et al. 1998). About a decade ago, anthocyanin pathway engineering led to the creation of the 'orange' petunia variety (Meyer et al. 1987). A delphinidin-producing carnation has been generated by the biotechnology company Florigene. Once the controlled manipulation of co-pigmentation and vacuolar pH is achieved, the dream of creating a blue rose may finally come true.

References

Alfenito MR et al. (1998) Plant Cell 10, in press
Chuck G et al. (1993) Plant Cell 5, 371-378.
de Vetten N et al. (1997) Genes Dev. 11, 1422-1434.
Elomaa P, Holton T (1994) Genet. Engin. Rev. 12, 63-88.
Holton TA, Cornish EC (1995) Plant Cell 7, 1071-1083.
Holton TA et al. (1993) Plant J. 4, 1003-1010.
Koes RE et al. (1994) BioEssays 16, 123-132.
Lloyd AM et al. (1992) Science 258, 1773-1775.
Ludwig SR et al. (1990) Science 247, 449-450.
Marrs KA (1996) Annu. Rev. Plant Phys. Plant Mol. Biol. 47, 127-158.
Marrs KA et al. (1995) Nature 375, 397-400.
Meyer P et al. (1987) Nature 330, 677-678.
Mol J et al. (1996) Crit. Rev. Plant Sci. 15, 525-557.
Mol J et al. (1998) TIPS 3, 212-217.
Moyano E et al. (1996) Plant Cell 8, 1519-1532.
Noda K et al. (1994) Nature 369, 661-664.
Quattrocchio F et al. (1998) Plant J. 13, 475-488.
Sainz M et al. (1997) Plant Cell 9, 611-625.
van Houwelingen A et al. (1998) Plant J. 13, 39-50.
Weisshaar B, Jenkins GI (1998) Current Opinion in Plant Biol. 1, 251-257.
Yoshida K et al. (1995) Nature 373, 291.

BIOSYNTHESIS OF SCENT AND FLAVOR COMPOUNDS

Eran Pichersky[1], Natalia Dudareva[1,2], Jihong Wang[1], Leland Cseke[1] and Efraim Lewinsohn[3]
[1]Department of Biology, University of Michigan, Ann Arbor MI 48109, [2]Horticulture Department, Purdue University, West Lafayette IN 47907, USA, and [3]Aromatic, Medicinal and Spice Crops Unit, Newe Ya'ar Research Center, Agricultural Research Organization, P.O. Box 1021, Ramat Yishay 30095, Israel

1. Introduction

Flowers of many plants emit scents, which are almost always a complex mixture of small volatile compounds. Floral scents may function as both long- and short-distance attractants, and as nectar guides to a variety of animal pollinators (Dobson 1993). Insects are able to distinguish between complex floral scents, and discriminatory visitation based on floral scent has important implications for plant population structure and reproductive isolation (Dodson et al. 1969; Galen 1985; Pellmyr 1986; Stanton et al. 1986).

Floral scent is an economically important trait in some crop plants, since the presence or absence of a scent appropriate to the locally available insect pollinators may have a large impact on the level of pollination and therefore on seed and fruit set (Buchmann, Nabhan 1996). Plants imported into a new environment by humans may be especially disadvantaged in this regard, as they have not coevolved with the local pollinators. Understanding how the production of scent components is regulated may allow us in the future to manipulate these traits and enhance the attractiveness of flowers to local pollinators. Genetically engineering floral scent may also enhance the value of cut flowers to consumers.

Floral scents are dominated by terpenoid, phenylpropanoid, and benzenoid compounds, with a smattering of other chemicals (Knudsen, Tollsten 1993; Knudsen et al. 1993), all of which are regularly defined as "secondary metabolites". Constituents of floral scents are often found also as flavorants in fruits or edible vegetative parts. For example, linalool and eugenol are found in basil leaves (Simon et al. 1990), many terpenes are found in citrus fruit (Kealey, Kinsella 1978), and volatile esters are important flavor compounds in melons, peaches, and many other ripe fruits (Olias et al. 1995). Thus, understanding the biosynthesis of these compounds on the biochemical and molecular levels will allow the manipulation of fruit flavors as well.

2. Results and Discussion

2.1. Pathways and enzymes, and their evolution: We have developed enzymatic assays for the production of five scent components in *Clarkia breweri* (Onagraceae), a plant

A. Altman et al. (eds.), Plant Biotechnology and In Vitro Biology in the 21st Century, 601–604.

native to California that serves as our model system. These compounds include one terpene and four pheylpropanoids/benzenoids (Fig. 1). So far, three of the four enzymes catalyzing these reactions have been purified, and the genes encoding them have been isolated and characterized. These enzymes and genes are described below.

Figure 1: Reactions catalyzed by LIS, IEMT, SAMT and BEAT

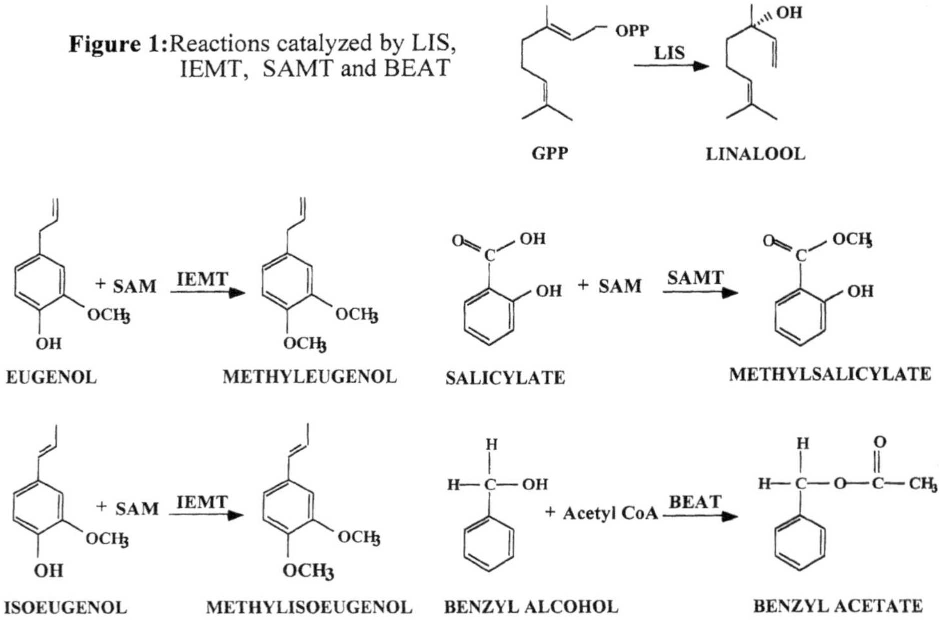

GPP —LIS→ LINALOOL

EUGENOL + SAM —IEMT→ METHYLEUGENOL

SALICYLATE + SAM —SAMT→ METHYLSALICYLATE

ISOEUGENOL + SAM —IEMT→ METHYLISOEUGENOL

BENZYL ALCOHOL + Acetyl CoA —BEAT→ BENZYL ACETATE

2.1.1 LIS: Linalool synthase, the enzyme that catalyzes the formation of linalool from geranylpyrophosphate (Fig. 1), was purified from stigmata (Pichersky et al. 1995). We next obtained peptide sequences from LIS that were used to construct oligonucleotides to isolate a LIS cDNA clone. The complete amino acid sequence of LIS, derived from the cDNA clone, showed it to be related to several enzymes involved in terpene synthesis (Dudareva et al. 1996). Both sequence comparisons and comparisons of the location of the introns in the genes suggest that LIS is actually a composite gene. It includes a portion of a copalyl pyrophosphate synthase-like sequence at its N-terminus coding region and a portion of another terpene synthase (a limonene synthase-like) in most of its second half. The sequence of the last 100 amino acids shows no homology to any sequence (Cseke et al., unpublished).

2.1.2. IEMT: We were also able to demonstrate the activity of an enzyme we designated (iso)eugenol O-methyltransferase (IEMT) which catalyzes the formation of methyleugenol and isomethyleugenol, two components of the floral scent of *C. breweri* (Fig. 1) (Wang et al. 1997). The sequence of IEMT is highly similar to that of caffeic acid O-methyltransferase (COMT). Interestingly, IEMT from *Clarkia* is 83% identical to

COMT from *C. breweri* (Wang, Pichersky 1997), and only 75% or less to any other COMT. The *C. breweri* COMT is 84% identical to COMT from a species in the same order, Myrtales, as *Clarkia* is, but less to other COMTs (Wang, Pichersky 1997). We were further able to show, by switching parts of the IEMT sequence with that of COMT, that the specificity of IEMT for (iso)eugenol rather than caffeic acid is due to no more than 16 amino acid substitutions in an 80-residue stretch of the protein (Wang, Pichersky 1998). These data indicate that *Clarkia* IEMT arose recently from COMT (presumably via gene duplication and divergence). This conclusion raises the possibility that IEMT in plants evolved independently more than once, since species not particularly close to *Clarkia*, such as basil, are also known to possess enzymatic activities that catalyze the formation of methyleugenol from eugenol and SAM (E. Lewinsohn, personal communication).

2.1.3. BEAT and other ester-forming enzymes: A major component of *C. breweri* scent is the ester benzylacetate, a common ingredient in many moth-pollinated floral aromas (Knudsen, Tollsten 1993). The two other benzenoid esters in the *C. breweri* scent, methylsalicylate and benzylbenzoate, are also widespread. We determined that benzylacetate is made from benzylalcohol and acetyl-CoA (Fig. 1) (Dudareva et al. 1998a), and we were able to purify and characterize the enzyme that catalyzes this reaction, benzylalcohol acetyltransferase (BEAT) and isolated a cDNA clone encoding BEAT (Dudareva et al. 1998b). BEAT sequence shows little overall similarity to any other known proteins, but a short segment of 35 amino acids in its middle has some similarity to other proteins believed to bind an acyl-CoA. Thus, we have identified a new family of acyltransferases distinct from those involved in fatty acid biosynthesis. Concerning the ester methylsalicylate, we showed that it is formed from salicylic acid and SAM in a reaction catalyzed by an enzyme designated salicylic acid carboxy methyltransferase (SAMT) (Dudareva et al. 1998a). We are now in the process of purifying this enzyme.

2.2. Genes and genetic control - *LIS*, *IEMT*, and *BEAT*: For all three genes, RNA blot analyses showed that mRNA accumulation began even before the flower opened, peaked on the first or second day after anthesis, and thereupon began to decline (Dudareva et al. 1996; Dudareva et al. 1998b; Wang et al. 1997). However, the level of decline varied among these genes. Overall, the conclusion was that levels of enzyme activities are regulated mainly at the mRNA level. F_2 segregation analyses (Dudareva et al., unpublished) showed that activity levels were inherited as quantitative traits, suggesting that additional loci influence the enzyme activity levels by affecting the rate of transcription of these genes, or the stability of their mRNAs.

2.2. Tissue and cellular localization: A unique aspect of scent production is that these small organic molecules, produced inside the cell but usually not water-soluble, have to come out of the cell and evaporate. Enzymatic assays showed that most of the IEMT, BEAT and SAMT enzyme activities are found in the petals, while high LIS activity levels are found both in the petals and in the stigma. Some activity of all four enzymes was also found in other flower parts. *In situ* hybridization experiments with LIS and IEMT probes indicated that in the petals, these genes are expressed in epidermal cells only, while in the stigma IEMT is expressed in epidermal cells but LIS is expressed in the cells of the transmitting tissue. Our *in situ* results indicate that expression of these genes is relatively uniform in all epidermal cells, with no evidence for specialized cells as found in flowers of orchids (Stern et al. 1987) and in vegetative tissue of some terpene-producing plants (Lewinsohn et al. 1998).

3. Conclusions

Several genes are already available for attempting to manipulate metabolic pathways in both flowers and fruits to make these organs produce secondary metabolites that will enhance floral scent and fruit flavor. Such experiments are now being carried out with several horticultural plants, and with tomatoes.

4. References

Buchmann S, Nabhan G (1996) The Forgotten Pollinators. Island Press, Washington DC.
Dobson HEM (1993) Insect-Plant Interactions, Vol V, pp 47-81, CRC Press, Boca Raton.
Dodson et al. (1969) Science 164,1243-1249.
Dudareva N et al. (1996) Plant Cell 8, 1137-1148.
Dudareva N et al. (1998a) Plant Physiol. 116, 599-604.
Dudareva N et al. (1998b) Plant J. 14, 297-304.
Galen C (1985) Ecology 66, 792-797.
Kealey KS, JE Kinsella (1978) CRC Critical Rev. Food Sci. Nutr. 11, 1-40.
Knudsen JT, L Tollsten (1993) Bot. J. Linn. Soc. 113, 263-284.
Knudsen JT et al. (1993) Phytochem. 33, 253-280.
Lewinsohn E et al. (1998) Annals Bot. 81, 35-39.
Olias JM et al. (1995) Fruit Flavors. Biogenesis, Chatacterization, and Authentication, pp 134-141, ACS Symposium Series 596.
Pellmyr O (1986) Oecologia 78, 304-307.
Pichersky E et al. (1995). Arch. Biochem. Biophys. 316, 803-807.
Simon JE et al. (1990) In: Basil: A source of essential oils. Advances in New Crops, Timber Press, Portland Oregon pp 484-489.
Stanton M et al. (1986) Science 232, 1625-1627.
Stern W et al. (1987) Am. J. Botany 74, 1323-1331.
Wang J et al. (1997) Plant Physiol. 114, 213-221.
Wang J, E Pichersky (1997) Plant Physiol. 114, 1567.
Wang, J and E Pichersky (1998) Arch. Biochem. Biophys. 349, 153-160.

Improving the Oxidative Stability of Corn and Soybean Oils by Genetic Modification

J. B. Shen, A. J. Kinney, and W. D. Hitz
E402/4255, Biotech Section, Global Technology, Ag Enterprise, DuPont Co., Wilmington, DE 19810-0402

Plant lipids have a variety of industrial and nutritional uses and are central to plant membrane function and climatic adaptation. These lipids represent a vast array of chemical structures, and these structures determine the physiological and industrial properties of the lipid. Many of these structures result either directly or indirectly from metabolic processes that alter the degree of unsaturation of the lipid.

The major use of plant lipids is as edible oils in the form of triacylglycerols. The specific performance and health attributes of edible oils are determined largely by their fatty acid composition. Most vegetable oils derived from commercial plant varieties are composed primarily of palmitic (16:0), stearic (18:0), oleic (18:1), linoleic (18:2) and linolenic (18:3) acids. Palmitic and stearic acids are, respectively, 16- and 18-carbon-long, saturated fatty acids. Oleic, linoleic, and linolenic acids are 18-carbon-long, unsaturated fatty acids containing one, two, and three double bonds, respectively. The relative amounts of saturated and unsaturated fatty acids in commonly used, edible vegetable oils are summarized below (Table 1).

Many recent research efforts have examined the role that saturated and unsaturated fatty acids play in reducing the risk of coronary heart disease. Several recent human clinical studies suggest that diets high in monounsaturated fatty acids and low in saturated fatty acids may reduce the "bad" (low-density lipoprotein) cholesterol while maintaining the "good" (high-density lipoprotein) cholesterol (Kinney, 1996).

It is generally thought that a vegetable oil low in total saturates and high in monounsaturated fatty acids would provide significant health benefits to consumers. For example, canola oil is considered a very healthy oil but its high polyunsaturated fatty acid content renders the oil unstable, easily oxidized and susceptible to development of disagreeable odors and flavors (Frankel, 1991). When exposed to air, unsaturated fatty acids are subject to oxidation which causes the oil to have a rancid odor (Frankel, 1980). Oxidation is accelerated by high temperatures, such as in frying conditions and the rate of oxidation is disproportionately increased by the addition of a second double

A. Altman et al. (eds.), Plant Biotechnology and In Vitro Biology in the 21st Century, 605–608.

most likely to involve several recessive genes. Thus it is desirable to produce oxidatively stable lines of corn and soybean by genetic engineering techniques.

Table 2. Fatty Acid composition of high oleate mutants.

	Line	16:0	18:0	18:1	18:2	18:3	20:0
Seeds	B73ol	11	1.1	57	29	1.4	-
	B73	12	1.5	22	62	1.3	-
	AEC272ol	9	1.8	63	25	0.4	-
	AEC272	10	2.3	36	51	0.5	-
	Mu/Mu	9	2.0	59	28	0.6	-
	-/-	10	2.5	28	58	0.8	-
Leaves	B73ol	14	2.8	1.3	22	47	12
	B73	14	2.0	0.7	16	54	13
	AEC272ol	18	2.2	1.6	21	46	10
	AEC272	14	2.0	2.5	18	41	20
	Mu/Mu	17	2.2	1.1	15	50	13
	-/-	17	2.1	1.1	16	50	12
Roots	B73ol	20	6.7	2.5	59	8.5	-
	B73	21	3.7	2.3	66	6.5	-
	AEC272ol	22	4.6	3.0	61	7.9	-
	AEC272	22	6.5	1.8	56	8.6	-
	Mu/Mu	20	5.3	5.3	61	5.0	-
	-/-	20	5.0	4.1	61	5.6	-

In soybeans, this has been achieved by suppressing a gene encoding an oleate desaturase using molecular biological techniques. This resulted in the production of a soybean oil with a total polyunsaturated content of less than 6% and oleic acid content of 84% (Figure 1).

Acetyl CoA C2 → **Palmitic acid C16:0** → **Stearic acid C18:0** → **Oleic acid C18:1** ⇸ **Linoleic acid C18:2** → **Linolenic acid C18:3**

Figure 1. Suppressing a gene encoding a fatty acid desaturase in soybean results in the production of soybean oil rich in oleic acid.

The high oleic acid phenotype in these transgenic soybeans was stable in a number of different environments over a number of years (this is usually not achievable in mutant soybean lines), and has been stable in the field for over ten generations. Yield trials indicated that the high oleic acid transgenic soybeans were indistinguishable from the elite parental lines, and thus the high oleic transgenic soybeans were competitive in terms of yield with the commercial line of soybean used in the transformation. These high oleic acid soybean lines are now in advanced stages of development and oil from bond to a monounsaturated fatty acid. Thus, linoleic acid with two double bonds is almost ten-fold more unstable than oleic acid which has only one double bond (Ibid.).

Table 1. Percentages of saturated and snsaturated fatty acids in the oils of selected oil crops.

	Saturated	Mono-unsaturated	Poly-unsaturated
Canola	6	58	36
Soybean	15	24	61
Corn	13	25	62
Peanut	18	48	34
Safflower	9	13	78
Sunflower	9	41	51
Cotton	30	19	51

The polyunsaturated fatty acid content of polyunsaturated vegetable oils may be reduced by hydrogenation, but this process results in the production of *trans* isomers of the remaining unsaturated fatty acids (Ratnayake, Pelletier, 1992), which have also been correlated with heart disease (Hu et al., 1997). As an alternative to hydrogenation, corn oil and products containing corn oil, are often packaged under nitrogen in special packaging materials such as plastic or laminated foil, or are stored under refrigeration to extend their shelf life. These extra measures to reduce oxidation and subsequent rancidity add considerable cost to products containing corn oil.

Thus, it is not surprising that serious efforts been made to improve the quality of vegetable oils through plant breeding, especially mutagenesis, and a wide range of fatty acid compositions have been discovered in experimental lines of many oil crops such as canola and soybean (Kinney, 1994).

In addition, we have recently identified a number of corn lines with reduced linoleate content. Table 2 shows the lipid profile of three high oleate mutants. The mutation was induced by EMS (B73ol and AEC272ol), or insertion of a transposable element, mutator (Mu). Their isogenic wild-types (B73 and AEC272), or wild-type segregant (-/-), were included for the comparison. The high oleate phenotype appears to be seed specific.

However, there are serious limitations of using mutagenesis to alter fatty acid composition. It is unlikely to discover mutations (a) that result in a dominant ("gain-of-function") phenotype, (b) in genes that are essential for plant growth, and (c) in an enzyme that is not rate-limiting or that is encoded by more than one gene. In addition, the introgression of desired mutant lines into elite lines by traditional breeding techniques will be slow and expensive, since the desired oil compositions in corn are these beans will be commercially available for the first time in late 1998. The oil from these transgenic beans has an oxidative stability of more than ten times that of RBD soybean oil and is similar in its storage and cooking stability to fully hydrogenated, heavy-duty frying shortening but does not contain any *trans* fatty acids.

Thus it is clear that efficient production of lines of both soybean and corn with modified fatty acid compositions will rely on the techniques of genetic engineering rather than mutagenesis or germplasm selection. Currently viable mutant lines are, however, useful for crossing with new transgenic lines to produce other useful phenotypes. In soybean examples of phenotypes from this type of cross include oils with a combined high oleic acid and high stearic acid content for use in margarine replacements, and oils with a very low saturated fatty acid content for healthier salad oils (Table 3).

Table 3. Soybean oils with novel phenotypes produced by combining transgenic and mutant lines

Trait	Fatty acid (% of total)					
	16:0	18:0	18:1	18:2	18:3	20:0
Commodity	12	4	24	54	7	-
High oleic	7	3	84	2	4	-
Low linolenic	11	3	40	41	4	-
Low saturate	4	3	22	63	8	-
Low sat./high oleic	4	3	88	1	3	-
High saturate	8	32	17	33	6	4
High sat./ high oleic	7	21	64	2	4	2
High palmitic	30	5	13	43	9	-
High linolenic	12	3	15	46	24	-

References

Frankel EN (1980) Prog. Lipid Res. 19,1-22.
Frankel EN (1991) J. Sci. Food Agric. 54:495-511.
Hu FB et al. (1997) N. Engl. J. Med. 337, 1491-1499.
Kinney AJ (1994) Curr. Opin. Biotechnol. 5:144-151.
Kinney AJ (1996) J. Food Lipids 3, 273-292.
Ratnayake WNM, Pelletier G (1992) J. Am. Oil Chem. Soc. 69, 95-105.

Thaumatin expression in transgenic cucumber plants

Maria Szwacka[1], Magdalena Krzymowska[2], Stefan Malepszy[1]
[1]Department of Plant Genetics, Breeding and Biotechnology, Warsaw Agricultural University, Poland
[2]Institute of Biochemistry and Biophysics PAS, Warsaw, Poland

1. Introduction

Thaumatin is produced in the African plant known *Thaumatococcus daniellii* as a family of five or more proteins [Higginbotham, Hough, 1977]. Thaumatins are nearly 100000 times sweeter than sucrose on a molar basis [Iyengar et al., 1979]. Taumatins: I and II are the most abundant in fruit, and the cDNA for thaumatin II has been cloned and sequenced [Edens et al., 1982]. So thaumatin is a potentially interesting target for genetic engineering studies and we have recently begun work in this direction developing *Agrobacterium*-mediated transformation of cucumber (*Cucumis sativus* L.) with thaumatin II gene. We used an efficient plant regeneration procedure described by Burza and Malepszy [1995]. Transformation was carried out using the binary plasmid pRUR528s bearing sense construct of *T. daniellii* thaumatin II cDNA under the control of constitutive promoter 35S CaMV [Szwacka et al., 1996]. Transformation status was confirmed by PCR amplification of *npt*II gene. Genomic DNA hybridization experiments proved that independent primary transformants carried one to three copies of the expression cassette. Here, we present data on thaumatin mRNA transcription in independent primary transformants (Fig. 1A). We observed marked differences in the level of thaumatin transcript. Western blot analysis (Fig. 1B) indicated the presence of mature thaumatin (22 kDa) [Vos de, 1985] in five from fifteen analysed primary transformants. In two of them (no. 214 and no. 224) high level of thaumatin mRNA correlated with high level of mature thaumatin (Fig. 1A and 1B). Preliminary sensory evaluation of ripe fruits of some T_1 cucumber plants showed the existence of phenotypes with improved tastes in comparison to the tasteless phenotype of control variety fruits.
Thaumatin displays sequence homology to proteins belonging to PR-5 family of so called PR proteins (pathogenesis-related) referred to as thaumatin-like (TL) [Cornelissen, et al. 1986]. Several studies suggest that enzymes from PR family may be components of a general defense system against pathogen invasion in several different plant species [Benhamou, et al. 1989, Jondie et al., 1989, Kauffmann et al., 1987]. Some of TL proteins display *in vitro* anti-fungal activity [Stintzi et al., 1993]. For this reason it was interesting for us to check if thaumatin expression in cucumber may enhance resistance to pathogen infection. Here, we demonstrated preliminary data on enhanced *in vitro* the

A. Altman et al. (eds.), Plant Biotechnology and In Vitro Biology in the 21st Century, 609–612.

resistance of some T_2 cucumber plants against downy mildew, disease of cucurbits caused by fungus *Pseudoperonospora cubensis* (Tab. 1).

2. Procedure

2.1. Materials and methods

2.1.1. Plant material and fungal infection assay

For transformation experiments *C. sativus* line (line B obtained from the Borszczaowski variety) was used. For fungal infection assays and seeds production cucumber plants were grown in the greenhouse in 16h/8h-day/night photoperiod (25°C-27°C in day/18°C-20°C in night). Light intensity in the greenhouse was about 1500 μmol m^{-2} s^{-1}.
P. cubensis was used for cucumber resistance assay. Fungus was maintained on leaves of susceptible cucumber cultivar Wisconsin SHR-18 placed on moist filter paper in the growth chamber with 100% humidity, to induce sporulation. Discs (5 cm in diameter) of fully expanded leaves of four weeks old cucumber plants were inoculated with single 20 μl - droplets of spore suspension deposited in the centre of the abaxial disc surface. Four leaf discs were cut from each of three to five T_2 plants tested and from control plants. Susceptible cv. Wisconsin SHR-18 and line B were used as negative and line WI4783 as positive control (Tab. 1). Only T_2 plants which T_1 progeny was more resistant to fungus infection were chosen. Spore titers were adjusted from $5x10^3$ /ml to 10^4/ml. Inoculated samples were incubated in moist chamber a 20 °C, the first 48h in the dark. Afterwards the incubation was carried out under 16h/8h light/dark regime. Seven days post-inoculation spore titer on inoculated discs were estimated in 0-9 scale, depending on the extent of infection (percentage of infected area):
1 several necrotic spots on the leaf disc
3 less than 30% of disc area infected
5 30% of disc area infected
7 from 60% to 70% of disc area infected
9 100% of disc area infected.

2.1.2. Vector plasmid

A vector pRUR528s was constructed by Dr. A Pałucha (Institute of Biochemistry and Biophysics PAS, Warsaw, Poland). This vector contains thaumatin II cDNA (2.1-kb, a gift from Dr. A.M. Ledeboer, Vlaardingen, The Netherlands) under the control of a 35S CaMV promoter and *npt*II gene driven by a nopaline synthase (nos) promoter.

2.1.3. RNA and protein gel blots

Independent primary transformants were submitted to molecular analysis at the RNA and protein level to confirm thaumatin gene expression. Total RNA was extracted from young leaves and in one case from ripe fruit samples using method described by Linthorst et al. [1993]. Twenty micrograms of total RNA per line was separated on 1%

agarose gel in 15 mM sodium phosphate, pH 6.5 and transferred to membrane Hybond N (Amersham). Then the filters were hybridized at 65°C in 250 mM sodium phosphate pH 7/1 mM EDTA/7% SDS/1% BSA with ^{32}P-labelled probe, 0.8-kb fragment of thaumatin II cDNA. Filter were washed with 2xSSC/ 0.1%SDS 20 min. at RT, an 2xSSC/0.1%SDS 30 min. at 45°C and exposed on autoradiographic film (Kodak) at -76°C for 29 to 30h. Protein leaf extracts and in one case (primary transformant no. 224) fruit extract, were prepared in buffer containing 50 mM Tris pH 8.0/1 mM EDTA/12 mM β-mercaptoethanol and 0.5% PMSF. Twenty five micrograms of protein extract, isolated from independent transformants were fractionated on 15% SDS-PAGE and subjected to immunoblot analysis using specific anti-thaumatin rabbit polyclonal antibodies and anti-rabbit IgG goat antibodies conjugated with alkaline phosphatase (AP). Immunoblot was developed using AP colorimetric detection kit (Boehringer Mannheim).

2.2. Illustrations

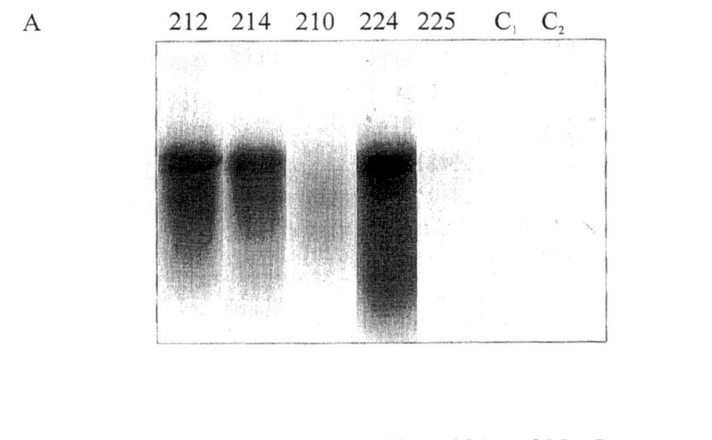

Figure 1. Expression of thaumatin gene in transgenic cucumber plants.
A. Northern blot analysis performed with thaumatin II cDNA as a probe. Total RNA was isolated from leaves of primary transformants. In case of primary transformant no. 224, RNA was isolated from fruit. As a control RNA from leaves (C_1) or ripe fruit (C_2) of untransformed cucumber (line B) was assayed.
B. Protein expression of thaumatin analysed by immunoblotting. 25 μg total protein extract prepared from leaves or from ripe fruit (224) was assayed using specific anti-thaumatin antibodies. A positive control (ST) was 0.25 μg thaumatin purchased from Sigma. As a negative control protein extract from untransformed plants was used (C_1). Bar indicates the expected size of mature thaumatin (22 kDa).

Table 1. Responses of T_2 progeny to *P. cubensis* infection in *in vitro* assays

Plant			Plant response
T_2 progeny		210 02	7
		03	5
		05	4
		212 04	4
		05	3
		06	3
		07	5
		08	3
		225 02	3
		03	4
		04	4
		05	5
		06	5
Control	line B		7
	cv. Wisconsin SMR-18		8
	line WI4783		1

This table shows that only one per thirteen T_2 plants (no. 210 02) exhibits the same response to fungal infection as the untransformed line (line B). The other T_2 plants have differentiated tolerance levels, from 3 to 5 according to 0-9 scale.

3. Acknowledgements

The authors would like to thank T. Gałecka for her help in downy mildew tests.
This work was supported by grant No. 227 PO6 96 11 from State Committee For Scientific Research.

4. References

Benhamou N. et al. (1989) Plant Cell 1, 1209-1221.
Burza W., Malepszy S. (1995) Plant Breed. 114, 341-345.
Cornelissen B.J.C. et al. (1986) Nature 321, 531-532.
Edens L. et al. (1982) Gene 18, 1-12.
Higginbotham J.D., Hough C.A.M. (1977) Sensory properties of Foods, Applied Sciences, London, pp 129-149.
Iyengar R. B. et al. (1979) Eur. J. Biochem. 96, 193-204.
Jondie D.J., Coors J.G., Duke S.H. (1989). Can.J.Bot. 67, 263-268.
Kauffmann S. et al. (1987) EMBO J. 6, 3209-3212.
Linthorst H.J.M. et al. (1993) Plant Mol. Biol. 21, 985-992.
Stintzi A. at al. (1993) Biochimie 75, 687-706.
Szwacka M. et al.(1996) Genet. Pol. 37A, 126-129.
Vos de A.M. et al. (1985) Proc. Natl. Acad. Sci. USA 82, 1406-1409.

Molecular breeding of flower color of *Torenia hybrida.*

Shinzo Tsuda, Ken-ichi Suzuki, Hui-min Xue, Yoshikazu Tanaka, Yuko Fukui, Masako Fukuchi-Mizutani, Yukihisa Katsumoto and Takaaki Kusumi

Institute for Fundamental Research, Research Center, Suntory Ltd., Osaka 618-8503, Japan. (Key words: *anthocyanin, chalcone synthase, dihydroflavonol 4-reductase, flavonoid 3'5'-hydroxylase, sense suppression, transgenic plants*)

Molecular breeding is a powerful method of plant breeding because it can change a specific characteristic of a plant without changing other desirable characteristics. Flower color is predominantly influenced by two types of pigments; flavonoids and carotenoids. The anthocyanin biosynthetic pathways of many plants have been well established (Holton et al, 1995) and conserved (Fig. 1). Chalcone synthase (CHS) and dihydroflavonol 4-reductase (DFR) are the first specific enzymes in flavonoid and anthocyanin biosynthesis, respectively. The presence of flavonoid 3'5'-hydroxylase (F3'5'H), cytochrome P-450 (Holton et al, 1993), is almost critical to the production of blue to purple anthocyanins. The flower color is reddish in its absence.

Torenia (*Torenia fournieri*), belonging to Scrophulariaceae, is one of the most important bedding plants. Suntory Ltd. successfully developed a new type of torenia cultivar Summerwave (*T. hybrida*), which has many characteristics superior to common torenia cultivars. Summerwave originally had one flower color (Blue, SWB). Because its male and female sterility predicted the difficulty of conventional breeding, a molecular approach was applied to widen its flower color variation.

Gene silencing is a common phenomenon in transgenic plants and it affects both transgenes and endogenous genes (Stam et al, 1997). Both constitutive expressions of a sense (Napoli et al, 1990) and antisense (Van der Krol et al, 1988) petunia CHS gene in transgenic petunia results in an altered flower pigmentation. The flower color patterns varied significantly among inter- and intra-transgenotes. The suppression pattern and stability were also varied. Although suppression of anthocyanin biosynthesis has been reported in some species (Gutterson et al, 1995), none of them were seemed to have commercial value. In this report, we made transgenic torenia plants harboring sense cDNAs encoded CHS, DFR and F3'5'H enzymes isolated from torenia SWB and observed flower color change with new and simple patterns.

Materials and methods

Plant material and genetic transformation. Torenia hybrida cv. Summerwave Blue, commercialized by Suntory Ltd., and *T. hybrida* inbred line T-33 were grown under standard greenhouse conditions. In vitro cultured shoots used for transformation were maintained on a MS agar medium. SWB and T-33 were transformed using the method of *Agrobacterium tumefaciens* mediated transformation of *T. fournieri* (Aida et al, 1995). *Pigment analysis.* Extraction and HPLC analysis of anthocyanin were used as described in Tanaka et al (1995). *Molecular analysis.* A SWB petal cDNA library constructed in the UNI-ZAP (Stratagene) was screened with rose CHS (unpublished result), DFR (D85102) and gentian F3'5'H (D85184) and their counterpart cDNAs were obtained.

A. Altman et al. (eds.), Plant Biotechnology and In Vitro Biology in the 21st Century, 613–616.

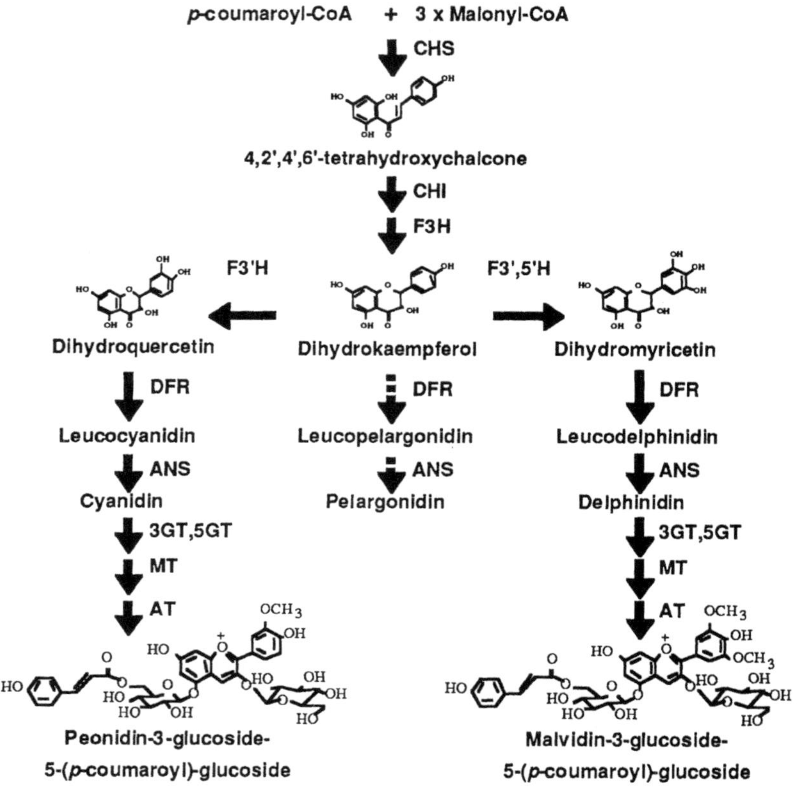

Figure 1. Schematic representation of the biosynthesis pathway for anthocyanin pigment production in *Torenia hybrida* cv. Summerwave Blue. The reactions shown by dotted arrows do not naturally occur in SWB. CHS: chalcone synthase; CHI: chalcone isomerase; F3H: flavanone 3-hydroxylase; F3'H: flavonoid 3'-hydroxylase; F3'5'H: flavonoid 3'5'-hydroxylase; DFR: dihydroflavonol 4-reductase; ANS: anthocyanidin synthase; 3GT: UDP-glucose:flavonoid-3-*O*-glucosyltransferase; 5GT: UDP-glucose:flavonoid-5-*O*-glucosyltransferase; MT: anthocyanin O-methyltransferase; AT: anthocyanin acyltransferase.

Results and discussion

Figure 1 shows the anthocyanin biosynthetic pathway which was determined on the basis of structural analysis of anthocyanins in the petals and molecular cloning of many of the genes involved in the pathway. It is known that the pathways to anthocyanidin-3-glucosides are well conserved and that further modification of them is variable among plant species. Malvidin-3-glucoside-5-(*p*-coumaroyl)-glucoside, derived from delphinidin, was the main product and occupied over 80 % of the total anthocyanins in the petals of SWB. Torenia CHS (AB012923), DFR (AB012924) and F3'5'H (AB012925) cDNAs were isolated and sequenced. For the reduction of these enzyme activities we generated transgenic plants with cosuppression binary constructs.

Torenia SWB was transformed with the binary vector containing whole torenia CHS sense cDNA or partial torenia DFR cDNA lacking 300 bp in the 5' region under the control of the enhanced CaMV35S promoter (Mitsuhara et al, 1996). Many of the transgenic plants had petals with reduced amounts of anthocyanins. Both transgenic plants of SWB CHS and SWB DFR showed the same phenotypic pattern. The degree of the reduction varied with each transgenic plant, and the petals had a range of colors from blue, to blue and white, and pure white. Interestingly, anthocyanin production was more consistently suppressed in the dorsal and ventral petal lobes, and the corolla tube, than in the lateral petal lobes. The degree of suppression of petal colors was classified into six groups, A to F described in Table 1. Jorgensen et al (1996) reported that 63% and 11 % of the 185 sense CHS transgenic plants studied had corollas with altered colors and white flowers, respectively. SWB CHS transgenic plants showed phenotypic change with a frequency similar to petunia (Table 1). Interestingly, variegated pigmentation was not observed in SWB. The frequency of CHS suppression with sense CHS gene was very low and no variegated phenotypes were observed in chrysanthemum (Gutterson et al, 1994), rose and carnation (Gutterson et al, 1995). Sense suppression phenomena may be different among plant species.

Table 1 Frequency of flower color change of transgenic plants. a: pure white; b: pale yellow; c: pure pink; d: white only in corolla tube; e: pale blue in whole corolla. In the case of SWB CHS and DFR transformants, phenotype A: a small region of pale blue or white color specifically in the inside of the dorsal and ventral petal lobes, B: a white color region was larger in the dorsal and ventral petal lobes, and the corolla tube was colored pale blue, C: a blue color in the lateral petal lobes and was almost completely white in other parts, except for a little pale blue color still visible in the corolla tube, D: the only lateral petal lobes showing blue color, E: a white color in all regions, with the exception of the tip of the lateral petal lobes, F: a pure white color, UC: unchanged flower color, IC: irregularly changed flower color.

Host	Transformed	No. of Transgenic Plants								
Plant	Gene	Total	UC	IC	A	B	C	D	E	F
SWB	CHS	121	91	0	4	0	3	0	4	19[a]
SWB	DFR	115	63	1[d]	9	4	8	12	10	8[a]
T33	CHS	96	64	0	10	9	10	2	0	1[b]
T33	DFR	80	46	0	12	4	9	3	4	2[b]
SWB	F3'5'H	105	88	1[e]	0	3	0	9	0	4[c]

In order to study the correlation between gene suppression and anthocyanin accumulation, CHS and DFR transcripts in the petals were analyzed. Petals in phenotype A-B had more mRNA than the control line expressing GUS protein. As with phenotype A to F, messages were decreased and finally phenotype F rarely showed the signal. In the flower of SWB, anthocyanin pigments accumulated only in the epidermal cell layer of the petal. Many genes of the flavonoid biosynthesis enzyme were expressed specifically in this cell layer (Helariutta et al, 1993). The transgene driven by the enhanced CaMV35S promoter might be expressed in all cell layers of the petals. Most of the message detected in our results may have been derived from the transgene. In phenotype F, expression of the endogenous CHS gene in epidermal cells should be very low. We concluded that the high level of transcription of the transgene caused the suppression of endogenous mRNA, as the result of cosuppression.

The binary vector contained the partial torenia F3'5'H cDNA containing only about 600 bp in the 3' region with sense direction was used to transform SWB. Some transgenic plants changed flower color from blue to pink. This change was consistently observed to be stronger in the dorsal and ventral petal lobes and the corolla tube compared to the lateral petal lobes seen in SWB CHS and SWB DFR plants. The plants showing intermediate phenotypes had pink colored flowers only in the corolla tube. No anthocyanins derived from delphinidin were produced in completely pink flowers. An increased amount of cyanidin and peonidin and a small amount of pelargonidin were observed. The suppression of the F3'5'H gene clearly caused a change from the delphinidin pathway to the cyanidin pathway in anthocyanin biosynthesis (Fig. 1).

Torenia T-33 petals, containing both anthocyanins and carotenoids, were dark brown and purple. To make yellow flower varieties from T-33, by specific suppression of biosynthesis of anthocyanins, T-33 was transformed with CHS or DFR cDNA with same methods used with SWB. About 40 % of the transgenic plants changed flower color (Table 1). The patterns of anthocyanin reduction in each construct were observed to be same as SWB/CHS and SWB/DFR, except for the pale yellow lines (phenotype F) which were isolated with a lower frequency than SWB (Table 1). Further molecular breeding to widen flower color of SWB is in progress.

References

Aida R et al (1995) Breed. Sci. 45, 71-74.
Gutterson N et al (1994) Bio/Technology 12, 268-271.
Guttrerson N et al (1995) HortScience 30, 964-966.
Helariutta Y et al (1993) Plant Mol. Biol. 22, 183-193.
Holton TA et al (1993) Nature 366, 276-279.
Holton TA et al (1995) Plant Cell 7, 1071-1083.
Jorgensen RA et al (1996) Plant Mol. Biol. 31, 957-973.
Mitsuhara I et al (1996) Plant Cell Physiol. 37, 45-59.
Napoli C et al (1990) Plant Cell 2, 279-289.
Stam M et al (1997) Plant J. 12, 63-82.
Tanaka Y et al (1995) Plant Cell Physiol. 36, 1023-1031.
Van der Krol AR et al (1988) Nature 333, 866-869.

NUTRITIONAL QUALITY IMPROVEMENT OF SORGHUM THROUGH GENETIC TRANSFORMATION

T.Yohannes[1], V. Frankard[1], L. Sagi[2], R. Swenen[2], M. Jacobs[1]

[1]Laboratory of Plant Genetics, Vrije Universiteit Brussel, Institute of Molecular Biology B-1640 Sint-Genesius Rode, Belgium

[2]Laboratorium voor Tropische Plantenteelt, Katholieke Universiteit Leuven; Kardinaal Mercierlaan 92, B-3001 Heverlee, Belgium

Sorghum is one of the primary cereals in many parts of the world, and ranks fifth among the world's cereals, following wheat, maize, rice, and barley. It is one of the main staples for the world's poorest and most food-insecure people, supporting more than 300 million lives. It performs well in the hot and dry agro-ecologies of the semi-arid tropics. In these areas, sorghum is a dual purpose crop, where it is used for human consumption and as feed for animals. However, although widely used and consumed, this crop is known to have low nutritional quality, as a result of its characteristic low content in the essential amino acid lysine. This low content results from end-product feedback inhibition of two key enzymes, aspartate kinase (AK) and dihydrodipicolinate synthase (DHDPS), in the aspartate family pathway (Karchi *et al.*, 1994; Galili, 1995). In particular, since lysine inhibits the activity of the enzyme (DHDPS) specific for its own synthesis, it is a major drawback that has to be alleviated to improve plant nutritional value. In this case, incorporation of a lysine overproducing trait in sorghum through foreign gene transfer could have considerable nutritional advantage for the food supply in several developing countries. Of the different gene transfer methods available, the utility of particle bombardment techniques has proved to be important in extending recombinant DNA methodology to the improvement of the more recalcitrant *Gramineae* species. Using this method, regeneration of stably transformed transgenic cereals expressing different antibiotic or herbicide resistant genes has been confirmed. These achievements to obtain monocot plants expressing different antibiotic or herbicide resistant genes can also be extended to the production of plants expressing nutritionally important genes. Our aim is thus to introduce a mutated gene (*dhdps-r1*) (Ghislain *et al.*, 1995) coding for a feedback insensitive dihydropicolinate synthase, the first enzyme of the lysine-specific pathway, with the goal to produce transgenic sorghum plants with increased lysine content.

1. Genotype screening

Based on the possibility of DNA delivery into organized tissues by high velocity microprojectile bombardment, different sorghum genotypes were screened for better *in vitro* performance using shoot tip and immature embryo tissues. Two out of eleven genotypes with high frequency in embryogenesis and regeneration were selected for further manipulation.

A. Altman et al. (eds.), Plant Biotechnology and In Vitro Biology in the 21st Century, 617–620.

2. Optimization of transformation conditions

To establish the conditions for sorghum transformation via particle bombardment, we have optimized various parameters, including the type and physiological condition of the explant, concentration of DNA and particles, particle accelerating pressure, target distance, and efficiency of selective agent. Different tissue types (immature embryos, mature embryos, shoot tips, proliferated pre-callus tissues, and organized calli) were bombarded at different velocities in relation to macrocarrier, microcarrier travel distance, gap width and chamber vacuum. The series of bombardment experiments were carried out using the gene construct pAHC25 (Christensen, H. Quail, 1996), with 0.83μg DNA per bombardment as described by Sanford *et al.*, (1993). Parameters were evaluated in relation to transient GUS expression levels that were quantified by histochemical GUS staining 48 hours after bombardment. Compared to other tissue types in replicated experimental observations, immature embryos and shoot tips gave adequate and consistent results at 6 cm for target distance and helium pressure of 1300 and 1100 psi, respectively.

3. Promoter analysis transient expression studies

In situations where transformation of a recalcitrant species like sorghum is difficult, and where effective expression of an introduced gene is not proved, it is important to test the expression level in combination with different promoter sequences. In this regard, transient expression studies using five gene constructs (Act1 D-BAR, McElroy *et al.*, 1990; pBAR-GUS, Fromm *et al.*, 1990; pAHC25 Christensen, H. Quail, 1996; and p35S-GUS (the GUS-uid A gene under the transcriptional control of CaMV 35S promoter), encoding the GUS (*uidA*) gene under the control of Act1, Adh1, CaMV 35S and Ubi1 promoter sequences were carried out by bombarding shoot tips and immature embryos. As a result of blue staining tests at 24, 48 and 72 hrs after bombardment, Adh1 activity was found to be about 1.5 and 2 or 3 times less than those of Act1 and Ubi1 promoters respectively, while CaMV 35S showed an activity about 1.5 times lower than the Adh1 promoter. The number of transient expression blue spots increases between culture periods of 24 and 48 hrs. However, the trend of expression level difference over time among the different promoters seems similar. Following the GUS staining test, other tissues were bombarded in similar conditions and after 48 hr culture period, protein was extracted and GUS activity was measured by fluorescence assay. This comparative promoter activity analysis also supports the result we obtained with the histochemical staining test.

4. Stable transformation studies

Based on the conditions we optimized, stable transformation studies were carried out using three different gene constructs (Table 1).
After bombardment, target tissues (immature embryos and shoot tips) were cultured on callus induction medium in the presence of S-aminoethyl cysteine (AEC, analogue of lysine), AEC and geneticin, and geneticin only, as selective agents wherever applicable. After a series of stepwise selections, a total of 9 plantlets derived from AEC resistant calli, 5 plantlets derived from AEC plus geneticin resistant calli and 3 plantlets derived from geneticin resistant calli survived transfer to soil. At flowering stage, Act D-neo and pGV4042 transformants showed normal growth and inflorescence formation, while p35TPM transformants exhibited branched shoots and panicles with partially or totally sterile spikelets.

Table 1. Gene constructs used in stable transformation of sorghum

Construct	Promoter	Selectable marker gene	Selective agent
pAct D-neo	Act1	*nptII*	geneticin
p35TPM	CaMV 35S	*dhdps-r1*	AEC
pGV4042	CaMV 35S	*dhdps-r1*	AEC
	NOS		geneticin

5. Molecular analysis

Total DNA was isolated from the leaf tissues of the R0 plants, and the presence and integration of the foreign genes were examined by Southern hybridization. With the genomic DNA samples from all three groups of transformants, PCR was carried out using the primers for *dhdps* and *nptII* genes. A band of approximately 0.7 kb from the *dhdps* gene in 8 out of 9 p35TPM and 5 pGV4042 transformants, and of around 0.5kb from the *nptII* gene in 5 pGV4042 and 3 pAct1 D-neo transformants were obtained showing positive amplifications as expected. No amplification was obtained from DNA samples isolated from nontransformed wild type control plants. The amplified samples were subjected to Southern hybridization using *dhdps* and *nptII* radioactive probes, and positive signals were obtained with respect to the positive PCR amplifications.

In the case of the pGV4042 transformants whose progenies showed total sensitivity to geneticin (Table 2), the presence of the *nptII* gene was shown by positive PCR amplification, and was later confirmed by Southern hybridization. In addition, using the *nptII* primers, another PCR test was done on DNA samples extracted from AEC-sensitive seedlings and AEC-resistant plus geneticin-sensitive seedlings. As a result, no amplification was obtained from the former samples, whereas the latter ones gave positive amplification confirming the presence of the introduced *nptII* gene.

Moreover genomic DNA samples isolated from the nine p35TPM transformants and one nontransformed control were digested with *Bam* HI and *Hind*III. Digested samples, together with one positive control (probe DNA) were hybridized with a 1.7 kb *Bam*HI-*Hind*III fragment of the *dhdps* probe. Eight out of the nine lines gave positive signals, among which four showed one band of 1.7 kb, and other four extra bands of higher molecular weight in addition to the expected band. No band was observed in one line which was also negative to PCR amplification, and as well as in the wild type control.

6. Biochemical analysis

6.1 Enzyme activity

Crude extracts of protein from leaf samples of the p35TPM and pGV4042 transformants and wild type control were tested for DHDPS activity. 2.5 to 4.5 times more activity was obtained from the different lines compared to the wild type. In the line where we did not get a positive signal for the presence of the foreign gene, the activity was comparable with that of the wild type.

6.2 Inhibition test

Inhibition of the DHDPS enzyme activity by increasing lysine concentrations (0-100μM) was also tested in a few lines. Clear inhibition in the wild type control was observed, while the enzyme from the transgene did not show any inhibition.

7. Acknowledgments

This work was supported by the EEC and a State-University concerted action.

8. References

Christensen AH, PH Quail (1996) Transgenic Research 5, 213-218
Fromm ME *et al.* (1990) Bio/Technology 8, 833-839.
Galili G (1995) The Plant Cell 7, 899-906.
Ghislain M. *et al.* (1995) Plant J. 8, 101-111.
Karchi H *et al.* (1994) Proc. Natl. Acad. Sci. USA 91: 2577-2581.
McElroy *et al.* (1990) Plant Cell 2, 163-171.
Sanford JC *et al.* (1991) Technique 3, 3-16.
Sanford JC *et al.* (1993) Meth. Enzymol. 217, 483-509.

SUPER-RUBISCO FOR IMPROVEMENT OF PHOTOSYNTHETIC PERFORMANCES OF PLANTS

A. Yokota[1,2], K. Uemura[2,4] and H. Kobayashi[3]

[1]Graduate School of Biological Sciences, Nara Institute of Science and Technology (NAIST), Takayama, Ikoma, Nara 630-0101, Japan

[2]Plant Molecular Physiology Laboratory, Reserach Institute of Innovative Technology for the Earth (RITE), Kizu, Kyoto 619-0225, Japan

[3]Graduate School of Nutritional and Envionmental Sviences, University of Shizuoka, Shizuoka 422-8526, Japan

[4]Present address: R & D Center, Yunichika Co., Ltd., Uji, Kyoto 611, Japan

1. Introduction

The CO_2-fixation step catalyzed by ribulose-1,5-bisphosphate carboxylase/ oxygenase (RuBisCO) in photosynthesis is the important rate-limiting step. Improving the enzymatic efficiency is a meaningful direction for improvements of plant water-use efficiency and crop productiviy. RuBisCO, even of higher land plants, has several disadvantages as an enzyme. The reaction turnover rate is up to 3/sec/reaction site; 1/100 to 1/1000 of enzymes found in nature. The affinity of the enzyme for CO_2 is 10 to 15 μM; just a quater of the enzyme in chloroplasts can participate in photosynthesis. Much worse is the unavoidable oxygenase reaction. Plant RuBisCO well adapted to the present oxygenic atmosphere still fixes O_2 once for every 2 to 3 CO_2 fixations in chloroplasts. A part of the reaction product is oxidized to CO_2. Totally, the oxygenase reaction reduces the productivity of crop plants up to 60%. We have been trying to improve the enzyme based on the molecular and biochemical mechanisms of the evolution and adaptation of the enzyme to the present atmosphere after the appearance of the enzyme in the nature. Information on the structure-function relationship of the RuBisCO evolution will be highly expected to give us various approaches useful for the improvement of the enzyme.

2. Results and Discussion

The biphasic reaction course, fallover, of carboxylation catalysed by ribulose 1,5-bisphosphate carboxylase/oxygenase (RuBisCO) has been known as a characteristic of

A. Altman et al. (eds.), Plant Biotechnology and In Vitro Biology in the 21st Century, 621–624.

the enzyme from higher land plants. Fallover consists of hysteresis in the reaction seen during the initial several minutes and a subsequent, very slow suicide inhibition byinhibitors formed from the substrate ribulose-1,5-bisphosphate (RuBP) (Yokota et al 1996). This study examined the relationship between occurrence of fallover, the putative hysteresis-inducible sites (Lys-21 and Lys-305 of the large subunit in spinach RuBisCO), and the relative specificity in the carboxylase and oxygenase reactions amongst RuBisCOs of a wide variety of photosynthetic organisms. Figure 1 shows the relationship between occurrence of fallover, amino acid residues at the hysteresis-inducible sites and the relative specificity of RuBisCOs (Uemura et al 1998).

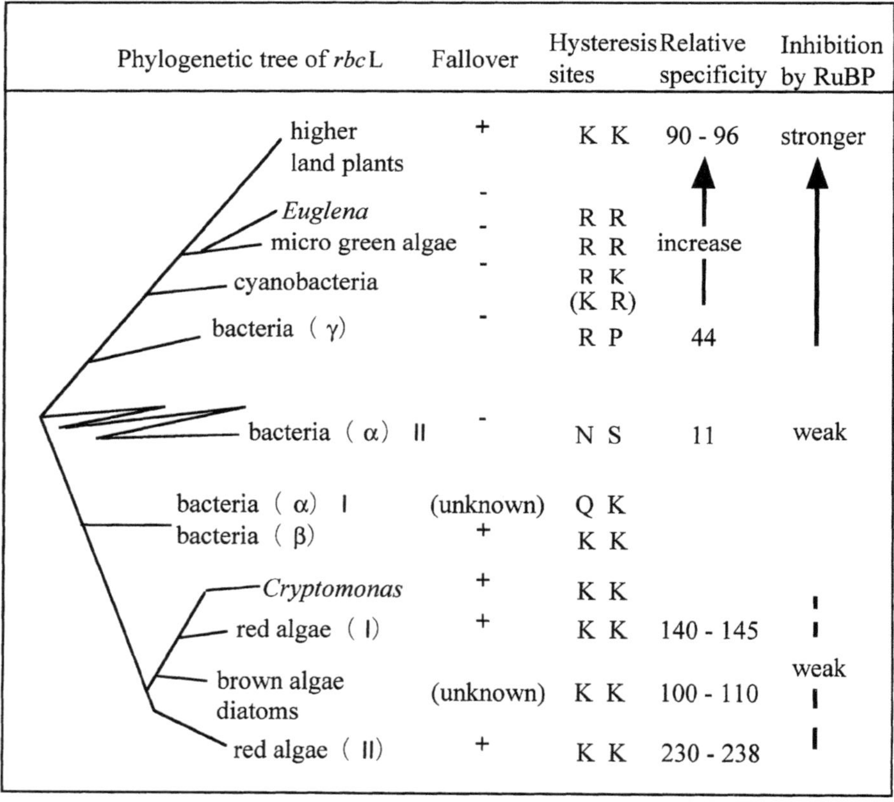

Fig. 1. Relationship between fallover, the hysteresis-iducible sites, the relative specificity and other enzymatic properties in RuBisCOs from various organisms

The phylogenetic tree for the evolution of the gene for the large subunits of RuBisCO, *rbc*L, has been well accepted (Ueda, Shibuya 1992, Zetsche, Valentin 1993). Occurrence of fallover and the hysteresis-inducible sites well followed the sequence of the adaptation of photosynthetic organisms to the terrestrial habitat or the increase in the relative specificity of RuBisCO.

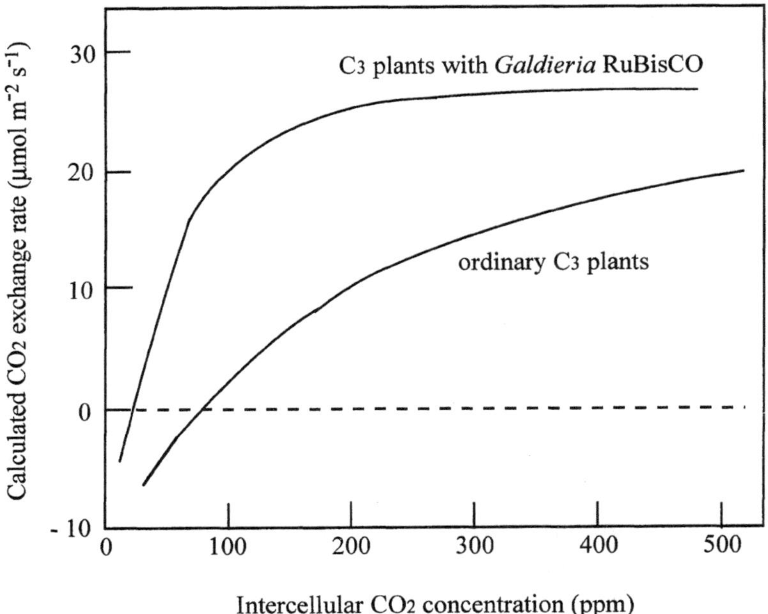

Fig. 2. Calculation of the photosynthetic CO_2 gas exchange rates of ordinary C_3-plants and of realizable C_3-plants in which *Galdieria* RuBisCO is functioning in place of the original enzyme. Calculations were done using the kinetic parameters for spinach and *Galdieria* RuBisCOs (Uemura et al 1997) and equations for photosynthetic gas exchange (Farquhar, von Caemmerer 1981).

From this line of our studies, we expected that introduction of the hysteresis-inducible sites into the photosynthetic bacterial (γ) enzyme would give rise to an increase in the relative specificity of the bacterial enzyme. However, this was not the case. The mutant *Chromatium vinosum* RuBisCO having lysine residues at R21K and P305K showed fallover, but its relative specificity was very similar to that of the wild enzyme.

Another interesting point is occurence of the hysteresis-inducible sites in ß-purple bacteria and non-green algae (Fig. 1). The relative specificity of the non-green algae was much higher than that of higher C_3-plants. Interestingly, red algae are dividied into two groups in the phylogenetic tree of *rbc*L. The group including *Porphiridium* and *Porphyra* live at moderate temperatures in the presence of salts. The other group contains *Cyanidium* and *Galdieria*, which grow at higher temperatures. The relative specificity of RuBisCOs from the latter group were the extremes of RuBisCOs examined so far (Uemura et al 1997). The higher specificity for CO_2 fixation in these RuBisCOs was partly due to their higher affinities for CO_2 (6.6 μM) and partly to an higher activation energy in the oxygenation reaction (28.6 kcal mol^{-1}).

Figure 2 shows the A/Ci curve for the ordinary C_3-plants, calculated by the equations of Farquhar and von Caemmerer (1981) with the reported kinetic values for plant RuBisCO. The CO_2 compensation point is 50 ppm intercellular CO_2 and the CO_2 fixation shifts from the RuBisCO-limiting phase to the RuBP-regeneration-limiting phase at 170 ppm CO_2. If RuBisCO is substituted for plant RuBisCO, the realizable transgenic plants will have the CO_2 compensation point at 16 ppm CO_2. The phase transition will occurr around 70 ppm CO_2. This predicts that the introduced *Galdieria* enzyme will utilyze the photosynthetic chemical energy efficiently even in the presence of low concentrations of CO_2.

These considerations teaches us that changing the enzymatic properties of RuBisCO of C_3-plants is the meaningful direction for improvement of plant productivity. Particularly, increasing the relative specificity and the affinity for CO2 of RuBisCO is the meaningful direction in plant biotechnology.

3. Acknowledgements

This work was partly supported by PEC/MITI of Japan.

4. References

Farquhar GD, von Caemmerer S (1981) Encyclopedia of Plant Physiology 12, pp 549-587, Elsevier, Amsterdam.

Ueda K, Shibuya H (1992) 5th International Colloquim on Endocytobiology and Symbiosis, pp 369-376, Tübingen University Press, Tübingen.

Uemura K et al (1997) Biochem. Biophys. Res. Commun. 233, 568-571.

Uemura K et al (1998) Plant Cell Physiol. 39, 212-219.

Yokota A et al (1996) J. Biochem. 119, 487-499.

Zetsche K, Valentin K (1993) Endocytobiosis Cell Res. 10, 107-127.

QUALITY ISSUES IN A TISSUE CULTURE LABORATORY

Pierre Debergh
Faculty of Agriculture and Applied Biological Sciences, University Gent, Department of Plant Production, Coupure links 653, 9000 Gent, Belgium

This contribution should not be considered as a research paper but as considerations of someone with approximately thirty years of experience in plant tissue culture. Anyone involved in this discipline has been confronted with dramatic situations of considerable losses of cultures due to unexpected problems. Therefore it might be interesting to learn from someone other problems in research and commercial tissue culture technology.

In food manufacturing there is a quality control program which is abbreviated as HACCP, which stands for Hazard Analysis Critical Control Points. It might be indicated to apply the same analytical procedure to the different steps involved in a successful tissue culture (TC) procedure. This contribution means to be a first draft of a list of critical control points, and is therefore non exhaustive.

The biggest challenge of a TC-laboratory is delivery of the correct amount of quality plant material of the cultivar which has been ordered in due time. All items discussed hereafter can interfere in this duty and therefore delay the planned date of delivery or create problems concerning the quality of the final product.

1. Mother plant

Starting with the correct plant material seems evident, but still too often mistakes are made. In our own laboratory we have made this mistake when we taught to initiate a bamboo species (*Phyllostachys*), using a grass (*Pogonantherum*) as mother plant (El Hassan, Debergh 1987). Also later on problems can arise due to mislabeling or mixing up of cultivars. Although a very simple problem, it was and still is a major complain of many customers of TC-laboratories.

Success or failure when initiating a culture can depend to a large extent on the physiological status, e.g.the floral or non floral status, the degree of maturity of the source of explants, the position of the explant, ... It is also possible to condition the explants by a pre-treatment [light conditions (intensity, photoperiod, spectrum); forcing solution, hormonal sprays, use of biocides, grafting on seedling rootstocks, ...].

A. Altman et al. (eds.), Plant Biotechnology and In Vitro Biology in the 21st Century, 625–628.

Stage 0 can help to yield "cleaner" explants, which allows to use milder sterilisation procedures, thereby minimising hypersensitivity reactions (Debergh & Read, 1991).

2. Explant

The size of the explant can influence the phytopathological status of the resulting off spring, and the success during initiation. Rather often the use of TC-technology is abused to claim that disease free material has been produced. TC is not a guarantee for this, only when appropriate tests have been run it is possible to speak about indexed plants for a specific contaminant. Therefore, terminology as disease free, bacteria free, free of contaminants should be banished, as it can not been proven.

Choosing the appropriate explant for subcultures requires well trained personnel, able to make the right interpretation. Important choices to be made are: axially or adventitious shoots, position of the explant.

The number of explants per container can have a dramatic influence on the final quality of the product or shorten the culture period, e.g. in orchids the duration of a subculture can be halved (3 months instead of 6) by halving the number of explants in a container.

3. Sterilisation

A good TC-procedure should be reliable; therefore regular initiation of new cultures is a must. This requires adequate sterilisation; for many plants stage 0 opens possibilities to use softer sterilisation procedures and this in turn can minimise browning problems.

Autoclaving is still the most used sterilisation device for TC-media. Appropriately operating an autoclave requires a double check: when the autoclave is not completely filled with steam, a specific pressure will not correspond to the temperature which should normally be reached.

The use of alcohol to flame the instruments can be problematic. Indeed specific bacteria can develop in alcohol and therefore can be a source for the systematic spread of contamination. So far no negative reports are available on the use of glass bead sterilisers.

4. The TC-container

Some types of container, especially disposable ones, can release specific substances, which can eventually be toxic.
The closure device determines the exchange capacity of the head space of the container with the surrounding and the chances of contamination due to insects and arthropods.

5. Medium ingredients

All ingredients of a TC-medium are important. Water is the major component, but is often the least suspected ingredient causing problems. In some laboratories tap water is used, and differences between summer and winter production have been observed. Possible causes are the use of different concentrations of chlorine in water purification stations depending on the season. Also the harvesting season of field crops can create problems with water quality: once the crop was harvested there was more leaching of herbicides (e.g. triazines) used, and this influenced the quality of the freatic water, with consequences for the quality of the well water.

It is well documented that the label and the concentration of the gelling agent have a tremendous impact on the physicochemical characteristics of a culture medium.

Changing the label of whatever ingredient can have far reaching consequences on the performance of the cultures. This is especially the case for hormones and charcoal, but also for sugar (beet sugar vs. cane sugar).

One of the crucial instruments in medium preparation is the pH-meter. A lower pH makes a softer medium. Therefore if the pH-meter is not well calibrated problems can arise. It is also important to specify when the pH is adjusted: cold or warm medium?; is the pH-meter equipped with a temperature compensation probe?

6. Medium preparation

The physical and chemical properties of a medium are altered by the way it is prepared and by the storing time before inoculation. In the traditional way a medium is heathen up for preparation and thereafter during the autoclaving procedure. When using a medium preparation device, there is only one heating cycle. This has consequences on the production of HMF (Maillard reaction).

The physical properties of a prepared medium poured in the container change dramatically during the first 48 h after preparation, before reaching a steady state. Therefore it is better to store the medium for a few days before use.

7. Culture room

Designing a culture room with a uniform ambience is an Utopia. As in a greenhouse the larger the volume the better the buffering capacity, and the more homogeneous conditions are created. One of the most challenging problems is monitoring the inner atmosphere of a container. Indeed the dissension is the need to have a tight container to avoid contamination and the need to have an aerated container to allow gas exchange with the environment. Bottom cooling is an alternative to keep at least one of the problems causing factors, namely relative humidity (a better terminology is water retention capacity of the surrounding head space), under control. Indeed, a high relative humidity is the major factor responsible for hyperhydricity.

Photoperiod and intensity are the most studied light parameters. The interest in producing photoautotrophic plants engendered a lot of research work with high light intensities, and much more attention was paid to PAR (Photosynthetic Active Radiation), instead of light intensity expressed in lux. Light quality (photomorphogenesis) has not been the subject of much research work, although it can influence the propagation ratio, plant height and leaf surface (Moreira da Silva, Debergh 1997), breakdown of hormones (Hangarter, Stasinopoulos 1991).

8. Propagation technique

Most commercial micropagation systems favour axillary branching. Adventitious shoot and embryo production are most often avoided so far, because of problems encountered with genetic stability. However, it is not always easy to distinguish between axillary and adventitious branching.

9. Quality of the final product

Too often the quality of the final product of micropropagation is evaluated by visual perception. However, a good looking plant is therefore not a guarantee for a good plant. Photoautotrophy is not a must, water retention capacity of the leaves is much more important.

10. Weaning procedure

High quality tissue culture material can be spoiled by inappropriate weaning procedures, as the preventive use of pesticides. Different fungicides have a hormonal effect (Werbrouck et al 1996).

REFERENCES

Debergh PC, Read PE (1991) Micropropagation - Technology and Application, pp. 1-13,. Kluwer Academic Publishers, Dordrecht, The Netherlands
El Hassan A, Debergh PC (1987) Plant Cell Tissue Organ Culture 10:73-77 + corrigendum 15:93
Hangarter RP, Stasinopoulos TC (1991) Plant Sci 79: 253-257
Moreira da Silva MH, Debergh PC (1997) Plant Cell Tissue Organ Culture 51:187-193
Werbrouck et al (1996) Plant growth regulation 15:87-93

TEMPORARY IMMERSION FOR PLANT TISSUE CULTURE

E. Etienne [2] C. Teisson [1], D.Alvard[1], M. Lartaud [1],, M. Berthouly [1], F. Georget [1], M. Escalona [3], JC Lorenzo [3]

[1]: CIRAD, BP 5035 Montpellier, France, [2]: CATIE Po B 104 Turrialba, Costa Rica and [3] CBP, Ciego de Avila, CP 69450,Cuba.

1. Introduction

Mass propagation of plants by tissue culture is labour-intensive and costly. Gelling agents have many drawbacks: they are not an inert medium component and do not enable easy automation. So liquid culture systems, provided that hyperhydricity is eliminated, are considered to be far better: culture conditions are much more uniform, media can be changed easily, sterilisation via ultrafiltration is possible and vessel cleaning after the culture period is greatly simplified. Bioreactors previously developed are not suitable as they are mainly adapted to bacterial culture and do not take into account the specific requirements of plant cells, such as sensitivity to shear forces, mechanical damages or foam formation in bubble aerated bioreactors. New systems have to be developed which must

- avoid continuous immersion which adversely affects growth and morphogenesis
- provide adequate oxygen transfer, sufficient mixing and limit shear levels
- enable sequential medium changes and automation
- reduce the risk of contamination
- be as cheap as possible

In a previous paper we indicated that temporary immersion in a liquid medium is one of the best ways of achieving all these goals (Teisson, Alvard 1994). We demonstrated too that the period of immersion must be very short and that both the frequency and length of these periods have a direct effect on morphogenetic events. We will now illustrate new results obtained either with a new specific apparatus or with quite simple home-made apparatus.

2 Material

The conceptual basis of the process lies in the periodic and pneumatic driven transfer of liquid medium from a specific tank to the container where the plants are. To avoid excess tubing these two volumes are preferably part of the same vessel. Overpressure is applied

A. Altman et al. (eds.), Plant Biotechnology and In Vitro Biology in the 21st Century, 629–632.

by a solenoid valve or a compressor connected to a programmable plug. This application determines the time and duration of floodings. All the air flows are sterilised through 0.2 µ hydrphobic filters.At the plant compartment inlet air flow must be broken to allow gentle stirring and bubbling.

2.1 Rita

The container we used in preliminary experiments was an autoclavable filtration unit on general sale which was modified for the purpose in our laboratrory. Despite the success we obtained, using this vessel resulted in some financial and technical drawbacks so we developed a new apparatus specifically intented for plant tissue culture. The vessel comprised two compartments, an upper one with the plants and a lower one with the medium. These two compartments were linked in such a manner that the overpressure applied in the lower compartment pushed the medium into the upper one. Plants were immersed as long as the overpressure was applied. During the immersion period a flow of air bubbled through the medium, stirred the plants and renewed the head space atmosphere inside the culture vessel as the overpressure escaped through outlets on the top of the apparatus. When pressure dropped, the medium returned to the lower compartment by gravity. Particular attention was paid to reduce the risk of contamination (apart from the outlets there was only one possible communication - through the screw of the lid- between the outside and inside) to facilitate handling (assembling the apparatus, changing the medium, manipulating the plants) and to reduce the cost. The size of the vessel was kept at one litre which we think is suitable for experimental purpose and for mass propagation:
- increasing the size may lead to an increase in the incidence of contaminated batches
- in the future, mass propagation will be mainly achieved via somatic embryos which are small

The resulting vessel has been patented, called Rita according to a French acronym meaning Recipient for Automatised Temporary Immersion. It is made of reusable and autoclavable plastic (polysulfone). Several thousand units have been manufactured and are working in experimental and preindustrial laboratories in Europe, Africa, Central America and Asia.

2.2 Twin flasks

The Rita system, as will be illustrated later, is mainly intended for mass propagation through somatic embryogenesis. For organogenesis the size -i.e the number- of propagules may require a larger volume and very cheap vessels.

The easiest way to carry out pneumatically-driven temporary immersion is to connect two glass flasks - any size, any shape- by tubing and to apply alternative overpressure to push the medium into the other recipient. This system is very simple, cheap, easy to built and has been used with vessels from 250 ml to 10 litres.

The Rita vessel can be easily adapted to this configuration.

3. Results

3.1 Meristem proliferation. Microcutting.

The first results we published on temporary immersion concerned banana meristem proliferation (Alvard"et al.". 1993). Later a Cuban team, in the Centro de Biopletas in Ciego de Avila, worked extensively on the topic. After achieving success with home made apparatusses they improved their technology using twin flasks of 10 l each. The multiplication rate sharply increased in comparison to a semi-solid medium: 3 to 4.5 times on pineapple and sugar cane, double on banana and 1.5 on malanga (*Xanthosomas caracu*). Up to 5000 pineapple plants can be harvested at the same time in one flask. The system is extensively used for industrial propagation.

3.2 Microtuberisation

The beneficial effect of partial but permanent immersion on microtuberisation was previously described by Akita, Takayama (1994) . Using a very simple procedure in a double Rita system, 3 microtubers were obtained per single node 10 weeks after inoculation. 50% of the microtubers exeeded 0.5 g and sprouting, still in a temporary immersion system, was very quick and efficient: three to four stems started to develop from one single tuber. Similar results were obtained with 3 different cultivars (Teisson, Alvard. 1998) in very rough experiments and the process seemed to be easily improvable with an ultimate target of direct transfer to the field.

3.3 Somatic embryogenesis

Temporary immersion can be used for somatic embryogenesis at different stages.
Development of Hevea somatic embryos is routinely performed in our lab. Around 150 embryos at a cotyledonnary stage are harvested 4 to 8 weeks after the transfer of 3 g of embryogenic callus to a Rita. These embryos are transferred to semi solid medium for germination. Until now the upright orientation of the plantlets seems to be necessary for germination, which was impossible to achieve in a Rita vessel, as a new orientation occurs every flooding.To date more than 5000 Hevea plantlets developed in Rita have made it possible to carry out the first fully clonal field experimentation.
Where applicable, the best way to perform mass propagation through somatic embryogenesis is to follow the process established for coffee and banana. Multiplication of embryogenic callus in a liquid medium is carried out in erlenmeyer flasks, which are the cheapest biotreactors, and then expression and germination (coffee) or germination only (after plating on a semi-solid medium) (banana) are achieved in a Rita vessel.In this way large quantities of callus, embryos and plantlets can be manipulated with minimum handling. Up to 5000 coffee somatic embryos and more that 500 banana plantlets can be harvested in one Rita. After germination in Rita the vigour and conformity of very young

embryos are such that direct transfer to in vivo conditions is possible. The quality of the propagules and the reduction in labour costs result in entirely new and valuable conditions for the economic exploitation of somatic embryogenesis. It can be applied to the mass propagation of other species.

4. Conclusion

Plant tissue culture can be carried out by pneumatic-driven temporary immersion in a wide range of vessels with a wide range of processes.
The keys to biological success appear to result from the physical conditions created in the culture vessel:
- short immersion in a bubbled, well oxygenated medium
- good supply of nutrients by renewed, direct contact with the medium
- low disruption of gas exchanges between the plant and the atmopsphere by very short immersion times
- desiccation prevented by the capillary film of the medium during emmersion
- complete renewal of the atmosphere inside the vessel at regular intervals
- splitting of plant tissues caused by gentle stiring during the immersion phase

The economic advantages are linked to:
-all the usual advantages of liquid medium: change of medium, sterilisation, cleaning...
- propagule quality
- wide range of simple, automatisable and cheap pneumatic-driven bioreactors

Although it sharply decreases in comparison with permanent immersion, hyperhydricity remains the major problem and requires an accurate definition of the frequency and duration of flooding periods, as well as possible modification of medium composition. Few teams seem to have been interested in this topic but we felt that temporary immersion created really new economic conditions for mass propagation, which may be valuable for several species.

5. References

Akita M, Takayama S (1994) Plant Cell Reports13, 184-187.
Alvard et al (1993) Plant Cell Tissue and Organ Culture 32, 55-60.
Teisson C, Alvard D (1994) VIII International Congress of Plant Tissue and Cell Culture, Florence, Italie, 12-17 juin.
Teisson C, Alvard D (1998) In Conference Potato seed production by tissue culture, Cost 822, Brussels, 25-28 February.

ROBOTIC MICROPROPAGATION FOR COMMERCIAL FORESTRY

Janet R. Gorst & Robert D. Teasdale
ForBio Research, PO Box 1729, Toowong, Queensland 4066, Australia

1. Introduction

With the predicted growth in the world population of over 80 million people per annum, the demand for timber in the form of wood and paper products is continuing to increase. By 2010, it is estimated that an additional 50 – 100 million hectares of plantation forests will be needed to meet the projected wood requirements for developing countries. The total global demand for planting stock of the top 5 tree species (Eucalypt, Acacia, Rubberwood, Radiata Pine and Teak) amounted to 1,650 million units in 1995 and is forecast to reach 2,475 million units in the year 2000. The commercial success of a robotic micropropagation system compared with manual micropropagation is dependent on the production of very large numbers of propagules; as the above numbers indicate, timber propagules represent an ideal niche for robotic propagation.

One of the problems associated with clonal propagation of elite trees is that, traditionally, there is a requirement to wait many years to identify the elite individuals and by this time the trees have passed into a mature stage that often acts as a barrier to vegetative propagation. In order to maximize propagation potential at a commercial level, it is judicious to combine biotechnology with robotics. The use of Marker Assisted Selection [in which DNA markers can be linked to the quantitative trait loci governing a particular valuable phenotype such as tree height or diameter at breast height] allows potentially elite individuals to be identified at the seedling stage so that clones can be propagated while the plant material is still responsive. This concept is the philosophy behind the development of the Vitron™ 501 robot and the associated automated tissue culture system at ForBio. It guarantees that crucial time is not lost in waiting to analyse trees for performance and that maximum financial benefits can be gained from the propagated material, since high quality is assured at the outset. The Vitron™ 501 robot was released in 1997 and was specifically designed to handle tree species.

A. Altman et al. (eds.), Plant Biotechnology and In Vitro Biology in the 21st Century, 633–636.

2. The Vitron™ 501 Robot

2.1. Construction

The robot (see Fig. 1) has an electro-pneumatic-mechanical construction with 11 axes (motor driven components) and tools (harvester, cutter, planter) with up to 5 degrees of freedom. The tool axes are all driven by open-loop stepper motors. The 'brains' of the robot are two 486 computers, one for the control system and one for the vision; they communicate over a serial line. Both computers run in DOS and are programmed in C/CC+. Although the robot is technically complex it is constructed from parts 90% of which are 'off-the-shelf'. The dimensions of the robot are such that it fits through standard doorways. Operation requires electrical power (240V single phase), compressed air, a vacuum system, and a supply of culture vessels (Vitron™ Tissue Culture Boxes) filled with an agar-based nutrient medium. The TC Boxes were designed at ForBio and are disposable (polyethylene terephthalate = PET) containers. All parts of the robot are made of biologically inactive materials. The frame, all of the panels, fasteners, tool posts and tool end effectors are made of stainless steel. The robot has a HEPA (High Efficiency Particulate Air) filter that maintains a constant positive pressure within the machine and forces clean air across the work area (in a manner similar to the principle used in laminar flow cabinets) thus preventing contaminants entering the Process Cell.

2.2. Operation

The robot has three separate compartments within each of which specific operations are performed -

i) The Loading Module

- Vitron™ boxes containing plants to be propagated, on the one hand, and media for propagation, on the other, are loaded manually into the chutes of the Loading Module which can take sufficient Boxes for continuous 2 h running before replenishment is required.
- The boxes are then automatically moved by conveyor into the Process Cell

ii) The Process Cell

- The Cell has three conveyors, the first of which carries the box of plants to be propagated, the second of which carries a box into which tips are planted and the third of which carries a box into which nodes are planted.
- Upon entry to the Process Cell, container lids are removed.

♦ Plantlets in the boxes are planted in defined rows. A harvesting tool moves sequentially along each row and cuts each plant which is then imaged and dissected into explants of designated length. The explants are then planted into target trays according to a pre-defined pattern (up to 40 plants/box). Any waste material is dealt with by a vacuum system.

♦ The Vitron™ Box lids are replaced just before exit from the module.

♦ Asepsis is maintained within the Process Cell by the robot automatically sterilizing the instruments at regular intervals and the operator manually cleaning the interior after pre-determined times. The operator is also reminded at pre-set times to exchange instruments for fresh ones and to drain and replace the disinfectant in the instrument baths.

iii) The Unloading Module

♦ The boxes of harvested plants and freshly planted tips and nodes are stored in the module until manually removed and taken to the incubation area or nursery.

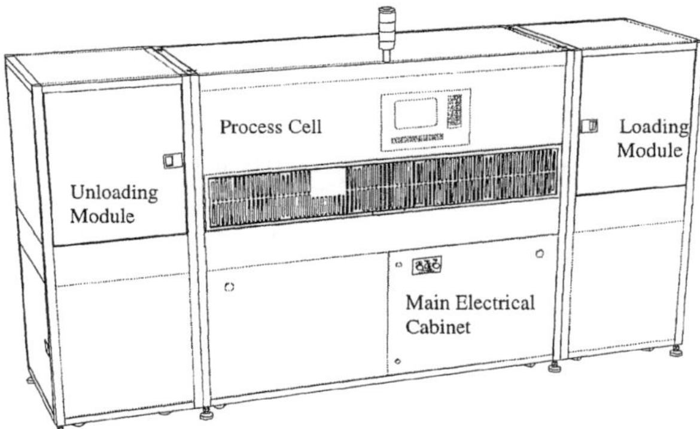

Figure 1. Sketch of the Vitron™ Robot

2.3. Robotically Produced Plants

The form of the plants must be such that there is growth of a single, apically dominant, rooted shoot (Fig. 2) . In order to achieve this, it is likely that there will be no cytokinin present in the culture medium and that there will be some type of auxin added to the medium. The robot is capable of processing input plants into new tip and nodal explants and leaving behind a rooted base. If, as will be the case in species like *Eucalyptus*, the nodal explants subsequently produce bifurcated shoots (Fig. 3), a Manual Replacement stage will be required since the robot is not currently capable of handling bifurcated shoots. This involves laminar flow operators manually processing the shoots and placing the products (tip and nodal explants) into Vitron Boxes.

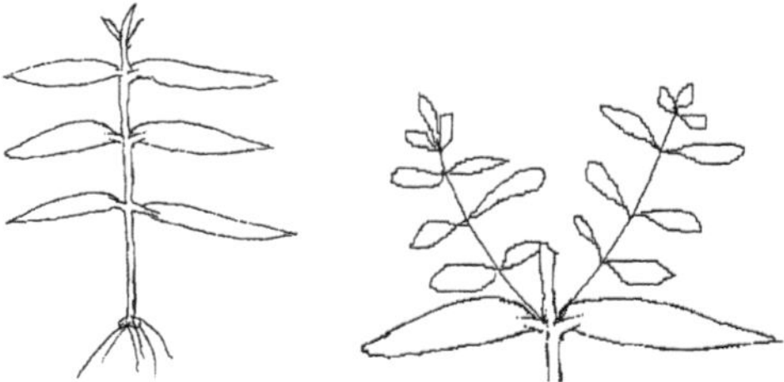

Fig. 2. Typical robotic form of eucalypts

Fig. 3 Development of a bifurcated shoot system in eucalypt after incubation of a nodal explant

The saleable products from this system will come after a suitable incubation period in the Growth Room following robotic harvesting and may be one or all of the following

- a plant such as represented in Fig. 2 that has come from a tip explant and has a single stem and roots.
- a plant that has come from a nodal explant and has a well developed axillary shoot (or shoots if the plant has a tendency to bifurcation) and roots.
- a plant that has come from a rooted base and has a well developed axillary shoot (or shoots if the plant has a tendency to bifurcation).

3. Commercial Application

Monsanto of the USA is one of the world's largest and leading biotechnology companies and, together with ForBio has established an Indonesian-based micropropagation facility – PT Monfori Nusantara – just outside Bogor (50 Km south of Jakarta). The facility has the potential for utilizing 4 Vitron™ Robots and producing around 15,000,000 plants/year. Currently, the species being propagated are elite eucalyptus, teak and acacia developed by ForBio. Monfori will market the trees to major forestry companies in Indonesia, with the expectation of future expansion into other plant types and consumer markets in the ASEAN region.

Acknowledgements: A large number of people (too many to name individually) within the ForBio Group in Australia and from ForBio's subsidiaries and joint venture partners overseas have contributed, and continue to contribute, to the development of the Vitron™ Robot.

Design Considerations for a Mechanized Micropropagation Laboratory

Amnon Gross[1] and Robert Levin[2]
[1]A. Gross Engineering, 46 Hahagana St. Raanana 43423, Israel, www.gross.co.it
[2] Osmotek Ltd. P.O.B. 550, Rehovot 76120. Israel. www.osmotek.com

A commercial micropropagation laboratory is designed to provide an efficient flow of material and personnel, reduce waste from contamination introduced by personnel and other factors and minimize construction and operation costs. Efficient flow of material and personnel is achieved by a design that places the equipment and personnel for each task in the appropriate location. Contamination losses are limited by controlling the air quality and the number of people and how they are dressed in areas which effect the cultures. Both of these factors have cost factors associated with them and therefore effect the cost of production of the facility. Balancing these factors to produce the lowest cost of production is the basis for the design considerations of the modern plant tissue.

Examples of these design considerations and their economic consequences are shown below for a conventional manual labor micropropagation laboratory and laboratories using mechanization procedures and equipment such as the Osmotek system (Levin et al., 1997). These laboratories are all designed based on the following assumptions:

1. Production of 20 million units per year distributed equally throughout the year.
2. One-month multiplication and growth cycles.
3. A multiplication rate of 4:1 less 5% contamination.
4. Where shelves are employed in the growth room, the shelves are in tiers of 8.
5. Where vessels are employed they are reusable, have 100cm^2 outside dimensions, hold 40 clusters, and use 200 ml of medium per vessel.
6. Where bioreactors are employed reactors have a 5-liter working volume, hold 6,000 clusters and a rack of ten reactors take up 1 m^2 of floor space.
7. Cool white fluorescent tubes provide illumination at 1,500 – 1,800 lux with a 16-hour photoperiod.
8. For manual labor clusters are transferred at the rate of 250 per hour.
9. For mechanized labor using reactors for multiplication and vessels for elongation the transfer rate is 3000 clusters per hour.
10. For mechanized labor using reactors for both multiplication and elongation the transfer rate is 4,500 clusters per hour.

Model 1is a laboratory designed for production with conventional technology using manual transfers and elongation and multiplication in vessels on semisolid medium. The laboratory is divided into four zones each with a different level of air handling to provide a different level of cleanliness (see Table 1). The white area has the highest level of air filtration (> than class 100,000) to protect the cultures were they may be exposed to contamination. This area is maintained at the highest air pressure and protected by air locks from other areas. Air pressure declines by 1 mm of mercury from zone to zone. Thus when there is airflow in the laboratory if flows from white to gray to black to green thus protecting the cultures.

The white area includes the changing rooms where personnel change to clean uniforms and head coverings to reduce contamination being transferred from the personnel to the cultures. The explant introduction room, the media storage area, the

A. Altman et al. (eds.), Plant Biotechnology and In Vitro Biology in the 21st Century, 637–642.

transfer rooms and growth rooms are also in the white area to protect the cultures. The gray area is a service area that communicates through air locks with the white area and includes services such as media preparation and storage, washing etc. The black areas are functionally peripheral areas such as packing of finished products. The green areas directly communicate with the outside environment and include storerooms, machinery rooms, a dining room and offices.

The costs of construction and operation of model 1, under Israeli conditions is shown below. The annualized cost of operation of model 1 is shown in Table 7. The annual cost of labor, energy and the annualized cost of construction and major services such as air conditioning and air handling for model 1 are $ 2,9 million. The annualized cost of equipment and furniture for Model 1 is $1.35 million. Taking into account the costs of media, vessels and miscellaneous expenses the cost of production for each unit of the 20 million units produced each year is $0.16.

Model 2 is a laboratory designed for mechanized production using a system such as the Osmotek Ltd. system with multiplication in bioreactors, mechanical separation of the multiplication culture and elongation in vessels on semisolid medium. There is a reduction in work places and space requirements in all zones of the laboratory. However, the cost of equipment and furniture remains unchanged. Larger spaces of lower cost clean benches and growth room shelving are replaced with fewer but higher cost workstations, fermenter racks, mechanical separators etc. This results in little or no change in the cost of equipment and furniture between model 1 and model 2.

The reduction in space and labor, however, does results in a decrease in the annual cost of labor, energy and the annualized cost of construction and major services such as air conditioning and air handling from $2.9 million to $.88 million in model 2. The costs of media, vessels and miscellaneous expenses remain unchanged, but the reduction of labor space and energy requirements of model 2 reduces the unit cost of production to $0.08.

Model 3 is a laboratory designed for mechanized production with multiplication and elongation in bioreactors (or ebb and flow bioreactors) and mechanical separation of the multiplication culture. There is a small further reduction in work places and space requirements in the laboratory. There is, however, a significant reduction in the area of the white zone, which results from shifting elongation to the Gray area. This shift also results in a reduction in energy costs. Taken together these factors result in a further reduction in the annual cost of labor, energy and the annualized cost of construction and major services to $0.66 million. The costs of media, vessels and miscellaneous expenses again remain unchanged, but the reduction of labor space and energy requirements moderately reduces the unit cost of production to $0.07 per unit.

Although a reduction in costs is the most outstanding feature of a mechanized laboratory, other benefits are also realized. As can be seen in table 2. the reduction in labor requirements is quite pronounced. Consequently very large laboratories can be efficiently managed. Further, the reduction in gray and white areas means simplified construction and maintenance.

The reduction of energy requirements for micropropagation not only reduces cost, but makes the facility less vulnerable to power outages. The reduction in energy requirements results from fermenters needing less light than shelves, reduction in cooling needed for personnel, a reduction in clean benches and an overall reduction in air conditioning. Autoclaving is also eliminated since he fermenters are gamma sterilized and the media is filter sterilized.

From the above it is apparent that mechanized micropropagation allows the design of a facility which produces larger volumes at less cost and in a less complicated facility.

References

Levin, R et al (1997). Acta Hotrt. 447, 659 – 663.

Table 1 Zones for Laboratory Models 1, 2 and 3

Green (30% filtration)	Black (60% filtration)	Gray (98% filtration)	100,000
Dining room	Corridors	Changing rooms	Changing rooms
Office	Raw material prep.(RMP)	Media prep	Corridors
Raw material storage (RMS)	Packaging material prep. (PMP)	Temporary storage area 1	Explant introduction room
Packaging material storage(PMS)	Packing area	Washing area	Temporary storage area 2
Finished product storage (FPS)	Quality Control (QC)	Corridors	Media storage
Changing rooms	Technical floor	Office corridor**	Transfer rooms
Media prep. area	Finished product rinsing area*	Rack washing area*	Stock culture growth room
Machinery room**	Package drying room**	Elongation growth room **	Multiplication growth room
Office building**			Elongation growth room
Corridors			Training and R&D**
Ramps			

* Model 3 only ** = excluded in the calculations

Table 2 Estimated Laboratory Labor Cost for Laboratory Models 1, 2 and 3
Model Number

	1		2		3	
	Shift 1	Shift 2	Shift 1	Shift 2	Shift 1	Shift 2
Number of Workers						
Cost per Month Per Person ($)	1,100	1,350	1,100	1,350	1,100	1,350
Total Labor Cost ($) Per Month Per Shift	85,800	91,000	28,600	24,300	25,300	18,900
Total Labor Cost Per Month ($) Per Shift	180,300		52,900		44,200	
Total Labor Cost Per Year ($) Per Shift	2,163,600		634,800		530,400	

Table 3 Estimated Construction Cost Per Zone for Models 1, 2 and 3
Model No.1

Zone	Green	Black	Grey	White	Total Lab
Filtration	30%	60%	98%	> class 100,000	
SQM	312	701	193	894	1,500
Cost $ per SQM	750	1,000	1,250	1,700	
Total ($)	234,000	701,000	241,000	1,520,000	2,696,000

Model No. 2

Zone	Green	Black	Grey	White	Total Lab
Filtration	30%	60%	98%	> class 100,000	
SQM	295	574	166	621	1183
Cost $ per SQM	750	1,000	1,250	1,700	
Total ($)	221,000	574,000	208,,000	1,056000	2,059,000

Model No. 3

Zone	Green	Black	Grey	White	Total Lab
Filtration	30%	60%	98%	> class 100,000	
SQM	295	512	175	405	990
Cost $ per SQM	750	1,000	1,250	1,700	
Total ($)	221,000	512,000	219,000	689,000	1,641,000

Table 4 Estimated Annual Cost of Major Investments for the Laboratory

	Building		Annual	Annual	Annual	Annual	Total
	Cost $ 1,000	e Years	Interest %	Cost of Building $ 1,000	Labor Cost $ 1,000	Energy Cost $ 1,000	Annual Cost $ 1,000
Model 1	2,696	20	4	140.19	2,163.6	209.16	2,912.95
Model 2	2,059	20	4	107.07	634.8	137.52	879.399
Model 3	1,641	20	4	85.33	530.4	48.96	664.683

5 Estimated Annual Cost of Major Equipment and Furniture for the Laboratory

	Equipment Cost $ 1,000	Furniture Cost $ 1,000	Amortize Years	Annual Interest %	Total Annual Cost $ 1,000
Model 1	1.000	300	10	4	135.2
Model 2	1,000	300	10	4	135.2
Model3	1,000	300	10	4	135.2

6 Estimated Energy Costs for Each Model for the Major Energy Use Factors

	Lighting Growth Room KW	Personnel KW	Clean Benches KW	Outside Air* KW	Aircond. R134a KW	Autoclave KW	Total KW	Energy+ Cost/hr $	Energy+ Cost/yr. $1,000
Model 1	92	33	15	66	175	200	581	34.86	6
Model 2	60	11	6.5	50.5	104	100	382	22.92	2
Model 3	24.5	8.5	2.5	39	61.5	-	136	8.16	48.96

* = 3 changes per hour + = $0.06 kW

Table 7 Estimated Cost of Production Per Unit for Each Model

	Total Cost $1 ,000 From Table 4	Total Cost $1,000 from Table 5	Unit Costs per unit $ for Tables 4 & 5	Unit Costs for Media $	Unit Costs for Vessels $	Unit Misc. Costs $	Total Unit Costs $
Model 1	2,912.95	135.2	0.13	0.01	0.01	0.01	0.16
Model 2	879.39	135.2	0.05	0.01	0.01	0.01	0.08
Model 3	664.683	135.2	0.04	0.01	0.01	0.01	0.07

Quality From A Young Plant Perspective

Gary Hennen, Oglesby Plants International, Inc.

When I was asked to make this presentation, I accepted assuming this subject was fairly straight forward, and the information easily assembled and presented. However, after considering the subject for some time, I realized that this is not the case. Quality, as an attribute, is very subjective and defies clear and concise definitions. Truly, quality is in the eye of the beholder and quality is a moving target. To use another cliché, "Quality perceived, is quality indeed."

Quality can be defined as a "distinguishing attribute that refers to a particular or essential characteristic of a product." Quality attributes of plants are easy to list but sometimes difficult to quantify. Each of us thinks we know what quality is when we see it, but describing or quantifying quality can be quite challenging. As an example, the size of a plant is easily defined using height and width measurements but we tend to prefer subjective terms such as "tall," "short," "compact" or "stretched." Sometimes plant size is even described by the size of the pot or the number of cells in a cavity tray. Fullness can be described quantitatively by the number of plants or branches in each pot or subjectively by terms such as "full" or "well-branched."

Quality as a dynamic concept varies considerably based on crop, producers, customers and sophistication of the market place. Economic pressures to produce lower cost plants within a market place can also have a dramatic effect on quality issues. Within each market there are quality ideals, whether written or unwritten, by design or by default. Each grower within a market has quality ideals, whether written or unwritten, by design or by default. These quality ideals are what competing products are judged by. Market competitors must try to meet or exceed these ideals to gain or maintain their market share. Exceeding quality ideals can help you gain a short-term marketing advantage. However, if widely accepted by the market, they become the new starting position for quality, essentially changing quality ideals to a new level.

A. Altman et al. (eds.), Plant Biotechnology and In Vitro Biology in the 21st Century, 643–645.

Achieving young plant quality, however it is defined, is directly related to numerous controllable inputs which I have placed into two categories, Organization and Art. Following is a brief discussion of individual components within each category.

It is the role of the Organization to set the stage for quality to be achieved. Quality must be defined and communicated to all involved in the production process. Defining, clearly communicating and periodic review of these definitions are crucial to success. Development and implementation of quality standards, such as ISO 9000 is a clear indication that the Organization has placed a high priority on quality.

The Organization is also responsible for allocating appropriate resources and trained personnel to perform the work. Utilizing the newest technologies and environmental controls is usually looked at on a cost, benefit basis. Is the newest technology and environmental structures necessary to achieve quality? Usually the answer is no. However, the answer may change when asked if you can achieve your quality goals at a competitive cost without utilizing new technologies.

We must constantly remind ourselves of the basic fundamental art and science of cultivating plants. Although I will not address many of these issues, a comprehensive understanding of nutrition, water, light and environment are all essential in consistently achieving your quality goals.

The use of "healthy," "uniform" starter plants allows for easier achievement of your quality goals. Starting with poor quality starter plants usually finishes with poor quality. Size grading of the starter plants is an important step in improving crop uniformity and reducing finished plant size grading. Our company routinely defines three size grades of microcuttings prior to planting. The more size grades defined prior to planting, the better crop uniformity achieved. Most of our customers accept our three grades but others further define grades within our size grades. Starter plant grading systems that utilize machine vision, weight or biomass may offer advantages over human grading. The choice to use microcuttings or plantlet clusters as a starter plant can have a large impact on the uniformity and loss percentages of the crop but may be somewhat dictated by your market competition. Experience has shown that proper planting technique can have a dramatic effect on the crop. Planting perfectly graded starter plants at different depths in the soil can have the same effect as using poorly graded starter plants. Proper spacing of multiple plants in a cavity tray cell can also affect the uniformity of the crop.

Growing medium selection is dependent on factors such as crop type and environmental conditions. Growers sometimes disregard growing medium as an important variable and many times it is the first target for cost cutting efforts. Our experience has shown it to be a critical element in the success of a program. The decision to use soil, soil-less mixes or synthetic medium can be crucial to your quality goals. Carefully test soil and other components on each delivery to ensure uniformity of the medium over time.

Cavity tray type and cell count used in young plant production is very market dependent. Florida's tropical foliage young plant producers have informally standardized to primarily 50, 72 and 98 cell count trays. The manufacturers and dimensions of these trays may differ but the cell counts are standard. Each market will have some formal or informal tray standards. The cavity size and shape, such as round, square, hexagonal or octagonal all have an affect on growth rates and root development and should be given consideration.

Grading of the finished young plant prior to delivery is your last opportunity to assure your product meets your and your customers' quality ideals. It gives the grower a final opportunity to improve size uniformity as well as remove off-types. However, the final grading is the most expensive place to add quality if quality controls are not used throughout the production process.

By consistently producing high quality plants, your customers will recognize your commitment to quality and help you achieve your future business goals.

DEVELOPMENT OF NEW PLANT TISSUE CULTURE MEDIA

M. Henry[1], C. Fulcheri[1, 2] and P. Morard[2]
[1]Faculty of Pharmacy, 54000 Nancy and [2] E.N.S.A. Toulouse, 31426 Castanet Tolosan, France.

1. Introduction

The setting-up of the composition of plant culture media is one of the first thankless but essential steps to perform successfully the plant cell, tissue or organ in vitro cultures. Different ways were employed to define the best constituent composition of culture media allowing the optimal growth of the plants (whole or in parts) cultivated in vitro. The simplest approach consisted in testing different well-known culture media found in the literature and in choosing the best one according to the growth of the plant tissue (Fujita et al 1981a; Rokem et al 1985). Conversely the most difficult means was developed by the first plant biotechnologists (Murashige, Skoog, 1962; Gamborg et al 1968) who created a new culture medium by studying the effect of each mineral and organic constituent well chosen at a precise concentration and in relationship with the others that improved the growth of the plant tissues chosen for their study. But the most used of these ways was a compromise between the two previous ones by the choice of one of the well-known media with the adjustment, element by element, of the concentration of the main constituents of the medium according to the reaction of the plant cultivated in vitro (Fujita et al 1981b; Rokem et al 1985).
Here we present a different strategy for developing new culture media allowing the optimal growth conditions by studying previously the mineral element needs and consumption of the plant during its growth cycle and we tested them also by the way for the production of secondary metabolites. The technique was applied to three different plant materials as suspension, root and multiple shoot cultures belonging to three different species (*Solanum paludosum*, a lignous plant and *Gypsophila paniculata* and *Saponaria officinalis* herbaceous plants) producing steroidal or triterpenoidal saponins.

2. Material and Methods

2.1. Plant mineral analysis

The whole plants were weighed and dried. Total water and mineral nutrient consumption by the plants were measured during the culture. Macronutrient or micronutrient contents were analysed by classical methods (Morard et al. 1990) after for

A. Altman et al. (eds.), Plant Biotechnology and In Vitro Biology in the 21st Century, 647–650.

example sulfuric acid mineralisation for total nitrogen and nitric acid for: P, Bo by automatic absorption spectrophotometry, S by turbidimetry and K, Ca, Mg, Fe, Mn, Zn, Cu by atomic absorption spectrophotometry as described in Fulcheri et al. 1998.

2.2. Establishment of plant tissue and organ cultures and secondary metabolite analysis

The conditions of achievement and maintain of the different plant cultures and respectively their secondary metabolite analysis were described previously: El Badaoui et al. 1996 for *Solanum paludosum* and Fulcheri et al. 1998 for *Gypsophila paniculata* and *Saponaria officinalis*. Batch cultures in liquid medium were performed in 250ml Erlenmeyer's flasks on a gyratory shaker (80 rpm) in a room culture chamber at 25°C in the light for *S. paludosum* and in the dark for *S. officinalis* and *G. paniculata*. The tissue growth and the saponin contents were measured on the total biomass of each flask at each harvest date.

2.3 Calculation of the mineral nutrient composition of the culture media

The new mineral formulation was performed using soilless culture methods. In hydroponics, the approach developed is based upon the calculation of the composition of the nutrient solution from the macronutrient and micronutrient consumption balance of the whole plant during all its growth cycle for herbaceous plants (Morard, 1995). For lignous plants, the mineral element composition of the whole plant after one year of growth is only taken in account since there is a good correlation between mineral consumption and mineral composition during the first year of growth. This permits the development of a new mineral solution adapted to the plant mineral requirements.

The mineral nutrient requirements measured in hydroponic conditions during the vegetative growth for each species were divided by the corresponding water coefficient of transpiration measured during the same period. The results expressed in milliequivalents per litre determined the composition of the macronutrient concentrations of the solution. The same calculation could be carried out for the micronutrients. Since there is no exact balance between the sum of the anions consumed and that of the cations due for example to precipitations and to losts of some elements during the culture of the plants in hydroponic conditions, the values of mineral nutrient consumption could not be used directly. Various adjustments were therefore necessary taking in account the chemical and physiological behaviours of some of the salts in relation to the plant culture studied. For example, calcium precipitated in water solution in presence of phosphate at pH 6.

The mineral nutrient formulation was performed by the use of a one to one mapping table in which the different values obtained for the anions were put in correspondence to the cations, taking in account there also of the solubility in water of the corresponding salts and their commercial availability. From these anions-cations associations the following salts were deducted successively for the conception of macronutrient solution: $MgSO_4$ $7H_2O$, KH_2PO_4, KNO_3, Ca $(NO_3)_2$ $4H_2O$. The use of NH_4NO_3 as nitrogen source is due to the non efficiency of the nitrate reductase step in the nitrogen metabolism of some plant tissues cultivated in vitro. The same protocol was applied to the micronutrient formulation. Sometimes when the mineral consumption of the main organs of the plant (e.g. roots) was quite different from the whole plant it was necessary to adapt the mineral solution to the corresponding organs.

3. Results

3.1 Mineral nutrient formulations for the three in vitro cultures

The mineral analysis of the plants of the three species chosen are reported in tables 1 and 2 (in meq/l):

Table 1	NO_3^-	$H_2PO_4^{--}$	SO_4^{--}
S. paludosum	20.6	1.8	3.2
G. paniculata	14.5	0.7	1.2
S. officinalis	10.8	0.6	1.0

Table 2	K^+	Ca^{++}	Mg^{++}
S. paludosum	4.5	15.2	3.4
G. paniculata	3.8	7.4	1.3
S. officinalis	3.2	3.7	1.0

One to one mapping tables in which anions and cations are in correspondence were then used. The filling in of these tables was begun by Mg^{++} and SO_4^{-} which concentration values were rather the same. The ions taken in account in a second step correspond to the following salts: KH_2PO_4, KNO_3 and Ca $(NO_3)_2$ $4H_2O$. Finally, for *S. officinalis* NH_4NO_3 as nitrogen source was used in addition of the nitrate salts since after a first assay of the basic mineral solution with only nitrates the suspensiones failed to grow. For example, the following composition was established after calculation (Table 3) for *S. officinalis*. Some small mathematical adjustments by rounding of the values were carried out to obtain exact correspondences between anions and cations.

Table 3: Example of macronutrient composition for *S. officinalis* suspension culture.

meq/l	NO_3^-	$H_2PO_4^{--}$	SO_4^{--}	cations
K^+	3	1	-	4
Ca^{++}	4.5	-	-	4.5
Mg^{++}	-	-	1.5	1.5
NH_4^+	1	-	-	1
Anions	8.5	1	1.5	11

The same tables were obtained for *S. paludosum* and *G. paniculata* (El Badaoui et al 1996 and Fulcheri et al 1998). The same protocol was also applied for the formulation of the micronutrient composition (Fulcheri et al 1998). Then, the organic constituents classically defined by Murashige and Skoog (1962) or Gamborg (1968) for example were added to the mineral composition with phytohormone balances and carbohydrare source and the new medium was tested with the corresponding organ or tissue cultures.

3.2 Improvement of the tissue growth and the secondary metabolite production

By comparison with the new formulation media the different organ or tissue cultures were also subcultured on classical media: on Murashige and Skoog's (MS) medium and half strength macronutrient MS medium for *S. paludosum* multiple shoot culture (El Badaoui et al 1996), on MS medium for *S. officinalis* suspension culture and on Gamborg's (B5) medium for *G. paniculata* which were considered as the best culture media prior this study.
The kinetics of growth and saponin production are reported previously (El Badaoui et al 1996 and Fulcheri et al 1998). For *S. paludosum* multiple shoot cultures and *Gypsophila paniculata* root cultures the saponin production followed rather exactly the growth of the biomass. So the largest differences between the different conditions occurred during the beginning of the stationary phase. For *S. officinalis* suspension cultures the maximum of the saponin production occurred in the first part of the exponential growth phase with only a slight decrease after. So we could compare the different values in the same conditions as the previous plant tissue cultures (Table 4). In each case the growth and the saponin production were enhanced at least by a factor two when the different cultures are performed the new mineral formulation media compared to the classical media even these last media are diluted half strength to keep the same total concentration of the different anions and cations. So this technique could bring a good improvement in the plant tissue cultures.

Table 4: Comparison of the effect of the different media on the batch growth and the saponin production of the different tissue cultures studied.

	S. paludosum multiple shoot cultures			S. officinalis suspensions		G. paniculata root cultures	
Media	MS	MS/2	SPOM	B_5	MH_2	MSA	MH_3
Growth FW g/l	260	300	420	340	740	340	670
Saponins Mg/gFW/L	4.1	4.5	12.6	11	31	2.3	4.3

4. References

Fujita Y et al (1981a) Plant Cell Rep. 1, 59-60.
Fujita Y et al (1981b) Plant Cell Rep. 1, 61-63.
Gamborg OL et al (1968) Exp. Cell Res. 50, 151-158.
El Badaoui H et al (1996) Plant Cell Tiss. Org. Cult. 45, 153-158.
Morard P et al (1990) J. Plant Nutr. 13, 249-268.
Morard P, Henry M (1998) J. Plant Nutr. accepted for publication.
Murashige T, Skoog F (1962) Plant Physiol. 15, 473-497.
Rokem JS (1985) J. Nat. Prod. 48, 210-222.

ENVIRONMENTAL CONTROL IN MICROPROPAGATION: EFFECTS OF CO_2 ENRICHMENT AND SUPPORTING MATERIAL ON GROWTH AND PHOTOSYNTHESIS OF EUCALYPTUS SHOOTS/PLANTLETS CULTURED PHOTOAUTOTROPHICALLY *IN VITRO*

C. Kirdmanee
National Science and Technology Development Agency, National Center for Genetic Engineering and Biotechnology, Rama VI Road, Bangkok 10400, Thailand

Abstract

Eucalyptus camaldulensis shoots were cultured photoautotrophically *in vitro* for 6 weeks with four different supporting materials (agar matrix, Gelrite matrix, plastic net, and vermiculite) under CO_2 enriched (1200 μmol mol^{-1}) and CO_2 non-enriched conditions. Plantlets from each treatment *in vitro* were then grown *ex vitro* in a greenhouse for 4 weeks. CO_2 enrichment significantly increased the dry weight, net photosynthetic rate, and rooting percentage of shoots/plantlets *in vitro*, as well as the survival percentage of plantlets *ex vitro* regardless of the type of supporting materials. The dry weight, net photosynthetic rate, and rooting percentage of shoots/plantlets *in vitro* were the highest in the vermiculite, followed by the plastic net, Gelrite matrix, and agar matrix (in descending order). Plantlets with a high survival percentage were obtained in response to the CO_2 enriched conditions and use of vermiculite as a supporting material.

1. Introduction

Recent research has revealed that chlorophyllous shoots/plantlets have a high photosynthetic ability. They grow faster, in many cases, under photoautotrophic conditions than under heterotrophic and photomixotrophic culture conditions. The physical/chemical environment affected the different photosynthesis and growth of shoots/plantlets *in vitro* (Kozai, 1992, Jeong 1993). However, there is little information relative to the influence of the environment on quality of shoots/plantlets *in vitro*. The environmental control for production of quality shoots/plantlets grown photoautotrophically *in vitro* are the key question to be addressed in this experiment.
The objective of this experiment was to increase the growth rate, net photosynthetic rate, dry matter percentage and rooting percentage of shoots/plantlets grown photoautotrophically *in vitro* with different types of supporting materials under CO_2

A. Altman et al. (eds.), Plant Biotechnology and In Vitro Biology in the 21st Century, 651–654.

enriched and CO_2 non-enriched conditions. The survival percentage of plantlets after transplanting was also evaluated.

2. Materials and methods

Actively growing shoots (initial shoot length: 2.2 ± 0.2 cm) of *Eucalyptus camaldulensis* were excised from axillary shoots, and inserted vertically to a depth of 0.8 ± 0.2 cm into 70 ml of sugar-free medium. The sugar-free medium consisted of quarter strength Murashige and Skoog (1962) inorganic salts and vitamins, 0.25 g l^{-1} casein hydrolysate (Gurung, Rajbhandari, 1989). There were eight treatments in this experiment. Four types of supporting materials were used: 8 g l^{-1} agar (Difco Bacto agar, Difco Laboratories, Detroil, MI), 2.5 g l^{-1} Gelrite (Merck & Co., San Diego, CA), plastic net (mesh size: 3 x 3 mm) or 180 g l^{-1} vermiculite under CO_2 enriched (1200 ± 100 μmol mol^{-1}) or CO_2 non-enriched (400 ± 50 μmol mol^{-1}) conditions. Four shoots were cultured in each 370-ml polycarbonate box. One filter membrane (Milli-Seal, Millipore, Japan; pore size: 0.5 μm) was attached over a hole (area: 0.8 cm^2) on each of four sides of culture box to increase the number of air exchanges. These culture boxes were estimated to have 6.8 h^{-1} air exchanges according to the method described by Kozai et al. (1986). The pH of medium was adjusted to 5.8 before autoclaving. All cultures were maintained for 6 weeks at 25 ± 1°C air temperature, 65 ± 5% relative humidity and 16 h d^{-1} photoperiod provided by cool-white fluorescent lamps at photosynthetic photon flux density of 80 ± 5 μmol m^{-2} s^{-1}.

At 6 weeks after culture *in vitro*, 160 plantlets were removed from the culture boxes, rinsed thoroughly of the substrate or medium, planted into 370-ml plastic pots with vermiculite, and placed immediately in the uncovered trays with water of 2-cm depth. Plantlets were grown for 4 weeks in an environmental control greenhouse at 35°C/31°C (photoperiod/dark period) air temperature, 50 ± 5% relative humidity and 450 kJ m^{-2} d^{-1} solar radiation. Plantlets were subirrigated once daily with tap water.

At 6 weeks after culture *in vitro*, 20 shoots/plantlets were destructively harvested from each treatment to determine dry weight, dry matter percentage, leaf area, and rooting percentage. At 4 weeks after transplanting, survival percentages were measured.

At 6 weeks after culture *in vitro*, CO_2 concentrations inside and outside the five culture boxes from each treatment were determined by analyzing gas samples with a gas chromatograph (GC-12A, Shimazu Co. Ltd., Japan). Net photosynthetic rates on a leaf area basis were calculated according to the method described by Fujiwara et al. (1987).

The experimental design was a 2^4 factorial with five replications and four shoots/plantlets per replication for dry weight, dry matter percentage, rooting percentage, net photosynthetic rate, and survival percentage. Data were tested by analysis of variance (ANOVA). Least significant difference (LSD) at P = 0.05 was used to determine significance of the type of supporting materials.

3. Results and discussion

The photoautotrophic micropropagation of *Eucalyptus* shoots/plantlets *in vitro* was established. At 6 weeks after culture *in vitro*, the dry weight was almost twice as great in the vermiculite compared with the agar matrix under either the CO_2-nonenriched or CO_2-enriched conditions (Table 1). CO_2 enrichment significantly increased the dry weight and dry matter percentage of shoots/plantlets *in vitro*, regardless of the type of supporting materials. It is possible to reduce hyperhydricity of shoots/plantlets *in vitro* by CO_2 enrichment. The rooting percentage was the highest in the vermiculite followed by the plastic net, Gelrite matrix, and agar matrix (in descending order) under either the CO_2-nonenriched or CO_2-enriched conditions.

Net photosynthetic rates of shoots/plantlets *in vitro* were almost twice as great in the vermiculite, compared with the agar matrix under either the CO_2-nonenriched or CO_2-enriched conditions (Table 1). The increased net photosynthetic rate *in vitro* promoted the development of root system (Kirdmanee, 1995).

Plantlets from the CO_2 enriched condition showed a significantly higher survival percentage 4 weeks after transplanting regardless of the type of supporting materials, when compared with those from the CO_2-nonenriched condition (Table 1). Plantlets with a high survival percentage were obtained in response to the CO_2 enriched conditions and use of vermiculite as a supporting material. Positive effects of CO_2 enrichment on survival percentage of plantlets after transplanting have been reported in grapevine (Lakso et al., 1986) and raspberry (Deng, Danielle, 1993). It is possible to increase survival percentage of plantlets by *in vitro* acclimatization.

Photoautotrophic growth of plantlets in the vermiculite under CO_2 enriched condition can substantially reduce costs of micropropagation and increase efficiency of the production schemes due to the increased rooting percentage and reduced hyperhydricity, reduced biological contamination risks *in vitro*, limited processes of washing off sugar and gelling agent during transplanting, reduced *ex vitro* acclimatization processes, and increased survival percentage of plantlets after transplanting. An automated system for transplanting would be easy to develop.

4. References

Deng R et al (1993) *In vitro* hardening of red raspberry by CO_2 enrichment and reduced medium sucrose concentration, Hort. Sci. 28, 1048-1051.

Fujiwara K et al (1987) Measurements of carbon dioxide gas concentration in closed vessels containing tissue cultured plantlets and estimates of net photosynthetic rates of the plantlets, J. Agric. Meteorol. 43, 21-30.

Gurung S et al (1989) Micropropagation of *Eucalyptus camaldulensis*-a fast growing essential oil-bearing tree, Proceedings of International WorkshopTissue Culture and Biotechnology of Medicinal and Aromatic Plants, pp.17-21, CIMAP, Lucknow, India.

Jeong BR et al (1993) Carbon dioxide enrichment in autotrophic micropropagation: methods and advantages. HortTechnology 3, 332-334.

Kirdmanee C et al (1995) Effect of CO_2 enrichment and supporting material *in vitro* on photoautotrophic growth of *Eucalyptus* plantlets *in vitro* and *ex vitro*: anatomical comparisons, Acta Hortic. 393, 111-118.

Kozai T et al (1986) Fundamental studies of environments in plant tissue culture vessels. II. Effects of stoppers and vessels on gas exchange rates between inside and outside of vessels closed with stoppers, J. Agric. Meteorol. 42, 119-127.

Kozai T et al (1992) Transplant production systems, pp. 247-282 Kluwer Academic Publishers, Dordrecht, The Netherlands.

Lakso AN et al (1986) Carbon dioxide enrichment for stimulation of growth of *in vitro*-propagated grapevines after transfer from culture, J. Amer. Soc. Hort. Sci. 111, 634-638.

Murashige T et al (1962) A revised medium for rapid growth and bioassays with tobacco tissue culture, Physiol. Plant 15, 473-497.

Table 1 Effects of CO_2 enrichment and supporting material *in vitro* on dry weight, dry matter percentage, rooting percentage, and net photosynthetic rate (NPR) of shoots/plantlets 6 weeks after culture *in vitro*, as well as survival percentage of plantlets 4 weeks after transplanting.

CO_2 condition	Supporting material	Dry weight (mg)	Dry matter (%)	Rooting (%)	NPR (μmol m^{-2} s^{-1})	Survival (%)
CO_2 nonenriched	Agar matrix	45	15	24	1.2	55
(400 ± 50	Gelrite matrix	49	15	48	1.3	60
μmol mol^{-1})	Plastic net	64	17	64	1.5	75
	Vermiculite	82	21	72	2.1	85
CO_2 enriched	Agar matrix	54	19	32	1.4	70
(1200 ± 100	Gelrite matrix	62	21	56	1.6	70
μmol mol^{-1})	Plastic net	76	18	80	2.1	90
	Vermiculite	103	22	100	2.7	100
LSD $p = 0.05$		8	1	12	0.4	9
ANOVA						
CO_2 condition (C)		**	**	**	**	**
Supporting material (S)		**	**	**	**	**
C × S		NS	**	**	**	NS

.. Significant at $P = 0.01$, respectively.
NS Nonsignificant at $P = 0.05$.

ENVIRONMENTAL CONTROL IN PHOTOAUTOTROPHIC MICROPROPAGATION

T. Kozai, Q.T. Nguyen and C. Chun
Laboratory of Environmental Control Engineering, Faculty of Horticulture, Chiba University, Matsudo, Chiba 271-8510, Japan

1. Introduction

The growth and development of chlorophyllous explants, shoots and plantlets in vitro are influenced by aerial and/or physical environmental factors more significantly in photoautotrophic micropropagation using sugar-free culture medium than in heterotrophic or photomixotrophic micropropagation using sugar-containing culture medium. It is because, in photoautotrophic micropropagation, carbon dioxide(CO_2) in air is a sole carbon source for increase in dry mass by photosynthesis. Thus, measurement and control of aerial and/or physical environmental factors are essential in photoautotrophic micropropagation to promote their photosynthesis, growth and development and to reproduce a number of quality plantlets in vitro at low production costs. In this article, methods of environmental measurement and control used in photoautotrophic micropropagation are briefly discussed. A detailed discussion has been given in Kozai et al. (1995).

2. CO_2 measurement and control

2.1. Infrared CO_2 analyzer/controller

Carbon dioxide (CO_2) concentration is most commonly measured with an infrared CO_2 analyzer providing continuous monitoring for controlling the CO_2 concentration in a culture room, a plant growth chamber, a box containing culture vessels and a greenhouse.

CO_2 enrichment can be conducted in a culture room, a plant growth chamber and a large culture box equipped with a CO_2 control unit consisting of a liquid CO_2 container, an electric solenoid valve, and an infrared CO_2 controller with a dehumidifier and an air pump for air sampling. A CO_2 control unit widely used for CO_2 enrichment in greenhouses can be adopted without any modification for CO_2 enrichment in photoautotrophic micropropagation.

2.2. Gas chromatography

Carbon dioxide concentration can be measured accurately by gas chromatography, but it

A. Altman et al. (eds.), Plant Biotechnology and In Vitro Biology in the 21st Century, 655–658.

cannot be readily automated for continuous monitoring and control. An advantage of gas chromatography is its small amount of sample air (half ml or less) required for CO_2 measurement. Air is sampled using a syringe and injected into the detector with a carrier gas .

2.3. Estimation of vessel natural ventilation rate or number of natural air exchanges

Natural ventilation rate of or the number of natural air exchanges of a culture vessel can be estimated based on the measurements of CO_2 concentration inside and outside the vessel. Net photosynthetic rate of plantlets in the vessel with natural ventilation during the photoperiod can be estimated by multiplying the natural ventilation rate and the difference in CO_2 concentration at a steady state between inside and outside the vessel.

3. Light measurement and control

3.1. Photosynthetic photon flux (PPF) meter

A photosynthetic photon flux (PPF) meter measures a photosynthetic (waveband: 400-700 nm) photon flux with a unit of mol $m^{-2}s^{-1}$, not an energy flux with a unit of W m^{-2} or J $m^{-2}s^{-1}$. PPF expresses the light environment with respect to photochemical reaction of plants. Energy per photon is about 1.5 times greater in blue than in red region. The sensor size is typically 1.5 cm in diameter and 2 cm in height.

3.2. Photosynthetically active radiation (PAR) flux meter

A photosynthetically active radiation (PAR) flux meter measures a photosynthetic radiation flux or photosynthetic irradiance with a unit of energy flux, W m^{-2} or J $m^{-2}s^{-1}$. PAR flux expresses the light environment with respect to photosynthetic energy flux. PPF and PAR flux sensors are similar in appearance to each other. The PPF is proportional to PAR flux only when they are measured under the identical light source.

3.3. Spectroradiometer

Spectral radiant energy or photon flux distribution, i.e., energy or photon flux at each wavelength, of light can be measured with a spectroradiometer with a unit of W m^{-2} nm^{-1} or mol $m^{-2}s^{-1}nm^{-1}$. Measurement of energy or photon flux in near-ultraviolet (300-400 nm), blue (400-500 nm), red (600-700 nm) and far-red (700-800 nm) waveband using a spectroradiometer are sometimes required because photomorphogenesis and/or development of plants are often influenced by the specific wavebands. Spectral transmittance of a leaf can be measured with this meter by placing the leaf above the sensor.

3.4. Light source

Fluorescent tubes are most commonly used as light source for micropropagation. The radiation emitted from the fluorescent tubes falls mostly in a photosynthetic waveband region, and the tube surface temperature is typically 35-40 ℃ at a room air temperature of 25 ℃. They give a relatively uniform distribution of PPF on the culture shelf when they are placed 35-40 cm above it. Light emitting diodes (LED) are recently used as small-size light source in micropropagation research. They emit blue, green, red or far-red light with a relatively narrow waveband (peak wavelength ± 20-30 nm). They are useful for controlling photomorphogenesis of plants. However, their price is currently still too high and light output is too low for use as light source for photosynthesis in commercial micropropagation. The conversion efficiency from electric to light energy is about a few times greater with a red LED than with a blue LED, and its efficiency of red LED is approximately equal to that of cool white fluorescent tubes.

3.5. Light environment

Light environment cannot always be expressed satisfactorily by measuring PPF or PAR flux on culture vessels or an empty culture shelf. The PPF or PAR flux sensor receives photosynthetic photons or energy with the correction of a cosine factor. In reality, however, plants in culture vessels often receive light from the sides, which is not sensed by the sensors. Thus, the light environment of plants in culture vessels is sometimes better measured with a spherical sensor than with a flat sensor used for a PPF or PAR flux meter. The photosynthetic photon flux measured with a spherical sensor is called "photosynthetic photon fluence"(Langhans et al., 1997). The photosynthetic photon fluence is about 4 times greater than PPF if measured in a uniform radiation field. Use of reflective material on walls in a culture room is effective to increase the PPF and photosynthetic photon fluence.

4. Relative humidity measurement and control

Relative humidity (RH) in a culture vessel can be measured using a high polymer humidity sensor, which operates on the basis of changing electric capacitance or resistance of the sensing part (5mm wide x 10 mm long x 2 mm thick or less) as a function of RH. This sensor is usually mounted together with a thermistor, a small temperature sensor, to estimate the absolute humidity, dewpoint temperature, wet bulb temperature or water vapor saturation deficit of the air. Ultrasonic humidifier can be operated with the RH sensor to control the RH in a culture vessel with a forced ventilation system and in a culture room.

4. Forced ventilation rate and air movement measurements

Forced air flow rate through tubes can be accurately and continuously measured and controlled with a thermal mass flow rate meter/controller. This meter/controller can be used to monitor and control the forced ventilation rate of a culture vessel with a forced ventilation unit. Net photosynthetic rate of plantlets in the vessel during the photoperiod can be estimated as a product of the forced ventilation rate and the difference in CO_2 concentration at the air inlet and air outlet of the vessel. Air movement in a vessel with natural or forced ventilation can be visualized by adding some fine particles in the vessel and taking pictures of the movements of the fine particles with a high resolution video camera under controlled environment conditions. Analysis of air movement in a vessel is critical for improving the uniform growth of plantlets in the vessel.

5. Other environmental and biological factors to be measured and/or controlled

Water potentials of the culture medium and plantlets can be measured with a psychrometer. Their water potentials together with those of the air inside and outside the vessel determine the direction of water flow in a culture medium-plantlet-inside air-outside air continum.
Chlorophyll fluorescence measured with a chlorophyll fluorescent analyzer is an interesting index showing the normality or abnormality of photosystem II of plantlets in vitro. Stomatal resistance to water vapor and CO_2 flow can be measured with a porometer. Recently several sophisticated portable equipment for measuring rates of net photosynthesis and transpiration and stomatal resistance at different CO_2 concentrations and PPFs have been available.
Leaf area can be measured with a CCD camera or an image scanner with a computer software if the leaves are detached in advance and placed separately on a flat plate under the camera.

6. References

Langhans RW et al (eds) (1997) Plant growth chamber handbook (North Central Region Research Publication 340). Iowa Agric. and Home Economics Exp. St. Iowa, U.S.A., 240pp.
Kozai T et al (1995) Environmental measurement and control systems. (in: Aitken-Christie J et al (eds), Automation and environmental control in plant tissue culture), Kluwer Academic Publishers, Dordrecht, The Netherlands, 539-574.

PHOTOAUTOTROPHIC MICROPROPAGATION OF TROPICAL PLANTS

Q.T. Nguyen[1, 2], T. Kozai[2], K.L. Nguyen[1] and U.V. Nguyen[1]
[1] Institute of Tropical Biology, NCST-VN, 1 Mac Dinh Chi Street, Ho Chi Minh City, Vietnam, and [2] Faculty of Horticulture, Chiba University, Matsudo, Chiba 271, Japan

1. Introduction

Recently, tissue culture technology has become one of the most effective ways to reproduce selected varieties of many food crops, ornamental and medicinal plants. Furthermore, micropropagation is currently a competitive approach with other vegetative propagation methods, such as seedling production, cutting propagation or grafting (I.Y.E. Chu, 1992). Micropropagation activities have been enhanced via practical commercialization. However, high production cost, resulted from high labor cost and other expenses due to biological contamination, low percent survival of in-vitro plantlets in the acclimatization, and formation of off-type products during the culture stages, is a major factor that limits the marketability of the method. Thus, cost-effectiveness becomes a challenging issue for plant tissue culturists and micropropagators.

Most research on chemical components (e.g., plant growth regulators, sugar types and concentrations, vitamins, etc.) of the culture medium seems unable to provide a satisfactory approach for solving the problem. Extensive studies on physically environmental factors (light intensity and quality, gaseous components, humidity, temperature, air movement, aeration of supporting materials, etc.) in the culture vessel have been recently conducted. All prove that plantlet growth and quality can be considerably improved when the control of those factors is achieved (T. Kozai, Y. Iwanami, 1988; R. Infante et al., 1989; T. Kozai et al., 1992; Y Desjardins, 1995). The quality improvement of the cultures and yield increase, thus, has brought about the importance of environmental measurement and control system in plant tissue culture methods for producing plants at lower costs. The studies have gone further in proving the high photoautotrophic potential of several species and cultivars cultured in vitro when the atmospheric CO_2 is used as a carbon source for plantlet growth (T. Kozai, 1991; M. Hayashi et al., 1993).

In recent years, micropropagation methods have provided a large number of pathogen-free plantlets of several tropically important industrial and fruit crops (R.E. Litz, V.S. Jaiswal, 1991). However, the effect of a photoautotrophic (sugar-free) culture system on the production of tropical qualified plantlets has not been focused. This study is mainly concerned with the response to changes of micro-environmental conditions of two tropical species, coffee (*Coffea arabusta*) and banana (*Musa spp.*) cultured photoautotrophically in vitro.

2. Growth of in-vitro plantlets under different sucrose concentration, supporting materials, CO_2 concentration, photoperiod, photosynthetic photon flux (PPF), and vessel ventilation rate conditions

2.1. Coffee

Single nodal cuttings of in-vitro coffee (*Coffea arabusta*) plantlets were used as plant material. Table 1 shows the effects of sucrose concentration on the dry weight, shoot length and leaf area of coffee plantlets cultured on half-strength Murashige and Skoog (1962) medium with either 0

A. Altman et al. (eds.), Plant Biotechnology and In Vitro Biology in the 21st Century, 659–662.

or 20 g l^{-1} sucrose in polycarbonate box-type vessels, and under PPF of 75 μmol m^{-2} s^{-1}, photoperiod at 10 h d^{-1} and an ambient atmospheric CO_2 concentration (350-450 μmol mol^{-1}). Shoot length and leaf area were significant greater when coffee plantlets were cultured on the medium with 0 g l^{-1} sucrose than on medium with 20 g l^{-1} sucrose. In addition, no callus formation was observed at the shoot base of plantlets cultured on the sugar-free medium, whereas all plantlets grown on sugar containing medium were found having callus. It is no doubt that there is a relationship between presence of sucrose in the medium and callus formation at shoot base of in-vitro coffee plantlets.

Table 1. Effects of presence/absence of sucrose in the medium on the growth of in-vitro coffee plantlets on day 45

Sucrose conc. (g l^{-1})	Dry weight (mg)	Shoot length (mm)	Leaf area (mm^2)
0	68	28^a	222^a
20	63	23^b	173^b

a,b within a column show a significant difference at $p = 0.01$

The ability to develop photoautotrophy of coffee plantlets was more emphasized when different CO_2 concentrations and PPFs, the two most important factors involved in the photosynthesis of chlorophyllous cultures, were investigated. Table 2 shows effects of CO_2 concentrations and PPFs on increased fresh weight (IFW), shoot length (SL), leaf area (LA) and net photosynthetic rate (P_n) of in-vitro coffee plantlets cultured on the sugar-free medium. The higher the CO_2 concentration, the greater the plantlet growth was. P_n measured *in situ* also increased with the increase of CO_2 concentration. However, under the same CO_2 concentration of the experiment, increasing PPF did not increase P_n significantly. The growth of in-vitro coffee plantlets was the greatest at 1450 μmol mol^{-1} CO_2 and 150 μmol m^{-2} s^{-1} PPF.

Table 2. Effects of high/low CO_2 concentration and PPFs on the photoautotrophic growth of in-vitro coffee plantlets on day 45

Treatment Code	IFW (mg)	SL (mm)	LA (cm^2)	P_n (μmol h^{-1} $plantlet^{-1}$)
HH	390^b	17^b	51^b	14
HL	760^a	31^a	81^a	15
LH	220^c	8^c	27^c	5
LL	200^c	10^c	24^c	4

a, b, c within a column show a significant difference at $p = 0.05$. As treament codes, H and L in the left represent for CO_2 concentration at 1450 and 450 μmol mol^{-1}, respectively. L and H in the right represent for PPF at 150 and 350 μmol m^{-2} s^{-1}, respectively.

Effects of two kinds of supporting materials – agar and Florilite (a mixture of vermiculite and cellulose fibers, Nissinbo Co., Tokyo) - on the photoautotrophic growth of in-vitro coffee plantlets were considered. Increased dry weight and shoot length of plantlets developed on Florilite were significantly greater than those of plantlets cultured on agar. (Q.T. Nguyen et al., unpublished). Data on leaf area, % dry matter and chlorophyll a/b ratio also proved in-vitro coffee plantlets had a better growth when agar was replaced by Florilite. The high air porosity of Florilite may enhance root growth, and this will help plantlets have a larger uptake of mineral salts and water in the medium. Cellulose plugs, rockwool and vermiculite, when replacing agar in the culture of carnation (Takazawa, Kozai, 1992) and Eucalyptus (Kirdmanee et al., 1995), also resulted in an increase in growth of these species.

2.2. Banana

Banana (*Musa spp.)* shoots cultured on half-strength Murashige and Skoog (1962) medium were used as explants. Figure 1 shows effects of sucrose concentration and photoperiod on the increased dry weight and shoot/root dry weight ratio of banana plantlets. There was no significant difference in development between plantlets growing either on sugar-free medium or on sugar containing (20 g l^{-1} sucrose) medium under the same photoperiod regime. However, plantlets grew better with increasing photoperiod regardless of sucrose concentration. Dry weight was about 3 times greater at photoperiod of 16 h d^{-1} than photoperiod 8 h d^{-1}. Shoot/root ratio, on the other hand, decreased with the increase of photoperiod. This showed that roots developed as better as shoots, especially when in-vitro banana plantlets were cultured on sugar-free medium.

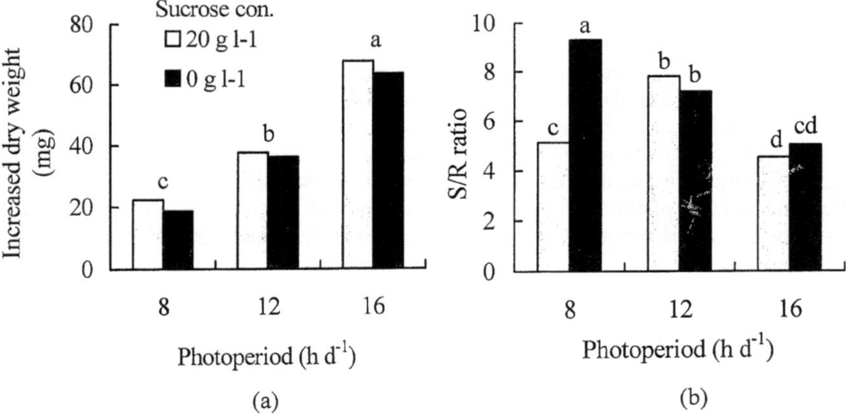

Figure 1. Increased dry weight (a) and shoot/root dry weight ratio (b) of in-vitro banana plantlets on day 28. [a, b, c, d] on top of columns show a significant difference at $p = 0.01$

Increased fresh weight and number of unfolded leaves of in-vitro banana plantlets as affected by different PPFs and air ventilation under photoautotrophic conditions are shown in Figure 2. Fresh weight increased nearly twice when PPF was raised from 50 to 100 μmol m^{-2} s^{-1} with increased ventilation rate. The number of unfolded leaves per plantlet was also significantly greater at higher PPF and air ventilation rate. However, under the ambient atmospheric CO_2 (350-450 μmol mol^{-1}) condition, no significant difference was found in fresh weight and number of unfolded leaves of in-vitro banana plantlets when PPF was increased to 200 μmol m^{-2} s^{-1}, and even at higher ventilation rate. The results implied that an increase in growth at higher PPFs could be obtained under CO_2- enriched condition as ever proved in the study of other plant species (Kozai, 1991).

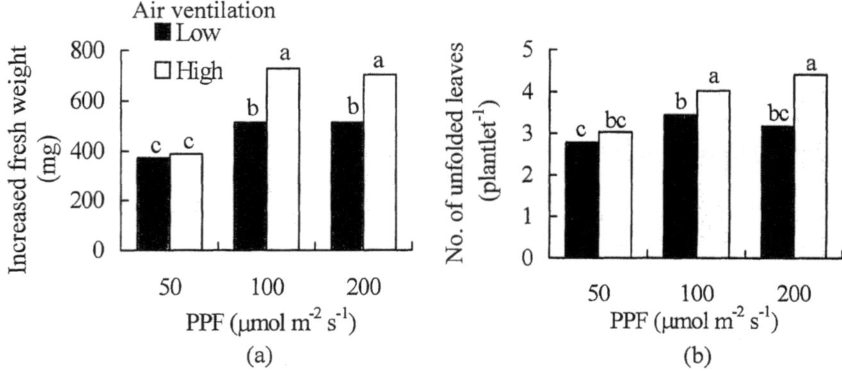

Figure 2. Increased fresh weight (a) and number of unfolded leaves (b) of in-vitro banana plantlets on day 28. [a, b, c] on top of columns show a significant difference at $p = 0.01$

3. Conclusion

In the present study, both coffee and banana plantlets cultured in vitro express an ability of photoautotrophic development. Under photoautotrophic culture conditions, a higher CO_2 concentration increased the growth of coffee plantlets significantly; whereas longer photoperiod, higher PPF and air ventilation rate resulted in a better growth of banana plantlets. Supporting materials with high air porosity, such as Florilite, also enhanced the development of coffee plantlets. The micro-environmental control is, thus, essential to improve the growth of plantlets cultured in vitro. The application of this approach is feasible to produce high quality in-vitro coffee and banana plantlets at a reduced cost. Further research on an environmental control system for mass photoautotrophic micropropagation of these and other tropical species should be conducted.

4. References

Chu IYE (1992) Transplant Production Systems, pp 137-150, Kluwer Academic Publishers, Dordrecht, The Netherlands.

Desjardin Y (1995) Ecophysiology and Photosynthetic *In vitro* Cultures, 145-160, Aix-en-Provence (France), 1-3 December 1993.

Hayashi M et al. (1993) SHITA J. 4(2), 107-110.

Infante R et al. (1989) Physiol. Plant. 77, 191-195.

Kirdmanee C et al. (1995) In Vitro Cell. Dev. Biol. – Plant, 31, 144-149.

Kozai T, Iwanami Y (1988) J. Jap. Soc. Hort. Sci. 57(2), 279-288.

Kozai T (1991) Biotechnology in Agriculture and Forestry, 17, pp 313-343, Springer-Verlag Berlin Heidelberg, Germany.

Kozai T et al. (1992) Transplant Production Systems, pp 247-282, Kluwer Academic Publishers, Dordrecht, The Netherlands.

Litz RE, Jaiswal VS (1991) Micropropagation, pp 247-263, Kluwer Academic Publishers, Dordrecht, The Netherlands.

Murashige T, Skoog F (1962) Physiol. Plant. 15, 473-497.

Takazawa A, Kozai T (1992) Environ. Control Biol. 30(2), 65-70.

Mechanization of Micropropagation

Abed A. Watad[1*], Yekutiel Alper[2], Ran Stav[1] and Robert Levin[3]

Departments of Ornamental Horticulture[1] and Agriculture Engineering[2], ARO, The Volcani Center, Bet Dagan 50250, Israel. Osmotek Ltd[3] POB 550, Kiryat Weizmann, Rehovot, Israel.*Corresponding author.

1. Introduction

Expansion of commercial micropropagation will require the implementation of scale up and mechanization technology (Vasil, 1991; Vasil, 1994). In developed countries labor cost have been reported to be 65–70 per cent of the cost of production (Chu, 1995; George, 1996). In these countries, mechanization would permit cost-effective micropropagation of transplant vegetable and forestry crops and production of ornamental crops at costs that are competitive with exports of low labor cost countries. In countries where labor costs are low, mechanization would allow standardized production and reduce operator-associated contamination.

A number of mechanized, robotic and automated systems have been described (Kurata, 1995), however, there are very few commercially available components for mechanization of micropropagation. We have, in association with other Israeli groups, developed methods and equipment for mechanization of micropropagation that have been applied to a number of crops. The equipment and methods have been commercialized by Osmotek Ltd., Israel.

2. Materials and Methods

The principal equipment components of the Osmotek system are the LifeReactor bioreactor vessel system, the LifeRaft floating raft support system, the VitroCut mechanical cutter and LifeLine vented vessels. Different components of the mechanization system were used in the experiments. These components are described below.

2.1. The floating membrane raft system

The LifeRaft™ system includes a microporous membrane raft frame and a buoyant float. The raft supports the plant tissue while the float maintains the membrane at the air/liquid interface.

2.2. The bubble reactor

The bubble reactor was 1 liter working volume air lift reactor similar to that reported by Ziv and Hadar (1991).

2.3. The plastic film bioreactor:

The LifeReactor™ vessels are 1.5 and 5 liter working volume presterilized plastic sleeve bioreactors which were adapted from the bubble reactor designed by M. Ziv, The Hebrew University of Jerusalem. They include an integral sparger, a large port for charging and harvesting plant material, an integral drain port with a screen to maintain culture in the reactor while removing medium, an air exhaust port and a medium inlet port.

2.4. The vented vessels

LifeLine™ vented vessels are a family of vessels having a microporous membrane vent. The polycarbonate or polypropylene vessels are with a vent size varies from 10 mm to 40 mm.

2.5. The mechanical separation device

The separation device is the VitroCut™ which is an autoclaveable stainless steel knife grid with a manually activated piston like head that engages the knife grid. Clusters of multiplying plant tissue culture placed on the knife grid. The head is then lowered to engage the knife grid and the separated clusters fall into a vessel below the knife grid. For the present experiments the knife grid was a 5mm or 7mm mesh.

A. Altman et al. (eds.), Plant Biotechnology and In Vitro Biology in the 21st Century, 663–666.

2.6. Micropropagation procedures using bioreactors and mechanical separation of multiplication cultures

A) Micropropagation of *Spathiphyllum* cv Petite

A bank of *Spathiphyllum* was maintained by monthly subcultures on solid multiplication medium composed from MS medium (Murashige and Skoog, 1962) supplemented with 2 mg/l kinetin, 3% sucrose and 0.8% agar. The cultures were incubated at 25±2°C under a 16-hour

photoperiod with a light intensity of 55 μmoles.m^{-2}.s^{-1}. At each subculture leaves were removed and the base divided in half giving a cluster multiplication rate of 2 every four weeks.

Clusters from the bank were divided, trimmed as above and transferred to 500 ml flasks (15 clusters per flask) with 200 ml of liquid multiplication medium (MS medium plus 2 mg/l kinetin, 3% sucrose and 2mg/l ancymidol). The flasks were incubated on an orbital shaker at 100 rpm under the same environmental conditions as the bank. After 4 weeks on a multiplication medium the clusters were cut with the VitroCut™ (5mm x 5mm mesh) and 150 ml of cut clusters per reactor were placed in a LifeReactor with 1 liter of liquid multiplication medium. The LifeReactors were incubated either under conditions similar to the flasks or in the dark and clusters were produced after 3–4 weeks.

When the light grown LifeReactors were harvested and cut about 250 – 300 ml of chopped clusters were produced. Clusters were either spread over solid multiplication medium in vessels to produce clusters for transplant to the greenhouse or returned to LifeReactors for further multiplication.

B) Micropropagation of potato cv Desiree

Liquid medium was employed for proliferation. The flask proliferation medium (PF) included MS basal salts and vitamins, 0.4 mg/l thiamine HCl, 0.01 mg/l NAA, 2 mg/l pantothenic acid, 0.25 mg/l GA3, and 20 g/l sucrose. The bioreactor proliferation medium (PB) included MS basal salts and vitamins, 0.4 mg/l thiamine, HCl, 20 mg/l pantothenic acid, 0.4 mg/l GA3, 20 g/l sucrose, and 1.5 mg/l ancymidol. The elongation medium was PF medium solidified with agar.

Twenty to thirty single node explants were removed from agar cultured potato stock plants and subcultured to 500 ml Erlenmeyer flasks containing 200 ml of PF medium. Flasks were incubated on an orbital shaker at 100 rpm for 21 days with a 16-hour photoperiod (60 μmoles.m^{-2}.s^{-1}). Bud clusters produced in the flasks were chopped for six seconds in a blender, washed in a sieve, and inoculated into reactors with 1 liter of PB medium. The reactors were incubated under constant light (70 μmoles.m^{-2}.s^{-1}) at 25±2°C for 21 days. Bud clusters harvested from the reactors were chopped in a blender again and 20 grams of chopped material was again inoculated into LifeReactor with 1 liter of PB medium and incubated as before. After 21 days the LifeReactors were harvested and the potato clusters were chopped in a blender once again. Between 4 and 4.5 grams of chopped clusters were inoculated on 100 ml of elongation medium in LifeGuard™ vessels. And incubated for 28 days with 16 hours of light (60 μmoles.m^{-2}.s^{-1}) at 25±2°C.

2.7. Micropropagation procedures using floating membrane rafts system with and without mechanical separation of multiplication cultures

A) Micropropagation of *Spathiphyllum* cv Petite using the raft system

Explants of *Spathiphyllum* were obtained from a bank as in the bioreactor studies above. Five uniform propagules were either cultured in Magenta GA7 vessels on 50 ml of agar solidified medium or on 50 ml of liquid medium with LifeRaft™ supports. The liquid medium composition was identical to the solid medium except that agar was omitted (MS basal medium and vitamins plus 2 mg/l kinetin, 3% sucrose). Each propagule was a 6-8 mm shoot with an average initial fresh weight of 125 mg.

Twenty-five propagules were used for each treatment (five propagules per vessel, five vessels per treatment) and experiment were repeated twice. All cultures were maintained in a growth room at 24±2°C under 16-hour photoperiod with a photon flux density of

(60 μmoles.m^{-2}.s^{-1}) provided by cool white fluorescent bulbs. Plant material was examined after 25 and 45 day.

B) Micropropagation of potato cv Desiree using the raft system

Single nodes of potato were subcultured from agar cultured potato stock plants to a microporous membrane raft (LifeRaft™) 70–90 nodes per raft. The rafts were placed on floats in LifeGuard™ vented vessels (40 mm vents) with 150 ml of liquid raft proliferation (LRP) medium. LRP medium consisted of MS basal salts and vitamins, 0.4 mg/l thiamine HCl, 0.01 mg/l NAA, 20mg/l pantothenic acid, 0.4 mg/l GA3, 1.5 mg/l ancymidol and 20 g/l sucrose. The vessels were incubated for 30 days under a 16-hour photoperiod (60 μmoles. m^{-2} .s^{-1}) at 25±2°C. During incubation, compact shoot cluster with very small leaves were produced. The shoot clusters were

mechanically chopped with the VitroCut (7mm x 7mm grid) and reinoculated into raft multiplication culture (MS basal salts and vitamins, 20 mg/l pantothenic , 1 mg/l ancymidol, 20 g/l sucrose and 1 mg/l kinetin) for two additional multiplication cycles. Culture produced on the third multiplication cycle was transferred to tuberization (TI) medium. Tuberization medium included MS basal salts and vitamins, 0.4 mg/l thiamine HCl, 20 mg/l pantothenic acid, 5 mg/l BA, 1 mg/l ancymidol, 1.3 mg/l CCC and 80 g/l sucrose. Moving the LifeRaft containing the culture to a LifeGuard vented vessel (40 mm vents) holding 150 ml of TI medium effected the transfer. The vessels were incubated at 19±1°C. For the first 10 days incubation was with 16-hours of light (60 μmoles.m^{-2}.s^{-1}) after which incubation was in the dark. After the first two weeks the medium was changed weekly.

3. Results

3.1. Results of *Spathiphyllum* cv Petite in bioreactors

When dark grown clusters were cut excessive phenols were produced, although, this problem was not seen with the light grown clusters. This procedure was transferred to several commercial laboratories. In one of these laboratories, 5 liter LifeReactor were routinely inoculated with 2 liters of chopped clusters. Over three subculture generations these reactors produced an average of 1.4 liters of unchopped clusters over and above the 2 liters of chopped clusters that were needed to restock the multiplication LifeReactor every 3–4 weeks. The unchopped clusters were spread over solid culture medium in vessels for elongation and produced, after three weeks incubation in the laboratory an average of 420 planted cells in the greenhouse. Harvesting, chopping, reinoculating the LifeReactor, and producing the culture vessels takes about 2 hours. When this data is extrapolated it can be calculated that one person produces 1,680 plantable clusters per 8-hour shift. In this laboratory, manual labor would require 34 hours for the same production.

3.2. Results of potato cv Desiree in bioreactors

Flasks prepared for the bioreactors experiments produced 10-12 g of potato bud clusters, after three weeks, twenty g of chopped and washed material of flask buds, increased in the bioreactor to 195±5 g after the second three week cycle. The 4-4.5 g of potato chopped and washed clusters, from the bubble reactors produced about 75±5 plants per vessel on 100 ml of elongation medium.

3.3. Results of *Spathiphyllum* cv Petite on floating rafts

On agar solidified medium only six shoots developed after 25 days compared to 14.4 shoots on the raft multiplied culture. There was no increase in the number of shoots at 45 days for the raft grown culture, but the culture grown on agar solidified medium increased to an average of 10 shoots. Fresh weight gain was also greater (about twice) in raft grown culture at 25 and 45 days. In general the propagules on the rafts developed more quickly and had better root development.

3.4. Results of potato cv Desiree on floating rafts

The single nodes placed on the LifeRaft produced a dense mat of shoot clusters after 3–4 weeks of incubation. Each raft produced shoot clusters which when divided by the VitroCut produced inoculum for three new rafts. After the third cycle, the rafts produced an average of 94 plants. When a raft with regenerated plants was placed on TI medium the raft produced an average of 140 microtubers in 9 weeks with 43 of the microtubers weighing more than 500 mg.

4. Discussion

The above results indicate that liquid medium mechanization procedure based on multiplication of bud or shoot clusters and using commercially available equipment can be applied to commercially relevant crops. This system can apparently reduce the labor and space cost of micropropagation and thus may be a significant step in expanding the market for commercial micropropagation. These results supports a number of observations (George, 1996; Watad, *et al.*, 1995, 1996), that liquid medium often results in faster rates of growth and multiplication than solid medium.

5. References

Chu I. (1995) In: Automation and environmental Control in Plant Tissue Culture (Aitken-Christie J., Kozai T. and Smith MAL, eds). p. 19-27.

George E.F. (1996) Plant propagation by tissue culture, Part 2. Exegetics, Ltd. Edington, Wilts., England.

Kurata K. (1995) In: Atiken-Christie J., Kozai T. and Smith MAL (Eds) Automation and environmental control in plant tissue culture. p. 257-272. Kluwer Academic Publishers, Dordrecht, Netherlands.

Murashige T. and Skoog F. (1962) Physiol. Plant. 15, 473-497.

Vasil I.K. (1994) Automation of plant propagation. Plant Cell Tiss. Org. Cult. 39, 105-108.

Vasil I.K. (1991) In: Vasil I.K. (Ed.) Scale up and automation in plant propagation (p. 1-6). Academic Press. New York, USA.

Watad A.A., Kochba M., Nissim A. and Gaba V. (1995) Plant Cell Reports 14, 345-348.

Watad A.A., Ahroni A., Zuker A., Shejtmen H., Nissim A. and Vainstein A. (1996) Scientia Hortic. 65, 313-320.

Ziv. and Hadar A. (1991) Isr. J. Botany 40, 7-16.

SURPRISING MICROPROPAGATION TOOLS

S Werbrouck[1], K Goethals [2], M Van Montagu[2], P Debergh[1]
Universiteit Gent, [1] Faculty of Agriculture and Applied Biological Sciences, Department of Plant Production, Coupure links 653, [2] Faculty of Sciences, Laboratory of Genetics, affiliated to VIB, KL Ledeganckstraat, 9000 Gent, Belgium

Abstract

Recently interesting organogenic and morphogenic effects on micropropagated plants have been achieved by applying pesticides to the culture medium or by infesting plants with *Rhodococcus fascians*. Both interventions result in cytokinin-like effects (abundant shoot and bud formation), but in neither case it could be proven that cytokinins are directly involved. Some data suggest that gibberellin biosynthesis plays a major role but this is not confirmed in all cases.

1. Introduction

Plant growth regulators are the medium ingredients which are most often manipulated to control axillary and adventitious organogenesis in a tissue culture system. Thereby the guideline is still contained in the now classical paper of Skoog and Miller from 1957, in which organogenesis is presented in terms of auxin - cytokinin ratio's. In many cases this approach has given very satisfactory results. However, it is also evident that only manipulating this ratio does certainly not allow to solve all the problems.

2. Hormonal effects of some imidazole pesticides

2.1. Organogenic and morphogenic effects

When testing fungicides to avoid fungal contaminations in tissue culture we observed a cytokinin-like effect of the imidazole fungicides, imazalil (IMA) and prochloraz, on *Spathiphyllum* and *Anthurium* (Werbrouck and Debergh, 1996). Around the same period we were confronted with a damage claim in greenhouse grown *Anthurium*., after a preventive treatment with prochloraz (Werbrouck et al., 1996). In both cases the major symptom was excessive shoot formation at the base of the plant.

In micropropagated *Ficus benjamina* and *F. elastica* chimeras with variegated leaves, 5 µM prochloraz subtly disturbed the integrity of the histogenic layers of the shoot meristem. This caused deformations and green leaf spots. At higher concentrations the

A. Altman et al. (eds.), Plant Biotechnology and In Vitro Biology in the 21st Century, 667–672.

meristems developed like witches' brooms. A number of shoots, which developed from these meristem clusters, had lost their chimerical character or were rearranged chimeras, among others we selected a GWW (L1 = green - L2 = white - L3 = white) chimera. Another damage case, with acclimatising micropropagated GWG chimeras of *F. benjamina* and *F. elastica*, draw our attention. Here too, analogous symptoms were observed a few weeks after spraying micropropagated plantlets with 995 μM prochloraz before planting. In both cases the integrity of the histogenic layers was disturbed and some uncontrolled cell divisions took place in the exterior tunica layer (L1) of the shoot meristem. These genetic green cells intruded in the albino interior tunica layer (L2), which caused deformations and green spots on the later developing leaves.

In all cases the use of an imidazole pesticide resulted in inhibition of shoot elongation and the development of thicker roots.

Bioreactor production of somatic embryos of *Citrus* yields a lot of abnormalities, among others fasciation and multiple shoot development from one embryo. By adding IMA to the culture medium these abnormalities could be suppressed to a large extent (R.M. Perez et al., pers. comm.).

2.2. Analysing the hormonal effects

Imazalil and prochloraz were only effective in the induction of abundant shoot formation in the presence of a cytokinin (Z, BA, 2iP, TDZ, meta-topoline), and not in se. They didn't change the metabolism of exogenous cytokinins, such as BA in *Spathiphyllum*. Comparing the endogenous cytokinin pool with or without imazalil did not allow to draw conclusions (Werbrouck et al., 1996).

The structural similarity of imidazole pesticides with triazole types of plant growth inhibitors suggested their interference in the gibberellin pathway. The shoot inducing effect of imazalil, but also of the cytokinins, could be counteracted by adding GA_3 to the medium (Werbrouck et al., 1996). Thus the hypothesis was formulated that gibberellins might act as a "brake" on the shoot inducing potential of cytokinins; a brake which is released by adding GA-inhibitors, such as imazalil or paclobutrazol.

The leaf spots on *Ficus* were not completely avoided by adding GA_3. Shoot induction experiments with GA-mutants (*gai*, insensitive and *ga1,* leaky) from *Arabidopsis thaliana* confirmed that the pesticide-effect cannot be completely explained by the presence or absence of GA. Nevertheless the hypothesis is also supported by the findings reported in a few papers. Koornneef et al. (1980) observed that GA deficient mutants of *A. thalianna* developed more basal shoots than the wild type *ex vitro*, and Ezura and Harberd (1995) reported that the same mutants when cultured *in vitro* regenerate shoot buds more readily than does root callus derived from wild-type controls.

2.3. Effects on cell division

The effect of GA_3 and imazalil was studied on genetically engineered *Arabidopsis thaliana* transformed with cycB1;1 promotor (activated during G2 - M transition of the cell cycle) fused to the ß-glucuronidase reporter gene (Ferreira et al., 1994). Root elongation was strongly inhibited by IMA, although intensive GUS-staining was observed. With no additives, or with only GA_3 or GA_3 + IMA almost no GUS-staining was observed. This is indicative that IMA slowed down mitosis in root tips and that GA_3 is not involved.

3. Hormonal effects of *Rhodococcus fascians*

3.1. Leafy gall formation by *Rhodococcus fascians*

The Gram positive bacterium *Rhodococcus fascians* can infect a wide array of plants. The symptoms that are provoked on the host plants are the result of alterations of apical dominance and of plant development. They range from deformed shoots, over witches' brooms, to fasciations and leafy galls depending on the host plant, the severity of infection or on the inoculation method used. Leafy galls are dense structures of meristematic tissues that are covered with small shoots of which the capacity to elongate is inhibited by the bacteria. Unlike *Agrobacterium tumefaciens,* the continuous presence of *Rhodococcus fascians* is required to sustain the symptoms (Jaziri et al., 1997).

Gall formation by *Rhodococcus fascians* is correlated with the presence in the bacteria of a linear plasmid pFiD188 (Fasciation Inducing) containing different virulence loci for synthesis and secretion of several signal molecules influencing plant development (Crespi et al., 1992). Although this bacterium produces several types of cytokinins, which are probably related to pathogenesis, the virulence genes coded by pFiD188 are responsible for the synthesis of signal molecules that differ in structure from the classical cytokinins (Vereecke et al., 1996). Infection of tobacco plants with *R. fascians* leads to the transcriptional induction of GA-20-oxidase and cytochrome P450 related plant genes that possibly act in changing the gibberellin spectrum in the leafy gall or that reflect a change in gibberellin metabolism in the infected tissues (Simon-Mateo et al., in preparation). Killing the bacteria by antibiotics allows the induced shoots to elongate. Thus, among other plant hormone families, gibberellins might also be involved in leafy gall formation and/or persistence.

3.2. Molecular mechanisms underlying leafy gall formation (fig.1)

Although the exact structure of the *R. fascians* produced signal molecules underlying leafy gall formation is still unknown, preliminary data suggest the involvement of signals with a unique structure and activity. Two loci on pFiD188, the *fas* and *att* locus, have been directly related to fasciation. The gene expression of the *fas* and *att* genes is subject to a tight control involving the presence of a leafy gall derived inducer molecule

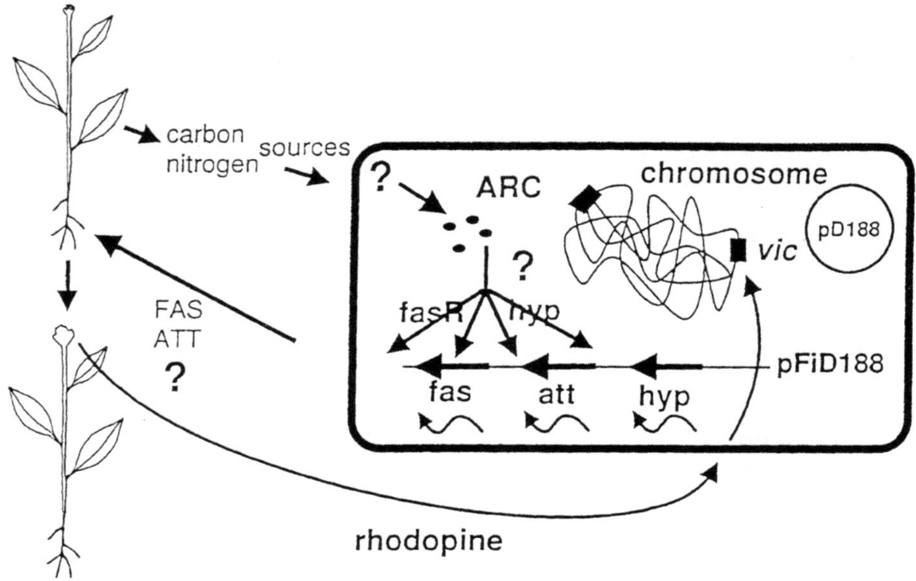

Fig. 1 Schematic overview of the interaction between *R.fascians* and its host plant. For details see text. ARC: autoregulatory compound; *sass, att, hyp*: pathogenicity loci on the linear plasmid; FAS, ATT: complex molecules deregulating normal plant development; *vic:* rhodopine catobolic locus; ?: unsolved issues such as the perception of the host, synthesis of ARC, complete regulatory cascade, possible other secreted bacterial compounds that alter plant development, way in which plant development is altered.

(Vereecke et al., 1996). The *fas* locus is crucial to leafy gall formation and potentially encodes genes that govern the biosynthesis and secretion of a complex signal molecule. One of these genes shows homology with the isopentenyl transferase (*ipt*) genes of different phytopathogenic bacteria implicated in the synthesis of isopentenyl adenine, a general precursor of cytokinins. Other *fas* genes encode functions such as cytochrome P450 and other redox-enzymes that are involved in other modifications (Crespi et al.. 1992, 1994; Goethals et al., 1995). Although these data suggest a role of the *fas* locus in the production of a cytokinin, preliminary mass spectrometry data indicate the *fas*-derived synthesis of a large complex molecule with a unique chemical structure (Temmerman et al., in preparation). Analysis of the *att* locus suggests its involvement in the synthesis and secretion of still another signal molecule that would be implicated in optimal leafy gall formation. Currently, the structure and activity of both the *fas*- and *att*- related signals is being determined.

3.3. Effects on cell division

Studying the effects of *R. fascians* on the behaviour of tobacco BY-2 cells that were synchronised for cell division using aphidicolin, showed that the *fas* locus interferes with the plant cell cycle. A consistent broadening of the mitotic index (MI)-peak was evident treating the plant cells with *R. fascians* wild type cells, pre-induced for *fas* gene expression. No such broadening could be seen in controls where the plant cells were treated with pre-induced *R. fascians* bacteria that carry a mutation in the *fas* locus. Using purified *fas*-related signal molecules, a similar MI-peak broadening was shown (Temmerman et al., in preparation). Although the molecular basis for these effects is still not clear, these data suggest an interference of the *fas*-signals with the machinery controlling plant cell division.

4. Parallelism between pesticides and *Rhodococcus*

Imidazole fungicides as well as *Rhodococcus fascians* induce the same kind of morphogenic and organogenic responses, characterised by abundant adventitious bud formation. In both cases the first working hypothesis was that a cytokinin-like effect is involved. Comparing the analytical data of the cytokinin content of IMA-treated plants and the different controls did not allow to conclude that the observed effects are due to cytokinins in se. Since up to 11 cytokinins have been isolated from culture supernatant of *R. fascians* (Eason et al., 1996) it was generally thought that this was the cause of the altered plant development, by interfering with the hormonal balance of the plant through secretion of vast amounts of cytokinins. Nevertheless, preliminary data on the structure of the complex secreted bacterial compounds and of the isolated plant genes that are induced during the interaction indicate that the mechanism used by *R. fascians* to induce leafy gall formation is much more refined and affects several hormone families.

The IMA-effect can be counteracted by adding GA_3 to the system. Therefore it was logical to investigate the hypothesis that GA_3 might be involved in the phenomenon. In the *R. fascians* system, shoot elongation is initiated after killing the bacteria, which also looks like a gibberellin effect. This is substantiated by the induction upon bacterial infection of plant genes that are correlated with gibberellin metabolism. Currently, experiments are designed to address the role of gibberellins in both interactions and to compare their physiological and molecular basis.

5. New tools

Both systems have potential as novel tools in micropropagation and in tissue culture applications in general:

- both can dramatically enhance the propagation ratio in some crops; *R.fascians* has a host range of 30 families, two of which are monocotyledons (*Iridaceae* and *Liliaceae*), and 67 genera (Bradbury 1986; Stange et al., 1996, Vereecke 1997); IMA has proven effective in this regard on the following plant: *Spathiphyllum, Anthurium, Ficus, Arabidopsis thaliana*;
- IMA can correct for abnormal developmental patterns, e.g. multiple apices and fasciation in *Citrus* embryos.

REFERENCES

Bradbury JF (1986) Guide to plant pathogenic bacteria, pp 185-187, CAB-International Mycological Institute, UK
Crespi et al. (1992) EMBO Journal 11: 795-804
Crespi et al. (1994) J. Bacteriol. 176: 2492-2501
Eason JR et al. (1996) Plant Pathol 45:323-331
Ezura H, Harberd NP (1995) Planta, addendum B 197:301-305
Ferreira PCG et al. (1994) The Plant Cell, **6,** 1763-1774
Goethals et al. (1995) Med. Fac. Landbouwwet. Univ. Gent 60/4a, 1553-1558
Jaziri et al. (1997) European patent application, February 1997, number: EP/21.02.1997/97200513
Koornneef M, Van der Veen JH (1980) Theor Appl Genet 58: 257-263
Skoog F, Miller CO (1957) Symposium Soc Exp Biol 11:118-131
Stange RR et al. (1996) Plant Pathol 45:407-417
Vereecke D (1997) PhD thesis University Gent, Lab Genetics
Vereecke et al. (1996) Med. Fac. Landbouww. Univ. Gent, 61/2a: 231-240
Werbrouck SPO, Debergh PC (1996) J Plant Growth Regul **15**:81-85
Werbrouck SPO et al. (1996) J Plant Growth Regul **15**:87-93

ORGANOGENIC PLANT REGENERATION IN BIOREACTORS

M. Ziv
Agricultural Botany and the Warburg Center for Biotechnology in Agriculture, The Hebrew University of Jerusalem, P.O. Box 12, Rehovot 76100, Israel

1. Introduction

Plant micropropagation is as yet encumbered by intensive manual manipulation and requires scaling-up and automation to provide low cost propagation (Aitken-Christie 1991). Scaling-up necessitates liquid media in bioreactors (Preil 1991; Takayama 1991), and measures must be taken to arrest shoot and leaf growth to prevent shoot malformation – hyperhydricity (Ziv 1991a). Plants developing in liquid cultures exhibit abnormal hyperhydrated leaves and cannot survive transplanting *ex vitro* (Ziv 1991a, Ziv 1994). Shoot morphogenesis *in vitro* can be controlled by culture condition which limit or prevent leaf expansion and shoot growth, while at the same time enhances proliferation. The formation of spherical organized structures, in which the shoots are reduced to buds or meristematic tissue was reported in several species. Protocorm-like bodies formed in liquid cultured gladiolus (Ziv 1989, 1990) and Brodiaea (Ilan et al. 1995). In woody species McCown et al. (1988) and Aitken-Christie et al. (1988) described nodular structures in poplar and meristematic tissue in radiata pine respectively. In bioreactor cultured fern shoots kinetin (KIN) and adenine sulphate induced the formation of meristematic clusters (Ziv and Hadar 1991). Levin et al. (1988) described a several fold biomass increase of organogenic clusters of several species. The clusters were separated, sorted in a controlled bioprocessor and dispensed to agar cultures for further plantlet development.

In several species treatment with growth retardants, inhibitors of gibberellin biosynthesis (Grossman 1990), was necessary to inhibit leaf expansion and enhance bud proliferation (Ziv 1989, 1990, 1992a, b; Ziv and Shemesh 1996; Ziv et al. 1998). The spherical organized meristematic structures that surrounded a central cavity developed into shoots, upon transfer to agar regeneration medium.

The present paper describes the development of dense meristematic cell masses or bud clusters into spherical structures; the effects of liquid media with plant growth regulators on leaf arrest, enhanced proliferation, cluster formation and plant development in bioreactors. The importance of meristematic clusters for controlled propagation of ornamental, vegetable and woody plants in bioreactors is discussed.

A. Altman et al. (eds.), Plant Biotechnology and In Vitro Biology in the 21st Century, 673–676.

2. Materials and Methods

Explants were isolated from various plant organs. Fern runner shoots (Ziv and Hadar 1991), Philodendron and potato axillary buds (Ziv and Ariel 1991; Ziv and Shemesh 1996); terminal and axillary buds from gladiolus corms (Ziv 1989, 1990) and *in vitro* banana shoots (Ziv et al. 1998) and from *in vitro* roots of poplar shoots (Carmi et al 1995).

The explants were established in full or ½ strength MS salts, vitamins and inositol (Murashige and Skoog 1962) agar medium, with various levels of growth regulators as detailed in the results. Aseptic explants developing buds or meristematic clusters after 2-3 weeks were subcultured to liquid media in Erlenmeyer flasks on a rotary shaker at 90-100 rpm.

The arrest of leaf growth was achieved by treatment with the growth retardants ancymidol (ANC, Elanco), paclobutrazol (PAC, ICI) or flurprimidol (FLP, FN9086 Dow Elenco) with the exception of fern which proliferated to form clusters in the presence of kinetin and adenine sulphate. The proliferating spherical structures were subcultured to 1 l bubble column, 1.5 l internal loop, 2 l central drought airlift (LH, England) or disposable presterilized ('Life-Reactor' Osmotek) bioreactors (for details see Ziv and Hadar 1991; Ziv and Ariel 1991; Ziv 1991b; Ziv et al. 1998).

Aeration and circulation was achieved by introducing humidified air through a sinter disc at the base of the bubble or plastic bioreactors or through glass spargers in the airlift bioreactors. Initial air flow was 0.5 and increased to 1.5 or 2.0 vvm (volume air/volume medium/min) according to biomass increase and the need for cluster circulation. All bioreactor cultures were kept at 25±2°C and illuminated continuously by cool white fluorescent lamps, 15-20 $\mu mols \cdot m^{-2} \cdot s^{-1}$ at the outer jacket of the vessel. Bud or meristematic clusters were separated mechanically in a Waering blender in which the blades were covered with silicon tubing (fern and gladiolus clusters) or by a grid of knives with 4x4 mm opening (potato, philodendron and banana bud clusters).

The macerated 3-4 mm propagule units were transferred to 10x10x6 cm polystyrene boxes with either a growth or a storage organ induction agar medium. The growth medium consisted of ½ strength MS salts with naphthalene acetic acid (NAA), KIN BA, depending on the species as detailed below. Storage organ induction medium contained elevated sucrose 6-8% and in potato and gladiolus included ANC or PAC. The cultures were grown in a growth chamber at 25/22°C day/night temperature under a 16/8 photoperiod, 60 $\mu moles \cdot m^{-2} \cdot s^{-1}$.

3. Results

Meristematic clusters developed on fern leaves in the presence of 9.3 μM kinetin, 0.5 μM NAA and 475 μM adenine sulphate in the initial stage of runner tips cultured in shaken flasks (Fig. 1). The subculture of fern clusters to bubble column or airlift bioreactor induced a 2-5-fold increase in the biomass, which was affected by both the size of the initial inoculum and the aeration rate. A smaller inoculum of 20 g/l resulted in a greater biomass production than a 40 g/l inoculum. Aeration rate had an effect on the morphogenetic quality of the clusters, which were compact, and without any leaves (Ziv and Hadar 1991).

Liquid cultured philodendron buds proliferated into clusters with arrested leaf and shoot growth in the presence of NAA, BA and 19.5 or 39 μM ANC. A short 3-day induction treatment with 19.5 μM ANC was as effective in reducing leaf growth as the continuous presence of ANC in the medium. The effect of 39 μM ANC given for 3 days was less inhibiting then its continuous presence on both leaf arrest and the biomass growth. In addition the transient treatment had a smaller carry over effect in retarding growth during the growth and development stage *ex-vitro* (Ziv and Ariel 1991).

Potato nodal explants including 8-10 mm of the stem tissue, a leaf and an axillary bud, developed bud clusters in liquid medium in the presence of 9.7 μM ANC after two subcultures over a period of 6-8 weeks. The clusters were chopped to sections with 3-4 buds and subcultured to disposable presterilized bioreactors. The bud clusters continued to proliferate, developing condensed shoot aggregates hyperhydricity symptoms. The subculture of the small propagule units to a growth medium induced shoot formation. On a tuber induction medium with 19.5 μM ANC and 6% sucrose the buds developed into tubers reaching 180-200 mg FW after 4-5 weeks.

Gladiolus buds proliferated in liquid shake cultures in the presence of 0.25 μM NAA, 10 μM BA and various growth retardants, the most effective were PAC and ANC (Ziv 1989; 1990) at 3.4 and 9.8 μM respectively. In the presence of PAC the clusters were compact with meristematic centers surrounding a central cavity. Biomass growth in the bioreactor was affected by the level of sucrose and PAC in the medium. The highest growth value was obtained in the 3% sucrose without PAC and the lowest in the presence of 6.8 μM PAC and 6% sucrose. However, dry weight was highest in the latter increasing from 8.8% in the control to 14.3%.

Upon the removal of PAC, leaf primodia developed from multiple shoot apices and in agar media these developed into plantlets. The addition of a second layer of liquid medium on top of the agar after 4 weeks, with ½ strength MS minerals 0.21 μM NAA, 2.22 μM BA, 6% sucrose and 17 μM PAC induced cormlet development after a period of 8-10 weeks (Fig. 1a).

Protocorm-like clusters were induced in bioreactor cultures in the presence of ANC in *Brodiaea* and, as can be seen in banana ANC induced bud clusters (Fig. 1b). Poplar roots cultures in disposable plastic bioreactor in the presence of 4.5 x 10^{-2}M TDZ regenerated buds, as many as 2500 per one g fresh weight of roots (Fig. 1c).

4. Discussion

The development of spheical meristematic or bud clusters in liquid cultured plants belonging to different genera, provides an advanced propagation system which can be utilized for *in vitro* scale-up, mechanized separation and automated inoculation (Levin et al. 1988). Cluster formation appears to be associated in most species with continuous circulation, agitation and submergence in a liquid medium as well as with the presence of growth retardants and/or cytokinins.

The clusters are made of densely packed organized cells, actively dividing and forming meristematic centers, similar to the nodules described by McCown et al.

(1988). Meristematic centers appeared in the outer peripheral layers and posses a distinct morphogenetic potential.

The advantage of using clusters for micropropagation are: suitability to scaled-up liquid cultures in bioreactors, mechanical homogenization, separation and uniform propagule sizing, automated subculture and transfer, controlled shoot hyperhydricity in liquid media, high rates of proliferation and regeneration.

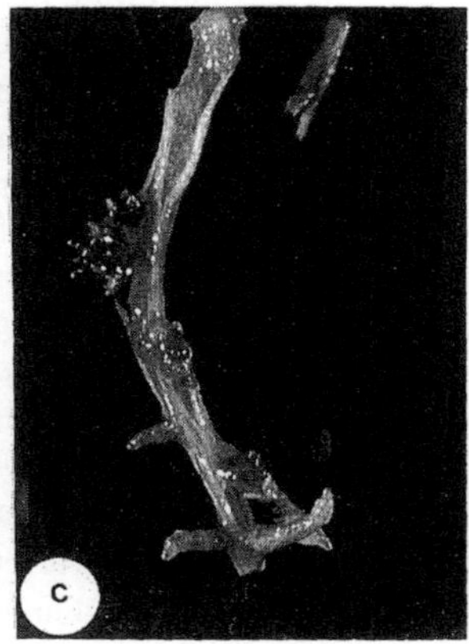

Fig. 1. Corm development on gladiolus clusters - a; ANC treated banana bud clusters - b; Bud formation in poplar roots treated with TDZ - c.

5. References

Aitken-Christie J, Singh AP (1988) Genetic Manipulation of Woody Plants. pp. 413-432 Plenum Publ. Corp. New York.

Ilan A et al. (1995) Sci. Hortic.63, 101-112.

Levin R et al. (1988) Biotechnol. 6, 1035-1040.

McCown BH et al. (1988) Genetic Manipulation of Woody Plants. pp. 149-166 Plenum Pub. Corp. New York.

Murashige T, Skoog F (1962) Physiol. Plant. 15, 473-497.

Preil W (1991) Micropropagation. pp. 425-455 Kluwer Acad. Dordrecht. Pub.

Takayama S (1991) Biotech. in Agr. and Forestry, pp. 1-46 Springer Verlag, Berlin.

Ziv, M. (1989) Pl. Cell Tiss. Org. Cul 17, 101-110.

Ziv M (1990) Acta Hortic. 280, 207-214.

Ziv M (1991)) Micropropagation. pp. 45-69. Kluwer Acad. Pub, Dordrecht.

Ziv M (1992a) Acta Hortic. 319, 119-124.

Ziv M (1992b) Progress in Plant Growth Regulation. pp. 809-817 Kluwer Acad. Pub, Dordrecht.

Ziv M, Ariel T (1991) J. Plant Growth Regul. 10, 53-57.

Ziv M, Hadar A (1991) Isr. J. Bot. 40, 7-16.

Ziv M, Shemesh D (1996) *In Vitro* Cell Dev Biol P. 32 31-36

Ziv M et al(1998) *In Vitro* Cell Dev Biol. P. 34, 152-158.

PLANT TISSUE CULTURE INDUSTRY IN THE MIDDLE EAST, AN OVERVIEW

Abo El-Nil[1] M. and Ilan A[2].
1. Kuwait Institute for Scientific Research, Kuwait.
2. Rahan Meristem. Rosh Hanikra 22825, Israel

The plant micropropagation industry in the Middle East was developed to meet the demand for clonally propagated crop plants. The opportunity presented by low production costs in some of the Middle Eastern countries is an incentive for the establishment of plant tissue culture laboratories. Currently, the scientific infrastructure is sufficient to provide adequate support to the industry. The ability of micropropagators to penetrate existing markets of plant production will be determined by prices, quality, quantity and novelty of the products, as well as regulation and legislation in each country.

1. Introduction

The use of plant tissue culture for commercial plant propagation (micropropagation and pathogen elimination) has begun in the late sixties. Micropropagation has since become an important commercial tool for plant propagation. High production cost and technological obstacles have limited the scope of this industry in comparison to other plant propagation industries (Ilan, Khayat 1996). First attempts to establish plant propagation methods for industrial purposes in different Middle Eastern countries were made in the late sixties and early seventies (Kochba et al. 1972, Reuveni, Lilien-Kipnis 1974). These attempts led to the establishment of the first micropropagation laboratories in the region in the mid seventies and early eighties.

In recent years, interest in the application of plant tissue culture for plant production in Middle Eastern countries has drawn the attention of public as well as private sectors. Training and establishment of research activities in many governmental organizations preceded the commercial application in the region. Fragmentation of the market for planting stock, diversity of crops, high front-end capital investment and high technical training required are some of the problems facing the application of tissue culture. For this reason, technology transfer and adaptation, cooperation and coordination of efforts in production and marketing are necessary. Labor cost is regarded a major component of the total production cost (Ilan, Khayat 1996, Standaert-de Metsenaere 1991). Among the incentives for the development of the industry in Middle Eastern countries was economic potential, due to low labor cost in some of the countries in the region. Another incentive

A. Altman et al. (eds.), Plant Biotechnology and In Vitro Biology in the 21st Century, 677–680.

for the development of plant tissue culture industry in the Middle East is the need for rapid multiplication and cleaning from diseases in the major crops for local and export markets such as: floriculture crops, banana and strawberry. In some cases, restrictions of massive plant import allowed the development of a local micropropagation industry as a tool for new introductions. The rapid development of irrigation systems in recent years is a vital factor for growing of tissue culture plants in arid and semi arid zones. The opportunity presented by the low production cost in some of the countries of the Middle East, combined with a gradual shift from traditional farming to intensive agriculture, promoted the establishment of plant tissue culture laboratories.

2. Current situation of micropropagation industry in the Middle East

The success of micropropagation industry depends upon strong research and development. Tissue culture research has been initiated in all countries of the region at different levels and for various objectives (Table 1). The economic activity of North African countries has an effect on Middle Eastern markets due to the close historical relation with the Middle East. These countries have a significant impact on Middle Eastern agricultural development (Table 1). The commercial application of micropropagation of banana, strawberry, date palm and ornamental plants is well established. Eight countries in the region have commercial production facilities with an estimated total annual production of 23 million plants, out of which 20 million are produced in Israel alone (Table 2). An extensive floriculture industry was the main drive behind the successful and early start of commercial tissue culture in Israel.

Date palms (*Phoenix dactylifera*) is one of the most important crops in the Middle East and neighboring desert areas. Novel technologies were developed and are being utilized for large-scale production (Sudhersan et al. 1993). Date palm is commercially propagated from tissue culture in Israel, Kuwait, Saudi Arabia and the United Arab Emirates (Table 2).

3. Potential cooperation in Research and Production

The basic requirements for the development of micropropagation are: market, production, knowledge and finance. The existence of some economical advantage i.e.: a large local market, low production cost, etc. is fundamental. Cooperation between micropropagators from different countries, preferably from the same region, is important to future development. In this regard, labor, finance and knowledge are limiting factors in some and available in other countries.
Many areas of applicable research can support the micropropagation industry. Among these areas are somatic embryogenesis, automation and mechanization.

Research on propagation of desert plants can support existing crops such as date palms and contribute to the introduction of desert plants from other parts of the world to the region. Scientists from the Middle East and neighboring areas were among the pioneers of date palm tissue culture (Reuveni, Lilien-Kipnis 1974; Ammar, Benbadis 1977; AboEl-Nil

Table 1. A list of tissue culture research and production laboratories in Middle Eastern and North African countries.

Country	Number of TC labs			
	Commercial	Products	R&D	Crops in research
Algeria	None		1	Date Palm
Bahrain	None		1	Date Palm
Egypt	5	Banana, Strawberry Potato, Ornamentals	7	Date Palm, Fruit trees, Gladiolus, Medicinal plants
Iran	None		1	Date P., Fruit trees, Pistachio
Iraq	None		4	Date Palm, Fruit trees
Israel	7	Banana, Strawberry, Orn., Date P., Flowers	5	Tomato, Garlic, Date Palm, Potato, Fruit trees
Jordan	1	Fruit trees, Strawberry	1	Fruit trees
Kuwait	1	Date Palm	1	Date P, Strawberry, Native plants, Ornamentals, Mangrove
Lebanon	None		1	Citrus, Grapes, Cedar
Morocco	2	Date P., Potato, Roses Banana, Caper, Pista.	4	Date P., Strawb, Fruit t., Tomato, Begonia, Avocado, Wheat
Oman	None		1	Date Palm
Qatar	None		1	Date Palm
Saudi Ar	1	Date Palm	3	Date P., Potato, Medicinal plants
Syria	None		2	Potato, Fruit trees, Pistachio
Tunisia	1	Potato, Strawberry	3	Date Palm, Potato
UAE	2	Date Palm	1	Date Palm
Total	21		37	

Table 2. A list of estimated numbers of commercially produced tissue culture plants in Middle Eastern countries.

Country	Total number	Crops ranked in decreasing order
Egypt	3 M	Banana, Strawberry, Potato, House plants.
Israel	20 M	Banana, Strawberry, House plants, Flowers, Date palm.
Jordan	????	Fruit trees, Strawberry.
Kuwait	.05 M	Date palm.
Morocco	.25 M	Banana, Date palm, Potato, Caper, Pistachio, Roses.
Saudi Arabia	.01 M	Date palm.
Tunisia	.05 M	Potato, Strawberry.
U. A. E.	.03 M	Date palm.
Total	23.390 M	Strawberry, Banana, Potato, House plants, Roses, Date P.

1989). Plant tissue culture techniques have been developed for rapid propagation of extinct plants for the restoration of vegetation in desert areas (Abo El-Nil 1997).

5. Potentials for the micropropagation industry in the era of recombinant DNA.

Recent advances in plant biotechnology are expected to greatly increase the demand for micropropagated plants. The transgenic approach is heavily dependent on the in vitro technology. Since a large number of transgenic crops are expected to be introduced in the forthcoming decade, the micropropagation industry will have new opportunities for development, as propagation of transgenic plants can provide an appropriate solution to the vast demand for these products. In addition, commercial tissue culture laboratories can support the development of transgenic crops. Countries of the Middle East can play an important role in the newly created opportunities due to a rare combination of technology, financial resources, inexpensive energy and large markets for the newly created products.

6. References

Abo El-Nil M. (1989) Refining methods for date palm micropropagation. Proceedings of the Second Symposium on Date Palm. King faisal University, Saudi Arabia, pp. 29-40, Mars Publishing House Riyadh, KSA.

AboEl-Nil (1997) Tissue culture of native plants in the developing countries. In: Hort. Biotech. In Vitro Culture and Breeding. Altman A. and Ziv M. (eds.) Acta Hort. 447: 507-514.

Ammar S. and Benbadis A (1977) Multiplication vegetative du palmier datier *(Phoenix dactylifera* L.) par la culture de tissus de jeunes plantes issues de semis. C. R. Acad. Sci. Paris 284. Serie D:1789-1792.

Ilan A. and Khayat E. (1996) An overview of commercial and technological limitations to marketing of micropropagated plants. Acta Hort. (in press)

Kochba J., Spiegel-Roy P. and Safran H. (1972) Adventitive plants from ovules and nucelli in citrus. Planta 106: 237-245.

Reuveni O. and Lilien-Kipnis H. (1974) Studies of the in vitro culture of date palm (*Phoenix dactylifera* L) tissues and organs. The Volcani Institute of Agricultural Research #145. pp. 40.

Sudhersan, C., M. AboEl-Nil and A. Al-baiz 1993."Occurrence of direct somatic embryogenesis on the sword leaf of in vitro plantlets of *Phoenix dactylifera* L. cultivar Barhee." Current Science 65:(11) 887-888.

Standaert-de Metsenaere, R.A.E. (1991). Economic considerations, pp. 123-140 In: P.C. Debergh and R.H. Zimmerman (eds.) Micropropagation. Kluwer Academic Publ., Dordrecht, The Netherlands.

INTEGRATION THROUGH IN VITRO CULTURE

M. Alonso[1], E. Gisbert[1] and M. Borja[1,2]
[1]Fundación PROMIVA, Departamento de I+D, Apdo Correos 35, Boadilla del Monte 28660-Madrid, and [2] Universidad Complutense de Madrid, Departamento de Bioquímica y Biología Molecular, 28040-Madrid, SPAIN

Introduction

Fundación Promiva is a non for profit organisation where we believe in the integration and normalisation of mentally handicapped persons in our society through having fair wages and adequate jobs for their capacities (laborious and repetitive tasks are preferred). In Fundación Promiva mentally handicapped persons have been working in our ornamental plants nursery for 16 years already. Since in vitro culture is a very laborious and repetitive task, two years ago we decided to set up an in vitro culture laboratory where most of the members would be mentally handicapped persons. We have done the necessary modifications in order for them to be able to perform all the usual tasks in an in vitro culture lab. They work on a sterile zone for the transplant of the explants. Some of them are also able to prepare media if the instructions are adapted for them. Currently we have 120 workers of which 10 of them have jobs in the laboratory. We suggest the possibility for other micropropagation companies of hiring mentally handicapped persons as means for their integration and normalisation.

Material and methods

The laboratory was set up with inside pressure in order to avoid contaminations and to facilitate the technicians job (Kyte, Kleyn, 1996). The following plants where chosen for production: Pelargonium, Petunia, Nephrolepis, Saintpaulia, Cordyline, Gerbera, Syngonium, Primula, Hydrangea, Kalanchoe, Ficus, Begonia, Spathiphillum, Dieffenbachia, Hedera and Chrysanthemum. The plant culture media where used as published (Kyte, Kleyn, 1996). Pelargonium meristem culture (Debergh, Maene, 1977; Beauchesne et al., 1997, Hakkaart, Hartel,1979) and callus regeneration (Pillai, Hilderbrandt, 1968; Desilets *et al.*, 1993; Abo El-Nil, 1971) were performed as already described. For Pelargonium phytopathogens detection the cultivable fungi where sporulated in a humid chamber. Fungi and bacteria where cultivated in Potato-Dextrose -Agar plates and for specific viruses (cucumber mosaic virus (CMV), pelargonium leaf curl virus (PLCV), pelargonium line pattern virus (PLPV), pelargonium flower break virus (PFBV), Tospoviruses (from groups I, II and III) (Tospo) and tobacco mosaic virus (TMV)) and Xhantomonas (Xcp) a triple sandwich ELISA was performed with Bioreba antibodies.

A. Altman et al. (eds.), Plant Biotechnology and In Vitro Biology in the 21st Century, 681–684.

Results

The mentally handicapped technicians where able to prepare the media, although not all of them are able to use a balance or the pH-meter. They are able to follow a recipe, that has been specially modified so even persons that can not read can prepare media. We have purchased an automatic media dispenser and a labelling machine. The label identifies the plant type the media and the date. Two are able to set up a non automatic autoclave. The transfer room has three large flow hoods and the technicians use glass beads sterilizers in order to avoid accidents with the alcohol. The ELISA tests where performed by the mentally handicapped technicians as well. Two technicians can analyse 200 plants in two day.

Figure1.

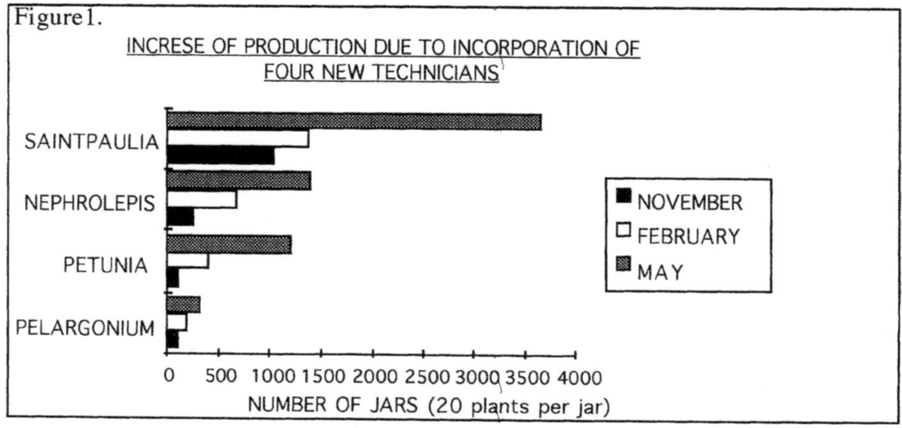

Pelargonium

Since Pelargonium is a traditional crop in Spain we are sampling Pelargonium varieties that have been in the same site for at least 25 years or longer. The samples are collected from different bioclimatic areas in Spain, from three different *Pelargonium* species (*P. zonale*, *P. grandiflora* or, *P. peltatum*). These old cultivars are interesting because we are trying to find cultivars that have developed resistance against some pathogens, and we wanted to know which are the most frequent diseases in Spain. So all of them have been analysed to detect the infections they could have before their establishment in culture. We used ELISA and agar mediated detection. In the figure 2. we can appreciate the percentage of infection caused by different geranium viruses in the analysed samples, including the appearance of multiple infections (2X, 3X or 4X). The 45% of the healthy varieties where micropropagated. The rest where evaluated for other traits (flower and leaf colour, original plant growing habit...) and if it's characteristics where of interest the sample was heat therapy treated and meristem cultured. Once the culture was successful it was evaluated again for disease. If it was healthy the cultivar was kept if not it was discarded.

Figure 2.

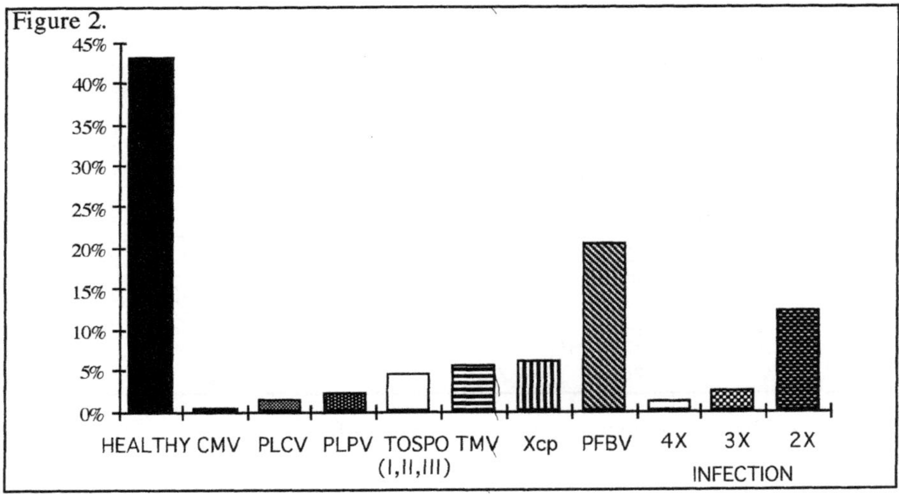

To prevent oxidation in the culture establishment we use cystein (2 mg/L) and initiate the cultures in a MS-medium free of hormones The biggest handicap is to develop the appropriate culture medium for each of the varieties, since we have observed that it has to be optimised in each case. A week after we transplant the samples to 12 different media consisting on MS-medium, supplemented with different combinations of auxins (10, 5, 2, 1 mg/l of kinetin) and citokinins (1; 0,5; 0,2 mg/l of IAA or 1; 0,5; 0,2 mg/l of NAA). Different explants (leaves, petioles and shoot tips) are evalueated from each of the samples. Table 1. represents the summary of the results of the percentage of regeneration of all the explants probed in the different media.

Table 1.

		KINETIN (mg/l)			
IAA (mg/l)	NAA (mg/l)	10	5	2	1
0,2	-	0%	0%	4%	0%
0,5	-	0%	0%	2%	2%
1	-	6%	4%	6%	6%
-	0,2	8%	10%	6%	0%
-	0,5	21%	4%	4%	2%
-	1	0%	8%	8%	2%

The samples where collected in different bioclimatic locations so we evaluated shoot formation at different temperatures to see if the in vitro cultured plants had the same temperature requirements as the original plants. The results are shown in figure 3 for three samples representative from three different regions: cold (Palencia), hot (Sevilla) and humid (Galicia).

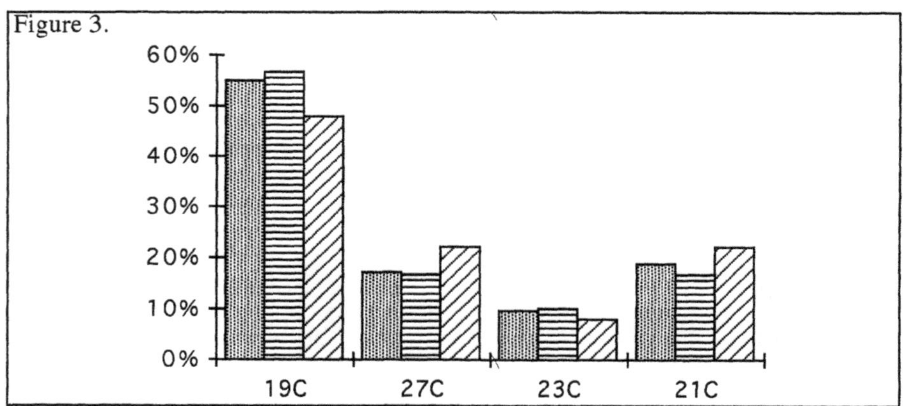

We observed that the optimum temperature for shoot formation was 19°C for all of the samples.

Discussion

We have trained 10 mentally handicapped (33%-66% of incapacity) technicians to perform most of the tasks needed in an in vitro culture lab. In previous experiences in our institution we had demonstrated that it is cheaper to train a mentally handicapped person to perform a useful job where he or she can earn a fair wage to cover their needs than to have them live out of our social security system. In the last year we have demonstrated that they are able to produce commercially in vitro culture plants. Mentally handicapped persons can be integrated in our society in a normalised environment if there is a social will. Although their training is longer and it is necessary to adapt certain protocols for the characteristics of each person they are usually very hard workers and they have very developed manual skills. We suggest the possibility for other micropropagation companies of hiring mentally handicapped persons as means for their integration and normalisation.

Acknowledgements

This work was supported by a grant from the MINER, ATYCA B5/1997. M.A. was a recipient of a MEC predoctoral fellowship. E.G. was a recipient of Fundación PROMIVA predoctoral fellowship.

References

Abo El-Nil N M, Hilderbrandt AC (1971) Plant Dis.Rept. 55, 1017-1020.
Beauchesne G et al (1997) Acta Hort. 78, 397-402.
Debergh H P, Maene L (1977). Acta. Hort. 78, 449-454.
Desilets H et al (1993) Can.J.Plant Sci. 73, 871-878.
Hakkaart F A, Hartel G (1979) Neth. J. Plant Path. 85, 39-46.
Kyte L, Kleyn (1996) Plants from Test Tubes, Timber Press Inc. Portland. Oregon.
Pillai S K, Hildebrandt A C (1968) Phyton 25, 81-87.

Commercial Needs and Creating New Markets in Eastern European Countries

Atanas Atanassov[1], Ludmil Antonov[1], Tomasz Twardowski[2]
[1] Institute of Genetic Engineering, 2232 Kostinbrod-2, Bulgaria and Norman Borlaug IPMBR, De Montford University, Leicester LE7 9SU, England; [2] Institute of Bioorganic Chemistry, Poznan, Poland

1. Introduction

Conventional plant breeding programmes alone can no longer increase food, feed and fibre productivity at a growth rate which could satisfy the population growth, and at the same time to respond to the negative changes in the environment and to maintain high level of sustainable agriculture. Nowadays biotechnology is regarded as a new technological improvement. It is becoming an essential element of the existing plant breeding programmes. There are real expectations that the biotechnology implementation can lead to food self-sufficiency and thus to alleviate the poverty and to conserve the national resources and to protect the environment (Borlaug, 1997). These challenges can be met if the public and private sectors work together (James, 1996). However, the successful collaboration is possible only if the problems related with the intellectual property rights (plant breeding rights) and the legislation and the regulations of genetically modified plants are properly settled. The Eastern European countries which are entering in the world of the market economy can serve as an experimental "field" for testing and implementation of new principles which the international organisations as UPOV, UNEP, OECD, UNIDO, FIS/ASSINSEL, EU can provide in the coming years. If this is the case, one can expect increasing commercial needs and faster creation of new markets in this changing and challenging part of the world called CEE.

2. Current state of the science in CEEC's (Central and Eastern European Countries)

During the transition period in CEEC's the capacity for scientific research has been decreased significantly. Nevertheless, science seems to have survived, especially in the Czech Republic, Hungary and Poland. Although in many cases diminished and reduced in tasks, the Academies of Sciences and their predecessors have played a crucial role in protecting the interests of the scientific community and in the survival of science as a whole. Many institutes dealing with applied research collapsed due to the disappearance of central planning and funding and also due to the privatisation of most of the state-owned companies (Coopers, Zybrand, 1997). At this moment, still most of CEEC's do not recognise the importance of well-balanced combination of basic and applied research. Competition for grants in research for both basic and applied has been introduced in all CEEC's.
For most of the countries needs are related with priority establishments, links between the science and industry (technology transfer) and active patenting and licensing of the research results.

3. Plant Biotechnology

3.1. General statement

Following the worldwide boom in the beginning of the '80s, biotechnology has been given priority among the national economies and research including CEEC's. Since 1985,

A. Altman et al. (eds.), Plant Biotechnology and In Vitro Biology in the 21st Century, 685–688.

significant investment has been made in both industry and academia. After the drastic cuts in the central funding for research, centres found themselves unable to compensate these by attracting alternative funds for their expensive research. Actually only those research centres which were capable to contract research and have some expertise in business and commercialising biotechnology before the changes started not only survive but develop their plant biotechnology programmes successfully. These are Biological Research Centre in Szeged and Agricultural Biotechnology Centre in Gödölö, Hungary, Institute of Genetic Engineering, Kostinbrod, Bulgaria, Institute of Bioorganic Chemistry, Poznan, Poland, Institute of Molecular Biology, Czeske Budovice, Czech Republic, Institute of Cell Biology and Genetic Engineering, Kiev, Ukraine, and Centre of Bioengineering, Moscow, Russia.

3.2. Plant Biotechnology Achievements

3.2.1. Tissue culture

This technology was developed almost equally well as in the Western Countries. However, except clonal micropropagation not in all other cases it proved to be a commercial success. The rest of the achievements have usually been forgotten after completing their R & D stage without their commercial potential to have been explored. The values of many of them exceed the local industry needs. Most of these original results achieved have not been patented and registered because of the lack of expertise and the settlement of appropriate property right (Wagner et al., 1997).

3.2.2. Gene manipulations

This technology was lagging far behind in comparison with the developed countries. Few exceptions that exist are related with the research organisations mentioned above. Because of the efforts of the Hungarian, Russian, and Bulgarian units, genetically engineered products have been locally developed. Transgenic crop field trials with them have been performed since 1991 in Bulgaria (tobacco and alfalfa); since 1993 in Hungary (alfalfa, corn, canola, eggplant, potato, and tobacco), and Russia (corn and potato). These three countries face the first transgenic variety release soon after the year 2000 (James, 1997).(Table 1)

3.2.3 Biotechnology legislation

From the survey performed several important characters can be stated:

a) All the CEEC's ratified the Environmental Program Convention on Biodiversity (Rio de Janeiro, 1992) – UNEP – CBD.
b) Full members of the EU and OECD (Organisation of Economical Co-operation and Development) are Czech Republic, Poland, and Hungary. The others except the NIS (National Independent States) are associates.
c) Full members of UPOV (Union for the Protection of New Varieties of Plants) are Poland, Hungary, Czech Republic, and since 24 of April 1998, Bulgaria and Russia.
d) National legislation
   - National "gene law" has been established so far only for Russia at the end of 1997 and in April 1998 by Hungary based on the Directive 90/219 EEC and 90/220 EEC regarding the Amendments issued in 1996. National guidelines for biosafety in genetically manipulated plants (GMP) have been put in action in Bulgaria since August 1996.
   - National Biotechnology Committees for GMP have been established in all CEEC's. Their role is to regulate the release of GMP to the environment. The important role which Biotechnology Committees in every CEEC plays is to see how to give an official permission for field trials tests which the multinational companies have begun to require since 1996. This year Monsanto, HiBred-Pioneer, Novartis, AgrEvo, and some others have provided field trials for transgenic corn, soybean, potato, rapeseed etc., in almost

all CEEC's.

4. Commercial needs and new markets

In the recent study of J. Burke and S. Thomas, 1997 "Agriculture is biotechnology's future in Europe" it is estimated that the potential market for biotechnology related products within the European Union is expected to be 250 billion ECU (US $278 billion) by the year 2005. Their prediction is that 70% of the biotechnology growth will come from the agriculture and food sector.

CEEC's face to enter into EU in the following decades. Hungary, Poland and the Czech Republic are already there. Basically speaking, plant biotechnology infrastructure in these countries still exists, although the physical equipment in most of the research centres is already very old and poor almost to the point of non-existence. On the other hand, its intellectual capacity based upon many years of excellent education and enthusiastic support for the scientific endeavours of the communist world has ensured that it has the intellectual and research capability to make a major contribution to Europe's future intellectual and economical assets.

The present lack of necessary effective capital system is a serious obstacle to the future commercialisation of R & D plant biotech projects. European Community, NATO, the other international organisations as IAEA, UNEP, UNIDO, FAO, ICGEB by their programs will not give the expected financial support to fill the gap which exists between the Western and Eastern European countries. Establishment of various contracts between the research and industry which already entered in a new era of consolidation and also the settlement of new private, venture capital for joint ventures seems to be right approaches to go further. Development financial facilities are becoming available from the World Bank, EBRD, JOPP, PHARE, Eureka, etc. Another critical point is providing marketing expertise and the settlement of appropriate intellectual property rights legislation that is already the case in some of the CEEC's. Positive changes are already observed in the laws and regulations for banking, tax, foreign investments and custom duties.

Improved intercommunications between the private and public actors in this field are also substantial. It is therefore essential that the communication between the various parties be improved, physically, socially, and on a business basis.

Only then one can expect industrial utilisation of the research results and as a return the basic science also to be developed in this part of Europe.

Conclusion

Plant biotechnology has been identified as one of the key technologies for the decades to come. In the CEEC's, the modern plant biotechnology is still waiting to be recognised. The human resources working in the field of plant cellular and molecular biology and genetics are of high value. However, the conversion of basic to applied research is very slow and inefficient. It needs capital and investment and legislation formulation harmonised with the European system. If this occurs soon, the CEEC's plant biotech can become an important player in the European and international arena in the upcoming 21st century.

References

- Burke JF, Thomas SM (1997). Agriculture is biotechnology's future in Europe. Nature

Biotechnology,15, 695-696

- Wagner W et al (1997). Europe ambivalent on biotechnology. Nature, 387, 845-847
- Coopers, Zybrand (1997). Thematic papers. Thematic Workshop Vienna, p. 48
- Borlaug N (1997). Feeding a world of 10 billion people: the miracle ahead. Biotechnology and Biotechnology Equipment, 11, 3-13
- James C (1996). Agricultural research and development: The need for public-private sector partnership. Issue in Agriculture 9. CGIAR, p. 48

Table 1. Transgenic Crop Field Trials in Eastern Europe and Russia: By Crop and Trial (1986 through 31 December 1995)

Country		**1987**	**1988**	**1989**	**1990**	**1991**	**1992**	**1993**	**1994**	**1995**	**Total**
Bulgaria	Alfalfa									M	**1**
	Tobacco									BR,VR	**2**
	Subtotal									3	**3**
Hungary	Alfalfa								IR	IR	**2**
	Corn								2HT,2IR	2HT,2IR	**8**
	Canola								HT	HT	**2**
	Eggplant								IR,VR	IR,VR	**4**
	Potato							VR	VR	VR	**3**
	Tobacco							VR	VR	VR	**3**
	Subtotal							2	10	10	**22**
Russia	Corn									3IR,3HT	**6**
	Potato								2VR	3VR	**5**
	Subtotal								2	9	**11**
Total EE/Russia								**2**	**12**	**22**	**36**

Source: Modified and updated from Krattinger (1994).

BR:	Bacterial Resistant	IR:	Insect Resistant	Q:	Quality Characteristics
FR:	Fungal Resistant	M:	Marker Gene	VR:	Virus Resistant
HR:	Herbicide Tolerant	MS:	Male Sterility		

EARLY RESEARCH AND COMMERCIAL USE OF *IN VITRO* PLANT PROPAGATION IN ISRAEL

J. Ben-Jaacov
Dept. of Ornamental Horticulture, A.R.O
The Volcani Center, P.O. Box 6, Bet Dagan, 50-250, Israel.

This paper describes the main research activities during the late sixties and early seventies, in the use of tissue culture technology for micro propagation of plants. These research activities lead to the development of the first commercial tissue culture laboratories in Israel.

International activities in the use of tissue culture for rapid, vegetative propagation, in the early seventies (Bhojwani et al 1986).

This subject was in its early stages during those years. Most of the research was on propagation of Orchids (a very special use of tissue culture - Sagawa et al 1970) and on the use of meristem culture for the development of Virus-Free plants (Quak 1972).

At the XVIII th International Horticultural Congress which was held in Tel-Aviv in March, 1970, the section on "Propagation by tissue and organ culture" included an introductory lecture by Murashige titled "Some factors in the application of tissue and organ cultures to plant propagation" but only very few additional papers on *in vitro* micro propagation were presented. (Gorn, Mendel 1972).

Tissue culture research in Israel in the early seventies.

Lavee, returning from Skoog's laboratory, in the USA, in the early sixties started the first tissue culture research laboratory, at the Volcani Institute. Lavee's laboratory was instrumental in helping others to initiate tissue culture research in Israel. Most investigators, who were active in the early seventies, were using tissue culture techniques as a tool for physiological, genetical and horticultural research (Altman, Goren, 1971, Altman, Goren, 1974, Kochba, et al.1972, Kochba, Spiegel-Roy 1977, Lavee, Messer, 1969 , Messer, Lavee, 1969, Raveh et al 1973, Umiel, Goldner 1976 and Vardi et al 1975).

Research in tissue culture of vegetative propagation, in Israel, during the early seventies.

Much of the early research, done in Israel, on the use of tissue culture for rapid vegetative propagation and for curing plants from virus diseases was documented only in local publications and in internal reports. The research was mainly done in three laboratories:

An early research on regeneration of *Gladiolus* through tissue culture, which was done at the Faculty of Agriculture of the Hebrew University by Meira Ziv and her coworkers, was later used as the basic information for a large scale "Gladiolus virus free project" which was carried out in the country (Ziv et al 1970).

As early as the late sixties Oppenheimer and Reuveni, of the Volcani Institute, started studying "Methods for quick propagation of new and superior date varieties- *In vitro*" (Reuveni et al 1972). Reuveni, who was a great horticulturist and plant propagator, tried many of the plants he investigated in plantations and in the

A. Altman et al. (eds.), Plant Biotechnology and In Vitro Biology in the 21st Century, 689–692.

greenhouse also in tissue culture. These included bananas, pineapples, papayas and others (Reuveni, 1988, Reuveni, et al, 1990). Reuveni devoted much of his time and effort to put the scientific discoveries into practical use.

The tissue culture laboratory at the Department of Floriculture of the Volcani Institute was started in 1971 by Ben-Jaacov who studied at Cornell University in F.C.Steward laboratory.With Bob Langhans of the Department of Floriculture of that University, he developed a method for rapid vegetative propagation of chrysanthemum (Ben-Jaacov, Langhans 1970a, Ben-Jaacov, Langhans 1972). He reported his finding also in the local, Israeli flower growers' magazine, a fact that simulated interest in the subject by local scintist and farmers. (Ben-Jaacov, Langhans 1970b). Soon after (1972), the new tissue culture laboratory at the Volcani Institute, become the center for the development of the " Selected, Virus-free Spray Carnation Stock"

Selected, virus free spray carnation project at the Volcani Institute (Ben-Jaacov, et al, 1975).

Spray carnation production in Israel was increasing rapidly in the early seventies, until a catastrophic virus infection threatened this new crop in 1973/74. In three years the crash program on the "Development of Selected, Virus-Free Stock of Spray Carnation" replaced all the diseased plants and thus lead to recuperation and further rapid growth of the carnation industry which expanded to over 300 hectares in the early eighties.

This program consisted of the following stages: Selection of superior plants, excision of their meristem tips and culturing them *in vitro* . Hardening off and farther growth of the resulting plants in individual containers, in an insect proof greenhouse. Repeated virus testing followed by horticultural testing and selections. Offspring of each meristem were kept under separate identity (sub- clone) for further testing and follow up.

The development and successful introduction of the superior healthy stock material was a well planned, integrated, national project. The main team for this project included Hannah Lilien- Kipnis, E. Ephrat and J. Ben-Jaacov of the Dept.of Floriculture in close collaboration with scientists from the Dept. of Virology (Smookler, Loebenstein, 1974). The cooperation with The Plant Protection and Inspection Services of the Israeli Ministry of Agriculture was crucial in the development of local Certification Scheme which included a well defined program of three nursery generations: Nuclear, Foundation and Certified stock (Elhanan, 1984). During the first year both the Nuclear and the Foundation stocks were kept at the Volcani Institute. Later the Foundation stocks were grown at a few well controlled private nurseries. Only several years later was the Nuclear stock also moved to private, specialized nurseries (Ben-Jaacov, et al, 1975).

The effect of the above described program on the carnation industry in Israel is presented in Table 1. When comparing the annual increase in acreage, the number of flowers exported and the export value before and after the introduction of the selected, virus-free, stock the effect of this superior stock material is very clear. The increase in planted area from 1972/73 to 1973/74 was 48%. The yearly increase in yield, however was only 10%. On the other hand, from 1975/76 to 1976/77 the increase in area planted was 100% the increase in flower yield was 119% and the increase in revenue was 126% (Table 1). It may be assumed that the relatively larger increase in yield and export values were due to the superior stock material that was introduced in 1976 (Anonymous, 1979).

Development of commercial tissue culture laboratories in Israel (Anonymous, 1978).

The commercial potential of the tissue culture technology become obvious. Kibbutz Rosh Hanikra and especially their dynamic nurseryman Shimen Zakai applied to the Volcani Institute for help in training their future tissue culture personnel. Uri Levanoni and Avihai Ilan (now Managing Director of Rahan Meristem) worked and obtained their first experience in tissue culture at Reuveni's laboratory in Sub-tropical Horticulture and at the laboratory of the Floriculture Department. Kibbutz Rosh Hanikra built, the first commercial tissue culture laboratory in Israel in 1975 and called it "Rahan Meristem". Soon, Kibbutz Bit Haemek sent their trainees to the Volcani Institute and built their tissue culture laboratory: "Biological Industries, Plant Propagation Ltd."

In Israel today there are three large and additional five smaller *in vitro* plant propagating and Biotechnology oriented companies. These laboratories produce about 20 million plants annually (Abo El-Nil, Ilan 1998).

Table number 1: Spray Carnation acreage, the number of flowers exported and their dollar value in the years 1969/70 - 1977/78*.
Note (bold-shaded figures) the % annual increase in: acreage, number of flowers exported and export value before (1973/74) and after (1977/78) the introduction of selected, virus-tested stock in Israel.

Year	Acreage		Flowers exported		Export value	
	Hectars	Annual increase (%)	Milions	Annual increase (%)	Million $	Annual increase (%)
1969/70	7.3	-	6.7	-	0.3	-
1970/71	16.0	119	10.7	60	0.6	100
1971/72	17.0	6	16.6	55	1.0	67
1972/73	25.0	47	31.4	89	2.3	130
1973/74	37.0	**48**	34.6	**10**	2.5	**9**
1974/75	47.0	27	40.5	17	4.0	60
1975/76	65.0	38	63.7	57	5.3	33
1976/77	130.0	**100**	139.2	**119**	12.0	**126**
1977/78	210.0	61	-	-	-	-

* Source: Annual Reports for the years 1970-1979. The production and marketing board of ornamental plants. Tel Aviv, Israel. (Anonymous 1979).

References

Abu El-Nil M, Ilan A (1998) Abst. pp. 22. IX International congress on plant tissue culture,Jerusalem, Israel, June 14-19, 1998.

Altman A, Goren R (1971) Plant Physiol. 47: 844-846.

Altman A Goren R (1974) Plant Physiol.30: 240-245.

Anonymous (1978) The Grower. January, 19, 1978: 121.

Anonymous (1979) Annual Reports for the years 1970-1979. The production and marketing board of ornamental plants. Tel Aviv, Israel.

Bhojwani S S, Dhawan V, Cocking, E W, .(1986) Plant Tissue Culture - a Classified Bibliography. Elsevier 1986.

Ben-Jaacov J, Langhans R W (1970a) Proceedings of the 18th Int. Hort. Congress. P. 211. Tel-Aviv, Israel.

Ben-Jaacov J, Langhans R W, (1970b) Hamamot V'Prachim Sept. 1970: 5-7 (in Hebrew).

Ben-Jaacov J, Langhans R W, (1972) HortScience 7 (3): 289-290.

Ben-Jaacov J, Ephrat E, Lilien-Kipnis Hannah,(1975) In: Institute of Field and Garden Crops, Scientific Activities 1971-1974. pp 189-190. ARO Bet Dagan, Israel.

Elhanan S, (1984) EPPO Bull 14 (3): 37-379.

Gorn Nurit, Mendel K, (1972) Proceedings of the 18th. International Horticultural Congress . Ministry of Agriculture, Tel-Aviv, Israel.

Kochba J, Spiegel-Roy P, Safran Hannah, (1972) Planta (Berl.) 106: 237-245.

Kochba, Spiegel-Roy P, (1977) HortScience 12 (2): 110-114.

Lavee S, Messer Glenda, (1969) J. of Experimental Botany, 20(64): 604-14.

Messer Glenda, Lavee S, (1969) J. hort. Sci. 44: 219-33.

Quak F, (1972). In: Gorn, Mendel K, (1972) Proceedings of the 18th. International Horticultural Congress, Vol. III : 12-25. Ministry of Agriculture, Tel-Aviv, Israel.

Raveh Dina, Huberman E, Galun E, 1973) In Vitro, 9 (3) : 216-222.

Reuveni O, Adato Y, Lilien-Kipnis Hannah,.(1972) Forty-ninth Annual Date Grower's Institute, 1972.

Reuveni O, (1988) INIBAP- Proceedings of an international workshop held at Los Banos, Philippines 5-10 September, 1988.

Reuveni O, Shlesinger D R, Lavi U, (1990) Plant Cell, Tissue and Organ Culture 20: 41-46.

Sagawa Y, Kim K K.Kunisaki J T, (1970) In: Proceedings of the 18th Int. Hort. Congress. P. 210. Tel-Aviv, Israel.

Smookler M, Loebenstein G, (1974) Phytopathology 64: 979-984.

Umiel N, Goldner Rina (1976) Protoplasma 89: 83-89.

Vardi Aliza, Spiegel-Roy P, Galun E, (1975) plant Science letters, 4 (1975) 231-236.

Ziv Meira , Halevy AH, Shilo Ruth, (1970) Ann. of Botany 34 (136) : 671-676.

THE APPLICATION OF PLANT BIOTECHNOLOGY IN DEVELOPING COUNTRIES ON THE AFRICAN CONTINENT

J.A. Brink[1], B.R. Woodward[1] and E.J. DaSilva[2]
[1]ARC-Roodeplaat, P/Bag X293, Pretoria, Rep of South Africa and [2]UNESCO, Paris, France

Introduction

The current world population is approximately 5,8 billion and is expected to double by the year 2050 (James, 1997). The population increase in developing countries constitutes 97% of the global increase (Swaminathan, 1995). It is estimated that by 2050, 90% of the population will reside in the developing countries of the South. The challenge for the future therefore lies in global food security which will require a doubling of food production in the next 50 years to meet the needs of the population (James, 1997).

Africa is an extremely poor continent and is also the most vulnerable in terms of food security. During mid 1995 the population of Africa, with the fastest growth rate in the world, was 720 million, which constitutes 13% of the world population. At the present rate of growth this number will double over the next 2 years (Nyira, 1995). This population expansion makes it very difficult to achieve economic growth and also decreases food security and environmental sustainability. To meet Africa's food requirements it will be necessary to increase the efficiency of food production. Africa has lagged behind in the benefits of the "Green Revolution" of the 1970's and 1980's due to the limited utilization of high yielding varieties of maize, wheat and rice. Five key factors are required for improved crop production i.e. appropriate agrochemicals, sustainable irrigation, high yielding (adapted) varieties, efficient crop management as well as the application of plant biotechnology. The major increase in agricultural productivity will however be achieved through the direct use of genetic improvement and biotechnology (Villalabos, 1995). On a global scale, plant biotechnology has impacted almost exclusively on crops of high economic importance such as maize, wheat, soybean, sunflower, rice and potato. Other species, important to developing countries in Africa due to their socio-economic contribution, do not attract the interest of multinational Seed- and Biotechnology Companies. The result of this is that genetic and biotechnological improvement of these "neglected" food species can only take place locally within interested countries. A serious limiting factor to this is that research levels and use of biotechnology in African countries is not adequate to successfully improve

A. Altman et al. (eds.), Plant Biotechnology and In Vitro Biology in the 21st Century, 693–698.

local crops (Villalobos, 1995). Africa therefore also risks failing the opportunity to take advantage of biotechnology or the so called "gene revolution".

Biotechnology in the broad sense can be defined as "any technique that uses living organisms, or substances from these organisms, to make or modify a product, to improve plants or animals, or to develop microorganisms for specific uses" (Persly, 1992). In plant biotechnology, 3 broad applications can be identified, i.e. plant tissue culture, genetic engineering as well as plant molecular markers. The applications range from very simple to very sophisticated but can in many cases still be very appropriate for Africa.

According to Lynam (1995), investments in, and development of a plant biotechnological research capacity in Africa is best accomplished in phases. The first phase should be plant tissue culture, which is appropriate for Africa as many of the important food crops such as cassava, sweet potato, yam and banana are vegetatively propagated. The second phase is the application of biotechnological tools, which can improve the efficiency of selection and breeding of varieties/cultivars. A prerequisite for this phase is to have an operational breeding programme in place. The third phase is the development of capacity to produce transgenic plants. To successfully produce transgenic plants requires an adequate infrastructure, expertise in tissue culture and molecular biology, and a critical mass of researchers and sustainable funding to cover the high cost of this research. Only a few laboratories in Africa have the capacity to produce transgenic plants, but may still not have the ability to "commercialize" the product, or to ensure that these plants reach the end user, the African-farmer. To bridge this gap it is necessary to form partnerships with either seed companies, producer organisations or government institutions which can ensure that the sophisticated technology be delivered in the most well known and accepted technology known to farmers – the seed (James, 1996).

Constraints to the Application of Plant Biotechnology in African countries

Constraints include the following:

1. <u>Lack of resources for Plant Biotechnology</u>
   - *Human resource component*: A serious deficit of skilled expertise in plant sciences and biotechnology is evident in Africa. Working opportunities in Africa are often inadequate as funds to pay salaries and running costs of projects may also be limiting, resulting in reduced personnel. In addition, training gained abroad is often not applicable to local needs.
   - *Infrastructure component*: In several African countries, basic facilities even for the simplest tissue culture techniques such as micropropagation are not available. Well developed communication systems, telephones, fax and access to e-mail and Internet is also lacking. Furthermore, unreliable power supply in many African countries as well as the availability of chemicals and consumables for research may also be serious constraints.
2. <u>Selection of crops for research</u>

- Crops utilized in Africa are often not important enough to attract foreign investment in research. Important African crops such as cassava and sweet potatoes attract only donor funding
- Most commercialized world crops are not well adapted to Africa

3. Lack of linkages and networks
   - Funding to run networks is limited
4. Political / Legislature component
   - A lack of (national/regional) priority setting in agricultural research is evident in many African countries.
   - Biosafety regulations and legislation's are in place only in a few countries of Africa, which impairs the evaluation and release of genetically modified organisms (GMO's).
   - The lack of protection of intellectual property rights (IPR) in Africa will also hamper the development of new technologies, profitable inventions and investments.
5. General constraints
   - In many countries, there is a total absence of a viable seed industry or private initiative, resulting in the absence of an appropriate channel for the transfer of products developed through plant biotechnology.
   - Africa's natural resources are often exploited with no returns to African countries.

The current status of Plant Biotechnology in Africa

The current status will be reviewed in different regions of Africa. Only major applications will be included.

North Africa	
Country	**Area of Research**
Egypt	- Genetic engineering of potatoes, maize, tomatoes
Morocco	- Micropropagation of forest trees, date palms - Development of disease and stress tolerant plants - Molecular biology of date palms and cereals - Field tests for transgenic tomato
Tunisia	- Abiotic stress tolerance and disease resistance - Genetic engineering of potatoes - Tissue culture of date palms, Prunus rootstocks and citrus - DNA markers for disease resistance
West Africa	
Cameroon	- Micropropagation of cassava, cocoyam, banana
Ghana	- Micropropagation of cassava, plantain, yam, cocoyam, pineapple and cocoa
Nigeria	- Micropropagation of ginger, cassava, yam, banana, tree species
Nigeria: IITA	- Genetic engineering of cowpea, yam, cassava and banana - Molecular marker studies, cassava, yam, banana

	- Virus elimination studies on banana/plantain, yam, cassava , sweet potato - Germplasm conservation of cassava, *Manihot* species, yam, sweet potato, cocoyam, banana / plantain - Training and transfer of technology
East / Central Africa	
Ethiopia	- Tissue culture research applied to tef - Micropropagation of forest trees
Kenya	- Production of disease free plants and micropropagation of pyrethrum, bananas, potatoes, strawberries, sweet potato, citrus, sugar cane - Micropropagation of ornamentals (carnation, alstromeria, gerbera, anthirium, leopard orchids) and forest trees - *In vitro* selection for salt tolerance in finger millet - Transformation of sweet potato with proteinase inhibitor gene - Transformation of sweet potato with Feathery Mottle Virus, Coat protein gene (Monsanto, ISAAA, USAID, ABSP, KARI) - *In vitro* long term storage of potato and sweet potato - Marker assisted selection in maize for drought tolerance and insect resistance
Uganda	- Micropropagation of banana, coffee, cassava, citrus, granadella, pineapple, sweet potato and potato - *In vitro* screening for disease resistance in banana - Production of disease free plants of potato, sweet potato and banana
Southern Africa	
Malawi	- Micropropagation of potato, sweet potato, yam, banana, trees (*Uapaca*), tropical woody species

South Africa	- *Tissue culture*
	  - Production of disease free plants – potato, sweet potato, cassava, dry beans, banana, ornamental bulbs
	  - Micropropagation of potato, ornamental bulbs, rose rootstocks chrysanthemum, strawberry, apple rootstocks, endangered species, coffee, banana, avocado, blueberry, date palm
	  - Embryo rescue of table grapes, sunflower and dry beans
	  - *In vitro* selection for disease resistance – tomato nematodes, guava wilting disease
	  - Long term *in vitro* storage – several crops
	- *Genetic engineering*
	  - <u>Cereals</u>: maize, wheat, barley, sorghum, millet, soybean, lupins, sunflower
	  - <u>Vegetables and ornamentals</u>: potato, tomato, cucurbits, ornamental bulbs, cassava and sweet potato
	  - <u>Fruits</u>: apricot, strawberry, peach, apple, table grapes, wine grapes
	  - <u>Other</u>: forestry crops, sugar cane

	- *Molecular marker applications* - Cultivar identification – potatoes, sweet potato, ornamentals, cereals - Seed-lot purity testing – cereals - Marker assisted selection – maize, tomato - Markers for disease resistance – wheat, forestry crops - Commercial production of forest trees, indigenous ornamentals - Commercial contracts with European firms - *Commercialized transgenic products* - Yieldgard maize, Bollgard cotton
Zimbabwe	- Genetic engineering of maize and sorghum - Micropropagation of potato, cassava, tobacco, sweet potato, ornamental plants, coffee - Marker assisted selection
Zambia	- Micropropagation of cassava, potato, trees (*Uapaca*), banana

The following recommendations can be made for the successful implementation of Plant Biotechnology activities in developing countries:

- A National Biotechnology Policy Strategy should be formulated by the NARS of each country.
- With limited resources available, demand driven plant tissue culture should be phased in.
- Opportunities in training of plant biotechnologists in Africa by the UNESCO/BAC BETCEN and IITA (short term and medium term training) as well as African Universities (long-term training) should be utilized.
- Cooperation between universities, research institutions, the private sector and government agencies should be encouraged
- Biosafety regulations should be put into place to be able to benefit from the "gene revolution"
- The natural resources of each country should be developed and conserved.
- A country's own intellectual property rights should be protected along with the rights of other countries.
- Linkages between African countries as well as with the developed world should be stimulated through networks and joint projects.

Conclusion

Africa can learn from previous experiences and can achieve its goal through proper planning, facilitating cooperation among and between countries, and forming networks. There are tremendous opportunities to conserve and developed the natural resources of Africa's wild relatives of commercial crops, neglected and underutilized crops, and plants with pharmaceutical applications. This should be the primary focus of African researchers. Household food and health security can be insured through breeding of disease free, higher yielding plants, mass propagation of better quality plants and crops with specific desirable characteristics. Biotechnology can therefore play a role in commercializing crops, which can create jobs, earn foreign exchange and ensure a better quality of life for all.

References

James C. (1996). Chairman's Commentary. The First Decade of crop biotechnology. In: Advancing altruism in Africa. ISAAA Annual Report 1996.

James C. (1997). Progressing public-private sector partnership in International Agriculture Research and Development. In: ISAAA Briefs No 4, p. 1-32.

Lynam J.K. (1995). Building biotechnology research capacity in African NARS. In: Turing priorities into feasible programs. Proc. Of a Regional Seminar on planning priorities and policies for Agricultural Biotechnology, South Africa, April 1995, p. 33-40.

Nyira, Z.M. (1995). Need challenges and objectives for Biotechnology and agriculture in Africa. In: Turing priorities into feasible programs. Proc. Of a Regional Seminar on planning priorities and policies for Agricultural Biotechnology, South Africa, April 1995, p. 17-23.

Persly G.J. (1992). Beyond Mendel's Garden: Biotechnology in agriculture. In: Biotechnology enhancing research on tropical crops in Africa. CTA/IITA co-publication, p. 11-19.

Swaminathan M.S. (1995). Population, environment and food security. Issues in Agriculture, No 7. CGAIR, Washington DC.

Villalabos V.M. (1995). Biotechnology in agriculture: How to obtain its benefits while limiting risks. In: Induced mutations and molecular techniques for crop improvement. IAEA Publication, Austria, p. 477-486.

COMMERCIAL MICROPROPAGATION IN THE UNITED STATES, 1965-1998

Robert D. Hartman, Twyford International, Inc., Sebring, Florida, United States and Richard H. Zimmerman, United States Department of Agriculture, Beltsville, Maryland, United States

Summary

Commercial micropropagation started in the U.S. with orchids about 1965. The techniques used were soon adapted to many other crops, with important commercial production first occurring with foliage plants. Application to woody plants took longer, but now very significant quantities of fruit, ornamental and forest species are produced. In addition, large quantities of other important crops are being produced, e.g. potatoes, and steady expansion is occurring in the production of herbaceous perennial plants. As the advantages of micropropagated plants become evident and production efficiencies improve, possibilities develop for additional species to be propagated using this technology. The current level of output, 120 million plants per year, will almost certainly continue to grow in the next decade. Labor remains the major cost component of production which limits the numbers of plant varieties which are commercially propagated. Significantly lower labor costing airlift and flooding bioreactor techniques are rapidly becoming a commercial reality. As such, the numbers and kinds of plants which will be commercially micropropagated is being greatly enhanced and the need for lower value commodity crops being produced in low labor costing areas of the world is being greatly lessened. Thus international market control though intellectual property rights for proprietary crops and focusing on regional rather than international market needs for commodity crops will dominate future marketing and production issues within micropropagation companies.

Introduction

Micropropagation is a relatively recent addition to the methods employed by horticulturists, foresters and agronomists to multiply plants of a specific genotype (clone). In the U.S., application of the method to commercial production of plants dates back only about 30 years. The principles and techniques of micropropagation have been described numerous times but a clear and concise account of these is presented by Hartmann et al. (1990) in Chapters 16 and 17 of their textbook. The primary uses of micropropagation as currently applied by commercial laboratories in the U.S are: (1) mass propagation of specific clones, sometimes used for rapid introduction of new cultivars, (2) maintenance of pathogen-free (indexed) germplasm, (3) use as the initial step in a nuclear stock crop production system, (4) clonal propagation of parental stocks for hybrid seed production, and (5) year-around production of plants.

Historical background

Micropropagation of orchids was the key event in the founding of the micropropagation industry (G. Morel, 1974; R.J. Griesbach, 1986). Orchid micropropagation developed as an offshoot of early successes in culturing meristem-tips for production of virus-free plants. Culture of larger meristem-tips yielded protocorms that could be cut up and recultured. The rate of multiplication was far more rapid than other vegetative means of propagation and the aseptic techniques required were already familiar to those producers germinating orchid seed *in vitro*. At the time, it was not clearly understood by many practioners of micropropagation that using tissue culture methods did not automatically insure that the plants produced would be virus free. Thus, an

A. Altman et al. (eds.), Plant Biotechnology and In Vitro Biology in the 21st Century, 699–707.

unintended consequence of some of the early production, particularly by small laboratories, was the spread of virus-infected orchids, rather than the production of disease-free plants.

The meristem-tip culture technique, originally shown to be effective for *Dahlia* and potato (G. Morel, C. Martin, 1952, 1955), was applied to numerous other crops. The techniques developed in this research, together with the work on micropropagation of orchids and the development of a widely adaptable tissue culture medium (T. Murashige, F. Skoog, 1962), led to successful *in vitro* propagation of numerous crops. These early results were summarized in a classic review by Murashige (1974). Most of the early commercial applications of micropropagation were to foliage plants widely used in the rapidly expanding area of interior landscaping (R.D. Hartman, F.W. Zettler, 1986). Considerable success was also achieved with flower crops of various types (W. Oglevee-O'Donovan, 1986; D.P. Stimart, 1986).

Dr. Toshio Murashige of the University of California - Riverside, was the key figure in the industry during this early stage. He did research on many commercially important crops, lectured widely, advised commercial growers and, in particular, taught many training courses. In addition, many of his graduate students went on to make significant contributions in micropropagation and other areas of plant research using *in vitro* technology.

Many researchers contributed to the efforts to adapt micropropagation technology to woody plants. A key contribution was the report by Jones (1976) that phloridzin and phloroglucinol would stimulate axillary shoot production on apple shoot tips *in vitro*. This information spurred much research on apple and other woody plants, which led to the development of protocols successful for numerous species and cultivars. In the U.S., early applications were made to small fruit crops (H.J. Swartz, J.T. Lindstrom, 1986). Anderson (1975, 1978, 1984) and later Lloyd and McCown (1981) did much of the developmental work on micropropagation of *Rhododendron* and other ericaceous plants; this work was the foundation for commercial micropropagation of these crops in the U.S. (B.A. Briggs, S.M. McCulloch, 1983).

The use of micropropagation has been limited to relatively few vegetable crops, but it is advantageous in certain situations, particularly for asparagus, potato and sweet potato (Seckinger, 1991 and references therein). It has proven very useful in the propagation of parent lines for hybrid seed production in such crops as *Brassica oleracea* (broccoli and cauliflower) and asparagus (R.D. Hartman, unpublished). In the U.S., it is currently applied on a large scale to production of certified potato plants to be used for minituber production in the greenhouse or seed tuber production in the field. The technology has been reviewed by Jones (1988).

The growth of commercial micropropagation has not been steady. Early successes with many crops through the early 1980s led to a rapid increase in number and size of commercial laboratories by 1985. This quickly resulted in over production of certain crops and economic competition became very severe. Some laboratories closed and others consolidated within the industry. Following several years of relative stability and growth, the industry, particularly the segment producing foliage plants, was faced in 1992 with a rapid increase in the number of imported tissue cultured plantlets that were selling for very low prices. Further consolidation of the industry resulted. Quality control and crop scheduling problems with some imported material has reduced the impact of these imports in the past few years. Production once again seems to be increasing as demand continues to grow (R.D. Hartman, 1995).

Newly developed protocols continue to broaden the range of plants for which it is economic to use micropropagation on a commercial scale. Efforts with forest tree species have lagged because of two reasons: the species of interest are often recalcitrant in culture and the cost per propagule must be very low to make the method economically feasible. Nevertheless, significant progress has been made and small scale production trials can be anticipated in the near future. Already multiple millions of certain tree species such as *Paulownia elongata* are being produced annually (R.D. Hartman, unpublished).

Description of the industry

Commercial micropropagation laboratories are located throughout the U.S., with labs documented in at least 26 states. Most of these laboratories are situated near important production areas of the horticultural industries that they service, e.g. the foliage plant industry in central Florida and potato production areas in several states. Florida is the leading state in laboratory capacity and plants produced. Florida will expand its lead with the addition of the newest of the central Florida laboratories (more than 6,000,000/yr capacity), which is due to begin operations in May, 1998 (R. Strode, personal communication). California is the second state in both categories. Washington state is third in production, Oregon fourth and Colorado fifth. Only California and Florida have more than 10 laboratories, whereas Hawaii, Minnesota, Colorado, Montana, Washington and Oregon each have between 5 and 10 laboratories. The remaining laboratories are well distributed around the country. Within states, the laboratories are often clustered in certain areas. In Florida, the heaviest concentration is near Apopka, where many large foliage plant production nurseries are located. Nevertheless, there are large laboratories in northern and south central Florida that serve regional, national and international markets. In Washington and Oregon, the labs are located west of the Cascade mountains stretching from near the Canadian border to just south of Portland, Oregon. California labs are clustered mainly near the Pacific coast north of Los Angeles and near San Diego.

Commercial laboratories can be classified as independent or as part of or owned by another horticultural enterprise, such as a nursery or a potato seed production farm. The independent laboratories can be further classified based on the type of additional facilities available to the lab. Thus, laboratories without greenhouse facilities sell only unrooted shoots ("micros" or "microcuttings") or rooted plantlets ("rooted microcuttings") that have not been acclimatized. This type of production is decreasing in comparison to the situation 5-10 years ago. It is still, however, the product form of choice when it comes to international distribution because of lower freight costs and the product is subject to fewer quarantine regulations since it is free of soil. Laboratories with greenhouse facilities may still sell non-acclimatized shoots or plantlets, but are more likely to sell acclimatized plants, some of which may have been grown on for a period of weeks or months to obtain a higher plant value.

Laboratories associated with nurseries or other production facilities may produce only for internal use, so that the micropropagated plants are not offered for sale until they have been grown for several months to several years. A more general case is where part of the laboratory production is used within the company and the rest is sold on the open market. It would be rare for a laboratory of this type to have sales only to other producers.

Both independent laboratories and those associated with nurseries may sell on contract. Some independent laboratories sell only on contract, and some specialize in doing very small quantities. Because of the capital, management and labor intensive aspects of building and operating a commercial laboratory, many companies with small and mid-size laboratories are choosing to contract with larger laboratories to produce plants, rather than expand, when their business needs exceed their own laboratory capacity. Most production is sold on the open market rather than on contract, however. Even open market sales may be booked months in advance for crops or cultivars in high demand, so laboratories often have only limited supplies of plants available to meet unanticipated orders.

Laboratory sizes vary from production of only a few thousand plants per year to tens of millions per year. Small laboratories (fewer than 500,000 units per year) account for about 60% of the slightly more than 100 laboratories identified; 24 of these small labs produce only 50,000 units per year or fewer. About 30% are medium size laboratories (500,000-2,500,000 units per year); large laboratories (2,500,000-6,000,000 units) and very large laboratories (more than 6,000,000 per year) account for the remaining 10%.

Laboratory capacity is difficult to evaluate and an earlier estimate of 150 million units per year (R.H. Zimmerman, J.B. Jones, 1991) may now be somewhat low. Laboratory capacity not only depends on physical size of the facility but also how it is managed, the crops it is producing and the production technology being used. At present, it appears that capacity would be closer to 175 million units per year. This would include some laboratories not currently propagating plants that could quickly re-open if the management so decides, therefore one needs to distinguish between capacity and occupancy. Not included are several large new laboratories planned to open in 1997 and beyond. The eleven laboratory firms currently having the largest production would account for about 57% of the total capacity. The laboratories classified as medium would account for about 27% and the small labs would have more than 10%, with the remainder in labs not currently in production.

Production of micropropagated plants

Total production of micropropagated plants is now estimated to be about 120 million plants per year based on a detailed survey conducted in March and April, 1996 (Table 1). This quantity is considerably higher than earlier estimates of 61 to 75 million plants (J.B. Jones, 1986; R.H. Zimmerman, J.B. Jones, 1991; R.D. Hartman, 1995). Part of the discrepancy can be accounted for by the much broader range of crops counted in this report and inclusion of crops not micropropagated 5-10 years ago. Another factor is the greater number of laboratories identified and contacted to obtain the figures reported here.

The plants produced in tissue culture laboratories can be grouped according to their uses or area of horticulture. General categories are foliage and flowering plants for indoor use, annual and perennial flowers for the landscape, woody ornamental plants, vegetable crops, fruit crops, **forest crops** plus some miscellaneous crops.

Foliage plants comprise the largest category of plants micropropagated in the U.S, accounting for almost 54% of the total production (Table 1). Ferns constitute the largest group of foliage plants, accounting for about 11% of all U.S. micropropagated plants. *Spathiphyllum* production is nearly as large and other important genera would include *Syngonium*, *Dieffenbachia*, *Ficus*, *Calathea* and *Philodendron*.

Orchids are about one-third of the production in the category of greenhouse flowers and orchids (Table 1). This figure does not include any *in vitro* orchid seed germination, which is done on a much larger scale than micropropagation. *Gerbera* makes up nearly half of the production in this category. Other important crops include *Anthurium* and bromeliads.

Herbaceous perennials are a rapidly increasing segment of the production (Table 1). *Hosta* and *Hemerocallis* are micropropagated in the largest quantities; other perennials being produced include *Stokesia*, *Gypsophila*, *Heuchera*, *Tiarella*, *Lobelia*, *Leucanthemum* and *Rudbeckia*. Aquatic plants and species used for aquariums and wetland restoration and other ecological purposes are a rapidly growing part and may account for as much as 30% of this category. Finally, some annual flowers are micropropagated to provide seed parents for hybrid seed production (e.g. *Tagetes*, *Petunia*, etc.).

In the category of trees and shrubs, ericaceous plants (*Rhododendron* and azalea, *Kalmia*, *Pieris*, *Leucothoe*) account for more than 23% of those now micropropagated (Table 1). Trees account for another 22% and these include both shade and ornamental trees (*Acer*, *Betula*, *Magnolia*, *Malus*, *Prunus*, *Ulmus* and other genera) and trees grown for paper pulp or biomass (*Eucalyptus*, *Populus*). The remainder of the plants in this category are shrubs such as *Nandina*, *Syringa*, *Fothergilla*, *Hydrangea*, *Photinia*, *Viburnum* and numerous other genera.

Potatoes account for about 90% of the micropropagated vegetable crops (Table 1). Most of the potato production is plantlets grown on in a greenhouse for minituber production with a smaller proportion planted in the field for elite seed potato production. A limited amount of *in vitro* microtuber production is done by several laboratories. Other crops propagated in tissue culture include asparagus, rhubarb (*Rheum*), garlic and sweet potato (*Ipomoea batatas*). There is some production of cabbage and cauliflower parent plants for hybrid seed production.

Fruit crop production is primarily that of small (soft) fruits: raspberry, blackberry, blueberry, strawberry and currant (Table 1). Blueberry and raspberry are the main crops here. Strawberry production is rather limited and is done primarily to obtain stock plants for runner production; only a few laboratories are producing plants. Some limited production of fruit tree rootstocks occurs, but this is primarily for establishing stool beds for layering.

The miscellaneous crops include a wide assortment of ornamental and tropical fruit crops and specialty crops (Table 1). Limited production of bulbs, corms and tubers occurs, e.g. *Lilium* and *Gladiolus*. Banana is one of the major tropical fruits produced, but most of this production is used in home gardens in southern states, especially Florida. A limited amount (± 50,000 plants) of *Citrus* rootstock has been micropropagated (R.D. Hartman, unpublished), however there may be a significant increase in this activity in the near future (Bowman et al, 1997). The integration of micropropagation into citrus breeding and release programs will greatly enhance the citrus rootstock base resulting in increased potential profits for citrus growers. Other specialty crops include peppermint, spearmint, wasabii or Japanese horseradish (*Wasabia japonica*), and sugarcane.

Other sources of micropropagated plants

In addition to the plants micropropagated in the U.S., significant quantities of tissue cultured plants are imported. These are primarily foliage plants (ferns, *Spathiphyllum*, *Syngonium*) and orchids. No attempt has been made to quantify these imports, but they probably exceed 20 million plants per year. With unused U.S. lab capacity one might question why the U.S would import micropropagated plants. The principal reason is cost. Average U.S. laboratory labor costs are in the range of \$5.50-\$8.00/hour range. Benefits add another 30%-35%. These costs put many plants with good micropropagation protocols and market potential out of the reach of domestic producers since labor costs for many Latin American and Asian producers are in the range of several dollars/day to $1.00-$2.00/hour range.

Costs

Production costs associated with applying current micropropagation technology limit its broader use in the U.S. and subjects it to lower cost imports. In his 1995 analysis, Chu (1995) gave a laboratory production cost breakdown for a microcutting of an average U.S. laboratory as follows:

Item	Est. Cost	% of Total
Direct Labor	$0.078	49%
Supervision	$0.024	15%
Materials	$0.015	9%
Utilities	$0.013	8%
Depreciation	$0.021	13%
Other	$0.009	6%
Total	$0.160	100%

With labor and supervision costs nearly 65% of total cost of production, U.S. laboratories have been very vulnerable to competitive imported product from lower labor costing areas of the world such as India, Malaysia, Thailand, Vietnam, China and certain areas of Central and South America. To deal with this competitive threat, U.S. laboratories began focusing on the development and marketing of proprietary varieties in recent years. Proprietary, patent protected, varieties not only tend to perform better, but they also afford U.S. producers market protection because, without a propagation license, proprietary varieties can be prevented from entering the U.S. This of course is only a temporary solution as breeding is done throughout the world, including areas of lower labor costs, and nothing can prevent non U.S. countries from inventing and protecting superior plant varieties and exporting them to the U.S. The only real long term solution for U.S. producers is to largely eliminate the labor component of cost of production, as current production technologies, though efficient compared to many other modes of propagation, are labor and management intensive. Systems which greatly limit these two cost components are being commercialized by several U.S. laboratories.

Future and Automation

Temporary immersion (flooding) and air lift bioreactors will play a significant role in the future of micropropagation. Alvard et al (1993) did a comparative analysis of five different propagation systems and demonstrated a 2-5 times greater accumulation of dry matter in bananas cultured in aerated liquid medium and those subjected to temporary immersion when compared to conventional gelled or liquid medium production techniques. They also documented a significant increase in multiplication rate, especially in the temporary immersion production system.

The use of bud (meristematic) clusters and somatic embryogenesis techniques in conjunction with air lift bioreactors has enormous potential to mitigate labor and supervision costs. These systems are being developed both privately and for commercial sales. Osmotek is leading the way in developing production systems for sales to commercial laboratories. Available from Osmotek (Osmotek Ltd. Israel) is a description of methodology which employs multiplication of meristematic or bud clusters on interfacial membrane rafts and in plastic film air lift bioreactors, separation of the clusters by a mechanical cutting device and subculture to multiplication or to a growth medium which they have proven useful for a number of crops. Comparative cost calculations indicate a savings of approximately 52% ($0.0767 versus $0.160) over traditional systems with 80% of the savings coming in the areas of direct labor (66%) and labor supervision (14%) (Osmotek Ltd, personal communication).

Because of the tremendous increase in yields and handling efficiencies, the use of these technologies is rapidly becoming a commercial reality and will have the effect of greatly leveling international competitiveness as far as labor costs are concerned. An excellent overview of the status of automation in micropropagation has been provided by J. Aitken-Christie et al (1995).

Conclusions

A significant commercial micropropagation industry has developed in the U.S. in the past 30 years. Currently more than 100 laboratories are producing at least 120 million plants per year in a wide range of ornamental and food crops. Despite past problems from excess laboratory capacity, over production of some crops, and strong competition from low-priced imports, the industry continues to grow. With emphasis on the production of proprietary varieties and the implementation of newer automated micropropagation technologies, growth is expected to substantially increase in the future.

Acknowledgments

Dr. Robert Griesbach is gratefully acknowledged for his significant assistance in obtaining the data on production of orchids and other flower crops and for many useful discussions. Dr. Mark Bridgen of the Department of Plant Science at the University of Connecticut, Storrs, kindly shared the information in his 1994 Directory of plant Tissue Culture Laboratories in the United States of America and Canada. The cooperation and assistance of the operators of more than 100 U.S. commercial micropropagation laboratories in providing production information is gratefully acknowledged.

References

Atiken-Christie J (1995) In: Atiken-Christie J et al. (eds.). Automation and environmental control in micropropagation, pp 1-18, Kluwer Achademic Publishers, Dordrecht, The Netherlands.
Alvard D et al. (1993) Plant Cell Tiss. Org. Cult. 32,55-60.
Anderson WC (1975) Comb. Proc. Int. Plant Prop. Soc. 25,129-135.
Anderson WC (1978) Comb. Proc. Int. Plant Prop. Soc. 28,135-139.
Anderson WC (1984) J. Amer. Soc. Hort. Sci. 109,343-347.
Bowman KD et al. (1997) In: Proceedings of the Florida State Horticultural Society, Vol. 110 (in press).
Briggs BA, SM McCulloch. (1983) Comb. Proc. Int. Plant Prop. Soc. 33,239-248.
Chu I (1995) In: Atiken-Christie J et al. (eds.). Automation and environmental control in micropropagation, pp 19-27, Kluwer Achademic Publishers, Dordrecht, The Netherlands.
Griesbach RJ (1986) In: Zimmerman RH et al. (eds.). Tissue culture as a plant production system for horticultural crops, pp 343-349, Martinus Nijhoff Publishers, Dordrecht, The Netherlands.
Hartman RD (1995) Greenhouse Product News 5(12),10-14.
Hartman RD, FW Zettler. (1986) In: Zimmerman RH et al. (eds.). Tissue culture as a plant production system for horticultural crops, pp 293-299, Martinus Nijhoff Publishers, Dordrecht. The Netherlands.
Hartmann HT et al. (1990) Plant propagation, principles and practices. 5th Edition. Prentice-Hall, Inc., Englewood Cliffs, NJ.
Jones ED (1988) Amer. Potato J. 65,209-220.
Jones JB (1986) In: Zimmerman RH et al. Tissue culture as a plant production system for horticultural crops, pp 175-182, Martinus Nijhoff Publishers, Dordrecht, The Netherlands.
Jones OP (1976) Nature 262,392-393.
Lloyd G, B McCown (1981) Comb. Proc. Int. Plant Prop. Soc. Proc. 30,421-427.
Morel G (1974) In: Withner C (ed.). The orchids: scientific studies, pp 169-222, John Wiley and Sons, New York.
Morel G, C Martin. (1952) C. R. Hebd. Seances Acad. Sci. 235,1324-1325.
Morel G, C Martin. (1955) C. R. Seances Acad. Agr. France 41,472-475.
Murashige T (1974) Annu. Rev. Plant Physiol. 25,135-166.
Murashige T, F Skoog. (1962) Physiol. Plant. 15,473-497.
Oglevee-O'Donovan W (1986) In: Zimmerman RH et al. (eds.). Tissue culture as a plant production system for horticultural crops, pp 119-123, Martinus Nijhoff Publishers, Dordrecht, The Netherlands.
Seckinger G (1991) In: Debergh PC, RH Zimmerman (eds.) Micropropagation: technology and application, pp 265-284, Kluwer Academic Publishers, Dordrecht, The Netherlands.
Stimart DP (1986) In: Zimmerman RH et al. (eds.). Tissue culture as a plant production system for horticultural crops, pp 301-315, Martinus Nijhoff Publishers, Dordrecht, The Netherlands.
Swartz HJ, JT Lindstrom. (1986) In: Zimmerman RH et al (eds.). Tissue culture as a plant production system for horticultural crops, pp 201-220, Martinus Nijhoff Publishers, Dordrecht, The Netherlands.
Zimmerman RH, JB Jones. (1991) In: Debergh PC, RH Zimmerman (eds.) Micropropagation: technology and application, pp 173-179, Kluwer Academic Publishers, Dordrecht, The Netherlands.

Table 1. Production of micropropagated plants in the United States of America by geographic region and type of crop. Data are based on a survey conducted in March, 1996. Numbers are thousands of plants.

Crops	Eastern U.S.	Florida	West Central U.S.	Pacific Northwest	California & Hawaii	Total
Foliage plants	1,200	47,975	0	0	14,520	**63,695**
Greenhouse flowers & orchids	278	5,178	495	0	5,346	**11,297**
Herbaceous annuals & perennials	4,388	1,030	1,320	2,080	630	**9,448**
Trees & shrubs	1,434	2,480	1,400	7,850	2,130	**15,294**
Vegetables	3,692	0	4,849	1,130	3,191	**12,862**
Fruits	1,431	10	50	1,970	260	**3,721**
Miscellaneous	25	975	1,715	1,270	560	**4,545**
Total	**12,448**	**57,648**	**9,829**	**14,300**	**26,637**	**120,862**

SECURING COST EFFECTIVE PRODUCT QUALITY, AN ESSENTIAL FOR SUSTAINABLE BUSINESS

D. P. Holdgate and E. A. Zandvoort
Integrated Plant Technology Applications, 53 Church Road, Tasburgh, Norfolk NR15 1ND, UK

1. Introduction.

The need to decrease crop losses, caused by viral infections, was the spur for commercial application of *in vitro* culture developments and four phases may be identified in progressing micropropagation applications: 1. establishment of technical capability and credibility, 2. widening of crop application and business establishment, 3. expansion and rationalisation, and 4. business maturation.

2. The establishment of technical capability and credibility.

The application of tissue culture procedures to crop productivity started in the mid 50's on the confirmation that virus-free plants could be secured through meristem tip culture (Morel and Martin, 1955). The procedure has been applied to many crops including potato and Chrysanthemum for production of certified propagules for commercial crops (Quak, 1977). The use of meristem tip culture for virus elimination in *Cymbidium* (Morel, 1960) coincidently gave rise to juvenile protocorm structures *in vitro*, which through subdivision gave rise to large numbers of healthy, true-to-type plants. Commercial micropropagation of selected clones of *Cymbidium* was started by Vacherot et Lecoufle in France in 1960, followed in the UK by McBeans Orchids Ltd. 1964 and in 1966 by Twyford Laboratories Ltd.(TLL) at which the first author DPH was managing director in the period 1965-82. TLL applied *in vitro* propagation techniques to pineapple for plantation trials in 1966 and to cut flower and pot plant species which were incorporated into the established orchid business in 1968 to create the first commercial multi-genera micropropagation company. Cooperation between TLL and breeders facilitated clonal cut flower productivity trials, leading to major expansion of for instance the lily bulb industry and gerbera businesses.

3. Expansion of crop application and establishment of the business.

A purpose built facility by TLL at the end of 1971 enabled further species to be introduced and sales increased from about 0.5 million to 3.5 million in 1975. Significant competitive operations started to appear in The Netherlands, Belgium, and Germany and also in France and Italy from 1973. Micropropagation in the fruit sectors led to the rapid international distribution by TLL of the virus-indexed East Malling bred dwarfing rootstocks of apple, cherry and plum from the mid 1970's; whilst potentials for strawberry were pursued in Belgium and Germany. France and Italy progressively developed high quality production centres for top, stone and soft fruit plants. Application to vegetables was less dramatic, but was used in breeding and seed production programmes.

By 1979 the total volumes sold approached 20 million plants, and the capability to produce healthy, true-to-type plants had been established in many species. This enabled genetic based gains of breeding programmes in some crops to be maximised and led to increased yields from reduced planting densities and improved quality and better prices.

A. Altman et al. (eds.), Plant Biotechnology and In Vitro Biology in the 21st Century, 709–712.

4. Commercial expansion and rationalisation.

Major expansion of the industry from 1980 onwards is illustrated by the production figures from The Netherlands (fig. 1). The decrease of production in the 90's is mainly due to a relocation of the activities to other countries (eg Poland, India, South Africa) and subsequent imports from those countries (fig. 3). Figure 2 illustrates performance in other countries. The output in France, based on large numbers of pot plants, roses and fruits has decreased dramatically. Production in Germany and Belgium has increased through niche markets (shrubs and perennials) and new technologies (bamboos). Italy and the UK appear constant in production levels, although new investments in the UK may have a significant effect in the near future. Total production in the main centres of Western Europe appears constant, with some changes in bulbs and orchids.

Figure 1
Micropropagation in The Netherlands 1980 - 1995 (Pierik and Ruibing, 1997)

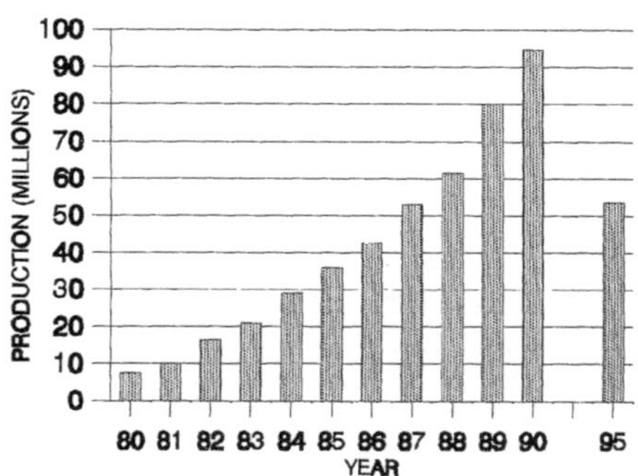

Figure 2
Micropropagation in main centres of W. Europe in 1988, 1993 and 1997

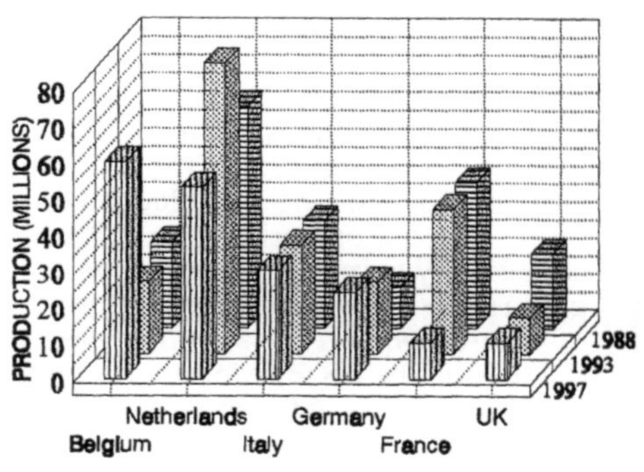

Figure 3
Production and imports of micropropagules in The Netherlands (Pierik and Ruibing, 1997)

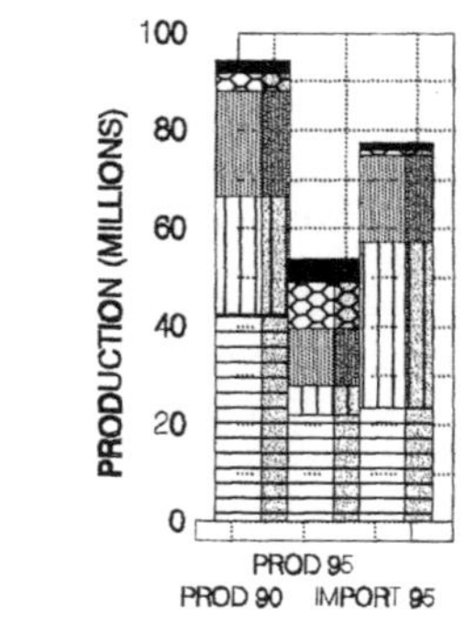

Figure 4
Micropropagation in Western Europe Production per type, 1993 and 1997

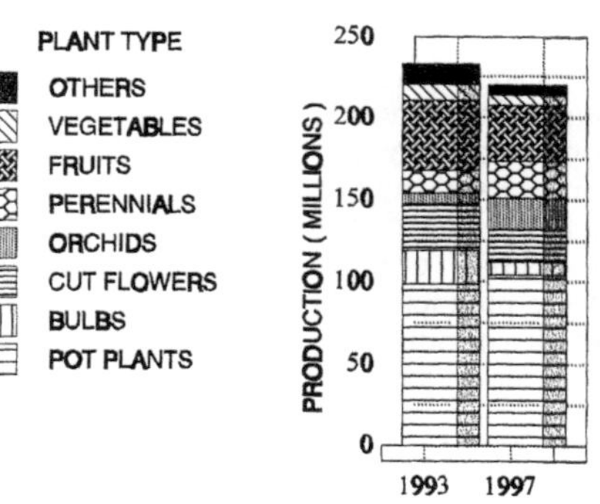

5. Threats and Opportunities

In many developing countries micropropagation has been identified as an opportunity to enter a labour intensive industry with local and relatively high value export markets. The practical realisation of this is limited. Experience with direct imports has been poor to very poor, and expensive. Common problems are weak cultures, poor health status, genetic and morphologic variations, failure to conform to delivery requirements, and incorrect labelling. The use of antibiotics or other biostats to mask systemic infections and, or, contamination levels has been frequent. Until quality standards are improved, entry to the important European market will be limited. Distribution of named varieties propagated to low standards, without owner approval, may cause reputation difficulties for European breeders, but the livelihood of many farmers in several countries can be destroyed.

During recent surveys on tissue culture needs, applications and requirements, serious misconceptions regarding the technology were found at official administrative and technical decision making levels, and by owners and managers of commercial operations. Micropropagation was considered as an old, well known technology, cheap to install and perform, and which automatically produces true-to-type disease free plants. Such beliefs have adverse effects on the infrastructure, regulatory matters, and viable capacities at research and production levels. Both commercial and official programmes, initiated to realise targets designated as of high priority, have failed to reach the objectives due to one or more basic errors. For example: poor initiation procedures, inappropriate reliance on published protocols, poor management of cultures, services and materials supply, basic errors in facility design, and failure to comprehend the reason, needs and nature of the products required by the farmer.

6. The future.

The principles and technologies for securing and maintaining healthy cultures (Holdgate 1977, Holdgate and Zandvoort 1997), should be implemented by all those using tissue culture for development and sustainable applications. This is fundamental and outweighs, on a simple cost basis, all other operational considerations.

Questions on mechanisation and automation frequently arise as an important issue. A number of routine procedures are mechanised and will be extended, for example to diagnostics, post culture selection and handling. Substantial automation at the culture cycle level using image analysis and laser cutting has been realised at the pilot scale level (Holdgate and Zandvoort 1992). Automation may appear highly attractive, but all biological procedures and their management must be of the highest standards for cost benefit realisation.

The use of liquid media has been the subject of much work over the last 30 years. Stationary vessel, rotating flask or bubble aeration systems were favoured by early micropropagators of orchids, bromeliads and ferns. Complex to simple airlift and gas diffusion bioreactors, ebb and flood, and raft systems have appeared and all claim better growth rates than in semi-solid medium. The various containers and wider range of culture products, including multiple nodules, must be assessed for ease of handling, maintenance of asepsis, and final plant quality. Embryogenesis and encapsulation to generate artificial seed has been a major topic of research. However, only when the product is appropriate for the habitat, planting and performance needs of the grower and capable of reaching crop maturation, can the application be viable.

Micropropagation is identified as of major importance in many countries, but in reality the performance is often at a low level of efficiency due to poor target control, inadequate culture management and high culture loss. This position is serious since cost effective output is essential irrespective of country, funding source or nature and purpose of the activity. The IPTA consultancy has introduced a management system, referred to as REPS (Real End Product System). REPS is designed to secure operational cost benefits for clients and to bridge the gap between academic training and application in product development and production. REPS is a pro-active management system covering:(i) facility design (ii) technical and operational procedures (iv) staff training and management (v) cost management (vi) customer relations. Although best introduced at the initial set up of a project REPS can be applied to existing operations to help secure short and long term cost-effective production improvements, and progression to a sustainable business.

Micropropagation operations face, and must address, all those production, financial, marketing, personnel and legal restraints, and customer and public awareness faced by other industries. Application of *in vitro* cultures must now be classified as commercially mature and have the responsibility to utilise technologies correctly to provide sustainable performance and quality products.

7. Conclusions.

The use of commercial micropropagation in Western Europe is now 30 years old and estimated to produce in the order of 220 million units, and utilise a further importation of not less than 100 million units. The greatest application remains in the ornamental sector (foliage plants 100 M, lily 40 M, gerbera 35 M, orchids 30 M). These exclude the enormous quantities of certified commercial propagules made from *in vitro* derived diseased indexed stocks of for example potato, chrysanthemum, carnation and strawberries. Experience has proven that the industry can support sustainable businesses, but only if cost effective product quality can be secured.

8. References.

Holdgate, D.P. (1977) In J. Reinert and Y. Bajaj (eds) Applied and Fundamental Aspects of Plant Cell, Tissue, and Organ Culture. pp 18-43 Springer-Verlag, Berlin

Holdgate D.P. and Zandvoort E.A. (1992) In K.Kurata and T.Kozai (eds). pp 297-311. Transplant Production Systems. Kluwer Academic Publishers, Dordrecht, The Netherlands.

Holdgate D. and Zandvoort E.A. (1997) In Pathogen and Microbial Contamination Management in Micropropagation. A. Cassells (ed) pp 15-22 Kluwer Academic Publishers, Dordrecht, The Netherlands.

Morel, G. (1960) Producing virus free cymbidiums. Am. Orchid Soc. Bull.29, 495-497.

Morel, G. and Martin, C. (1955) Guérison de pommes de terre atteintes de maladies à virus. Compt. Rend. 41, 472 - 474.

Pierik R.M.L. and Ruibing, M.A. (1997) Developments in the Micropropagation Industry in The Netherlands. Plant Tissue and Biotechnology 3, 152-156.

Quak, F. (1977) In J.Reinert and Y.Bajaj Meristem Culture and Virus-Free Plants. (eds) Applied and Fundamental Aspects of Plant Cell, Tissue, and Organ Culture. pp 598- 615 Springer-Verlag, Berlin.

SCOPE OF THE TISSUE CULTURE INDUSTRY IN INDIA

A.F.Mascarenhas

Tissue Culture Pilot Plant, National Chemical Laboratory, Pune 411 008, India.

INTRODUCTION AND HISTORY

Commercial tissue culture was born in India in 1987 when A.V.Thomas and Company Kerala (AVT) established their first production unit in Cochin for clonal propagation of superior genotypes of selected cardamom plants. This was based on a small scale laboratory technology indigenously developed and released to the company by the National Chemical Laboratory (NCL) Pune, India. This technology was later refined and scaled up by the company in collaboration with a UK based firm to make the method more production oriented so that the process is cost and quality effective and could guarantee an efficient delivery system. This pioneering effort in the application of biotechnology research by AVT was followed by several other entrepreneurs, who entered the field. In 1988, a second company Indo-American Hybrid Seeds at Bangalore, Karnataka, who were in the nursery business in hybrid flowers and vegetables, imported a tissue culture laboratory and green houses with a capacity of 10 million plants/ annum.

By 1988 the number of units increased to 4, and later showed a rapid rise till 1996 to around 50 labs (Mascarenhas 1991). According to a compilation made by David Constantine (1997) the number now stands at 105. However, many of these labs have closed their shutters or are in the process of doing so, leaving barely 40-50 units which are still operational. This is similar to the situation that prevailed in the early nineties in the Netherlands and USA (Pierik, Ruibing 1997; Zimmerman 1996). The main reasons for the closure is that entrepreneurs have now realised the high risks involved in this business. The major difficulties being in maintaining sterility and quality of propagules and in obtaining a steady market. This paper describes the status of the tissue culture industry in India.

SOURCE OF INFORMATION AND LOCATION OF LABORATORIES

The data represents a cross section of over 40 known micropropagation units who were contacted for first hand information by, direct visits and discussions, telephonically, or by other means of communication. Additionally a survey was made of existing literature published on this subject. The Directory of tissue culture laboratories in India by Constantine (1997) and the paper by Govil and Gupta (1997) was used as a base for compiling the distribution of units in different states of India and is presented in Table 1. Commercial micropropagation laboratories are located throughout India with labs documented in at least 14 States. Most of these units are situated in locations close to international airports. The state of Maharashtra has the largest number of labs followed by Karnataka, Andhra Pradesh, Tamil Nadu, and New Delhi. A smaller number of units are also situated in other states.

A. Altman et al. (eds.), Plant Biotechnology and In Vitro Biology in the 21st Century, 713–720.

Table 1 :
Geographical Distribution and Production Capacities of Some Commercial Micropropagation Units (1996)

State	Units	Annual Production Capacity (in millions)
Maharashtra	25	31
Karnataka	9	31
West Bengal	9	5
Kalimpong (Darjeeling)	4	1
Andhra Pradesh	6	10
Delhi	4	7
Tamil Nadu	4	27
Kerala	4	25
Harayana	3	13
Gujarat	3	12
Uttar Pradesh	3	25
Punjab	1	2
Himachal Pradesh	1	5
Orissa	1	1
Rajasthan	1	1

Ref. : Govil & Gupta, 1994, 1997; Constantine 1997.

CAPACITY OF LABORATORIES

Laboratory sizes (Table 2) vary from production capacity of less than 0.1 million plants per year to a few with over 20 million. The laboratories have been categorized depending on their size and production capacity as small, medium and large laboratories. The smaller laboratories (less than 1.0 million) account for about 70% of the total laboratories listed. Medium size laboratories which are around 20% produce between 1 to 10 million plants. A small percentage of laboratories have capacities of 10-20 million. Laboratory capacity has been difficult to evaluate and to correlate with their actual output. At present it appears that the production in India is not over 50 million plants annually. Some of the facilities compare with the best laboratories in the world. However, a majority of these laboratories have not been put to their full capacity utilisation with the result that actual production is much lower than what is reported (Govil and Gupta 1997.)

Table 2
Production Capacitywise Distribution of Labs

Production Capacity (M)	No. of Units
15-20	4
10-15	3
5-10	12
3-5	4
1-3	7
Below 1	Most labs (70%)

FUNDING AGENCIES

Inspite of several important breakthroughs achieved in the area of plant tissue culture by scientists in India, there are very few commercial units in India set up with indigenous know how. Although NCL, Pune has helped entrepreneurs to set up laboratories as early as mid 1980's and even sold protocols , many entrepreneurs obtain the entire technology package consisting of plans for a lab, greenhouse, protocols for plant production etc. on a turn key basis from foreign collaborators. Earlier foreign collaborators also promised a market. This is now discontinued because of many failures due to packaging, transportation and quality of plants when received by the foreign buyers.

Realising these difficulties and recognising the fact that for success of a tissue culture venture a complete package for production, processing, transportation and marketing, is essential; several funding agencies (Banks, Department of Biotechnology, Council of Scientific and Industrial Research, Department of Atomic Energy, Indian Council of Agricultural Research and others came forward to finance research projects in different laboratories in the country.

ROLE OF DEPARTMENT OF BIOTECHNOLOGY, NEW DELHI -

The Department of Biotechnology (DBT), has funded programmes on *in vitro* studies for agricultural crops and forest trees in a big way. In view of their commitment and interest, their programmes will be covered in greater detail. All the information which will be presented is through the courtesy of DBT. DBT funding in micropropagation has focussed on the following areas.

1. Research and Development of protocols
2. Setting up Pilot plant units
3. Scale up of protocols already available with the laboratories and field evaluations.

Considering the need for improved high yielding forest trees in large numbers for afforestation, and also of horticultural and plantation crops of desirable characteristics for the domestic and export markets, DBT initiated and funded tissue culture programmes in many laboratories located in different states of the country. (Table 3).

Table 3:
Number of Projects Funded – Statewise

State	No. of Projects Funded
Karnataka	6
Kerala	15
Tamil Nadu	8
Assam	4
Himachal Pradesh	1
Delhi	4
Uttar Pradesh	10
Orissa	3
Gujarat	2
Rajasthan	5
Madhya Pradesh	4
West Bengal	4
Bihar	1
Maharashtra	3
Jammu & Kashmir	2
Chandigarh	1
Andaman & Nicobar Islands	1

Initially most of the projects were restricted to laboratory scale production. Presently a few protocols have been extended to field evaluation studies particularly in the two tissue culture pilot plants, at NCL and Tata Energy Research Institute (TERI), where over 5 million plants of forest trees have been produced and field tested in many locations in different states of India. As a result of funding, successful protocols for tissue culture propagation have been developed by some laboratories which are available for large scale production. The plants cover forest and fruit trees, plantation crops, mangroves, orchids, and others. The main objective of the research is to increase productivity per unit area, be it biomass, yield and quality of fruits, or other traits.

In forest trees the main species for propagation have been *Eucalyptus tereticornis*, *E. camaldulensis*, Bamboo species, *Tectona grandis*, *Anogeissus* spp, *Populus deltoides* etc. Micropropagation protocols have also been developed for horticultural and plantation crops.

DBT have also identified some priority crops based on their economic importance for the country and multi institutional network programmes are being conducted. Some of the crops identified are, mango, cashew, cotton, grapes, forest trees, foliage plants and others.

In addition India is in a unique position because of its diverse agroclimatic zones and rich heritage of germplasm. This has permitted setting up of tissue culture laboratories near the growing areas for selective crops. For instance the Rubber Research Institute and the Plantation Crops Research Institute are situated in Kerala the home of these plants. The Indian Institute of Horticultural Research is at Bangalore. For forest trees there are many laboratories located in different parts of India.

HIGHLIGHTS

As a result of DBT funding of about 150 R&D projects, protocols for about 20 species have been developed for regeneration of elite planting material of forest trees, horticultural and plantation crops (Table 4). Two tissue culture pilot plant facilities have been set up for forest trees, one at NCL, Pune and the other at TERI, Delhi, where over 5 million plants have been produced and are being field evaluated at many locations all over the country in approximately 3500 Ha. These Tissue Culture Pilot Plants are now converted to Micropropagation Technology Parks for transfer of proven technologies to entrepreneurs and for generation of skilled manpower. Funding by DBT since 1989 has been to the tune of Rs.45 crores , (11.2 million US $) for R & D programmes for tissue culture which also includes the two pilot plants.

Table 4:
Plants for which Protocols have been Developed (DBT or other Funding)

Species	Contact Scientist	Institute
I . Forest Trees		
Populus deltoides	H.C. Chaturvedi Vibha Dhavan	NBRI, Lucknow TERI, New Delhi
Alnus nepalensis	D.R. Sharma	YSP University, Solan
Dendrocalamus hamiltonii	LMS. Palni	Almora
Boswellia serrata	S.D. Purohit	MLS University, Udaipur
Pinus kesiya	Promod Tandon	NEHU, Shillong
Anogeissus pendula *Anogeissus latifolia*	Vibha Dhavan N.S. Shekhawat	TERI, New Delhi JNV University, Jodhpur
Dendrocalamus strictus *Tectona grandis* *Eucalyptus camaldulensis* *Salvadora persica* *E. tereticornis*	R.S. Nadgauda R.S. Nadgauda Laxmi Sita	NCL, Pune NCL, Pune IISC, Bangalore
Sapindus trifoliatus	A.R. Mehta	GAU, Anand

II. Fruit Trees		
Apple CVS Golden delicious Tykeman early Worcester	D.R. Sharma	YSP University, Solan
Pomegranate (*Punica granatum*)	R.S. Nadgauda	NCL, Pune
III. Orchid species (nine species)	S.P. Vij	Punjab University, Chandigarh
IV. Plantation Crops		
Piper nigrum	P.A. Nazeem K.V. Peter V. J. Philip	KAU, Trichur NRCS, Calicut Calicut University, Kerala
Elettaria cardamom *Zingiber officinalis* *Curcuma Longa*	R.S. Nadgauda	NCL, Pune
Crocus sativum	B.L. Kaul	RRL, Jammu
Bunium persicum	A.K. Waklu	Jammu University, Jammu
V. Desert plants		
Ceropegia bulbosa	M.S. Shekhawat	JNV University, Jodhpur

Abbreviations :

1) GAU : Gujarat Agricultural University
2) IISC : Indian Institute of Science
3) JNV : J. N. Vyas University
4) KAU : Kerala Agricultural University
5) MSSRF : M.S. Swaminathan Research Foundation
6) MLS : M L. Sukhadia University
7) NBRI : National Botanic Research Institute
8) NEHU : North Eastern Hill University
9) NRCS : National Research Centre for Spices
10) RRL : Regional Research Laboratory
11) SPIC : Southern Petrochemicals Industries Corporation Ltd.

CONSTRAINTS IN INDIA FOR THE TISSUE CULTURE BUSINESS –

Some of the constraints that have been identified are given below.

1. India's trade in tissue culture plants is with free sale cultivars, which are mainly foliage plants such as cultures of *Ficus*, *Syngonium*, *Spathyphyllum* and others which are the easiest type of product with which to enter the market but are the least profitable.
2. It has been difficult to fully quantify the quality of micropropagated plants. Features such as the height of shoots, stem callipers, number of leaves, number of shoots per clump, number of roots per plant, diameter and weight of bulbs have not been used to define quality by most companies.
3. Timely delivery - not adhered to.
4. Careful costing has not aimed at being competitive and still getting the best quality product.
5. Establishment of professional credibility, has not been developed by acquiring a reliable track record in the market.
6. Breeding is the slowest and most expensive method of developing a proprietary product. Companies have not developed breeders proprietary

cultivars by breeding or selection of a chance mutation in a large production unit or finding new cultivars from the wild.

Corrective measures are underway.

PRESENT STATUS AND FUTURE SCOPE -

1. Banana is the major crop presently produced by tissue culture for the domestic market and will remain the most reliable assuring a continuous market. Tissue culture plants have been tested in several laboratories and found to give higher yields. The total area under cultivation in India is estimated at 390000 Ha. (Singh and Chadha, 1997). In a single state of Maharashtra, the area under banana cutivation is 52,000 Ha. Even if only 50% of this area is planted with tissue culture plants it will require 125 million plants annually .
2. Sugarcane also has a great potential for the domestic market. New varieties need to be rapidly multiplied. Earlier studies have clearly indicated increase in cane yields using tissue culture produced plants . Research Institutes in different states in the country, in Tamil Nadu, Uttar Pradesh, Maharashtra and others are carrying out breeding studies.
3. Forest trees - Work is largely experimental although present studies from extensive field evaluation data indicate its potential for commercialisation. For details see paper of Nadgauda et al. in same volume.
4. Fruit crops - India presently accounts for 39% of mango fruits exported which is projected to grow annually. The area under mangoes is over 1 million Ha, and is covered by different varieties, some of very poor quality which need to be replaced with high quality varieties. Work on tissue culture of mango is at R & D level.
   Grapes - The export of grapes is also expected to grow by 2000 AD. Many new varieties by Indian Agricultural Research Institute, Delhi (IARI) have been released for wine, juice, table purposes and raisin making. The latest is "Arka Neelamani" a black seedless hybrid, which has been released from the Indian Institute of Horticultural Research, Bangalore. It has a high market potential. However, commercial protocols are still to be developed.
5. India's herb export has increased more than three times within last few years, giving new and better opportunities for the herbal industry. Many of these herbs are found to be very effective in local medicine resulting in overharvesting and bringing many species to the brink of extinction. In view of this it is necessary to create a favourable economic environment for commercial cultivation, propagation and multiplication of these herbs for which tissue culture can be a boon.
   As an example a wonder plant similar to ginseng, *Trichopus zelayanicus* has recently been located in the mountainous pockets of Kerala. It is an energy and vitality enhancer, viewed as a potential export earner, and marketed by a company in Coimbatore. However, it is difficult to propagate, and has scope for expanding cultivation.

In a similar manner there are other herbs, shrubs and trees which have been located in the forests which need to be conserved.

SECONDARY METABOLITE PRODUCTION

Several laboratories in India are conducting research on medicinal and aromatic plants for production of secondary metabolites, but no product has been commercialised. The main plants are those containing high value products like *Taxus* spp, *Vinca rosea, Capsicum annuum*, plants with anti-cancer, antiviral and anti-AIDS properties used in some local medicines.

The names of some labs are listed in Table 6.

Table 6 :

Some Laboratories Involved in Research on Secondary Metabolites (Mainly Medicinal and Aromatic Plants)

Contact Scientist	Laboratory
P.S. Ahuja	Institute of Himalayan Bio Resource Technology, Palampur, H.P.
S. Badari Narayan	Dabur Research Foundation, Ghaziabad, U.P.
M.R. Heble	Kelkar Research Foundation, Mumbai
K.V. Krishnamurthy	National Chemical Laboratory, Pune
A.K. Kukreja	Central Institute of Medicinal and Aromatic Plants, Lucknow
Usha Mukundan	Mumbai University, Mumbai
P.S. Rao	Bhabha Atomic Research Centre, Mumbai
P.S. Srivastava	Hamdard University, New Delhi
G.A. Ravi Shankar	Central Food Technology Research Institute, Mysore

CONCLUSION

In general labs in India are producing plantation and horticultural crops, ornamentals (foliage and flowering plants) and forest trees species. The actual number of plants produced have not been given as the figures have been very conflicting and the numbers very erratic and contradictory. Many commercial labs are not willing to divulge this information for purposes of confidentiality. A major proportion of the market is in foliage and flowering plants followed by plantation and horticultural crops and forest trees.

The major crop produced for the domestic market by several laboratories is banana estimated to be around 10-15 million plants annually. Some laboratories restrict their production to one or a few crops. A variety of spice and floriculture crops are also offered for the domestic market. Micropropagated forest trees are gradually making progress with some species like teak, poplars, eucalypts, *Anogeissus* spp. etc. being produced and field evaluated prior to large scale production for afforestation. A major limitation will be in the cost of plants for which automation or robotics can be a solution.

The tissue culture industry in India has come to stay. By integrating the tissue culture industry with other safe and profitable business ventures, many companies are able to sustain. This seems to be the present strategy, which some companies are following to ensure their survival.

ACKNOWLEDGEMENTS

The help of the Department of Biotechnology for kindly sending some of the information reported in this manuscript is gratefully acknowledged, particularly to Dr.(Ms) Manju Sharma, Secretary and Dr.(Ms) Renu Swarup. The author is also thankful to Dr. David Constantine, UK, and to commercial micropropagation units particularly Drs. R. Doreswamy, D. Shreedhar, B.Vatsya, Jitendra Prakash, Manish Jain, Anil Kush, Joe Thomas and several others for providing valuable information. The help of NCL staff members, Dr. R.S. Nadgauda, Dr. C. K. John and Shuchishweta Kendurkar in preparation of the manuscript, Swati Bhalerao in computer work, Sajita Nair for typing and Parag Akkadkar for photography is acknowledged.

If any names have been left out this has been done inadvertently.

REFERENCES -

1. Constantine D (1998) Directory of Tissue Culture Laboratories in India, U.K.
2. Department of Biotechnology, Government of India,Annual Report (1997-1998).
3. Govil Suman, S.C.Gupta. (1997) Plant Cell, Tissue and Organ Culture, 65 -73, 1997.
4. Govil Suman, S.C.Gupta (1994) J.Sci.Ind.Res. 53, 768-777.
5. Mascarenhas AF (1991) Current Science 60, 547-551.
6. Pierik RLM, Ruibing MA (1997) Developments in Micropropagation industry in the Netherlands, 152-156.
7. Singh HP, Chadha KL (1997) Banana and Plantain in India. Infomusa 5, 22-25.
8. Zimmerman RH (1996) J.Korean Society of Horticultural Science, 486-490.

Commercialisation of Tissue Culture for Forest Tree Species in India

Rajani S. Nadgauda, S.V. Kendurkar and A.F. Mascarenhas
Tissue Culture Pilot Plant,
National Chemical Laboratory, Pune – 411 008

1 INTRODUCTION

Tissue culture of tree species was initiated in India in the 1960's (Konar et al., 1963). The possibility of plant production using the meristem tissue from a 100 year old teak tree was first demonstrated in 1980 at National Chemical Laboratory (NCL), Pune, (Gupta et al.1980) and is considered a land mark. Since then NCL has carried out pioneering research in this area (Gupta et al., 1983, Mascarenhas et al.,1988,1989, Nadgir et al.1984). Simultaneously micropropagation of eucalyptus (Lakshmita et al. 1979,1985) and somatic embryogenesis in sandalwood (Bapat, Rao 1979, Lakshmisita et al., 1979) was also reported. Studies were also reported on protoplast culture (Rao, Ozias Akins, 1985) and artificial seeds in sandalwood (Bapat, Rao, 1988) . The status of tissue culture in India has been reviewed (Kendurkar, Mascarenhas 1987, Muralidharan, Mascarenhas 1989). Currently, many research groups are engaged on forest tree tissue culture which includes desert species (Shekhavat et al., 1998). At present, micropropagation protocols are available for more than 30 tree species and studies are underway on more than 20 tree species (courtesy : Department of biotechnology, DBT, New Delhi). Although methods were developed for micropropagation of several forest tree spp. and small scale field trials were conducted (Mascarenhas et al. 1988, Gupta et al. 1991) for commercial utilization the main challenges were (1) To develop, refine and upscale the protocols to make them cost effective and (2) To evaluate growth of the plants in field. To answer both these questions extensive research work for upscaling and extension of the technology to the field level was necessary. These are described in this paper.

In this context, following events have played a major role in commercialisation of tree tissue culture in India.

1. The pioneering efforts of NCL, in forest tree tissue culture in the early eighties.
2. Research in different laboratories all over India on development of micropropagation protocols for forest trees.
3. Funding by National Bank for Agriculture and Rural Development (NABARD), Mumbai, for developing bench scale protocols and preliminary field trials at NCL, Pune in mid eighties.
4. Setting up of Pilot Plants by DBT at NCL, Pune and at Tata Energy Research Institute (TERI), New Delhi
5. Refinement of the protocols and extensive field verificatory trials using tissue culture raised plants produced in these Pilot Plants.
6. Setting up of demonstration plantations to assess the cost benefits and realisation of benefits incurred for sustainable economic and ecological development by the farmer and forest community.
7. Generation of awareness and confidence in the industry.
8. Transfer of technology to industry.

A. Altman et al. (eds.), Plant Biotechnology and In Vitro Biology in the 21st Century, 721–724.

2 TISSUE CULTURE PILOT PLANTS (TCPPs)

In 1989, DBT decided to set up two Pilot Plants one at NCL, Pune and other at TERI, New Delhi, with the major objectives of refinement and upscaling of the laboratory scale protocols, for mass propagation of forest tree species and to assess the technocommercial feasibility of tissue culture propagules by conducting multilocation field trials. The plant species selected for NCL were *Tectona grandis, Eucalyptus camaldulensis, Eucalyptus tereticornis, Dendrocalamus strictus* and for TERI were *Anogeissus pendula, Anogeissus latifolia, Populus deltoides and E. tereticornis.*

2.1 TISSUE CULTURE PILOT PLANT FACILITY AT NCL – A CASE STUDY

Tissue Culture Pilot Plant Facility with a production capacity of 1 million plants per year with green house was designed at NCL, Pune and commissioned in 1992. This facility has a semiautomated, fully airconditioned laboratory and a greenhouse, monitored by an advanced computer system and is at par with most sophisticated labs in the world. The forest tree species being multiplied in this unit (mentioned above) were selected on the basis of their economic importance.

The techniques for all the these plant species reported earlier in 80's were refined considerably. The extensive research and development work carried out was focussed on the following lines.

- Development, refinement and upscaling of protocols by reduction in the number of stages, use of minimal media, rapid multiplication rate and improved survival rates.
- Methods for successful transportation of plants from the greenhouse to the field.
- Field verificatory trials.

Fresh green (apical and axillary) buds were collected from the identified 'plus' trees located in far flung forest areas. Season of collection, location and methods for transport were standardized for each species. Protocols were developed to control contamination and phenolics in the explants and cultures were established. These shoot cultures were multiplied by meristem proliferation laying emphasis on increasing the number of shoots per explant on minimal media. The protocols were tested and extended to several genotypes distributed in different forest regions in India.

A simple efficient method for rooting of microshoots directly in soil i.e. rooting *ex vitro* was developed for Eucalyptus spp. and teak. The method involves, a) harvesting of microshoots from culture bottles b) quick dip of microshoots in a root induction solution and c) transfer to rooting substrate. These shoots were initially maintained under 80-90% humidity for 10-15 days. Rooting and hardening took place simultaneously, resulting in reduction in number of stages, labour and time (Nadgauda et al.1997).

In case of bamboo (*Dendrocalamus strictus L.*) a simple, method for rapid in vitro multiplication and rhizome formation on minimal media devoid of any growth regulators has been developed using seedling cultures (Shirgurkar et al.1997). *In vitro* rhizome formation ensured 99% survival of bamboo plants in the soil.

2.2 FIELD EVALUATION OF TISSUE CULTURE RAISED PLANTS

Using these refined protocols for eucalyptus, teak and bamboo so far over a million plants have been produced and field planted at 278 different locations in different states of India and Nepal covering an area of 1000 ha. Field trials were undertaken in collaboration with forest departments/ corporations, agricultural universities and private sector organizations. These trials were designed as progeny trials, provenance trials and clonal trials with the objectives :

- To test clonal homogeneity of the micropropagated plants.
- To assess the performance of micropropagated plants over seed raised progeny.
- To identify suitable provenance/s for different clones so as to realise the full potential of individual genotypes.

The plant survival in the field was more than 93%. The growth data collected from different sites and analysed revealed that the tissue culture plants exhibit high uniformity, higher biomass leading to early rotation.

The clonal trials conducted using different clones at different locations have shown very interesting observations. In one of the field trials on eucalyptus conducted in 1994 in

Maharashtra State using different tissue culture raised clones of ERK-4; SR-5; R-3 and R-2, plants of ERK-4 after 40 months showed a growth of 9 meters in height compared to 7 meters attained by other clones and 0.5 meters by seed raised progeny. This growth was observed inspite of the fact that the soil pH was 8.0 as against normal pH of 6.8 to 7.
Similarly, teak clones exhibited varied performance at different provenances. Among the teak clones numbered NC-21, NE, TD, DN, KLS, the clones NC-21 and NE proved to be best performers in the area of Maharashtra and Gujarat. In one of the trials conducted at Gujarat in 1997 the recorded height was 281 cm for clone NE over other clones where the average height noted was between 200-215 cms in 10 months.

In a similar programme carried out at the pilot plant by TERI, New Delhi, India tissue culture raised plants of *Anogeissus*, *Populus* and *Eucalyptus* planted at various locations. The available data of the field performance of the different species clearly suggests that tissue culture plants of *Anogeissus pendula* , have survived well and have performed better in the field with respect to height and girth measurements. Some clones of the other species were also found to be superior to controls (Vibha Dhawan Personal Communication).

2.3 MOLECULAR MARKERS FOR TESTING CLONAL HOMOGENEITY

Genetic analysis of tissue culture raised plants of *Eucalyptus tereticornis, E. camaldulensis* and *Tectona grandis* was carried out by Dr. S.N. Raina and his group at Delhi University, New Delhi, using cytochemical and molecular assays. Their results indicated that there is no change in total DNA content among the progeny and mother clone. The genetic fidelity was further confirmed using RAPD techniques . The amplified genomic DNA fragments from *in vivo* and *in vitro* materials were monomorphic across the plants tested (Rani and Raina 1997).

2.4 DEMONSTRATION PLOT AT FARMERS LAND

In order that the lab technology reaches the land a demonstration plot was planted using tissue culture raised eucalyptus plants on farmers fields with the help of Cooperative Agroforestry Federation, Nashik, Maharashtra, India.
Cooperative Agroforestry Federation is an apex body of Cooperative tree growers Society, Maharashtra, India and has been successful in developing a system for eucalyptus marketing. In this programme the production is decentralised and marketing is centralised. Where the price is linked with quality. This builds quality – consciousness, sets product evaluation standards and gears production in accordance with market needs. (FAO 1996).

These trials demonstrated higher growth rates, early rotation leading to a higher cost : benefit ratio (1:3). Wood volume of tissue culture raised plants was 45 cu.m. / Ha higher and fetched 38.91% more value for wood which resulted in higher net profit of 42%.

Realizing the potential economic benefits of this technology the Federation has encouraged farmers to undertake large scale plantations at their fields using micropropagated eucalyptus plants of selected genotypes. This has created a market demand leading to the establishment of a commercial arrangement with, EPC irrigation at Nasik having a production facility of 0.5 million plants per annum to cater the need of the farmers.

3 CONCLUSION

These projects have contributed in building up of a strong knowledge base and have been put upgraded to "Micropropagation Technology Park" to act as a knowledge based catalyst for the development of tissue culture technologies for subsequent transfer and adoption to industries. These parks are housed in the premises of existing tissue culture pilot plants, operated with the assistance of expertise and staff trained under Pilot Plants.

Extensive efforts in the country including establishment of the two pilot plant units for taking the technology from laboratory to land has resulted in the development of commercially viable indigenous technologies. The field verificatory trials and testing at molecular level have confirmed the feasibility of these tissue culture propagules for clonal forestry. Commercial plantations have resulted in generation of confidence and awareness among foresters, farmers and industries on the benefits of micropropagation.

This has opened up avenues for commercialization of forest tree tissue culture in India and abroad. At present more than seven commercial units in the country are engaged in the production of forest tree propagules using tissue culture (Govil, Gupta 1997). The technology on teak micropropagation has already been transferred to laboratories in India and also U.K. and have been successfully adopted. The protocol developed on micropropagation of teak is based on segmentation of shoots into nodal segments which in turn multiply by axillary proliferation. This system has scope for robotic operations and would have an impact in automation of the process by facilitating the production of plants on a very large scale at the levels required for afforestation programmes.

ACKNOWLEDGEMENTS

The authors gratefully acknowledge funding agencies to support this activity; TCPP, NCL staff; Forest Departments/ Corporations, Agriculture Universities; Mr. Vinayakrao Patil, Chairman, Co-operative Agroforestry Federation, Nasik for field trials; Mr. G.M. Abhangkar for field data analysis and Mrs. Swati Bhalerao, Mrs. Shobha Bhondve, Ms Sajita Nair for typing the manuscript.

REFERENCES

1. Bapat VA, Rao PS (1979) Ann Bot. 44 : 629-630.
2. Bapat VA, Rao PS (1988) Plant Cell Rep. 7 : 434-436.
3. Govil S, Gupta SC (1997) Plant Cell Tiss. Org. Cult. 51 :65-73.
4. Gupta PK et al (1983) Plant Cell Rep. 2 :296-299.
5. Gupta PK et al (1980) Plant Sci. Lett. 17 :259-268.
6. Gupta PK et al (1991) In vitro Cell. Dev. Biol. 27P : 159-164.
7. Kendurkar SV and Mascarenhas AF (1987) Proc. Symp. Plant Cell and Tiss. Cult. Of Economically Important Plants, Hyderabad, India, 181-190.
8. Konar RN (1963) Phytomorphology, 13: 65.
9. Lakshmisita G, Shobharani B (1985) Plant Cell Rep. 4 :63-65.
10. Lakshmisita G et al (1979) Plant Sci. Lett. 15 : 265-270.
11. Mascarenhas AF et al (1988) Genetic manipulation of Woody Plants, ed. By JM Hanover and DE Keathly, 391-412.New York, Plenum.
12. Mascarenhas AF, Muralidharan EM (1989) Curr. Sci. 58:606-613.
13. Mascarenhas et al.(1993)
14. Nadgir AL et al (1984) Silvae Genetica 33 :219-223.
15. Nadgauda RS et al (1997) IUFRO Symp. Innovations in Forest Tree Seed Science and Nursery Technology, Raipur, India.
16. Rani V, Raina SN (1997) Plant Cell Rep. 16 : 1-7.
17. Rao PS, Ozias A (1985) Protoplasma. 124 : 80-86.
18. Shekhawat NS et al (1998) National Symp. Commercial Aspects of Plant Tissue Culture, Molecular Biology and Medicinal Plant Biotechnology.
19. Shirgurkar MV et al. (1996). Current Science. 70 : 940-943.

INVOLVEMENT OF BIOTECHNOLOGY IN COMMERCIAL PLANT TISSUE CULTURE

B. N. Okole[1], C. Memela[1], V. Chibba[1], S. Rademan[2], K. Kunert[2] and C. Cullis[2,3]
[1]Research and Development Department, AECI, Modderfontein 1645, South Africa, [2]NovoMark Technologies, Cleveland, OH, US and [3]Department of Biology, Case Western Reserve University, Cleveland, OH 44106 USA

1. Introduction

Production of plants via in vitro micropropagation has developed into a multibillion market. This market is, however, mainly covered by small to medium-sized companies producing from thousand to several million plants and only a small number of companies are producing up to 20 to 30 million commodity-type plants mainly in the ornamental sector. Since non-automated plant micropropagation is very labor-intensive, there has recently been a significant shift for low cost production to developing countries. With their lower wage scales, companies in developing countries have mostly buy-back arrangements with companies from developed countries, which might include both technology and equipment transfer.

Simple multiplication of low cost plants using non-sophisticated in vitro technologies easily results in the flooding of the plant market with identical products. Consequently, only the price with a low profit margin and limited cash resources might determine the success or failure of a business. Since value-added products would significantly improve competitiveness and might guarantee survival and healthy growth of a micropropagation business, the question might be asked: How can recent advances in plant biotechnology/molecular biology help to add value to tissue culture plants and therefore improving the competitiveness of a tissue culture business?

In this paper, we report about our experience with three biotechnology tools to improve the competitiveness of a banana tissue culture facility. These tools include a molecular marker system for early monitoring of off-type plants in a production system and both genetic engineering and in vitro selection of plant material as non-conventional breeding tools for disease resistance.

2. Experiences

2.1 Molecular Markers

A. Altman et al. (eds.), Plant Biotechnology and In Vitro Biology in the 21st Century, 725–728.

Somaclonal variation is a common problem in tissue culture plants. Genetic changes in the nuclear, mitochondrial or chloroplast genomes, produce inferior off-types with little commercial value. Traditionally, an experienced examiner is required to identify off-types by a morphological description and visual monitoring of plant characteristics. Further, many characteristics are only expressed in a more mature stage of a plant and not in in vitro plants or very young plants. For example in banana, off-types, such as dwarfs, mosaics, which have irregular bright yellow spots or stripes on the leaf, and Masadas, which have abnormal foliage showing depressions and thicker leaves, are difficult to identify at both the tissue culture and nursery stages (Israeli et al. 1991).
To improve genetic conformity of in vitro propagated plants, we have investigated the possibility of introduce a molecular marker system for off-type detection. Such a system should have the potential to systematically monitor and control the material and to determine if the genetic information has changed during the micropropagation process. The RAPD (Random Amplified Polymorphic DNA) technique is currently widely used by several groups for off-type detection and recently a RAPD marker has been identified for a dwarf banana off-type (Damasco et al. 1996). The RAPD technique has, however, several disadvantages, which includes a lack of reproducibility of results and the use of a large number of different primers to detect variation in only a small portion of the genome.
As an alternative to RAPDs, we have evaluated the technology of Representational Difference Analysis (RDA), which belongs to a general class of DNA subtractive technologies (Lisistyn et al. 1993). This technology offers the advantage to compare a much larger portion of the genome when trying to find differences between closely-related material which is essentially uncharacterised at the molecular level. A further advantage is that RDA might be executed in a commercial laboratory by a cost-effective dot blot system.
The basic concept of the RDA technology is to compare two DNAs by using one of the DNAs in excess to remove (subtract) all of the sequences held in common between the two DNA samples. Therefore, what are left are those sequences, which vary between the two. When a representation is used, as is necessary for complex DNAs, only some of the differences between the genomes being compared will be found. DNA fragments from Williams mosaic and Masada, off-type plants have been isolated using this technique. The final products are currently beeing evaluated for their potential as a diagnostic tool to identify off-types in banana tissue culture plants.

a. In vitro Selection

Non-conventional breeding techniques can be used to improve the value of plant material by adding useful characteristics, such as increased resistance to plant diseases. For banana improvement, we have investigated the two technologies, in vitro selection and genetic engineering. We selected somaclonal variants in the presence of fungal toxins to obtain black Sigatoka resistant banana plants. The isolated non-host specific Mycosphaerella fijiensis toxin, juglone, and the host-specific fungal toxin 2,4,8-trihydroxytetralone were applied for selection, and hand-cut micro-cross sections from the corm of Grande Naine and Williams tissue culture plants were used as explants (Okole, Schulz 1996). This toxin-based procedure has the advantage of easy translocation of the metabolite through the tissue and regeneration of shoot buds directly from the micro-cross section. We found a positive correlation between in vitro studies with the toxin 2,4,8-trihydroxytetralone and in vivo studies with the pathogen. The in vitro selection of micro-cross sections using a two-step selection against a crude fungal filtrate, followed by selection with the purified toxin 2,4,8-trihydroxytetralone gave the highest degree of tolerance against the fungus. Using this technique over the last 2 years, we have created sufficient plants necessary to carry out a first field trial with selected variants in two fungal infested areas. The trials include both Williams and Grande Naine in Cameroon to test for increased black Sigatoka resistance, and in South Africa to test if any cross-protection against Fusarium wilt can be achieved by our selected material.

2.3 Genetic Engineering

Much attention is paid to genetic engineering of plants to increase disease resistance. This route is currently followed by several multinational companies, which are involved in plant biotechnology. Such companies have made over the last several years extensive investments into the plant transformation technology and the isolation of useful genes. Engineering of banana for disease resistance has recently also become an attractive idea and significant progress has been made by several groups to engineer banana (Sagi et al. 1995; May et al. 1995).

As a banana micropropagation company, we were also attracted by the idea to use genetic engineering of banana plants as a biotechnology tool to add value to our existing products. In the last two years, we have evaluated the strategy to increase resistance against nematodes and fungal diseases, such as black Sigatoka and Fusarium wilt (Panama disease), by using either proteinase inhibitors or designed antifungal peptides. However, progress of our genetic engineering program was primarily not determined by our technical ability but rather by economic factors, relevant to a micropropagation company operating in a developing country.

When engineering of a commodity-type, and therefore a relatively inexpensive product, like banana with a low profit margin is envisaged, a number of cost factors can be prohibitive in executing successfully an extensive and costly genetic engineering program. Representing a company with a very small in-house research in plant biotechnology and a lack of extensive local expertise, we had to assume that we have to obtain various technology licenses for useful genes, plant specific expression systems and also for the gene transformation system. We considered these licenses for "enabling technologies", which in most cases are only available from companies or institutes in developed countries, as vital for efficient manufacturing, cultivating and making economic use of the corresponding plants. These licenses are generally very expensive for a company operating in a developing country. Also, by focusing on production of a commodity-type and therefore inexpensive product like banana, we had to consider a short-term return on the company's investment delivering large volumes of plants, which has finally prohibited any long-term development program for a costly biotechnology product.

3. Conclusions

Plant biotechnology is widely seen as one of the key technologies for generating innovations and new products. But in this respect, biotechnology is highly capital intensive and demands from a company a long-term commitment to the technology with sufficient financial back up and possibly substantial in-house research executed by skilled personal. From our experience, these requirements are difficult to fulfill by most companies in developing countries. However, certain biotechnology segments might be affordable and we identified two biotechnology tools, which might help a micropropagation company in a developing country to improve competitiveness by adding value to existing plant products.
Firstly, a molecular marker system developed as a diagnostic tool will have the potential to identify very early somaclonal variants in the tissue culture process. In general, plants being non-true-to-type have severe financial implications for both growers and plant producers, who may only discover this fact after some period of time in the field. Such markers, which are currently being developed and tested by us, might be ideal to define optimal culture conditions and to optimize plant multiplication rates, which are economically more favorable, without producing a significant number of off-types.
Secondly, in contrast to genetic engineering, where the financial outlay for a company might severely limit the available technology, in vitro selection represents a very cost-effective non-conventional breeding technique. This technique avoids an extensive investment into both expensive "high tech" equipment and enabling licenses. Since lines with useful variations might also contain other undesirable changes in important plant characteristics, we are currently carrying out in Africa the required very careful and extremely stringent long-term selection process of variant plant material.

4. References

Damasco O et al (1996) Plant Cell Rep. 16, 118-122.
Israeli Y et al (1991) Scientia Hortic. 48, 71-88.
Lisistyn NA et al (1993) Science 259, 946-951.
May GD et al (1995) Biotechnology13, 486-492.
Okole BN, Schulz FA (1996) Plant Cell Rep. 16, 39-343.
Sagi L et al (1995) Biotechnology 13, 481-485.

The possibilities of commercialization of Indian orchids: application of tissue culture techniques

S.P. Vij

Orchid Laboratory, Department of Botany, Panjab University, Chandigarh- 160 014, India.

The orchids represent a group of botanically interesting and commercially significant plants which have outnumbered and outsmarted their counterparts by evolving ingenuity and higher levels of specializations. They comprise nearly 20,000 species in 800 genera, culminate one of the evolutionary lines of monocots, and are still in an active state of speciation due to poorly developed barriers of reproductive isolation. Suppressed development of endosperm, microscopic seeds with undifferentiated embryos, and dependence upon a variety of factors including specific pollinators and fungi for continued reproduction in nature are some of their distinguishing features. The orchids possess extremely beautiful, intricately fabricated, highly colourful, and long-lasting flowers of myriad sizes and have significantly contributed to the development of international trade in cut-flowers and pot-plants. Possibilities of producing new and novel varieties through successful wide matings have added new dimensions to the significance of these plants in the trade; they rank among the top ten sought after plants for cut-flower production. Many of them are rich in phytochemical contents as well and are extensively used in local medicines for their curative and aphrodisiac properties. Thailand, Singapore, Malaysia, Taiwan, Korea, and Sri Lanka are the major asian players in the orchid trade with Japan and U.S.A. as the biggest export destinations.

India with a vast geographic expanse and varied climatic regions is rich in plant genetic resources. The orchids have naturalized here in great profusion; they are represented by nearly 1200 species (Sathish Kumar, Manilal, 1994). The rich diversity of Indian orchids includes several with proven ornamental and/or therapeutic values. Many of them particularly in the genera *Aerides, Arachnis, Calanthe, Cymbidium, Cypripedium, Dendrobium, Paphiopedilum, Phalaenopsis, Pleione, Renanthera, Rhynchostylis, Vanda* etc. are source of incalculable aesthetic pleasure because of their floral excellence. Besides adding to the prized collections of almost all the famed botanical gardens, the world over, they have been extensively used to progenate internationally acclaimed hybrids with heavier texture, exotic colour combinations, and long keeping qualities of flowers (Bose, Bhattacharjee, 1980). Unfortunately, the orchid genetic resource remained unexploited in

A. Altman et al. (eds.), Plant Biotechnology and In Vitro Biology in the 21st Century, 729–736.

the country due to different national priorities and lack of awareness about its commercial significance. The scenario has now changed with the increased level of awareness about these plants and changed national policies which favour orchid based floriculture, but we are faced with conservation related problems. A sizeable number of Indian orchids have succumbed to unregulated collection and habitat destruction pressures; their natural populations are declining rapidly. While the Government of India has already banned the export of orchids collected in wild and has taken steps to protect orchid rich habitats, the efforts are not as yet commensurate with the dimensions of the problem. The orchids are still being collected stealthily from their natural and foster homes and shall continue to be collected as long as they enjoy an economic status and demand. Mass propagation using conventional and tissue culture techniques thus seems to be the only strategy to commercialize orchids and save their natural populations from collection pressures.

The orchids are primarily sexual but they reproduce a great deal through vegetative means as well. Their vegetative propagation through conventional means i.e. *keikis*, back-bulbs, division of shoots, etc., is rather slow. Even their natural regeneration through seeds is limited. Only 0.2-0.3% seeds germinate because they are poorly organized (the development of endosperm is suppressed and that of embryo remains arrested at the 'globular'/'pre-heart-shaped' stage), lack an appropriate metabolic machinery to utilize their own lipidaceous food reserves, and require a suitable mycorrhizal association for germination in nature. The fungus is believed to provide the necessary stimulus by aiding carbohydrate/auxin and/or vitamin transport during germination (Hayes, 1969). The tissue culture techniques have, on the other hand, opened new possibilities in conservation and commercialization of orchids. Round the year propagation of genetically uniform, disease-free, fast maturing and high yielding plants is now within the realms of reality. The technique exploits the regenerative potential of plants much more effectively than the conventional methods of vegetative reproduction. and can easily satisfy the large scale requirements for genetically identical planting materials. Moreover, the micropropagated plants grow more vigorously due to their better health status. Even to keep pace with the ever changing consumer preferences, the possibility of using tissue culture techniques for mass multiplication of new and better developed plant varieties (through desired matings and/or genetic engineering) is tremendous. It is also possible to enlarge the genetic base by inducing somaclonal variations. Incidentally, the orchids represent the first floricultural crop successfully propagated through tissue culture techniques. The immature embryos and shoot meristems are amongst the most commonly used explants for *in vitro* propagation of orchids. The embryo culture helps in enlarging the genetic diversity and shoot meristems in maintaining genetic purity.

Seed/Embryo Culture

Ever since Knudson (1922) demonstrated the possibility of bypassing the fungal requirement of orchid seeds during germination *in vitro*, asymbiotic seed germination has been accepted as an important procedure for propagating orchids (Arditti et al., 1982). The orchid seeds can also germinate prior to reaching maturity. The technique of culturing immature seeds is variously referred to as ovule/embryo/green pod/green fruit culture technique (Sagawa, 1963). It is easy to follow, ensures better germination frequency, and favours rapid production of virus-free seedlings. Its additional utility in recovering the progenies of desired matings, propagating rare and endangered species, and cloning apomictic (obligate) genotypes is also well documented (Vij, 1992). The asymbiotic germination potential of seeds, representing different developmental stages, has been positively tested in several commercially viable and/or threatened Indian taxa (cf. Vij, 1995). In orchids, very young ovules do not form suitable explants because the embryo sac development is a post-pollination phenomenon and fertilization a pre-requisite for obtaining seedlings. However, as the ovules can be used for raising cultures immediately after fertilization, the importance of information on time interval between pollination and fertilization has often been stressed (Valmayor, Sagawa, 1967). Though such studies are yet to be initiated in the Indian orchids, the time (in wks) after pollination when the ovules can be successfully germinated seems to vary with species (Vij, 1995). *Eulophia hormusjii* embryos, procured between 8 and 16 wks after pollination, germinate readily but their germination frequency declines sharply with further passage of time; only a few of them respond when obtained later than 16 wks of pollination. Likewise in *Satyrium nepalense*, *Nephalaphyllum cordifolium*, *Phaius tankervillieae*, and cymbidiums, the germination frequency shows a sharp decline when the embryos are collected 3-4 wks prior to fruit dehiscence; the mature embryos either fail to germinate or germinate very poorly. Since they can be induced to germinate by treating with cold temperature, 20% sucrose, and/or KCl, it appears that dormancy rather than the non-viability factors are responsible for their poor response *in vitro*. The impaired germination of the mature seeds has also been attributed to the quality of their food reserves. Harvais (1974) reported that the food reserves comprise starch in the immature and lipids in the mature seeds of *Corallorhiza maculata*, and suggested that conversion of starch and other simpler carbohydrates into lipids, during seed maturation may be a common feature of orchids. The inability of the mature seeds to germinate, with ease, was correlated with lack of an appropriate metabolic machinery (glyoxysome bodies) to utilize their own lipid reserves. Raghavan (1976) recorded changes in the enzyme complements at different stages of seed maturation in orchids but the critical stage at which they acquire dormancy is yet to be identified. Moreover, as the dormant (mature) and metabolically active (immature) seeds are morphologically more or less indistinct, Mitra (1986) stressed the importance of information on histochemical and biochemical features for selecting the fruits for the right type of immature ovules (seeds). In this connection, the ablility of embryos from fruits which develop prominent ridges along the valves and cease to grow in diameter to respond better (Vij, 1995) may be useful in selecting the right stage of fruits for embryo culture.

Meristem Culture

Resident meristems:

Since the orchids are outbreeders and generate a great deal of heterozygosity in their progenies, their propagation through embryo culture appears to be a disadvantageous proposition in cut-flower industry where pure lines of desired genotypes are preferred.

Morel (1960) demonstrated the possibility of using excised shoot-meristems for regenerating complete plants of *Cymbidium, in vitro*, and Wimber (1963) formulated, described, and published a procedure for the purpose. The technique of resident meristem (shoot-tip, axillary bud) culture has opened new vistas in orchid micropropagation (cf. Arditti, Ernst, 1993). Upto 200,000 plants can be regenerated, within a year, from a single resident meristem. The technique has, however, a limited utility in monopodial taxa as it requires the sacrifice of the entire new growth or the only growing point and endangers the survival of the mother plant.

Adventitious meristems:
Several efforts have been made to develop an equally effective multiplication system by inducing *de novo* formation of adventitious meristems in organs whose excision is not detrimental to the survival of mother plant. The proliferative potential of explants from leaf (Chaturvedi, Sharma, 1986; Mathews, Rao, 1985; Seeni, 1988; Seeni, Latha, 1992; Vij et al., 1984, 1986a, 1988a; Vij, Pathak, 1988, 1990), root (Chaturvedi, Sharma, 1986; Sood, Vij, 1986; Vij, et al., 1987; Vij, Pathak, 1988b; Vij, 1993, 1994) and flower stalks (Singh, Prakash, 1984; Kaur, Vij, 1995; Vij, et al., 1986b; Vij et al., 1997) has been positively tested in some Indian orchids. The source, genetic constitution, and physiological age of the explants are, however, some of the important factors for regeneration. The juvenile tissues from greenhouse grown plants respond better than the mature ones from plants grown outdoors. The proliferative loci, in general, get activated in the dermal cells and soon develop into somatic embryos (Protocorm Like Bodies: PLBs). Direct or callus mediated development, multiplication, and differentiation of the PLBs is influenced by the chemical stimulus in the nutrient pool (Vij, Pathak, 1990; Seeni, Latha, 1992). Most orchid tissues require a treatment with plant growth regulators for growth and development. This requirement, however, varies with the species and source of explant. The use of NAA and either of BAP and KN, in the medium, yields a rich crop of PLBs in *Luisia trichorhiza, Satyrium nepalense, Vanda cristata,* and *V. testacea* leaf segment culture (Vij, 1995). Similarly, a synergistic action of KN and IAA/ NAA, in peptone enriched medium, favours enhanced production of PLBs in *Rhynchostylis retusa* cultures. Yeast extract is obligatory for regeneration in *Aerides multiflorum, Papilionanthe teres,* and *Satyrium nepalense* foliar cultures and peptone in those of *Vanda*. However, as the juvenile leaves can proliferate all along and the relatively older ones only in their basal regions (Vij et al., 1986) and the tip region regenerates infrequently when excised and more frequently in an entire leaf (Mathews, Rao, 1985), it was suggested that the inherent (latent!) meristematic potential in orchid leaves is perhaps controlled by some factors emanating from the leaf base and it diminishes with maturity (Vij, Pathak, 1990).

In the root explants, the effect of an exogenous supply of plant growth regulators is species specific and it varies during initiation, multiplication, and differentiation of cultures. It is possible to induce etiolated growth in the explants or produce plantlets therefrom (through direct or callus mediated development of either of PLBs and shoot buds) by varying the quality of the growth adjunct(s) in the medium. Incidentally, as the explants from juvenile roots (with poorly developed root- caps) regenerate with ease, and the ones from mature roots with well developed root-caps fail to do so, it appears that the root-cap plays an important role in controlling the regenerative potential of roots. In this connection, it is worthwhile to mention that root-cap is an active site of IAA accumulation and the transformation of a root meristem into a shoot meristem is positively influenced by the endo- and/or exo-genous level of the auxin; its supra-optimal level maintains root meristem and sub-optimal level brings about a switch in the activities of the meristematic cells with the result that the quiescent centre and other derivative cells are induced to develop into a shoot meristem (Philip, Nainar, 1988).

The advantages of leaf and root segment culture are apparent for more than one reason; they are easy to obtain, easier to disinfect, and their excision does not endanger the mother plant. Moreover, as the regeneration occurs in the dermal cells which are cytologically more stable, mass production of genetically uniform plants from these is a distinct possibility. A two-step technique involving culture initiation from flower stalk cuttings and subsequent use of foliar explants is expected to yield better results for cloning desired genotypes.

The release of brownish exudates (phenolics!) by the explants into medium is a serious problem in orchid cultures as these significantly impair the growth of the germinating entities or regenerants. We have successfully alleviated the harmful effect of these exudates by using activated charcoal (AC) in the medium. According to Yam et al. (1989), AC favours better health of the cultures because of its ability to adsorb exudates/growth inhibitors, enhance medium aeration and, absorb light and provide enhanced quantum of energy per unit plant material.

Precocious flowering

Free gene-flow across the taxonomic limits favours the development of new genotypes in orchids; their evaluation and subsequent selection for floricultural purposes is, however, a time consuming process because these plants take several years to reach maturity. The recent possibilities of inducing precocius flowers *in vitro*, however, offer tremendous opportunities to the breeder to select suitable clones from a hybrid progeny for commercial purposes within a short period of time. We have successfully induced *in vitro* florigenesis in *Dendrobium denudans*, *Dendrobium* Snow Fire, *Vanda coerulea* , *Eulophia dabia*, *Spiranthes sinensis*, and *Zeuxine strateumatica* cultures. While a cytokinin (BAP/KN) treatment proved useful for flower induction in most of these taxa, a treatment with IAA was obligatory for the purpose in *Dend. denudans*. In *Vanda*, the *in vitro* induced floral buds failed to open into complete flowers despite repeated subcultures as is also true in *Phalaenopsis* and *Dendrobium* cultures; frequent subcultures were required for normal development of floral buds in *Doriella* cultures (Duan, Yazawa, 1994).

Culture transplant and transport

Incidentally, regeneration in the explants, proliferation of the regenerants, differentiation of the plantlets (complete with roots and shoots) and their transplantation to soil/field are the 4 important stages in micropropagation process. The micropropagants have poorly developed cuticle, stomatal apparatus, photosynthetic ability, and conducting tissues, and they fail to withstand direct exposure to harsher climates outside the *in vitro* regimes until or unless they are properly acclimatized prior to transplanting. The procedure, known as hardening, helps cutting down their mortality rate, during lab to land transfers, by correcting the above mentioned morpho-physiological aberrations. Storage and long distance transport of the micropropagants are among other problematic aspects. The possibility of overcoming these problems by preparing somatic/artificial 'seeds' or beads (encapsulation of propagules in a nutritive gel) is being increasingly realized. Such 'seeds' have so far been prepared in nearly 20 orchid species by encapsulating their meristems/embryoids in nutrient supplemented calcium alginate gel. The quality of the beads varies with the concentration of calcium chloride and sodium alginate in the medium; they are best formed by using 2.5% sodium alginate and 100 mM calcium chloride with a complexation period of 30-40 min (Vij et al.,1992). They can germinate readily on a variety of substrata including agar, sand, vermiculite, etc. and retain their viability for about 4 months when stored at 4°C. The synthetic 'seeds' are easy to handle and ensure economy of space, medium, and time during storage, lab to land transfer, and long distance transport of tissue culture raised genotypes. The nutritive gel, in them, acts like an artificial endosperm and provides necessary growth stimulus to the enclosed embryoids/meristems besides saving them from mechanical injury. Since the encapsulated propagules invariably proliferate during germination, the synthetic 'seeds' represent a novel system for exploiting the inherent polyembryonate potential of orchids.

Conclusions

The international trade in floriculture is developing rapidly and the prevalent 'orchid mania', the world over augurs well for orchid based floriculture. India is a rich repository of orchids (1,200 species). Many an Indian species have been used to progenate internationally acclaimed hybrids; they are even known for their curative and aphrodisiac properties. Unfortunately, till very recently, the orchid based floriculture could not pick up in the country due primarily to different national priorities, conservation related problems, and lack of appropriate planting materials. Application of tissue culture techniques has, however, opened up new possibilities to conserve and commercialize Indian orchids. The immature embryos and shoot meristems are amongst the most commonly used explants for *in vitro* propagation of orchids; large scale propagation of elite clones and disease-free propagules, and development of new and novel varieties are within the realm of reality. Embryo culture helps in enlarging genetic diversity and shoot meristems in maintaining genetic purity. Since the shoot meristem culture requires the sacrifice of entire new growth or the mother plant itself, the possibility of using leaves (whole/segments/dermal peels), roots, and flower stalks as alternate but equally effective donor tissues has been positivley tested. Their efficacy, however, varies with the species, maturity level of the donor tissues, and chemical stimulus in the nutrient pool. Since orchid plants raised form seeds require several years to reach maturity, the technique of inducing precocious flowering in the protocorms provides the breeder with an opportunity for early selection of desirable hybrid genotypes for commercial propagation. Proper development of orchid based floriculture, however, calls constant efforts to improve the quality of planting material through desired matings and by inducing somaclonal variations and to develop cost effective reproducible micropropagtion protocols for commercially viable genotypes particularly the hardy ones with a wider ecological amplitude.

References

Arditti J, Ernst R (1993) Micropropagation of Orchids, John Wiley and Sons, Inc., New York.

Arditti J et al (1982) Orchid Biology, Reviews and Perspectives, Cornell University Press, Ithaca and London.

Bose TK, SK Bhattacharjee (1980) Orchids of India, Naya Prokash, Calcutta.

Chaturvedi HC, AK Sharma (1986) Biology, Conservation, and Culture of Orchids, Affiliated East-West press, New Delhi.

Duan JX, S Yazawa (1994) Proc. Nagoya International Orchid Show, pp. 87-96, The Organizing Committee, NIOS'94, Nagoya, Japan.

Harvais G (1974) Can. J. Bot. 52, 451-460.

Hayes AB (1969) Am. Orchid Soc. Bull. 38, 597-600.

Hey GL, MG Hey (1966) Proc. Fifth World Orchid Conf., pp. 35-39, Long Beach, California.

Kaur P, SP Vij (1995) J. Orchid Soc. India 9, 85-90.

Knudson L (1922) Bot. Gaz. 73, 1-25.

Mathews VH, PS Rao (1985) Proc. Indian Natl. Sci. Acad. 51(4), 496-504.

Mitra GC (1986) Biology, Conservation, and Culture of Orchids, Affiliated East-West press, New Delhi.

Morel GM (1960) Am. Orchid Soc. Bull. 29, 495-497.

Philip VJ, SAZ Nainar (1988) Ann. Bot. 61, 143-149.

Raghavan V (1976) Experimental Embryogenesis in Vascular Plants, Academic Press, London.

Sagawa Y (1963) The Florida Orchidist 6, 296-297.

Sathish Kumar C, KS Manilal (1994) A Catalogue of Indian Orchids, Bishen Singh Mahendra Pal Singh, New Delhi.

Seeni S (1988) Proc. National Seminar on Current Research Trends in Indian Orchids with a special reference to tissue culture technology, NEHU, Shillong.

Seeni S, PG Latha (1992) Plant Cell Tiss. Org. Cult. 29, 167-172.

Singh F (1992) J. Orchid Soc. India 6, 11-14.

Singh F, D Prakash (1984) Sci. Hortic. (Amsterdam) 24, 385-390.

Sood A, SP Vij (1986) Biology, Conservation, and Culture of Orchids, Affiliated East-West press, New Delhi.

Valmayor HL, Y Sagawa (1967) Am. Orchid Soc. Bull. 36, 766-769.
Vij SP (1992) Proc. Fourth Asia Pacific Orchid Conference, Chiang Mai, Thailand.
Vij SP (1993) J. Orchid Soc. India 7, 61-72.
Vij SP (1994) Proc. The Nagoya International Orchid Show Nagoya, pp. 72-80, Japan.
Vij SP (1995) Advances in Horticulture Vol. 12- Ornamental Plants I., Malhotra Publishing House, New Delhi, India.
Vij SP, P Kaur (1992) J. Orchid Soc. India 6, 71-74.
Vij SP, P Pathak (1988a) Proc. National Seminar on Current Research Trends in Indian Orchids with a special reference to tissue culture technology, NEHU, Shillong.
Vij SP, P Pathak (1988b) Proc. National Seminar on Current Research Trends in Indian Orchids with a special reference to tissue culture technology, NEHU, Shillong.
Vij SP, P Pathak (1990) J. Orchid Soc. India, 4, 69-88.
Vij SP et al (1997) J. Orchid Soc. India 11, 75-78.
Vij SP et al (1987) J. Orchid Soc. India 1, 71-74.
Vij SP et al (1988) Proc. National Symposium on Regulation of growth and differentiation in Plants, Chandigarh.
Vij SP et al (1986a) Curr. Sci. 55(21), 1100-1101.
Vij SP et al (1986b) Biology, Conservation, and Culture of Orchids, Affiliated East-West press, New Delhi.
Vij SP et al (1992) Orchid News 8-9, 12-13.
Vij SP et al (1984) Bot. Gaz. 145(2), 210-214.
Wimber DE (1963) Am. Orchid Soc. Bull. 32, 105-107.
Withner CL (1955) Am. Orchid Soc. Bull. 24, 380-392.

TRANSGENIC RICE: DEVELOPMENT, PRODUCTS, ACCEPTANCE AND ECONOMIC IMPACT

Swapan K. Datta
International Rice Research Institute, P.O. Box 933, 1099 Manila, Philippines, e-mail: SDATTA@IRRI.CGNET.COM

1. Introduction

Rice (*Orzya sativa*) is the major staple food for about 2.5 billion people, almost all of whom live in developing countries. By the year 2020, the number of rice consumers will almost double, to about 4 billion. To maintain an adequate supply of rice for the tremendous annual increase in population between now and 2020, and beyond, presents a formidable challenge to the scientific community which must achieve this increase in the face of declining arable land and water supplies, and in a manner that protects the soil, water and biotic resource base from which all food must come. A key research tool in the global effort to meet this challenge is genetic engineering, whose utilization in plant improvement through the development of transgenic products is increasing.

The research and development of transgenic products are moving very fast. These developments are being driven by private firms and public research institutions. Private firms (Monsanto, Du Pont, AgroEvo, Novartis, etc.) are now spending several billions of dollars annually on R&D of plant biotechnology. The payoff for this is expected to be seen over the next several years when biologists will be able to design the blueprint, for the insertion of appropriate single/multiple genes in plants for the commercial products for human, and animal consumption. This could result in transgenic plants taking a share of industrial and food markets worth $500 billion per year (John Pierce, Du Pont). However, the benefit of such knowledge-based intensive technology to developing countries depends on greater access by them to the technological advances of agro-industry with acceptable agreements of mutual benefit. Decisions in the CGIAR system in general, and at IRRI in particular, will have an insignificant effect on the "march of biotechnology". The non-aggressive stance on biotechnology by the system has already delayed access to developing-country biotechnology by a number of years (Evenson, 1998). Nevertheless the progress in transgenic rice development has been enormous and I will sum up the critical developments and the possible economic impacts for the future.

A. Altman et al. (eds.), Plant Biotechnology and In Vitro Biology in the 21st Century, 737–740.

2. Materials and Methods

Protocols for rice transformation which allow transfer of foreign genes from diverse biological systems into rice have been developed. Direct DNA transfer via protoplast (Datta et al. 1990), biolistic (Christou et al. 1991) and *Agrobacterium* system (Hiei et al. 1994) are routinely used for rice transformation at IRRI (Datta et al. 1997). The gene technology development and introduction of agronomically important genes into rice are summarized in Table 1.

3. Biosafety regulations and public acceptance

Field testing of transgenic crops in USA, Canada, Australia and China is now a routine practice. The progress in Asia is slower, although biosafety guidelines are now in place in China, India, Indonesia, Japan, South Korea, Malaysia, Singapore, Taiwan and Thailand.

3.1 NGOs: Approtech Asia, the principal NGO partner in relation to biosafety, is the umbrella organization of about 42 NGOs that share an interest in the application of appropriate technology to agriculture in the Asian regions. It operates in 16 Asian countries. The International Service for the Acquisition of Agri-biotech Application (ISAAA) has its Southeast Asia office at IRRI. ISAAA is sharing its expertise in intellectual progress acquisition, technology transfer and biosafety with IRRI and member countries of the Asian Rice Biotechnology Network (ARBN).

3.2 Risks and benefits: Transgenic plants are designed for use in agriculture and in industrial production and will be set available in commercial scale. The public and parts of the scientific community, have raised concerns on the possible ecological impacts of transgenic plants once they are released in the field. Most of the risks cited are speculative and merely concerns in many cases. For example, the concern that herbicide-resistant rice may promote the development of a herbicide-resistant weed. Horizontal gene flow is a natural phenomenon that has occurred for centuries, but has not produced unwanted weeds in the cultivated land. It depends more on our agricultural practice. In most well-studied cases, a post-emergent herbicide with no or very little residual soil activity greatly reduces the chance that resistant weeds could appear in a weed population over time. A few such concerns and scientific knowledge-based possible solutions have been explained in a recent publication by the World Bank (Kendall et al. 1997).

4. Economic impact

Transgenic rice developed by genetic engineering has opened vast opportunities for the improvement of the plant. It ensures higher yields or decreases losses due to biotic and abiotic stress. Improved transgenic products (seeds) provide the best opportunity for such use of biotechnology and transferring the benefits from laboratory to farmers land.

During the 12 year period (1986-1997), approximately 25,000 transgenic crop field trials of more than 60 crops with 10 traits have been conducted in 45 countries. China was the first Asian country to commercialize virus-resistant tobacco, in the early

1990s. Use of Bt cotton in the USA alone in 1996 resulted in a total national benefit of $60 million; the figure was $190 million for Bt corn in1997.
Attacks by insect pests, sheath blight, bacterial blight and abiotic stresses etc. can cause considerable annual yield losses in rice equivalent to 200 million tons (Herdt 1991). However, genetic engineering can meet the challenge and provide insect-resistant (Datta et al. 1998) sheath-blight-resistant (Lin et al. 1995), and BB-resistant rice (Tu et al. 1998b). The first biotic stress-related field testing of transgenic rice for bacterial blight resistance developed at IRRI has been conducted successfully in China (unpublished). The transgenic elite IRRI variety, IR72, has been used and thus can be grown extensively in most Asian countries.

Table 1. Gene technology and important traits introduced in rice*

Rice variety	Method used	Genes transferred	Traits conferred	References
Japonica	P	*hph*	Antibiotic resistance	Shimamoto et al. 1989
Indica	P	*hph*	Antibiotic resistance	Datta et al. 1990
I/J	B	*bar/gus*	Resistance to herbicide	Christou et al. 1991
IR72 (I)	P	*bar*	Resistance to herbicide	Datta et al. 1992
Japonica	*Ag*	*hph*	Antibiotic resistance	Hiei et al 1994
Indica	P	*chi11*	Resistance to SB	Lin et al.1995
Japonica	B	*Xa-21*	BB resistance	Song et al. 1996
Japonica	B	*HVA1*	Osmoprotectant	Xu et al. 1996
Japonica	B	*Psy*	Phytoene synthesis	Burkhardt et al. 1997
IR72 (Indica)	B	hybrid *Bt*	Stem borer resistance	Tu et al. 1998a
I/J	P/B	*cryIA(b)*	Stem borer resistance	Datta et al.1998
IR72 (Indica)	B	*Xa-21*	BB resistance	Tu et al. 1998b
I/J	P/B	*ENOD-12*/gus	Early nodulin gene exp.	Reddy et al. 1998
Japonica	*Ag*	*Ferritin*	Higher iron synthesis	Yoshihara et al. 1998
I (MHR)	B	cryIA(b)	Insect resistance	Alam et al. 1998a
India (DWR)	B	*cryIA(b)*	Stem borer resistance	Alam et al. 1998b

*Selected publication; I/J = indica/japonia; I(HR) = indica/hybrid rice; P = protoplast; B = biolistic; *Ag.* = *Agrobacterium*; SB = sheath blight; BB = bacterial blight; MHR = maintainer line of hybrid rice

5. Conclusion

Agribusiness based on biotechnology, biodiversity and value-added characteristics can provide significant avenues for sustainable agriculture and economic returns. Patent protection via IPR (mostly by private companies) and farmers rights, along with the future role of CGIAR centers, particularly IRRI, in making the technology accessible to and transferable to NARS is of great interest. Success of such technology transfer and partnership with private sector and NARS will bring about greater utilization of transgenic products.

Acknowledgment

I gratefully thank NOVARTIS, USA, CSIRO, Australia, Dr. S. Muthukrishnan, Dr. I. Altosaar, Dr. P. Ronald, Dr. H. Uchimiya, Dr. T. Okita, Dr. R. Wu, Dr. Y. Fan for providing gene constructs. Thanks are also due to Dr. Karabi Datta, L. Torrizo, E. Abrigo, N. Oliva, M. Viray for research and technical help and Robert D. Huggan for valuable comments. Financial support from the BMZ (Germany) and the Rockefeller Foundation (USA) are gratefully acknowledged.

References

Alam MF et al. (1998a) Plant Cell Rep. (in press)
Alam MF et al. (1998b) Plant Science 135, 25-30
Burkhardt P et al (1997) The Plant Journal 11(5), 1071-1078.
Christou P et al (1991) Bio/Technology 9, 957-962.
Datta K et al (1998) Theor. Appl. Genet. 97, 20-30.
Datta SK et al (1992) Plant Mol. Biol. 20, 619-629.
Datta SK et al (1990 Bio/Technology. 8, 736-740.
Datta SK et al (1997) IRRI Discussion paper series #21, IRRI, Los Baños.
Evenson RE (1998) International workshop on prioritization of Rice Research in Asia at IRRI, 20-22 April 1998, Los Baños.
Hiei Y et al (1994) The Plant J. 6, 271-282.
Herdt RW (1991) Rice biotechnology. pp. 19-54, CAB/ IRRI, Los Baños.
Kendall et al. (1997) Bioengineering of crops. The World Bank, Washington.
Lin E et al (1995) Bio/Technology. 13:686-691.
Reddy PM et al (1998) Plant Journal 14: 693-702.
Shimamoto K et al (1989) Nature 337:274-276.
Song WY et al (1996) Science 270:1804-1806.
Tu J et al (1998a) Theor. Appl.. Genet. 97:31-36.
Tu J et al (1998b) Plant Biotech (in press)
Xu D et al (1996) Plant Physiol 110:249-257.
Yoshihara S et al (1997) in 5th Int. Congress of Plant Mol. Biol. Abs. 303., Singapore.

Intellectual Property Protection for Plant Technologies

E. Anthony Figg
Rothwell, Figg, Ernst & Kurz
Washington, D.C.

Agricultural biotechnology and modern plant-breeding techniques offer enormous benefits to farmers and society as a whole. These technologies provide the potential for superior products of food and fiber, increased agricultural efficiencies, improved yields, disease resistance, decreased use of chemical herbicides and pesticides and the use of plants to make products never before produced by agricultural means.

The patent system has been proven to provide the incentives necessary to stimulate investments in high-risk, high-cost research, and it is already doing so in the relatively new field of plant biotechnology. The fruits of this patent-incentivized research will inure to the benefit of all mankind.

The United States has led the world in providing flexible and meaningful intellectual property protection for plant technology. While patenting of plants and other life forms remains a controversial issue in Europe, it now appears that the European Union is poised to provide protections similar to those available in the United States. Other developed nations, like Japan and Australia, similarly provide broad protection for novel plants and plant technologies, while Canada and numerous other countries prohibits patenting of higher life forms.

In the United States, patent protection for novel plants and plant parts began with the 1985 Patent and Trademark Office decision, In re Hibberd, which recognized the patentability of a novel plants and its seeds.[1] Prior to that case, it was uncertain that traditional systems for protecting intellectual property had a role in the emerging science of plant biotechnology.

Three federal statutes provide systems for protecting intellectual property relating to plants in the U.S.[2] These statutes are (1) the Plant Patent Act,[3] (2) the Plant

A. Altman et al. (eds.), Plant Biotechnology and In Vitro Biology in the 21st Century, 741–745.

Variety Protection Act[4] and (3) the general utility patent statute.[5] As a result of these diverse statutory schemes, both traditional plant breeders as well as plant molecular biologists have available an unusual selection of forms of protection for their inventions and discoveries.

Most other countries provide similar forms of protection (*i.e.*, patents and some form of plant breeders' rights), although policies regarding the patenting of plants and other higher life forms vary significantly. The so-called TRIPs agreement of GATT gives WTO countries the option of including or excluding patent protection for plants and animals. This paper addresses the use of patents for the protection of inventions and discoveries arising out of plant research, with emphasis on the protections available in the United States.

One aspect of U.S. patent law that is unique is a special patent statute for plants. The Plant Patent Act, enacted in 1930, is the oldest form of protection for plants in the United States. This Act was established to recognize the contributions of notable plant breeders, such as Luther Burbank. Special legislation was thought to be necessary because of the perceived impossibility of providing an enabling written description of a plant -- a requirement for patent protection for other types of inventions.

A plant patent is granted to one who "invents or discovers and asexually reproduces any distinct and new variety of plant, including cultivated spores, mutants, hybrids, and newly found seedlings, other than a tuber propagated plant or a plant found in an uncultivated state . . ." At the time of enactment of the Plant Patent Act, tuber propagated plants, such as potatoes and Jerusalem artichokes, were excluded from protection because of a concern that tubers are propagated by the same part of the plant that is sold as food.

In addition to the requirements for distinctness and asexual reproduction, the statute imposes the same standards of novelty and non-obviousness that are required for other types of inventions. "Distinctness" has been defined as having characteristics clearly distinguishable from those of existing plant varieties, whether such characteristics are inferior or superior to the existing varieties.[6]

The description requirements for a plant patent application are much less onerous than those for a conventional utility patent application. The application need only include a description which is "as complete as reasonably possible." Generally, the description includes a photograph or drawing along with a discussion of important distinguishing morphological and agronomic characteristics.

It has recently been held that, due to the asexual reproduction requirement of the Plant Patent Act, plant patents cover only a single plant and its asexually reproduced progeny. Arguments that the protection extended to other varieties having the same essential characteristics as the patented plant have been rejected.[7]

There has been considerable interest in extending the protection of plant patents to include plant parts. Such a change would allow a patentee to exclude others from growing a patented plant outside of the United States and importing and selling the fruit, flowers or other parts of the plant. Those who depend upon plant patents to protect their plant-breeding programs have long advocated amendment of the statute to close what they see as an obvious loophole.

Until the 1985 In re Hibberd decision, it was the position of the United States Patent and Trademark Office that plants, seeds and plant tissue cultures did not constitute patentable subject matter. Relying on the landmark U.S. Supreme Court decision of Diamond v. Chakrabarty,[8] which held that a genetically engineered microorganism constituted patentable subject matter, the Patent Office held in the Hibberd case that claims to a maize plant and to seeds and tissue cultures of such plant were patentable under the general utility patent statute. Since the Hibberd decision, the Patent Office has recognized the patentability of novel and unobvious plants, whether created through traditional plant-breeding procedures or through genetic engineering.

The utility patent statute provides much greater flexibility in protecting inventions of new plants and related subject matter than is available under the Plant Patent Act or the Plant Variety Protection Act. This greater flexibility does, however, have a price, which is the substantially more detailed disclosure that is required. Through a utility patent, an inventor can claim the novel plant, seeds and other propagating material, and even fruits and flowers. For transgenic plants, claims to genes and other nucleic acid molecules, vectors, transformed plant cells, intact plants, seeds and processes for making and using such materials are typically included in a utility patent application.

The disclosure of a U.S. utility patent application must contain a written description of the invention as claimed, a description of how to make and use the claimed invention (*i.e.*, an enabling disclosure) and the best mode contemplated by the applicant for carrying out the invention at the time of filing. Frequently, compliance with the disclosure requirement necessitates the deposit of seeds, propagating material, or other biological specimens with a public depository. The specification must conclude with one or more claims distinctly claiming and specifically pointing out what the applicant regards as the invention.

Because of the disclosure requirements, utility patent applications are much more complicated and lengthy (and accordingly, more expensive) than applications for plant patents or plant variety protection certificates. This additional complexity and expense is usually justified by the scope of protection that is provided.

The availability of utility patent protection for plant biotechnology in Europe has vacillated over the past decade. From the late 1980's until 1995, patents for transgenic plants were allowed under the European Patent Convention. Such patents were permitted, notwithstanding the prohibition in Article 53(b) against patenting of "plant or animal varieties or processes for the production of plants or animals." The bases for such patents arose out of the European Patent Office Board of Appeals decision in the Harvard Oncomouse case[9], in which it was held that a claim to a transgenic animal was not directed to a "variety."

In 1995, the EPO Board of Appeals reversed this policy in the case of Plant Genetics Systems/Glutamine Synthetase Inhibitors.[10] In that case, the Board upheld an opposition to claims to transgenic plants and seeds on the ground that those claims embraced new varieties and were therefore prohibited by Article 53(b). Examination of patent claims directed to plants and seeds has been suspended in view of this case.

On May 7, 1998 in Strasbourg, France, the European Parliament cleared the European Union's Biotechnology Patenting Directive. This directive recognizes a right to patent life forms, including transgenic plants, tissue cultures and seeds. This directive reportedly will allow patenting of plants of a scope similar to that available in the United States and Japan. The directive will now go the the fifteen European Union member states for final clearance.

Unless the European Parliament's biotechnology directive is approved, the value of European patent protection for plant technologies is questionable. While claims to genetic materials, transformed cells and cell cultures and processes for making transgenic plants are allowed, the extent to which these claims will protect progeny plants separated from the original transgenic plant by several generations is not clear.

Patent claims to plants and seeds *per se* provide the most direct and valuable protection for novel plants. Such claims provide the incentives necessary to stimulate investment in the expensive and risky research needed to develop beneficial new plant technologies.

Endnotes

1. In re Hibberd 227 U.S.P.Q. 443 (P.T.O. B.P.A.I. 1985).

2. A fourth form of intellectual property protection arises from the state laws of trade secrets and unfair competition.

3. Title 35, United States Code, §161 *et seq.*

4. Title 7, United States Code, §2401 *et seq.*

5. Title 35, United States Code, §100 *et seq.*

6. Yoder Brothers, Inc. v. California-Florida Plant Corporation, 537 F.2d 1347, 1378 (5th Cir. 1976).

7. Imazio Nursery, Inc. v. Dania Greenhouses, 69 F.3d 1560 (Fed. Cir. 1995).

8. Diamond v. Chakrabarty, 447 U.S, 303 (1980).

9. T 19/90, OJ. E.P.O. 1990, 476.

10. T 356/93, OJ. E.P.O. 1995, 545.

Plant Breeders' Rights and Plant Biotechnology

Robert J. Jondle, Ph.D, J.D.
Patent Attorney
Rothwell, Figg, Ernst & Kurz
Omaha, Nebraska, United States

The development of new plant innovations by either private or public plant breeders and biotechnologists involves considerable research and development expenses. Each year, private companies invest hundreds of millions of dollars in biotechnology and plant breeding research. Effective protection of these new plant products is necessary to provide an incentive to make this large research investment. Universities also are protecting certain plant research products to encourage technology transfer from the university to private industry or to increase resources for research.

Plant protection options vary from country to country, depending on the type of plant invention, such as a new biotechnology method, a gene, plant cultivar, or hybrid. For example, in the United States there are three main types of protection including: 1) utility patents; 2) Plant Variety Protection (UPOV); and 3) plant patents for asexually reproduced plants. Plant science, plant breeding, and biotechnology are interconnected and international in scope. Therefore, it is important to consider international plant protection options available for plant innovations.

This article focuses on breeders' rights under the UPOV conventions. The purpose of the UPOV convention and seed registration laws of certain countries is to recognize and ensure to the breeder of a new plant cultivar certain rights regarding the use by others of the protected cultivar.

The International Convention for the Protection of New Varieties of Plants (the UPOV) was concluded in 1961, and was revised in 1972, 1978, and 1991. Article 2(1) of the original 1961 UPOV Convention permitted each member state the option to provide protection rights to new cultivars by grant of either a special title of protection or a patent, but required a member state to provide only one of these forms for one botanical genus or species. Over the next two decades, countries that enacted breeders' rights legislation

A. Altman et al. (eds.), Plant Biotechnology and In Vitro Biology in the 21st Century, 747–749.

implemented the UPOV Convention, which prohibited having protection by both patent and breeders' rights. At present, there are 37 member states in the UPOV.

On March 19, 1991, the UPOV adopted a new convention, with significant changes from the 1978 Convention. Compared to the 1978 Convention, the 1991 UPOV Convention offers important modifications in the areas of:

1) Rights over harvested material and their direct products
2) Allowing multiple types of protection
3) Adopting a concept of "essentially derived" varieties.

The 1991 Convention provides breeders' rights over harvested material when 1) the seed is obtained without authorization of the breeder, and 2) the breeder has no reasonable opportunity to exercise his or her right. Also, as an optional provision, a breeder may have rights over products made directly from harvested material of protected lines. For example, seed of a protected wheat variety may be produced without authorization of the breeder and then used to produce flour and bread. The owner of the breeders' rights certificate could enforce his or her proprietary rights against the flour or the bread in a country that has enacted legislation containing the 1991 UPOV Convention.

Multiple types of protection allowed. The 1991 Convention allows multiple types of plant protection where each country decides for itself which types of plant protection it will use. The new Convention abandons the prohibition of "double protection," which is present in the 1978 Convention. The United States was excepted from this provision because patents for asexually reproduced plants under the 1930 Plant Patent Act were available in the United States at the time the Convention was finalized.

"Essentially derived" varieties. The concept of an essentially derived variety in the 1991 UPOV Convention states that in order for a new variety to be an essentially derived variety, the variety must be derived from at least one protected variety. If one of the parents of a new line are protected, then any new line derived from these unprotected parents will be essentially derived from the initial protected variety when:

1) the new cultivar has been predominately derived from the initial cultivar
2) the new cultivar is clearly distinguishable from the initial cultivar
3) the new cultivar conforms to the initial variety in the expression of the essential characteristics.

Examples of methods that may result in an essentially derived cultivars, which are listed in the 1991 UPOV Convention, are:

1) selection (from plants of the initial variety) of:
   a) a natural or induced mutant
   b) a somaclonal variant, or
   c) a variant individual
2) backcrossing
3) transformation by genetic engineering.

Genetically modified crops in the European Union the regulatory framework and public acceptance

A. Koening, Monsanto Europe S.A., Av. de Tervuren 270-272, 1150 Brussels, Belgium

Substantial equivalence - a principle for the safety assessment of genetically modified plants

In both the United States and the European Union the fundamental concept for the food and environmental safety assessment of products derived from modern biotechnology is the concept of substantial equivalence, where the novel product is compared to a closely related product that has an accepted standard of safety. The concept was initially introduced by the World Health Organization (WHO) and the United Nations Food and Agricultural Organization (FAO) (WHO, 1991). In 1992, the Organisation for Economic Cooperation and Development (OECD) elaborated the underlying concept and introduced the term "substantial equivalence" (OECD, 1993). The application of substantial equivalence has since been reinforced by international expert bodies (WHO, 1995; FAO, 1996) and has been adopted by regulatory authorities in most countries.

The assessment of substantial equivalence occurs in two steps. First, agronomic and compositional parameters of the novel product and the traditional counterpart are compared to assess equivalence taking into account the limits of natural variation. Secondly, it is demonstrated that the introduced change is well characterised and of no harm to human health and the environment; this involves a molecular characterisation and the safety assessment of the introduced proteins. A novel plant variety that is established to be substantially equivalent to its traditional counterpart should be treated in the same way with respect to safety as the traditional counterpart.

Although the fundamental principle for assessing the safety of the genetically modified plants and plant products in the US and European regulatory systems is similar the scope of the legislation, risk perception, data requirements and time required for completion of the regulatory processes within these systems differ considerably. These differences have the potential to cause trade conflicts for commodity crop products.

The regulatory system in the United States

The United States Office of Science and Technology Policy (OSTP) published the "Coordinated Framework for Regulation of Biotechnology" in 1992 which recommended maintaining the regulatory authority for the products of plant biotechnology within the three existing regulatory agencies and emphasized that regulatory assessments must be science-based and risk-based (OSTP, 1986). Based on these guidelines, the regulatory authority for plant biotechnology products in the United States resides in all three US regulatory agencies, the Food and Drug Agency (FDA), the Environmental Protection Agency (EPA) and the United States Department of Agriculture (USDA). EPA focuses

A. Altman et al. (eds.), Plant Biotechnology and In Vitro Biology in the 21st Century, 751–756.

on the safety of pesticidal components, USDA focuses on the assurance that the new plants do not present any plant pest risks and FDA assesses the food and feed safety aspects of the products. EPA also regulates the use of any pesticides used on herbicide tolerant plants. Genetically modified plants that produce a pesticidal substance such as proteins from *Bacillus thuringiensis* are regulated under the jurisdiction of all three agencies. Genetically modified plants that are non-pesticidal are under the regulatory jurisdiction of the FDA and USDA, but not EPA.

USDA regulates the movement and release of genetically modified plants under the Federal Plant Pest Act and the Plant Quarantine Act, which is administered by the Animal and Plant Health Inspection Service (APHIS). USDA issued regulations on June 16, 1987 (7 CFR Part 340), with revisions on March 31, 1992 (Federal Register 58:17044-17059) and a second revision proposed August 11, 1995 (Federal Register 60:4356743573). Prior to unconfined released (e.g. commercialization), the USDA/APHIS must make a determination that the genetically modified plant is not a plant pest and, therefore, is not a regulated article under 7 CFR Part 340, using the types of information specified in these guidance documents.

EPA has joint responsibility along with USDA and FDA for the regulatory oversight for any genetically modified plant that produces a pesticide. EPA derives its authority from the Federal Insecticide, Fungicide and Rodenticide Act (FIFRA). EPA has provided guidance for data requirements to support registration and tolerance exemption under their 1994 proposed policy (Federal Register 59:60496-60547).

FDA maintains its authority over the human food and animal feed safety and wholesomeness of all plant products, including those produced via genetic modification under the Federal Food Drug and Cosmetic Act (FFDCA, 1988). The FDA 1992 "Statement of Policy: Foods Derived from New Plant Varieties" (FDA, 1992; Kessler et al., 1992) described the decision-tree safety assessment approach that FDA recommended for the safety assessment of food or feed derived for new plant varieties, including those derived via genetic modification. FDA concluded that food and feed derived from genetically modified plants pose no unique safety concerns and, therefore, that the food and feed products derived from these plants should be regulated no differently than food or feed derived from traditional plant breeding or any other genetic modification approach. The FDA food policy reiterates that it is the responsibility of the producer to assure the safety of respective foods and provided a consultation process between the producer and FDA to discuss the appropriate safety assessment and review the results.

The regulatory system in the European Union

The regulatory system in the European Union (EU) is currently undergoing significant change with the recent adoption of the Novel Foods Regulation administered by Directorate General (DG) III Industry and Internal Affairs of the European Commission. This introduces an EU-wide regulatory approval system for novel foods, including those foods derived from genetically modified crop plants. The EU implemented two EU-wide environmental assessment Directives 90/219/EEC and 90/220/EEC in 1990. These two directives regulate all activities with all genetically modified organisms (GMOs). The Directive 90/219/EEC covers contained use operations in laboratories, growth rooms and glass houses, while the Directive 90/220/EEC covers the deliberate release of genetically modified organisms into the environment. The Directive 90/220/EEC is divided into major two parts: Part B, which applies to research and development work and Part C, which applies to the approval to place a genetically modified product on the market. Specific risk assessments, and in certain cases simplified procedures, have been introduced for the deliberate release of genetically modified plants in field trials. Revisions of Directives 90/219/EEC and 90/220/EEC are currently in progress.

The Directives 90/220/EEC and 90/219/EEC and the Novel Food Regulation regulate activities broadly across all genetically modified organisms, including plants, microorganisms and humans, including gene therapy, since they are based on the process of genetic modification and not the products produced. The legislations require the appointment of national competent authorities, who have the responsibility to consider applications for release and marketing and to coordinate national activities at the European level.

In addition to gaining the necessary environmental and food approvals, an EU-wide approval for animal feed use is also being considered. Presently, animal feed safety assessments are being addressed within the 90/220/EEC review process. There are also specific national and EU-wide procedures for varietal certification of new plant varieties that also cover new plant varieties that are derived by genetic modification. The U.S. does not have a varietal certification or registration process.

Existing national and EU authorities are responsible for the authorisation of pesticidal substances are also responsible for the registration of pesticidal substances on genetically modified plants.

Comparison of the EU and the U.S. systems

The unpredictable nature of requirements and timelines to gain an EU-wide approvals for imports and cultivation through the 90/220/EEC process continues to be a limitation to the commercialisation of genetically modified crop plants in Europe and elsewhere. The time required to obtain a determination that the genetically modified plant is not a plant pest from the USDA is an average of 6 months for 22 of the 23 genetically modified crops in the U.S. In comparison, the time required for the 90/220/EEC procedure has ranged from 17 months for Roundup Ready™ soybean to over two years for the approvals of the canola product submitted by Plant Genetic Systems and the corn product of Ciba Geigy. (See Table 1 Timelines of product approvals under Directive 90/220/EEC).

Risk perception and data requirements that affect the timelines and the predictability of the approval procedure continue to differ between the U.S. and the EU. Harmonisation of the regulation for transgenic crops across the world would help to reduce uncertainties for companies preparing to launch genetically modified crop plants, for farmers and the crop commodity chain, including international traders.

Public acceptance of plant biotechnology in the European Union

The majority of the scientific community in the EU agrees that there are no specific risks associated with plant biotechnology. Environmental and consumer organisations that target plant biotechnology in particular are however successful in raising concerns. The confidence in the food chain has been undermined as European consumers were faced with a wide range of food scares since the mid 1980's including salmonella in eggs, the use of growth hormones and Bovine Spongiform Encephalopathy. Furthermore, philosophic concerns exist that hereditary material should not be tampered with, and that if this occurs consumers should at least have the freedom to choose to avoid such 'unnatural' products. Due to a lack of accurate information and oversimplification of complex scientific scenarios, scientifically unfounded issues such as the generation of 'super weeds' and the undermining the value of B.t. in organic farming are reported in the press.

It has however been observed that consumer concerns voiced in opinion polls do not necessarily translate to an altered buying behaviour. The food industry has started to voluntarily label products as "produced with modern biotechnology". More than 40 such products are available in the Netherlands and more than 90 in the United Kingdom, ranging from soup to frozen pizza and meat balls. The market share of the products did not change upon the introduction of the biotechnology label.

In response to consumer concerns industries involved in the discourse ranging from food processors to companies directly involved in plant biotechnology in 1997 joined together in the Food and Biotechnology Communication Initiative to make more information available on the safety assessments of products derived from plant biotechnology, and on the benefits derived from the technology. A wide variety of tools are used including hot lines, web sites, fact sheets, television programs, and a broad advertising campaign to shed light on to the myths generated by environmental organisations. It has been shown in Switzerland that the provision of information to the general public in the context of the referendum helped to convince 66% of the voters, a significant share of the consumers that biotechnology is safe and should not be banned.

Conclusion

A regulatory system that is predictable and is harmonised across the world, will lower the barriers to market for plant biotechnology products and will help to reduce political uncertainties.

Industry, scientists and the governments providing sufficient information on products developed with these new technologies that will increasingly contribute to improvements in the food chain will help to rebuild confidence of consumers in the food chain and more specifically in plant biotechnology.

Table 1. Genetically modified plant products approved according to European Directive 90/220/EEC

Company/Institution	Crop	Trait	Time Required For Determination*
Rhone-Poulenc	Tobacco	Herbicide Tolerant	~ 9 months
Plant Genetic Systems	Corn	Male Sterility/Herbicide Tol.	~ 24 months
Monsanto	Soybean	Herbicide Tolerance	17 months
Bejo-Zaden	Chicory	Male Sterility/Herbicide Tol.	~ 18 months
Ciba-Geigy	Corn	Insect Protection	~ 25 months

* Approximate average time from submission to rapportuer country to publication in the European Official Journal.

References

FDA. 1992. Statement of policy: Foods derived from new plant varieties: Notice. Federal Register 1992;57: 22984-23005.

FAO/WHO. 1996. Biotechnology and food safety. Report of a Joint JAO/WHO Consultation. FAO, Food and Nutrition Paper 61.

IFBC (International Food Biotechnology Council). 1991. Biotechnologies and Food: Assuring the Safety of Foods Produced by Genetic Modification. Regul. Toxicol. Pharmacol. 12: SI-SI96.

OECD (Organization for Economic Cooperation and Development). 1993. Safety Evaluation of Foods Produced by Modem Biotechnology: Concepts and Principles. OECD, Paris.

OECD. 1996. OECD Documents: Food Safety Evaluation. OECD, Paris.

OSTP (Office of Science and Technology Policy). 1986. Coordinated framework for regulation of biotechnology. Federal register 1986;51: 23302-23350.

WHO, 1991. Strategies for Assessing the Safety of Foods Produced by Biotechnology. Report of a Joint FAO/WHO Consultation. World Health Organization, Geneva.

WHO. 1993. Health Aspects of Marker Genes in Genetically Modified Plants. Report of a WHO Workshop. World Health Organization, Geneva. WHO/FNU/FOS/93.6.

WHO. 1995. Application of the Principles of Substantial Equivalence to the Safety

Evaluation of Foods and Food Components from Plants Derived by Modern Biotechnology. Report of a WHO Workshop. World Health Organization, Geneva. WHO/FNU/FOS/95. 1.

DEMYSTIFYING THE GENETIC ENGINEERING OF PLANTS: TALKING BIOTECHNOLOGY IN THE COMMUNITY

Peggy G. Lemaux
University of California, Berkeley

Although the potential for risk in genetically engineered foods can be fairly accurately assessed scientifically, the perception of risk is quite different. Public concern, can be inconsistent with risk measurement, depending on the familiarity, "friendliness" and voluntariness of the risk. With biotechnology public perception of risk varies widely. To help consumers, end-users, public opinion leaders and governmental regulators make informed decisions based on real, not perceived, risks and benefits, information must be communicated in easy-to-understand terms. As scientists, we are often not prepared to do this and therefore do not play a role in educating the public. Depending on target audience needs, different communication strategies can be required.

- **Strictly informational**; people just want to know what products might arise from the technology and how they will differ from what they see now.
- **Strategic**; government agencies or state or national organizations need information to make appropriate public policy decisions.
- **Enabling**; an industry needs information to determine whether biotechnology provides appropriate tools to address a problem and wants to know how they can utilize these technologies to solve the problem.

General concepts of biotechnology are well-understood by scientists, but it can be difficult to identify the information and issues of importance and to relay this information in a jargon-free manner. As a public-sector communicator, I have developed aids that help me communicate with varied audiences; these include slides containing easy-to-understand analogies and a generic talk written in modules, allowing the presenter to pick and choose aspects appropriate to the intended audience. The slides and text, available at **http://plantbio.berkeley.edu/~outreach**, are divided into the following modules.

- A **general introduction** to the history of foods and agriculture,
- An explanation of the methods using easy-to-understand **analogies,**
- An **example** contrasting classical breeding technologies and biotechnology,
- A look at how biotechnology is already **impacting agriculture** and
- A look at the **biotechnology pipeline** highlighting products on the market and those being developed.

In discussing biotechnology, the tone and content of the talk can be changed, but certain basic concepts need to be conveyed. These concepts are presented below.

General Introduction: What is Biotechnology? Biotechnology is not the first technology to impact the food supply in the 10,000-year history of farming. Other technologies have been used through the years to change our food supply, *e.g.* domestication of plants and animals, mechanization, chemical fertilizers, pesticides and these too have raised issues involving risk and benefit assessment. Most people in North America don't understand how the food in restaurants and groceries came to look and taste like it does. This general lack of understanding about how foods have been modified by human intervention in the past makes it extremely difficult to communicate about how the foods for tomorrow's table will be generated.

A. Altman et al. (eds.), Plant Biotechnology and In Vitro Biology in the 21st Century, 757–760.

Since the beginnings of agriculture, humans have been involved in the natural exchange of information between plants and animals, a process called breeding. Humans have been actively involved in this process by selecting the parents that participate in genetic exchange and have thus changed the way foods look, taste and yield. The principle of human involvement can be exemplified by showing an ancient ancestor of modern hybrid corn and a new vegetable that resulted from a wide species cross, the broccoflower. This type of human involvement has contributed to the dramatic improvements in agricultural productivity. Around 1930, U.S. farmers needed 379 million acres to feed roughly 100 million people. By 1984, we farmed only 312 million acres to feed 2.5 times the population that existed in 1930.

What happens to the genetic information when we cross, for example, two corn plants, one with increased disease resistance and one with increased yield. Using an analogy, the genetic information in a cell is the recipe which determines what the cells will do and imparts on the plant or animal its characteristics - like whether the tomato fruit is yellow or red. That recipe, the genome, is made up of chemical units. If each chemical unit in the cell of a wheat plant (or a human being) is represented by an alphabetic letter, it takes 1700 books, each of 1000 pages, to hold all that information. Stacked on top of one another, the books would be as high as a 20-story building!

How does biotechnology differ from classical methods of genetic manipulation? So what happens when we do breeding? This question can be effectively addressed with an example that contrasts a classical breeding approach with one using biotechnology, highlighting the similarities and differences in the two approaches. The example chosen, based on work at the University of California (1), was aimed at enhancing the flavor of the commercial tomato by increasing its sugar content. This was made possible because certain wild tomato relatives, although not like the domesticated tomato in appearance, have one desirable quality, a higher sugar content. The idea was to transfer only the higher sugar characteristic to the domesticated tomato, leaving behind the small size, bitter taste and lower yield of the wild relative.

In the classical breeding approach the wild tomato was crossed with the domesticated tomato and over many years, a higher sugar tomato was achieved. In terms of the book analogy, two different stacks of books were combined; the information in those books was randomly mixed. Genetic rules dictate that the final literary piece contain only 1.7 million pages, not 3.4 million, the combined total of the two stacks of books. After continuing to cross the resulting progeny with the original domesticated tomato, the "editors of these volumes" ended up with a final stack of books that had mostly volumes from the domesticated species, with about 100-200 pages from the wild species. In that 100 pages was the information for higher sugar content and a lot of other information they hadn't "read". Part of the additional information from the wild species turned out to cause reduced fertility in the resulting tomato plants. With classical breeding, the breeder does not have complete control over the outcome. Sometimes you get characteristics that you don't want. Breeders have little control over the final outcome since they are observers not orchestrators.

In the second approach the goal was the same, to increase the sugar content of the commercial tomato. This time they attempted to achieve the goal by looking at the "recipe" for the tomato fruit. They identified a single piece of genetic information (one responsible for the breakdown of sugar) which, if removed, would result in a sweeter tomato. Through modern molecular technologies they turned off the machinery that makes the sugar-degrading enzyme. Instead of exchanging 100 pages, they changed only a half page of information in our 1.7 million-page series. In the classical breeding approach they knew the information from the wild species contained the information for higher sugar, but weren't expecting the other information that interfered with fertility. In the second case where "genetic engineering" methods were used, they knew everything about that new half-page of information they were adding; they had "read it". For this reason, the outcome was more predictable because it was more precise. Another important difference between classical and molecular approaches is that with classical methods, gene exchange is limited to interchange between very close relatives. With the newer technologies, the source of information can vary.

The book analogy can also be used to address issues of risk/benefit. With classical breeding, information exchange occurs only between closely related species. This is functionally equivalent to exchanging hundreds of pages from two 1.7 million-page mystery novels written by the same author about the same subject. The likelihood that a romantic novel will be created is negligible; however, it is possible that an outcome or a minor character in the original mystery novel might be changed. But the editors for these changes (breeders) have a history of making good editing changes. In the case of genetic engineering, less than a page of information in the 1.7 million-page mystery novel is changed; however, the new information can be from the same mystery novel, a different mystery novel or a romantic novel. In this case the impact of the change will depend on the precise exchange and each exchange must be judged individually. Most changes of this magnitude will not appreciably change the novel; however, such exchanges, just as with changes resulting from the classical method, are not without the potential for change. Each exchange (literary or genetic) must be analyzed for its potential for marked change.

How is biotechnology already impacting agriculture? Some applications of the technology are already in the commercial marketplace. Some applications don't involve the actual modification of the food but simply involve the use of the technology to, for example, make better diagnostics to facilitate breeding or to assess food safety. Other applications involve changes in the way food is processed, such as cheese made from an enzyme identified in a calf and moved into a bacterium. Another product that entered the U.S. market was the Flavr Savr or Endless Summer tomato aimed at delivering a tastier, less spoilage-prone tomato to the consumer. Virus-resistant squash and papaya are also available. The prediction for 1998 production is that 50% of the cotton acreage, 30% of the soybean and 20% of the maize acreage in the U.S. will be genetically engineered (2). The products of this technology are therefore no longer just an idea waiting in the research laboratory; they are real products being planted and harvested by farmers and appearing in our groceries.

The new product pipeline for "biotechnology" It is important for consumers and end-users to understand that these are just the first products of the technology; there is much more on the horizon. In order to determine the fate of the technology and to help end-users understand the breadth of its impact. I provide a large number of examples covering various applications of the technology to give people a better idea of its potential for the future. These include plant pest protection strategies, *i.e.* fungal, bacterial, viral and insect resistance and food and processing quality traits, including the new areas of neutraceuticals or functional foods, where certain properties of food are changed to make the food more digestible, more nutritious or less allergenic. I also include some non-food applications, such as changes in the appearance of ornamental crops, the utility of plants as sources of animal and plant vaccines and the use of plants as "pharms" to make products to replace those now being made from nonrenewable resources, such as industrial oils, replacements for gasoline and thermostable, biodegradable plastics. Plants are also being used along with microbes as bioremediators to clean up existing contamination of metals and other organic and inorganic pollutants and to prevent future contamination.

SUMMARY

In summary, we expect the products of genetic engineering to have a major impact in agriculture, but we need to be able to communicate to consumers and clientele about these impacts and how they might be realized. We need to have an informed public and informed policy makers who make decisions based on "scientific risk" and not on perception of risk issues mandated by a wary public.

REFERENCES

1) Bennett AB, Chetelat RT and Klann EM (1994). Exotic germplasm or engineered genes: Comparison of genetic strategies to improve tomato fruit quality. In: 208th National Meeting of the American Chemical Society, Washington, D.C., USA, August 21-25, 1994. Abstracts of Papers American Chemical Society, v.208, n.1-2: AGFD 12.
2) Kilman S and Warren S (1998) Old Rivals Fight for New Turf. The Wall Street Journal, pp. B1-B2.

LEGISLATION AND PUBLIC PERCEPTION OF BIOTECHNOLOGY IN POLAND

Tomasz Twardowski
Institute of Bioorganic Chemistry, Polish Academy of Sciences, Poznań, Poland

So far, biotechnology has been of small economic importance in Poland. However, Poland represents a market and potential of almost 40 mln people and R+D sectors of agrobiotechnology, pharmaceutical industry and environment protection are focused on modern technologies.

1. Legislation

Poland will join the EU and will have to accept the rules coming from Brussels. That also means the acceptance of the European legislation, e.g. EC Directives (90/219 and 90/220) as the examples strictly related to biotechnology. In the 1990s, we have seen international conventions significantly influencing national legislation, e.g. Convention for Biological Diversion (1995), membership of OECD (1996), the Budapest Treaty and protection of intellectual property rights (1993). Over the last three years, there has been a tendency to implement a new Polish law, which would be a 'national copy' of the EU directives implemented with the assistance of the international organisations (OECD, UNEP, UNIDO). The new law on the „protection of environment", consistent with the „Biodiversity Convention", was passed by the Parliament and signed by President (Sept. 1997).
1997 saw the first releases of GMP to the environment (potato, maize, common-beet) carried out with the permission of the Ministry of Agriculture under strict

A. Altman et al. (eds.), Plant Biotechnology and In Vitro Biology in the 21st Century, 761–764.

supervision of experts. In 1998, we expect to run about 20 experiments with GMO in the environment.

2. Public Perception

The Polish society is oriented towards the Western Hemisphere. Modern technology, e.g. genetic engineering, has a significant effect on the public opinion and reflects the public perception of the future prospects and the conversion of attitudes. The public in general takes biotechnology to be a key technology for the future. The future prospects of biotechnology in Poland are connected with agrobiotechnology, health industry and services.

Until November 1995, the mass media demonstrated hardly any interest in biotechnology in general, or genetic engineering in particular. Unfortunately, public debates rarely concern scientific issues, including those related to biotechnology. The exceptional situation concerned soybeans (Nov. 1996), transgenic carp (Christmas 1996) and the Dolly sheep story which raised concern about cloning of people (March 1997). In November 1995, the Polish Green Federation (Department Kraków) published the report „Playing the God (transgenic food in Central and Eastern Europe)". We observed real „explosion" of the mass media interest in the problems of modern genetic technology and, as a result, in molecular biology. Many articles have been published in the press. However, already at the beginning of the debate, the insufficient knowledge of the subject both by the journalists and the public could be observed. Today, the quality and quantity of the articles and radio and TV broadcasts on biotechnology have significantly improved.

Mass media in general cover the progress of science and technology and present related risks. After 1990, we can also observe a change of policy; following the priority of national politics, i.e. the membership of the EU, Polish legislation has been a copy of EU directives. It can also be seen in the mass media. Polish press is more quiet, more solid and less sensational in comparison to the press of Western Europe, e.g. Germany or UK. It seems not to be influenced by the Parliament or significantly affected by the Greens.

3. Perspectives

The scientific society have recognized three key components of the „value added chain" represented by modern biotechnology: science with technology + legislation with intellectual property rights + market with public perception. However, in the light of common interpretation we can expect that the future of biotechnology depends much more on moral dogmas and legislation than on science and technology. Reports in the mass media supported by the Greens' reports on „patenting a life" and „cloning of the human being" etc. produced mixed feelings.

International cooperation is of special importance. For example, the United Nations Environment Programme (UNEP) plays a key role in the sustainable development of the environment, particularly in the situation of the dynamic development of modern technologies of economic significance, such as biotechnology. In compliance with Agenda 21 of the UN and following the Biodiversity Convention Poland as a Member State has decided to strengthen its cooperation on biosafety. We cannot compare a well established western economy (e.g. of Germany or Britain) with a country going through a transition period. The progress is tremendous and it goes in the right direction. Time and money are the most important factors; then comes capacity building. In my opinion, Poland together with Hungary and the Czech Republic is a leading country in Central Europe .

References:

„Rozwój biotechnologii. Projekt rozwiązań prawnych dotyczących stosowania genetycznie modyfikowanych organizmów" [„Biotechnology and a proposal for the Polish gene law"], Ed. T . Twardowski, Poznań 1997; [in Polish with extended English summary].

T. Twardowski „Społeczne i prawne aspekty biotechnologii", Politechnika Łódzka, Łódź, 1996 [in Polish].

T. Twardowski, B. Suchocki, A. Przestalski „Public perception and legal aspects of biotechnology during the transformation period in Central Europe; case Poland", a lecture presented in Chicago „Public Understanding of Science", USA, 15 - 17.08.1997, Chicago Academy of Sciences.

Acknowledgement:

This work was partially supported by UE grant „Biotechnology and the European Public", nr IC20-CT96-0030.

AUTHOR INDEX

www.ingramcontent.com/pod-product-compliance
Ingram Content Group UK Ltd.
Pitfield, Milton Keynes, MK11 3LW, UK
UKHW021332070726
13610UKWH00011B/44

* 9 7 8 9 4 0 1 1 4 6 6 2 3 *